Springer Proceedings in Physics

Volume 203

The series Springer Proceedings in Physics, founded in 1984, is devoted to timely reports of state-of-the-art developments in physics and related sciences. Typically based on material presented at conferences, workshops and similar scientific meetings, volumes published in this series will constitute a comprehensive up-to-date source of reference on a field or subfield of relevance in contemporary physics. Proposals must include the following:

- name, place and date of the scientific meeting
- a link to the committees (local organization, international advisors etc.)
- scientific description of the meeting
- list of invited/plenary speakers
- an estimate of the planned proceedings book parameters (number of pages/ articles, requested number of bulk copies, submission deadline).

More information about this series at http://www.springer.com/series/361

Md. Naimuddin
Editor

XXII DAE High Energy Physics Symposium

Proceedings, Delhi, India, December 12–16, 2016

Springer

Editor
Md. Naimuddin
Department of Physics
University of Delhi
New Delhi
India

ISSN 0930-8989 ISSN 1867-4941 (electronic)
Springer Proceedings in Physics
ISBN 978-3-030-10328-6 ISBN 978-3-319-73171-1 (eBook)
https://doi.org/10.1007/978-3-319-73171-1

Printed on acid-free paper

This Springer imprint is published by the registered company Springer International Publishing AG part of Springer Nature
The registered company address is: Gewerbestrasse 11, 6330 Cham, Switzerland

Preface

The research in the area of high energy physics remains as active as ever. With the turning on of Large Hadron Collider (LHC) in 2010, the field has received an enormous push in terms of its reach and breadth. The LHC opens up new horizons and opportunities for the physicists to discover and measure the things that have not been done so far. In the sector of neutrino physics, there are plenty of activities going on, both theoretical and experimental. With the neutrino physics experiments spread across three continents, the field is emerging even stronger. Similar is the case with particle astrophysics and cosmology where new horizons are being explored. As a result of ever-expanding reach of high energy experiments, the demand for better and more sophisticated detector and electronics has also been growing. This has brought a significant push in the detector and instrumentation research. Although the high energy physics research falls under the category of fundamental science and deals with the basic building blocks of universe, its outcome has much diverse consequences. The technology that is developed for pursuing this basic research has found applications in many areas of immediate human benefit.

The activity of high energy physics is spread across all over the world, but the high energy physicist maintains a very close working relationship among themselves. This stimulates the research collaborations spanning people across many countries and continents. In order to exchange ideas, and share and discuss results in person, the physicists meet with each other at various occasions. One such regular occasion is the DAE-BRNS high energy physics symposium. The DAE-BRNS high energy physics (HEP) symposium is a conference series held every other year in India. The DAE-BRNS HEP symposium is the premiere event organized in the field of high energy physics and covers almost all aspects relevant to Particle Physics. The deliberations and discussions cover both theory and experiment with variety of physics interests like particle, neutrino, flavor, heavy ion, astroparticle, physics beyond standard model, societal applications. The last symposium in the series took place at Department of Physics and Astrophysics, University of Delhi, New Delhi, from December 12 to 16, 2016. The symposium saw participation of approximately 400 physicists, postdocs, engineers, and students. The symposium academic program consisted of invited plenary talks, invited parallel talks or mini-review talks, and contributed parallel talks. In order to encourage and motivate the young and amateur physicists, a poster session with more than 100 posters was also organized.

The symposium also drew a very enthusiastic participation from Industrial partners and engineering institutions. During the symposium, results from running experiments like CMS, ATLAS, ALICE, LHCb, Belle were discussed in detail. The prospects of planned experiments like DUNE, INO-ICAL were also discussed. For the first time, the symposium had a dedicated theme of societal applications: medical physics, imaging, etc. A good number of presentations were made on this important topic, and future collaborations were formed.

This book contains the selected papers that were presented during the symposium. Each of these papers has been peer-reviewed before accepted for the final publication. The papers cover a wide variety of results and ideas on topics like standard model physics, neutrino physics, beyond standard model physics, heavy ion and QCD physics, Particle Astrophysics and Cosmology. The book also contains papers presented outlining the future experiments and recent developments in detector development and instrumentation. Recent progress in formal theory has also been discussed. Finally, the ideas and efforts in harnessing the high energy physics research spin-offs in terms of societal applications are also included.

New Delhi, India Md. Naimuddin

Acknowledgements

Any event having hundreds of participants coming from all over the world is not an easy one to organize successfully. The XXII DAE-BRNS high energy physics symposium had about 400 participants coming from all over the world. During the preparation and organization of the symposium, many individuals and organizations provided their support and help because of which the symposium could be organized flawlessly. After the symposium was over, the next task was to review hundreds of manuscripts submitted for publication as proceedings. Many individuals agreed and devoted their valuable time in reviewing the manuscripts in order to maintain the quality and scientific value of these papers. While it is not possible to mention the contributions of all the individuals by their name but this event would not have been possible without their unending support, so I would like to thank everyone who lent their support in any way during the entire process. In particular, I would first like to show my heartfelt gratitude and thanks to the members of the National Organizing Committee and Local Organizing Committee for their unconditional support throughout. The event would not have been possible without the support of the Delhi University administration and staff. My sincere thanks go to the Delhi University administration and support staff. I would also like to thank the Head of the Department of Physics and Astrophysics for their support during the symposium. An event of such a scale requires lot of resources and funding to meet all the expenses. I am extremely thankful to Department of Atomic Energy—Board of Research in Nuclear Sciences for their support for the symposium. The contributions and participation of our sponsors CAEN, Hamamatsu, Tektronix–Keithley, Nucleonix, AMETEK, and KAF International are duly acknowledged. We would also like to thank Tata Institute of Fundamental Research, Mumbai, and Institute of Mathematical Science, Chennai, for their support. The symposium and the review process of proceedings would not have been possible without the support of my colleagues in the high energy physics group of Delhi University. My heartfelt thanks go to Prof. Brajesh Choudhary, Prof. Kirti Ranjan, Dr. Ashok Kumar, Dr. Ashutosh Bhardwaj, and Dr. Sanjeev Kumar Verma. Thanks also go to my Ph.D. students and postdocs. A special thanks to Dr. Shivali Malhotra and

Dr. Daljeet Kaur for their valuable support in the organization of the symposium and also in the review process of proceedings. Thanks also go to all the Ph.D. students, postdocs, and support staff of our high energy physics group. And in the end, I would like to thank my family, my father, my wife, my son, my daughter, and my niece for their valuable support and sacrifices during the run-up to the symposium, during the symposium and all through for keeping up with me.

Contents

Contributors

S. P. Adhya Experimental High Energy Physics and Applications Group, Variable Energy Cyclotron Centre, Kolkata, India

Sanjib Kumar Agarwalla Institute of Physics, Bhubaneswar, India; Homi Bhabha National Institute, Mumbai, India

Zubayer Ahammed EHEP&A Group, Variable Energy Cyclotron Centre, HBNI, Kolkata, India

Rizwan Ahmad Department of Physics and Astrophysics, University of Delhi, Delhi, India

Zafar Ahmed Nuclear Physics Division, Bhabha Atomic Research Centre, Trombay, India

F. Akbar Department of Physics, Aligarh Muslim University, Aligarh, India

M. Rafi Alam Department of Physics, Aligarh Muslim University, Aligarh, India

Ashutosh Kumar Alok Indian Institute of Technology, Jodhpur, India

Balasubramanian Ananthanarayan Centre for High Energy Physics, Indian Institute of Science, Bangalore, India

Sankaran Aniruddhan Indian Institute of Technology Madras, Chennai, India

Mathew Thomas Arun Department of Physics, Mar Thoma College, Thiruvalla, Kerala, India; Department of Physics and Astrophysics, University of Delhi, New Delhi, India

Richa Arya Physical Research Laboratory, Ahmedabad, India

M. Sajjad Athar Department of Physics, Aligarh Muslim University, Aligarh, India

David Attié CEA Saclay, Gif Sur Yvette, France

Anjali Attri Panjab University, Chandigarh, India

T. Aziz Tata Institute of Fundamental Research, Mumbai, India

Partha Bagchi Variable Energy Cyclotron Centre, Kolkata, India

Seema Bahinipati Indian Institute of Technology Bhubaneswar, Bhubaneswar, India

Triparno Bandyopadhyay Department of Physics, University of Calcutta, Kolkata, India

Debika Banerjee Department of Physics, National Institute of Technology, Durgapur, West Bengal, India

Debika K. Banik Department of Physics, Dibrugarh University, Dibrugarh, Assam, India

Sebika K. Banik Department of Physics, Dibrugarh University, Dibrugarh, Assam, India

Anshika Bansal University of Delhi, New Delhi, India

S. Bansal Akal University, Talwandi Sabo, India

R. C. Baral Institute of Physics, HBNI, Bhubaneswar, India

Nirbhay K. Behera Indian Institute of Technology Bombay, Mumbai, India

P. K. Behera Physics Department, Indian Institute of Technology Madras, Chennai, Tamil Nadu, India

Prafulla Kumar Behera Department of Physics, Indian Institute of Technology Madras, Chennai, India

Suman B. Beri Panjab University, Chandigarh, India

Arunava Bhadra HECRRC, University of North Bengal, Siliguri, WB, India

Akanksha Bhardwaj Physical Research Laboratory (PRL), Ahmedabad, Gujarat, India; IIT Gandhinagar, Palaj, Gandhinagar, India

Ashutosh Bhardwaj Centre for Detector and Related Software Technology, Department of Physics and Astrophysics, University of Delhi, New Delhi, India

Shankita Bhardwaj Department of Physics and Astronomical Science, School of Physical and Material Sciences, Central University of Himachal Pradesh (CUHP), Dharamshala, Kangra, Himachal Pradesh, India

M. A. Bhat Tata Institute of Fundamental Research, Mumbai, India

A. D. Bhatt Homi Bhabha National Institute, Mumbai, India; Tata Institute of Fundamental Research, Mumbai, India

Disha Bhatia Tata Institute of Fundamental Research, Mumbai, India

Neha Bhatnagar Department of Physics, Banaras Hindu University, Varanasi, India

Sonali Bhatnagar Dayalbagh Educational Institute, Agra, India

V. Bhatnagar Department of Physics, Panjab University, Chandigarh, India; Akal University, Talwandi Sabo, India

Vipin Bhatnagar Physics Department, Panjab University, Chandigarh, India

Aparajita Bhattacharya Department of Physics, Jadavpur University, Kolkata, India

Deb Sankar Bhattacharya Institute of Physics, Bhubaneswar, India

Purba Bhattacharya Weizmann Institute of Science, Rehovot, Israel

Rajarshi Bhattacharya Saha Institute of Nuclear Physics, Kolkata, India

Sudeb Bhattacharya Saha Institute of Nuclear Physics, Kolkata, India

Gautam Bhattacharyya Saha Institute of Nuclear Physics, Kolkata, India

Tanvi Bhavsar Department of Physics, Sardar Patel University, Vallabh Vidyanagar, India

Kalyan Bhuyan Department of Physics, Dibrugarh University, Dibrugarh, Assam, India

Anirban Biswas Harish-Chandra Research Institute, Jhunsi, Allahabad, India; HBNI, Mumbai, India

Aritra Biswas The Institute of Mathematical Sciences (affiliated to HBNI), Chennai, India

Rathijit Biswas Bose Institute, Centre for Astroparticle Physics and Space Science (CAPSS), Kolkata, India

S. Biswas Department of Physics, Rishi Bankimchandra College, Naihati, India; Bose Institute, Department of Physics and CAPSS, Sector V, Kolkata, India; Centre for Astroparticle Physics and Space Science, Bose Institute, Kolkata, India

Kalpana Bora Physics Department, Gauhati University, Guwahati, Assam, India

Happy Borgohain Department of Physics, Tezpur University, Tezpur, India

Gerhard Brandt II. Physikalisches Institut, Georg-August-Universität Göttingen, Göttingen, Germany

Ankita Budhraja Indian Institute of Science Education and Research, Bhopal, Madhya Pradesh, India

Dipankar Chakrabarti Department of Physics, Indian Institute of Technology Kanpur, Kanpur, India

Mainak Chakraborty Centre of Excellence in Theoretical and Mathematical Sciences, SOA University, Bhubaneswar, India

H. C. Chandola Department of Physics (UGC-Centre of Advanced Study), DSB Campus, Kumaun University, Nainital, India

Vinod Chandra Indian Institute of Technology Gandhinagar, Gandhinagar, Gujarat, India

Kanwaljeet S. Channey University of Delhi, New Delhi, India

Sandeep Chatterjee School of Physical Sciences, National Institute of Science Education and Research Bhubaneswar, HBNI, Jatni, India

Suman Chatterjee Tata Institute of Fundamental Research, Mumbai, India

S. Chattopadhyay VECC, Kolkata, India

Subhasis Chattopadhyay Variable Energy Cyclotron Centre, HBNI, Kolkata, India

Utpal Chattopadhyay Department of Theoretical Physics, Indian Association for the Cultivation of Science, Jadavpur, Kolkata, India

R. R. Chaturvedi Applied Physics Department, Sardar Vallabhbhai National Institute of Technology, Gujarat, India

Bhavesh Chauhan Physical Research Laboratory, Ahmedabad, India; Indian Institute of Technology, Gandhinagar, India

B. C. Chauhan Department of Physics and Astronomical Science, School of Physical and Material Sciences, Central University of Himachal Pradesh (CUHP), Dharamshala, Kangra, Himachal Pradesh, India

S. Chauhan Department of Physics, Aligarh Muslim University, Aligarh, India

Satendra Kumar Chauhan Department of Physics, University of Lucknow, Lucknow, India

Ridhi Chawla Panjab University, Chandigarh, India

Rahul Chhabra National Institute of Technology, Jalandhar, India

Sandhya Choubey Homi Bhabha National Institute, Mumbai, India; Harish-Chandra Research Institute, Allahabad, India; Department of Theoretical Physics, School of Engineering Sciences, KTH Royal Institute of Technology, Stockholm, Sweden; AlbaNova University Center, Stockholm, Sweden

Brajesh C. Choudhary Department of Physics and Astrophysics, University of Delhi, New Delhi, India

Subikash Choudhury Bose Institute, Centre for Astroparticle Physics and Space Science (CAPSS), Kolkata, India; Variable Energy Cyclotron Centre, HBNI, Kolkata, India

Sudipto Paul Chowdhury Institute of Physics, Sachivalaya Marg, Bhubaneswar, India

Sabya Sachi Chatterjee Institute of Physics, Bhubaneswar, India; Homi Bhabha National Institute, Mumbai, India

Ruchi Chudasama Nuclear Physics Division, Bhabha Atomic Research Center, Mumbai, India

Paul Colas CEA Saclay, Gif Sur Yvette, France

S. I. Cooper University of Alabama, Tuscaloosa, AL, USA

Harleen Dahiya Department of Physics, Dr. B. R. Ambedkar National Institute of Technology, Jalandhar, India

Ranjeet Dalal Department of Physics and Astrophysics, Centre for Detector and Related Software Technology, University of Delhi, New Delhi, India

Zubair Ahmad Dar Aligarh Muslim University, Aligarh, Uttar Pradesh, India

Diganta Das Physical Research Laboratory, Navrangpura, Ahmedabad, India

Mrinal Kumar Das Department of Physics, Tezpur University, Tezpur, Assam, India

Pallabi Das Tata Institute of Fundamental Research, Mumbai, India

S. Das Centre for Astroparticle Physics and Space Science, Bose Institute, Kolkata, India

Sabita Das Institute of Particle Physics and Key Laboratory of Quark and Lepton Physics (MOE), Central China Normal University, Wuhan, China

Supriya Das Bose Institute, Centre for Astroparticle Physics and Space Science (CAPSS), Kolkata, India

Ajay Kumar Dash School of Physical Sciences, NISER, HBNI, Jatni, India

N. Dash Indian Institute of Technology, Bhubaneswar, India

Sadhana Dash Department of Physics, Indian Insititute of Technology Bombay, Powai, Mumbai, India

V. M. Datar Tata Institute of Fundamental Research, Mumbai, India; India-based Neutrino Observatory, Tata Institute of Fundamental Research, Mumbai, India

Debabrata Deb Department of Physics, Indian Institute of Engineering Science and Technology, Howrah, West Bengal, India

K. N. Deepthi Physical Research Laboratory, Ahmedabad, India

Frank F. Deppisch Department of Physics and Astronomy, University College London, London, UK

S. Dev Department of Physics School of Sciences, HNBG Central University, Srinagar, Uttarakhand, India

Ram Krishna Dewanjee Laboratory of High Energy and Computational Physics, KBFI, Tallinn, Estonia

Abhishek Dey Maulana Azad College, Government of West Bengal, Kolkata, India

Rajat K. Dey Department of Physics, University of North Bengal, Siliguri, West Bengal, India

Sourav Dey Saha Institute of Nuclear Physics, Kolkata, India

Preeti Dhankher Indian Institute of Technology Bombay, Mumbai, India

Lobsang Dhargyal Institute of Mathematical Sciences, Chennai, India

Nitish Dhingra Panjab University, Chandigarh, India; G.H.G. Khalsa College, Ludhiana, India

Nisha Dhiman Department of Physics, Dr. B. R. Ambedkar National Institute of Technology, Jalandhar, India

Shashi K. Dhiman Department of Physics, Himachal Pradesh University, Shimla, India

Alexander Dierlamm Karlsruhe Institute of Technology, Eggenstein-Leopoldshafen, Germany

Brian Dorney CERN, Geneva, Switzerland

A. K. Dubey VECC, Kolkata, India

Sunil Dutt Department of Physics, Government College for Women, Jammu, J&K, India

D. Dutta Tata Institute of Fundamental Research, Mumbai, India

Nirupam Dutta Variable Energy Cyclotron Centre, Kolkata, India

Suchandra Dutta Saha Institute of Nuclear Physics, Kolkata, India

Andreas Ekstedt Department of Physics and Astronomy, Uppsala University, Uppsala, Sweden

Rikard Enberg Department of Physics and Astronomy, Uppsala University, Uppsala, Sweden

S. Ganesh Department of Physics, Birla Institute of Technology and Science, Pilani, India

Avijit K. Ganguly Department of Physics (MMV), Banaras Hindu University, Varanasi, India

Serguei Ganjour CEA Saclay, Gif Sur Yvette, France

Rajesh Ganai EHEP & A Group, Variable Energy Cyclotron Centre, Kolkata, India; Bose Institute, Centre for Astroparticle Physics and Space Sciences, Kolkata, West Bengal, India; GSI Helmholtzzentrum für Schwerionenforschung GmbH, Darmstadt, Germany; Training School Complex, Homi Bhabha National Institute, Mumbai, India

Kajal Garg Dayalbagh Educational Institute, Agra, India

Rocky Bala Garg Department of Physics and Astrophysics, University of Delhi, New Delhi, India

Ankit Gaur Department of Physics and Astrophysics, University of Delhi, New Delhi, India

Radha Raman Gautam Department of Physics, Himachal Pradesh University, Shimla, India

C. Ghosh VECC, Kolkata, India

Gayatri Ghosh Physics Department, Gauhati University, Guwahati, Assam, India

Kaushik Ghosh Vivekananda College, University of Calcutta, Kolkata, India

Monojit Ghosh Department of Physics, Tokyo Metropolitan University, Hachioji, Tokyo, Japan

S. Ghosh PG Department of Physics, Vijaya College, Bengaluru, India

Sabyasachi Ghosh Department of Physics, University of Calcutta, Kolkata, India; School of Physical Sciences, National Institute of Science Education and Research Bhubaneswar, HBNI, Jatni, India

Shamik Ghosh Saha Institute of Nuclear Physics, HBNI, Kolkata, India

Rohini Godbole Centre for High Energy Physics, Indian Institute of Science, Bengaluru, India

Mohit Gola Department of Physics and Astrophysics, University of Delhi, New Delhi, India

Srubabati Goswami Physical Research Laboratory, Ahmedabad, India

Marek Michal Gruchala Ghent University, Ghent, Belgium

B. K. Guha Department of Physics, Indian Institute of Engineering Science and Technology, Howrah, West Bengal, India

Anshu Gupta GD Goenka University, Gurugram, Haryana, India

K. K. Gupta Department of Physics, Ramjas College, University of Delhi, New Delhi, India

Pallavi Gupta School of Physics and Material Science, Thapar University, Patiala, Punjab, India

Rajat Gupta Panjab University, Chandigarh, India

Shashikant Gupta GD Goenka University, Gurugram, Haryana, India

H. Haider Department of Physics, Aligarh Muslim University, Aligarh, India

Chandan Hati Theoretical Physics, Physical Research Laboratory, Ahmedabad, India

Chandan Gupta Physical Research Laboratory, Ahmedabad, India; Homi Bhabha National Institute, Mumbai, India

P. V. Hunagund Department of Applied Electronics and Research, Gulbarga University, Gulbarga, India

S. S. R. Inbanathan Department of physics, The American College, Madurai, Tamilnadu, India

D. Indumathi The Institute of Mathematical Sciences, CIT Campus, Chennai, India; Homi Bhabha National Institute, Mumbai, India

Gunnar Ingelman Department of Physics and Astronomy, Uppsala University, Uppsala, Sweden

Mohammad Irfan Department of Physics, Indian Institute of Science Education and Research, Bhopal, India

Manoj Jadhav Indian Institute of Technology Bombay, Mumbai, India

Ambar Jain Indian Institute of Science Education and Research, Bhopal, Madhya Pradesh, India

Chakresh Jain Centre for Detector and Related Software Technology, Department of Physics and Astrophysics, University of Delhi, New Delhi, India

Geetika Jain Centre for Detector and Related Software Technology, Department of Physics and Astrophysics, University of Delhi, New Delhi, India

Manoj K. Jaiswal Department of Physics (MMV), Banaras Hindu University, Varanasi, India

Bharti Jarwal University of Delhi, New Delhi, India

Abhik Jash Saha Institute of Nuclear Physics, Kolkata, India; HBNI, Mumbai, India

Agam K. Jha Department of Physics, Kirori Mal College, University of Delhi, New Delhi, India

T. K. Nayak Variable Energy Cyclotron Centre, HBNI, Kolkata, India

Rupam Kalita Department of Physics, Tezpur University, Tezpur, India

S. D. Kalmani Tata Institute of Fundamental Research, Mumbai, India

Amandeep Kaur Kalsi Panjab University, Chandigarh, India

R. Kanishka Physics Department, Panjab University, Chandigarh, India

Supriya Kar Department of Physics & Astrophysics, University of Delhi, Delhi, India

Siddhartha Karmakar Discipline of Physics, Indian Institute of Technology Indore, Simrol, Indore, India

Anterpreet Kaur Panjab University, Chandigarh, India

Ashpreet Kaur Department of Physics, Panjab University, Chandigarh, India

Daljeet Kaur S.G.T.B. Khalsa College, University of Delhi, New Delhi, India

M. Kaur Department of Physics, Panjab University, Chandigarh, India

Navdeep Kaur Department of Physics, Dr. B. R. Ambedkar National Institute of Technology, Jalandhar, India

P. Kaur Tata Institute of Fundamental Research, Mumbai, India

Puneet Kanwar Kaur Tata Institute of Fundamental Research, Mumbai, India

Sandeep Kaur Panjab University, Chandigarh, India

Abhiram Kaushik Centre for High Energy Physics, Indian Institute of Science, Banglore, India

Kiyotomo Kawagoe Kyushu University, Nishi-ku Fukuoka, Japan

Shuaib Ahmad Khan Variable Energy Cyclotron Centre, HBNI, Kolkata, India

Amina Khatun Institute of Physics, Bhubaneswar, India; Homi Bhabha National Institute, Training School Complex, Mumbai, India

Sarif Khan Harish-Chandra Research Institute, Jhunsi, Allahabad, India; HBNI, Mumbai, India

Bharti Kindra Physical Research Laboratory, Ahmedabad, India; Indian Institute of Technology, Gandhinagar, India

Partha Konar Physical Research Laboratory (PRL), Ahmedabad, Gujarat, India

A. Kumar Department of Physics, Panjab University, Chandigarh, India; VECC, Kolkata, India

Achint Kumar Department of Physics, Birla Institute of Technology and Science, Pilani, Goa, India

Ajay Kumar Indian Institute of Technology Indore, Indore, India

Vineet Kumar Nuclear Physics Division, Bhabha Atomic Research Center, Mumbai, India

Arvind Kumar National Institute of Technology, Jalandhar, India

Ashok Kumar Department of Physics and Astrophysics, University of Delhi, New Delhi, India; Panjab University, Chandigarh, India

Dinesh Kumar Indian Institute of Technology, Bombay, India

Girish Kumar Physical Research Laboratory, Ahmedabad, India

Jitendra Kumar Indian Institute of Technology Bombay, Mumbai, India

Lokesh Kumar Department of Physics, Panjab University, Chandigarh, India

Narinder Kumar Department of Physics, Dr. B. R. Ambedkar National Institute of Technology, Jalandhar, India

Purnendu Kumar Indian Institute of Technology Madras, Chennai, India

Rajeev Kumar Punjab Agricultural University, Ludhiana, Punjab, India

Sachin Kumar Theoretical Physics Section, Bhabha Atomic Research Centre, Trombay, India

Sanjeev Kumar Department of Physics and Astrophysics, University of Delhi, New Delhi, India

Shyam Kumar Department of Physics, Indian Institute of Technology Bombay, Mumbai, India

Sumit Kumar Department of Physics, Panjab University, Chandigarh, India

Sushil Kumar Department of Physics and Electronics, Hans Raj College, University of Delhi, New Delhi, India

Yogesh Kumar Department of Physics, Deshbandhu College, University of Delhi, New Delhi, India

Ajay Kumar Rai Department of Applied Physics, S.V.N.I.T., Surat, Gujarat, India

Priyanka Kumari Panjab University, Chandigarh, India

Suman Kumbhakar Indian Institute of Technology, Bombay, India

Sourav Kundu National Institute of Science Education and Research, HBNI, Jatni, India

S. M. Lakshmi Physical Research Laboratory, Ahmedabad, India

A. Lokapure Tata Institute of Fundamental Research, Mumbai, India

Johan Löfgren Department of Physics and Astronomy, Uppsala University, Uppsala, Sweden

Namit Mahajan Physical Research Laboratory, Ahmedabad, India

Neelu Mahajan Department of Physics, G.G.D.S.D. College, Chandigarh, India

B. Mahakud Tata Institute of Fundamental Research, Mumbai, India

Anwesha Mahapatra Department of Physics, Central University of Karnataka, Kalaburagi, India

Priya Maji Department of Physics, National Institute of Technology, Durgapur, West Bengal, India

Tanmay Maji Department of Physics, Indian Institute of Technology Kanpur, Kanpur, India

Nayana Majumdar Saha Institute of Nuclear Physics, HBNI, Kolkata, India

G. Majumder Department of High Energy Physics, Tata Institute of Fundamental Research, Mumbai, India

Gobinda Majumder Tata Institute of Fundamental Research, Mumbai, India

Shivali Malhotra Department of Physics and Astrophysics, University of Delhi, New Delhi, India

Chandrima Mallick Department of Physics, Central University of Karnataka, Kalaburagi, India

Bhabani Prasad Mandal Banaras Hindu University, Varanasi, India

M. Mandal Department of Physics, Government General Degree College, Burdwan, India

Rusa Mandal The Institute of Mathematical Sciences, Tharamani, Chennai, India

Sanjoy Mandal The Institute of Mathematical Sciences, Tharamani, Chennai, India

Tanumoy Mandal Department of Physics and Astronomy, Uppsala University, Uppsala, Sweden

Manisha Department of Physics, Panjab University, Chandigarh, India

A. Manna Electronics Division, Bhabha Atomic Research Centre, Trombay, Mumbai, India

N. Marimuthu Physics Department, Institute of Science, Banaras Hindu University, Varanasi, India; Department of Physics, The American College, Madurai, Tamilnadu, India

Mehedi Masud AHEP Group, IFIC—CSIC/Universitat de València, Parc Cientific de Paterna, Paterna, València, Spain; Harish-Chandra Research Institute, Allahabad, India

Barilang Mawlong School of Physics, University of Hyderabad, Hyderabad, India

K. Mazumdar Tata Institute of Fundamental Research, Mumbai, India

Kajari Mazumdar Tata Institute of Fundamental Research, Mumbai, India

Greeshma Koyithatta Meethaleveedu Indian Institute of Technology Bombay, Mumbai, India

K. K. Meghna School of Physical Sciences, National Institute of Science Education and Research, HBNI, Jatni, Odisha, India

Ankita Mehta Panjab University, Chandigarh, India

Poonam Mehta School of Physical Sciences, Jawaharlal Nehru University, New Delhi, India

Shaifali Mehta School of Physics and Material Sciences, Thapar University, Patiala, Punjab, India

Jeremie Alexandre Merlin CERN, Geneva, Switzerland

Debadeepti Mishra National Institute of Science Education and Research, HBNI, Jatni, Odisha, India

M. Mishra Department of Physics, Birla Institute of Technology and Science, Pilani, India

S. Mishra Department of Physics, Central University of Karnataka, Kalaburagi, India

Anuradha Misra Department of Physics, University of Mumbai, Mumbai, India

Jubin Mitra Variable Energy Cyclotron Centre, HBNI, Kolkata, India

S. Mitra Indian Institute of Technology Gandhinagar, Gandhinagar, Gujarat, India

Sukanya Mitra Indian Institute of Technology Gandhinagar, Gandhinagar, Gujarat, India

Tanmoy Modak NTU, Taipei, Taiwan

Lakshmi S. Mohan Physical Research Laboratory, Ahmedabad, India

R. Mohanta University of Hyderabad, Hyderabad, India

Rukmani Mohanta School of Physics, University of Hyderabad, Hyderabad, India

B. P. Mohanty Department of Biophysics, Panjab University, Chandigarh, India

Bedangadas Mohanty School of Physical Sciences, National Institute of Science Education and Research Bhubaneswar, HBNI, Jatni, India

G. B. Mohanty Tata Institute of Fundamental Research, Mumbai, India

Subhendra Mohanty Physical Research Laboratory, Ahmedabad, India

Swetalin Mohapatra Department of Physics, Central University of Karnataka, Kalaburagi, India

Kuntal Mondal Saha Institute of Nuclear Physics, Kolkata, India

Mitali Mondal Variable Energy Cyclotron Centre, Kolkata, India

N. K. Mondal Tata Institute of Fundamental Research, Mumbai, India; Saha Institute of Nuclear Physics, Bidhannagar, Kolkata, India

S. Mondal Tata Institute of Fundamental Research, Mumbai, India

Surya Mondal Tata Institute of Fundamental Research, Mumbai, India

Suryanarayan Mondal Homi Bhabha National Institute, Mumbai, India

Tanmoy Mondal Physical Research Laboratory (PRL), Ahmedabad, Gujarat, India; Regional Centre for Accelerator-based Particle Physics, Harish-Chandra Research Institute, HBNI, Jhusi, Allahabad, India

Ananya Mukherjee Department of Physics, Tezpur University, Tezpur, Assam, India

Supratik Mukhopadhyay Saha Institute of Nuclear Physics, HBNI, Kolkata, India

D. Nag Centre for Astroparticle Physics and Space Science, Bose Institute, Kolkata, India

P. Nagaraj Tata Institute of Fundamental Research, Mumbai, India

Bharati Naik Indian Institute of Technology Bombay, Mumbai, India

Z. Naik Sambalpur University, Burla, India

Md. Naimuddin Department of Physics and Astrophysics, University of Delhi, Delhi, India

Basanta. K. Nandi Department of Physics, Indian Insititute of Technology Bombay, Powai, Mumbai, India

Basanta Kumar Nandi Indian Institute of Technology Bombay, Mumbai, Powai, India

Ekata Nandy Variable Energy Cyclotron Centre, Kolkata, India

Newton Nath Physical Research Laboratory, Ahmedabad, India; Indian Institute of Technology, Ahmedabad, India

Kishora Nayak National Institute of Science Education and Research, HBNI, Jatni, India

Lopamudra Nayak Department of Physics, Central University of Karnataka, Kalaburagi, India

Ranjit Nayak Department of Physics, Indian Insititute of Technology Bombay, Powai, Mumbai, India

Varun Neelamana Department of Physics, Central University of Karnataka, Kalaburagi, India

Neha India-based Neutrino Observatory, Tata Institute of Fundamental Research, Mumbai, India; Homi Bhabha National Institute, Mumbai, India

M. Nizam Homi Bhabha National Institute, Mumbai, India; Tata Institute of Fundamental Research, Mumbai, India

Antonio Palazzo Dipartimento Interateneo di Fisica "Michelangelo Merlin", Bari, Italy; Istituto Nazionale di Fisica Nucleare, Sezione di Bari, Bari, Italy

Ashutosh Kumar Pandey Department of Physics, Indian Insititute of Technology Bombay, Powai, Mumbai, India

H. C. Pandey Department of Physics, Birla Institute of Applied Sciences, Bhimtal, India

Sharda Pandey Department of Physics, University of Lucknow, Lucknow, India

Vipul Kumar Pandey Banaras Hindu University, Varanasi, India

J. N. Pandya Applied Physics Department, Faculty of Technology and Engineering, The Maharaja Sayajirao University of Baroda, Gujarat, India

Tanmay Pani Indian Institute of Technology Bombay, Mumbai, India

Sonia Parmar Panjab University, Chandigarh, India

Bhavin Patel Scarborough, ON, Canada

Shesha D. Patel Department of Physics, Faculty of Science, The Maharaja Sayajirao University of Baroda, Vadodara, Gujarat, India

E. Pathaleswar India-based Neutrino Observatory, Tata Institute of Fundamental Research, Mumbai, India

Pathaleswar Tata Institute of Fundamental Research, Mumbai, India

Rajendra Nath Patra Variable Energy Cyclotron Centre, HBNI, Kolkata, India

Sudhanwa Patra Center of Excellence in Theoretical and Mathematical Sciences, Siksha 'O' Anusandhan University, Bhubaneswar, India

Riley Patrick Center of Excellence for Particle Physics at the Terascale, University of Adelaide, Adelaide, SA, Australia

Mario Pelliccioni Istituto Nazionale di Fisica Nucleare (INFN), Turin, Italy

S. Pethuraj Homi Bhabha National Institute, Mumbai, India

Aman Phogat Department of Physics and Astrophysics, University of Delhi, New Delhi, India

M. Pooja Department of Physics, Central University of Karnataka, Kalaburagi, India

P. Poulose Department of Physics, Indian Institute of Technology Guwahati, Guwahati, Assam, India

Anil Prabhakar Indian Institute of Technology Madras, Chennai, India

S. K. Prasad Bose Institute, Centre for Astroparticle Physics and Space Science (CAPSS), Kolkata, India

K. Priyabrat Pandey Inderaprastha University, New Delhi, India

Arnab Purohit Saha Institute of Nuclear Physics, HBNI, Kolkata, West Bengal, India

Moh. Rafik Department of Physics and Astrophysics, University of Delhi, New Delhi, India

Farook Rahaman Department of Mathematics, Jadavpur University, Kolkata, West Bengal, India

A. K. Rai Applied Physics Department, Sardar Vallabhbhai National Institute of Technology, Gujarat, India

Desh Raj Department of Physics, Himachal Pradesh University, Shimla, India; Department of Applied Science and Humanities, J. N. Government Engineering College, Sundernagar, India

Sonia Rajput University of Jammu, Jammu, India

S. Kalyana Rama The Institute of Mathematical Sciences, Chennai, India

R. Ramanathan Department of Physics and Astrophysics, University of Delhi, New Delhi, India

Aditee Rane Indian Institute of Science, Education and Research, Pune, India

Raghavan Rangarajan Physical Research Laboratory, Ahmedabad, India

Kirti Ranjan Centre for Detector and Related Software Technology, Department of Physics and Astrophysics, University of Delhi, New Delhi, India

K. C. Ravindran Tata Institute of Fundamental Research, Mumbai, India

K. Raveendrababu Physics Department, Indian Institute of Technology Madras, Chennai, Tamil Nadu, India; Physical Science Division, Homi Bhabha National Institute, Mumbai, India

Deependra Singh Rawat Department of Physics (UGC-Centre of Advanced Study), DSB Campus, Kumaun University, Nainital, India

Vaibhav Rawoot Department of Physics, University of Mumbai, Mumbai, India

Saibal Ray Department of Physics, Government College of Engineering and Ceramic Technology, Kolkata, West Bengal, India

Amitava Raychaudhuri Department of Physics, University of Calcutta, Kolkata, India

Jorge C. Romão CFTP, Lisboa, Portugal

Jogesh Rout School of Physical Sciences, Jawaharlal Nehru University, New Delhi, India

Prasant Kumar Rout Saha Institute of Nuclear Physics, HBNI, Kolkata, India

Sourav Roy Chowdhury Department of Physics, Indian Institute of Engineering Science and Technology, Howrah, West Bengal, India

Suvankar Roy Chowdhury Saha Institute of Nuclear Physics, Kolkata, India

Ankhi Roy Indian Institute of Technology Indore, Indore, India

Ashim Roy Saha Institute of Nuclear Physics, HBNI, Kolkata, India

P. K. Roy High Energy Nuclear and Particle Physics Division, Saha Institute of Nuclear Physics, Kolkata, India

Probir Roy Centre for Astroparticle Physics and Space Science, Bose Institute, Kolkata, India

Victor Roy National Institute of Science Education and Research, Odisha, India

P. Rumerio University of Alabama, Tuscaloosa, AL, USA

Divya Sachdeva Department of Physics and Astrophysics, University of Delhi, New Delhi, India

Soumya Sadhukhan PRL, Ahmedabad, India

Jafar Sadiq Department of Physics, Indian Institute of Technology Madras, Chennai, India

Abhijit Kumar Saha Department of Physics, Indian Institute of Technology Guwahati, Guwahati, Assam, India

Satyajit Saha Saha Institute of Nuclear Physics, Kolkata, India

A. R. Sahasransu IISER, Mohanpur, West Bengal, India

Sarita Sahoo Institute of Physics, HBNI, Bhubaneswar, India

Shibananda Sahoo Department of Physics, Indian Institute of Technology Guwahati, Guwahati, Assam, India

Suchismita Sahoo School of Physics, University of Hyderabad, Hyderabad, India

Sukadev Sahoo Department of Physics, National Institute of Technology, Durgapur, West Bengal, India

P. K. Sahu Institute of Physics, HBNI, Bhubaneswar, India

J. Saini VECC, Kolkata, India

Lalit Saini DAV University, Jalandhar, India

Rome Samanta Saha Institute of Nuclear Physics, HBNI, Kolkata, India

Deepak Samuel Department of Physics, Central University of Karnataka, Kalaburagi, India

S. Sandilya University of Cincinnati, Cincinnati, OH, USA

M. N. Saraf Department of High Energy Physics, Tata Institute of Fundamental Research, Mumbai, India

Kapil Saraswat Department of Physics, Banaras Hindu University, Varanasi, India

Debojit Sarkar Variable Energy Cyclotron Centre, HBNI, Kolkata, India

Sandip Sarkar Saha Institute of Nuclear Physics, Kolkata, India

Subir Sarkar Saha Institute of Nuclear Physics, Kolkata, India

Swarnendu Sarkar Department of Physics and Astrophysics, University of Delhi, New Delhi, India

Tamal Sarkar USIC, University of North Bengal, Siliguri, WB, India

Utpal Sarkar Department of Physics, Indian Institute of Technology Kharagpur, Kharagpur, India

Balachandran Sathiapalan The Institute of Mathematical Sciences, Chennai, India

B. Satyanarayana Department of High Energy Physics, Tata Institute of Fundamental Research, Mumbai, India

Manibrata Sen Tata Institute of Fundamental Research, Mumbai, India

Varun Sethi Department of Physics and Astrophysics, University of Delhi, New Delhi, India

Priyanka Sett Department of Physics, Indian Insititute of Technology Bombay, Powai, Mumbai, India

Aashaq Shah Department of Physics and Astrophysics, University of Delhi, New Delhi, India

Manan Shah P. D. Patel Institute of Applied Sciences, Charusat, Changa, India

Zalak Shah Department of Applied Physics, S.V.N.I.T., Surat, Gujarat, India

J. S. Shahi Department of Physics, Panjab University, Chandigarh, India

Amit Sharma Amity University Gurugram, Gurugram, Haryana, India

Ankita Sharma University of Jammu, Jammu, India

Archana Sharma CERN, Geneva, Switzerland

Ashish Sharma Department of Physics and Astronomical Science, School of Physical and Material Sciences, Central University of Himachal Pradesh (CUHP), Dharamshala, Kangra, Himachal Pradesh, India

Gazal Sharma Department of Physics and Astronomical Science, School of Physical and Material Sciences, Central University of Himachal Pradesh (CUHP), Dharamshala, Kangra, Himachal Pradesh, India

Natasha Sharma Department of Physics, Panjab University, Chandigarh, India

Pankaj Sharma Department of Physics, ARC Center of Excellence for Particle Physics at the Terascale, University of Adelaide, Adelaide, SA, Australia

Ram Krishna Sharma University of Delhi, New Delhi, India

S. Sharma Department of Physics, Panjab University, Chandigarh, India

Satendra Sharma Yobe State University, Damaturu, Nigeria

Varun Sharma Department of Physics and Astrophysics, University of Delhi, New Delhi, India

R. R. Shinde Department of High Energy Physics, Tata Institute of Fundamental Research, Mumbai, India

Mehulkumar Shiroya Sardar Vallabhbhai National Institute of Technology, Surat, Gujarat, India

Prashant Shukla Nuclear Physics Division, Bhabha Atomic Research Centre, Mumbai, India; Homi Bhabha National Institute, Mumbai, India

Arunansu Sil Department of Physics, Indian Institute of Technology Guwahati, Guwahati, Assam, India

D. Sil Department of High Energy Physics, Tata Institute of Fundamental Research, Mumbai, India

Dipankar Sil Tata Institute of Fundamental Research, Mumbai, India

João P. Silva CFTP, Lisboa, Portugal

R. N. Singaraju Variable Energy Cyclotron Centre, HBNI, Kolkata, India

Abhishek K. Singh Rajdhani College, University of Delhi, New Delhi, India

Amandeep Singh Panjab University, Chandigarh, India

Captain R. Singh Department of Physics, Birla Institute of Technology and Science, Pilani, India

Govind Singh Department of Physics and Astronomical Science, School of Physical and Material Sciences, Central University of Himachal Pradesh (CUHP), Dharamshala, Kangra, Himachal Pradesh, India

J. B. Singh Akal University, Talwandi Sabo, India

Janardan P. Singh Faculty of Science, Department of Physics, The Maharaja Sayajirao University of Baroda, Vadodara, Gujarat, India

Jaydip Singh Department of Physics, Lucknow University, Lucknow, India

Jyotsna Singh Department of Physics, University of Lucknow, Lucknow, India

Lakhwinder Singh Institute of Physics, Academia Sinica, Taipei, Taiwan

Meghendra Singh Dr. A.P.J. Abdul Kalam Technical University, Lucknow, Uttar Pradesh, India

R. B. Singh Department of Physics, University of Lucknow, Lucknow, India

S. K. Singh Department of Physics, Aligarh Muslim University, Aligarh, India

S. S. Singh Department of Physics and Astrophysics, University of Delhi, New Delhi, India

S. Somorendro Singh Department of Physics and Astrophysics, University of Delhi, New Delhi, India

Soram Robertson Singh School of Physics, University of Hyderabad, Hyderabad, India

Sunita Singh Miranda House College, University of Delhi, New Delhi, India

V. Singh Physics Department, Institute of Science, Banaras Hindu University, Varanasi, India

Venktesh Singh Department of Physics, Institute of Science, Banaras Hindu University, Varanasi, India

Shivaramakrishna Singirala School of Physics, University of Hyderabad, Hyderabad, India

Roopam Sinha Saha Institute of Nuclear Physics, Kolkata, India

Nilanjan Sircar School of Physics and Centre for Theoretical Physics, National Institute for Theoretical Physics, University of the Witwatersrand, Wits, Johannesburg, South Africa

Bipin Sonawane Department of Physics, University of Mumbai, Mumbai, India

N. R. Soni Applied Physics Department, Faculty of Technology and Engineering, The Maharaja Sayajirao University of Baroda, Gujarat, India

C. Soumya University of Hyderabad, Hyderabad, India

K. Sridhar Department of Theoretical Physics, Tata Institute of Fundamental Research, Mumbai, India

Ajit M. Srivastava Institute of Physics, Bhubaneswar, India

Rahul Srivastava IFIC, Valencia, Spain

M. Sruthilaya University of Hyderabad, Hyderabad, India

Abhaya Kumar Swain Physical Research Laboratory (PRL), Ahmedabad, Gujarat, India

Tarak Thakore Louisiana State University, Baton Rouge, LA, USA

S. Thakur Department of Physics, Panjab University, Chandigarh, India

Suman Thakur Department of Physics, Himachal Pradesh University, Shimla, India

R. Thiru Senthil The Institute of Mathematical Sciences, Chennai, India

S. H. Thoker University of Kashmir, Srinagar, India

Tobias Toll Shiv Nadar University, Greater Noida, UP, India

Sridhar Tripathy Saha Institute of Nuclear Physics, Kolkata, India

S. K. Tripathy Institute of Physics, Bhubaneswar, India; Sambalpur University, Burla, India

S. Uma Sankar Indian Institute of Technology Bombay, Mumbai, India

S. S. Upadhya Tata Institute of Fundamental Research, Mumbai, India

A. Upadhyay School of Physics and Material Science, Thapar University, Patiala, Punjab, India

José W. F. Valle AHEP Group, Institut de Fisica Corpuscular—C.S.I.C./Universitat de Valencia, Parc Cientific de Paterna, Paterna (Valencia), Spain

Raghava Varma Department of Physics, Indian Institute of Technology Bombay, Mumbai, India

Massimo Venaruzzo CAEN S.p.A., Viareggio (LU), Italy

P. Verma Tata Institute of Fundamental Research, Mumbai, India

S. Verma Department of Physics, Panjab University, Chandigarh, India

Surender Verma Department of Physics and Astronomical Science, School of Physical and Material Sciences, Central University of Himachal Pradesh (CUHP), Dharamshala, Kangra, Himachal Pradesh, India

P. C. Vinodkumar Department of Physics, Sardar Patel University, Vallabh Vidyanagar, India

Y. P. Viyogi Variable Energy Cyclotron Centre, HBNI, Kolkata, India

Genius Walia Panjab University, Chandigarh, India

Anthony G. Williams Center of Excellence for Particle Physics at the Terascale, University of Adelaide, Adelaide, SA, Australia

H. T. Wong Institute of Physics, Academia Sinica, Taipei, Taiwan

Nu Xu Key Laboratory for Quark and Lepton Physics (MOE) and Institute of Particle Physics, Central China Normal University, Wuhan, China; Nuclear Science Division, Lawrence Berkeley National Laboratory, Berkeley, CA, USA

D. Yadav Department of Physics (UGC-Centre of Advanced Study), DSB Campus, Kumaun University, Nainital, India

M. Younus Institute of Physics, HBNI, Bhubaneswar, India

M. Yousuf Jamal Indian Institute of Technology Gandhinagar, Gandhinagar, Gujarat, India

E. Yuvaraj Department of High Energy Physics, Tata Institute of Fundamental Research, Mumbai, India

Chapter 1
Exploration of the QCD Phase Diagram at Finite Baryon Density Region: Recent Results from RHIC Beam Energy Scan-I

Nu Xu

1. Introduction: The goal of high-energy nuclear collisions is to study the properties of the QCD phase diagram. Specifically one wants to first discover the new form of matter Quark-Gluon Plasma (QGP) and the transition from QGP to normal hadronic region. Early experimental results from RHIC $\sqrt{s_{NN}} = 200\,\text{GeV}$ Au+Au and LHC $\sqrt{s_{NN}} = 2.76 - 5.02\,\text{TeV}$ Pb+Pb collisions have led to the conclusion of the formation of strongly-coupled Quark-Gluon Plasma (sQGP) at vanishing net-baryon density in such collisions.[1] Regarding the nature of the transition, Lattice Gauge Theory calculations have demonstrated that the transition from sQGP to hadronic region at $\mu_B \sim 0$ is a smooth-crossover [1], indicated as thin-dashed-line in Fig. 1.1. Note that so far no direct experimental evidence for the smooth-crossover transition. On the other hand, in large net-baryon density region, it has been speculated that the transition from the sQGP is of the first-order [4] shown as black-line in Fig. 1.1. If both the smooth-crossover at $\mu_B \sim 0$ and the first-order phase transition at large

[1]In high-energy nuclear collisions at LHC $\sqrt{s_{NN}} = 2.76 - 5.02\,\text{TeV}$ and at RHIC $\sqrt{s_{NN}} = 200\,\text{GeV}$, the values of baryonic chemical potential are found to be $\mu_B \sim 0$ and 25 MeV, respectively. All are much smaller than that of the transition temperature $T \approx 155\,\text{MeV}$ [2].

N. Xu (✉)
Key Laboratory for Quark and Lepton Physics (MOE) and Institute of Particle Physics, Central China Normal University, Wuhan 430079, China
e-mail: nxu@lbl.gov

N. Xu
Nuclear Science Division, Lawrence Berkeley National Laboratory, Berkeley, CA 94720, USA

Md. Naimuddin (ed.), *XXII DAE High Energy Physics Symposium*, Springer Proceedings in Physics 203, https://doi.org/10.1007/978-3-319-73171-1_1

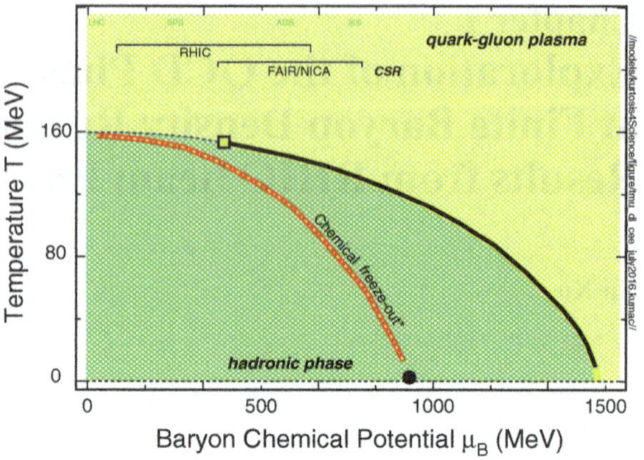

Fig. 1.1 Schematic QCD phase diagram: temperature as a function of baryonic chemical potential. The corresponding energy ranges for different accelerator facilities are indicated in the figure. Red-line: chemical freeze-out [5]; Black-dot at $T = 0$: mass of nucleon; Black-line: speculated first-order phase boundary

net-baryon density region are true, there must be a QCD critical end point,[2] shown as square in Fig. 1.1. Ever since the discover of the sQGP in 2005, scientists have asked: What is the properties and structures of the QCD phase diagram at large net-baryon region? Should the QCD critical point exist, where is it? Aimed at the exciting goal, RHIC carried out its first beam energy scan (BES-I) program that covers the center of mass energy from $\sqrt{s_{NN}} = 7.7$ to 200 GeV corresponding to the baryonic chemical potential from $\mu_B \approx 420$ to 20 MeV in central Au+Au collisions, respectively.

In this short review, I will first discuss selected recent results from RHIC BES-I followed by the discussions on the opportunities in future physics program in order to address the key questions regarding the QCD phase structure including the illusive critical point. I would stress that adequate detector upgrades, focused at the large baryon density region, are essential for the future physics program.

2. Freeze-out Dynamics: In order to set the landscape for future discussion, we first discuss the freeze-out dynamics observed in RHIC BES-I. STAR's recent results of hadron spectra from the BES-I has been submitted for publication [6]. Assuming that the thermal equilibrium is reached in the heavy-ion collisions at RHIC, the extracted chemical freeze-out parameters temperature and baryonic chemical potential are shown in Fig. 1.2. Filled-circles and open-squares are form central (top 5%) and peripheral (60–80%) Au+Au collisions, respectively. The yellow-line represents the systematics of hadron resonance gas (HRG) fit of chemical freeze-out [5] and the green-band is the LGT results [2]. The BES-I alone covers a wide range of the baryon density region: $20 \le \mu_B \le 420$ MeV. At a given collision energy, it appears that the temperature parameter are very similar but a clear centrality dependence of the chemical potential parameter: the more the central collision the larger the potential.

[2]In high-energy nuclear collisions, due to the finite sizes and finite evolution time, precisely speaking there is no critical point but rather a critical region. All experimental efforts should look for the critical 'region' [3].

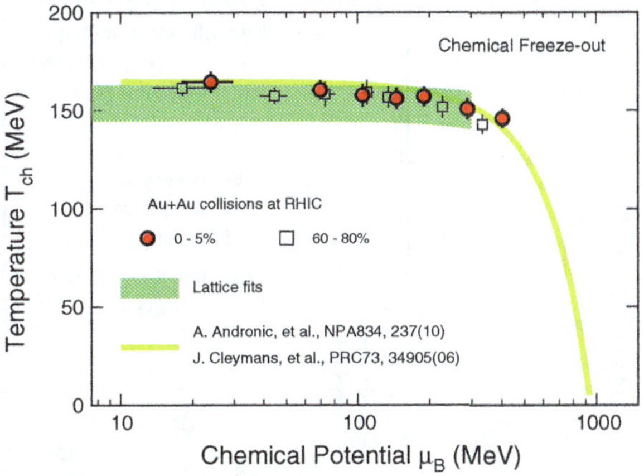

Fig. 1.2 Experimental results of chemical freeze-out temperature as a function of the baryonic chemical potentials from the RHIC BES-I [7]. Red-circles and black-squares represent results from the top 10% and 60–80% Au+Au collisions, respectively. The hatched green-band represents the Lattice results [2]. The yellow-line shows the empirical thermal fits results [5]

At the baryon density $\mu_B \leq 300\,\text{MeV}$, results from LGT and HRG are consistent and they are relatively flat as a function of the μ_B. Starting about $\mu_B \sim 300\,\text{MeV}$, the temperature seems decrease.

3. Search for the QCD Critical Region: Fluctuation of conserved quantities are useful tool for criticality analysis. Especially in high-energy heavy-ion collisions, the size and evolution duration of the created medium are finite so the higher the order of the fluctuation the better sensitivity is expected [9, 10]. The net-proton, net-Kaon and net-charge multiplicity distributions have been used to search for the QCD critical region in RHIC BES-I [7, 11, 12]. Figure 1.3 shows STAR preliminary results of the fourth-order fluctuations $\kappa\sigma^2$ of net-protons (filled-circles), anti-protons (open-triangles) and protons (open-squares) from the most top 5% central Au+Au collisions. As the energy decreases or in another words as the baryon density increases, one can see that the net-proton $\kappa\sigma^2$ starts to decrease and shows a dip at about $\sqrt{s_{NN}} \sim 20\,\text{GeV}$, followed by a sharp increase above unity for collision energy below 10 GeV. Such an oscillation pattern was predicated in a model calculation where the criticality is expected [3]. On the other hand, the UrQMD results show a continuous monotonic decrease, see yellow-band in the figure, presumably mainly due to the baryon number conservation.

Note that when $\sqrt{s_{NN}} \leq 15\,\text{GeV}$ all data points are above unity indicating that the interactions among these protons are attractive. But all known transport models have failed in reproducing the attractive behavior in the high net-baryon density region $\mu_B \geq 300\,\text{MeV}$, even the attractive interaction was turned on [13]. A similar sharp increase of net-proton directed-flow slope parameter $(dv_1/dy|_{y=0})$ has also been observed in the same energy region [14]. Unlike the fluctuation case, the observation indicates a repulsive interaction instead of the attractive one. In [15], the authors

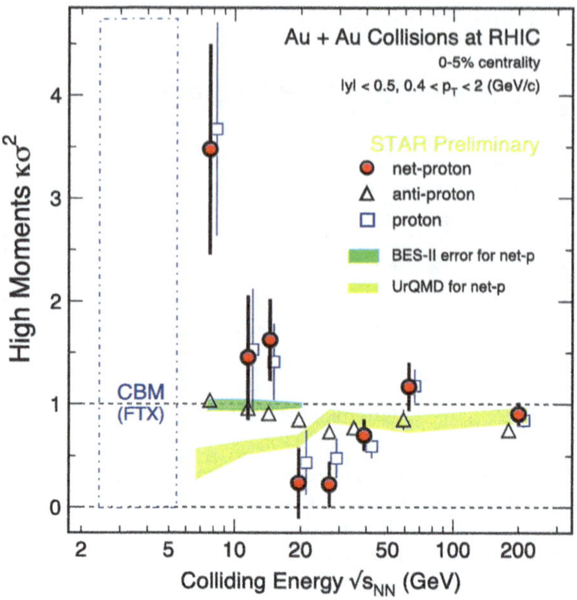

Fig. 1.3 STAR preliminary results [8] on the energy dependence of the fourth-order fluctuations $\kappa\sigma^2$ of net-protons (filled-circles), anti-protons (open-triangles) and protons (open-squares) from the most top 5% central Au+Au collisions. The yellow-band represents the results from UrQMD model calculations while the green-band shows the estimated statistical errors for the net-proton $\kappa\sigma^2$ from RHIC BES-II. Future fixed-target experiment at FAIR will cover the energy region of $\sqrt{s_{NN}} = 2.5$ to 5.5 GeV, shown as dashed-blue-box

have argued that multi-nucleon interactions might be responsible to the observed enhancement in net-proton $\kappa\sigma^2$ at $\sqrt{s_{NN}} \leq 15$ GeV, but no convincing explanations for the data of both net-proton fluctuation and directed-flow yet.

Summary and Outlook: The successful RHIC BES-I program has yielded several interesting results especially the non-monotonical energy dependence of the net-proton fluctuation as well as the directed-flow. In the BES-II, we will focus on the region $7.7 \leq \sqrt{s_{NN}} \sim 20$ GeV and the experimental uncertainties will be reduced dramatically [16]. In order to extract meaningful information on the critical region, we have to understand the baryon dynamic at the even higher density region. There, homogeneous and large acceptance (in both p_T and rapidity y) and high efficiency for proton measurements will be essential [16]. In this energy region, compared to the collider mode of the fixed-target collisions is more efficient. The next generation experiment CBM at FAIR is designed for the purpose [17]. It will cover the crucial energy region: $2 \leq \sqrt{s_{NN}} \sim 5$ GeV ($800 \leq \mu_B \leq 500$ MeV), as indicated by the dashed-blue-box in Fig. 1.3.

Acknowledgements I thank Drs. P. Braun-Munzinger, V. Koch, X.F. Luo, B. Mohanty, J. Stroth and J. Xu, for inspiring discussions. This work was supported in part by the China MoST No.2015CB856901 and US DE-AC03-76SF00098.

References

1. Y. Aoki et al., Nature **443**, 675 (2006)
2. A. Bazavov et al., Phys. Rev. **D85**, 054503 (2012); O. Kaczmarek et al., Phys. Rev. **D83**, 014504 (2011); M. Swagato, private communications (2012)
3. M.A. Stephanov, Phys. Rev. Lett. **107**, 052301 (2011)
4. P. De Forcrand, O. Philipsen, Nucl. Phys. **B642**, 290 (2006)
5. A. Andronic et al., Nucl. Phys. **A834**, 237 (2010); J. Cleymans et al., Phys. Rev. **C73**, 34905 (2006)
6. L. Adamczyk et al., (STAR Collaboration), arXiv:1701.07065
7. N. Xu, (STAR Collaboration), Nucl. Phys. **A931**, 1c (2015)
8. X.F. Luo, (STAR Collaboration), PoS(CPOD2014)019, 2015, arXiv:1503.02558
9. X.F. Luo, N. Xu, arXiv:1701.02105; S. Gupta et al., Science, **332**, 1525 (2011)
10. M.A. Stephanov, Phys. Rev. Lett. **102**, 032301 (2009)
11. L. Adamczyk et al., (STAR Collaboration), Phys. Rev. Lett. **112**, 032302 (2014); ibid. Phys. Rev. Lett. **113**, 092301 (2014)
12. J. Xu et al., (STAR Collaboration), arXiv:1611.07132
13. S. He et al., Phys. Lett. **B762**, 296 (2016)
14. L. Adamczyk et al., (STAR Collaboration), Phys. Rev. Lett. **112**, 162301 (2014)
15. A. Bzdak, V. Koch, V. Skokov, arXiv:1612.05128
16. B. Ling, M.A. Stephanov, Phys. Rev. **C93**, 034915 (2016); L. Adamczyk et al., (STAR Collaboration), arXiv:1609.05102 and references therein
17. T. Ablyazimov et al., (CBM Collaboration), EJPA, arXiv:1607.01487

Chapter 2
Silicon Sensors for Experiments in High Energy Physics

Alexander Dierlamm

2.1 Introduction

High energy physics (HEP) is probing the nature of matter. With large accelerators new particles are generated and huge detection systems identify the type and properties of those to challenge predictions of particle physics theories. The measurements agree very well with current theories, but more accurate measurements or detection of very rare events can help to identify the best descriptions of our universe. That pushes the acceleration systems to higher energy and luminosity. High luminosity comes with a large number of simultaneous tracks to be distinguished and a very harsh radiation environment. The detection systems closest to the interaction points are tracking and vertexing systems, which mainly rely on silicon sensors as sensitive elements and are strongly affected by increasing luminosity.

The upgrade of the LHC to the high-luminosity LHC will push the peak luminosity to $5 - 7.5 \times 10^{34}\,\mathrm{cm}^{-2}\,\mathrm{s}^{-1}$ and the integrated luminosity to $3000\,\mathrm{fb}^{-1}$ [1] for the CMS and ATLAS experiments. This results in an equivalent fluence of up to $2 \times 10^{16} \mathrm{n_{eq}/cm^2}$ for pixel sensors and $1 \times 10^{15} \mathrm{n_{eq}/cm^2}$ for strip sensors (Fig. 2.1). The total ionizing dose reaches 10 MGy and 700 kGy, respectively.

The silicon sensors will suffer from radiation damage due to displacement damage in the bulk and ionization in the silicon oxide at the surface of the bulk silicon. Displacement damage generates vacancies and interstitials, which can form higher order defects combining with themselves or with impurities (e.g. oxygen, carbon, phosphorus, boron) in the crystal lattice. These defects show up as energy levels in the band gap of silicon [2]. Depending on the energy level and the interaction

Alexander Dierlamm for the CMS Collaboration.

A. Dierlamm (✉)
Karlsruhe Institute of Technology, Hermann-von-Helmholtz-Platz 1, 76344 Eggenstein-Leopoldshafen, Germany
e-mail: alexander.dierlamm@kit.edu

Md. Naimuddin (ed.), *XXII DAE High Energy Physics Symposium*, Springer Proceedings in Physics 203, https://doi.org/10.1007/978-3-319-73171-1_2

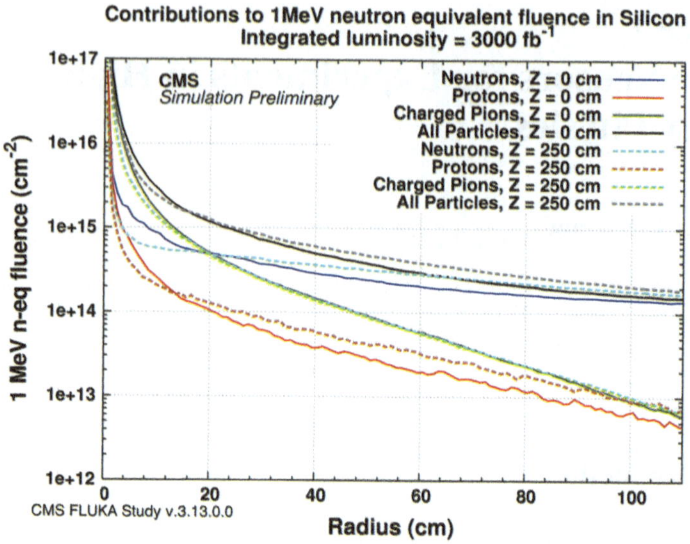

Fig. 2.1 Expected radiation environment for the CMS tracker at the HL-LHC [6]

cross sections these defect levels can act as generation centers for volume generated leakage current, as trapping centers to fix mobile charges for a short period of time (long enough to get lost for fast electronics) and as acceptors/donors modifying the effective space charge and the shape of the electric field. Ionizing radiation leads to a positive charge within the silicon oxide and generation/activation of interface traps which both alter the electric properties close to the surface [3, 4]. One effect for example is the attraction of mobile electrons to the positively charged oxide creating an accumulation layer of electrons, which could short circuit n-doped strips/pixels. This has to be prevented by measures to isolate the strips by p-stop or p-spray techniques [5].

2.2 Silicon Sensor Types

This section discusses silicon sensor types of current interest mainly for the upgrades of the LHC experiments.

2.2.1 Planar Silicon Sensors

Manufacturing of planar silicon sensors is a mature technology, which is being developed and applied for more than 35 years [7]. The sensors are based on structured

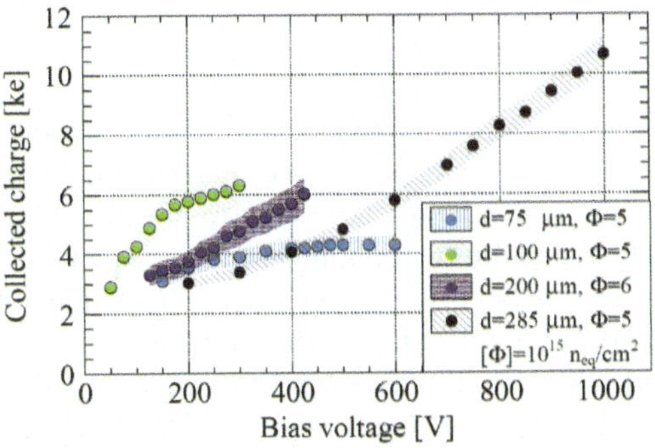

(a) Collected charge vs. bias voltage for pixel sensors of different thicknesses. [22]

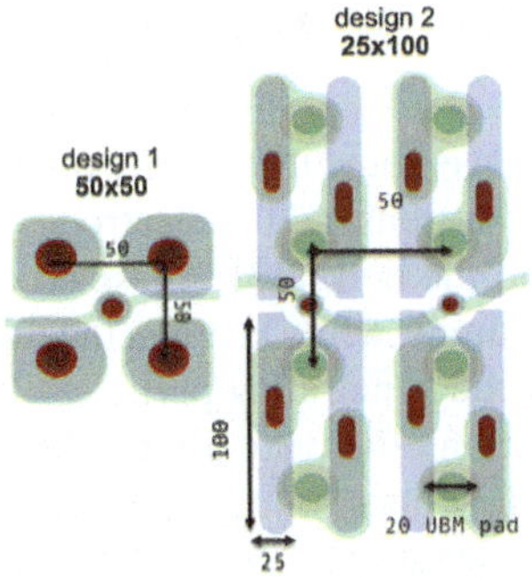

(b) Design of very small pixel cells. Dimensions in μm [23].

Fig. 2.2 Illustration of the main developments for planar silicon sensors: thin devices and small pixel cells (here with punch-through connections for biasing without bump-bonded read-out chip)

pn-junctions operated under reverse bias. Sensors with structured p-type strips/pixels on an n-type bulk (p-in-n) collect holes and have been used in large scale detectors up to now. Trapping is much more severe for this kind of sensor due to the slower drift of holes that results in higher trapping probability. To gain higher radiation resistance (as required for the strip trackers at HL-LHC and vertex detectors) electrons need to be read out [8] and n-in-p or n-in-n sensors are used. For even higher radiation one needs to increase the electric field strength in the sensors. Increasing the operation voltage has a practical limit, therefore thinner sensors are being developed that collect greater charge at moderate voltages compared to thick sensors after high irradiation (Fig. 2.2a).

The initially smaller charge compared to thick sensors is accepted in view of increased longevity, reduced leakage current and lower material budget.

For vertex detectors the high density of tracks requires higher granularity of the sensors and therefore smaller pixel cells (e.g. $2500\,\mu m^2$ cells in Fig. 2.2b). Small cells reduce the individual readout channel noise, which enables the detection of smaller signals after severe radiation damage.

The two general purpose experiments at LHC, ATLAS and CMS, will receive new tracking systems for the HL-LHC phase each equipped with $200\,m^2$ of n-in-p type silicon sensors [9, 10]. Also the vertex detectors will be replaced and equipped with several square meters of planar silicon pixel sensors. These big systems will be exceeded in size by the CMS High Granularity Calorimeter, which will be equipped with $600\,m^2$ of planar silicon pad sensors [11]. LHCb will build a new tracking system with $10\,m^2$ of silicon sensors (upstream tracker) utilizing double-metal layers for signal distribution and front-side biasing for eased module production [12].

2.2.2 3D Sensors

To overcome the penalty of less generated charge when reducing the drift distance in planar sensors, the electrodes were proposed to be manufactured in form of columns into the bulk [13]. These so called 3D sensors allow lateral drift to closely spaced columns, while keeping the thickness high and thus maintaining the charge generated by the penetrating particles. The devices typically show more than 50% charge collection efficiency after severe irradiation ($>5 \times 10^{15} n_{eq}/cm^2$) even when operated at low bias voltages between 150 and 200 V. This sensor type requires specialized and lengthy process steps and is currently offered by three vendors (FBK, CNM and Sintef [14]). This sensor type exhibits an inefficiency for particle detection when the particle traverses one of the columns, which can be overcome by placing the sensors at an angle to the incoming particles. Three-dimensional sensors were first deployed in the ATLAS Inner B-Layer, for which production yields of about 65% were reached [15, 16].

2.2.3 HV-CMOS Sensors

Monolithic Active Pixel Sensors (MAPS) combine sensitive elements and readout electronics and can overcome the limitations of hybrid systems such as connectivity (costly bump-bonding) and very low mass. The largest application of such MAPS is the future ALICE Inner Tracking System [17], which will be equipped with $10\,m^2$ of these detectors. Although MAPS are widely used (from CMOS cameras to the STAR experiment [18]), they suffer a lot from radiation damage and the readout speed is slow compared to the requirements for proton collisions at LHC (40 MHz). A fairly new development is the usage of industrially available HV-CMOS processes to allow biasing of the substrate up to about 150 V leading to a collection of the charge carriers by fast drift rather than slow diffusion as for conventional CMOS processes. Therefore, HV-CMOS sensors are much more radiation hard and can be considered for applications in a much wider field including LHC experiments [19]. Still, there are several challenges to overcome like time resolution, power consumption and production of large sensors, which requires stitching for sizes larger than about $5\,cm^2$. The first application of HV-CMOS sensors in a particle physics experiment is at Mu3e [20, 21].

References

1. http://hilumilhc.web.cern.ch/about/hl-lhc-project
2. M. Moll, Ph.D. thesis, DESY-THESIS-1999-040 (1999)
3. T.R. Oldham, F.B. McLean, IEEE Trans. Nucl. Sci. **50**, 483–499 (2003)
4. J. Zhang, Ph.D. thesis, DESY-THESIS-2013-018 (2013)

5. M. Printz, NIM A **831**, 38–43 (2016) ; R. Dalal et al., JINST **9**, P04007 (2014)
6. CMS Collaboration, CMS-DP-2015-022, http://cds.cern.ch/record/2039908
7. B. Hyams et al., NIM Phys. Res. **205**, 99105 (1983)
8. G. Casse, NIM A **612**, 464–469 (2010)
9. D. Contardo et al., CERN-LHCC-2015-010
10. ATLAS Collaboration, CERN-LHCC-2012-022
11. A.-M. Magnan, JINST **12**, C01042 (2017)
12. LHCb Tracker Upgrade Technical Design Report, CERN-LHCC-2014-001
13. S.I. Parker et al., NIM A **395**, 328–343 (1997)
14. https://www.fbk.eu ; http://www.imb-cnm.csic.es; https://www.sintef.no/en/
15. The ATLAS IBL Collaboration, JINST **7**, P11010 (2012)
16. M. Backhaus, NIM A **831**, 65–70 (2016)
17. P. Yang et al., JINST **10**, C03030 (2015)
18. G. Contin et al., JINST **10**, C03026 (2015)
19. I. Perić et al., NIM A **731**, 131–136 (2013)
20. The Mu3e Experiment, http://www.psi.ch/mu3e/
21. H. Augustin et al., JINST **11**, C11029 (2016)
22. S. Terzo et al., JINST **9**, C12029 (2014)
23. N. Savic et al., NIM A **845**, 154–158 (2017)

Chapter 3
Physics and Detectors at Future Linear Colliders

Kiyotomo Kawagoe

3.1 Future Linear Colliders

The history of particle physics tells us that hadron and lepton colliders are complementary and synergetic. While hadron colliders are good at producing heavy particles with their very high energies, lepton colliders are good at their precision measurements with simple and clean experimental environments. The ATLAS and CMS experiments at the LHC run 1 with center of mass energies 7 and 8 GeV have discovered the Higgs boson during 2012 [1, 2]. They have been searching for new physics with the LHC run 2 that started in 2015 with center of mass energy of 13 TeV, but with no success. The diphoton mass peak around 750 GeV observed in the 2015 data, which generated widespread interest, was not confirmed by the 2016 data with much more statistics [3, 4].

In this situation, where we don't have any apparent signature of new physics at the LHC, precision measurements of the Higgs boson and the top quark at future e^+e^- colliders gain more importance than ever. Through precision measurements one can reveal symmetries and basic principles behind the Higgs boson and the top quark, and possibly obtain information of as-yet undiscovered particles. The hadron and e^+e^- colliders are complementary to each other and the synergy of the two will open a new vista of particle physics.

The future energy-frontier e^+e^- collider will be necessarily linear, not to suffer from energy loss due to synchrotron radiation. The energy loss per turn of a charged particle is proportional to the fourth power of the energy over mass, and inverse of the radius of a circular accelerator. This leads to two possible ways to reduce the energy loss. One is to circulate heavy charged particles, like protons as done at the LHC, and the other is to use an accelerator with a large radius. A linear collider is an accelerator with an infinitely large radius, i.e. with no synchrotron radiation. It is

K. Kawagoe (✉)
Kyushu University, 744 Motooka, Nishi-ku Fukuoka 819-0395, Japan
e-mail: kawagoe@phys.kyushu-u.ac.jp
URL: http://epp.phys.kyushu-u.ac.jp

Md. Naimuddin (ed.), *XXII DAE High Energy Physics Symposium*, Springer Proceedings in Physics 203, https://doi.org/10.1007/978-3-319-73171-1_3

recognized that this is the only option to realize an e^+e^- collider of center-of-mass energy beyond 500 GeV. In addition, linear colliders have two excellent advantages. Firstly, the collision energy can be extended by simply adding LINAC if the tunnel is long enough. Secondly, polarization of electron and/or positron beams can be controlled to select initial states and hence to select interactions of e^+e^- collisions.

There are two future linear colliders in the planning stage. The first one is the ILC (the International Linear Collider) that has been is developed via a global effort. The cold technology was chosen by an international panel in 2004. Since then an international team called GDE, the Global Design Effort, worked on its R&D, and completed ILC TDR (Technical Design Report) back in 2013 [5]. The center of mass energy of the ILC is designed to be 250–500 GeV, which is extendable to more than 1 TeV. As described in the TDR, the key technologies, namely the superconducting RF and beam focusing are well matured. Therefore, the ILC is now in a phase moving from the design to the reality, led by a new organization LCC (the Linear Collider Collaboration) [6].

The other is the CLIC, the Compact Linear Collider, being developed by CERN. This collider is based on two-beam acceleration concept and has a potential energy reach up to 3 TeV or more. Its Conceptual Design Report (CDR) was published in 2012 [7], while the baseline was updated in 2016 with a staging scenario, starting at the center-of-mass energy of 380 GeV [8]. It is understood that the CLIC needs more R&D efforts than the ILC. The CLIC stands among the future programs of CERN, after completion of the HL-LHC (High Luminosity LHC) project [9].

In the following, we focus on the ILC. After the Higgs discovery and completion of the ILC TDR, the Japan Association of High Energy Physicists (JAHEP) proposed to host the ILC in Japan. The Kitakami mountain area in the north part of Japan's main island was evaluated as the most possible candidate site, as it has long and stable granite bedrocks without faults, being ideal for the construction and operation of the ILC. Whether to host the ILC in Japan is now officially discussed by the ILC advisory panel established in 2014 in MEXT (Ministry of Education, Culture, Sports, Science and Technology) of the Japanese Government.

3.2 Physics at the ILC

The physics case for the ILC was described in the ILC TDR [5] Volume 2 as well as in [10]. First of all, the ILC is recognized as a Higgs factory, where the Higgs boson will be copiously produced with small background contamination. The precise Higgs measurement at the ILC is aimed at finding the direction of physics beyond the standard model through it. The measurement will be performed in three steps. The Higgsstrahlung process is dominant at $\sqrt{s} = 250\,\text{GeV}$, where various decay modes, including the invisible one, can be measured very precisely. At higher center of mass energies, we can measure the Higgs boson production via vector-boson fusion processes. This measurement will enable us to determine the total decay width of the Higgs boson, and to normalize relative branching ratios, resulting in determination of absolute Higgs couplings. At center of mass energy of 500 GeV, it would be possible to measure the Higgs self-coupling, which is key to understand the Higgs potential.

Precisions of the Higgs measurements at the ILC with an operating scenario of 20 years [11] are typically by factor 5–10 better than those at the HL-LHC for most of the decay modes, while synergy between the HL-LHC and ILC is remarkable for the measurement of $H\gamma\gamma$ coupling. We will be able to not only discover new physics via the deviation from the standard model, but also to distinguish models of new physics with specific patterns of deviations.

Discovery of the triple Higgs coupling is very important, as this leads to a direct evidence for vacuum condensation. The measurement is very tough even at the ILC, because of the low cross section and multi-jet final states, coupled with presence of interfering diagrams. Results of previous simulation studies show precisions of the cross sections and coupling typically of several 10%. Efforts are being made to improve the analysis for a 10% level measurement.

The top mass can be very precisely measured, at a level of 50 MeV, by the $t\bar{t}$ threshold scan. The mass measured in this way will provide an important input to check the stability of vacuum, or the fate of our universe. Polarized beams are essential to precisely measure the $Zt\bar{t}$ left-right couplings, which would enable us to distinguish various composite models.

If the dark matter particle χ is lighter than the half Higgs mass, it can be searched for by studying the process $e^+e^- \rightarrow ZH$ followed by an invisible decay $H \rightarrow \chi\chi$, using the recoil mass calculated only from the Z decay products. Model independent search for the dark matter can also be performed with mono-photon events, namely an initial state photon plus nothing ($e^+e^- \rightarrow \gamma Z^* \rightarrow \gamma\chi\chi$). If the Lightest Supersymmetric Particle (LSP) is Wino-like or Higgsino-like, the mass difference between the Next Lightest Supersymmetric Particle (NLSP) and the LSP is very small in most cases, and the LHC may not be able to explore this region. In such a case, the ILC is very powerful, where one can look for events with ISR photon and soft particles coming from the decay of NLSP to LSP.

3.3 Detectors for the ILC

Driving requirements for linear collider detectors are listed bellow. Unprecedented performances are needed to take full advantage of a linear e^+e^- collider.

1. An excellent vertex system with pixel detectors and low material budget is necessary to achieve desired impact parameter resolution for jet-flavor tagging.
2. Excellent momentum resolution is required for the precise reconstruction of the recoil mass of the Higgs boson using lepton pairs from the Z boson. The momentum resolution goal is $\sigma_{1/p_T} \approx 2 \times 10^{-5}\,\mathrm{GeV}^{-1}$.
3. Jet energy resolution should be good enough to identify W and Z bosons in multi-jet final states.

The particle flow algorithm (PFA) [12–14] is the leading philosophy of detector optimization, which has been developed to realize an excellent jet energy resolution. The PFA optimally combines measurement of following types of particles in jets to reconstruct the latter energy;

- charged particles, about 60% on average, are very precisely measured by the tracker with good precision,
- photons (~30%) are measured by the electromagnetic calorimeter with a relative resolution of about $15\%/\sqrt{E(\text{GeV})}$ or better, and
- neutral hadrons (~10%) are measured by the hadron calorimeter with a relative resolution of about $55\%/\sqrt{E(\text{GeV})}$ or better.

In this algorithm it is essential to efficiently separate particles in the calorimeter. In the ideal case, where each particle is resolved, a jet energy resolution of $19\%/\sqrt{E(\text{GeV})}$ could be obtained [12–14]. This calls for high granularity both for the electromagnetic and hadron calorimeters [15].

Two detector concepts, ILD and SiD, are validated as experiments at the ILC, both optimized for PFA with highly granular calorimeters. The main difference between the two detectors is their inner tracking system; SiD uses all-silicon tracking system, while ILD uses a combination of TPC and silicon tracking detectors. SiD has smaller inner radius of the calorimeter system, and stronger magnetic field of the solenoid. Many R&D studies for components of the linear collider detectors are ongoing. They are carried out by large groups such as CALICE and LC-TPC, as well as by individual small groups [16].

References

1. G. Aad et al., [ATLAS Collaboration], Phys. Lett. B **716**, 1 (2012)
2. S. Chatrchyan et al., [CMS Collaboration], Phys. Lett. B **716**, 30 (2012)
3. The ATLAS Collaboration [ATLAS Collaboration], ATLAS-CONF-2016-059
4. V. Khachatryan et al., [CMS Collaboration], Phys. Rev. Lett. **117**(5), 051802 (2016)
5. ILC Technical Design Report, https://www.linearcollider.org/ILC/Publications/Technical-Design-Report
6. Linear Collider Collaboration, http://www.linearcollider.org
7. CLIC Conceptual Design Report, https://lcd.web.cern.ch/lcd/CDR/CDR.html
8. Updated baseline for a staged Compact Linear Collider, http://clicdp.web.cern.ch/content/updated-staging-baseline
9. The HL-LHC Project, http://hilumilhc.web.cern.ch
10. K. Fujii et al., arXiv:1506.05992 [hep-ex]
11. T. Barklow, J. Brau, K. Fujii, J. Gao, J. List, N. Walker, K. Yokoya, arXiv:1506.07830 [hep-ex]
12. J.C. Brient, H. Videau, eConf C **010630**, E3047 (2001)
13. V.L. Morgunov, eConf C **010630**, E3041 (2001)
14. M.A. Thomson, Nucl. Instrum. Method A **611**, 25 (2009)
15. F. Sefkow, A. White, K. Kawagoe, R. Pöschl, J. Repond, Rev. Mod. Phys. **88**, 015003 (2016)
16. ILC Detector R&D Liaisons, http://www.linearcollider.org/P-D/Working-groups/Detector-R-D-liaison

Chapter 4
The Hierarchy Problem and Physics Beyond the Standard Model

Gautam Bhattacharyya

4.1 Introduction

In quantum field theory, the mass of a spin-zero elementary particle receives uncontrollably large quantum corrections. Then why the Higgs boson mass is only around 125 GeV? Why it has not been jacked up to the grand unification theory (GUT) scale (10^{16} GeV) or even beyond? Does it mean that the SM needs to be supplemented by a more fundamental beyond the Standard Model (BSM) theory controlling dynamics at a more microscopic level? What should be the properties of such theories to have a satisfactory solution to this so-called 'hierarchy problem' [1–4]? In this talk, I shall highlight how 'supersymmetry' [5] and 'composite Higgs' [6], as two broad scenarios, address this problem.

4.2 Hierarchy Problem

The Higgs boson receives one-loop quantum corrections from fermion (f) and scalar boson (s) as: $\Delta m_h^2(f) = -(y_f^2 2\Lambda^2)/16\pi^2$ and $\Delta m_h^2(s) = +(\lambda_S^2 \Lambda^2)/16\pi^2$, where Λ is the highest scale of the theory. These are not divergences which can be removed by regularizations. We are talking about large finite corrections arising from the Λ-scale fermions and bosons which may couple to the Higgs. Even if Λ is very large, the sign difference between fermion and boson loops can be tuned to keep the net contribution to the Higgs mass in 100 GeV range. However the couplings need to be tuned order by order in perturbation theory to keep the net result same. This is what we call the hierarchy problem. It implies that the electroweak vacuum expectation value (vev) $v(=246)$ GeV is not protected either ($m_h \sim \lambda v$, where λ is

G. Bhattacharyya (✉)
Saha Institute of Nuclear Physics, 1/AF Bidhan Nagar, Kolkata 700064, India
e-mail: gautam.bhattacharyya@saha.ac.in

Md. Naimuddin (ed.), *XXII DAE High Energy Physics Symposium*, Springer Proceedings in Physics 203, https://doi.org/10.1007/978-3-319-73171-1_4

the quartic coupling). Thus the entire SM is plagued by the hierarchy problem. Do we then have to accept a miraculous cancellation of one part in 10^{28} between unrelated quantities to get $m_h^2 = \mathcal{O}\,(100\,\text{GeV})^2$? In QED, electron mass is protected by chiral symmetry, and the photon always remains massless thanks to gauge symmetry. Can we associate the Higgs boson with a similar symmetry that can protect its mass? The SM does not have an answer. Can this be a guiding principle when we look for BSM physics?

4.3 Supersymmetry

Supersymmetry relates matter particles with force particles. For every fermion there is a bosonic partner and vice versa. Non-observation of superparticles means that supersymmetry is very badly broken (in masses, but not in couplings). Still, the quadratically divergent contributions to the Higgs mass cancel between loops containing virtual fermions and bosons because the couplings are related by supersymmetry. Minimal supersymmetry contains two Higgs doublets, and hence three neutral (two CP-even and one CP-odd) scalars and one pair of charged scalars. The lightest CP even neutral scalar is identified with the observed Higgs boson. Today, weak scale supersymmetric model building is suffering from the following tension:

(1) The quartic couplings are functions of gauge couplings. As a result, at tree level the Higgs mass is at most the Z boson mass. However, large top Yukawa coupling induced loop corrections push it to higher values. At one loop, the approximate expression for the Higgs mass is given by:

$$m_h^2 \simeq M_Z^2 \cos^2 2\beta + 3m_t^4 \ln(m_{\tilde{t}}^2/m_t^2)/\left(\sqrt{2}\pi^2 v^2\right), \tag{4.1}$$

where $\tan\beta = v_2/v_1$, the ratio of the two vevs. For moderate to large $\tan\beta$, m_h (tree) reaches its maximum value of $M_Z \simeq 91\,\text{GeV}$. So to reach up to 125 GeV, a really heavy stop is necessary ($m_{\tilde{t}} \sim 10\,\text{TeV}$). Adding higher loop effects brings the required stop mass down to 3 GeV. Still, the overall message is that one needs very heavy stop to generate $m_h = 125\,\text{GeV}$.

(2) One of the Higgs mass-square, $m_{H_u}^2$, starting from a positive value at high scale after RG running, due to large top Yukawa coupling, flips sign at the weak scale triggering electroweak breaking. An over-simplified one-loop expression reads

$$0.5\, M_Z^2 \simeq -|\mu|^2 - M_{H_u}^2 \simeq -|\mu^2| + \mathcal{O}(1)\, m_{\tilde{t}}^2. \tag{4.2}$$

Note that $m_{H_u}^2$ receives large corrections from the stop mass. But the loop suppression factor is neutralized by a large log (of GUT scale to weak scale). It implies a cancellation between μ (supersymmetry preserving higgsino mass) and the the stop mass so as to generate $M_Z = 91\,\text{GeV}$. But since $m_{\tilde{t}} \sim$TeV to reproduce $m_h = 125\,\text{GeV}$, what is required is a cancellation between $m_{\tilde{t}}$ and μ to the level of one part in 10^3, which constitutes the 'little hierarchy' problem. Although the big hierarchy is successfully addressed by supersymmetry, this little hierarchy is getting increasingly worrisome. This is basically the tension! It is not so much the non-observation of the superparticles, but the requirement of m_h =125 GeV, that created this tension. Supersymmetric model building is all about addressing this issue.

4.4 Composite Higgs

Can Higgs be some sort of a pseudo-Goldstone boson like a pion? If so, the big hierarchy is solved because the Higgs would be a composite object, and beyond the compositeness scale (not very large) the Higgs would dissolve. This is precisely why no one bothers about GUT scale correction to the pion mass. Can we employ the same mechanism (chiral symmetry breaking) to trigger electroweak breaking? The answer is 'No', since $M_W = g f_\pi/2 = 29\,\text{MeV}$ – a disaster!

Technicolor was invented to address this issue, but the simple versions didn't survive due to many issues, most notably, unacceptably large flavor changing neutral currents. In technicolor, $f_\pi \equiv f \sim$ few hundred GeV, so that the correct W mass is reproduced. However, the oblique parameter $S \sim v^2/f^2 \sim 1$, which fails to satisfy the LEP constraint $S \leq 0.01$. This was a big blow for technicolor.

To treat v and f as independent parameters, the strongly interacting sector is prohibited from directly participating in electroweak breaking (unlike in technicolor). Its rôle is limited to providing us with a set of pseudo Nambu-Goldstone bosons (pNGBs). Then we construct the SM scalar doublet out of those states. As an example, we assume that the strong sector has a global group G = SO(5) which spontaneously breaks to H = SO(4), giving 4 Goldstone bosons. Since G is also *explicitly* broken by gauge and Yukawa interactions, those Goldstone bosons would generate a potential. The Higgs is one such pNGB, and its Yukawa and gauge couplings are modulated by strong sector dynamics. For example, the $hVV(V = W, Z)$ coupling is modified from g_{ew} to $g_{\text{ew}}\sqrt{1-\xi}$, where $\xi = v^2/f^2$. The Yukawa couplings too get modified. Extremely precise measurements of Higgs branching ratios might sense these variations.

4.5 Conclusion

Supersymmetry and Composite Higgs are two broad scenarios containing many interesting features that can be tested at the LHC (and then at the ILC). These models provide useful guidelines for a more general category of BSM search.

References

1. G. Bhattacharyya, Rep. Prog. Phys. **74**, 026201 (2011), arXiv:0910.5095 [hep-ph]
2. R.K. Kaul, arXiv:0803.0381 [hep-ph]
3. G. Bhattacharyya, Pramana **72**, 37 (2009), arXiv:0807.3883 [hep-ph]
4. G.F. Giudice, *Perspectives on LHC Physics*, ed. by G. Kane, P. Aaron, arXiv:0801.2562 [hep-ph], pp. 155–178
5. R.N. Mohapatra, *Unification and Supersymmetry: The Frontiers of Quark-Lepton Physics* (Springer, New York, 1992; M. Drees, R. Godbole, P. Roy, *Theory and Phenomenology of Sparticles: An Account of Four-Dimensional N=1 Supersymmetry in High Energy Physics* (World Scientific, Singapore, 2004); H. Baer, X. Tata, *Weak Scale Supersymmetry: From Superfields to Scattering Events* (Cambridge University Press, Cambridge, 2006)
6. G. Panico, A. Wulzer, Lect. Notes Phys. **913**, 1 (2016), arXiv:1506.01961 [hep-ph]

Chapter 5
Electroweak Physics at the LHC

Gerhard Brandt

5.1 Introduction

The electroweak sector is a central part of the Standard Model (SM). It is intricately connected to the mechanism of electroweak symmetry-breaking (EWSB). Many electroweak processes also represent important background to the study of the Higgs boson, searches for new physics, or are sensitive to contributions from new physics themselves. Consequently a large number of related results have been produced by the ATLAS [1] and CMS [2] collaborations at the LHC [3]. With excellent progress on the detector understanding in recent years, the data from the LHC Run-1 with $\sqrt{s} = 7, 8$ TeV can now be exploited to their fullest potential and provide precision measurements of electroweak processes at that energy. First results from LHC Run-2 with $\sqrt{s} = 13$ TeV data are also becoming available. They provide unprecedented opportunities for precision measurements of single and di-boson production and for the first time provide access to rare processes such as vector boson fusion, vector boson scattering, exclusive production of vector bosons and tri-boson production. The processes are also sensitive to anomalous triple gauge couplings (aTGC) and quartic gauge couplings (aQGC) and allow to constrain them using a traditional chiral Lagrangian approach or a more modern approach based on effective field theory [4]. This talk provides an overview of selected recent results.

On behalf of the ATLAS and CMS Collaborations.

G. Brandt (✉)
II. Physikalisches Institut, Georg-August-Universität Göttingen,
37077 Göttingen, Germany
e-mail: gbrandt@cern.ch

Md. Naimuddin (ed.), *XXII DAE High Energy Physics Symposium*, Springer Proceedings in Physics 203, https://doi.org/10.1007/978-3-319-73171-1_5

5.2 Inclusive Vector Boson Production

High precision measurements of W and Z production cross sections are now available [5]. They allow to obtain a new PDF set `ATLAS-epWZ16` from a QCD analysis of LHC and HERA data, and provide confirmation that the strange to light sea quark density in the proton is close to unity at low x_{Bj}. A new CKM parameter V_{cs} measurement from a fit to these data is competitive with previous results. First inclusive W, Z cross sections from 13 TeV data are already sensitive to different PDF sets [6–8]. For the first time a measurement of the W mass, m_W, has been made at the LHC [9]. The result $m_W = 80370 \pm 7(\text{stat.}) \pm 11(\text{exp. syst.}) \pm 14(\text{mod. syst.})\ \text{MeV} = 80370 \pm 19$ MeV is competitive to current m_W measurements from the TeVatron and compatible with the PDG world average, as well as the SM prediction from a global electroweak fit. Measurements of vector boson production in association with jets at 13 TeV are also available [10, 11] in addition to new results at 7, 8 TeV [12, 13]. The cross section has been measured as function of jet multiplicity and in a wide range of differential distributions. Comparisons to various fixed order calculations at (N)NLO are in reasonable agreement with the data.

5.3 Di-boson Production and Limits on Anomalous Triple Gauge Couplings

A large number of di-boson cross sections has been measured by ATLAS and CMS in 7, 8 and 13 TeV data. The cleanest final state, but smallest rate compared to other di-boson processes, is available in ZZ production which has been measured in 8 TeV [14–17] and 13 TeV [18, 19] in various decay channels. Measurements of WZ production at 7, 8 TeV [20, 21] and 13 TeV [22, 23] combine the low background of Z final states and higher rate of W processes. Template fits of aTGC models to the high-p_T tails of distributions sensitive to aTGC contributions, such as the transverse mass of the $W - Z$ system, m_T^{WZ}, provide some of the most stringent limits on aTGCs connected to a WWZ coupling to date. Measurements of W^+W^- production provide the largest cross section of di-boson processes but also face the most challenging background situation. A number of results in 8 TeV [24–26] and 13 TeV [27, 28] data are available. In such opposite charged W-pair production measurements it is essential to suppress contributions from top-pair production. This requires the precise understanding of the application of jet vetoes which has seen recent progress. All measurements are compatible with SM predictions, often involving state-of-the-art pQCD predictions at NNLO. For example [28] measures $\sigma_{W^+W^-} = 115 \pm 11\ \text{pb}^{-1}$ for $118.7 \pm 3.6\ \text{pb}^{-1}$ expected in the SM. As mentioned the high-p_T tails of the di-boson measurements are sensitive to aTGC contributions and have been used to derive a large number of limits on charged (WWZ and $WW\gamma$ couplings) and neutral ($Z\gamma\gamma$, $ZZ\gamma$, γZZ and ZZZ) couplings. A review of all results can be found in [29], providing summary tables for all charged and neutral aTGC limits from the LHC and previous experiments. Results from LEP and TeVatron are no longer competitive to the high-precision LHC limits. A first combination of ATLAS and CMS of the respective *ZZ* production measurements provides a further 20% increase in precision on the corresponding aTGC limits [30, 31].

5.4 Rare Processes and Limits on Anomalous Quartic Gauge Couplings

Vector boson scattering (VBS) is intimately connected to the EWSB mechanism and remains an essential probe of the SM even after discovery of the Higgs boson. The measurements are sensitive to aQGCs and spectacular signatures are expected for vector boson fusion and VBS with two high p_T forward jets, one or two high-p_T central leptons and rapidity gaps in-between. In nature these turn out to be challenging to observe due large reducible and irreducible backgrounds. Nevertheless both electroweak production of Z bosons [32] and W bosons [33] have now been observed in fits to the electroweak component in enriched selections. Exclusive W-pair production $\gamma\gamma \rightarrow W^+W^-$ is important to understand as background to Higgs production. The process has now been observed by both collaborations at 3σ significance [34, 35]. Analysis of same sign $W^{\pm}W^{\pm} + jj$ final states remains the seminal analysis in VBS studies and has been measured by both experiments [36, 37]. To complete our picture also studies of $W\gamma + jj$ [38], $Z\gamma + jj$ [39] production have been made. Low statistics still dominate results, but the processes have been observed. Searches for $W^{\pm}V \rightarrow \ell^{\pm}\nu + jj$ [40–42] with hadronic decays of $V = W, Z$ use novel techniques such as large-R jet reconstruction at high-p_T. These results are aimed at the most stringent limits on aTGCs and aQGCs, for example on the aQGC parameters $-0.024 < \alpha_4 < 0.030$ and $-0.028 < \alpha_5 < 0.033$ [42]. First tri-boson searches for $W\gamma\gamma$ [43, 44] and $Z\gamma\gamma$ [45] and even WWW [46] production have been done. Since no deviations from the SM are found, all these results have been used to derive a large number of limits on aQGC, also summarised in [29]. A variety of competitive limits has been produced where generally VBS data are more sensitive than tri-boson processes.

5.5 Summary

A large number of electroweak processes has already been explored at the LHC. No significant deviations from the SM predictions are observed anywhere reinforcing our trust in the description of the electroweak sector of the SM. High rate processes allow precision tests of NNLO SM predictions. This precision will increase even further with more luminosity and detector understanding. Limits on aTGCs and aQGCs have been derived and are now testing the sensitivity of our experiments. The planned LHC

Run-3 and upgrade programme should bring a big increase in available integrated luminosity, especially for the High Luminosity LHC. If we do not find new physics elsewhere (for example new resonances from SUSY), the electroweak sector is the best place to probe. Sensitivity projection studies, for example for longitudinal VBS, motivate a strong electroweak programme in the future [47]. And we are just at the beginning of exploration of not-yet-observed rare electroweak processes like VBS and triboson production.

References

1. ATLAS Collaboration, JINST **3**, S08003 (2008)
2. CMS Collaboration, JINST **3**, S08004 (2008)
3. L. Evans, P. Bryant (eds.), JINST **3**, S08001 (2008)
4. C. Degrande et al., Ann. Phys. **335**, 21–32 (2013)
5. ATLAS Collaboration, CERN-EP-2016-272 (Subm. to EPJC)
6. CMS Collaboration, CMS-PAS-SMP-15-011
7. CMS Collaboration, CMS-PAS-SMP-15-004
8. ATLAS Collaboration, Phys. Lett. B **759**, 601 (2016)
9. ATLAS Collaboration, CERN-EP-2016-305 (Subm. to EPJC)
10. CMS Collaboration, CMS-PAS-SMP-16-005
11. ATLAS Collaboration, ATLAS-CONF-2016-046
12. CMS Collaboration, CMS-SMP-14-012 (Subm. to JHEP)
13. CMS Collaboration, CMS-SMP-14-023 (Subm. to Phys. Rev. D)
14. CMS Collaboration, PLB **740**, 250 (2015)
15. ATLAS Collaboration, PLB **753**, 552–572 (2016)
16. CMS Collaboration, EPJC **75**, 511 (2015)
17. ATLAS Collaboration, CERN-EP-2016-194 (Subm. to JHEP)
18. ATLAS Collaboration, Phys. Rev. Lett. **116**, 101801
19. CMS Collaboration, PLB **763**, 280 (2016)
20. ATLAS Collaboration, EPJC **72**, 2173 (2012)
21. ATLAS Collaboration, Phys. Rev. D **93**, 092004 (2016)
22. CMS Collaboration, CMS-SMP-14-014
23. ATLAS Collaboration, ATLAS-CONF-2016-043
24. CMS Collaboration, EPJC **76**, 401 (2016)
25. ATLAS Collaboration, PLB **763**, 114 (2016)
26. ATLAS Collaboration, JHEP **09**, 029 (2016)
27. ATLAS Collaboration, ATLAS-CONF-2016-090
28. CMS Collaboration, CMS-PAS-SMP-16-006
29. D. Green, P. Meade, M. Pleier (Subm. to RMP), arXiv:1610.07572
30. ATLAS Collaboration, ATLAS-CONF-2016-036
31. CMS Collaboration, CMS-PAS-SMP-15-001
32. CMS Collaboration, EPJC **75**, 66 (2015)
33. CMS Collaboration, JHEP **11**, 147 (2016)
34. ATLAS Collaboration, Phys. Rev. D **94**, 032011 (2016)
35. CMS Collaboration, JHEP **08**, 119 (2016)
36. CMS Collaboration, Phys. Rev. Lett. **114**, 051801 (2015)
37. ATLAS Collaboration, CERN-EP-2016-167 (Subm. to Phys. Rev. D)
38. CMS Collaboration, CMS-PAS-SMP-14-011
39. CMS Collaboration, CMS-PAS-SMP-14-018
40. CMS Collaboration, CMS-PAS-SMP-13-008

41. CMS Collaboration, CMS-PAS-SMP-16-012
42. ATLAS Collaboration, Phys. Rev. D **95**, 032001 (2017)
43. CMS Collaboration, CMS-PAS-SMP-15-008
44. PRL **115**, 031802 (2015)
45. Phys. Rev. D **93**, 112002 (2016)
46. ATLAS Collaboration, CERN-EP-2016-172 (subm. to EPJC)
47. Phys. Rev. D **90**, 032008 (2014)

Chapter 6
Exotics in Flavor Factories

Seema Bahinipati

6.1 Introduction

Although the Standard Model (SM) allows any color-neutral combinations, conventional hadrons exist as bound states of either three quarks, qqq (baryon) or a quark and an anti-quark, $q\bar{q}$ (meson). The exotic color-neutral combinations, that are also allowed in SM as proposed by Gell-Mann [1] and Zweig [2], include tetraquarks ($q\bar{q}q\bar{q}$), pentaquarks ($qqqq\bar{q}$), glueballs (gg), and so on. The first exotic appeared in the charmonium sector in 2003 , and since then, several exotics have been discovered.

6.2 Flavor Factories

The two asymmetric flavor factories, KEK-B, in Tsukuba, Japan and PEP-II, at SLAC, California, USA and their companion detectors, Belle [3] and BaBar [4] have produced several landmark results in flavor physics results, providing stringent tests of the SM. These flavor factories were instrumental in discovery of several exotic states as well [5–10]. Other collaborations contributing to the exotic search include CMS [11] and LHCb [12], BES III [13], CDF [14] and D0 [15]. The exotic searches in all these collaborations is performed by fitting the invariant mass spectrum of the selected candidates within the mass window of interest.

Seema Bahinipati on behalf of the Belle Collaboration.

S. Bahinipati (✉)
Indian Institute of Technology Bhubaneswar, Bhubaneswar, India
e-mail: seema.bahinipati@iitbbs.ac.in

Md. Naimuddin (ed.), *XXII DAE High Energy Physics Symposium*, Springer Proceedings in Physics 203, https://doi.org/10.1007/978-3-319-73171-1_6

6.3 Properties of $X(3872)$

The first exotic observed, $X(3872)$ was discovered by Belle in 2003 as a resonance of $\pi^+\pi^- J/\psi$ in the decay $B \to K^+(\pi^+\pi^- J/\psi)$ [5]. In 2004, the Belle collaboration reported the evidence of $X(3872)$ in $B^+ \to K^+\pi^+\pi^-\pi^0 J/\psi$, which was the first observation of an $X(3872)$ decay mode other than $\pi^+\pi^- J/\psi$ [16]. The collaboration observed $X(3872)$ state in $B \to X(3872)K$ decays, where $X(3872)$ decays to $J/\psi\gamma$ [17]. The existence of $X(3872)$ has since been confirmed by several experiments such as CDF [18, 19], D0 [20], BaBar [6, 21], and CMS [22].

Based on the angular correlations in $B^+ \to X(3872)K^+$, where $X(3872)$ decays to $\pi^+\pi^- J/\psi$, the quantum numbers of the $X(3872)$ meson are determined to be $J^{PC} = 1^{++}$ by the LHCb collaboration [23], which ruled out the only alternative assignment allowed by previous measurements from CDF, $J^{PC} = 2^{-+}$ with a confidence level exceeding 8σ. This result favors the exotic explanations of the $X(3872)$ state. The BES III collaboration presented the first observation of $e^+e^- \to \gamma X(3872)$ with a statistical significance of 6.3σ [24] and measured the production rate $\sigma[e^+e^- \to \gamma X(3872)] \times \mathcal{B}[X(3872) \to \pi^+\pi^- J/\psi]$ to be $0.27 \pm 0.09 \pm 0.02$ pb.

6.4 Exotic Nature of $Y(4260)$

The second charmonium-like exotic, $Y(4260)$ was discovered in 2005 by the BaBar collaboration [7] using the initial-state radiation (ISR) technique in $e^+e^- \to \gamma_{ISR}\pi^+\pi^- J/\psi$. Such states have $J^{PC} = 1^{--}$ (same quantum numbers as γ), and are called Y. It was confirmed by the Belle collaboration in 2007 [8], which also observed an additional structure Y(4008). Later, BaBar made a more precise measurement of the $Y(4260)$ parameters [25]. However, they did not confirm the new structure seen by Belle at $\approx$4 GeV.

The BES III collaboration also reported the study of the process $e^+e^- \to \pi^+\pi^- J/\psi$ [26] at a center-of-mass energy of 4.260 GeV using a 525 pb^{-1} data sample. The Born cross-section was measured to be $(62.9 \pm 1.9 \pm 3.7)$ pb, consistent with the production of the $Y(4260)$. They also observed a structure at around 3.9 GeV in the $\pi^+\pi^- J/\psi$ mass spectrum, which is referred to as the $Z_c(3900)$. The $Y(4260)$ state was used in experiments by both the BES III [26] and Belle [27] collaborations for the first confirmation of a charged charmonium-like state, $Z_c^+(3900)$, in 2013. It appears as a resonance in the decay $Y(4260) \to \pi^-(\pi^+ J/\psi)$, and its quantum numbers have been determined to be $J^P = 1^+$. As a charged state, its minimal flavor content is $c\bar{c}u\bar{d}$, meaning that it was the first manifestly exotic state ever confirmed beyond 5σ by the two experiments.

$Y(4260)$ has been confirmed by other experiments, and is charmonium-like. This is inferred both by its mass and decay modes, such as $\pi^+\pi^- J/\psi$. Since all of the expected conventional $c\bar{c}$ states in the 1^{--} channel up to the $\psi(4415)(4S)$ have been seen, $Y(4260)$ is referred to as an exotic state.

6.5 Observation of $Z(4430)$

$Z^+(4430)$ state was first claimed by Belle in $B \to K[\pi^-\psi(2S])$ decays [9]. It was assumed to have a minimal quark content of $c\bar{c}d\bar{u}$. This state was explained by BaBar through reflections of $K\pi$ system [10]. Later, Belle performed an amplitude analysis confirming the new state. Recently, the LHCb collaboration performed a 4-dimensional amplitude analysis of $B \to K[\pi^-\psi(2S)]$ [28] and performed a model-independent extraction of the resonant phase.

The $Z(4430)$ state has a component with $J^P = 1^+$ (hereafter, referred as Z_1^-). The resonant nature of $Z(4430)$ is determined from $B^0 \to K[\pi^-\psi(2S)]$ decays. This observation was confirmed by a model independent analysis. In 2014, LHCb not only confirmed [28] the existence (at 13.9σ) of a charged charmonium-like state, $Z^-(4430) \to \pi^-\psi^{'}$ (first seen by Belle in 2008 [29]) with $J^P = 1^+$, but also measured the phase shift in the resonant region that it found to be compatible with the expected looping behavior. While nature could be unkind and generate this behavior even in the absence of a resonance, such an accident seems unlikely, and moreover, further tests by LHCb [30] support that $Z^-(4430)$ is not merely a kinematical reflection.

6.6 Observation of $P_c^+(4380)$ and $P_c^+(4450)$

In 2015, the LHCb collaboration observed exotic structures [31] in the $\Lambda^0 \to J/\psi K^- p$ channel, which we refer to as candidates of pentaquark states using an amplitude analysis of the three-body final state. The two baryonic charmonium-like resonances, P_c^+ (4380) (at 9σ) and P_c^+ (4450) (at 12σ) decay into J/ψp, and are therefore expected to possess pentaquark $c\bar{c}uud$ quantum numbers.

To obtain a satisfactory fit of the structures seen in the $J/\psi K^- p$ mass spectrum, it is necessary to include two Breit–Wigner amplitudes each describing a resonant state. The significance of each of these resonances is more than 9 standard deviations. The first one has a mass of $4380 \pm 8 \pm 29$ MeV and a width of $205 \pm 18 \pm 86$ MeV, while the second is narrower, with a mass of $4449.8 \pm 1.7 \pm 2.5$ MeV and a width of $39 \pm 5 \pm 19$ MeV. The preferred J^P assignments are of opposite parity, with one state having spin $3/2$ and the other $5/2$.

6.7 Conclusions

Several exotic hadrons have been discovered since 2003. The X, Y, Z exotic states have been observed at the e^+e^- flavor factories namely Belle and BaBar and other collaborations including CDF, BES III, CMS and LHCb have also observed several of these states. Recent additions to the family of exotic particles are pentaquarks, which were discovered by the LHCb collaboration. Theoretical work [32] in order to understand the structure of these states (such as, whether they are "molecular charmonium" [33], "hybrid charmonium" [34], or other possibilities allowed within the framework of QCD) has been performed as well.

Acknowledgements We thank the KEKB group, all institutes and agencies that have supported the work of the members of the Belle Collaboration.

References

1. M. Gell-Mann, A schematic model of baryons and mesons. Phys. Lett. **8**, 214 (1964)
2. G. Zweig, Report No. CERN-TH-401
3. Belle Collaboration, M. Staric et al., Phys. Rev. Lett. **98**, 211803 (2007)
4. BaBar Collaboration, B. Aubert et al., Phys. Rev. Lett. **98**, 211802 (2007)
5. S.-K. Choi et al., (Belle Collaboration), Phys. Rev. Lett. **91**, 262001 (2003)
6. B. Aubert et al., (BaBar Collaboration), Phys. Rev. **D71**, 071103 (2005)
7. B. Aubert et al., (BaBar Collaboration), Phys. Rev. Lett. **95**, 142001 (2005), arXiv:hep-ex/0506081
8. C.Z. Yuan et al., (Belle Collaboration), PRL **99**, 182004 (2007)
9. S.-K. Choi et al., (Belle Collaboration), PRL **100**, 142001 (2008)
10. A.G. Mokhtar (BaBar Collaboration) (2008), arXiv:0810.1073 [hep-ex]
11. CMS Collaboration, JINST **3**, S08004 (2008)
12. LHCb Collaboration, A.A. Alves Jr. et al., The LHCb detector at the LHC, JINST **3**, S08005 (2008)
13. M. Ablikim et al., (BESIII Collaboration), Nucl. Instrum. Methods Phys. Res. Sect. A **614**, 345 (2010)
14. R. Blair et al., (CDF II Collaboration). The CDF II Detector, Technical design report. FERMILAB-Pub-96/390- E CDF (1996)
15. V.M. Abazov et al., (D0 Collaboration), Nucl. Instrum. Methods Phys. Res. A **565**, 463 (2006)
16. K. Abe et al., (Belle Collaboration) (2004), arXiV:0408116 [hep-ex]
17. V. Bharadwaj et al., (Belle Collaboration), Phys. Rev. Lett. **107**, 091803 (2011)
18. D. Acosta et al., (CDF II Collaboration), Phys. Rev. Lett. **93**, 072001 (2004)
19. A. Abulencia et al., (CDF II Collaboration), Phys. Rev. Lett. **98**, 132002 (2007)
20. V.M. Abazov et al., (D0 Collaboration), Phys. Rev. Lett. **93**, 162002 (2004)
21. B. Aubert et al., (BaBar Collaboration), Phys. Rev. **D77**, 111101 (2008)
22. S. Chatrchyan et al., (CMS Collaboration), JHEP **04**, 154 (2013)
23. R. Aaij et al., (LHCb Collaboration), Phys. Rev. Lett. **110**, 222001 (2013)
24. M. Ablikim et al., (BESIII Collaboration), Phys. Rev. Lett. **112**, 092001 (2014)
25. J.P. Lees et al., (BaBar Collaboration), PRD **86**, 051102(R) (2012)
26. M. Ablikim et al., (BESIII Collaboration), Phys. Rev. Lett. **110**, 252001 (2013), arXiv:1303.5949 [hep-ex]
27. Z.Q. Liu et al., (Belle Collaboration), Phys. Rev. Lett. **110**, 252002 (2013), arXiv:1304.0121 [hep-ex]

28. R. Aaij et al., [LHCb Collaboration], Phys. Rev. Lett. **112**, 222002 (2014), arXiv:1404.1903 [hep-ex]
29. S.K. Choi et al., [Belle Collaboration], Phys. Rev. Lett. **100**, 142001 (2008), arXiv:0708.1790 [hep-ex]
30. R. Aaij et al., [LHCb Collaboration], Phys. Rev. D **92**, 112009 (2015), arXiv:1510.01951 [hep-ex]
31. R. Aaij et al., (LHCb Collaboration), Phys. Rev. Lett. **115**, 072001 (2015), arXiv:1507.03414 [hep-ex]
32. R.L. Jaffe, K. Johnson, Phys. Lett. **60B**, 201 (1976); M.B. Voloshin, L.B. Okun, Pisma Zh. Eksp. Teor. Fiz. **23**, 369 (1976) [JETP Lett. **23**, 333 (1976)]
33. See, for example, M. Bander, G.L. Shaw, P. Thomas, Phys. Rev. Lett. **36**, 695 (1977); M.B. Voloshin, L.B. Okun, JETP Lett. **23**, 333 (1976); A. De Rujula, H. Georgi, S.L. Glashow, Phys. Rev. Lett. **38**, 317 (1977); N.A. Tornqvist, Z. Phys. **C61**, 525 (1994); A.V. Manohar, M.B. Wise, Nucl. Phys. **B339**, 17 (1993)
34. See, for example, S. Godfrey, J. Napolitano, Rev. Mod. Phys. **71**, 1411 (1999), and reference cited therein

Chapter 7
Mini Review on Transport Coefficients of Quark-Gluon-Plasma

Victor Roy

7.1 Introduction

Today it is widely accepted that the hot and dense QGP formed in high energy heavy ion collisions behave as almost perfect fluid (shear viscosity over entropy density ratio $\frac{\eta}{s}$ very close to the lower bound predicted by string theory and uncertainty principle). It is worthwhile to mention that the the fluid evolution is governed by the kinematic viscosity $\frac{\eta}{s}$ which is very small for QGP but the dynamic viscosity η is not. The large entropy density s of QGP makes the η/s small. Relativistic viscous hydrodynamics has been proven to be exceptionally effective in describing the space-time evolution of QGP by treating it as a locally equilibrated drop of fluid. By using appropriate algorithm (Cooper–Frye freezeout prescription) one can calculate experimentally measured quantities such as invariant yield and flow harmonics of charged hadron from viscous hydrodynamics simulation for a given equation of state (usually constructed from lattice QCD data and hadronic resonance gas model) and initial conditions (usually Glauber/Color Glass Condensate). One of the bulk experimental observable proven to be very useful for extracting transport coefficient of QGP is the flow harmonics v_n where n $= 1, 2, 3, \ldots$ corresponds to directed, elliptic, triangular flow and so on. In principle the η and ζ (and other transport coefficients) of QGP should be a function of temperature and they can be calculated from fundamental theory of strong interactions i.e, QCD (more specifically lattice QCD). However, all the available lattice results of QGP transport coefficients contains large error and there is no direct way to extract the transport coefficients of QGP from the direct comparison of experimental data and lQCD results. We can, however, extract η and ζ of QGP by comparing the result from relativistic viscous hydrodynamics simulation where η and ζ are treated as a free parameter to the corresponding experimental data. This method is in use since year 2007 and even today we are essentially

V. Roy (✉)
National Institute of Science Education and Research, Odisha 752050, India
e-mail: victor@niser.ac.in

Md. Naimuddin (ed.), *XXII DAE High Energy Physics Symposium*, Springer Proceedings in Physics 203, https://doi.org/10.1007/978-3-319-73171-1_7

doing the same thing but with more sophisticated hydrodynamics model. In this mini review we will discuss some of the important developments made in the last decade which led us to the present precision measurement of η/s and ζ/s of QGP, and the uncertainties related to it.

7.1.1 Essential Ingredients of Relativistic Viscous Hydrodynamics Model

Before we proceed any further let us very briefly discuss the essential ingredients of relativistic viscous hydrodynamics (details of which can be found elsewhere). The fluid evolution is governed by the conservation of energy-momentum $\partial_\mu T^{\mu\nu} = 0$ and net conserve charges $\partial_\mu N^\mu = 0$, where $T^{\mu\nu}$ and N^μ are energy-momentum tensor and conserved four current respectively. In general the energy momentum tensor contains dissipative fluxes and in the second order causal theory (due to Israel-Stewart and Mueller) of relativistic viscous hydrodynamics these dissipative fluxes ($\pi^{\mu\nu}$ and Π) are treated as an independent quantity and are governed by the following relaxation equations (or some variant of it) with finite relaxation time τ_π, τ_Π.

$$D\pi^{\mu\nu} = -\frac{1}{\tau_\pi}\left(\pi^{\mu\nu} - 2\eta\nabla^{<\mu}u^{\nu>}\right) - \left[u^\mu\pi^{\nu\lambda} + u^\nu\pi^{\mu\lambda}\right]Du_\lambda \tag{7.1}$$

$$D\Pi = -\frac{1}{\tau_\Pi}\left[\Pi + \zeta\nabla_\mu u^\mu + \frac{1}{2}\zeta T\Pi\partial_\mu\left(\frac{\tau_\Pi u^\mu}{\zeta T}\right)\right] \tag{7.2}$$

where $D = u^\mu\partial_\mu$ is the convective time derivative, $\nabla^{<\mu}u^{\nu>} = \frac{1}{2}\left(\nabla^\mu u^\nu + \nabla^\nu u^\mu\right) - \frac{1}{3}\partial_\mu u^\mu\left(g^{\mu\nu} - u^\mu u^\nu\right)$ is a symmetric traceless tensor, u^μ is fluid four velocity. η and ζ are treated here as a free parameter in the numerical simulation. For vanishing net baryon numbers we can ignore the charge conservation equation. The rest of the equations are closed with an Equation of State (EoS) which relates pressure (p) to the energy density (ε), p = p(ε). Another important input to viscous hydrodynamics model is the initial entropy/energy density profile at the thermalisation time (τ_i). Two most popular and successful models in use are Glauber and Color Glass Condensate and their variants (from now on we use the abbreviation MCG for Monte Carlo Glauber, OG for optical Glauber and MC KLN for Monte Carlo Kharzeev–Levin–Nardi). In the next section we shall mention initial conditions, EoS, and other important informations which are found to be most influential for the extracted value of transport coefficients of QGP.

7.1.2 Chronological Development in Extracted Value of η/s and ζ/s

Experience shows us that the most sensitive experimental observable for extracting η/s and ζ/s are the transverse momentum (p_T) dependence of flow harmonics $v_n(p_T)$. In the early studies mostly the second harmonics $v_2(p_T)$ was used to extract η/s. Here we shall report the chronological development starting with the very first paper by Luzum et al. which appears in the year 2007. Due to the limitation of space

Table 7.1 Viscous hydrodynamics models and extracted value of η/s, ζ/s

Year	Reference	Model (system)	Initial condition	EoS	$\eta/s \leq$	$\zeta/s \leq$
2007	Luzum et al. [1]	2+1D viscous hydro (Au+Au 200 GeV)	OG CGC	HRG+PQCD	0.23	0.0
2010	Song et al. [2]	2+1 D viscous hydro + URQMD (Au+Au 200 GeV)	MCG MC KLN	HRG+LQCD s95p-PCE	0.08 0.23	0.0
2011	Schenke et al. [3]	3+1 D viscous hydro e-by-e (Au+Au 200 GeV)	MCG	HRG+LQCD s95pv1	0.16	0.0
2011	Roy et al. [4]	2+1 D viscous hydro (Pb+Pb 2.76 TeV)	OG	HRG+LQCD	0.16	0.0
2014	Novak et al. [6]	2+1 D viscous hydro+hadron cascade (Au+Au 200 GeV) multi-parameter Markov Monte Carlo method	OG	LQCD	0.15	0.0
2015	Ryu et al. [7]	3+1 D viscous hydro + URQMD (Pb+Pb 2.76 TeV) non-zero bulk viscosity	IP Glasma + MUSIC	HRG+LQCD	0.09	0.3
2016	Nopoush et al. [8]	3+1 D viscous anisotropic hydro + Therminator (Pb+Pb 2.76 TeV)	OG	LQCD	0.23	0.0

we will not be able to discuss in details all the references mentioned here, instead we show comparison of various model studies (arranged chronologically) and their main differences in Table 7.1. We would like to point out that the result of temperature dependence shear viscosity (see [5]) is not included in the Table 7.1 because it does not provide a single value of η/s like all other studies where η/s is constant. In [5] it was found that the hadronic viscosity is important for Au+Au collisions at RHIC and QGP viscosity at LHC collisions. From Table 7.1 it is clear that as we improve the hydrodynamics model η/s approaches the lower bound $1/4\pi$ (in unit of $\hbar$ and c). However, it is probably not legitimate to claim that we have achieved such high precision in the extracted value of η/s or ζ/s as there are some less unexplored field which may introduce uncertainty in the present known value. In [8] more relevant anisotropic initial condition is used and it is clear that in this case there is a rise in the value of η/s. However, it should also be noted that the study [8] is in a preliminary stage and it needs to be improved before we draw any definitive conclusion. Before we discuss the open problems and outlook it is worthwhile to mention some important observation: (a) QGP is one of the most perfect fluid ever created in any experiment. (b) p_T differential v_n's are most sensitive experimental observable to the QGP viscosity.

We would like to point out the following less studied topics which are important for a precision measurement of QGP transport coefficients.

- Use of a realistic 3+1D initial condition with longitudinal fluctuation
- More study is needed for non-zero bulk viscosity.
- Effect of strong electric and magnetic field in the initial stage of QGP evolution might also be important.
- Arguably the best method for extraction of η/s or ζ/s will require a full 3+1 d viscous hydrodynamics model with the following features: realistic anisotropic initial condition which contains transverse and longitudinal fluctuation + hadronic cascade + electric and magnetic field + proper freezeout distribution function with correction for shear and bulk viscosity and electromagnetic field. One should use this model and extract the η/s and ζ/s by multiparameter fit as was done in [6].

References

1. M. Luzum, P. Romatschke, Conformal relativistic viscous hydrodynamics: applications to RHIC results at s(NN)**(1/2) = 200-GeV. Phys. Rev. C **78**, 034915 (2008), Erratum: [Phys. Rev. C **79**, 039903 (2009)]. https://doi.org/10.1103/PhysRevC.78.034915, https://doi.org/10.1103/PhysRevC.79.039903, arXiv:0804.4015 [nucl-th]
2. H. Song, S.A. Bass, U. Heinz, T. Hirano, C. Shen, 200 A GeV Au+Au collisions serve a nearly perfect quark-gluon liquid. Phys. Rev. Lett. **106**, 192301 (2011), Erratum: [Phys. Rev. Lett. **109**, 139904 (2012)]. https://doi.org/10.1103/PhysRevLett.106.192301, https://doi.org/10.1103/PhysRevLett.109.139904, arXiv:1011.2783 [nucl-th]
3. B. Schenke, S. Jeon, C. Gale, Elliptic and triangular flow in event-by-event (3+1)D viscous hydrodynamics. Phys. Rev. Lett. **106**, 042301 (2011). https://doi.org/10.1103/PhysRevLett.106.042301, arXiv:1009.3244 [hep-ph]

4. V. Roy, A.K. Chaudhuri, Charged particle's elliptic flow in 2+1D viscous hydrodynamics at LHC ($\sqrt{s} = 2.76$ TeV) energy in Pb+Pb collision. Phys. Lett. B **703**, 313 (2011). https://doi.org/10.1016/j.physletb.2011.08.006, arXiv:1103.2870 [nucl-th]
5. H. Niemi, G.S. Denicol, P. Huovinen, E. Molnar, D.H. Rischke, Influence of the shear viscosity of the quark-gluon plasma on elliptic flow in ultrarelativistic heavy-ion collisions. Phys. Rev. Lett. **106**, 212302 (2011). https://doi.org/10.1103/PhysRevLett.106.212302, arXiv:1101.2442 [nucl-th]
6. J. Novak, K. Novak, S. Pratt, J. Vredevoogd, C. Coleman-Smith, R. Wolpert, Phys. Rev. C **89**(3), 034917 (2014). https://doi.org/10.1103/PhysRevC.89.034917, arXiv:1303.5769 [nucl-th]
7. S. Ryu, J.-F. Paquet, C. Shen, G.S. Denicol, B. Schenke, S. Jeon, C. Gale, Importance of the bulk viscosity of QCD in ultrarelativistic heavy-ion collisions. Phys. Rev. Lett. **115**(13), 132301 (2015). https://doi.org/10.1103/PhysRevLett.115.132301, arXiv:1502.01675 [nucl-th]
8. M. Nopoush, M. Strickland, R. Ryblewski, Phenomenological predictions of 3+1d anisotropic hydrodynamics, arXiv:1610.10055 [nucl-th]

Chapter 8
Results on Quarkonium and Heavy Meson Production in PbPb Collisions by CMS Experiment

Prashant Shukla

Introduction

The aim of colliding two heavy nuclei at high energies is to create and study Quark Gluon Plasma (QGP), a phase of strongly interacting deconfined matter. The quarkonium states have been one of the most interesting probes of QGP. Since, the LHC started PbPb collisions at $\sqrt{s_{NN}} = 2.76\,\text{TeV}$, a plethora of data have become available on quarkonium production [1–7]. The CMS experiment [8] with its muon detection capabilities has enabled several measurements on quarkonia (both charmonia as well as bottomonia) via dimuon channel in PbPb collisions at $\sqrt{s_{NN}} = 2.76\,\text{TeV}$. Detailed measurements of J/ψ and $\psi(2S)$ have been made in different kinematic ranges [2–4]. The excellent mass resolution in dimuon channel allows precise measurement of the three Υ states and their relative yields in pp, pPb as well as in PbPb system [5–7]. The LHC results on quarkonia have been interpreted using a combination of suppression and regeneration models [9] although the regeneration phenomena needs to be confirmed by other observables. Recently, quarkonium measurements in Pb+Pb collisions at $\sqrt{s_{NN}} = 5.02\,\text{TeV}$ have been made which are described in this report. CMS has excellent capability of displaced vertex measurements which enables measurement of D an B mesons containing heavy quarks. These measurements can constrain various energy loss mechanisms [10].

Prashant Shukla on behalf of CMS collaboration.

P. Shukla (✉)
Nuclear Physics Division, Bhabha Atomic Research Center, Mumbai, India
e-mail: pshukla@barc.gov.in

P. Shukla
Homi Bhabha National Institute, Anushakti Nagar, Mumbai, India

Md. Naimuddin (ed.), *XXII DAE High Energy Physics Symposium*, Springer Proceedings in Physics 203, https://doi.org/10.1007/978-3-319-73171-1_8

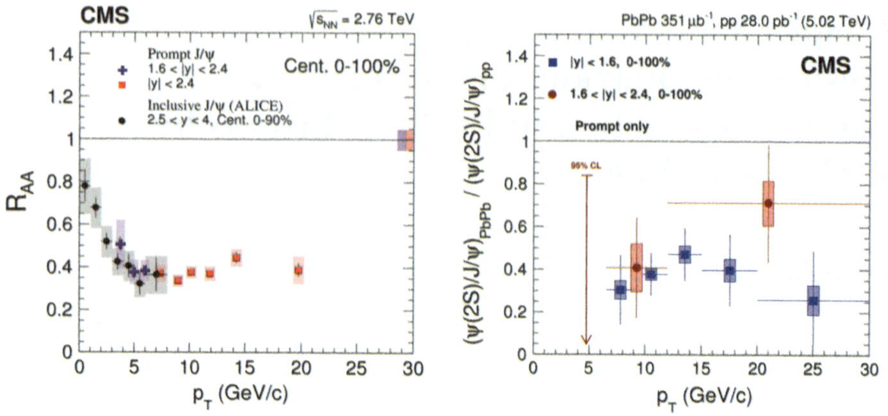

Fig. 8.1 (Left panel) The nuclear modification factor (R_{AA}) of J/ψ as a function of p_T measured by CMS [3] and ALICE [1] experiments. (Right panel) Double ratio of $\psi(2S)$ and J/ψ as a function of p_T measured by CMS in two rapidity regions [11]

Charmonia Measurements

The CMS experiment carries out J/ψ measurements in $|y| < 2.4$. In the forward rapidity range ($1.6 < |y| < 2.4$) it can measure J/ψ from $p_T > 3.0\,\text{GeV}/c$ for PbPb analysis which require tighter muon selection.

Figure 8.1 (left panel) shows R_{AA} of J/ψ in PbPb collisions at $\sqrt{s_{NN}} = 2.76\,\text{TeV}$ as a function of p_T measured by CMS [3] and ALICE [1] experiments. The three low p_T points of CMS are consistent with the ALICE data. The R_{AA} shows slightly upward trend for p_T above $5\,\text{GeV}/c$. The ALICE J/ψ data shows that R_{AA} increases with decreasing p_T below $4\,\text{GeV}/c$. This could be suggestive of regeneration of J/ψ at low p_T. Figure 8.1 (right panel) shows the double ratio of $\psi(2S)$ and J/ψ as a function of p_T in PbPb collisions at $\sqrt{s_{NN}} = 5.02\,\text{TeV}$ [11]. The circles are for forward rapidity region ($1.6 < |y| < 2.4$) and the squares are for central rapidity region ($|y| < 1.6$). The figure shows that $\psi(2S)$'s are more suppressed than J/ψ in all kinematic regions at 5.02 TeV.

Bottomonia Measurements

CMS measurements reveal that the higher Υ states are more suppressed relative to the ground state. This phenomenon is called sequential suppression. Figure 8.2 (Left panel) shows the double ratio of Υ(2S)/Υ(1S) as a function of N_{part} in PbPb collisions at $\sqrt{s_{NN}} = 5.02\,\text{TeV}$ and the right panel shows the same as a function of p_T [12]. These measurements show that the excited states Υ(3S) and Υ(2S) are more suppressed than Υ(1S). The double ratio Υ(2S)/Υ(1S) is flat over measured p_T and rapidity range.

Heavy Flavour Measurements

The heavy flavoured mesons are tagged through a displaced vertex. The detection of D^0 and B^+ mesons are made via $K + \pi$ and $J/\psi + K$ channels, respectively.

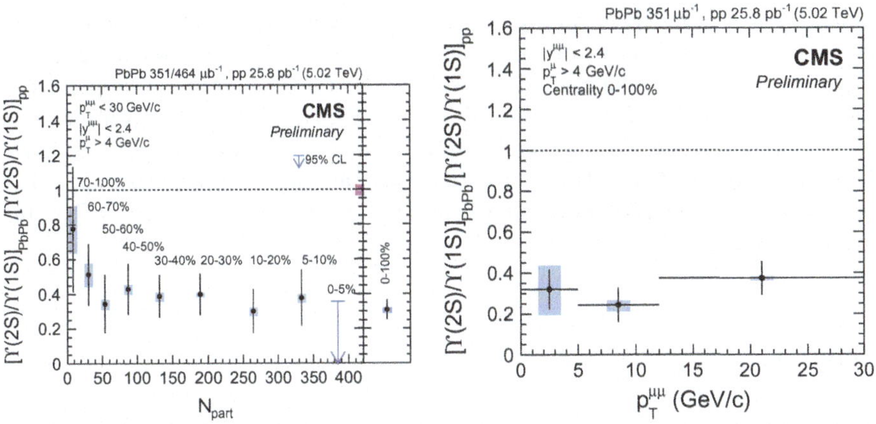

Fig. 8.2 (Left panel) The double ratio of $\Upsilon(2S)/\Upsilon(1S)$ as a function of N_{part} in PbPb collisions at $\sqrt{s_{NN}} = 5.02\,\text{TeV}$. (Right Panel) Same as left panel but as a function of p_T [12]

Fig. 8.3 Nuclear modification factor (R_{AA}) of D and B mesons compared to R_{AA} of light charged hadrons [13, 14]

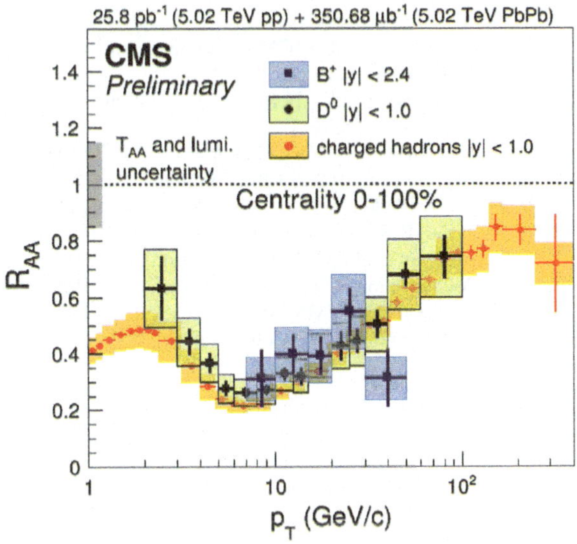

Figure 8.3 shows the R_{AA} of D^0 and B^+ compared to R_{AA} of light hadrons [13, 14]. The measurements show that at $p_T > 8\,\text{GeV}/c$ the suppressions of both D and B mesons are consistent with that of light hadrons.

Summary

With the recent results from PbPb collisions at $\sqrt{s_{NN}} = 2.6$ and $5.02\,\text{TeV}$, an overall understanding of quarkonium and heavy flavour production in heavy ion collisions is emerging. The R_{AA} of J/ψ shows that it remains suppressed even at very high p_T upto $20\,\text{GeV}/c$. The ALICE J/ψ data suggest regeneration of J/ψ at low p_T due to recombination of independently produced charm pairs. The double ratio of $\psi(2S)$ and J/ψ shows that $\psi(2S)$ are more suppressed than J/ψ in all kinematic regions at $\sqrt{s_{NN}} = 5.02$ TeV. The excited states $\Upsilon(3S)$ and $\Upsilon(2S)$ are more suppressed as compared to $\Upsilon(1S)$. The double ratio $\Upsilon(2S)/\Upsilon(1S)$ is found to be flat over measured p_T and rapidity range. The measurements of D^0 and B^+ show that at $p_T > 8$ GeV/c, the suppressions of D and B mesons are consistent with light hadrons at 5.02 TeV. Detailed measurements of R_{AA} of quarkonia and heavy flavours at $\sqrt{s_{NN}} = 5.02$ TeV are reported in Quark Matter 2017.

References

1. B.B. Abelev et al., [ALICE Collaboration], Phys. Lett. B **734**, 314 (2014), arXiv:1311.0214 [nucl-ex]
2. S. Chatrchyan et al., [CMS Collaboration], JHEP **1205**, 063 (2012), arXiv:1201.5069 [nucl-ex]
3. V. Khachatryan et al., [CMS Collaboration], arXiv:1610.00613 [nucl-ex]
4. V. Khachatryan et al., [CMS Collaboration], Phys. Rev. Lett. **113**, 262301 (2014), arXiv:1410.1804 [nucl-ex]
5. S. Chatrchyan et al., [CMS Collaboration], Phys. Rev. Lett. **107**, 052302 (2011), arXiv:1105.4894 [nucl-ex]
6. S. Chatrchyan et al., [CMS Collaboration], Phys. Rev. Lett. **109**, 222301 (2012), arXiv:1208.2826 [nucl-ex]
7. S. Chatrchyan et al., [CMS Collaboration], JHEP **1404**, 103 (2014), arXiv:1312.6300 [nucl-ex]
8. S. Chatrchyan et al., [CMS Collaboration], JINST **3**, S08004 (2008)
9. V. Kumar, P. Shukla, R. Vogt, Phys. Rev. C **92**, 024908 (2015), arXiv:1410.3299 [hep-ph]
10. K. Saraswat, P. Shukla, V. Singh, Nucl. Phys. A **943**, 83 (2015), arXiv:1507.06742 [nucl-th]
11. A.M. Sirunyan et al., [CMS Collaboration], arXiv:1611.01438 [nucl-ex]
12. CMS Collaboration [CMS Collaboration], CMS-PAS-HIN-16-008
13. CMS Collaboration [CMS Collaboration], CMS-PAS-HIN-16-001
14. CMS Collaboration [CMS Collaboration], CMS-PAS-HIN-16-011

Chapter 9
Recent Status of the Understanding of Neutrino-Nucleus Cross Section

H. Haider, M. Sajjad Athar and S. K. Singh

9.1 Introduction

The neutrino-nucleus (ν-A) cross sections are important input to the systematics of analysing ν oscillation experiments. For precise measurements, there are two challenging tasks: (i) good knowledge of neutrino fluxes and (ii) well understood nuclear medium effects (NME) in the entire region of neutrino energy spectrum. In this presentation, we have considered the following neutrino interactions:

$$
\begin{array}{rlrl}
\mathrm{QE}: & {}^{\circ}{}_{\mathrm{l}} + \mathrm{n} \rightarrow l^- + p & \pi: & \nu_l + N \rightarrow l^- + \pi^i + N' \\
\mathrm{K}: & {}^{\circ}{}_{\mathrm{l}} + \mathrm{N} \rightarrow l^- + K^i + N'; & \eta: & \nu_l + N \rightarrow l^- + \eta + N' \\
\mathrm{AP}: & {}^{\circ}{}_{\mathrm{l}} + \mathrm{N} \rightarrow l^- + Y + K^i; & \mathrm{DIS}: & {}^{\circ}{}_{\mathrm{l}} + \mathrm{N} \rightarrow \mathrm{l}^- + \mathrm{X}
\end{array}
$$

$$
i = +,\ 0\ or\ -; \quad N, N' = p\ or\ n; \quad Y = \Lambda, \Sigma; \quad X = jet\ of\ hadrons.
$$

on nucleon and nuclear targets. Similar reactions take place with $\bar{\nu}_l$. It is estimated that due to NME the cross sections have an overall uncertainty of 20–25% [1, 2].

9.2 Quasielastic (QE) Reactions

Most of the present Monte Carlo generators use Fermi gas model (FGM) of Smith and Moniz [3] to analyse the experimental results on $\nu - A$ cross section. However, in the literature there are other variants of FGM like the model used by Aligarh [4, 5] and Valencia [6] groups, where local FGM with long range correlations of the particle-hole excitations in the nuclear medium were included or the calculations where nucleon spectral functions are used [7]. The data from MiniBooNE, K2K and

H. Haider (✉) · M. Sajjad Athar · S. K. Singh
Department of Physics, Aligarh Muslim University, Aligarh 202002, India
e-mail: huma.haider8@gmail.com

Md. Naimuddin (ed.), *XXII DAE High Energy Physics Symposium*, Springer Proceedings in Physics 203, https://doi.org/10.1007/978-3-319-73171-1_9

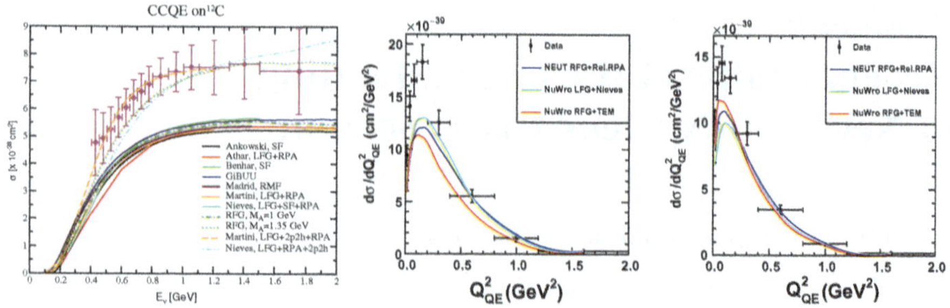

Fig. 9.1 (Left) σ_{CCQE} versus E_{ν_μ} in ^{12}C calculated within several models along with the MiniBooNE data [1]. (Mid and Right) Comparison of the best fit MC predictions in NEUT and NuWro to MINERvA CCQE data. The clear difference in the normalisation between the MC and the experimental data arises from the MINERvA data placing a much stronger constraint on the cross-section shape than its normalisation [10]

MINOS could not be explained by using Smith and Moniz model [3] of QE reactions using the world average value of axial dipole mass $M_A(=1.026\,\text{GeV})$. Martini et al. [8] and later Nieves et al. [9] showed the importance of two particle-two hole (2p-2h) contributions, the inclusion of 2p-2h along with the FGM with RPA effect gives satisfactory explanation of the data. Their results are also consistent with the values reported from NOMAD and MINERvA. The results are shown in Fig. 9.1 (left). Recently Stowell et al. [10] compared the MINERvA CCQE data with two different MC generators NEUT and NuWro, the results of which have been shown in Fig. 9.1 (mid and right). It may be seen that the data and the results from the MC generators are not in agreement and it may be concluded that better theoretical models are still needed.

9.3 Single π Production

The data on the weak pion production in ν-A scattering in the experiments performed at MiniBooNE, SciBooNE and more recently from MINERνA have highlighted the inadequacy of our present understanding of nuclear medium and final state interaction (FSI) effects. Furthermore, there is lack of consensus in the theoretical modeling of basic reaction mechanism of $\nu(\bar{\nu})$ induced 1π production from the free nucleon, specially concerning the role of background terms and higher resonances. By including the contribution of the background terms and higher resonances along with the dominant Δ resonance, Aligarh group [11] has studied weak charged and neutral current induced 1π production from nucleons and fitted the reanalysed ANL and BNL data [12], and the results are shown in Fig. 9.2 (left). It was concluded that the best description of the reanalyzed experimental data of ANL and BNL experiments is obtained when we take $C_5^A(0) = 1.0$ and $M_A = 1.026\,\text{GeV}$ for $\text{N} - \Delta$ axial vector transition current form factor $C_5^A(Q^2)$. Recently Valencia group [13]

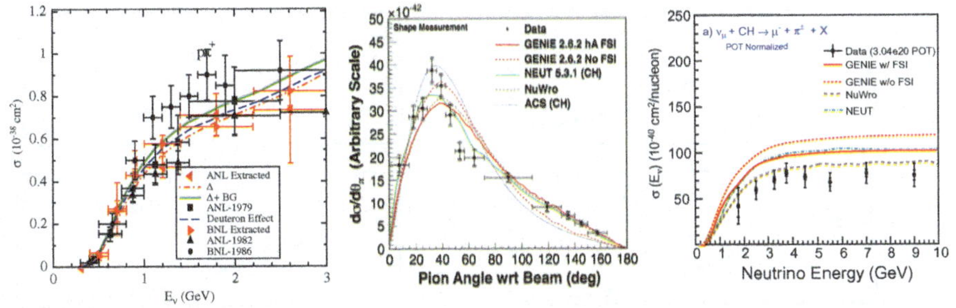

Fig. 9.2 (Left) $\sigma_{CC1\pi}$ versus E_{ν_μ} for $\nu_\mu p \rightarrow \mu^- p \pi^+$ process. Experimental results are reanalyzed data points of ANL and BNL experiments. For details please see [11]. (Mid) Pion angular distribution in several models along with the MINERvA data [15]. (Right) MINERνA data of CC1π production and comparison with the different MC generators [16]

implemented unitarity using Watson's theorem and applying it to fit old and reanalysed ANL and BNL data, the best fit was found with $C_5^A(0) = 1.12 \pm 0.11$ and $C_5^A(0) = 1.14 \pm 0.07$ respectively, and $M_A = 0.954 \pm 0.063$ GeV. In the case of 1π production from nuclear targets, the dominant contribution from the Δ resonance is suppressed due to modifications of mass and width of the Δ propagator in the nuclear medium [14]. Moreover, the produced pions undergo FSI with the residual nucleus which further reduces the pion production. In Fig. 9.2 (mid), the results for CC1π production in ^{12}C obtained by different theoretical groups are shown along with the MINERνA data [15]. Recently McGivern et al. [16] have published MINERνA data of CC1π production and compared the results with different theoretical approach incorporated in MC generators (Fig. 9.2 right). It may be clearly observed that there is a discrepancy between the theoretical and experimental results which need further study. There is another mode of pion production from nuclear target known as coherent pion production where a nucleus after the interaction remains in the ground state and all the energy is transferred to the outgoing pion. The work by the Aligarh group [17] and the Valencia group [18] have shown that the contribution from coherent pion production is highly suppressed due to NME and is 2–3% of the total pion production. Besides 1π production there can be 2π or multipion production for which details can be found in [19].

9.4 Deep-Inelastic Scattering (DIS)

There is no sharp kinetic region to distinguish the onset of the DIS region from the resonance region but the region $W \geq 2.0$ GeV and $Q^2 \geq 1.0$ GeV^2 is considered to be the safe DIS region. At high energy and Q^2, the inclusive DIS cross sections are usually expressed in terms of the structure functions which are derived in terms of quark PDFs using the methods of perturbative QCD. These structure functions are experimentally determined from DIS experiments on nucleon and nuclear targets.

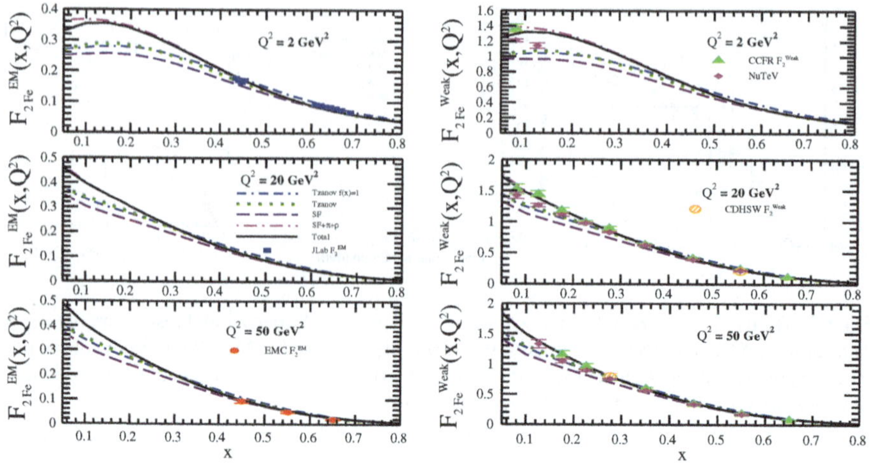

Fig. 9.3 Results of EM (Left panel) and Weak (Right panel) nuclear structure functions in ^{56}Fe (isoscalar) obtained using spectral function (long dashed line), including mesonic contribution (dashed-dotted line), full model (solid line)

The observation of EMC effect has led to the presence of strong NME in DIS region. Recently a comparative study of NME in $F_{2A}^{EM}(x, Q^2)$ and $F_{2A}^{Weak}(x, Q^2)$ has been done by Kulagin and Petti [20] and the Aligarh group [21]. In Fig. 9.3, we show the results [21] for $F_{2A}^{EM,Weak}(x, Q^2)$ obtained using the spectral function only, also including the mesonic contribution, and finally using the full model which also includes shadowing. It may be observed that NME in EM and Weak interactions are not exactly the same.

9.5 $|\Delta S| = 0$ and $|\Delta S| = 1$ Processes off Nucleon

The single hyperon (by $\bar{\nu}$) or single kaon ($K/\bar{K}$) are produced by $|\Delta S| = 1$ reaction mechanism (Cabibbo suppressed) while η meson and associated production of kaon are induced by the $|\Delta S| = 0$ weak currents. It has been shown by Aligarh group [22–24] that for the precise determination of neutrino oscillation parameters these reactions are important.

References

1. T. Katori, M. Martini, arXiv:1611.07770 [hep-ph]
2. J.G. Morfin et al., Adv. High Energy Phys. **2012**, 934597 (2012)
3. R.A. Smith, E.J. Moniz, Nucl. Phys. B **43**, 605 (1972)
4. S.K. Singh, E. Oset, Phys. Rev. C **48**, 1246 (1993)

5. M. Sajjad Athar, S.K. Singh, Phys. Rev. C **61**, 028501 (2000); F. Akbar et al., Int. J. Mod. Phys. E **24**, 1550079 (2015)
6. J. Nieves et al., Phys. Rev. C **70**, 055503 (2004)
7. O. Benhar, Int. J. Mod. Phys. E **23**, 1430005 (2014)
8. M. Martini, M. Ericson, Phys. Rev. C **87**, 065501 (2013)
9. J. Nieves et al., Phys. Lett. B **721**, 90 (2013)
10. P. Stowell et al., arXiv:1611.03275 [hep-ex]
11. M. Rafi Alam et al., Int. J. Mod. Phys. E **25**, 1650010 (2016)
12. C. Wilkinson et al., Phys. Rev. D **90**, 112017 (2014)
13. L. Alvarez-Ruso et al., Phys. Rev. D **93**, 014016 (2016)
14. M. Sajjad Athar et al., Phys. Rev. D **75**, 093003 (2007)
15. B. Eberly et al., Phys. Rev. D **92**, 092008 (2015)
16. C.L. McGivern et al., Phys. Rev. D **94**, 052005 (2016)
17. S.K. Singh et al., Phys. Rev. Lett. **96**, 241801 (2006)
18. L. Alvarez-Ruso et al., Phys. Rev. C **75**, 055501 (2007)
19. E. Hernandez et al., Phys. Rev. D **77**, 053009 (2008)
20. S.A. Kulagin, R. Petti, Phys. Rev. D **76**, 094023 (2007)
21. H. Haider et al., Phys. Rev. C **84**, 054610 (2011); Nucl. Phys. A **955**, 58 (2016)
22. M. Rafi Alam et al., Phys. Rev. D **82**, 033001 (2010); AIP Conf. Proc. **1680**, 020001 (2015)
23. S.K. Singh et al., AIP Conf. Proc. **1680**, 020012 (2015)
24. F. Akbar et al., Phys. Rev. D **94**, 114031 (2016)

Chapter 10
Review of Latest RHIC Results and Future Perspectives

Lokesh Kumar

10.1 Introduction

The RHIC experiments aim to study the formation of Quark Gluon Plasma (QGP), its properties, and the Quantum Chromodynamics (QCD) phase diagram [1]. These aspects are explored by colliding heavy-ions at the relativistic speed and analyzing the data produced. The study of particle production under these extreme conditions shed light on properties of the medium produced in such collisions. In 2010, RHIC experiments started the energy scan to explore the QCD phase diagram called the BES program [2]. The motivation was to search for the QCD phase boundary, first-order phase transition, and the critical point [3]. The first phase started with collection of data for Au+Au collisions at three energies $\sqrt{s_{NN}} = 7.7$, 11.5, and 39 GeV. In 2011, data were collected for two more energies $\sqrt{s_{NN}} = 19.6$ and 27 GeV. This enabled the RHIC to cover large portion of the phase diagram that included data from 7.7 to 200 GeV. The results presented here are from both Solenoidal Tracker at RHIC (STAR) and PHENIX experiments [1].

10.2 Results from the Beam Energy Scan

Figure 10.1 shows the chemical freeze-out temperature T_{ch} versus the baryon chemical potential μ_B for three different centralities at different energies [4]. The curves represent the model predictions [5]. For lower energies, the T_{ch} versus μ_B exhibits the centrality dependence. Higher RHIC energies show little centrality dependence. As can be seen, the coverage of phase diagram is quite large at RHIC and it is pos-

L. Kumar (✉)
Department of Physics, Panjab University, Chandigarh 160014, India
e-mail: lokesh@pu.ac.in

Md. Naimuddin (ed.), *XXII DAE High Energy Physics Symposium*, Springer Proceedings in Physics 203, https://doi.org/10.1007/978-3-319-73171-1_10

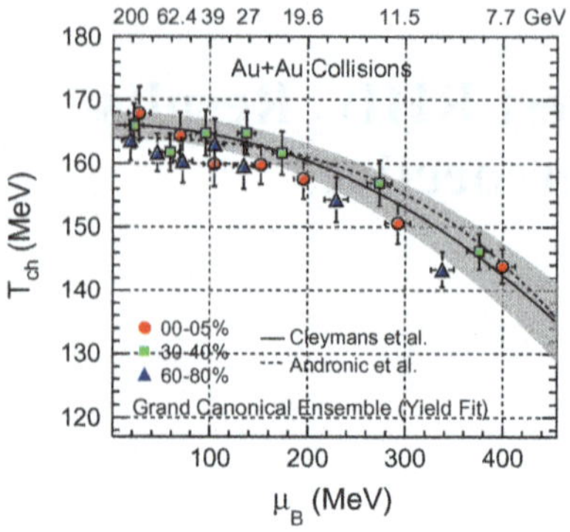

Fig. 10.1 Chemical freeze-out temperature versus baryon chemical potential for different centralities

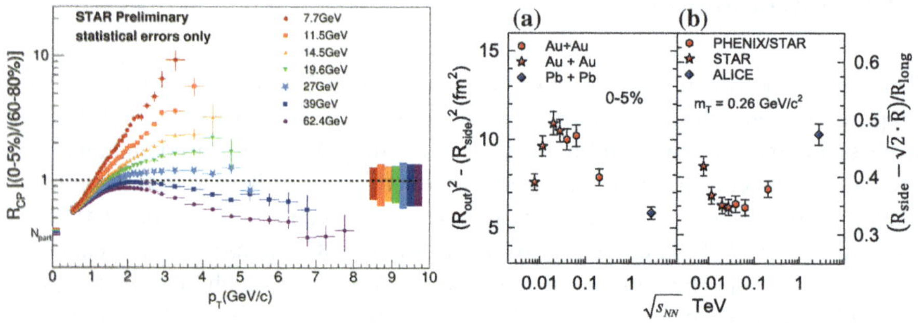

Fig. 10.2 Left: R_{CP} versus p_T at different energies.; Right: Energy dependence of **a** $R_{\mathrm{out}}^2 - R_{\mathrm{side}}^2$ and **b** $(R_{\mathrm{side}} - \sqrt{2}\bar{R})/R_{\mathrm{long}}$

sible to look for signatures of critical point, phase boundary and first-order phase transition at RHIC energies.

Figure 10.2 (left) shows the nuclear modification factor R_{CP} versus p_T at different energies [6]. Suppression of high-p_T particle at higher RHIC energies reflects the partonic energy loss while traversing QGP dense medium. Thus, $R_{\mathrm{CP}} < 1$ at high-p_T is considered as one of the signatures of the QGP formation at higher energies. The figure shows that at lower energies, the $R_{\mathrm{CP}} \geq 1$ suggesting that QGP might not have formed at lower energies. There are other results which are consistent with disappearance of QGP at lower energies such as baryon-meson splitting, deviation of ϕ-meson v_2 from other particles [7], disappearance of splitting between same and opposite sign charges [8].

Figure 10.2 (right) shows the energy dependence of the observables constructed using the Hanbury Brown and Twiss (HBT) radii for central collisions [9]. The observable $R_{\mathrm{out}}^2 - R_{\mathrm{side}}^2$ reflects the emission duration of the source, while $(R_{\mathrm{side}} - \sqrt{2}\bar{R})/R_{\mathrm{long}}$, reflects its expansion speed [10]. The energy dependence of these observables show a non-monotonic trend i.e a maximum is observed for emission duration and a minimum is observed for the expansion speed. This might indicate the passage of the system through the softest region in the equation of state [10]. Another observable that can provide information related to first-order phase transition is the slope of proton directed flow v_1 as a function of rapidity. The energy dependence of proton v_1 slope also shows a non-monotonic behavior [11]. A simple UrQMD model which does not have the phase transition cannot explain this trend.

In the vicinity of the critical point, the correlation length is expected to diverge. The higher moments (skewness S and kurtosis κ) of distributions of conserved quantities such as net- baryons, net-charge, and net-strangeness, have a good sensitivity to the correlation length [12]. The moment products such as $\kappa\sigma^2$ and $S\sigma$ are equal to the ratios of various order susceptibilities calculated in Lattice QCD and HRG model. Figure 10.3 shows the energy dependence of various products of moments of net-proton distributions at different centralities [13]. The bottom panel shows $\kappa\sigma^2$ compared with poisson expectation and UrQMD model results. Both the poisson distribution and UrQMD model do not include any critical behavior and hence show a flat/monotonic dependence with energy. The $\kappa\sigma^2$ seems to exhibit a non-monotonic behavior. This may be an indication of possible critical point around this region. However, the error bars at lower energy data are large and it is difficult to make conclusive statement.

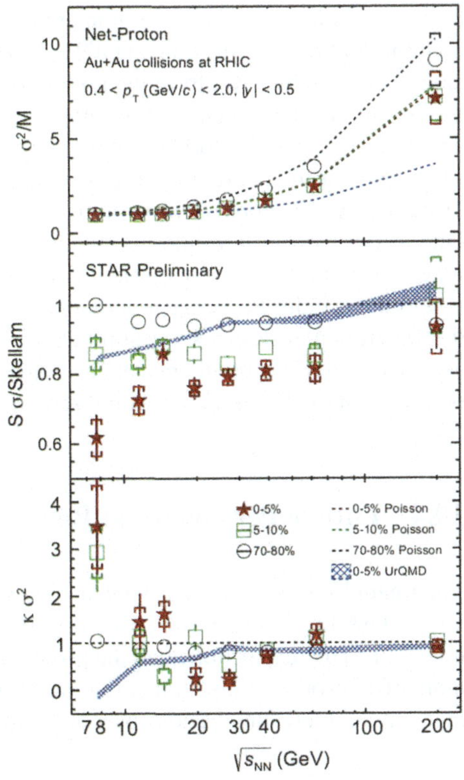

Fig. 10.3 Energy dependence of various products of moments of net-proton distributions at different centralities

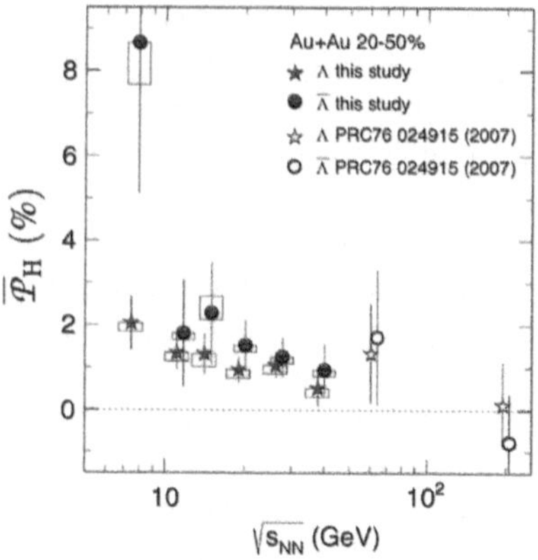

Fig. 10.4 Energy dependence of Λ ($\bar{\Lambda}$) polarization

10.3 Observation of Global Λ Polarization

The non-central heavy-ion collisions can lead to a large angular momentum of the order of $1000\hbar$ and large magnetic field. The shear forces generated in heavy-ion collisions may result in vorticity to the fluid [14]. The spin-orbit coupling in QCD can generate a spin alignment or polarization along the direction of vorticity which is on average parallel to total angular momentum [14]. Thus, polarization measurements of hadrons emitted from the fluid or QGP can be used to determine the vorticity. Here, the spin of emitted hyperon from the midrapidity fireball is aligned with the net angular momentum of the system and is called global polarization. The global polarization arising from QCD spin-orbit coupling would align both Λ and $\bar{\Lambda}$ along the direction of net angular momentum with same magnitude. The global polarization signal may also arise due to coupling of hadronic dipole magnetic moment with magnetic field. Due to this effect, Λ would be aligned anti-parallel and $\bar{\Lambda}$ along the direction of magnetic field which is same as of net angular momentum. Thus, overall effect would be to induce a larger polarization for $\bar{\Lambda}$ than Λ baryons. Figure 10.4 shows the experimental results suggesting positive polarization at each energy for both Λ and $\bar{\Lambda}$ [14]. The experimental measurements are also not inconsistent with larger polarization of $\bar{\Lambda}$ than Λ but would require more data to quantify the difference.

10.4 Future Perspectives: Beam Energy Scan-II

The Beam Energy Scan-I has been a great success. It covers a large portion of QCD phase diagram to search for critical point, phase boundary, and first-order phase transition. The results indicated the disappearance of QGP signals at lower energies, hints of first-order phase transition, and hints of possible critical point. There are many more interesting trends observed at lower energies. These include observation of difference between particle and anti-particle v_2 at lower energies [7], different behavior of Ω/ϕ ratio as a function of p_T at lower energies [15], almost constant behavior of $\langle m_T - m \rangle$ for light hadrons as a function of collision energy [4], and non-monotonic behavior of third-harmonic azimuthal correlations as a function of collision energy [16].

However, most of the important results need higher statistics data to make conclusive statements. In view of this, RHIC has planned the BES-II which will consolidate the BES findings. The RHIC will collect high statistics data below 20 GeV during 2019–2020. From collider side, there will be improvement (about a factor of 6–7 near 20 GeV) in luminosity due to the use of electron cooling device and longer bunches. From experimental side, there will be upgrades of detectors to improve the important measurements by improving acceptances in momentum and rapidity. In addition, there is a plan in STAR to run the Fixed Target program (FXT) to extend the reach of the BES program up to μ_B 700 MeV [17].

Acknowledgements We thank Prof. B. Mohanty and Dr. N. Sharma for the discussions. This work was supported by the UGC start-up grant of LK.

References

1. J. Adams et al., (STAR Collaboration), Nucl. Phys. A **757**, 102 (2005); K. Adcox et al., (PHENIX Collaboration), Nucl. Phys. A **757**, 184 (2005)
2. B. Mohanty, Nucl. Phys. A **830**, 899c (2009); M. Aggarwal et al., (STAR Collaboration) (2010), arXiv:1007.2613; L. Kumar, Mod. Phys. Lett. A **28**, 1330033 (2013)
3. S. Ejiri, Phys. Rev. D **78**, 074507 (2008); E.S. Bowman, J.I. Kapusta, Phys. Rev. C **79**, 015202 (2009); M.A. Stephanov, Int. J. Mod. Phys. A **20**, 4387 (2005)
4. L. Adamczyk et al., (STAR Collaboration) (2017), arXiv:1701.07065
5. J. Cleymans et al., Phys. Rev. C **73**, 034905 (2006); A. Andronic et al., Nucl. Phys. A **834**, 237c (2010)
6. S. Horvat, (STAR Collaboration), Nucl. Phys. A **956**, 838 (2016)
7. L. Adamczyk et al., (STAR Collaboration), Phys. Rev. Lett. **110**, 0142301 (2013)
8. L. Adamczyk et al., (STAR Collaboration), Phys. Rev. Lett. **113**, 052302 (2014)
9. R.H. Brown, R.Q. Twiss, Nature **178**, 1046 (1956)
10. A. Adare et al., (PHENIX Collaboration) (2014), arXiv:1410.2559
11. L. Adamczyk et al., (STAR Collaboration), Phys. Rev. Lett. **112**, 162301 (2014)
12. M.M. Aggarwal et al., (STAR Collaboration), Phys. Rev. Lett. **105**, 022302 (2010); F. Karsch, K. Redlich, Phys. Lett. B **695**, 136 (2011); S. Gupta et al., Science **332**, 1525 (2011)
13. J. Thäder, (STAR Collaboration), Nucl. Phys. A **956**, 320 (2016)
14. L. Adamczyk et al., (STAR Collaboration) (2017), arXiv: 1701.06657
15. L. Adamczyk et al., (STAR Collaboration), Phys. Rev. C **93**, 21903 (2016)
16. L. Adamczyk et al., (STAR Collaboration), Phys. Rev. Lett. **116**, 112302 (2016)
17. K. Meehan, (STAR Collaboration), Quark Matter 2017

Chapter 11
Status of INO-ICAL Detector

Venktesh Singh

Introduction

The existence of non-zero neutrino masses has profound implications on various fields apart from being of fundamental interest to particle physics [1]. Although we know a few properties of neutrino, still large number of properties is unknown to us. The experimental field of neutrino is now moving into a phase where high precision experiments are needed to play a leading role [2]. INO is a major step in this direction. The Iron Calorimeter (ICAL) detector, which consists of a massive electromagnet (50 kt) having magnetic field of 1.3 T, with Resistive Plate Chambers (RPC) as active detector elements, expected to be placed inside a mountain at Pottipuram, Theni near Madurai in South India. INO Collaboration consists of nearly 100 scientists from 23 research institutes and universities all over India [3]. It is one of the largest basic science projects in India in terms of work force. INO-ICAL detector with dimensions of 49m $\times$ 16m $\times$ 14.5m will consist of three modules of 16.2m $\times$ 16m $\times$ 14.5m each. The details of the INO-ICAL are mentioned in Ref. [3]. Two prototype detector stacks were built, first one at TIFR, Mumbai and still taking data since very long time and the second one is magnetized at VECC, Kolkata, which took data for a short time. A third prototype detector stack of 2m $\times$ 2m dimension glass RPC was recently built and is currently operational at IICHEP Madurai. This article focuses on the short over view of the INO-ICAL detector and basic tools of quality control.

Status of ICAL Detector

ICAL will consist of around 29,000 RPC detectors made of 1.95m $\times$ 1.91m $\times$ 0.024m glass/bakelite electrodes with a gas gap of 2 mm. The performance and characteristics study in terms of various parameters of the RPC detectors have been done [4]. ICAL

For INO Collaboration.

V. Singh (✉)
Department of Physics, Institute of Science, Banaras Hindu University,
Varanasi 221005, India
e-mail: venkaz@yahoo.com

Md. Naimuddin (ed.), *XXII DAE High Energy Physics Symposium*, Springer Proceedings in Physics 203, https://doi.org/10.1007/978-3-319-73171-1_11

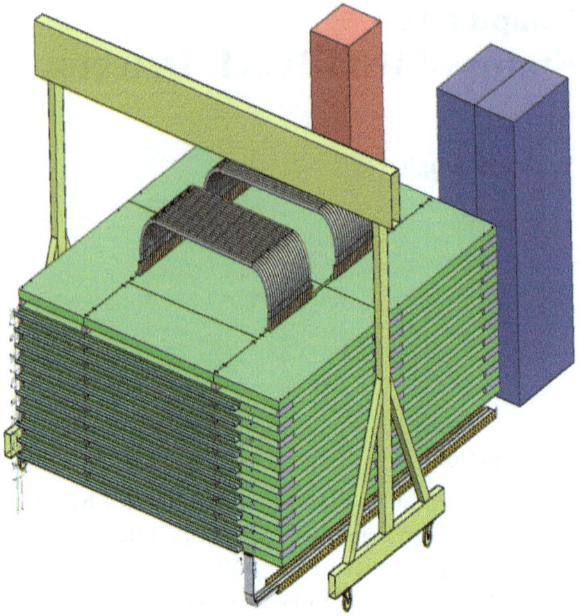

Fig. 11.1 Schematic design of mini-ICAL

Table 11.1 The mini-ICAL detector dissimilar physical parameter values with full ICAL are listed

Parameter	ICAL (Full scale)	mini-ICAL	Parameter	ICAL (Full scale)	mini-CAL
No. of modules	03	01	No. of RPCs/Road/Layer	08	02
Number of layers	150	10	No. of Road/Layer/Module	08	01
No. of RPC units/Layer	192	02	No. of RPC units	28,800 (1,07,266 m^2)	20 (74.5 m^2)
Gap for RPC trays	40 mm	45 mm	No. of readout strips	3,686,400	2560

module calibration, electronics, data acquisition etc protocols are developed. Now, with a fixed target date, collaboration is building a mini-ICAL as shown in Fig. 11.1, with components identical to those to be used in the future ICAL at IICHEP in Madurai. The RPC detector and readout channels are tabulated in Table 11.1. In this context, more than 400 gas gaps of full dimension are under production at Saint Gobain in Chennai out of which ~150 have been produced. All the related processes are implemented by automatic systems. Slow control parameters needed to monitor the RPC detectors and their environmental conditions such as the leak test, ambient temperature and pressure measuring devices have been more advanced and precise. A few dielectric materials such as silicon fiber sheet, PVC foam with Aluminum foil lamination, etc have been selected for the readout panel and the final choice is expected shortly. The performance and characteristic study of silicon fiber sheet based readout panel shows satisfactory results [5]. A magnetic field of 1.5 T will be used to discriminate between neutrino and anti-neutrino interactions, giving ICAL the capability to determining the neutrino mass hierarchy. Magnet design and its parameters have been tuned and finalized [6]. Required number/amount of soft iron plates has been procured. The design of the trigger scheme for the ICAL detector consists of a distributed and hierarchical architecture. The detector module is logically subdivided into identical segments for the purpose of trigger generation. The segment dimensions are chosen considering the expected hit pattern of the events of interest, the associated trigger rates and the feasibility of implementation. The ICAL trigger system is finalized and all the boards are fabricated [7]. Related firmware development and programming are in progress. Tools related to simulation and events reconstruction have been developed and validated. The closed-loop gas-recycling system has been designed, developed and validated [8]. Almost all requirements for INO-ICAL seem to be ready and the remaining ones will be completed in due course of time, as final experiment will proceed.

Development of Quality Control Tools

Various automatic tools have been developed for the testing of different parameters of the ICAL. Collection of the quality data requires that each connection of signal transmission must be properly ensured and long lasting. However, soldering on aluminum surface is not easy. Therefore, application of the Ansol Aluminium soldering flux provides satisfactory results [9]. An automatic prototype continuity checking system based on the Operational Amplifier i.e. an amplifier with high gain and high input impedance used especially in circuits for performing mathematical operations on an input voltage, has been developed. Connectivity between copper strip and the aluminum sheet of the readout panel has to be checked before installation. The connectivity scanning mechanism will scan/check all connection of the readout panel. A fast and accurate automatic connectivity scanning mechanism may affect the appraisal of the scientific results. Therefore, a detail study of the performance of upgraded version of the earlier system has been done [10]. Study of surface resistivity of the conductive paint coated glass electrode is equally important. However, for testing the uniformity of the coating on the glass surface require automatic system. Currently such automatic system is available at IIT-Chennai, Saint Gobain (Chennai) and Banaras Hindu University (Varanasi). At BHU, the performance of prototype automatic system has been demonstrated and testing of full-scale (2.25m $\times$ 2.25 m); system is under progress [11]. Such system is also useful for the calibration and efficiency study of the RPC detector using radioactive source.

A need for suitable material, having fire proof/flame resistant, moisture proof, flexible, lightweight, cost effective, locally available and most importantly with proper impedance, may act as dielectric material in the readout panel, is necessary and important. Study on Silicon Fiber Sheet (SFS) has shown promising result [5]. Chemical and elemental composition of SFS has been done using X-ray florescence techniques [12]. Development of various other quality control tools is under way.

Summary and Outlook

INO-ICAL collaboration has prepared all ingredients for the construction of the mini-ICAL at the transit campus of IICHEP in Madurai. RPC detector related research and development work is completed. However, R & D is a continuous process; it will mature, and be polished with time. Presently, two B = 0 prototype RPC detector stacks are operational, one using 1m × 1m RPCs at TIFR-Mumbai running since long time and the other using 2m × 2m RPCs operational at IICHEP in Madurai. The existing B = 0 IICHEP detector will continue to operate with the mini-ICAL. INO collaboration is progressing on various fronts such as study of various required minor modifications in electronics, magnet, DAQ, gas system etc. Meanwhile RPC production at Saint Gobain industry is progressing well.

Acknowledgements We acknowledge the financial support from Department of Science and Technology (DST), New Delhi. We would also like to thank the INO Collaboration and INO group at Tata Institute of Fundamental Research (TIFR), Mumbai for providing help and support in terms of the raw materials for detector construction.

References

1. A. Ghosh, Ph.D. thesis, Homi Bhabha National Institute, Bhabha Atomic Research Center, Mumbai (2014); A. Ghosh et al., JHEP **2013**, 09 (2013)
2. D. Kaur et al., Eur. Phys. J. **C75**, 156 (2015)
3. S. Ahmed et al., (INO Collaboration) (2015), arXiv:1505.07380
4. B. Satyanarayan, Ph.D. thesis, IIT Bombay (2009); Md. Naimuddin et al., JINST **9**, C10039 (2014); S. Biswas, Ph.D. thesis, University of Calcutta (2010)
5. M.K. Singh et al., JINST **12**, T01006 (2017)
6. S.P. Behera et al., IEEE Trans. Magn. **51**, 7300409 (2015)
7. S. Dasgupta et al., NIM **A678**, 105 (2012); S. Dasgupta, Ph.D. thesis, Homi Bhabha National Institute, Bhabha Atomic Research Center, Mumbai (2013)
8. S.D. Kalmani et al., JINST **11**, C11026 (2016)
9. M.K. Singh et al., Proc. DAE NP **59**, 950 (2014); A. Kumar et al., IJIREST **04**(11), 10766 (2015)
10. N. Marimuthu et al., IJIREST **05**(3), 3807 (2016); N. Marimuthu et al., Proc. DAE NP **61**, 1094 (2016)
11. S. Shree et al., Proc. DAE NP **59**, 854 (2014); A. Kumar et al., Proc. DAE NP **60**, 1008 (2015); A. Pandey et al., Proc. DAE NP **61**, 1060 (2016)
12. M.K. Singh et al., Proc. DAE NP **59**, 976 (2014); M.K. Singh et al., IJIREST **04**(10), 10035 (2015); M.K. Singh et al., Proc. DAE NP **60**, 950 (2015)

Chapter 12
Search for Sterile Neutrino Signal in the ^7Be Solar Neutrino Measurement with KamLAND

Ashish Sharma, Govind Singh, Gazal Sharma, Shankita Bhardwaj, Surender Verma and B. C. Chauhan

12.1 Introduction

The perspective of solar neutrino physics changed due to important and unprecedented achievements of KamLAND [1] and Borexino [2]. The first experimental observation [3] of solar neutrinos gave rise to Solar Neutrino Problem (SNP). The Sudbury Neutrino Observatory (SNO) [4] and Super-Kamiokande (SK) [5] experiment confirmed that there are at least three types of active solar neutrino flavors. The flavor conversion between different active neutrinos does not change the total neutrino fluxNeutrino flux observed by various experiments, unless some of the neutrinos are converting to the sterile ones. The conversion of active to sterile neutrino flavor is possible via Resonant Spin Flavor Precession (RSFP) [6] caused by interaction of solar neutrinos with transverse magnetic field of Sun through their magnetic moment on their propagation towards the surface. This conversion will be time dependent if

A. Sharma · G. Singh · G. Sharma · S. Bhardwaj · S. Verma
B. C. Chauhan (✉)
Department of Physics and Astronomical Science, School of Physical and Material Sciences, Central University of Himachal Pradesh (CUHP), Dharamshala, Kangra 176215, Himachal Pradesh, India
e-mail: chauhan@associate.iucaa.in

A. Sharma
e-mail: sharma.ashish2615@gmail.com

G. Singh
e-mail: gsgovind22@gmail.com

G. Sharma
e-mail: gazzal.sharma555@gmail.com

S. Bhardwaj
e-mail: shankita.bhardwaj982@gmail.com

S. Verma
e-mail: s_7verma@yahoo.co.in

Md. Naimuddin (ed.), *XXII DAE High Energy Physics Symposium*, Springer Proceedings in Physics 203, https://doi.org/10.1007/978-3-319-73171-1_12

the magnetic field of Sun is varying with time. Thus a significant time modulation in the solar neutrino fluxNeutrino flux can be seen as the existence of sterile neutrinos.

In this paper we worked on the existence of sterile neutrinos in relation to RSFP mechanism for the ^7Be solar neutrino measurement with KamLAND in the light of our model [7] and predictions done for ^7Be signal variation at KamLAND [8]. We re-investigate the time variation of solar magnetic field affect on the solar neutrino flux and presence of sterile neutrino signal thereof, in the KamLAND ^7Be solar neutrino data [9].

12.2 2+1 Model

The model was built with the motivation to explain the certain experimental observations which were not being explained by the accepted solution for SNP i.e. Large Mixing Angle (LMA) [10]. In the model two active neutrinos and one sterile neutrino flavor mix in mass eigenstates as [7, 11]

$$\begin{pmatrix} \nu_s \\ \nu_e \\ \nu_x \end{pmatrix} = \begin{pmatrix} c_\alpha & s_\alpha & 0 \\ -s_\alpha c_\theta & c_\alpha c_\theta & s_\theta \\ s_\alpha s_\theta & -c_\alpha s_\theta & c_\theta \end{pmatrix} \begin{pmatrix} \nu_0 \\ \nu_1 \\ \nu_2 \end{pmatrix}, \tag{12.1}$$

where θ and α denote the LMA vacuum mixing angle and vacuum-sterile mixing angle, respectively. We consider $\alpha = 0$, so that active states ν_e and ν_μ communicate to the sterile neutrino through transition magnetic moment only. The matter Hamiltonian thus obtained is

$$H_M = \begin{pmatrix} \frac{-\Delta m_{10}^2}{2E} & \mu_1 B & \mu_2 B \\ \mu_1 B & \frac{\Delta m_{21}^2}{2E} s_\theta^2 + V_e & \frac{\Delta m_{21}^2}{4E} s_{2\theta} \\ \mu_2 B & \frac{\Delta m_{21}^2}{4E} s_{2\theta} & \frac{\Delta m_{21}^2}{2E} c_\theta^2 + V_\mu \end{pmatrix}. \tag{12.2}$$

Here V_e and V_ν are the matter induced potentials for active neutrinos ν_e and ν_μ, respectively, and μ_1 and μ_2 are their transition magnetic moments to the sterile neutrinos. We consider the magnetic field of sun to be concentrated around the bottom of convective zone to have strong time modulation effect on solar neutrinos [12]. As a consequence both processes LMA and RSFP occur separately at very different solar radii. An efficient conversion to sterile neutrino requires that the transition moment associated with this order of magnitude difference dominates over the other so we choose $\Delta m_{10}^2 = \mathcal{O}(10^{-8}\,\text{eV}^2)$ and $\mu_2 = 0$. The neutrino survival probability (P_{ee}) was calculated by assuming the propagation inside the sun to be purely adiabatic except in the region where adiabaticity parameter reaches its minimum. So the survival and conversion probabilities became

$$P_{ee} = |(\nu_e(t), \nu_e)|^2 = c_\theta^2 c_{12i}^2 P_C + s_\theta^2 s_{12i}^2 , \quad (12.3)$$
$$P_{e\mu} = |(\nu_\mu(t), \nu_e)|^2 = s_\theta^2 c_{12i}^2 P_C + c_\theta^2 s_{12i}^2 . \quad (12.4)$$

Subscript i refers to the mixing angle at production point. The P_C represent jump probability given by Landau Zener approximation

$$P_C = exp\left(\frac{\pi}{2}\gamma_c\right) , \quad (12.5)$$

where γ_c is the adiabaticity parameter given by ratio between the eigenvalues difference and rate of mixing angle

$$\gamma_c = \left|\frac{\lambda_0 - \lambda_1}{2\dot{\theta}_{01}}\right| . \quad (12.6)$$

Here λ_0 and λ_1 are the mass matter eigenvalues corresponding to sterile neutrinos and solar ν_e. Our model predicts the clear time variational effect only in the low energy sector of solar neutrinos. For low energy solar neutrinos to have spin flavour precession resonating near the base of convective zone where the solar field is maximum, we choose the magnitude of mass square difference between one of the active state and sterile state as of $\mathcal{O}(10^{-8}\,\text{eV}^2)$.

12.3 Result and Discussion

The low energy threshold of KamLAND detector made it possible to measure the real time detection of ^7Be solar neutrinos. These neutrinos at KamLAND are detected through the electron elastic scattering process. The main problem with the ^7Be solar neutrino detection at KamLAND was the low energy background due to Kr, Pb and external gamma rays which makes it hard to analyse the signal. There are several methods developed by the KamLAND collaborators for the effective removal of these backgrounds [13]. The visible electron kinetic energy spectrum is

$$R_{ES} = Q_0 \int_0^{T'_M} dT' f(T', T)\phi(E)\left[P_1(E)\frac{d\sigma_W}{dT'} + P_2(E)\frac{d\sigma_{W'}}{dT'}\right] \quad (12.7)$$

Here Q_0 is normalization factor, $\phi(E)$ is ^7Be solar neutrinos flux observed by KamLAND corresponding to energy E. $P_1(E)$ is the survival probability of ν_e , while $P_2(E)$ is the production probability of ν_μ such that $P_1(E) + P_2(E) \leq 1$. The term $f(T', T)$ is the energy resolution function of the KamLAND detector in terms of physical (T') and measured (T) electron kinetic energy. However $\frac{d\sigma_W}{dT'}$ and $\frac{d\sigma_{W'}}{dT'}$ are the differential scattering cross-section corresponding to the ν_e and ν_μ scattering, respectively.

The ^7Be solar neutrinos are monoenergetic and consist of single line spectrum, therefore by fixing the value of Δm^2_{10}, all the neutrinos will resonate exactly at same position and experience the same solar magnetic field. As such these neutrinos are blind to solar magnetic field profile and as a consequence the ^7Be solar neutrino signal in KamLAND are solar field profile independent. At present the knowledge regarding the Sun's magnetic field is unknown. The KamLAND data taking phase for ^7Be solar neutrinos lies in the starting of current solar cycle (from April 2009 to 2011 June) and the corresponding observation of solar magnetic field variation is given in [14] (Fig. 12.1).

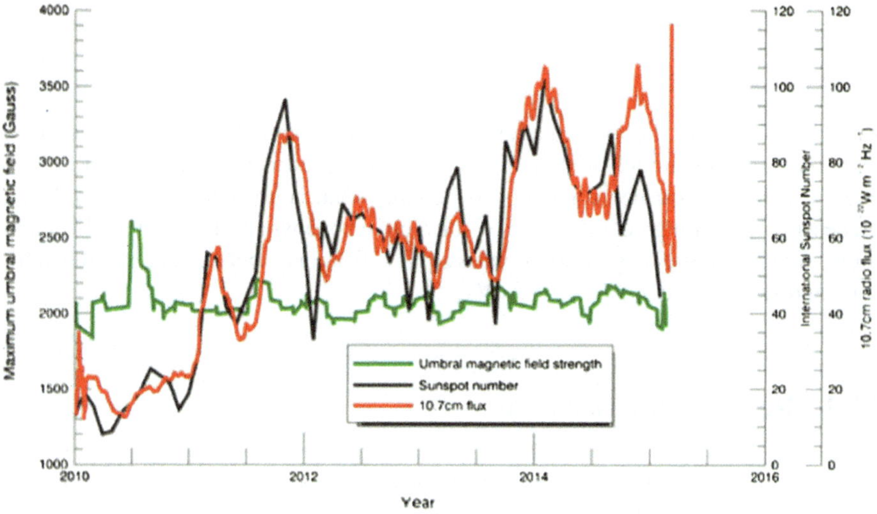

Fig. 12.1 Depiction of magnetic field with sunspot number for solar cycle 24 taken from [14]

Fig. 12.2 Background subtracted spectrum for ^7Be solar neutrinos at KamLAND

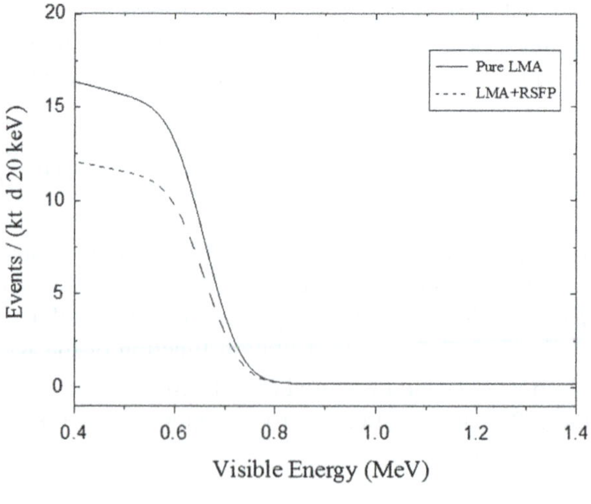

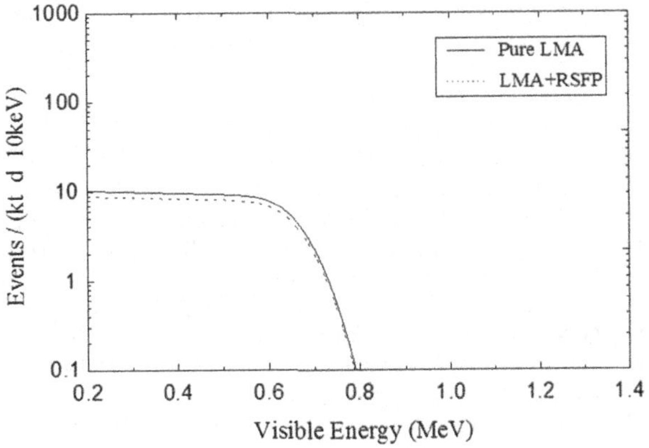

Fig. 12.3 Energy spectrum of selected ^7Be solar neutrinos at KamLAND

The KamLAND experiment determined the first direct spectroscopic value of ^7Be solar neutrino energy, $E_\nu = 862 \pm 16$ keV and observed solar neutrino interaction rate, $R_{KL} = 582 \pm 94(ktd)^{-1}$ which corresponds to ^7Be solar neutrino flux Neutrino flux of $(3.26 \pm 0.52) \times 10^9\,\text{cm}^{-2}\,\text{s}^{-1}$ with survival probability $P_{ee} = 0.66 \pm 0.15$ [9]. We use the KamLAND mixing parameters [1] consistent to explain the solar data and SNP and the value of neutrino magnetic moment $\mu_\nu = 2.9 \times 10^{-11}$ [15].

Figure 12.2 shows the background subtracted spectrum of KamLAND in the energy range of 0.4–1.4 MeV . The time modulation of the ^7Be solar neutrino spectrum with in the energy range 0.2–1.4 MeV shown in Fig. 12.3.

12.4 Conclusion

In conclusion, on re-investigation of our model predictions done for ^7Be solar neutrino of KamLAND, we found the flux variation and sterile neutrino signal thereof, are not conclusive. This is because the KamLAND data is not time binned and data available is not falling in the region of solar field peak as the solar activity during the exposure time was on the rising side of the current solar cycle (2014–2015).

Acknowledgements B.C. Chauhan and all the other authors acknowledge the financial support provided by University Grant Commission (UGC), Govt. of India vide Grant No. UGC-MRP-MAJOR-PHYS-2013-12281 and CUHP.

References

1. A. Gando et al., (KamLAND Collaboration), Phys. Rev. D **88**, 033001 (2013)
2. G. Bellini et al., (Borexino Collaboration), Phys. Rev. D **89**, 112007 (2014)
3. R. Davis, Phys. Rev. Lett. **12**, 303–305 (1964)

4. Q.R. Ahmad et al., (SNO Collaboration), Phys. Rev. Lett. **87**, 071301 (2001)
5. S. Fukuda et al., (Super-Kamiokande Collaboration), Phys. Rev. Lett. **86**, 5651 (2001)
6. C.S. Lim, W.J. Marciano, Phys. Rev. D **37**, 1368 (1988); E.Kh. Akhmedov, Yad. Fiz **48**, 599 (1988) [Sov. J. Phys. **48**, 382 (1988)]; Phys. Lett. B **213**, 64 (1988)
7. B.C. Chauhan, J. Pulido, JHEP **06**, 008 (2004)
8. B.C. Chauhan, JHEP **02**, 035 (2006)
9. A. Gando et al., (KamLAND Collaboration), Phys. Rev. C **92**, 055808 (2015)
10. K. Eguchi et al., (KamLAND Collaboration), arXiv:hep-ex/0212021
11. P.C. de Holanda, A.Yu. Smirnov, Phys. Rev. D **69**, 113002 (2004)
12. H.M. Antia, S.M. Chitre, M.J. Thompson, arXiv:astro-ph/0005587; S. Couvidat, S. Turck-Chieze, A.G. Kosovichev, Astrophys. J. **599**, 1434 (2003)
13. G. Keefer et al., Nucl. Instrum. Methods A **769**, 79 (2015)
14. W. Livingston, F. Watson, Geophys. Res. Lett. **42**, 9185–9189 (2015); C. Jin, J. Wang, Astrophys. J. **806**, 174 (2015)
15. A.G. Beda et al., Adv. High Energy Phys. **350150**, 12 (2012)

Chapter 13
A Model for Anisotropic Strange Stars

Debabrata Deb, Sourav Roy Chowdhury, Saibal Ray, Farook Rahaman and B. K. Guha

13.1 Introduction

The interior of the compact stars may have differential fluid pressures which are directionally dependent. Ruderman [1] argued that the nuclear matter may have anisotropic features with very high density ranges ($>10^{15}$ gm/cm^3), where the nuclear interaction must be treated relativistically. On the contrary, Bowers and Liang [2] investigated that anisotropy might have non-negligible effects on different physical parameters. However, one can note the recent literature regarding anisotropic matter distribution where several authors have considered anisotropy in connection to compact stars [3–7].

Our motivation in the present investigation is to construct a model for strange star by assuming that (i) the compact star is made up of anisotropic fluid, and (ii) the

D. Deb (✉) · S. Roy Chowdhury · B. K. Guha
Department of Physics, Indian Institute of Engineering Science and Technology, Shibpur, Howrah 711103, West Bengal, India
e mail: d.deb32@gmail.com

S. Roy Chowdhury
e-mail: b.itsme88@gmail.com

B. K. Guha
e-mail: bkguhaphys@gmail.com

S. Ray
Department of Physics, Government College of Engineering and Ceramic Technology, Kolkata 700010, West Bengal, India
e-mail: saibal@associates.iucaa.in

F. Rahaman
Department of Mathematics, Jadavpur University, Kolkata 700032, West Bengal, India
e-mail: rahaman@associates.iucaa.in

Md. Naimuddin (ed.), *XXII DAE High Energy Physics Symposium*, Springer Proceedings in Physics 203, https://doi.org/10.1007/978-3-319-73171-1_13

MIT Bag model is applicable in this environment. We have studied several physical parameters, viz., the radius, redshift and anisotropy among which it can be observed that anisotropy is increasing with the radial coordinate inside the stellar system from zero at the center to it's maximum at the surface for all the strange stars.

13.2 The Einstein Field Equations and Their Solutions

We consider the metric of the stellar system as follows

$$ds^2 = -e^{\nu(r)}dt^2 + e^{\lambda(r)}dr^2 + r^2(d\theta^2 + sin^2\theta d\phi^2), \tag{13.1}$$

where the metric potentials ν and λ are functions of the radial coordinate r only.

The energy-momentum tensor can be provided as

$$T_\nu^\mu = (\rho + p_r)u^\mu u_\nu - p_t g_\nu^\mu + (p_r - p_t)\eta^\mu \eta_\nu, \tag{13.2}$$

where ρ, p_r and p_t of (13.2) represents the energy density, radial and tangential pressures respectively of the fluid sphere.

According to the phenomenological MIT bag model, the EOS of strange matter can be written in a linear form

$$p_r = \frac{1}{3}(\rho - 4\,B_g), \tag{13.3}$$

where ρ, p_r and B_g are the energy density, the radial pressure and the bag constant respectively.

Again, assuming that the matter within the spherical system follows the density profile proposed by Mak and Harko [8]:

$$\rho(r) = \rho_c\left[1 - \left(1 - \frac{\rho_0}{\rho_c}\right)\frac{r^2}{R^2}\right], \tag{13.4}$$

where ρ_c and ρ_0 are the central and surface densities respectively.

Now, from the well known Einstein field equations one can find out expression for the anisotropy as

$$\Delta(r) = p_t - p_r = \frac{[(12\,{c_1}^2 r^4 - \frac{1}{5}\,c_1 c_5 r^2 + c_6)\,\pi - \frac{9}{2}\,c_1 R^2]r^2}{36R^2[r^2\left(\frac{3}{5}\,c_1 r^2 - \rho_c R^2\right)\pi + \frac{3}{8}\,R^2]}, \tag{13.5}$$

where $c_1 = (\rho_c - \rho_0)$, $c_5 = \frac{1}{5}\,(156\,\rho_c - 84\,\rho_0)\,R^2$, $c_6 = (6\,\rho_c - 3\,\rho_0)\,(4\,\rho_c - \rho_0)R^4$.

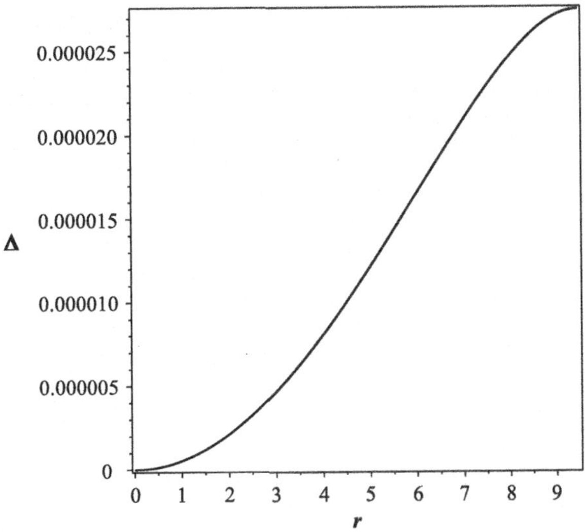

Fig. 13.1 Variation of anisotropy as a function of the radial coordinate r for the strange star $LMC\ X-4$

Table 13.1 Physical parameters as derived from the proposed model for $LMC\ X-4$ [9]

Observed mass (in $M_\odot$)	Predicted radius (km)	Redshift (Z)	Δ (0) (in km^{-2})	Δ (R) (in km^{-2})	Δ'' (R) (in km^{-2})
1.29 ± 0.05	9.48	0.29	0	0.000027	−0.0000031

13.3 Discussion and Conclusion

Based on the present model for compact stars let us consider a particular candidate $LMC\ X-4$ which is a representative of the strange stars. For this star, we have found out numerical values of the several physical parameters with the Bag constant 83 $\mathrm{MeV}/(\mathrm{fm})^3$. The model yields the values of the maximum mass and maximum radius as $M_{max} = 3.54 M_\odot$ and $R_{max} = 11.811\,km$ respectively. We also find that anisotropy is minimum at the center and maximum at the surface, which is clear from Fig. 13.1 as well as Table 13.1.

References

1. R. Ruderman, Annu. Rev. Astron. Astrophys. **10**, 427 (1972)
2. R. Bowers, E. Liang, Astrophys. J. **188**, 657 (1974)
3. V. Varela, F. Rahaman, S. Ray, K. Chakraborty, M. Kalam, Phys. Rev. D **82**, 044052 (2010)
4. F. Rahaman, S. Ray, A.K. Jafry, K. Chakraborty, Phys. Rev. D **82**, 104055 (2010)
5. F. Rahaman, P.K.F. Kuhfittig, M. Kalam, A.A. Usmani, S. Ray, Class. Quantum Gravity **28**, 155021 (2011)
6. F. Rahaman, R. Maulick, A.K. Yadav, S. Ray, R. Sharma, Gen. Relativ. Gravit. **44**, 107 (2012)

7. M. Kalam, F. Rahaman, S. Ray, S.M. Hossein, I. Karar, J. Naskar, Eur. Phys. J. C **72**, 2248 (2012)
8. M.K. Mak, T. Harko, Chin. J. Astron. Astrophys. **2**, 248 (2002)
9. M.L. Rawls, J.A. Orosz, J.E. McClintock, M.A.P. Torres, C.D. Bailyn, M.M. Buxton, Astrophys. J. **730**, 25 (2011)

Chapter 14
Re-discovery of the SM Higgs Boson in Diphoton Channel at $\sqrt{s} = 13$ TeV at CMS in LHC

Arnab Purohit

14.1 Introduction

After almost half a centuary from Higgs boson's theoretical developments, in July 2012, in the two large particle detectors, ATLAS and CMS, of LHC, a new particle whose properties are most consistent with Standard Model (SM) Higgs boson, with mass around 125 GeV, was announced to be discovered. In this report mass measurement of the Higgs boson in it's diphoton decay channel, at the CMS detector, is presented. The analysis [1] is based on $12.9\,\text{fb}^{-1}$ of p-p collision data collected at center-of-mass energy of 13 TeV. The SM Higgs bosons are produced in the LHC environment primarily through gluon-gluon fusion (ggH), vector boson fusion (VBF), and associated production with a W or Z (VH) or $t\bar{t}$ pairs (ttH). The clean final state topology of the diphoton decay channel allows the mass of the Higgs boson to be reconstructed with high precision. However there is large background originating from irreducible background processes producing two prompt γ and reducible background processes producing one prompt photon and a neutral meson, where the meson is misidentified as a γ and two misidentified neutral mesons, which are reconstructed as γ. The Higgs boson decays into two photons primarily via either a top quark loop or a W boson loop. But the amplitudes of those modes interfere destructively resulting in a branching fraction of around 0.2% for a SM Higgs boson at 125 GeV. Despite the small branching fraction and the presence of a large diphoton continuum background, the diphoton decay mode provides an expected signal significance for the 125 GeV SM Higgs boson that is one of the highest among all the decay modes.

On behalf of the CMS Collaboration.

A. Purohit (✉)
Saha Institute of Nuclear Physics, HBNI, Kolkata, West Bengal, India
e-mail: arnabpurohit@cern.ch

Md. Naimuddin (ed.), *XXII DAE High Energy Physics Symposium*, Springer Proceedings in Physics 203, https://doi.org/10.1007/978-3-319-73171-1_14

14.2 Analysis Strategy

A multivariate likelihood regression technique is used to correct individual photons energies and provide an estimate of the per-photon energy resolution. In order to select the primary vertex from those of other interactions in the same bunch crossing a boosted decision tree (BDT) is deployed. An additional BDT is trained to know the probability that the vertex choice was correct. A Boosted Decision Tree (BDT) is trained to distinguish prompt photons from non-prompt photons. Figure 14.1 shows the BDT output of the photon ID.

To achieve the best performance, after the selection of the diphoton system, the events are categorized in eight mutually exclusive classes depending on the production modes and the different signal over background ratio. Figure 14.2 shows the distribution of the diphoton MVA output for simulated events (stacked histograms) and data (black dots). The variable is built to have a flat distribution on signal events (red shades histograms). The vertical dashed lines show the boundaries for the definition of the four Untagged categories. Several systematic uncertainties have been evaluated: the scale and resolution for the photon's energy, photon identification (preselection and BDT), vertex finding efficiency, trigger efficiency, theoretical production cross sections and uncertainty due to the additional objects in the exclusive categories (jet, leptons and missing energy). Parametric models are built both for signal and background components in order to perform the signal extraction in the different categories. Extraction of the signal model is done from simulation, using a sum of gaussian functions to describe the different components contributing to the diphoton invariant mass. The model is fit to several Higgs mass hypotheses in a

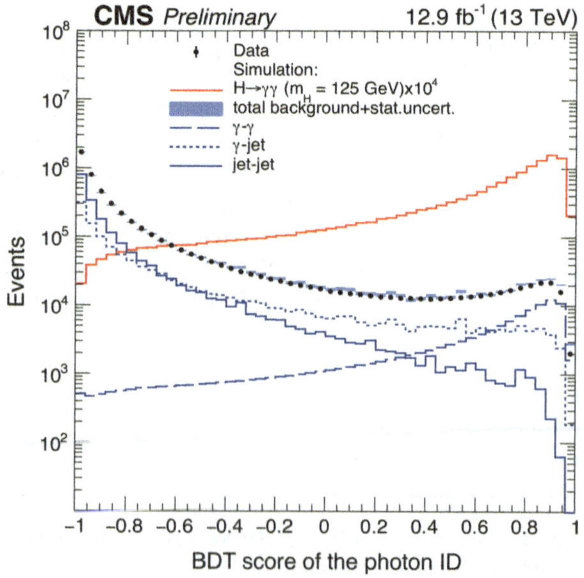

Fig. 14.1 Photon Id BDT output [1]

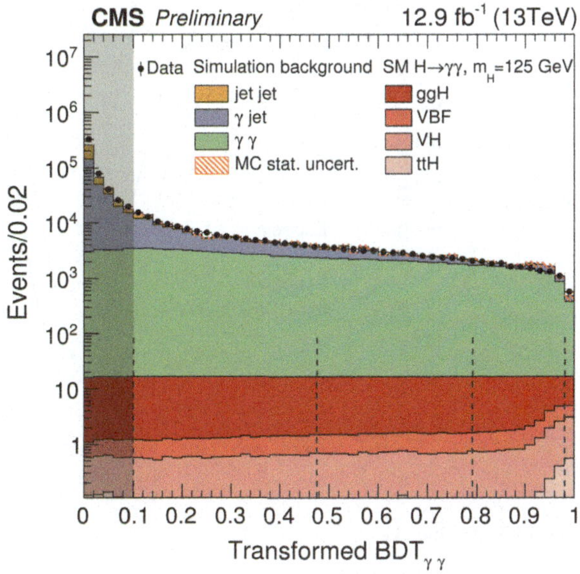

Fig. 14.2 Diphoton MVA output [1]

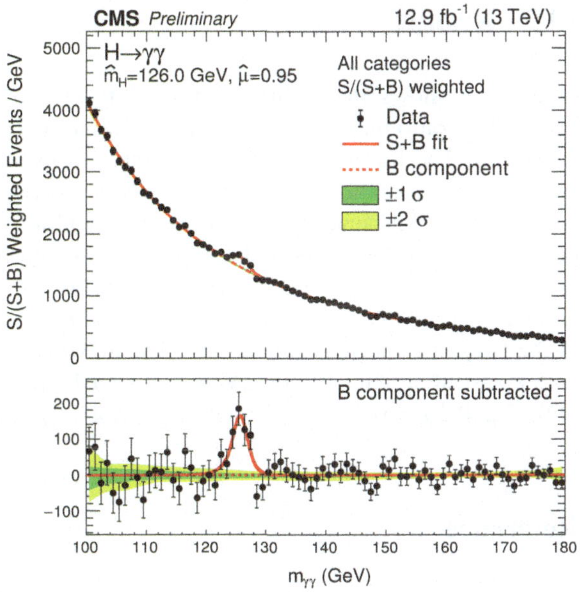

Fig. 14.3 Diphoton invariant mass distribution in data [1]

mass range between 120 and 130 GeV. The extraction of the background model is completely data driven. The choice of the functional form for the description of the background diphoton mass distribution in each category is treated as a discrete nuisance parameter floating over an initial set of possible functional forms. The analysis technique is described in detail in [1].

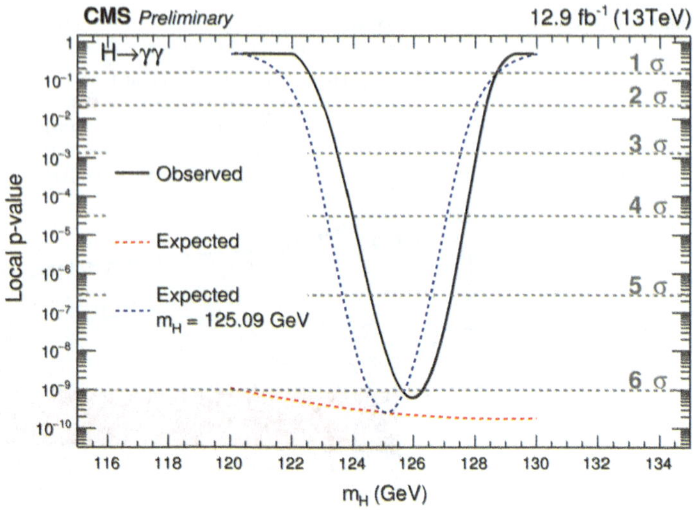

Fig. 14.4 p value [1]

14.3 Results

Figure 14.3 shows the invariant mass weighted for their signal to signal-plus-background ratio of the diphoton system for all the events selected by the different categories. The lower plot shows the excess of events due to the presence of the Higgs boson, after the subtraction of the background. The solid red line shows the signal-plus-background fit, while the dashed red line indicates the background component of the fit. Figure 14.4 shows the local p-value observed in data (solid black line), expected local p-value for a Higgs boson with a mass of 125.09 GeV (blue dashed line) and minimum expected local p-value as function of the mass hypothesis of the Higgs boson (red dashed line). The maximum observed significance is 6.1σ with a best fit at mass 126.0 GeV, with a signal strength relative to standard model of $0.95 \pm 0.20 = 0.95 \pm 0.17$ (stat)$^{+0.10}_{-0.07}$ (syst) $^{+0.08}_{-0.05}$ (theo).

Reference

1. Updated measurements of Higgs boson production in the diphoton decay channel at $\sqrt{s} = 13$ TeV in pp collisions at CMS. CMS-PAS-HIG-16-020, CERN, Geneva (2016)

Chapter 15
Neutrino Phenomenology Within Inverse and Type II Seesaw Based on S_4 Flavor Symmetry

Ananya Mukherjee and Mrinal Kumar Das

15.1 Introduction

Neutrinos have been shown to have tiny but nonzero mass by several neutrino oscillation experiments which remain massless in the SM. To explain the origin of tiny neutrino mass various BSM frameworks have already been there in the literature. At the same time some recent neutrino oscillation experiments **T2K**, **RENO** have also predicted non-zero value of reactor mixing angle. Here we present a TeV scale seesaw mechanism, the well known ISS [1] which offers the explanation for tiny neutrino mass at the cost of proposing the right handed neutrino mass at a scale much lower than that involved in the canonical seesaw. The leading order ISS neutrino mass is accompanied by type II seesaw to produce non-zero θ_{13} [2]. Non Abelian discrete symmetries have shown a promising role in explaining the origin of neutrino mass since long [3]. We present both of the seesaw mechanism within the framework of S_4 flavor symmetry. Z_2 symmetry helped us to construct the desired Yukawa coupling matrices from which we obtain various mass matrices of ISS and type II seesaw. We have extended the SM scalar sector by the inclusion of extra flavons whose charge assignments are given in Table 15.1. The Dirac Yukawa coupling is shown to play a crucial role in achieving the neutrino oscillation parameters within the correct 3σ range. The type II perturbation generates the non vanishing reactor mixing angle breaking the $\mu - \tau$ symmetric nature of light neutrino mass matrix obtained from ISS. We fit the model parameters with the help of recent neutrino oscillation data [4]. The Yukawa coupling is varied from $0.998 - 1$ to see whether the oscillation parameters fall within the correct 3σ or not. Standard and new physics contribution to effective mass prediction have been calculated. The model also shows some

A. Mukherjee (✉) · M. K. Das
Department of Physics, Tezpur University, Tezpur 784 028, Assam, India
e-mail: ananyam@tezu.ernet.in

M. K. Das
e-mail: mkdas@tezu.ernet.in

Md. Naimuddin (ed.), *XXII DAE High Energy Physics Symposium*, Springer Proceedings in Physics 203, https://doi.org/10.1007/978-3-319-73171-1_15

Table 15.1 Fields and their transformation properties under various symmetry group

	$\bar{L}$	ν_R	l_R	h	η	S	Φ_R	Φ_s	ζ	ξ	Δ
$SU(2)_L$	2	1	1	2	2	1	1	1	1	1	3
S_4	3_1	3_1	3_1	1_1	3_1	3_1	1_1	1_1	3_1	2	1
Z_2	+	−	+	−	−	−	+	+	−	−	−
Z_3	ω^2	ω	1	1	1	1	ω^2	1	ω	ω	ω
$U(1)_L$	−1	1	1	0	0	−1	0	2	0	0	0

correlations in obtaining the branching ratio governing LFV decays like $\mu \rightarrow 3e$ and $\mu \rightarrow e\gamma$. The Yukawa Lagrangian for the particle content of the presented model reads,

$$\mathcal{L} = y_D \bar{L}\nu_R\eta + y_D{}'\bar{L}\nu_R h + y_R\nu_R S\Phi_R + y_s SS\Phi_s + f_\nu LL\frac{\zeta\Delta}{\Lambda} + f_\nu LL\frac{\xi\Delta}{\Lambda}. \tag{15.1}$$

S_4 product rules allow us to have the ISS and type II seesaw mass matrices as follows

$$m_D = y\begin{pmatrix} v_h & 0 & 0 \\ 0 & v_h & v_\eta \\ 0 & v_\eta & v_h \end{pmatrix}, \mu = y_s\begin{pmatrix} 1 & 0 & 0 \\ 0 & 1 & 0 \\ 0 & 0 & 1 \end{pmatrix} v_s, M = y_R\begin{pmatrix} 1 & 0 & 0 \\ 0 & 1 & 0 \\ 0 & 0 & 1 \end{pmatrix} v_R. \tag{15.2}$$

with the vev alignments $\langle\Phi_R\rangle = v_R$, $\langle\Phi_s\rangle = v_s$, $\langle h\rangle = v_h$, $\langle\eta\rangle = v_\eta(1, 0, 0)$, $\Delta \sim v_\Delta$, $\langle\zeta\rangle \sim \sqrt{2}v_\zeta(1, 0)$, $\langle\xi\rangle \sim v_\xi(0, 1, -1)$. ζ and ξ are assumed to take the VEV in the same scale $v_\zeta = v_\xi = \Lambda$. Since, the charged lepton mass matrix is nondiagonal, U_ν is chosen in a basis, where charged lepton is diagonal, which implies $U_\nu \rightarrow U_{TBM} = U_\omega^\dagger U_\nu$. The ISS mass matrix and type II seesaw mass matrix are given by,

$$m_\nu = y^2\frac{y_s v_v}{y_R^2 v_R^2}\begin{pmatrix} v_h^2 & 0 & 0 \\ 0 & v_h^2 + v_\eta^2 & 2v_h v_\eta \\ 0 & 2v_h v_\eta & v_h^2 + v_\eta^2 \end{pmatrix}, m_{LL}^{II} = \begin{pmatrix} 0 & -w & w \\ -w & w & 0 \\ w & 0 & -w \end{pmatrix}. \tag{15.3}$$

15.2 Summery and Conclusion

It is worth mentioning that, the chosen Yukawa coupling range, which gives proper neutrino phenomenology also evinces $Sin^2\theta_{13}$ in agreement with experimental data. For the type II perturbation strength from 0.004295 to 0.005402 eV $Sin^2\theta_{13}$ falls around 0.017–0.0259 in case of Normal Hierarchy and in case of Inverted Hierarchy for type II strength from 0.00057 to 0.000681 eV we get $Sin^2\theta_{13}$ around

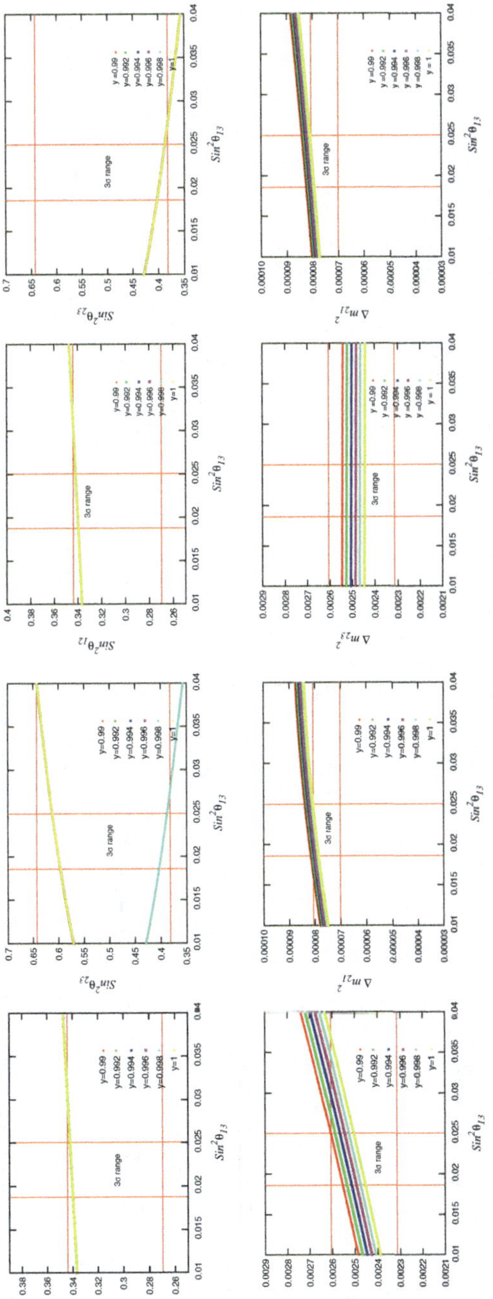

Fig. 15.1 Oscillation parameters as a function of $Sin^2\theta_{13}$ for NH (left panel) and IH (right panel)

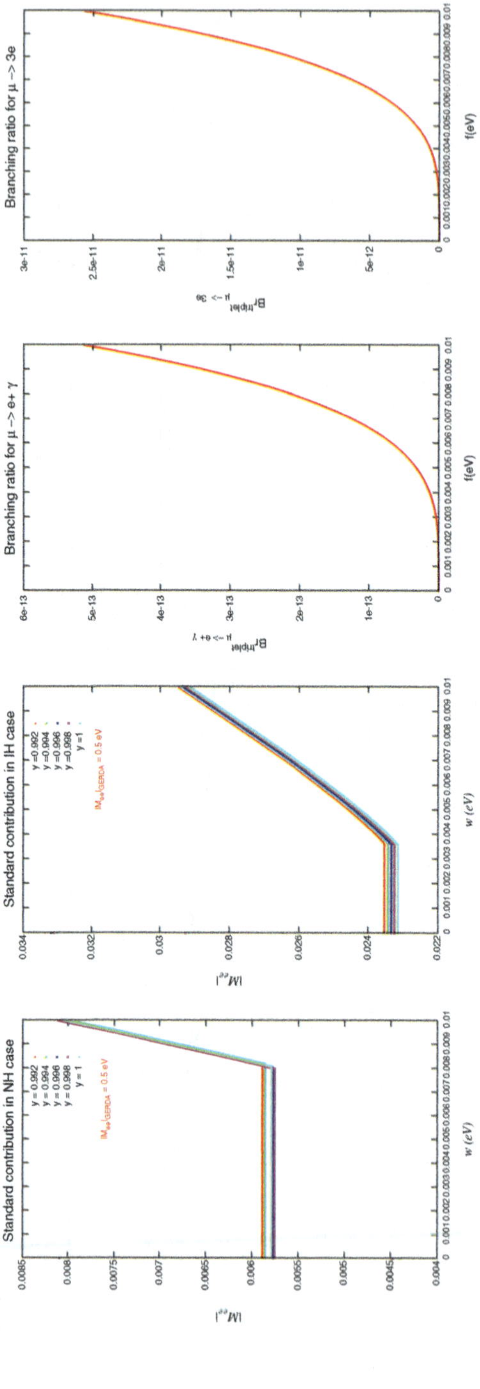

Fig. 15.2 Effective mass (left panel), Branching ratio (right panel) as a function of type II seesaw strength

0.017–0.0259. Figure 15.1 shows that the oscillation parameters are within the frame for Normal Hierarchy framework while taking into consideration the lower bound of 3σ deviation only. The proposed model also shows evidences for correct neutrino phenomenology using the best fit and lower 3σ bound in case of Inverted Hierarchy mass pattern. Both the standard and new physics contribution to $0\nu\beta\beta$ decay are obtained in the vicinity of experimental results. Figure 15.2. The model forecast for effective mass prediction fall within the GERDA bound. The model prediction for $Br^{triplet}_{\mu\rightarrow e\gamma}$ is found to be $<10^{-12}$ and for $Br^{triplet}_{\mu\rightarrow 3e}$ is 10^{-16}–10^{-12}. The presented model may give some notions in finding Baryon Asymmetry of the Universe through the process of a TeV scale leptogenesis which is natural in the proposed Inverse seesaw model.

References

1. A.G. Dias, C.A. de S. Pires, P.S. Rodrigues da Silva, A. Sampieri, A simple realization of the inverse seesaw mechanism. https://doi.org/10.1103/PhysRevD.86.035007
2. A. Mukherjee, M.K. Das, Neutrino phenomenology and scalar dark matter with A_4 flavor symmetry in inverse and type II seesaw. https://doi.org/10.1016/j.nuclphysb.2016.10.008
3. H. Ishimori et al., Non-Abelian discrete symmetries in particle physics. https://doi.org/10.1143/PTPS.183.1
4. J. Bergstrom, M.C. Gonzalez-Garcia, M. Maltoni, T. Schwetze, Bayesian global analysis of neutrino oscillation data. https://doi.org/10.1007/JHEP09(2015)200

Chapter 16
Twist-Six Corrections to $\eta\gamma$ and $\eta'\gamma$ Transition Form Factors in QCD

Janardan P. Singh and Shesha D. Patel

16.1 Introduction

The hard exclusive processes involving light mesons (η, η', π^0) can be described using the factorization, in the sense of convolution, of the short and long distance dynamics. The short-distance part is process dependent and calculable perturbatively as parton-level subprocess amplitude in which the mesons are replaced by their valance Fock components. The long-distance part is described by process-independent meson distribution amplitudes (DAs). The electromagnetic transition form factors (TFFs) of light pseudo scalar mesons are important ingredients to understand their structure. The aim of theoretical investigations is to explain the experimental data within the framework of QCD using factorization. The leading hard contribution to TFF starts at tree-level itself. For large photon virtualities,the full expression can be expanded as a sum of contributions of increasing twist:

$$F(q^2, (q-p)^2)_{\gamma^*\gamma \to P} = T_H(u, Q^2, \mu_F) \otimes \Phi(u, \mu_F) \tag{16.1}$$

Here μ_F represents the boundary between low and high energy parts. η and η' mesons posses $SU(3)_F$ singlet and octet valance Fock components and also two gluons. There is a DA corresponding to each of these three components. This feature leads to the flavor mixing for η and η' system, and to mixing of singlet quark and gluon distribution amplitudes under evolution. One of the key objectives of analysis of TFF of mesons is to model the meson DAs and, in the η–η' case, to extract information about their gluon contents. The gluonic parts of the η–η' meson DAs, deduced from the

J. P. Singh (✉) · S. D. Patel
Faculty of Science, Department of Physics, The Maharaja Sayajirao University of Baroda, Vadodara 390001, Gujarat, India
e-mail: jp.singh-phy@msubaroda.ac.in

S. D. Patel
e-mail: sheshapatel30@gmail.com

Md. Naimuddin (ed.), *XXII DAE High Energy Physics Symposium*, Springer Proceedings in Physics 203, https://doi.org/10.1007/978-3-319-73171-1_16

$\eta\gamma$, $\eta^{'}\gamma$ data, are important ingredients in studying a wide range of exclusive processes, such as B meson two body non-leptonic exclusive and semi-inclusive decays involving η-$\eta^{'}$ mesons. The form factors describing the mesons (η-$\eta^{'}$) transition in two (in general virtual) photons are defined as,

$$i\int d^4xe^{-iqx} < P(p)|T j_{\mu}^{em}(x) j_{\nu}^{em}(0)|0 >= e^2\epsilon_{\mu\nu\rho\sigma}q^{\rho}p^{\sigma}F(q^2,(q-p)^2) \quad (16.2)$$

$$j_{\mu}^{em}(x) = e_u\bar{u}(x)\gamma_{\mu}u(x) + e_d\bar{d}(x)\gamma_{\mu}d(x) + e_s\bar{s}(x)\gamma_{\mu}s(x) + \cdots \quad (16.3)$$

$$< 0|J_{\mu5}^{i}(0)|P(p) >= if_P^i p_{\mu}(i = 1, 8;\ P = \eta, \eta') \quad (16.4)$$

Measurements of the electromagnetic TFFs of $\gamma^*\gamma \rightarrow \eta$ and $\gamma^*\gamma \rightarrow \eta^{'}$ at space-like momentum transfers in the interval 4–40 GeV^2 [1] and at very large time-like momentum transfer 112 GeV^2 [2] by the BaBar collaboration have opened the question of the role of higher order power corrections coming from higher twist contributions on the one hand, and from finite quark and meson masses on the other hand. These measurements and their comparison to the space-like data for $\gamma^*\gamma \rightarrow \pi^0$ FF in the similar range by BaBar and Belle collaborations [3, 4] have started debate on whether hard exclusive hadronic reactions are under theoretical control [5, 6].

The $\eta\gamma$ and $\eta^{'}\gamma$ TFFs have been analyzed to leading-twist, twist-two accuracy and next-to-leading order (NLO) accuracy in [5, 6]. In [6], twist -four contribution has also been calculated. However, considerable difference between the data point and the calculated result for $\gamma^*\gamma \rightarrow \eta$ for large momenta has been found [5]. In view of this discrepancy,we have embarked upon the calculation of twist-six correction to TFF for $\gamma^*\gamma \rightarrow \eta$ in this work. This requires taking into account finite meson-mass as well as finite quark-mass.

16.2 The Method and Results

The light cone distribution amplitude (DAs) gives the momentum fraction distribution of valance quarks of a meson. The twist-two two-quark DAs $\phi_{2,P}^i$ of $P = \eta^{(')}$ meson are defined as

$$< 0|\bar{\psi}(z)C_i\not{z}\gamma_5[z,-z]\psi(-z)|P(p) >= ipzf_P^i\int_0^1 due^{i(2u-1)pz}\phi_{2,P}^i(u) \quad (16.5)$$

where z_{μ} is the light like vector, $[z, -z]$ is the path ordered gauge connection and u is the momentum fraction of a valance quark, $C_1 = 1/\sqrt{3}$ and $C_8 = \lambda^8/\sqrt{2}$. DAs of $\eta^{(')}$ are expanded in terms of Gegenbauer polynomials:

$$\phi_{2,P}^i(u,\mu_F) = 6u(1-u)\left[1 + \sum_{n=2,4,..} a_{P,n}^i(\mu_F)C_n^{3/2}(2u-1)\right] \quad (16.6)$$

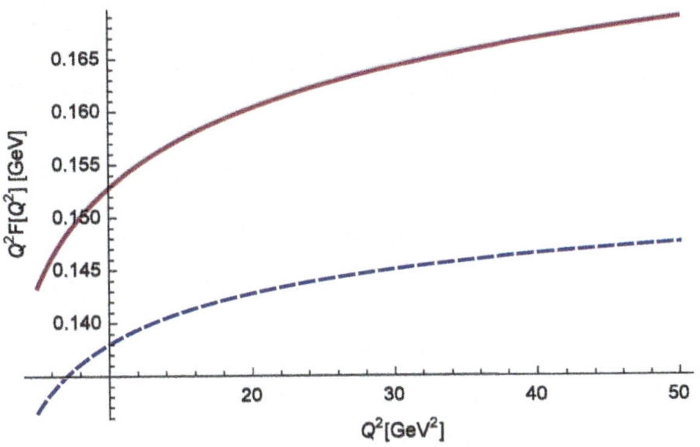

Fig. 16.1 Plots of $Q^2F(Q^2)$ for a typical set of parameters from second reference of [6] (broken curve) and our results on twist-six corrections added to the result from this set of parameters (solid curve)

The gluonic twist-2 DA $\phi^g_{2,P}$ of $P = \eta^{(')}$ mesons are defined as,

$$< 0|A^a_\alpha(z)A^b_\beta(-z)|P(p) >= \frac{1}{4}\epsilon_{\alpha\beta\rho\sigma}\frac{z^\rho p^\sigma}{pz}\frac{C_F}{\sqrt{3}}\frac{f^1_P\delta^{ab}}{8}\int_0^1 e^{i(2u-1)pz}\frac{\Phi^g_{2,P}(u)}{u(1-u)} \tag{16.7}$$

$$\Phi^g_{2,P}(u) = u^2(1-u)^2\left(\sum_{n=2,4,...} b^g_{n,P}(\mu_F)C^{5/2}_{n-1}(2u-1)\right) \tag{16.8}$$

As done by several authors, we are going to truncate the series at n = 2 in the present analysis. Hence the coefficients can be viewed as effective parameters. Our theoretical calculations involve IR regularization through dynamical gluon mass and through ρ-meson mass using vector-meson dominance hypothesis (since it arises due to on-shell photon). Evolution of parameters has not been considered yet. A similar analysis can be done for η' meson.

Parameters used: $m_q = 3.5\,\text{MeV}$, $m_s = 100\,\text{MeV}$, $f_{\eta 8} = 161\,\text{MeV}$, $f_{\eta 1} = 20\,\text{MeV}$, $a^8_{2,\eta} = 0.045$, $a^1_{2,\eta} = -0.12$, $b^g_{2,\eta} = 5.8$, $\Lambda_g = 400\,\text{MeV}$, $\alpha_s = 0.45$ (Fig. 16.1).

In summary, twist-six corrections can make the theoretical results fit the data better.

Acknowledgements Authors gratefully acknowledge the financial assistance from DST-SERB, New Delhi.

References

1. P. del Amo Sanchez et al., (BaBar Collaboration), Phys. Rev. D **84**, 052001 (2011)
2. B. Aubert et al., (BaBar Collaboration), Phys. Rev. D **74**, 012002 (2006)
3. B. Aubert et al., (BaBar Collaboration), Phys. Rev. D **80**, 052002 (2009)

4. S. Uehara et al., (Belle Collaboration), Phys. Rev. D **86**, 092007 (2012)
5. S.S. Agaev et al., Phys. Rev. D **90**, 074019 (2014), arXiv:hep-ph/1409.4311
6. P. Kroll, K. Passek-Kumericki, Phys. Rev. D **67**, 054017 (2003); J. Phys. G **40**, 075005 (2013)

Chapter 17
Recent Underlying Event Measurements at $\sqrt{s} = 13$ TeV

Rajat Gupta

17.1 Introduction

A measurement of the UE activity at a centre-of-mass energy of 13 TeV is presented using events with a Z-boson decaying into $\mu^+\mu^-$ [1]. The Z-boson production is experimentally clean and theoretically well understood, allowing a clear identification of the UE activity. The UE measurement with Z-boson events is complementary to measurements with leading jet and leading track, which will corroborate the universality of the UE. Another advantage of using Z-boson events is the possibility of partial separation of MPI from the radiation contribution [2, 3], which can further improve the optimization of model parameters.

The analysis is performed with a data sample of pp collisions corresponding to an integrated luminosity of 2.1 fb^{-1} at a centre-of-mass energy of 13 TeV, collected using the CMS detector at the LHC [4]. Events are triggered with the requirement of at least two isolated muon candidates with $p_T > 17$ and 8 GeV/c for leading and subleading muons respectively.

The offline selection criteria requires each event to have at least one well reconstructed primary vertex. A recommended tight muon selection is used for selecting the muons, which are required to have a p_T larger than 20 and 10 GeV/c respectively, to match the trigger selection. Both muons are required to lie within the pseudorapidity (η) range of $|\eta| < 2.4$. They are further required to be associated with the primary vertex that has the largest value of p_T squared sum of the associated tracks, designated as the signal vertex. The events with two oppositely charged muons are further required to have an invariant mass ($M_{\mu\mu}$) in the window of 81–101 GeV/c^2. After all the selections, there are about 1.3 million Z candidate events in the data, which is in

On behalf of the CMS collaboration.

R. Gupta (✉)
Panjab University, Chandigarh, India
e-mail: rajat.gupta@cern.ch

Md. Naimuddin (ed.), *XXII DAE High Energy Physics Symposium*, Springer Proceedings in Physics 203, https://doi.org/10.1007/978-3-319-73171-1_17

Table 17.1 Summary of the systematic uncertainties in the particle and Σp_T density [1]

Observable	Uncertainty (%)
Model dependency	2–5
Tracking efficiency	4–6
Pileup	0.5
Trigger	0.1
Physics background	0.5–1
Rochester momentum correction	0.4–0.7

agreement with simulated samples within 1–2%. Selected events have background contributions, mainly from top-quark and diboson decays, of about 0.3%.

In the selected Z-boson events, all the tracks with $p_T > 0.5$ GeV/c and $|\eta| < 2$ are considered for the UE measurements. The track selection efficiencies for data and simulated samples agree within 4–5%. These tracks are used to construct UE observables, namely the particle density and Σp_T density, which are calculated and defined as follows.

The UE activity is quantified in terms of the total number of charged particles (particle density) and the scalar sum of their p_T (Σp_T density). These observables are calculated in different phase-space regions defined with respect to the resultant azimuthal direction of the two selected muons. These regions are classified as *towards* region ($|\Delta\phi| < 60°$), *transverse* region ($60° < |\Delta\phi| < 120°$), and *away* region ($|\Delta\phi| > 120°$). The UE observables measured in different regions are studied as a function of p_T of the dimuon system ($p_T^{\mu\mu}$).

For the comparison with predictions from different simulations and tuning of model parameters, the UE distributions are corrected to the stable charged particle level using the iterative D'Agostini method [5], which properly considers the bin-to-bin migrations. At particle level, muons and charged particles are required to have the same kinematic selections as at the detector level. The 4-dimensional response matrix is constructed using MADGRAPH + PYTHIA8 events, and is used to unfold the detector level distributions. The unfolded measured distributions can get biased due to the selection criteria and simulated samples used for the unfolding. Table 17.1 summarizes the dominant systematic uncertainties in the particle and Σp_T density.

17.2 Results and Discussions

The unfolded distributions of the UE activity as a function of $p_T^{\mu\mu}$ are compared with predictions from various simulations. In order to understand the MPI evolution with centre-of-mass energies, measurements are also compared with previous results from the Tevatron and LHC.

Figure 17.1 shows the UE activity as a function of $p_T^{\mu\mu}$ at the centre-of-mass energy of 1.96, 7, and 13 TeV. The predictions of the POWHEG event generator with

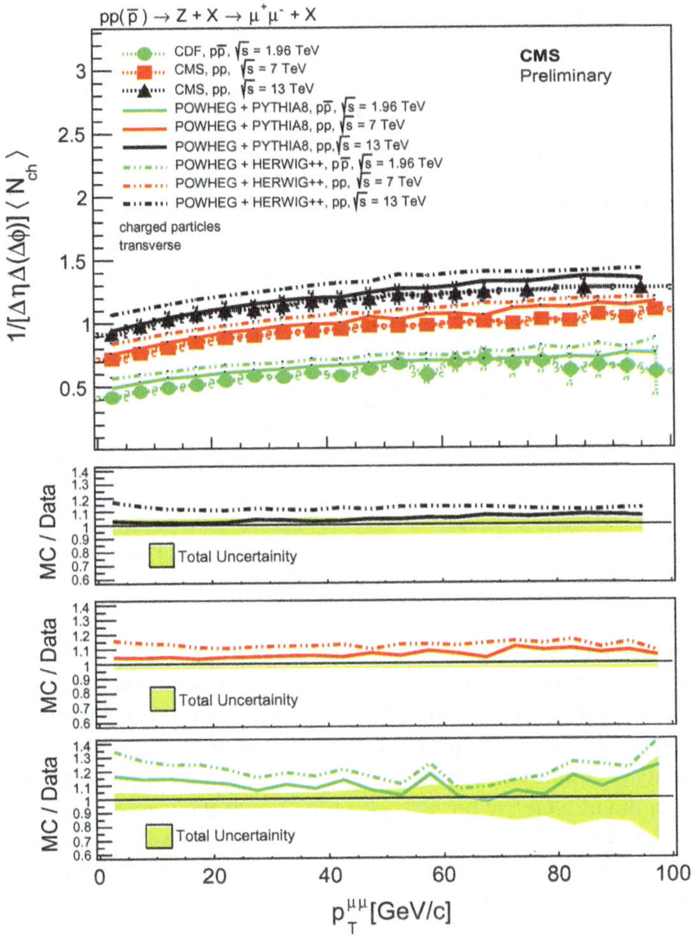

Fig. 17.1 Comparison of UE activity measured at $\sqrt{s} = 13$ TeV with the measurements at 7 and 1.96 TeV, by the CMS and CDF experiments, for particle density in the *transverse* region as a function of $p_T^{\mu\mu}$ [1]. The measured data distributions are also compared with predictions of POWHEG + PYTHIA8 and POWHEG + HERWIG++. The bottom panel of each plot reports the ratios of model predictions to the measurements. The error band in the bottom panel represents the statistical and systematic uncertainties added in quadrature

PYTHIA8, as well as with HERWIG++, are also shown. The ratios of simulations to the measurements are shown in the bottom panel of each plot. The POWHEG + PYTHIA8 predictions describe the measurements within 10% at centre-of mass energies of 1.96 and 7 TeV, and within 5% at 13 TeV. The combination of POWHEG and HERWIG++ describes the measurements within 10–15%, 10–20%, and 20–40% at a centre-of-mass energy of 13 TeV, 7 TeV, and 1.96 TeV respectively.

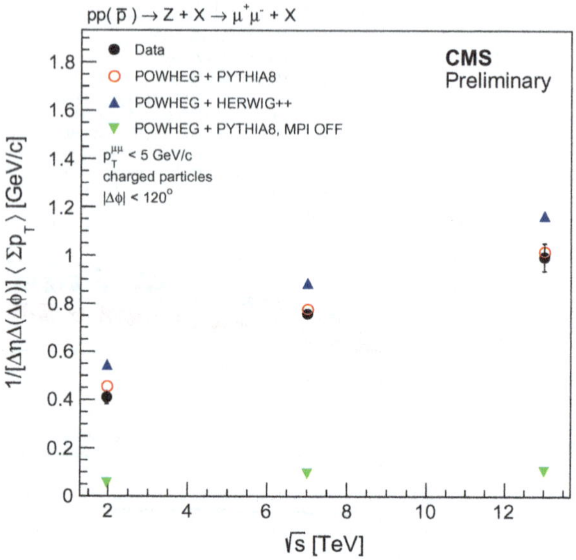

Fig. 17.2 Σp_T density, with $p_T^{\mu\mu}$ < 5 GeV/c as a function of centre-of-mass energy for data, and predictions from simulations by POWHEG + PYTHIA8, and POWHEG + HERWIG++ [1]. The predictions of POWHEG + PYTHIA8 without MPI are also shown. Error bars represent the statistical and systematic uncertainties added in quadrature.

To quantify the energy dependence of the UE activity, events with a $p_T^{\mu\mu}$ smaller than 5 GeV/c are considered. An upper cut on $p_T^{\mu\mu}$ reduces the radiation contribution and the resulting UE activity is mainly from MPI. With a requirement of $p_T^{\mu\mu} < 5$ GeV/c, the UE activity is similar in the *towards* and *transverse* regions. Therefore, the UE activity is combined in these two regions. Figure 17.2 shows the UE activity, after an upper cut of 5 GeV/c on $p_T^{\mu\mu}$, as a function of the centre-of-mass energy for data and the simulated samples. The predictions from POWHEG + PYTHIA8, without MPI, are also shown. It is clear from the comparison of the distributions, with and without MPI, that there is very small contribution from radiation, which increases very slowly with centre-of-mass energy. There is a logarithmic increase in UE activity with centre-of-mass energy which is well reproduced, qualitatively, by POWHEG. The energy evolution is better described by POWHEG events hadronized with PYTHIA8, whereas hadronization with HERWIG++ overestimates the UE activity at all energies.

17.3 Summary

A measurement of the UE activity using inclusive Z-boson events at a centre-of-mass energy of 13 TeV is presented. Analyzed data corresponds to an integrated luminosity of 2.1 fb^{-1}. The UE activity, quantified in terms of particle and Σp_T densities, is measured as a function of the resultant p_T of the two muons coming from the decay of the Z-boson. The distributions are corrected for detector effects and compared with various predictions. It is observed that MADGRAPH and POWHEG, hadronized with PYTHIA8 and CUET8PM1 tune, describe the measurements within 5%. The combination of POWHEG + HERWIG++ with tune EE5C, overestimates the measurements by 10–15%. The results obtained at 13 TeV are also compared with previous measurements at 1.96 and 7 TeV. The UE activity almost doubles, with logarithmic increase, as the collision energy increases from 1.96 to 13 TeV. Simulations nicely describe the increase in UE activity as the collision energy increases from 7 to 13 TeV but they underestimate the UE evolution from 1.96 to 7 TeV. Thus there is scope of further improvements in the UE modeling, especially in the energy dependence. The present measurement, in combination with previous results, will be important for the further optimization of the model parameters in various simulations.

References

1. CMS Collaboration, Measurement of the underlying event using the Drell-Yan process in proton-proton collisions at $\sqrt{s} = 13$ TeV, CMS-PAS-FSQ-16-008 (2016)
2. C.M.S. Collaboration, Measurement of the underlying event in the Drell-Yan process in proton-proton collisions at $\sqrt{s} = 7$ TeV. Eur. Phys. J. C **72**, 2080 (2012)
3. R. Kumar, M. Bansal, S. Bansal, J.B. Singh, New observables for multiple-parton interactions measurements using Z+jets processes at the LHC. Phys. Rev. D **93**(5), 054019 (2016)
4. C.M.S. Collaboration, The CMS experiment at the CERN LHC. JINST **3**, S08004 (2008)
5. G. D'Agostini, A Multidimensional unfolding method based on Bayes' theorem. Nucl. Instrum. Meth. A **362**, 487 (1995)

Chapter 18
Pion Mass Modification in Presence of External Magnetic Field

S. P. Adhya, M. Mandal , S. Biswas and P. K. Roy

18.1 Introduction

The modifications due to the magnetic field on various theoretical estimates in heavy ion collisions as well as neutron star have drawn significant attention in recent years [1–4]. As for example, in heavy ion collisions for off central collisions, magnetic field can be $\sim$0.02 GeV^2 (at RHIC) and $\sim$0.3 GeV^2 (at LHC). On the other hand, field strengths of $\sim 10^{15}$ Gauss can be found in some neutron stars. Thus, it will be worthwhile to calculate the mass modification of pions in such scenarios because of the comparable mass and field strengths. For our purpose, we use the Schwinger's proper time approach for calculation of the magnetic field modified Fermionic propagators [5]. A consistent calculation of the pion self energy has been presented for both the pseudo-scalar and pseudo-vector coupling of the pion-nucleon Lagrangian. However, we present the results in absence of any medium. The modification of the dispersion relations in presence of the medium and magnetic field will be reported shortly.

S. P. Adhya (✉)
Experimental High Energy Physics and Applications Group,
Variable Energy Cyclotron Cente, 1/AF Bidhannagar, Kolkata 700 064, India
e-mail: sp.adhya@vecc.gov.in

M. Mandal
Department of Physics, Government General Degree College,
Kalna-I, Burdwan 713 405, India

S. Biswas
Department of Physics, Rishi Bankimchandra College,
Naihati 743 165, India

P. K. Roy
High Energy Nuclear and Particle Physics Division, Saha Institute
of Nuclear Physics, 1/AF Bidhannagar, Kolkata 700 064, India

Md. Naimuddin (ed.), *XXII DAE High Energy Physics Symposium*, Springer Proceedings in Physics 203, https://doi.org/10.1007/978-3-319-73171-1_18

18.2 Formalism

The pion-nucleon phenomenological Lagrangian is written as,

$$\mathcal{L}_{\text{int}}^{\text{PS}} = -ig_\pi \bar{\Psi}\gamma_5(\phi \cdot \mathbf{\Phi}_\pi)\Psi \qquad (pseudo-scalar\ coupling) \tag{18.1}$$

$$\mathcal{L}_{\text{int}}^{\text{PV}} = -\frac{f_\pi}{m_\pi}\bar{\Psi}'\gamma_5\gamma^\mu\partial_\mu(\phi.\Phi'_\pi)\Psi' \qquad (pseudo-vector\ coupling) \tag{18.2}$$

where $\mathcal{L}_{\text{int}}^{\text{PS}}$ and $\mathcal{L}_{\text{int}}^{\text{PV}}$ are the Lagrangians corresponding pseudo-scalar and pseudo-vector couplings respectively. In the above equation, (Ψ, Ψ') and (Φ, Φ') are the nucleon and pion fields with the coupling constants g_π and f_π accordingly. Therefore, the pion self energy at one loop order is given as,

$$\Pi_\pi(q) = -i\int \frac{d^4k}{(2\pi)^4}\text{Tr}[\{i\Gamma(q)\}iS_a(k)\{i\Gamma(-q)\}iS_b(k+q)] \tag{18.3}$$

We consider the weak field approximation of the magnetic field on the propagators according to the condition $(eB << m_\pi^2)$ .Thus up to order $(eB)^2$, the Fermion propagators can be written as [6],

$$S(k) = S^{(0)}(k) + eB\, S^{(1)}(k) + (eB)^2\, S^{(2)}(k) + \mathcal{O}((eB)^3) \tag{18.4}$$

The form of $S^{(1)}(k)$ and $S^{(2)}(k)$ can be found in the [6]. Thus, using the modified propagator, we arrive at [6],

$$\begin{aligned}
\Pi_{\pi^0}(q)^{PS} &= -\frac{g_\pi^2}{2\pi^2}\Bigg[\int_0^1 dx\Bigg[\frac{(q^2-m_{\pi^0}^2)x(1-x)(m^2-3q^2x(1-x))}{\Delta_R} \\
&+ (m^2-3m_{\pi^0}^2x(1-x))\log\frac{\Delta_R}{m^2-m_{\pi^0}^2x(1-x)}\Bigg] \\
&+\frac{(eB)^2}{2}\int_0^1 dx\, x(1-x)\left(\frac{1}{\Delta_R}+\frac{m^2+x(1-x)q_{\|}^2}{\Delta_R^2}\right) \\
&+(eB)^2\Bigg\{\int_0^1 dx\,(1-x)^3\Bigg[\frac{1}{\Delta_R}+\frac{q^2x(1-x)+q_\perp^2x(4x-1)+m^2}{3\Delta_R^2} \\
&+\frac{2x^2q_\perp^2[q^2x(1-x)+m^2]}{3\Delta_R^3}\Bigg]+\int_0^1 dx(1-x)^2\left[\frac{1}{\Delta_R}-\frac{q_\perp^2x(1-x)}{\Delta_R^2}\right]\Bigg\}\Bigg].
\end{aligned} \tag{18.5}$$

where $\Delta_R = m_p^2 - q^2x(1-x)$. Similarly, $\Pi_{\pi^\pm}(q)^{PS}$, $\Pi_{\pi^0}(q)^{PV}$ and $\Pi_{\pi^\pm}(q)^{PV}$ can be derived likewise (for detailed discussion please refer to [6]). Finally, defining the effective pion masses by the positions of the pole of the propagator,

$$m_\pi^{*2} = m_\pi^2 + \mathrm{Re}\,\Pi(m_\pi^{*2}, \mathbf{q} = 0, B), \tag{18.6}$$

we calculate the variation of the mass with respect to the weak magnetic field.

18.3 Results

We present the variation of charged and uncharged pion masses with respect to the strength of the external magnetic field (according to the condition $(eB) \ll m_\pi^2$ for the weak field regime). For our case, we show the results considering both the pseudo-scalarand pseudo-vector interaction of the $\pi - N$ Lagrangian in Figs. 18.1 and 18.2 respectively. We observed that the effective pion mass decreases for the PS interaction, whereas there is a slight increase for the PV interaction.

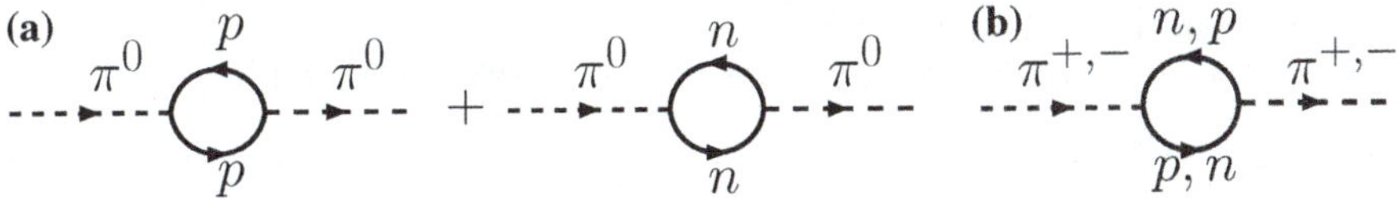

Fig. 18.1 **a** represents the one-loop self-energy diagram for π^0 and **b** represents the same for $\pi^\pm$

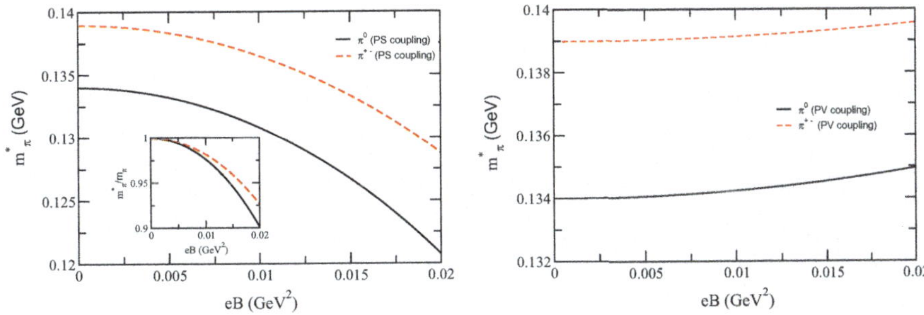

Fig. 18.2 Effective pion mass as a function of the magnetic field for PS and PV coupling

18.4 Summary and Discussions

The motivation of the present work has been to analyse the contribution of the weak magnetic field on the pion mass. We have introduced the weak magnetic field correction at one loop order through Schwinger's proper time method in the pion-nucleon Lagrangian. Our results show correction at $(eB)^2$ order for the field effects on the pion mass over the vacuum results. Finally, we have observed that the pion mass decreases for the PS coupling, whereas for the PV coupling, there is a marginal enhancement in mass for both the charged and uncharged pions [6]. The mass modification in medium due to magnetic field in realistic scenarios like neutron star or heavy ion collisions will be reported shortly.

References

1. K. Tuchin, Phys. Rev. C **88**, 024910 (2013)
2. A. Ayala, C.A. Dominguez, L.A. Hernandez, M. Loewe, R. Zamora, Phys. Rev. D **92**, 096011 (2015)
3. G. Colucci, E.S. Fraga, A. Sedrakian, Phys. Lett. B **728**, 19 (2014)
4. S. Ghosh, A. Mukherjee, M. Mandal, S. Sarkar, P. Roy, Phys. Rev. D **94**, 094043 (2016)
5. J. Schwinger, Phys. Rev. **82**, 664 (1951)
6. S.P. Adhya, M. Mandal, S. Biswas, P.K. Roy, Phys. Rev. D **93**, 074033 (2016)

Chapter 19
Can Stopped Cosmic Muons Be Used to Estimate the Magnetic Field in the Prototype ICAL Detector?

Neha, G. Majumder, B. Satyanarayana, E. Pathaleswar and V. M. Datar

19.1 Introduction

The ICAL detector consists of alternate layers of 56 mm thick soft iron plates with 40 mm gaps housing Resistive Plate Chamber (RPC) providing lateral position and fast timing information. Iron is the neutrino target material and is also an integral part of an electromagnet. Measuring the magnetic field seen by a passing muon in the iron plate is still an open problem. The Muon Spin Rotation (μSR) technique using stopped cosmic muons is one possibility of addressing this problem. Towards this end we have used the cosmic muon stack of 12 nos. of 1 m $\times$ 1 m glass RPCs at Tata Institute of Fundamental Research, Mumbai.

Neha (✉) · E. Pathaleswar · V. M. Datar
India-based Neutrino Observatory, Tata Institute of Fundamental Research, Mumbai 400 005, India
e-mail: neha_005@tifr.res.in

E. Pathaleswar
e-mail: Pathaleswar@tifr.res.in

V. M. Datar
e-mail: vivek.datar@tifr.res.in

Neha
Homi Bhabha National Institute, Mumbai 400 094, India

G. Majumder · B. Satyanarayana
Department of High Energy Physics, Tata Institute of Fundamental Research, Mumbai 400 005, India
e-mail: gobinda@tifr.res.in

B. Satyanarayana
e-mail: bsn@tifr.res.in

Md. Naimuddin (ed.), *XXII DAE High Energy Physics Symposium*, Springer Proceedings in Physics 203, https://doi.org/10.1007/978-3-319-73171-1_19

19.2 Muon Spin Rotation at ICAL

μSR is a standard condensed matter technique to study the internal magnetic field in small samples using secondary beams of polarized muons [1]. Parity violation in the $\pi \rightarrow \mu + \nu$ decay together with the steep energy dependence of cosmic pion energy spectrum [2] is responsible for partial longitudinal polarization of the muon [3]. The precession of μ-spin with B modifies the angular distribution of the decay e^-. The time spectra in the detectors above and below the magnetized iron are expected to show an oscillatory behavior superimposed on an exponential decay. The oscillations will have a 180° relative phase difference, and an oscillation frequency proportional to B. The asymmetry factor given as $2A = (N_T - N_B)/(N_T + N_B)$, where N_T (N_B) are the counts in top (bottom) detector.

19.3 Small Scintillator Set-Up

As a first step towards this goal we tried to measure the asymmetry in the counts of the top and bottom detector. A stack of 5 scintillators ($0.96 \times 0.32 \times 0.01$ m^3) sandwiching an Iron plate ($0.9 \times 0.4 \times 0.02$ m^3), as shown in Fig. 19.1, has been set up. If a μ passing through the stack stops in the Iron plate then the decay e^- could be detected in the top or bottom scintillator after a certain delay (~2μs). The trigger was generated by $A \cdot B \cdot C \cdot \overline{D}$ and the time differences between (a) trigger and delayed co-incidence of $B \cdot C$, (b) trigger and delayed co-incidence of $D \cdot E$ were analyzed and the histogram plots are shown in Fig. 19.2, left and right respectively. The time spectra were fit to $N = Aexp(-t/\tau) + C$. For top(bottom) scintillator, the measured lifetime (τ) is $1.9 \pm 0.2(2.2 \pm 0.2)$ μs which, within errors, is consistent with known value of lifetime. The measurement of decay asymmetry is under investigation.

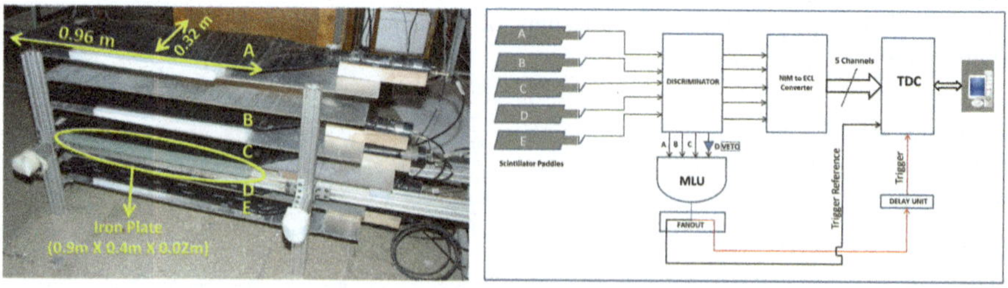

Fig. 19.1 Stack of 5 scintillators (*Left*) and electronics block diagram (*Right*)

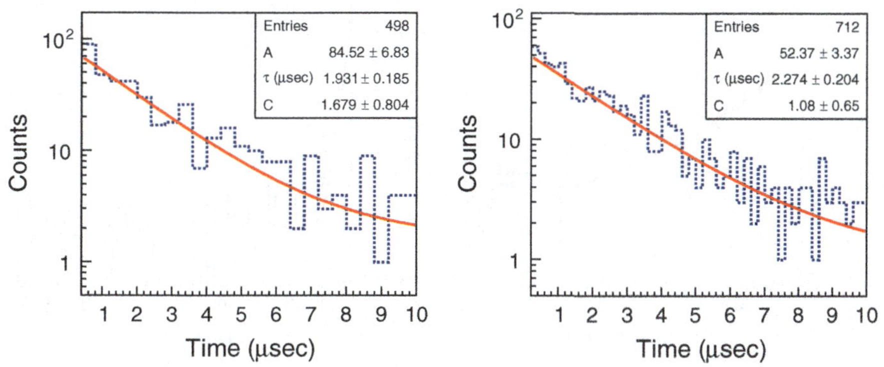

Fig. 19.2 Time distribution of top (*Left*) and bottom (*Right*) scintillators

19.4 Feasibility of Using RPCs for the μSR measurement using the TIFR 1 m x 1 m RPC stack

An Iron plate (1 m $\times$ 1 m $\times$ 0.02 m) was placed between the 2nd and 3rd layers in RPC stack of 12 layers at TIFR, as shown in Fig. 19.3. To check the feasibility of the experiment, a GEANT4 based simulation was performed by incorporating the full detector geometry of RPC stack. The detection efficiency of μ-decay is 31 and 16%, for 2 and 4 cm thick Iron plate respectively. In the experiment, the trigger was generated by $(L6 \cdot L5 \cdot L4)$ which ensures that the μ has passed through L3. For L2(L3), the measured lifetime is $2.09 \pm 0.18(0.148 \pm 0.007)$ μs. The measured μ lifetime is in good agreement with the known value for layer 2 but not for layer 3. This is due to the fact that the passage of a muon induces an avalanche in the RPC and reduces the electric field locally, making it "*dead*", for few ms. However, in the case of magnetized iron the B-field can be extracted from the time spectra of detector below the Iron plate. The magnetic field measurements using μSR technique will be carried out at a proposed prototype ICAL detector (4 m $\times$ 4 m $\times$ 1.2 m) consisting of 10 layers of RPCs and 11 layers of Iron plates magnetized by current carrying coils (Fig. 19.4).

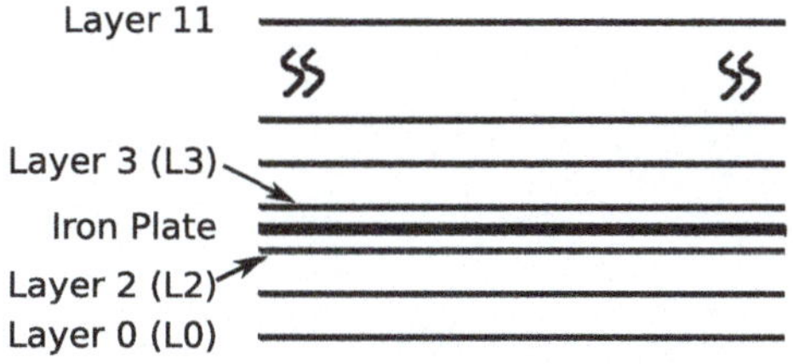

Fig. 19.3 Schematic diagram of the experimental setup

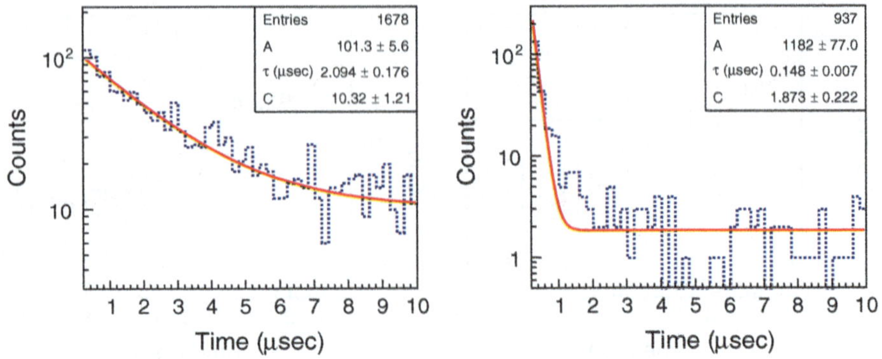

Fig. 19.4 Time distribution of layer 2 (*Left*) and layer 3 (*Right*)

References

1. S.J. Blundell, Contemp. Phys. **40**, 175 (1999)
2. Bruno Rossi, Rev. Mod. Phys. **20**, 537 (1948)
3. C. Scott Johnson, Phys. Rev. **122**, 1883 (1961)

Chapter 20
Techniques to Improve Time Resolution of Large Area RPCs

A. D. Bhatt, V. M. Datar, G. Majumder, N. K. Mondal, Pathaleswar and B. Satyanarayana

20.1 Introduction

The overall time resolution of the RPC chambers [1, 2] was estimated to be about 1.5 ns. The principal aim of the present work is to develop an offline calibration procedure using which time resolution of large area single gap RPC can be improved. The technique developed here will be used for ICAL detector to improve the time resolution of the RPCs which in turn will improve its capability to identify the up/down going muons.

20.2 Experimental Setup and Data Analysis

The detector setup and DAQ has been described in detail in [3, 4]. Event data consist of two informations: (i) cosmic ray muon hits per layer and (ii) the corresponding time of arrival. Hit information is used to get the angular distribution of muons and is described in detail in [3]. A preliminary analysis of time of arrival was also performed and a time resolution of 1.3 ± 0.2 ns was achieved [4]. In the present work, first the TDC time (which used leading edge threshold discriminator) is corrected for electronic offsets as explained in G. Majumder et al. [4].

In Fig. 20.1a, b, the efficiency maps for a typical large area RPC are shown. The efficiency plots have 32×32 pixels, where each pixel has an area of $3 \times 3\,\text{cm}^2$, given

A. D. Bhatt (✉)
Homi Bhabha National Institute, Mumbai 400094, India
e-mail: apoorva.bhatt@tifr.res.in

A. D. Bhatt · V. M. Datar · G. Majumder · Pathaleswar · B. Satyanarayana
Tata Institute of Fundamental Research, Mumbai 400005, India

N. K. Mondal
Saha Institute of Nuclear Physics, Bidhannagar, Kolkata 700064, India

Md. Naimuddin (ed.), *XXII DAE High Energy Physics Symposium*, Springer Proceedings in Physics 203, https://doi.org/10.1007/978-3-319-73171-1_20

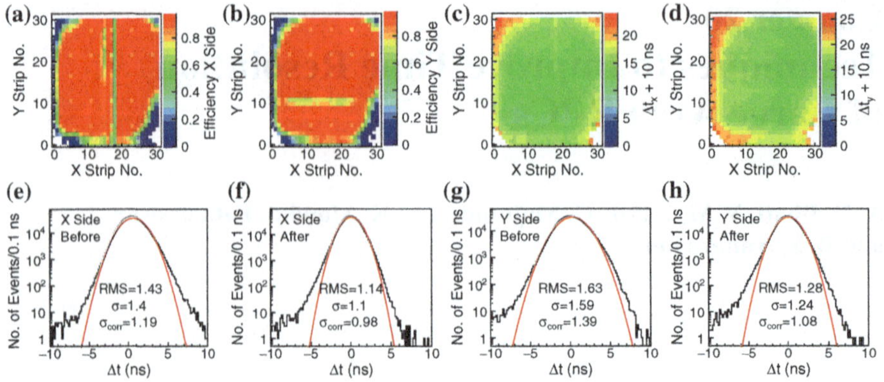

Fig. 20.1 **a, b** Efficiency for muon detection, and **c, d** average time delay of muon signal in the 1st RPC layer, for strips in the X-plane (**a, c**) and Y-plane (**b, d**). (Left two) Time resolution in layer-1 and (right two) measurement of muon direction with and without pixel-wise time offset correction. **e, g** Time resolution for the X and Y-plane in layer-1 before the pixelwise offset correction and **f, h** after the offset correction. Here, σ is the gaussian fitted standard deviation and σ_{corr} is σ corrected for the extrapolation error

the width of pickup strips. The first feature observed in the efficiency plot shown in Fig. 20.1a, b is that the efficiency map is similar in the X- and Y-planes of the RPC. Also, it can seen that there is a non-uniformity of efficiency and inefficiency in a large region near the corners. The main source for this position dependence of signal efficiency is the nonuniform thicknesses of glass electrodes and gas gap spacers. The thickness of glass used for RPC fabrication is 3.1 ± 0.1 mm, where as the thickness of spacers is 2.0 ± 0.2 mm, which results in variation in the gas gap affecting the electric field and consequently the gas gain of the RPC. This is indicated in the observed efficiency maps in Fig. 20.1a, b. It is difficult to improve the tolerance in glass thickness, as commercially produced glasses are used for the RPC fabrication.

Figure 20.1c, d shows the average time delay of the avalanche signal in the X- and Y-planes with respect to the expected time in that layer. There is a perfect correlation between the inefficient regions in Fig. 20.1a, b, and the time delay in Fig. 20.1c, d. This shows that there is a variation in time delay, due to the variation of the gain in different regions of the RPC. This is very prominent in the regions with lower efficiency. The lower efficiency in a region implies that the average gain is less, which results as a delay in the measured time. Also, in regions with nearly 100% efficiency, a small variation in the time delay is observed.

20.3 Position Dependent Time Offset Correction

The average time delay between the strip charge induction to the front-end electronics input varies with the position of the particle interaction in the RPC and more specifically with the position of the signal pickup strip with respect to the front-end electronics. Here, the time measurement itself is used to make this offset correction. The average time delay is measured in 32×32 pixels for both X- and Y-plane of the RPC. For each layer, the time is corrected using the offset corrections of the nearby four pixels.

20.4 Results

The measured time resolution for one of the RPCs in the detector stack is shown in Fig. 20.1e, g without any pixel-wise offset correction and in Fig. 20.1f, h after applying position dependent offset corrections. A substantial improvement can be observed in the time measurements after using the corrections.

20.5 Conclusion

The position dependent gain of a RPC has been identified as the principal factor affecting its intrinsic time resolution. A technique of offline correction to improve the time measurement has been discussed and applied to a stack detector for a cosmic ray data sample. After correcting for this along with correction for time of flight of the signal in strips as well as for the delay in the electronic chains, most of the RPCs show a time resolution better than 1 ns in any part of the detector. For INO experiment, all RPCs need be calibrated in a setup at surface, because muon flux in INO cavern will be very low, will not have sufficient statistics for this calibration. This technique can also be applied in a test beam setup.

References

1. Yu.N. Pestov, G.V. Fedotovich, A picosecond Time-of-flight spectrometer For The VEPP-2M based on local-discharge spark counter. SLAC-TRANS 0184, IYF 77–78 (1978). http://www-public.slac.stanford.edu/sciDoc/docMeta.aspx?slacPubNumber=SLAC-TRANS-0184
2. R. Santonico, R. Cardarelli, Development of resistive plate counters. Nucl. Instrum. Methods Phys. Res. **187**, 377–380 (1981). https://doi.org/10.1016/0029-554X(81)90363-3
3. S. Pal et al., Measurement of integrated flux of cosmic ray muons at sea level using the INO-ICAL prototype detector. JCAP 07, 033 (2012). https://doi.org/10.1088/1475-7516/2012/07/033
4. G. Majumder et al., Study of the directionality of cosmic muons using the INO-ICAL prototype detector. Nucl. Instrum. Meth. A **735**, 88–93 (2014). https://doi.org/10.1016/j.nima.2013.09.025

Chapter 21
Graviton Portal to Dark Matter in Universal Extra Dimensions

Mathew Thomas Arun and Divya Sachdeva

Introduction: Though the minimal Universal Extra Dimension scenario (UED) is well known to be phenomenologically less stringent (from flavour and electro-weak precision measurements), over their warped counterparts, the stability and Higgs mass hierarchy remain unanswered. The UED scenario is, also, known to posses a natural Dark Matter (DM) candidate, unlike minimal warped scenarios. Hence, various search strategies were envisaged at LHC. And, the current experiments rule out a compactification mass scale of below 1150 GeV [1].

Another constraint for UED comes from the dark matter relic density. For a compactification scale above 1400 GeV [2], the dark matter density become larger than the observed amount $\Omega_{dm} \sim 0.119$ [3]. This is because of the fact that lightest KK particle is absolutely stable and there are not enough channels through which they could annihilate. Since LHC has already put a lower bound for the compactification scale very close to this scale and in the next few years, with more data coming, the minimal UED scenario will get highly disfavored.

In the light of these new developments, we discuss a new model [4] in which such Universal Extra Dimension models could naturally emerge in a six dimensional double warped scenario on a $S^1/Z_2 \otimes S^1/Z_2$ orbifolded 6D manifold with vanishing 5-dimensional induced cosmological constant. The line element for this space time is given as,

$$ds^2 \sim e^{k(x_5-\pi)}\left(\eta_{\mu\nu}dx^\mu dx^\nu + R_y^2 dx_4^2\right) + r_z^2 dx_5^2\,, \quad \epsilon = \frac{k}{r_z M_6}\,,$$

where k is the warping along x_5 and M_6 the six-dimensional Planck scale.

M. T. Arun (✉)
Department of Physics, Mar Thoma College, Thiruvalla 689 103, Kerala, India
e-mail: thomas.mathewarun@gmail.com

M. T. Arun · D. Sachdeva
Department of Physics and Astrophysics, University of Delhi, Delhi 110 007, India

Md. Naimuddin (ed.), *XXII DAE High Energy Physics Symposium*, Springer Proceedings in Physics 203, https://doi.org/10.1007/978-3-319-73171-1_21

The presence of warped dimension provides the electro weak mass hierarchy required by the Standard Model starting from a Planck scale Higgs mass, and also the stability to the bulk moduli. Unlike a Minimal UED, here the graviton couplings are substantial and will introduce new processes involving the 5-dimensional brane localized Standard Model particles.

In this work we will show that the introduction of new co-annihilation and annihilation channels for dark matter decrease the tight bounds 5-dimensional Universal Extra dimensions have. And thus making *Natural* UED a viable model than a simple 5-dimensional UED construction.

Dark Matter Density: In the model prescribed above, vanishing of cosmological constant induced on 4-brane would mean that there warping is substantial only along the x_5−direction. On localizing the Higgs field at $x_5 = 0$ 4-brane, we get the usual protection for Higgs mass from large radiative corrections. The new Higgs mass is given as $m_h = \lambda \frac{1}{r_z} e^{-k\pi}$, where $\lambda \sim 1$.

The Standard Model localized on the 4-brane at $x_5 = 0$ will mimic the minimal UED scenario with a slightly broken KK-parity. Just like in UED, the orbifolding inherent to the system would not only eliminate unwanted modes, but also introduce a KK-parity that, in turn, provides for a Dark Matter candidate, namely the LKP in this natural UED scenario with mass $m_{B_1} \sim 1/R_y$. It might be argued, though, that with the warping along x_4 being different from zero, the KK-parity is not exact. This is indeed so, but with the extent of Z_2-breaking being determined by the (vanishingly small) induced cosmological constant $\tilde{\Omega}$, the lifetime of such DM-candidates could be greater than the age of the universe.

The graviton excitation in the model, by virtue of them not being localized on any brane, will have two modes. One, the quantization of the momentum in the x_4-direction and the other along x_5. The couplings of those graviton modes that are excited only along the x_4 direction, is Planck Scale suppressed. The only unsuppressed coupling of graviton excitations exists for those which have non-trivial wave profile in the x_5−direction.

The mass of the higher excitations of gravitons M_g is given as $M_g = \sqrt{\left(\frac{n}{R_y}\right)^2 + m_p^2}$, where $m_p = x_p \frac{k}{r_z} e^{-k\pi}$, x_p being the pth root of $J_{3/2}$.

In this project, we look at the Dark Matter annihilation through resonant production of the first non-trivial excited mode of the graviton. The coupling of LKP with the first excitation of the graviton with non-negligible coupling in the bulk is given as $C_{01} = \frac{1}{M_6^2 \sqrt{\pi R_y r_z}} \frac{1}{N_1} e^{5k\pi/2} J_{5/2}(x_1)$, where N_1 is the normalization of graviton wave function in x_5−direction and x_1 is the first root of $J_{3/2}$.

In MUED, Dark Matter candidate is the Kaluza Klein level 1 partner of hypercharge gauge boson $B_\mu^{(1)}$. Considering all others KK modes heavier than $B_\mu^{(1)}$, the relevant self annihilation of $B_\mu^{(1)}$ is into fermions and higgs bosons. Self annihilation alone sets an upper bound for $m_{B_1} \sim 900$ GeV [5]. As dark matter is non-relativistic at freeze out temperature, the center of mass energy is almost double of $B_\mu^{(1)}$ mass. Therefore, the resonance process plays major role in calculation of relic abundance.

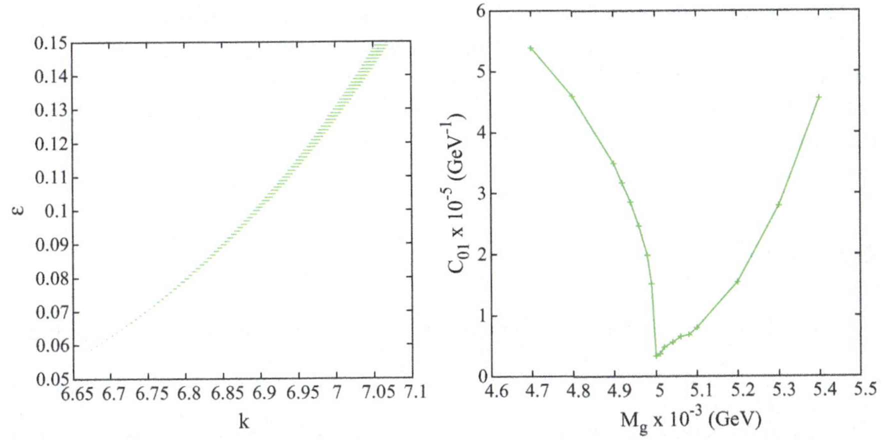

Fig. 21.1 Plot showing the k vs ϵ on left and mass vs coupling on right for $R_y^{-1} = 2500$ GeV. The green represents the region allowed by Dark Matter density limits (or the region above the curve in the mass vs coupling graph) with the inclusion of the new process mediated by graviton. C_{01} is the coupling of graviton to the LKP

Here, graviton in resonance increases the self annihilation rate and helps in satisfying relic density for $m_{B_1} \sim R_y^{-1}$ above 1.5 TeV. In particular, we have shown this for $m_{B_1} \sim R_y^{-1}$ values 2.5 TeV in Fig. 21.1.

Cross-section for $B_\mu^{(1)}$ (with mass m_v) annihilation to fermions (with mass m_f) via graviton (with mass M_g) is given by $\sigma = \frac{N_c C^4 s^2 \left(15(2-\beta^2)m_v^2 + (-9\beta^4+5\beta^2+30)s\right)}{270\pi\beta|(s-M_g^2-iM_g\Gamma)^2|}$, where $\beta = \sqrt{1 - m_f^2/s}$ and N_c is the colour factor. C and Γ are the coupling and decay width of graviton to SM particles. We have used MUED FeynRules [6] file in [7] after modifying it to include graviton couplings with all KK as well as SM particles. To calculate the relic density, we have implemented model in version 2.0 of MadDM [8]. As we are dealing with resonance, we have kept $iter\_W_{ij} = 10$ in MadDM as parameters. The green region in Fig. 21.1 satisfies the constraint $\Omega h^2 \leq 0.1199 \pm 0.0027$ [3].

Conclusion: In addition to co-annihilation processes involving KK level 1 particles, the s-channel exchange of KK level 2 particles near resonance and production of level 2 particles in final state contributes significantly to the total Relic density. Due to this contribution, the upper limit on m_{B_1} increases to 1.4 TeV. The graviton resonance process augments the level 2 resonance processes because mass of graviton is closer to the level 2 KK particles. It can be seen that there is a competition between large cross-sections in co-annihilation channels but with large Boltzmann suppressions, and large cross-section in self annihilation due to the addition of the new s-channel process involving gravitons.

Also, the graviton resonance mode discussed here does not affect the LUX limits as WIMP-nucleon scattering via graviton is a t-channel process and $B_\mu^{(1)}$ graviton coupling falls off drastically away from resonance. Hence, like in MUED, cross-sections for WIMP-nucleon scattering are small [9] and their magnitudes are below the limit.

References

1. D. Choudhury, K. Ghosh, Bounds on universal extra dimension from LHC Run I and II data. Phys. Lett. B **763**, 155–160 (2016)
2. G. Belanger, M. Kakizaki, A. Pukhov, Dark matter in UED: the role of the second KK level. JCAP **1102**, 009 (2011)
3. D. Paoletti, P. Collaboration, Planck 2015 cosmological results. In: Proceedings, Magellan Workshop: Connecting Neutrino Physics and Astronomy: Hamburg, Germany, March 17–18, 2016, pp. 71–86 (2016), [71(2016)]
4. M.T. Arun, D. Choudhury, Stabilization of moduli in spacetime with nested warping (2016)
5. G. Servant, T.M.P. Tait, Is the lightest Kaluza-Klein particle a viable dark matter candidate? Nucl. Phys. B **650**, 391–419 (2003)
6. N.D. Christensen, C. Duhr, FeynRules - Feynman rules made easy. Comput. Phys. Commun. **180**, 1614–1641 (2009)
7. A. Datta, K. Kong, K.T. Matchev, Minimal universal extra dimensions in CalcHEP/CompHEP. New J. Phys. **12**, 075017 (2010)
8. Backovic, M., Kong, K., McCaskey, M.: MadDM v.1.0: computation of dark matter relic abundance using MadGraph5. Phys. Dark Universe **5**, **6**, 18–28 (2014)
9. J.M. Cornell, S. Profumo, W. Shepherd, Dark matter in minimal universal extra dimensions with a stable vacuum and the right Higgs boson. Phys. Rev. D **89**(5), 056005 (2014)

Chapter 22
Various Studies with Gas Electron Multiplier (GEM) Detectors

Shivali Malhotra, Md. Naimuddin, Ashok Kumar, Mohit Gola, Anshika Bansal and Aashaq Shah

22.1 Introduction

The Micro-Pattern Gaseous Detectors (MPGDs) were developed in the continuity of the Micro Strip Gas Chambers (MSGCs). The most common MPGDs is the Gas Electron Multiplier (GEM) introduced by Fabio Sauli in 1996 [1]. GEM Foil is a thin layer of an insulating polymer, usually a 50 μm thick polyamide, coated on both sides with 5 μm of copper and chemically perforated with a high density of microscopic holes. There are two kind of geometries of GEM holes: Single Mask (or Conical) and Double Mask (or Bi-conical). A Triple GEM detector consists of three GEM Foils stacked one over another. Potential difference between the two GEM foils creates electric field in between which in general are known as: Drift Field, Transfer Field(s), and Induction Field.

22.2 Various Studies with GEM Detector

Studies were performed at University of Delhi (DU) with the triple GEM detector having gap configuration of 3:2:2:2 with an active area of $10 \times 10\,\text{cm}^2$. The following characteristics were measured:

- **Leakage Current**: It is the current that flows from the top of a GEM foil to the bottom along the dielectric surface of the walls of the holes in the presence of an applied voltage (~500 V). Figure 22.1 on left shows the leakage current for 3 GEM Foils used. We have also measured the Electric field at nominal voltage of 3600 V as shown in Fig. 22.1 (Right).

S. Malhotra (✉) · Md. Naimuddin · A. Kumar · M. Gola · A. Bansal · A. Shah
University of Delhi, Delhi 110 007, India
e-mail: shivali.malhotra@cern.ch

Md. Naimuddin (ed.), *XXII DAE High Energy Physics Symposium*, Springer Proceedings in Physics 203, https://doi.org/10.1007/978-3-319-73171-1_22

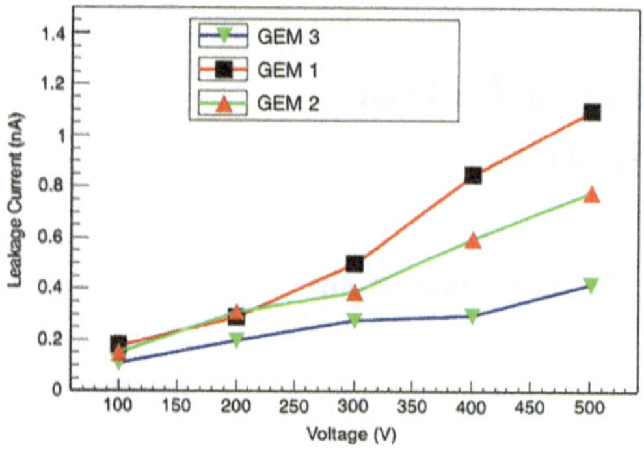

Region	GAP (mm)	Electric Field (kV/cm)
Drift	3	2.88
Transfer	2	3.35
Transfer	2	3.35
Induction	2	4.79

Fig. 22.1 Left: showing the Leakage Current of 10 cm × 10 cm GEM detector; Right: showing the Electric Field across the gaps of GEM detector

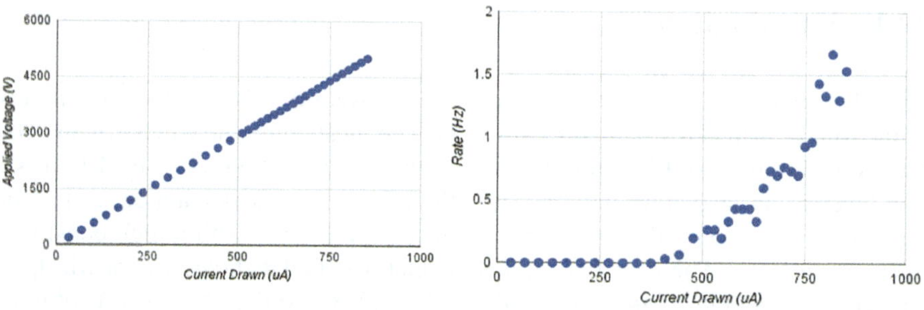

Fig. 22.2 Left: showing the HV characteristics of 10 cm × 10 cm GEM detector; Right: showing the Spurious signal rate

- **HV Characteristics and Spurious Signals**: Behaviour of the high voltage distribution circuit, which uses a resistive divider to ground. Detector is flushed with CO_2 for 5 h at 2.5 L/h prior to powering. The HV characteristics for the detector is shown in Fig. 22.2 (Left) and the total HV resistance was found to be 5.43MΩ. The signals observed with non-amplifying gas are termed spurious since they cannot originate from particle ionization. Figure 22.2 on right shows the rate of Spurious signal as a function of divider current.
- **Energy Spectrum**: Energy Resolution is calculated as $\frac{FWHM}{Peak\ Energy\ Value}$. Figure 22.3 (Left) shows MCA Spectrum obtained using X-ray at operating voltage 4150 V of GEM detector and its energy resolution was found to be ~30%.
- **Effective Gain**: Gain is the ratio of output current to the input current. The Effective gain is defined as: $G_{eff} = \frac{I_{ro}}{n_p.e.R_s}$, where I_{ro} is current collected on the readout, n_p is the average number of primary electrons with charge e created in the drift gap due to incident X-Rays, and R_S is the rate of the source at V_{Drift}. Figure 22.3 on right shows the variation of gain with the increasing voltage at drift.
- **Gas Purity**: Gas Chromatograph (GC) checks for different compounds (or any impurities) present in a gas mixture. GC provides the same signal for same amount

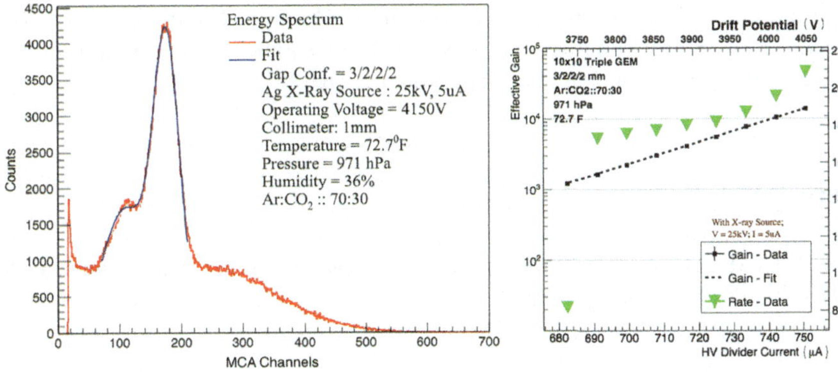

Fig. 22.3 Left: showing the MCA Spectrum for GEM detector using X-ray; Right: showing the rate of the source (green triangles) and effective gain (black dots) as a function of drift potential (upper *x-axis*) and divider current (lower *x-axis*) of detector with X-ray operated at 25 kV and 5 μA

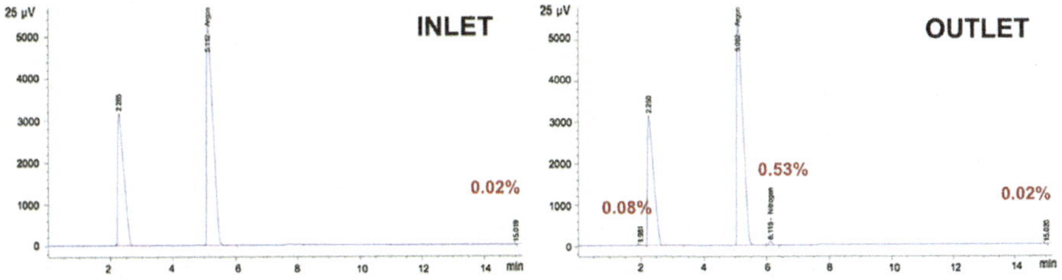

Fig. 22.4 Left: showing the gas content entering GEM detector; Right: showing the gas content after flushing the GEM detector with Argon (right peak) and CO_2 (left peak)

of material introduced to it, if all operating conditions are kept same. Different compounds are identified by the time required for them to travel through a column, called the *Retention Time*. Figure 22.4 on left shows the gas content (i.e. Argon and CO_2) at inlet of the GEM detector which is provided through the Gas Mixing Cylinder and Fig. 22.4 right shows the gas content at outlet of detector after flushing the gas mixture. We observed that few impurities ($<1\%$) were introduced after flushing the detector with the gas mixture.

22.3 Conclusions

We have performed various measurements (Leakage Current, HV Characteristics, Rate of Spurious Signals, Energy Spectrum, Effective Gain) with 10 cm × 10 cm triple GEM detector at DU. We have successfully repeated these measurements on large triple GEM detector prototype (~100 cm × 45 cm × 22 cm). We have also installed the Gas Chromatograph and saw that 99.98% pure gas was being used for the gaseous detectors which are built and tested at DU.

Reference

1. F. Sauli, GEM: a new concept for electron amplification in gas detectors. Nucl. Instrum. Meth. **A386**, 531 (1997)

Chapter 23
Z_2 Odd Sector Leading to Left-Right Symmetric Unification

Triparno Bandyopadhyay and Amitava Raychaudhuri

23.1 Introduction

Left-right symmetric (LRS) models [1] with gauge coupling unification address the issues of parity asymmetry and arbitrariness of hypercharge assignment in the electroweak (EW) sector of the standard model (SM), while automatically providing a mechanism for generation of neutrino masses. We extend the canonical LRS model to add suitable dark matter candidates leading to gauge coupling unification, without intermediate scales. The model is falsifiable with both the LRS breaking and unification scale testable at ongoing or upcoming experiments.

23.2 Model and Gauge Coupling Unification

The LRS symmetry is defined as: $SU(3)_C \otimes SU(2)_L \otimes SU(2)_R \otimes U(1)_{(B-L)}$. $SU(2)_R$ is a right-handed (RH) analogue of the standard $SU(2)_L$ and 'B', and 'L' are baryon and lepton numbers respectively. A discrete $L \leftrightarrow R$ symmetry, P_D, makes the left- and corresponding right-handed couplings equal, and is spontaneously broken by the vev, v_η, of a gauge singlet scalar, η, odd under P_D [2]. In addition to the SM fermions, there is an RH neutrino, N_l, per generation. The scalars in the model transform under the LRS symmetry as, $\Phi \equiv (1, 2, 2, 0)$, $\Delta_L \equiv (1, 3, 1, 2)$, $\Delta_R \equiv (1, 1, 3, 2)$, $\eta \equiv (1, 1, 1, 0)$. EW symmetry is broken by Φ, while Δ_R breaks LRS at some scale M_R.

T. Bandyopadhyay—Presenter.

T. Bandyopadhyay (✉) · A. Raychaudhuri
Department of Physics, University of Calcutta, 92 APC Road, Kolkata 700009, India
e-mail: gondogolegogol@gmail.com

A. Raychaudhuri
e-mail: palitprof@gmail.com

Md. Naimuddin (ed.), *XXII DAE High Energy Physics Symposium*, Springer Proceedings in Physics 203, https://doi.org/10.1007/978-3-319-73171-1_23

To this standard set of particles we add a couple of fermionic multiplets, $\chi_L^{1,2} + \chi_R^{1,2}$, in the representation $(1, 3, 1, 0) + (1, 1, 3, 0)$. $U(1)_{(B-L)}$ being broken by a $(B-L) = 2$ scalar, leaves a remnant $\mathbb{Z}_2$ symmetry [3]. Under this $\chi_{L,R}^{1,2}$ are even, while SM fermions are odd. This stabilises the lightest states of the multiplets. The mass lagrangian for $\chi_{L,R}^{1,2}$, is:

$$\mathcal{L}_m = \frac{M^i}{2}\left(\overline{\chi_L^{i\,c}}\chi_L^i + L \leftrightarrow R\right) + h^i(v_\eta + \eta)\left(\overline{\chi_L^{i\,c}}\chi_L^i - L \leftrightarrow R\right) + h.c.\,, i \in 1, 2. \tag{23.1}$$

Summing over 'i' is implied. The relative $-ve$ sign between the left- and right-handed Yukawa terms in the second parentheses is because $\eta \leftrightarrow -\eta$, under $L \leftrightarrow R$. The lagrangian although explicitly left-right symmetric to begin with, becomes asymmetric after η acquires a vev. The masses of the left- and right-handed multiplets are then: $M_{\chi_L^{1,2}} = M^{1,2}/2 + h^{1,2}v_\eta$, and $M_{\chi_R^{1,2}} = M^{1,2}/2 - h^{1,2}v_\eta$. With, $M^{1,2} \sim v_\eta$, the LH multiplets stay massive at the v_η scale.

The masses of the $SU(2)_R$ multiplets could be anything below v_η. However, for $M_R^{1,2} \sim \mathcal{O}$(TeV) there are multiple benefits. Firstly, the gauge couplings unify at a scale $M_U \sim 10^{16}$ GeV, with $M_P = M_U$, without any scales between M_R and M_U, implying minimal fine tuning. The unified coupling is given by, $g_U \sim 0.53$. Secondly, the right-handed symmetry breaking occurs at ~6.5 TeV, which can be probed at LHC-II. Also, the relic abundance of the multiplets matches that of dark matter in the Universe [5].

The model, when embedded in an $SO(10)$ symmetry, predicts a proton decay lifetime $\sim 2 \times 10^{34}$ years, clear of present bounds [4], and testable at the upcoming Hyper-Kamiokande experiment. The SM fermions reside in the 16 dim. representation of $SO(10)$, Φ in the 10 dim. fundamental, and $\Delta_{L,R}$, in the 126. $\chi_{L,R}^{1,2}$ are part of the adjoint 45 dimensional representation. η comes from the 210 dim. representation. $SO(10)$ itself is broken by the 45 dim. representation.

Each triplet contributes a pair of singly charged Dirac fermions and a Majorana fermion. At tree level the states are mass degenerate, however there is a radiative mass splitting. The charged states, $\chi_\pm^{1,2}$, are heavier than the corresponding Majorana $\chi_0^{1,2}$, the splitting being $\mathcal{O}$(GeV). Hence, the individual lightest states are neutral and can be dark matter (DM) candidates. $\chi_\pm^{1,2}$ decay to $\chi_0^{1,2}$ before big bang nucleosynthesis, with a lifetime about $\mathcal{O}$(ns).

23.3 Dark Matter Phenomenology

In Fig. 23.1, we show the contours in the $M_{\chi_R^2} \geq M_{\chi_R^1}$ plane for which the relic density of $\chi_{L,R}^{1,2}$ matches that of dark matter of the Universe. The dark matter candidates, $\chi_0^{1,2}$ do not couple to the Z, Z', or γ and as a result DM–DM annihilation channels are highly suppressed, the only possible channel being a t-channel annihilation to a W_R pair, mediated by $\chi^\pm$. Consequently there is no tree level coupling to nucleons,

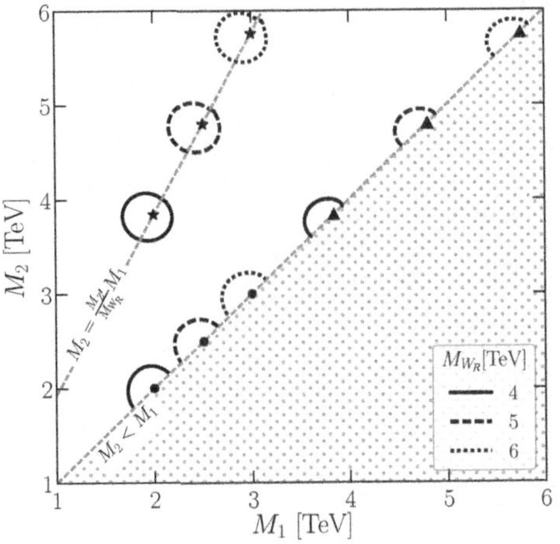

Fig. 23.1 The contours in the $M_{\chi^2} \geq M_{\chi^1}$ plane satisfying DM relic abundance, for different M_{W_R}. The contours center at the $(M_{\chi^1}, M_{\chi^2}) \equiv (M_{W_R}/2, M_{Z'}/2)$, (stars) $(M_{W_R}/2, M_{W_R}/2)$, (dots) and $(M_{Z'}/2, M_{Z'}/2)$ (triangles), with $M_{Z'} \sim 1.95 \times M_{W_R}$

and we do not expect the detection of the DM particles in current direct detection experiments.

23.4 Collider Prospects

The golden-channel for detection of the W_R, is $pp \to W_R \to lN_l^* \to lljj$. If the right-handed neutrino, N_l, is a Majorana particle as predicted in LRS models, equal numbers of same sign and opposite sign lepton pairs should be detected in the final state. In our analysis we find that at the LHC, with $\sqrt{s} = 14$ TeV, and an integrated luminosity, $\mathcal{L}_{\text{int}} \sim 500\,\text{fb}^{-1}$, a W_R as heavy as 6.5 TeV can be discovered. The Z' mass in our model is almost twice that of the W_R mass, hence its detection prospects are bleak, for $M_{W_R} \gtrsim 4$ TeV. The production cross section of $\chi_{\pm}^{1,2}\chi_{\mp}^{1,2}$ is too small, for $M_{W_R} \gtrsim 3.5$ TeV, to be detected at the LHC-II. However, $\chi_{\pm}^{1,2}\chi_0^{1,2}$ has a sizable cross section. The $\chi_0^{1,2}$ particles will be missed by the detector, while the $\chi_{\pm}^{1,2}$ particles, with a lifetime of $\mathcal{O}$(ns) and decay lengths of 0.1–1m, will manifest as disappearing tracks in the tracking chambers of CMS, ATLAS, and the MoEDAL detector at LHC.

23.5 Conclusion

In conclusion, we have presented a modified LRS model with gauge coupling unification and parity restoration at the GUT scale, and with suitable dark matter candidates, that can be falsified at ongoing and proposed experiments.

References

1. J.C. Pati, A. Salam, Phys. Rev. Lett. 31, 661 (1973); R. N. Mohapatra, J.C. Pati, Phys. Rev. D **11**, 566 (1975)
2. D. Chang, R.N. Mohapatra, M.K. Parida, Phys. Rev. Lett. **52**, 1072 (1984)
3. L.M. Krauss, F. Wilczek, Phys. Rev. Lett. **62**, 1221 (1989); L.E. Ibanez, G.G. Ross, Phys. Lett. B **260**, 291 (1991)
4. K. Abe et al., Phys. Rev. D **95**, 012004 (2017)
5. P.A.R. Ade et al., [Planck Collaboration], Astron. Astrophys. **594**, A13 (2016)

Chapter 24
Momentum Anisotropy, Chromo-Weibel Instability and QGP Phenomenology

Vinod Chandra

24.1 Introduction and Motivation

Space-time evolution of hot QCD matter–the Quark-Gluon Plasma (QGP), created during ultra-relativistic heavy-ion collisions is mainly governed by second order dissipative hydrodynamics. The initial spatial anisotropy in the geometry in the presence of anisotropic pressure gradients helps the system to develop collective flow through momentum anisotropy during the evolution. To capture the physics of such anisotropies (momentum) within semi-classical transport theory, we must have appropriate mathematical forms of momentum distributions of the gluons and quarks that constitute QGP before we study important phenomena such as heavy-quark dynamics and dilepton production, in this exotic medium. This sets the prime motivation for this work. The main focus is on the momentum anisotropy present in later stages of the heavy-ion collisions.

24.2 Momentum Anisotropy in Heavy-Ion Collisions and Near Equilibrium Quasi-parton Distributions

Momentum anisotropy in heavy-ion collisions can be captured in the anisotropic momentum distribution functions of quarks/anti-quark and gluonic degrees of freedom (up to linear order) as,

$$f_{g,q}\,(\mathbf{p}) = f^0_{g,q/\bar{q}} \;+ f^0_{g,q}\;(1 \pm f^0_{g,q})f_1^{g,q}(\mathbf{p}) + O(f_1^2) \tag{24.1}$$

V. Chandra (✉)
Indian Institute of Technology Gandhinagar, Gujarat 382355, India
e-mail: vchandra@iitgn.ac.in

Md. Naimuddin (ed.), *XXII DAE High Energy Physics Symposium*, Springer Proceedings in Physics 203, https://doi.org/10.1007/978-3-319-73171-1_24

where $f^0_{g,q}$ are isotropic distribution in the local thermal equilibrium of the QGP. Their forms are motivated by a recently proposed effective quasi-particle model of hot QCD medium [1] ($f^0_{g,q} = z_{g,q}\exp(-\beta u^\mu p_\mu)/(1 \mp z_{g,q}\exp(-\beta u^\mu p_\mu))$, here z_g is the gluon effective fugacity and z_q is the quark effective fugacity that lead to dispersion relations, $\omega_{g,q} = p + T^2\partial_p \ln(z_{g,q})$, and u^μ is the 4-velocity of the QGP fluid. The quantity $f_1^{g,q}(\mathbf{p})$ encodes the anisotropy.

As the medium is expanding in RHIC, the anisotropy while couples with expansion leads to Chromo-Weibel instability [2] in a similar way as Weibel instability in QED plasmas. The instability could lead to turbulent soft color fields which, in turn, contribute to the anomalous transport processes in the hot QCD plasma [3] and eventually modulates the transport coefficients of the medium quite substantially. This idea has been utilized to set-up an effective transport equation while considering the ensemble of turbulent soft color modes and solution of its linearized version determines the mathematical form of $f_1(\mathbf{p})$ [3, 4].

The effective transport equation that determines the form of the $f_1(\mathbf{p})$ in the local rest frame of the expanding medium (considering $(1+1)$- dimensional boost invariant expansion) reads [4]

$$\frac{p_i p_j (\nabla u)_{i,j}}{\omega_{g,q} T} f^0_{g,q}(1 \pm f^0_{g,q}) = \frac{g^2(T) C_2 < E^2 + B^2 > \tau_m}{3(N_c^2 - 1)\omega^2_{g,q}} \mathcal{L}^2 f_1^{g,q}(1 \pm f^0_{g,q}), \tag{24.2}$$

where $(\nabla u)_{i,j}$ is the symmetrized velocity gradients, $\mathcal{L}^2 := (\mathbf{p} \times \partial_{\mathbf{p}})^2 - [(\mathbf{p} \times \partial_{\mathbf{p}})^2]_z$, τ_m is the time scale of the instability. The quantities E and B denotes the color electric and magnetic fields respectively and the bracket $< ... >$ denotes the average with respect to ensemble of turbulent color fields [3].

The unknowns in the right hand side can be related to the phenomenologically known gluon-quenching parameter, $\hat{q} = \frac{2g^2 T C_2}{2(N_c^2-1)} < E^2 + B^2 > \tau_m$ [6], $C_2 = N_c$ for gluons and $(N_c^2 - 1)/2N_c$ for quarks (the respective Casimirs of $SU(N_c)$).

The near equilibrium anisotropic distribution functions thus obtained reads,

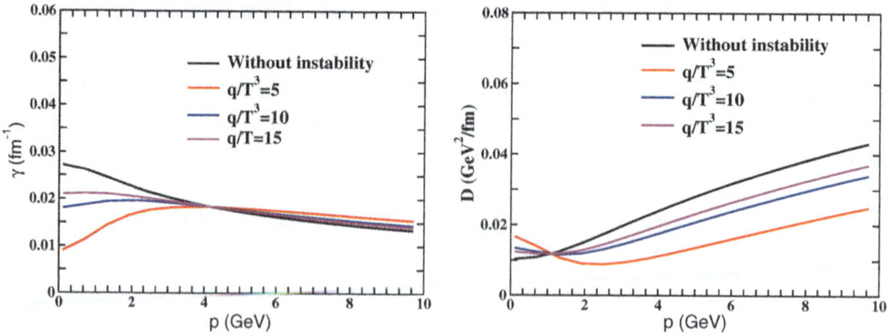

Fig. 24.1 Heavy quark drag and diffusion coefficients for, $Q = \hat{q}/T^3$

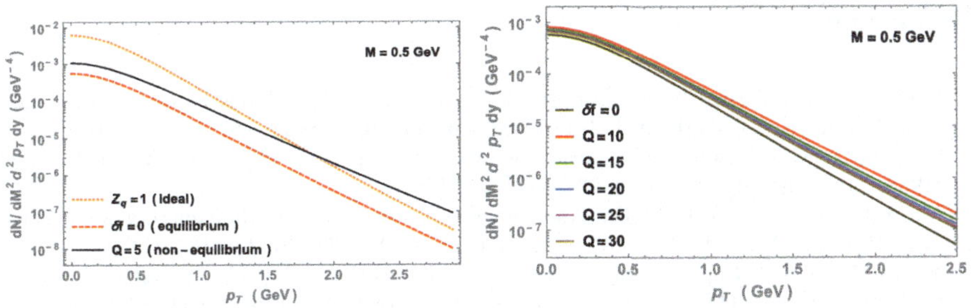

Fig. 24.2 Dilepton rate and yield: EOS and anisotropy effects, $Q = \hat{q}/T^3$

$$f_{g,q}(\mathbf{p}) = f^0_{g,q} - f^0_{g,q}(1 \pm f^0_{g,q})\frac{4\omega_g}{9\hat{q}_{g,q}T\tau}\cdot\left(p_z^2 - p^2/3\right) \tag{24.3}$$

Now, (24.3) is utilized to study heavy quark (charm quark) drag, γ and diffusion, D coefficients in the above anisotropic hot QCD medium while studying their scattering with light quarks and gluons in the medium. In addition, the dilepton production via $q\bar{q}$-annihilation in the anisotropic hot QCD/QGP medium is another implication of the anisotropic momentum distributions in (24.3). The impact of anisotropy on the heavy-quark dynamics is depicted in Fig. 24.1 and dilepton production in Fig. 24.2 for different values of the scaled jet quenching parameter, $\hat{q}/T^3$. Heavy-quark dynamics gets appreciably influenced by the momentum anisotropy. Clearly, from Fig. 24.2, the dilepton production rate notice significant changes in the presence of anisotropy [5]. Thus, while comparing these observables with experimental predictions, the effects from the momentum anisotropy can not be ignored. The validity of (24.3) is ensured whenever the second term in right-hand-side is much less in the magnitude as compared to the first one. In both the cases of dilepton production and heavy-quark transport the value $\hat{q}/T^3 = 5$ sets the limit of the approximation as it fails for the values of the same that smaller five.

24.3 Conclusions and Outlook

Momentum anisotropy that develops with the space-time evolution of the hot QCD matter seen to play crucial role while understanding the bulk and transport properties of the QGP. This fact has been realized while investigating heavy-quark dynamics as well as dilepton production in the anisotropic QGP medium. In both the cases, anisotropy is seen to play prominent role. A more quantitative understanding will require coupling the analysis with full (3 + 1)-hydrodynamical evolution of the QGP and be a matter of future investigations.

References

1. V. Chandra, V. Ravishankar, A quasi-particle description of (2+1)- flavor lattice QCD equation of state. Phys. Rev. D 84 074013 (2011). https://doi.org/10.1103/PhysRevD.84.074013
2. S. Mrowczynski, Stream instabilities of the quark-gluon plasma. Phys. Lett. B **214**, 587–590 (1988). https://doi.org/10.1016/0370-2693(88)90124-4
3. M. Asakawa, S.A. Bass, B. Müller, Anomalous viscosity of an expanding quark-gluon plasma. Phys. Rev. Lett. **96**, 252301 (2006). https://doi.org/10.1103/PhysRevLett.96.252301
4. V. Chandra, K. Das, Impact of momentum space anisotropy on heavy quark dynamics in a QGP medium. Phys. Rev. D **93**, 094036 (2016). https://doi.org/10.1103/PhysRevD.93.094036
5. V. Chandra, V. Sreekanth, Impact of momentum anisotropy and turbulent chromo-fields on thermal particle production in quark-gluon plasma medium. arXiv:1602.07142 [nucl-th]
6. A. Majumdar, B. Müller, X-.N. Wang, Small shear viscosity of a quark-gluon plasma implies strong jet quenching. Phys. Rev. Lett. **99**, 192301 (2007). https://doi.org/10.1103/PhysRevLett.99.192301

Chapter 25
Response Functions and Collective Modes of Hot QCD Medium

M. Yousuf Jamal, S. Mitra and V. Chandra

25.1 Introduction

Like QED plasma, hot QCD plasma (in the abelian limit) do possess similar collective excitation [1]. There have been various attempts to study collective plasma excitation in the isotropic and anisotropic hot QCD medium [2]. The collective excitations, that are studied by looking at the modes of the dispersion equations in the hot QCD medium, basically obtained from the polarization tensor in the medium. More precisely, the Polarization tensor that enters into the propagator, contains the dynamical information about the medium and the dispersion equation for the collective modes, are obtained in terms of poles of the propagator.

As the bulk of hot QCD medium that is created in ultra-relativistic heavy-ion collisions made up of light quarks/anti-quarks and gluons, therefore, there will be both gluonic and quarks collective modes in the medium. The main focus here is on the gluonic transverse and longitudinal collective modes which alternatively can be termed as plasmons. Another intriguing aspect of the hot QCD medium is the presence of momentum anisotropy throughout its space-time evolution. Therefore, while performing such investigations on collective excitations, it is desired to include the effects of anisotropy. Since, our investigations are within the framework of semi-classical transport theory, we require to invoke the appropriate momentum distributions of quarks and gluonic degrees of freedom in the medium. This is done within a recently proposed quasi-particle model for various hot QCD Equations of state ($\mathcal{O}(g^5)$ and $\mathcal{O}(g^6 log(\frac{1}{g}))$ pQCD, (2+1)-flavor lattice QCD equation of state [3]). In addition, refractive index(RI) of hot QCD medium has also been investigated.

M. Yousuf Jamal (✉) · S. Mitra · V. Chandra
Indian Institute of Technology Gandhinagar, Gandhinagar 382355, Gujarat, India
e-mail: mohammad.yousuf@iitgn.ac.in
S. Mitra
e-mail: sukanyam@iitgn.ac.in
V. Chandra
e-mail: vchandra@iitgn.ac.in

Md. Naimuddin (ed.), *XXII DAE High Energy Physics Symposium*, Springer Proceedings in Physics 203, https://doi.org/10.1007/978-3-319-73171-1_25

25.2 Collective Gluonic Modes

In the temporal gauge, the dispersion equations in the hot QCD medium could be obtained in terms of the poles of gluon propagator: $\Delta^{ij}(k) = ((k^2 - \omega^2)\delta^{ij} - k^i k^j + \Pi^{ij}(k))^{-1}$, where, the gluon polarization tensor reads:

$$\Pi^{ij}(k) = -g^2 \int \frac{d^3p}{(2\pi)^3} u^i \partial^l f_\xi(\tilde{\mathbf{p}}) \left(\delta^{jl} + \frac{u^j k^l}{k \cdot u + i0^+} \right) \tag{25.1}$$

The anisotropic distribution, f_ξ can be obtained as, $f_\xi(\tilde{\mathbf{p}}) = C_\xi f(\sqrt{\mathbf{p}^2 + \xi(\mathbf{p}.\hat{\mathbf{n}})^2})$. There is one more degree of freedom, the direction of anisotropy, $\hat{\mathbf{n}}$ with $\hat{\mathbf{n}}^2 = 1$. $\xi > 0$ correspond to contraction, $-1 < \xi < 0$ correspond to stretching of isotropic distribution function and $\xi = 0$ will take back to the case of isotropic medium with the distributions [4]: $f(\mathbf{p}) = 2N_c f_g(\mathbf{p}) + N_f[f_q(\mathbf{p}) + f_{\bar{q}}(\mathbf{p})]$, $f_{g/q} = z_{g/q}\exp[-\beta E_p]/(1 \mp z_{g/q}\exp[-\beta E_p])$. The tensor decomposition of Π^{ij}: $\Pi^{ij} = \alpha A^{ij} + \beta B^{ij} + \gamma C^{ij} + \delta D^{ij}$, with, $A^{ij} = \delta^{ij} - k^i k^j/k^2$, $B^{ij} = k^i k^j/k^2$, $C^{ij} = \tilde{n}^i \tilde{n}^j/\tilde{n}^2$, $D^{ij} = k^i \tilde{n}^j + \tilde{n}^i k^j$. The structure functions, $\alpha, \beta, \gamma, \delta$ could be obtained with the appropriate contractions. Next, the dispersion equation is obtained as, $\Delta(k) = \Delta_A[\mathbf{A} - \mathbf{C}] + \Delta_G[(k^2 - \omega^2 + \alpha + \gamma)\mathbf{B} + (\beta - \omega^2)\mathbf{C} - \delta\mathbf{D}]$ with, $\Delta_A^{-1}(k) = k^2 - \omega^2 + \alpha$ and $\Delta_G^{-1}(k) = \Delta_{G1}^{-1}(k)\,\Delta_{G2}^{-1}(k)$, where, $\Delta_{G1}^{-1}(k) = k^2 - \omega^2 + \alpha + \gamma$, $\Delta_{G2}^{-1}(k) = \beta - \omega^2$. We symbolically denote the collective modes as A and $G = (G1, G2)$ modes as done in [5]. These modes can take both real and imaginary ($\omega = i\Gamma$) values depicted in Fig. 25.1, for various interacting QCD/QGP EOSs along with the leading order (LO). In the case of real ω, we have found three modes in which two (ω_A and ω_{G1}) correspond to transverse mode while the third mode ($G2$) correspond to the

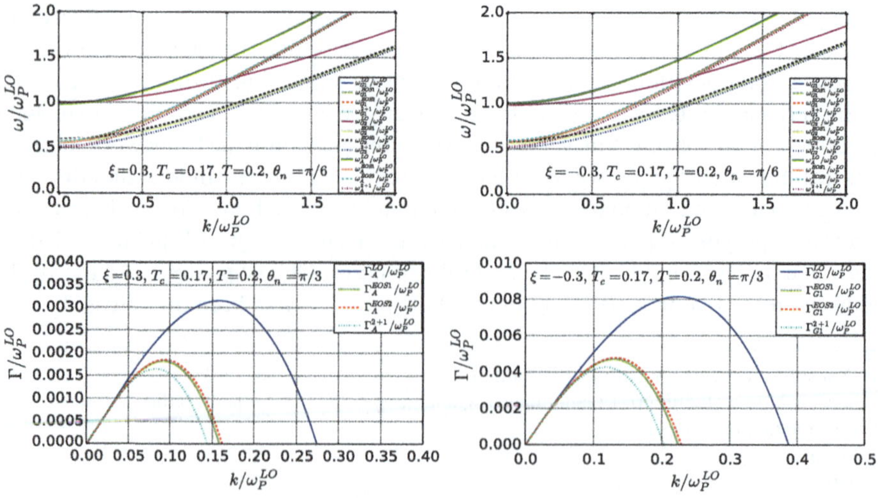

Fig. 25.1 Real and Imaginary modes for various EOSs along with LO

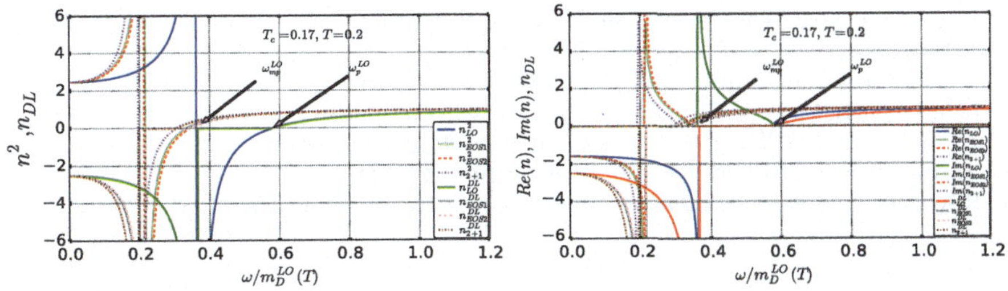

Fig. 25.2 n^2(Left) and n(right) along with n_{DL} for various EOSs along with LO

longitudinal one. In the case of imaginary frequency (i.e., $\omega = i\Gamma$), we found only two imaginary modes (Γ_A and Γ_{G1}), among them, Γ_{G1}, appears only for certain value of θ_n for $\xi > 0$ and for $\xi < 0$, we found only Γ_{G1} mode.

25.3 The Refractive Index

The RI is defined as, $n := \sqrt{\epsilon\,\mu}$, where in the small k-limit [6], $\epsilon(\omega) \approx 1 - m_D^2/3\omega^2 + \mathcal{O}(k^2)$, $\mu(\omega) \approx (1 - 2m_D^2/15\omega^2)^{-1} + \mathcal{O}(k^2)$. Considering, $\sqrt{2/15}\, m_D(T) = \omega_{mp}$ and $m_D(T)/\sqrt{3} = \omega_p$, where at ω_{mp}, there is a pole in μ which was absent in the holographic treatment of strongly coupled plasma [7] and ω_p is the plasma frequency. The refractive index is considered to be negative if Depine-Lakhtakia index [8], $n_{DL} = |\epsilon|\text{Re}(\mu) + |\mu|\text{Re}(\epsilon) < 0$.

In Fig. 25.2, one can observe that for the frequency $\omega < \omega_{mp}$, the RI is negative. Since, in this range, n^2 is positive, only Re(n) exist. In the range $\omega_{mp} \leq \omega \leq \omega_p$, n^2 is negative and so, only Im(n) is contributing while the $n_{DL} = 0$. In this range EM waves can not propagate and the medium is opaque. For the frequency $\omega > \omega_p$, again n^2 is positive and so, only Re(n) contributes and approaching towards unity.

Finally, we observe that hot QCD medium effects induce sizable modifications to collective modes and refractive index of hot QCD medium.

Acknowledgements VC would like to sincerely acknowledge DST, Govt. of India for Inspire Faculty Award -IFA13/PH-15 and Early Carrier Research Award(ECRA/2016) Grant.

References

1. E.S. Weibel, Phys. Rev. Lett. **2**, 83 (1959)
2. P. Romatschke, M. Strickland, Phys. Rev. D **68**, 036004 (2003); Phys. Rev. D **70**, 116006 (2004), S. Mrowczynski, Acta Phys. Polon. B **37**, 427 (2006); P.B. Arnold, J. Lenaghan, G.D. Moore, JHEP **0308**, 002 (2003), M.E. Carrington, K. Deja, S. Mrowczynski, Phys. Rev. C **90**(3), 034913 (2014)

3. P. Arnold, C. Zhai, Phys. Rev. D **50**, 7603 (1994); M. Cheng et al., Phys. Rev. D **77**, 014511 (2008); K. Kajantie et. al., Phys. Rev. D **67**, 105008 (2003)
4. V. Chandra, V. Ravishankar, Phys. Rev. D **84**, 074013 (2011)
5. M.Y. Jamal, S. Mitra, V. Chandra, Phys. Rev. D **95**, 094022 (2017)
6. J. Liu, M.j. Luo, Q. Wang, H.j. Xu, Phys. Rev. D **84**, 125027 (2011)
7. A. Amariti et al., JHEP **04**, 036 (2011); JHEP **10**, 104 (2011)
8. R.A. Depine, A. Lakhtakia, Microwave and Optical Technology Letters **41**, 315 (2004)

Chapter 26
Electromagnetic Response of a Hot QCD Medium in Heavy Ion Collisions

Sukanya Mitra and Vinod Chandra

26.1 Introduction

It is well established that in the early stages of the non-central heavy ion collisions an electromagnetic (EM) field is produced with considerable strength ($eB \sim m_\pi^2 \sim 10^{12}$ Gauss) [1], which is reflected on the observable properties of the extracted signals at the later stages of the collisions. In response of the concerned electric field, an induced current is generated that relates the electric field by the quantity called electrical conductivity σ_{el}. The sensitivity of the charge-dependent directed flow of the observed hadrons on initial stage charge asymmetry via σ_{el} and the dependence of soft photon and dilepton emission rates on σ_{el}, provide the motivation of investigating the values of σ_{el} along with its temperature behavior. Viewing this scenario, in this article the electrical conductivity and charge diffusion coefficient(D) have been estimated giving their explicit temperature dependences for a strongly interacting hot QCD (quantum chromodynamic) system which is likely to be created in heavy ion collisions.

26.2 Formalism

The effective quasi particle model (EQPM) employed here, is based on the idea of mapping the hot QCD medium effects present in the equations of states (EOSs) either computed within improved perturbative QCD ($O(g^5)$ and $O(g^6 \ln(1/g))$ or $(2+1)$-flavor lattice QCD simulations. Under the EQPM scheme the quasi-parton equilibrium distribution functions and the non-trivial dispersion relations are respectively the followings,

S. Mitra (✉) · V. Chandra
Indian Institute of Technology Gandhinagar, Gandhinagar 382355, Gujarat, India
e-mail: sukanyam@iitgn.ac.in
URL: http://www.iitgn.ac.in/pdrf/sukanya.htm

Md. Naimuddin (ed.), *XXII DAE High Energy Physics Symposium*, Springer Proceedings in Physics 203, https://doi.org/10.1007/978-3-319-73171-1_26

$$f_{g/q} = \frac{z_{g/q}\exp[-\beta E_p]}{\left(1 \mp z_{g/q}\exp[-\beta E_p]\right)}, \qquad \omega_{g/q} = E_p + T^2\partial_T ln(z_{g/q}), \tag{26.1}$$

where $z_{g/q}$'s are effective fugacities of the quasi partons that encode the effects of a strongly interacting thermal medium. With this definitions we finally lead to the effective coupling of QCD interaction as,

$$\alpha_{eff} = \alpha_s(T)\frac{\frac{2N_c}{\pi^2}PolyLog[2, z_g] - \frac{2N_f}{\pi^2}PolyLog[2, -z_q]}{\frac{N_c}{3}+\frac{N_f}{6}}. \tag{26.2}$$

The thermal relaxation times (τ) of the constituent quarks and gluons can be obtained from the relativistic transport equation as an explicit function of their mutual interactions. In terms of the temperature and effective coupling, the relaxation times for quasi gluons and quarks take the following form,

$$\tau \sim T\alpha_{eff}^2 ln\left\{\frac{1}{\alpha_{eff}}\right\}. \tag{26.3}$$

In the Chapman–Enskog method from kinetic theory application of a multi component quark gluon plasma (QGP) system, the expression of the electric conductivity and diffusion coefficient are,

$$\sigma_{el} = q_q^2\frac{l_{11}h_g - l_{1q}x_q}{h}, \qquad D = \frac{Tl_{11}}{nx_q x_g}, \tag{26.4}$$

with, $$l_{11} = \frac{1}{T}[x_q^2\tau_g\int\frac{d^3p_g}{(2\pi)^3}f_g^0(1+f_g^0) + x_g^2\tau_q\int\frac{d^3p_q}{(2\pi)^3}f_q^0(1-f_q^0)],$$

$$l_{1q} = \frac{1}{T}[-x_q\tau_g\int\frac{d^3p_g}{(2\pi)^3}f_g^0(1+f_g^0)(\omega_g - h_g) + x_g\tau_q\int\frac{d^3p_q}{(2\pi)^3}f_q^0(1-f_q^0)(\omega_q - h_q)].$$

The $q_q^2 = \sum_k \nu_k q_{qk}^2$ is the square of the fractional quark charges taking sum over quark degeneracy. For the detail derivations and the definitions of the thermodynamic quantities, [2] is being referred.

26.3 Results and Discussions

The temperature dependences of σ_{el} and D are distinctly showing the effects of different EOSs under EQPM. The leading log over coupling term in τ's are appearing to be majorly responsible for the considerably higher magnitude of σ_{el}, which however shows a good agreement with Lattice results from [3] and dynamical quasi particle

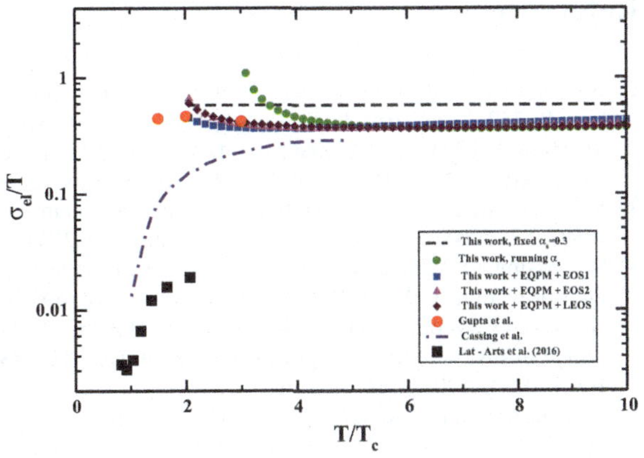

Fig. 26.1 σ_{el} for $N_f = 3$ as a function of T

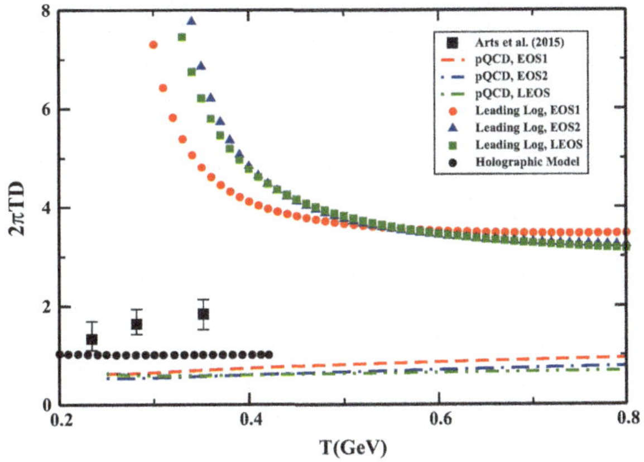

Fig. 26.2 D for $N_f = 3$ as a function of T

results from [4]. For D, the estimations with perturbative QCD cross section are comparable with Lattice results from [5] and holographic results from [6], whence the leading log results of D are showing larger magnitude due to the intrinsic nature of the leading log term itself (Figs. 26.1 and 26.2).

26.4 Conclusion and Outlook

In this current work we have estimated the temperature behavior of electrical conductivity and diffusion coefficients for a strongly interacting QCD system using an effective quasi particle model in order to describe the thermal medium. The obtained results are showing the different EOS effects through the quasi particle model noticeably. The results are observed to be in quite sensible agreement with other recent estimations such as lattice, holographic models and dynamical quasi particle models.

References

1. B.G. Zakharov, Electromagnetic response of quark–gluon plasma in heavy-ion collisions. Phys. Lett. B. **737**, 262 (2014). https://doi.org/10.1016/j.physletb.2014.08.068
2. M. Mitra, V. Chandra, Thermal relaxation, electrical conductivity, and charge diffusion in a hot QCD medium. Phys. Rev. D **94**, 034025 (2016). https://doi.org/10.1103/PhysRevD.94.034025
3. S. Gupta, The electrical conductivity and soft photon emissivity of the QCD plasma. Phys. Lett. B **597**, 57 (2004). https://doi.org/10.1016/j.physletb.2004.05.079
4. W. Cassing, O. Linnyk, T. Steinert, O. Ozvenchuk, Electrical conductivity of hot QCD matter. Phys. Rev. Lett. **110**(18), 182301 (2013). https://doi.org/10.1103/PhysRevLett.110.182301
5. G. Aarts, C. Allton, A. Amato, P. Giudice, S. Hands, J.I. Skullerud, Electrical conductivity and charge diffusion in thermal QCD from the lattice. JHEP **1502**, 186 (2015). https://doi.org/10.1007/JHEP02(2015)186
6. S.I. Finazzo, R. Rougemont, Thermal photon, dilepton production, and electric charge transport in a baryon rich strongly coupled QGP from holography. Phys. Rev. D **93**(3), 034017 (2016). https://doi.org/10.1103/PhysRevD.93.034017

Chapter 27
A Proof- of-Principle for Time of Flight-Positron Emission Tomography Imaging

Rajesh Ganai, Shaifali Mehta, Mehulkumar Shiroya, Mitali Mondal, Zubayer Ahammed and Subhasis Chattopadhyay

27.1 Introduction

Positron Emission Tomography (PET) [1], is a radio-tracer, nuclear medicine imaging technique. PET is used to observe metabolic processes in the body. The basic principle of PET is detecting a pair of back to back 511 keV photons created by the annihilation of a positron with an electron. The positron emitter, Fludeoxyglucose (^{18}F) (FDG) which is a radio-tracer, administered in the body annihilates into a pair of 511 keV photons, flying in opposite directions. Time of Flight (TOF) technique has found its application in PET imaging. The two gamma-ray interaction points define a so-called line-of-response (LOR) on which the annihilation must have taken place. A precise measurement of the arrival times of the coincident photons along with the time difference in flight-time of the two photons helps to localize the annihilation event on the LOR. Figure 27.1 illustrates the basic working and detection principle of any TOF-PET system.

R. Ganai (✉) · M. Mondal · Z. Ahammed · S. Chattopadhyay
EHEP&A Group, Variable Energy Cyclotron Centre, Kolkata 700064, India
e-mail: rajesh.ganai.physics@gmail.com

R. Ganai · M. Mondal · Z. Ahammed · S. Chattopadhyay
Training School Complex, Homi Bhabha National Institute,
Anushakti Nagar, Mumbai 400 085, India

R. Ganai
Bose Institute, Centre for Astroparticle Physics and Space Sciences,
P-1/12, CIT Road, Scheme VII-M, Kolkata 700054, West Bengal, India

R. Ganai
GSI Helmholtzzentrum für Schwerionenforschung GmbH, Planckstrasse 1,
64291 Darmstadt, Germany

S. Mehta
School of Physics and Material Sciences, Thapar University, Patiala 147004, Punjab, India

M. Shiroya
Sardar Vallabhbhai National Institute of Technology, Surat 395007, Gujarat, India

Md. Naimuddin (ed.), *XXII DAE High Energy Physics Symposium*, Springer Proceedings in Physics 203, https://doi.org/10.1007/978-3-319-73171-1_27

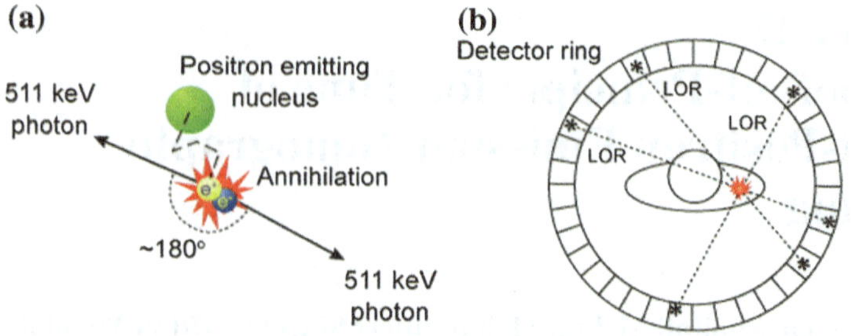

Fig. 27.1 Imaging principle of PET: **a** Two 511 keV photons are emitted in (almost) opposite directions after e^+ annihilates with e^-; **b** The two photons are simultaneously detected within a ring of detectors surrounding the patient. The line-of-response (LOR) connecting the two photons is generated. By recording many LORs the activity distribution can be tomographically reconstructed [5]

The Multi-gap Resistive Plate Chambers (MRPCs) [2] are advanced form of RPC [3] detectors where the gas gap of a single gap RPC is divided into multiple gas subgaps with highly resistive electrodes to improve the time resolution of the later without sacrificing its other good qualities like efficiency. The best time resolution from MRPC is 15.8 ps [4]. Better time resolution of MRPCs over scintillators can determine the LOR, hence the annihilation point more precisely. The major disadvantage of MRPCs is that they are low efficient in detecting photons.

27.2 Experimental Set up and Test Results

The aim of this work was to detect the two back to back gammas created by the annihilation of positron emitted from ^{22}Na with an electron with the developed prototype 5-gap glass MRPCs [6, 7] and also to sense a change in the position of the ^{22}Na source. In order to do so, the major challenge was to eliminate the cosmic muon background as MRPCs are known to have very good charged particle detection efficiency and was successfully achieved by veto method. The schematic of the experimental set up and the actual set up has been shown in Fig. 27.2. The two prototype MRPCs were kept horizontally and separated by a known distance of 17 cm. A ^{22}Na source was kept in between the MRPCs. Two aluminium plates of dimensions $\sim$30 cm $\times$ 30 cm $\times$ 0.5 cm ensured that the back to back photons does not reach the scintillators. Two plastic scintillators each of dimension $\sim$50 cm $\times$ 25 cm were used.

A suitable trigger $(\overline{(Scintillator - I) \cdot (Scintillator - II)} \cdot$ MRPC - I) was also chosen which initiated the START of the TDC module. The first set of TDC spectra was taken when the source was placed 3 cm away from the bottom or MRPC - II and the second set was taken when the source was kept at 14 cm away from the bottom MRPC which have been shown in Fig. 27.3. The mean channel of the TDC spectra after a Gaussian fit was obtained to be 1479 for Fig. 27.3a and 1514 for Fig. 27.3b. Clearly there is a shift in the mean of the TDC spectra, specially in the mean of the spectra by 35 TDC channels as the source was moved from position-I to position-II by 11 cm. Assuming the velocity of photons to be 30 cm/ns, a change in source

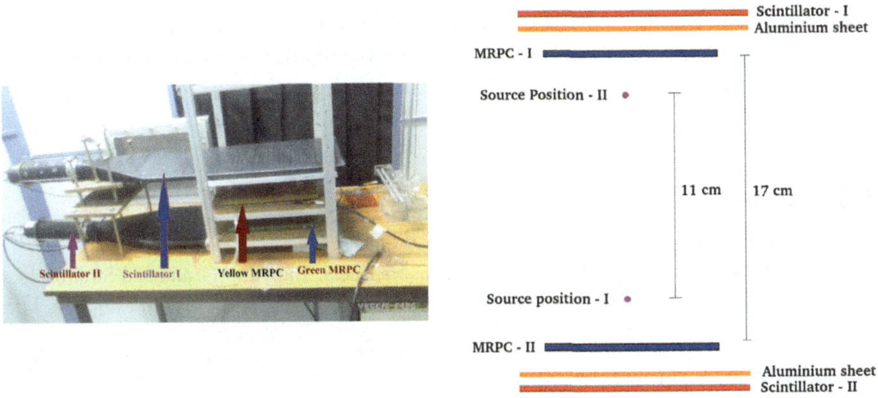

Fig. 27.2 Actual experimental set up and schematic of the experimental set up for TOF-PET with prototype MRPCs

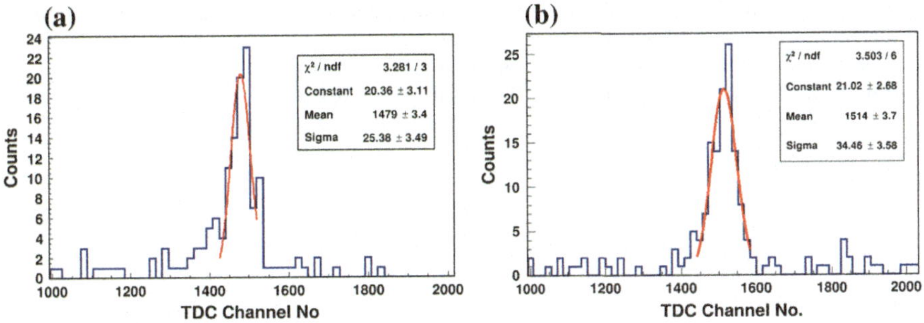

Fig. 27.3 **a** The TDC spectra of MRPC-II when the ^{22}Na source was kept at position-I. **b** The TDC spectra of MRPC-II when the ^{22}Na source was kept at position-II. The resolution of TDC is ~25 ps/channel

position by 11 cm should give TDC channel difference of 30 channels which is close to to the obtained value of 35 channels. From another way of looking at it, a shift in the mean of the TDC spectra of 35 channels should yield a change in the source position by 12.8 cm which is close to the actual change in source position by 11 cm.

27.3 Summary and Discussion

Excellent time resolution of MRPCs make them potential candidate to replace the scintillators in existing PET systems. If successful, the cost per scan of PET imaging will reduce drastically as MRPCs are relatively low cost detectors. As a first step towards this noble work, two prototype MRPCs have been tested in a two-MRPC coincidence set-up for the detection of back to back photons created by the annihilation of positron (emitted from Na^{22} source) with a nearby electron. The change in distance in the source position was successfully estimated from the time spectra obtained by using both the MRPCs. In order to bring down the estimated change in distance in the source position by MRPCs, the time resolution of the detectors needs to be improved.

References

1. J.A. Sorenson, M.E. Phelps, *Physics in Nuclear Medicine*, 2nd edn. (Grune and Stratton Inc., Orlando, 1987); M.M. Ter-Pogossian et al., Positron emission tomography. Sci. Am. **243**, 170–181 (1980)
2. E. Cerron Zeballos et al., A new type of resistive plate chamber: the multigap RPC. Nucl. Instrum. Meth. A **374**, 132–135 (1996)
3. R. Santonico, R. Cardarelli, Development of resistive plate counters. Nucl. Instrum. Meth. **187**, 377 (1981)
4. Crispin Williams, *Latest results from MRPC time resolution*. http://www.pas.va/content/dam/accademia/pdf/sv119/sv119-williams.pdf
5. https://www.tudelft.nl/en/faculty-of-applied-sciences/about-faculty/departments/radiation-science-technology/denkova-group/research/tof-pet/
6. Rajesh Ganai, Proc. DAE Symp. Nucl. Physics **61**, G18 (2016)
7. Rajesh Ganai, Proc. DAE Symp. Nucl. Physics **61**, G23 (2016)

Chapter 28
Development of 6-Gap Bakelite Multi-gap Resistive Plate Chamber (MRPC)

Rajesh Ganai, Mitali Mondal, Shaifali Mehta, Zubayer Ahammed and Subhasis Chattopadhyay

28.1 Introduction

Multi-Gap Resistive Plate Chamber (MRPC) [1] being a parallel plate avalanche gas detector has been used in various nulcear and high energy physics experiments, due to their excellent time resolution (~20 ps) [2], high detection efficiency, reliability and ability to cover a large area. MRPC works on the principle of gas-ionization and consists of small sub-gaps. The small gas-gaps are maintained by inserting thin resistive plates between two outer resistive plates. Formation of an avalanche of charge carriers in one or more than one gas gap induces a signal on metallic pick up strips. A schematic of an MRPC is shown in Fig. 28.1. Mostly glass MRPCs had been used in various high energy physics experiments until now. We have made an attempt to develop a six gap bakelite MRPC. Bakelite MRPCs have several advantages over glass MRPC such as it is more flexible, easy to handle, can be used in both avalanche and streamer mode gas-mixture.

The fabrication procedure of a six gap bakelite MRPC of total dimension $15\,\mathrm{cm} \times 15\,\mathrm{cm} \times 1\,\mathrm{cm}$ with each gap thickness of ~250 μm and preliminary test results of this MRPC using cosmic muons are discussed in this paper.

R. Ganai (✉) · M. Mondal · Z. Ahammed · S. Chattopadhyay
Variable Energy Cyclotron Centre, 1/AF Bidhan Nagar, Kolkata 700064, India
e-mail: rajesh.ganai.physics@gmail.com

R. Ganai
Bose Institute, Centre for Astroparticle Physics and Space Sciences, P-1/12, CIT Road, Scheme VII-M, Kolkata 700054, West Bengal, India

R. Ganai
GSI Helmholtzzentrum für Schwerionenforschung GmbH, Planckstrasse 1, 64291 Darmstadt, Germany

S. Mehta
School of Physics and Material and Sciences, Thapar University, Patiala 147004, Punjab, India

Md. Naimuddin (ed.), *XXII DAE High Energy Physics Symposium*, Springer Proceedings in Physics 203, https://doi.org/10.1007/978-3-319-73171-1_28

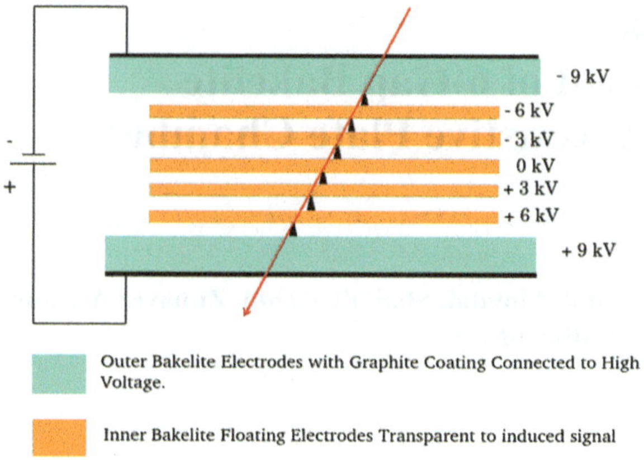

Fig. 28.1 Schematic diagram of a six gap MRPC

Development of 6-gap Bakelite MRPC

We have developed a six-gap bakelite MRPC of dimension 15 cm × 15 cm with a total gas gap of 0.15 cm. The outer two bakelite electrode plates have dimensions of 15 cm × 15 cm × 0.3 cm while the inner bakelite plates are thin (14 cm × 14 cm × 0.05 cm). Each gas gap has a thickness of 250 μm which is maintained by five button spacers made of G10 sheet. A perspex frame of 4 mm thick and 1 cm width serves as a side spacer between two outer plates. Two gas nozzles made of polycarbonate material, one for gas inlet and one for gas outlet, have been attached diagonally to the chamber for proper gas flow. We have applied a conductive layer prepared by mixing a conductive paint and special dry thinner in a ratio of 1:1 on the two outer plates for uniform application of high voltage. The different steps of developing the MRPC has been shown in Fig. 28.2.

Experimental Test Results using Cosmic Ray

The bakelite MRPC was characterized in standard cosmic ray test setup to measure efficiency and noise rate in avalanche mode with a gas mixture ratio of R134a:Isobutane:SF_6::94.3:5.3:0.4. The flow rate of the gas was 0.75 lit/hr. The temperature, pressure and relative humidity of the lab were ~15 °C, ~1 atm and ~50 – 55 % respectively throughout the experiment. The trigger was given by the coincidence from three scintillators, 2 paddle scintillators (Active area ~20 cm × 8.5 cm) and a finger scintillator (5 cm × 1.5 cm). Figure 28.3 shows the measured efficiency and the noise rate values as a function of the applied voltage. Efficiency plateau of ~90% was reached at high voltage values beyond 15.4 kV.

As expected it was found that the noise rate of the MRPC decreases by increasing threshold. The maximum noise rates of the MRPC are ~2.8 Hz/cm^2, ~1.8 Hz/cm^2 and ~2.2 Hz/cm^2 at negative discriminator thresholds of 20 mV, 50 mV and 100 mV respectively.

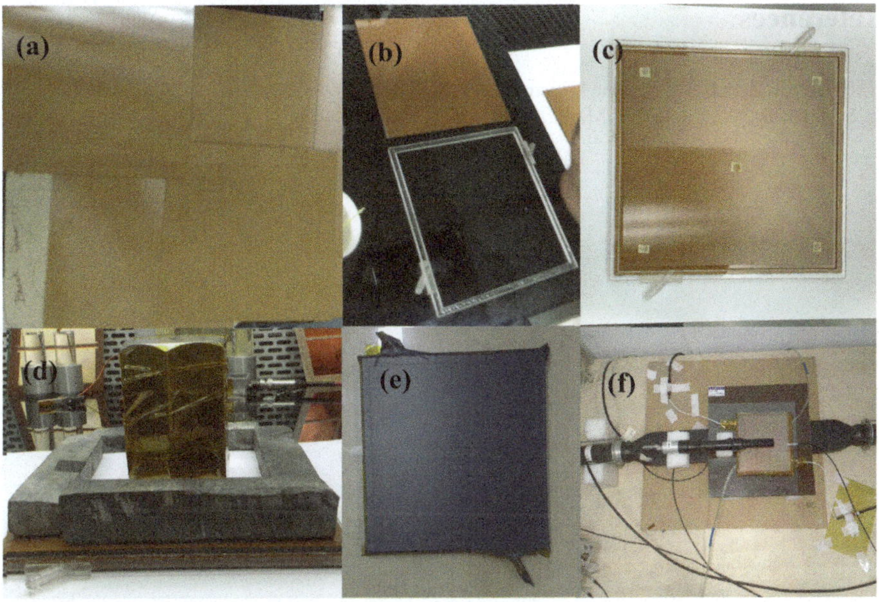

Fig. 28.2 **a** Inner bakelite plates of 500 μm thick, **b** the bottom electrode and perspex frame kept side by side, **c** Inner electrodes glued on bottom electrode spacing them by button spacer, **d** weights placed over developed MRPC, **e** Outer surface of the outer electrodes painted with black conduction paint, **f** Testing the developed MRPC with cosmic ray Setup

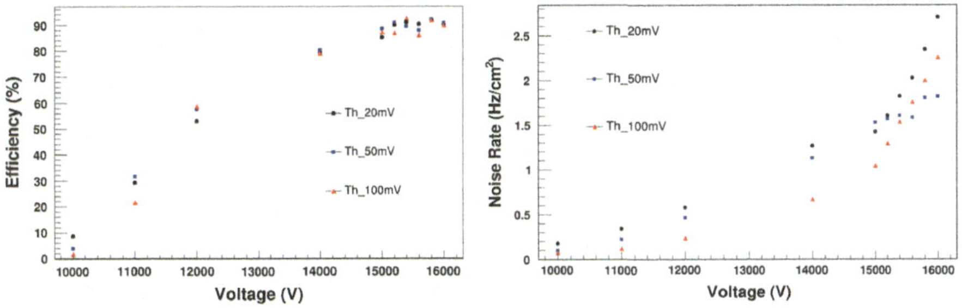

Fig. 28.3 Efficiency and Noise Rate of the MRPC as a function of applied HV

Conclusion

In this paper we have presented the development procedure of six gap bakelite MRPC and it has been characterized in avalanche mode gas mixture R134a (94.3%), isobutane (5.3%), SF_6 (0.4%) using cosmic ray flux. It has an efficiency of ~90% at noise rate 1.5 Hz/cm^2. As a future plan, we planned to measure the time and charge spectra of the MRPC.

References

1. E. Cerron Zeballos et al., Nucl. Instrum. Meth. A **373**, 132–135 (1996)
2. S. An et al., Nucl. Instrum. Meth. A **594**, 39–43 (2008)

Chapter 29
Conductivity of Holographic Superconductor in Soft-Wall Model

Neha Bhatnagar

29.1 Introduction

Experimentally observed novel features of condensed matter systems are intriguing but difficult to grasp as conventional non-perturbative approach comes with constraints. However advent of AdS/CFT duality(holography) [1] results in remarkable development in the area. Using this duality dynamics of strongly coupled systems are described from weakly coupled dual gravity theory residing in one higher dimension. High T_c superconductor is one such example which can be observed in laboratory but still needs a valid theoretical description. Conventional superconductivity with $T_c = 30K$ can be explained using BCS theory considering weak coupling of electrons. Recently various holographic models have been developed to investigate the phenomenon of high T_c superconductor where coupling of complex scalar field with gauge field is required in gravity action. In this work, we have applied soft-wall model [2] to study the transport properties of $2 + 1$ dimensional holographic superconductor. It was argued [3] that different dilaton profile resembles with different condensate in superconducting phase. Although phase transition could not be studied explicitly in present form but we can match our results qualitatively with experimentally observed transport properties. This approach provide us scope to work further and explore strongly coupled system in phenomenological simple way [4].

29.2 Holographic Model

We consider soft wall model in $(3 + 1)$ dimensional Einstein-Maxwell system.

$$S = \frac{1}{16\pi G_4} \int d^4x \sqrt{-g} e^{-2\phi} \left(R + \frac{6}{L^2} - F_{MN} F^{MN} \right) \tag{29.1}$$

N. Bhatnagar (✉)
Department of Physics, Banaras Hindu University, Varanasi 221005, India
e-mail: bhtngr.neha@gmail.com

Md. Naimuddin (ed.), *XXII DAE High Energy Physics Symposium*, Springer Proceedings in Physics 203, https://doi.org/10.1007/978-3-319-73171-1_29

where F^2 is U(1) gauge field strength with $F_{\mu\nu} = \partial_\mu A_\nu - \partial_\nu A_\mu, \kappa^2 = 8\pi G_4 = L = 1$ (G_4 is Newton's constant) and Λ(cosmological constant). Equations of motion from above action are given as,

$$R_{\mu\nu} - \frac{1}{2} g_{\mu\nu} R - \Lambda g_{\mu\nu} = 8\pi G_4 T_{\mu\nu} \quad and \quad \nabla_\mu (e^{-2\phi} F^{\mu\nu}) = 0 \tag{29.2}$$

considering metric ansatz as, (taking $z = \frac{r_+^2}{r^2}$)

$$ds^2 = \frac{L^2}{z^2} \left(-f(z)dt^2 + dx^2 + dy^2 + \frac{dz^2}{f(z)} \right) \tag{29.3}$$

where

$$f(z) = 1 + Q^2 \left(\frac{z}{z_h} \right)^6 - (1 + Q^2) \left(\frac{z}{z_h} \right)^4 \tag{29.4}$$

and z_h is the black hole horizon, $Q = \frac{\mu}{z_h}$(Q is the charge of the black hole) and Hawking temperature of system is given as, $T_H = \frac{f'(z_h)}{4\pi}$.

29.2.1 AC Conductivity

AC conductivity is calculated by introducing perturbations in matter and gravitational fields and studying linear response of the system. Using Fourier transformation as,

$$\delta A_x(t, z) = \int_{-\infty}^{\infty} \frac{d\omega}{2\pi} \tilde{a}_x(\omega, z) e^{-i\omega t}, \qquad \delta g_{tx}(t, z) = \int_{-\infty}^{\infty} \frac{d\omega}{2\pi} h_{tx}(\omega, z) e^{-i\omega t} \tag{29.5}$$

we get the gauge field equation used to study frequency response of conductivity (scaling $\tilde{a}_x = e^\phi a_x$).

$$a_x''(z) + \frac{f'(z)}{f(z)} a_x'(z) + \left(\phi''(z) - \phi'(z)^2 + \frac{f'(z)}{f(z)} \phi'(z) - \frac{A_t'^2 z^2}{f(z)} + \frac{\omega^2}{f(z)^2} \right) a_x(z) = 0 \tag{29.6}$$

Here we have fixed the dilaton profile as $\phi = z^2$ and using solution of A_t from (29.2) we study response of system. From (29.6) behavior of optical conductivity with varying chemical potential is shown in Fig. 29.1. We observe formation of gap in ac conductivity with increase in chemical potential. Thus using soft wall model we can study normal phase of holographic superconductor in 2 + 1 dimensions at finite charge density.

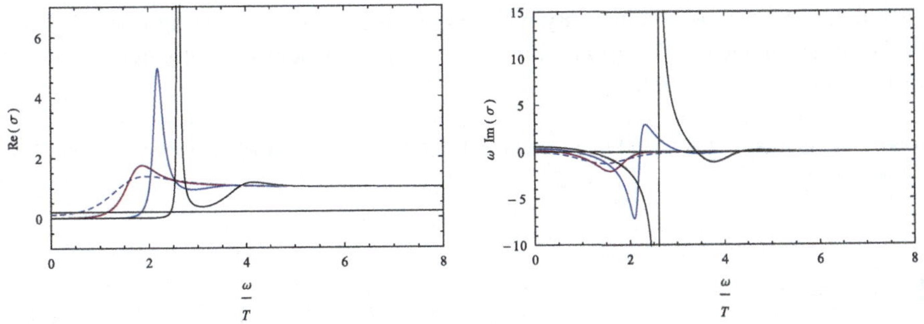

Fig. 29.1 Frequency dependent conductivity with varying chemical potential $\mu = 0$ (dashed), 0.5 (purple), 1.0 (blue) and 1.5 (black)

Fig. 29.2 Temperature dependent DC conductivity with varying chemical potential $\mu = 0.5$(blue), 1(black) and 1.5(brown)

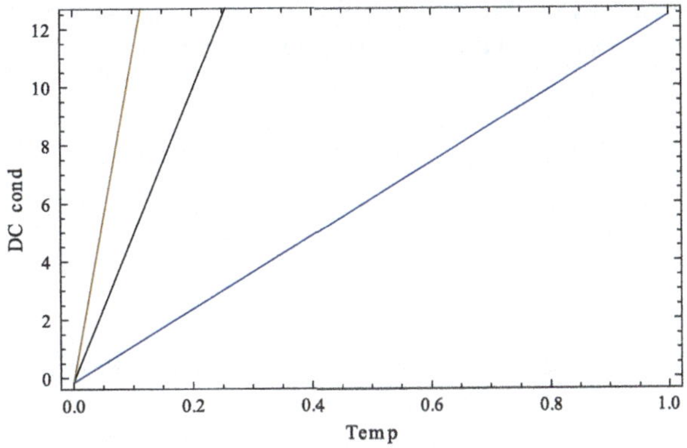

29.2.2 *DC Conductivity*

DC conductivity can be calculated using zero frequency limit of the above equation. But for simplification we have introduced perturbations in the form of time independent electric field as in [5], $\delta g_{tx} = h_{tx}$ and $A_x = -Et + a_x$. Coupled equations are solved considering regularity condition at horizon. DC electric conductivity can be obtained from Ohm's law as $\sigma = \frac{\partial J}{\partial E}$($J$ is the current density of the system). Dependence of dc conductivity on temperature with varying chemical potential has been shown in Fig. 29.2.

29.3 Results and Discussions

We study ac and dc conductivity of holographic superconductor using soft wall model. Soft wall model is a modified form of hard wall model with dilaton field to introduce the confinement in the system. For charged black hole system we have shown variation of ac conductivity with chemical potential indicating normal phase of the superconductor. This model can be developed further to investigate other novel features of strongly coupled systems.

Acknowledgements NB would like to acknowledge Dr. Sanjay Siwach and Prof. Debanand Sa for discussions.

References

1. J.M. Maldacena, Int. J. Theor. Phys. **38**, 1113 (1999) [Adv. Theor. Math. Phys. **2**, 231 (1998)]
2. A. Karch, E. Katz, D.T. Son, M.A. Stephanov, Phys. Rev. D **74**, 015005 (2006)
3. S.S. Afonin, I.V. Pusenkov, Eur. Phys. J. C **76**(6), 342 (2016)
4. N. Bhatnagar, S. Siwach, arXiv:1611.09996 [hep-th]
5. A. Donos, J.P. Gauntlett, JHEP **1411**, 081 (2014)

Chapter 30
Why T2K Should Run in Dominant Neutrino Mode to Discover CP Violation?

Monojit Ghosh

30.1 Introduction

Neutrino oscillation in standard three flavour is defined by three mixing angles i.e. θ_{12}, θ_{13}, θ_{23}, two mass squared differences i.e. Δm^2_{21}, Δm^2_{31} and the phase δ_{CP}. Among these six parameters at these moments the unknowns are: (i) neutrino mass hierarchy i.e. normal or inverted (NH: $\Delta m^2_{31} > 0$ or IH: $\Delta m^2_{31} < 0$), (ii) the octant of the mixing angle θ_{23} i.e. lower or higher (LO: $\theta_{23} < 45°$ or HO: $\theta_{23} > 45°$) and the leptonic CP phase δ_{CP}. The first hint of CP violation in the leptonic sector is believed to come from the currently running long-baseline experiment T2K [1] in Japan which has already indicated a mild preference towards $\delta_{CP} = -90°$. This hint came from the neutrino data of the T2K experiment [1] which is currently running in the antineutrino mode. In this work we ask the question that what should be the proportion of neutrino and antineutrino run to extract the best sensitivity from the T2K experiment regarding the discovery of leptonic CP violation. The capability of the T2K experiment to determine the phase δ_{CP} is limited by parameter degeneracies [2] which are (i) hierarchy-δ_{CP} degeneracy and (ii) octant - δ_{CP} degeneracy. It is already shown that the hierarchy - δ_{CP} degeneracy is same for neutrinos and antineutrinos but octant - δ_{CP} degeneracy behaves differently for neutrinos and antineutrinos. Thus a combination of neutrino and antineutrino can resolve the octant - δ_{CP} degeneracy but not the hierarchy- δ_{CP} degeneracy. In this work we will show that for the T2K experiment, if the parameter space is free from octant degeneracy then best CP sensitivity comes from the pure neutrino run of T2K experiment. On the other hand antineutrinos are required for the parameter space where octant degeneracy is present. To overcome this problem we also study the possibility of adding data from other experiments namely the ongoing accelerator base long-baseline experiment NOνA [3] in Fermilab and the proposed

---

M. Ghosh (✉)
Department of Physics, Tokyo Metropolitan University, Hachioji, Tokyo 192-0397, Japan
e-mail: mghosh@phys.se.tmu.ac.jp

Md. Naimuddin (ed.), *XXII DAE High Energy Physics Symposium*, Springer Proceedings in Physics 203, https://doi.org/10.1007/978-3-319-73171-1_30

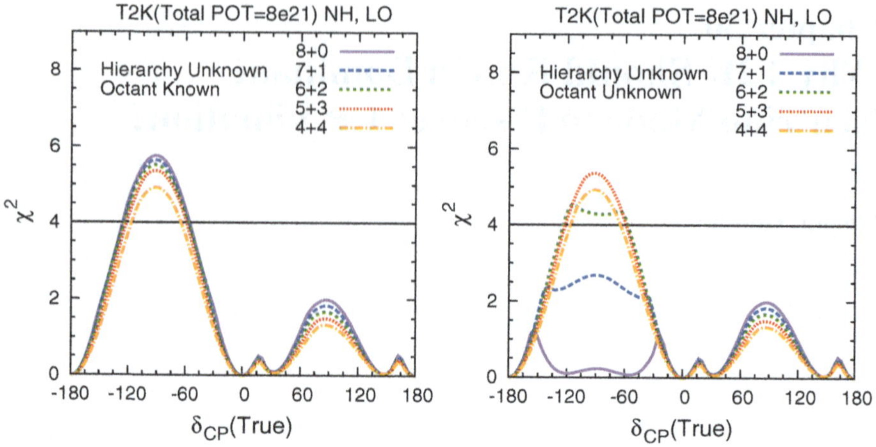

Fig. 30.1 CP Sensitivity of T2K for NH-LO

atmospheric neutrino experiment ICAL@INO [4] in India to show that the maximum CP sensitivity of the T2K experiment comes from the dominant neutrino run.

30.2 Results and Discussions

For the simulation of T2K experiment we consider a total exposure of 8×10^{21} protons of target (pot). We have divided this exposure in different proportion of "neutrino + antineutrino" running in units of 10^{21} pot.

In Fig. 30.1, we have plotted the CP violation discovery potential of T2K for NH ($\Delta m^2_{31} = 0.0024$ eV^2) - LO ($\theta_{23} = 39°$). From the figure we see that when octant is known (left panel), the best sensitivity comes from the pure neutrino run i.e. $8 + 0$ configuration. But when octant is unknown (right panel), $8 + 0$ gives the worst sensitivity. As the proportion of antineutrino increases, CP sensitivity gets improved. The maximum sensitivity is obtained for $5 + 3$ and further increase of antineutrino run decreases the sensitivity. This is because for $5 + 3$, the wrong-octant solution is completely removed and further addition of antineutrino reduces the statistics and hence the sensitivity gets decreased.

In Fig. 30.2 we plotted the same but for all the four combinations of hierarchy and octant assuming octant is unknown. IH corresponds to $\Delta m^2_{31} = -0.0024$ eV^2 and HO corresponds to $\theta_{23} = 51°$. From this figure we see that apart from $-90°$ - NH - LO and $+90°$ - IH - HO, $8 + 0$ configuration of T2K gives the best CP sensitivity. Thus to get a handle over these two situations, in Fig. 30.3 we plotted the same as Fig. 30.2 but for the combination of T2K+NOνA+ICAL.

For NOνA we assume a three years running in both neutrino and antineutrino mode and for ICAL we consider a 50kt iron calorimeter detector running for 10

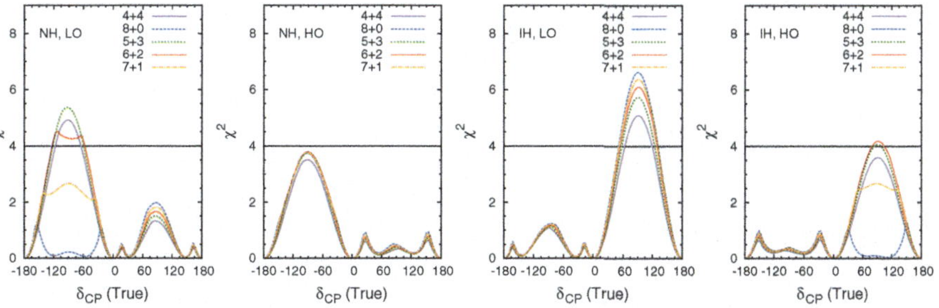

Fig. 30.2 CP Sensitivity of T2K for all the four combinations of hierarchy and octant

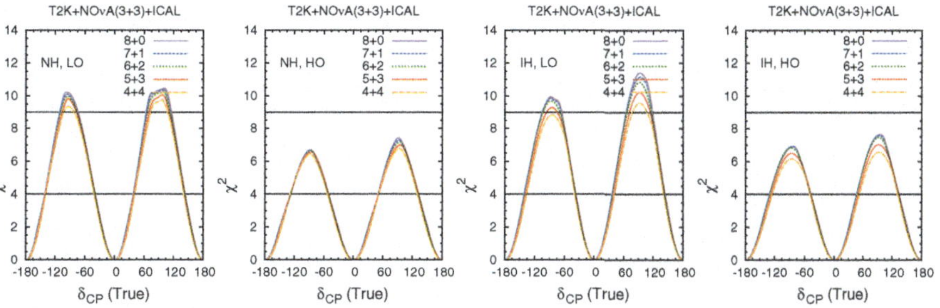

Fig. 30.3 CP Sensitivity of T2K, NOνA and ICAL for all the four combinations of hierarchy and octant

years. From the figure we see that when NOνA and ICAL are combined with the T2K data then the best CP sensitivity comes from the $7 + 1$ configuration of T2K.

Note that Figs. 30.1 and 30.3 of this work is published in [5] and some of the results of this work is discussed in [6].

Acknowledgements This work is partly supported by the Grant-in-Aid for Scientific Research of the Ministry of Education, Science and Culture, Japan, under Grant No. 25105009.

References

1. K. Abe et al., [T2K Collaboration], Phys. Rev. D **91**(7), 072010 (2015), arXiv:1502.01550 [hep-ex]
2. M. Ghosh, P. Ghoshal, S. Goswami, N. Nath, S.K. Raut, Phys. Rev. D **93**(1), 013013 (2016), arXiv:1504.06283 [hep-ph]
3. P. Adamson et al., [NOvA Collaboration], Phys. Rev. D **93**(5), 051104 (2016), arXiv:1601.05037 [hep-ex]
4. S. Ahmed et al., [ICAL Collaboration], arXiv:1505.07380 [physics.ins-det]
5. M. Ghosh, Phys. Rev. D **93**(7), 073003 (2016), arXiv:1512.02226 [hep-ph]
6. M. Ghosh, arXiv:1611.00220 [hep-ph]

Chapter 31
Near-Horizon Geometry and the Entropy of a Minimally Coupled Scalar Field in the Asymptotically Flat Black Holes

Kaushik Ghosh

31.1 Introduction

Since the four laws of black hole mechanics were formulated [1], there has been much effort to relate the laws of black hole mechanics to those of thermodynamics. The area theorem led Bekenstein to assign an entropy to a black hole [2]. Hawking established the thermodynamical aspects of black holes by showing that a black hole can radiate like a hot body at a temperature equal to a multiple of the surface gravity of the horizon [3]. The entropy of a black hole, considered as a thermodynamical system, was found to be $\frac{A}{4}$. Here A is the area of the horizon.

We have to consider quantum field theory in curved spaces in most of the works related to black hole thermodynamics. In flat space, a uniformly accelerated observer detects a thermal spectrum when the field is in the Minkowski vacuum [4, 5]. The temperature is dependent on the proper acceleration of the observer. A corresponding situation arises in the Schwarzschild black hole for a static observer outside the event horizon. The temperature is the same as that of the black hole if the static observer is at a large distance from the horizon. The corresponding vacuum is the Hartle–Hawking vacuum [6]. In this article we will discuss the expression of the entropy of a scalar field in a Kerr black hole and in thermal equilibrium with the black hole. The expression is derived by us in a previous article [8]. The corresponding expression for the Schwarzschild case can be obtained by considering an appropriate slow rotation limit [7, 8]. We have used a model proposed by G. 't Hooft known as the brick wall model [9] and used the Wentzel–Kramers–Brillouin (WKB) approximation to count the states. We can define a thermal equilibrium between a black hole and black body radiation of a massless scalar field surrounding the black hole at the same temperature as that of the black hole. The amount of radiation absorbed and emitted by the black hole is the same. Divergences appear when we try to calculate the entropy

K. Ghosh (✉)
Vivekananda College, University of Calcutta, 269 Diamond Harbour Road,
Kolkata 700063, India
e-mail: ghosh_kaushik06@yahoo.co.in

Md. Naimuddin (ed.), *XXII DAE High Energy Physics Symposium*, Springer Proceedings in Physics 203, https://doi.org/10.1007/978-3-319-73171-1_31

of the scalar field [7, 9]. This is associated with the continuous energy spectrum and unboundedness of the allowed angular quantum numbers in the near-horizon region. To regulate the divergence, t Hooft proposed a boundary condition on the scalar field near the horizon [9]. He assumed the scalar field to be vanishing at a small distance away from the horizon. This radial parameter is known as the brick wall cut-off parameter. This boundary condition is a good model since in thermodynamic equilibrium there is no net interchange of particles between the black hole and the surrounding matter. The details can be found in [7, 8]. The expression of the entropy is found to be logarithmically divergent in the brick wall cut-off parameter. The metric component g_{rr} has a simple pole at the horizon. The WKB quantization rule indicates that the divergence is expected to be logarithmic. In the Schwarzschild black hole, we can also use the blueshift factor together with the form of the proper distance in the near-horizon region to understand this logarithmic divergence. The thickness of the thin shell takes care of the unbounded nature of the allowed angular quantum numbers. The expression of the entropy obtained by us is proportional to the area of the horizon when we take the temperature to be same as the Hawking temperature. The entropy is also found to be a decreasing function of the thickness of the thin shell. These aspects may be taken to be the semiclassical signs of a holography principle and entropy bound. Here the semiclassical theory refers to quantum field theory in a fixed background.

The Lorentzian sector calculation of the scalar field entropy in the Kerr black hole is more significant compared to the Schwarzschild case since the corresponding Euclidean sector literature is not well formulated [10]. The expressions obtained in this article are also significant in the context of entanglement entropy and the Euclidean approach to calculate the matter field entropy in a black hole background. We will discuss these aspects and the significance of the results obtained in this article in Sect. 31.2.

31.2 Entropy of a Thin Shell of Scalar Field in the Kerr Black Hole

In this section we will consider the entropy of a minimally coupled scalar field in the Kerr black hole. We will discuss the case of a massless field but the results can be easily extended for a massive case [7, 8]. We will follow the method discussed in [7–9] and consider a thin shell of scalar field surrounding the horizon. We found earlier that in the spherical polar coordinates, the WKB approximation is suitable for a thin shell of scalar field [7, 8]. Since the near-horizon region gives us interesting physics, we proceed to discuss the entropy of a thin shell of scalar field confined in the near-horizon region. We impose the brick wall boundary condition:

$$\psi = 0, \quad for \quad r \leq r_+ + h \, . \tag{31.1}$$

Where r_+ is the black hole event horizon and h is a small radial parameter. We can consider the scalar field to be confined in a half-infinite potential well in the near-horizon region [7, 8]. The metric of the Kerr black hole is given by:

$$ds^2 = -\frac{\Delta}{\rho^2}[dt - a(\sin^2\theta)d\phi]^2 + \frac{\sin^2\theta}{\rho^2}[(r^2 + a^2)d\phi - a\,dt]^2 + \frac{\rho^2}{\Delta}dr^2 + \rho^2 d\theta^2 . \tag{31.2}$$

Where, $\rho^2 = r^2 + a^2\cos^2\theta$, $\Delta = r^2 - 2Mr + a^2$, M is the mass and a is the angular momentum per unit mass of the black hole. The scalar field satisfies the following wave equation: $(-g)^{-\frac{1}{2}}\partial_\mu[(-g)^{\frac{1}{2}}g^{\mu\nu}\partial_\nu\psi] = 0$. Here $[g_{\mu\nu}]$ is the metric tensor and g is it's determinant. The entropy of a scalar field confined within the near horizon region and as predicted by an observer at infinity is given by [8]:

$$S = \frac{1}{60}\,ln\left(\frac{\epsilon_d}{\epsilon_h}\right)\frac{A}{4\pi{\epsilon_d}^2} . \tag{31.3}$$

The corresponding expression in the Schwarzschild black hole is given by the following expression [7]:

$$S' = \frac{1}{30}\,ln\left(\frac{\epsilon_d}{\epsilon_h}\right)\frac{A}{4\pi{\epsilon_d}^2} . \tag{31.4}$$

Here ϵ_x represents the proper radial variable: $\epsilon_x = \int_{r_+}^{r_+ + x}\sqrt{g_{rr}}\,dr$, r_+ is the event horizon. 'A' represents the area of the horizon. The entropy of the scalar field in the Kerr black hole is half of the corresponding value in the Schwarzschild black hole. This is due to rotation and is consistent with the preferential emission of particles in the Kerr black hole with azimuthal angular momentum in the same direction as that of the black hole itself. We can obtain the Schwarzschild case expression by including a subleading term and taking the appropriate slow rotation limit [8]. The temperature in both the cases are chosen to be the Hawking temperature of the black holes. The expressions of the scalar field entropy contain two free parameters, the proper thickness of the thin shell and the proper brick wall cut-off parameter. We take the proper thickness of the thin shell to be of the order of atomic lengths. The other undetermined parameter is the proper brick wall cut-off. 't Hooft wanted to explain the black hole entropy in terms of the near horizon part of the matter field entropy and equated the two [9]. If we do so, the internal energy turns out to be finite: $U = \frac{3}{8}M$. This also determines the proper brick wall cut-off parameter. In the present case we then have:

$$\epsilon_h = (\epsilon_d)\exp\left[-60\pi\left(\frac{\epsilon_d}{l_p}\right)^2\right] . \tag{31.5}$$

Where l_p is the Plack length. If we take $\epsilon_d = 10^{-10}\ cm$, ϵ_h is given by the following expression in terms of the Planck length: $\epsilon_h \sim [10^{24}]\exp(-10^{50})(l_p)$. Thus, the brick wall is almost coincident with the horizon although the free energy is finite.

This is expected since there exists solutions vanishing on the horizon [7, 8]. These solutions are consistent with the brick wall boundary condition. We note that even in the extreme case of a thin shell of thickness comparable to the Planck length, the proper brick wall cut-off remains small compared to the thickness. However, in this case the semiclassical theory of quantum fields in a fixed background may not be appropriate. The back reaction problem is discussed in detail by Candelas and is not much important in the present case [11]. These discussions may become significant in the context of an entropy bound discussed later.

The expression of the scalar field entropy in an asymptotically flat black hole is given by (31.3) or (31.4). This expressions have some important characteristics. The scalar field entropy in a Kerr black hole with off mass shell temperature is given by:

$$S = \frac{16\pi^3}{15}\frac{1}{\beta^3}\frac{({r_+}^2 + a^2)^4}{(r_+ - r_-)^3}\frac{1}{{\epsilon_d}^2}\, ln\left(\frac{\epsilon_d}{\epsilon_h}\right). \tag{31.6}$$

Where a is the angular momentum per unit mass and r_- is the inner horizon of the black hole. We find that the entropy is a decreasing function of the proper thickness even when the temperature is not given by the Hawking temperature. Thus, we may conjecture that the entropy of a scalar field in an asymptotically flat black hole background which is correlated with the horizon through the brick wall boundary condition has an upper bound and this upper bound is proportional to the horizon surface area when the temperature is taken to be the Hawking temperature. The magnitude of the upper bound is to be fixed by choosing a canonical value of the proper thickness of the thin shells considered in this article. This requires comparison with a theory of quantum gravity. We will discuss these issues later. The conjecture is expected to remain valid for the semiclassical theory of quantum fields in curved spaces where back reaction problem is not severe. This is unlike a previous expression of entropy considered in [7] which contains an infrared divergent part and can not have an upper bound. We could increase the entropy indefinitely by increasing the radial dimension and there was no upper bound for the entropy. We can verify the validity of the present conjecture by considering the spectrum of the solutions of the wave equation defined throughout the manifold and satisfying the brick wall boundary condition. We can use numerical methods for this purpose. These discussions are also important to the holographic principle in asymptotically flat spaces [12].

The logarithmic divergence in (31.3) and (31.4) is important to the Euclidean sector calculation of the scalar field entropy [10]. It reduces a quadratic and a quartic divergent expression to logarithmically divergent expressions. The logarithmic divergence is expected on physical grounds as discussed briefly in the introduction and in detail in [7, 8]. The Euclidean sector method contains a parameter which can be fixed by comparing with the Lorentzian sector expression obtained in the present article. This will remove the problem of unphysical divergences. Lastly, we can consider the scalar field entropy to give quantum correction to the black hole entropy and use (31.3) and (31.4) to renormalize the gravitational constant [13].

References

1. J.M. Bardeen, B. Carter, S.W. Hawking, Commun. Math. Phys. **31**, 161 (1973)
2. J. Bekenstein, Phys. Rev. D **7**, 2333 (1973)
3. S.W. Hawking, Commun. Math. Phys. **43**, 199 (1975)
4. S.A. Fulling, Phys. Rev. D **7**, 2850 (1973)
5. W.G. Unruh, Phys. Rev. D **14**, 870 (1976)
6. J.B. Hartle, S.W. Hawking, Phys. Rev. D **13**, 2188 (1976)
7. K. Ghosh, J. Phys. Soc. Jpn. **85**, 014101 (2016)
8. K. Ghosh, Prog. Theor. Exp. **2016**, 093E03
9. G. 't Hooft, Nucl. Phys. B **256**, 727 (1985)
10. S.N. Solodukhin, Living Rev. Relativ. **14**, 8 (2011)
11. P. Candelas, Phys. Rev. D **21**, 2185 (1980)
12. R. Bousso, Rev. Mod. Phys. **74**, 825 (2002)
13. L. Susskind, J. Uglum, Phys. Rev. D **50**, 2700 (1994)

Chapter 32
PeV Neutrinos from Local Magnetars

Rajat K. Dey

32.1 Introduction

The detection of PeV astrophysical neutrinos with IceCube detector has recently opened up a whole new window on the energetic Universe [1]. Here, we propose a viable physical model of the interactions taking place in source regions of these neutrinos.

The possible origins of them have been discussed by many in recent times [2], but still remains a topic of much speculation. A fundamental question is therefore being raised: what classes of astronomical objects could accelerate hadrons, and in which types of interactions are neutrinos then produced?

Acceleration of protons in the vicinity of young local neutron star surface with magnetic field ($B \sim 10^{14-17}$ G), widely known as magnetars, is in fact supposed to be a possibility. The subsequent production of photomesons in radiative background proceeds, and was studied by many. Here, we have proposed an additional contribution to the target photon fields from the PS mechanism. The PS process is expected to modify the radiative background.

The $P - B$ diagram, attains a phase when the star could pass a neutrino-loud regime [3]. If the spin-down power of a magnetar for a favorable set of P and B is consumed to accelerate protons, the object might emit PeV muon neutrinos (ν_{μ}). The radiative background is believed to be filled with soft ultraviolet that are in turn produced from the effect of PS on magnetar's unmodified radiative background.

R. K. Dey (✉)
Department of Physics, University of North Bengal, Siliguri, West Bengal, India
e-mail: rkdey2007phy@rediffmail.com

Md. Naimuddin (ed.), *XXII DAE High Energy Physics Symposium*, Springer Proceedings in Physics 203, https://doi.org/10.1007/978-3-319-73171-1_32

32.2 Photon Splitting Process in the Magnetosphere

The PS, a QED process that splits a high-energy photon into a pair of low energy photons in presence of a pure magnetic field and/or magnetized plasma. The environment affects the rate by changing photo dispersion properties of the region.

The PS may occur via various possible channels which are determined by the electric field vector (**X**), momentum vector (**q**) associated with a photon, and the magnetic field vector (**B**). The state '1' refers to a configuration where **X** is perpendicular to the plane containing **q** and **B** while '2' corresponds the **X** being parallel to their plane. For low energy background target UV-C photons ($T_{\text{kin.}} \cong 0.1 - 0.2$ keV $<< m_e c^2$) the only physical mode, $\gamma_1 \to \gamma_2 + \gamma_2$ is responsible for splitting and, thereby soften background photon spectra in the magnetized plasma. But in the absence of plasma, the physical mode, $\gamma_1 \to \gamma_1 + \gamma_2$ is the significant one for PS over the channel, $\gamma_1 \to \gamma_2 + \gamma_2$.

32.3 PeV Neutrinos from Magnetars: Physical Model

It has been presumed that the protons and/or ions are injected, and accelerated in surrounding regions of cosmic accelerators. They interact with the radiative background and subsequently PeV neutrinos and gamma-rays are generated following (32.1),

$$p + \gamma \to \Delta^+ \to \begin{cases} p + \pi^0 \to p + 2\gamma \\ n\pi^+ \to n + e^+ + \nu_e + \nu_\mu + \bar{\nu_\mu} \end{cases} \tag{32.1}$$

The final products of all neutrino flavours turn into a ratio of $\nu_e : \nu_\mu : \nu_\tau = 1 : 1 : 1$ at earth.

The neutron star (NS) remnant or merger that appears from massive binary NSs ($M \sim 2\text{M}_\odot$) coalescence, may form a millisecond magnetar with thin ejecta walls across polar caps in its very early phase. As the star receives huge angular momentum from the binary, it possesses a rapid rotation at the moment of its birth. These magnetars also have strong magnetic fields. Here, we suggest that the star may transit a state, when spin-down power is comparable with its magnetic power.

The energetic protons will interact with soft UV photons close to the magnetar's polar caps, the Δ resonance state may form. The energy of the modified target photons is $2.8kT_\infty(1 + z_g) \sim 0.01$ keV, where $z_g \sim 0.4$ being the gravitational red shift. Thus the proton threshold energy $\varepsilon_{\text{p,Th}}$ for the Δ^+ resonance state ranges $\geq 3 \times 10^{16}$ eV. The proton flux emitted from the polar cap region would therefore be

$$\Phi_{\text{PC}} \simeq c f_{\text{d}}(1 - f_{\text{d}}) n_{\text{o}} A_{\text{PC}}, \tag{32.2}$$

where A_{PC} denotes polar cap area, and it is $\eta_{\text{A}}(4\pi R^2)$ with η_{A} accounts the ratio of polar cap area to the magnetar surface area. Some earlier calculations for estimating proton/ion flux in pulsar's polar caps took the parameter η_A as unity. The characteristic polar cap radius can be given by, $r_{PC} = R(\Omega R/c)^{1/2}$, and hence η_A takes the form $\Omega R/(4c)$.

The charge-changing reaction goes on just $\frac{1}{3}$rd of the reaction time, about three high-energy neutrinos (or a pair of ν_μ, $\bar{\nu_\mu}$) will accompany with four high-energy gamma-rays on the average. The total flux of neutrinos then will be

$$\Phi_\nu(r \simeq 1.2R) = 2cf_\xi A_{pc} f_d(1 - f_d) n_o P_c, \tag{32.3}$$

with f_ξ is 2/3. If now the duty cycle factor f_{dc} of the muon neutrino is taken into account, the phase averaged ν_μ flux on the Earth from a magnetar at a distance D is given by

$$\Phi_{\nu_\mu,\bar{\nu_\mu}} \simeq 2cf_\xi f_\zeta \eta_A f_{dc} f_s f_d(1 - f_d) n_o \left(\frac{R}{D}\right)^2 P_c \tag{32.4}$$

The effect of neutrino oscillations is represented by the parameter f_ζ (here, it is 1/2). The factor f_s is set equal to 1 for ν_μ.

We now calculate numerical values for ν_μ flux using the formula in the (32.4) for a typical galactic magnetar with $D \sim 2$ kpc, $P \sim 350$ ms, $B_{15} \sim 1.5$, $T_{0.1\,\mathrm{keV}} \sim 0.0255$, and $f_{dc} \leq 0.10$ in cases when $\eta_A = 1$ and $\eta_A = \Omega R/(4c)$. We have used here $R = 10$ km and choose $Z = 1$ and $f_d = 1/2$.

The ν_μ flux $(E^2\phi_{\nu_\mu})$ is 6.03×10^{-10} in GeVcm^{-2}s^{-1} for $\eta_A = 1$. This value is 0.0009×10^{-10} according to $\eta_A \neq 1$ in GeVcm^{-2}s^{-1}. When compared with IceCube results, i.e., $\sim 2.4 \times 10^{-9}$ GeVcm^{-2}s^{-1}, these predicted values look quite low, particularly for $\eta_A \neq 1$. IceCube has measured neutrino flux as $E^2\phi_{\nu_\mu+\bar{\nu}_\mu} \sim 3 \times 10^{-8}$ GeV cm^{-2} s^{-1} sr^{-1} corresponding to neutrino energy range 0.2–2 PeV.

32.4 Conclusions

The model suggests no possible indication of any statistically significant excess of PeV neutrino flux from the direction of any local magnetar to be observed by IceCube in near future.

References

1. M.G. Aartsen et al., (IceCube Coll.): Phys. Rev. Lett. **111**, 021103 (2013)
2. R. Laha et al., Phys. Rev. D **88**, 043009 (2013)
3. B. Zhang et al., ApJ **595**, 346 (2003)

Chapter 33
Parton Distribution Functions of Proton in a Light-Front Quark-Diquark Model

Tanmay Maji and Dipankar Chakrabarti

In recent years, there have been a lot of activities to investigate the polarised PDFs and three dimensional structure of nucleons. Different model investigations gave many interesting insights into the nucleon structure [1, 2]. Light front AdS/QCD predicts a particular form of wave function for a two body bound state [3] which can predict many interesting nucleon properties in quark-diquark model. Light front AdS/QCD wave function also encodes the non-perturbative ingredients of nucleons. Recently, the light front wave functions (LFWFs) for the nucleons in quark-scalar-diquark models [4, 5] have been constructed from the light-front AdS/QCD prediction. In [6], the contribution of vector diquark is included considering the spin-flavour SU(4) structure and different properties of nucleons are studied in detail. In this paper, we present the important predictions of the model of [6] and compared with available data.

PDFs in Light-Front Quark-Diquark Model

In the quark-diquark model, the nucleon is treated as a bound state of a quark and a diquark. The proton state is written in the spin-flavour SU(4) structure as

$$|P;\pm\rangle = C_S|u\ S^0\rangle^{\pm} + C_V|u\ A^0\rangle^{\pm} + C_{VV}|d\ A^1\rangle^{\pm}. \tag{33.1}$$

where $|u\ S^0\rangle$, $|u\ A^0\rangle$ and $|u\ A^0\rangle$ are two particle state with isoscalar-scalar, isoscalar-axial vector and isovector-axial vector diquarks respectively [1]. The states are written by the two particle Fock state expansion with $J^z = \pm 1/2$ for both the scalar and the axial vector diquarks. We adopt the LFWFs $\varphi_i^{(\nu)}(x, \mathbf{p}_\perp)$ from soft-wall AdS/QCD prediction [3] and modify as

T. Maji (✉) · D. Chakrabarti
Department of Physics, Indian Institute of Technology Kanpur, Kanpur 208016, India
e-mail: tanmay@iitk.ac.in

D. Chakrabarti
e-mail: dipankar@iitk.ac.in

Md. Naimuddin (ed.), *XXII DAE High Energy Physics Symposium*, Springer Proceedings in Physics 203, https://doi.org/10.1007/978-3-319-73171-1_33

$$\varphi_i^{(\nu)}(x, \mathbf{p}_\perp) = \frac{4\pi}{\kappa}\sqrt{\frac{\log(1/x)}{1-x}}x^{a_i^\nu}(1-x)^{b_i^\nu}\exp\left[-\delta^\nu\frac{\mathbf{p}_\perp^2}{2\kappa^2}\frac{\log(1/x)}{(1-x)^2}\right]. \quad (33.2)$$

The wave functions φ_i^ν $(i = 1, 2)$ reduce to the AdS/QCD prediction for the parameters $a_i^\nu = b_i^\nu = 0$ and $\delta^\nu = 1.0$. We use the AdS/QCD scale parameter $\kappa = 0.4$ GeV as determined in [4] and the quarks are assumed to be massless. The parameters are fixed by fitting the Dirac and Pauli form factors. More detail of this formalism is given in [6].

The parton distributions encodes the probability of finding a quark having light-cone momentum fraction $x = p^+/P^+$ in a proton. The PDFs are defined, at equal light-front time $z^+ = 0$, as

$$\Phi^{\Gamma(\nu)}(x) = \frac{1}{2}\int\frac{dz^-}{2(2\pi)}e^{ip^+z^-/2}\langle P;S|\bar{\psi}^{(\nu)}(0)\Gamma\mathcal{W}_{[0,z]}\psi^{(\nu)}(z^-)|P;S\rangle\bigg|_{z^+=z_T=0}, \quad (33.3)$$

where different Dirac structures e.g., $\Gamma = \gamma^+, \gamma^+\gamma^5$ and $i\sigma^{j+}\gamma^5$ gives unpolarized PDF $f_1(x)$, helicity distribution $g_1(x)$ and transversity distribution $h_1(x)$ respectively. $|P;S\rangle$ is the proton state having spin S and momentum $\mathbf{P}$. The quark fields ψ and $\bar{\psi}$, at two different points z^- and 0, are connected by a Wilson line $\mathcal{W}_{[0,z]}$. At the light-cone gauge $A^+ = 0$ the Wilson line becomes unity.

Since the different experiments have different energy scales, scale evolutions of PDFs are needed to compare with the experimental data. We simulate the DGLAP evolution of the PDF by making the parameters in the unpolarized PDF scale dependent where the values of the parameters at μ_0 are the same as in the LFWFs. The scale evolution in this model is discussed in [6]. The explicit expression of unpolarised PDFs are given as

$$f_1^\nu(x,\mu) = N_{f_1}^\nu(\mu)\left[\frac{1}{\delta^\nu(\mu)}x^{2a_1^\nu(\mu)}(1-x)^{2b_1^\nu(\mu)+1} + \frac{\kappa^2 x^{2a_2^\nu(\mu)-2}(1-x)^{2b_2^\nu(\mu)+3}}{(\delta^\nu(\mu))^2 M^2 \ln(1/x)}\right],$$

with a flavour ν and scale μ. $N_{f_1}^\nu(\mu)$ is the normalization constant. The model prediction for unpolarized PDFs, at the scale $\mu^2 = 10$ and 10^4 GeV^2, are shown in Fig. 33.1(left column) for u and d quarks. We compare our result with the experimental data and phenomenological extractions e.g., HERAPDF15(nnlo), NNPDF21(nnlo) and MSTW2008(nnlo). The PDFs are measured by the H1 and the ZEUS collaborations from the QCD analysis in exclusive $e^\pm p$ scattering at the HERA [7]. The NNPDF provides unpolarised PDFs at the next-to-next leading order (NNLO) using the Neural Network approach to the deep-inelastic data [8]. The MSTW2008 proposed phenomenological PDFs from the global fit analysis of hadron scattering data at LHC [9].

Helicity PDF $g_1(x)$ provides the information about longitudinal polarized quark. The model prediction for the helicity PDFs are shown in Fig. 33.1(middle column) at $\mu^2 = 1$ GeV^2. The error bands in this model (yellow) come from uncertainties in

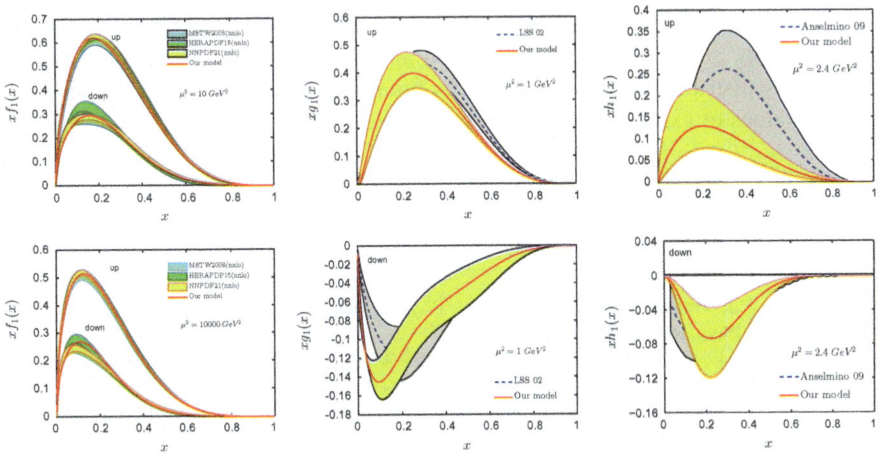

Fig. 33.1 Model prediction of $xf_1(x)$, $xg_1(x)$ and $xh_1(x)$ are shown by the continuous lines (red) for both u and d quarks. Left column: unpolarized PDF at $\mu^2 = 10$ and 10,000 GeV^2 and compared with experimental data and the phenomenological extractions [7–9]. Middle column: helicity distribution at $\mu^2 = 1$ GeV^2. Last column: transversity distribution at $\mu^2 = 2.4$ GeV^2

Table 33.1 Axial charge g_A^ν at $\mu^2 = 1$ GeV^2 and tensor charge g_T^ν at $\mu^2 = 0.8$ GeV^2 are listed and compared with the measurements [11, 12]

	g_A^u	g_A^d	g_A	g_T^u	g_T^d	g_T
LFQDM	0.71 ± 0.09	$-0.54^{+0.19}_{-0.13}$	$1.25^{+0.28}_{-0.22}$	$0.37^{+0.06}_{-0.05}$	$-0.14^{+0.05}_{-0.06}$	$0.51^{+0.12}_{-0.11}$
Measurement	0.82 ± 0.07	-0.45 ± 0.07	1.27 ± 0.14	$0.59^{+0.14}_{-0.13}$	$-0.20^{+0.05}_{-0.07}$	$0.79^{+0.19}_{-0.20}$

the model parameters and error bands in phenomenological extractions (gray) are from a constant relative error of 10% to g_1^u and 25% to g_1^d in the data taken from [10]. The axial charges, $g_A = g_A^u - g_A^d$, which are obtained from the first moment of the helicity distributions are given in Table.33.1 and compared with the measured data [11]. The model predictions agree with the experimental data within the error bars. The probability of finding a transversely polarized quark is given by the transversity PDF $h_1(x)$. The model predictions are shown to agree with the experimental data [12], see Fig. 33.1(last column), at $\mu^2 = 2.4$ GeV^2 for both u and d quarks. The first moment of the transversity distribution gives the tensor charge denoted by g_T. The model again predicts the tensor charges quite accurately as shown in Table.33.1 and fall within the uncertainty bands of the phenomenological fits for both u and d quark [12]. We observe $\mid g_T^\nu \mid < \mid g_A^\nu \mid$. The scale independent ratio $\mid g_T^d / g_T^u \mid = 0.38$ is very close to the phenomenological prediction [12]. This model also satisfies the Soffer bound $|h_1^\nu(x, \mu)| \leq \frac{1}{2}\big[f_1^\nu(x, \mu) + g_1^\nu(x, \mu)\big]$ at an arbitrary scale [13].

We show the prediction of the unpolarised and polarised PDFs in the quark-diquark model where the wave function is constructed from the soft-wall AdS/QCD prediction. Light-front AdS/QCD can accurately predict the PDF evolution up to a very high scale ($\mu^2 = 10^4$ GeV^2). The helicity and transversity PDFs are calculated as predictions of the model and have good agreement with the available data. The transversity distribution satisfies the Soffer bound. Our model reproduces the experimental values of axial and tensor charges quite well.

References

1. R. Jakob, P.J. Mulders, J. Rodrigues, Nucl. Phys. A **626**, 937 (1997); A. Bacchetta, F. Conti, M. Radici, Phys. Rev. D **78**, 074010 (2008)
2. C. Lorce, B. Pasquini, M. Vanderhaeghen, JHEP **1105**, 041 (2011); B. Pasquini, M. Pincetti, S. Boffi, Phys. Rev. D **72**, 094029 (2005)
3. S.J. Brodsky, G.F. de Teramond, Phys. Rev. D **77**, 056007 (2008); G.F. de Teramond, S.J Brodsky, arXiv:1203.4025 [hep-ph]
4. D. Chakrabarti, C. Mondal, Phys. Rev. D **88**(7), 073006 (2013); Eur. Phys. J. C **73**, 2671 (2013)
5. T. Gutsche, V.E. Lyubovitskij, I. Schmidt, A. Vega, Phys. Rev. D **89**(5), 054033 (2014); Erratum: [Phys. Rev. D **92**(1), 019902 (2015)]
6. T. Maji, D. Chakrabarti, Phys. Rev. D **94**(9), 094020 (2016)
7. F.D. Aaron et al., [H1 and ZEUS Collaborations], JHEP **1001**, 109 (2010)
8. L. Del Debbio et al., [NNPDF Collaboration], JHEP **0703**, 039 (2007)
9. A.D. Martin, W.J. Stirling, R.S. Thorne, G. Watt, Eur. Phys. J. C **63**, 189 (2009)
10. E. Leader, D.B. Stamenov, Phys. Rev. D **67**, 037503 (2003)
11. E. Leader, A.V. Sidorov, D.B. Stamenov, Phys. Rev. D **82**, 114018 (2010)
12. M. Anselmino, M. Boglione, U. D'Alesio, A. Kotzinian, F. Murgia, A. Prokudin, S. Melis, Nucl. Phys. Proc. Suppl. **191**, 98 (2009)
13. J. Soffer, Phys. Rev. Lett. **74**, 1292 (1995)

Chapter 34
Extended Scaling in Neutrino Majorana Mass Matrix

Probir Roy

34.1 Introduction

We focus on the observed three neutrino mixing and not so much on the dynamics of $\mathcal{O}$(eV) neutrino Majorana masses. In standard notation

$$-\mathcal{L}^{\nu}_{mass} = \frac{1}{2}\bar{\nu}^C_l (M_\nu)_{lm}\nu_m + h.c., \; U^T M_\nu U = M^d_\nu \equiv \text{diag}\,(\text{m}_1, \text{m}_2, \text{m}_3), \tag{34.1}$$

with m_i assumed to be real, positive and nondegenerate. The charged lepton mass matrix is taken as diagonal. U_{PMNS} has the standard form [1]

$$\begin{pmatrix} c_{12}c_{13} & e^{i\frac{\alpha}{2}} s_{12}c_{13} & s_{13}e^{-i(\delta-\frac{\beta}{2})} \\ -s_{12}c_{23} - c_{12}s_{23}s_{13}e^{i\delta} & e^{i\frac{\alpha}{2}}(c_{12}c_{23} - s_{12}s_{13}s_{23}e^{i\delta}) & c_{13}s_{23}e^{i\frac{\beta}{2}} \\ s_{12}s_{23} - c_{12}s_{13}c_{23}e^{i\delta} & e^{i\frac{\alpha}{2}}(-c_{12}s_{23} - s_{12}s_{13}c_{23}e^{i\delta}) & c_{13}c_{23}e^{i\frac{\beta}{2}} \end{pmatrix} \tag{34.2}$$

with $c_{ij} \equiv \cos\theta_{ij}, s_{ij} \equiv \sin\theta_{ij}, \theta_{ij} = [0, \pi/2], \delta, \alpha, \beta = [0, 2\pi]$. The latest relevant 3σ ranges [2] are: solar: $\Delta m^2_{21} \equiv m^2_2 - m^2_1$: (7.02–8.09) $\times 10^{-5}$ eV^2, atmospheric: $|\Delta m^2_{31}| \equiv |m^2_3 - m^2_1|$: (2.32–2.59) $\times 10^{-3}$ eV^2, Planck: $\Sigma_i m_i < 0.23$ eV. The 3σ intervals of the mixing angles are [2]: θ_{12}: 31.29°–35.91°, θ_{23}: 38.3°–53.3°, θ_{13}: 7.87°–9.11°.

The Simple Real Scaling (SRS) was first proposed [3] as an ansatz. Within our sign convention, this means $\frac{(M^{SRS}_\nu)_{e\mu}}{(-M^{SRS}_\nu)_{e\tau}} = \frac{(M^{SRS}_\nu)_{\mu\mu}}{(-M^{SRS}_\nu)_{\mu\tau}} = \frac{(M^{SRS}_\nu)_{\tau\mu}}{(-M^{SRS}_\nu)_{\tau\tau}} = k$, k being a real, positive, dimensionless scaling factor. It follows that [3] one physical neutrino is massless and the corresponding eigenvector, i.e. the third column of U_{PMNS},

P. Roy (✉)
Centre for Astroparticle Physics and Space Science, Bose Institute, Kolkata, India
e-mail: probirrana@gmail.com

Md. Naimuddin (ed.), *XXII DAE High Energy Physics Symposium*, Springer Proceedings in Physics 203, https://doi.org/10.1007/978-3-319-73171-1_34

is $(C_3^{SRS})^T = (0, (1+k^2)^{-\frac{1}{2}}e^{i\frac{\beta}{2}}, k(1+k^2)^{-\frac{1}{2}}e^{i\frac{\beta}{2}})$ implying that $m_3^{SRS} = 0 = \theta_{13}^{SRS}$ while θ_{23}^{SRS} is left undetermined. This SRS proposal got knocked out by the experimental exclusion of a vanishing θ_{13} at the 10σ level [4].

34.2 Residual Flavor Symmetry and Complex Extended Scaling

M_ν has a residual $\mathbb{Z}_2 \times \mathbb{Z}_2'$ flavor symmetry [5]. $\mathbb{Z}_2 : G_2^T M_\nu G_2 = M_\nu$, $\mathbb{Z}_2' : G_3^T M_\nu G_3 = M_\nu$, $G_{2,3}$ being unitary. Identify $\mathbb{Z}_2'$ as $\mathbb{Z}_2^{scaling}$ and deduce the form of G_3^{scaling} as given in [6]. The complex extension of this is [6]: $G_3^{\text{scaling}} M_\nu G_3^{\text{scaling}} = M_\nu^*$. Now the most general Complex Extended Scaling (CES) invariant form of M_ν is

$$M_\nu^{CES} = \begin{pmatrix} x & -y_1 k + i\frac{y_2}{k} & y_1 + iy_2 \\ -y_1 k + i\frac{y_2}{k} & z_1 - w\frac{k^2-1}{k} - iz_2 & w - i\frac{k^2-1}{2k}z_2 \\ y_1 + iy_2 & w - i\frac{k^2-1}{2k}z_2 & z_1 + iz_2 \end{pmatrix} \quad (34.3)$$

with six real mass dimensional parameters x, $y_{1,2}$, $z_{1,2}$ and w apart from the scaling factor k. It follows that [6]

$$G_3^{scaling}(U^{CES})^* = U^{CES}\tilde{d}, \tilde{d} = \text{diag}\,(\tilde{d}_1, \tilde{d}_2, \tilde{d}_3) \quad (34.4)$$

with $\tilde{d}_{1,2,3} = \pm 1$. The exploitation of (34.5) and the element by element identification of U^{CES} with the U_{PMNS} of (34.2) yield $\tilde{d}_1 = -1$, $\tilde{d}_2 = e^{-i\alpha}$ and $\tilde{d}_3 = e^{i(2\delta-\beta)}$. Furthermore, the input $c_{12} \neq 0$ leads to the results [6]

$$0 = \sin\alpha = \sin\beta = \cos\delta, \ \tan\theta_{23} = k^{-1}. \quad (34.5)$$

Thus Majorana CP violation is absent, Dirac CP violation is maximal and the deviation from maximality in θ_{23} is controlled by the deviation of the scaling factor k from unity.

34.3 Phenomenological Discussion

By inputting the observed 3σ ranges of Δm_{21}^2, $|\Delta m_{31}^2|$, θ_{12}, θ_{23} and θ_{13} as well as the cosmological upper bound on $\Sigma_i m_i$, the allowed mass ranges for a normal ($m_3 > m_2 > m_1$) and an inverted ($m_2 > m_2 > m_3$) mass ordering of the light neutrinos have been worked out [6]. For a normal mass ordering, we obtain 9.2×10^{-5} eV $< m_1 < 0.071$ eV, 0.01 eV $< m_2 < 0.077$ eV and 0.051 eV $< m_3 < 0.082$ eV. For the inverted case we have 0.051 eV $< m_1 < 0.085$ eV, 0.049 eV $< m_2 < 0.079$ eV and 8.2×10^{-5} eV $< m_3 < 0.068$ eV. Next, the relevant quantity for $0\nu\beta\beta$ decay is

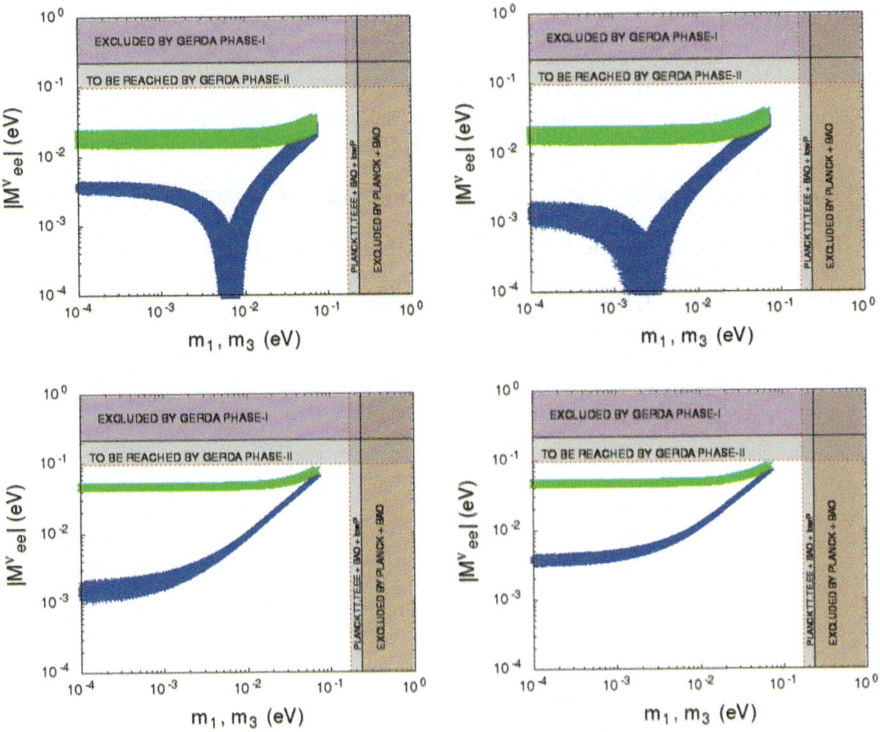

Fig. 34.1 Plot of $|M^{\nu}_{ee}|$ versus the lightest neutrino mass: the top two figures are for $(\tilde{d}_1, \tilde{d}_2, \tilde{d}_3) = (-1, 1, 1)$ (left) and (-1,1,-1) (right) while the figures in the lower panel are for $(\tilde{d}_1, \tilde{d}_2, \tilde{d}_3) = (-1, -1, 1)$ (left) and $(-1, -1, -1)$ (right)

given by

$$(M_{\nu})_{ee} = c_{12}^2 c_{13}^2 m_1 + s_{12}^2 c_{13}^2 m_2 e^{i\alpha} + s_{13}^2 m_3 e^{i(\beta - 2\delta)}. \tag{34.6}$$

There exist [6] four possibilities in our model. The corresponding mass bands are shown in Fig. 34.1. The upper and lower bands are for a normal (plotted against m_1) and for an inverted (plotted against m_3) mass ordering respectively. Only the extreme right corners of these plots will be accessible to forthcoming experiments such GERDA.

Acknowledgements My collaborators in this work have been A. Ghosal and R. Samanta of Saha Institute.

References

1. K.A. Olive et al., [Particle Data Group] Chin. Phys. **C38**, 090001 (2014)
2. M.C. Gonzalez-Garcia, M. Maltoni, T. Schwetz, arXiv:1512.06856 [hep-ph]
3. L. Lavoura, Phys. Rev. D **62**, 093011 (2000); R.N Mohapatra, W. Rodejohann, Phys. Lett. B **644**, 59 (2007)
4. B.Z Hu et al., (Daya Bay Collaboration), Phys. Rev. Lett. **115**, 111802 (2015) and references therein
5. C.S. Lam, Phys. Lett B **656**, 193 (2007); Phys. Rev. Lett. **101**, 121602 (2008); Phys. Rev. D **78**, 073015 (2008)
6. R. Samanta, P. Roy, A. Ghosal, Eur. Phys. J. C **76**, 662 (2016)

Chapter 35
Non-universal Z' Boson Effects in Rare Radiative Decay $B_s^0 \to \mu^+\mu^-\gamma$

Debika Banerjee, Priya Maji and Sukadev Sahoo

35.1 Introduction

In the current scenario, several experimental results from different accelerators have reported several anomalies around 3σ in B meson sector [1, 2]. These anomalies indicate the presence of new physics (NP) effects which need precise theoretical as well as experimental analysis. The rare decays of B mesons such as $B^0_{s,d} \to l^+l^-$ are important to probe the flavour sector of the standard model (SM) and to search NP beyond the SM. In particular, the rare radiative decays $B_s^0 \to l^+l^-\gamma$ are of special interest due to the presence of additional photon with the lepton pair in the final state removes the helicity suppression and promotes a large branching ratio (BR). $B_s^0 \to l^+l^-\gamma$ decays can be obtained from $b \to sl^+l^-$ transition by attaching a photon to any charged internal and/or external fermion lines in it. In case of photon emitted from initial quarks, the structure dependent (SD) part [3, 4] contributes more for the transition amplitude of $B_s^0 \to l^+l^-\gamma$ decays because of the absence of helicity suppression. Whereas for other situation, when the photon is emitted from external charged leptons, the internal Bremsstrahlung (IB) part [3, 5] adds small contribution to the total transition amplitude because it is proportional to lepton mass. To calculate the total transition amplitude we include both parts. The SM prediction for the branching ratio of $B_s^0 \to \mu^+\mu^-\gamma$ decay is 1.9×10^{-9} [6]. In this paper, we investigate the effects of non-universal Z' boson on the $B_s^0 \to \mu^+\mu^-\gamma$ decay to dig out the NP contributions. Non-universal Z' bosons exist in many theories beyond the SM [7] such as grand unified theories (GUTs), superstring theories and theories with large extra dimensions. The mass of Z' boson is not known as it is not detected in experiments so far. However, there are strict limits on the mass of Z' boson and the $Z - Z'$ mixing angle θ from the non-observation of direct production at the CDF [8] and indirect constraints from the precision data [9]. Tree level flavour changing

D. Banerjee (✉) · P. Maji · S. Sahoo
Department of Physics, National Institute of Technology,
Durgapur 713209, West Bengal, India
e-mail: rumidebika@gmail.com

Md. Naimuddin (ed.), *XXII DAE High Energy Physics Symposium*, Springer Proceedings in Physics 203, https://doi.org/10.1007/978-3-319-73171-1_35

neutral current (FCNC) interactions can be induced by an additional Z' boson as a result of the non-diagonal chiral coupling matrix. Here, we study the coupling of Z' boson with leptons and quarks. In the next section, we present the mathematical formalism to calculate the branching ratio of $B_s^0 \to \mu^+\mu^-\gamma$ decay in the Z' model.

35.2 Mathematical Formalism

The total matrix element for $B_s^0 \to \mu^+\mu^-\gamma$ decay can be obtained by adding the SD [3, 4] and IB [3, 5] part together, i.e. $M = M_{SD} + M_{IB}$ and hence squared matrix element is,

$$| M |^2 = | M_{SD} |^2 + | M_{IB} |^2 + 2Re(M_{SD} M_{IB}^*). \tag{35.1}$$

The corresponding differential decay width of the $B_s^0 \to \mu^+\mu^-\gamma$ is given as [3],

$$\frac{d\Gamma}{d\hat{s}} = \frac{G_F^2 \alpha^3}{2^{10}\pi^4} | V_{tb} V_{ts}^* |^2 m_{B_s}^3 \Delta, \tag{35.2}$$

where,

$$\Delta = \frac{4}{3} m_{B_s}^2 (1-\hat{s})^2 v_l ((\hat{s} + 2r_l)(| A_1 |^2 + | B_1 |^2) + (\hat{s} - 4r_l)(| A_2 |^2 + | B_2 |^2))$$
$$- 64[(f_{B_s}^2 r_l)/(m_{B_s}^2 (1-\hat{s}))] C_{10}^2 \left[(4r_l - \hat{s}^2 - 1) ln \frac{1+v_l}{1-v_l} + 2\hat{s} v_l \right] - 32 r_l (1-\hat{s})^2 f_{B_s} Re(C_{10} A_1^*) \tag{35.3}$$

Let us consider the $B_s^0 \to \mu^+\mu^-\gamma$ decay process in non-universal Z' model [7].The effective Hamiltonian for $b \to s l^+ l^-$ transition corresponding to Z' part can be written as [10]:

$$H_{eff}(Z') = -\frac{4G_F}{\sqrt{2}} V_{tb} V_{ts}^* [\Lambda_{sb} C_9^{Z'} O_9 + \Lambda_{sb} C_{10}^{Z'} O_{10}], \tag{35.4}$$

$$\text{where, } \Lambda_{sb} = \frac{4\pi e^{-i\phi_{sb}}}{\alpha V_{tb} V_{ts}^*}, C_9^{Z'} = | B_{sb} | S_{ll}, C_{10}^{Z'} = | B_{sb} | D_{ll}. \tag{35.5}$$

In the above expressions, the S_{ll} and D_{ll} are associated with the couplings of Z' boson with the left- and right-handed leptons respectively, B_{sb} corresponds to the off-diagonal left handed coupling of quarks with Z' boson and ϕ_{sb} is a new weak phase. Hence, the effect of non-universal Z' boson can be obtained by modifying the SM Wilson coefficients C_9 and C_{10} as:

$$C_9^{SM+Z'} = C_9^{eff} + \Lambda_{sb} \mid B_{sb} \mid S_{ll}, \tag{35.6}$$

$$C_{10}^{SM+Z'} = C_{10} + \Lambda_{sb} \mid B_{sb} \mid D_{ll}, \tag{35.7}$$

35.3 Numerical Analysis and Discussion

In this section, we calculate the branching ratio for $B_s^0 \to \mu^+\mu^-\gamma$ decay in non-universal Z' model. The values of Z' boson coupling parameters are obtained from the analysis of $B \to X_s\mu^+\mu^-$ [11], $B \to K^*\mu^+\mu^-$ [12], and $B \to \mu^+\mu^-$ [13] decays. The numerical values of all other parameters are taken from Particle Data Group [14]. Using all the required input parameters with three scenarios of Z' parameters we get: for Scenario 1, i.e. $\mid B_{sb} \mid = 5.5 \times 10^{-3}$, $\phi_{sb} = \pm 160°$, $S_{ll} = 0.8 \times 10^{-2}$ and $D_{ll} = -2.6 \times 10^{-2}$, the estimated BR ($B_s^0 \to \mu^+\mu^-\gamma$)= 1.0×10^{-9}; for Scenario 2, i.e. $\mid B_{sb} \mid = 4 \times 10^{-3}$, $\phi_{sb} = \pm 150°$, $S_{ll} = 0.8 \times 10^{-3}$ and $D_{ll} = -2.6 \times 10^{-3}$, the estimated BR ($B_s^0 \to \mu^+\mu^-\gamma$)= 1.8×10^{-9}; for Scenario 3, i.e. $\mid B_{sb} \mid = 3.5 \times 10^{-3}$, $\phi_{sb} = \pm 140°$, $S_{ll} = 0.8 \times 10^{-2}$ and $D_{ll} = -2.6 \times 10^{-2}$, the estimated BR ($B_s^0 \to \mu^+\mu^-\gamma$)= 3.1×10^{-9}. The estimated BR for $B \to \mu^+\mu^-\gamma$ decay depends on the Z' coupling parameters severely. Scenario 1 and 3 clearly distinguish between NP contributions and SM result. Therefore, the detection of $B \to \mu^+\mu^-\gamma$ decay in present/future colliders would explain the presence of NP predictions explicitly.

References

1. R. Aaij et al., (LHCb Collaboration), Angular analysis of the $B^0 \to K^{*0}\mu^+\mu^-$ decay using $3fb^{-1}$ of integrated luminosity. JHEP **02**, 104 (2016)
2. W. Altmannshofer, D.M. Straub, New physics in $b \to s$ transitions after LHC run 1. Eur. Phys. J. C **75**, 382 (2015)
3. R. Mohanta, Implications of the non-universal Z boson in FCNC mediated rare decays. Phys. Rev. D **71**, 114013 (2005)
4. F. Krüger, D. Melikhov, Gauge invariance and form factors for the decay $B \to \gamma l^+l^-$. Phys. Rev. D **67**, 034002 (2003)
5. S. Rai Choudhury, A.S. Cornell, N. Gaur, G.C. Joshi, Signatures of new Physics in dileptonic B-decays. Int. J. Mod. Phys. A **21**, 2617 (2006)
6. T.M. Aliev, A. Özpineci, M. Savci, Rare $B \to l^+l^-\gamma$ decay and new physics effects. Phys. Lett. B **520**, 69 (2001)
7. P. Langacker, M. Plumacher, Flavor changing effects in theories with a heavy Z' boson with family nonuniversal couplings. Phys. Rev. D **62**, 013006 (2000)
8. A. Abulencia et al., (CDF Collaboration), Search for $Z' \to e^+e^-$ using dielectron mass and angular distribution. Phys. Rev. Lett. **96**, 211801 (2006)
9. J. Erler, P. Langacker, Indications for an extra neutral Gauge Boson in electroweak precision data. Phys. Rev. Lett. **84**, 212 (2000)

10. M.A. Paracha, I. Ahmed, M.J. Aslam, Imprints of CP-violation asymmetries in rare $\Lambda_b \to \Lambda l^+ l^-$ decay in a family non-universal Z' model. Prog. Theor. Exp. Phys. **2015**, 033B04 (2015)
11. M. Iwasaki et al., (Belle Collaboration), Improved measurement of the electroweak penguin process $B \to X_s l^+ l^-$. Phys. Rev. D **72**, 092005 (2005)
12. R. Aaij et al., (LHCb Collaboration), Differential branching fraction and angular analysis of the decay $B^0 \to K^{*0} \mu^+ \mu^-$. Phys. Rev. Lett. **108**, 181806 (2012)
13. R. Aaij et al., (LHCb Collaboration), Strong constraints on the rare decays $B_s^0 \to \mu^+ \mu^-$ and $B^0 \to \mu^+ \mu^-$. Phys. Rev. Lett. **108**, 231801 (2012)
14. C. Patrignani et al., [Particle Data Group], Review of particle physics. Chin. Phys. C **40**, 100001 (2016)

Chapter 36
Bulk Viscosity Coefficient of Hadronic Matter

Sabyasachi Ghosh, Sandeep Chatterjee and Bedangadas Mohanty

36.1 Introduction

Recent research of heavy ion physics has concluded that the medium, formed in relativistic heavy ion collisions, must have very small shear viscosity, which is in contrast to the weak coupling picture, described by the standard finite temperature calculation of quantum chromo dynamics (QCD). Owing to this motivation, several microscopic calculations of shear viscosity have been done in recent times. Estimation of other transport coefficient like bulk viscosity of this strongly interacting matter is also becoming a contemporary research interest. In this context, this present article has addressed the bulk viscosity calculation for hadronic matter, where equilibrium thermodynamics for all hadrons in medium are described by Hadron Resonance Gas (HRG) model. The non-equilibrium part involving the thermal widths of medium constituents have been calculated by using effective Lagrangian densities for the hadronic medium. Here we have assumed pions and nucleons as the most abundant constituents of the hadronic medium and we have calculated their thermal widths, which basically reflect their in-medium scattering with different mesonic and baryonic resonances.

S. Ghosh (✉)
Department of Physics, University of Calcutta, 92, A. P. C. Road, Kolkata 700009, India
e-mail: sabyaphy@gmail.com

S. Ghosh · S. Chatterjee · B. Mohanty
School of Physical Sciences, National Institute of Science Education and Research Bhubaneswar, HBNI, Jatni 752050, India

Md. Naimuddin (ed.), *XXII DAE High Energy Physics Symposium*, Springer Proceedings in Physics 203, https://doi.org/10.1007/978-3-319-73171-1_36

36.2 Formalism

For the equilibrium part of the medium, we have used the ideal HRG model, where the hadrons and resonances with masses up-to 2 GeV have been taken from the Particle Data Book. Constructing total partition function of hadronic medium, one can easily derived all thermodynamical quantities like pressure (P), energy density (ϵ), entropy density (s). Their temperature dependence at zero baryon chemical potential ($\mu_B = 0$) are quite close to the corresponding results in the hadronic temperature domain, obtained by the lattice quantum chromo dynamics (LQCD) calculations [1]. In terms of P and ϵ, the square of the speed of sound for constant baryon density (n_B) is defined as $v_n^2 = \left(\frac{\partial P}{\partial \epsilon}\right)_{n_B}$, plays an important role in bulk viscosity calculation. Either from the Relaxation Time Approximation (RTA) in kinetic theory approach or from the one-loop diagram representation in Kubo framework, one can get standard expressions of bulk viscosity coefficient for pion and nucleon components [2]:

$$\zeta_\pi = \left(\frac{g_\pi}{T}\right) \int \frac{d^3\mathbf{k}}{(2\pi)^3} \frac{n_\pi [1 + n_\pi]}{\omega_\pi^2 \Gamma_\pi} \left\{\left(\frac{1}{3} - v_n^2\right) \mathbf{k}^2 - v_n^2 m_\pi^2\right\}^2 \tag{36.1}$$

and

$$\zeta_N = \left(\frac{g_N}{T}\right) \int \frac{d^3\mathbf{k}}{(2\pi)^3} \frac{1}{\omega_N^2 \Gamma_N} \left[\left\{\left(\frac{1}{3} - v_n^2\right) \mathbf{k}^2 - v_n^2 m_N^2 - \omega_N \left(\frac{\partial P}{\partial n_B}\right)_\epsilon\right\}^2 n_N^+ (1 - n_N^+) \right.$$
$$\left. + \left\{\left(\frac{1}{3} - v_n^2\right) \mathbf{k}^2 - v_n^2 m_N^2 + \omega_N \left(\frac{\partial P}{\partial n_B}\right)_\epsilon\right\}^2 n_N^- (1 - n_N^-)\right], \tag{36.2}$$

where $n_\pi = 1/\{e^{\omega_\pi/T} - 1\}$ with $\omega_\pi = \{\mathbf{k}^2 + m_\pi^2\}^{1/2}$ for pion, $n_N^\pm = 1/\{e^{(\omega_N \mp \mu_B)/T} + 1\}$ with $\omega_N = \{\mathbf{k}^2 + m_N^2\}^{1/2}$ for nucleon and their respective degeneracy factors are $g_\pi = 3$ and $g_N = 2 \times 2$. The m_π, m_N stand for masses of pion, nucleon and their momenta are denoted by $\mathbf{k}$. The Γ_π and Γ_N are thermal widths of pion and nucleon, which have been calculated from the imaginary part of their self-energies (on-shell) at finite temperature. We have considered different mesonic and baryonic loops for pion self-energy, whereas, pion-baryon intermediate states are taken in nucleon self-energy [3, 4].

36.3 Results and Conclusion

Using the explicit momentum $\mathbf{k}$, T and μ_B dependent thermal widths of pion and nucleon in the integrands of (36.1) and (36.2), we can calculate bulk viscosity of pion and nucleon components respectively. Adding these two components, we get total bulk viscosity, which is plotted in the left panel of Fig. 36.1a. The entropy density s, obtained from HRG and the ratio ζ/s are also drawn in the left panel of Fig. 36.1b and c, which reveal their temperature dependence at $\mu_B = 0$ (solid line), 0.250 (dotted

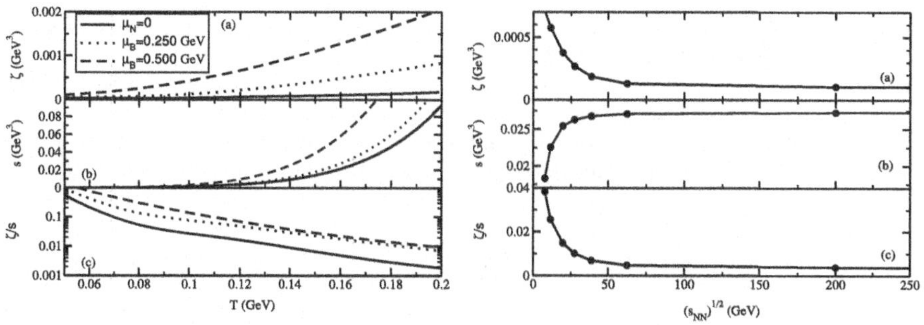

Fig. 36.1 Left: T dependence of bulk viscosity (**a**), entropy density (**b**) and their ratio ζ/s (**c**) at $\mu_B = 0$ (solid line), 0.250 GeV (dotted line) and 0.500 GeV (dash line). Right: Variation of same quantities ζ (**a**), s (**b**) and ζ/s (**c**) with the variation of center of mass energy ($\sqrt{s_{NN}}$)

line) and 0.500 GeV (dashed line). We notice that ζ and s both increase with T and μ_B. Whereas, ζ/s is a decreasing function of T (but still increase in μ_B) because the increment of $s(T)$ is larger than the increment of $\zeta(T)$.

Next, right panel of Fig. 36.1a–c show respectively the variation of ζ, s and ζ/s with the variation of center of mass energy $\sqrt{s_{NN}}$ We have used the parameterization from [5], where beam energy dependence of T and μ_B used in computation are those obtained from fits to hadron yields. We notice in the right panel of Fig. 36.1 that ζ (a) as well as ζ/s (c) are decreasing with $\sqrt{s_{NN}}$, which is qualitatively agreeing with the results of earlier [6]. The decreasing trend of ζ and ζ/s with $\sqrt{s}$ can be understood from the fact that μ_B decreases with $\sqrt{s}$ while T remains fairly constant in the range of $\sqrt{s}$ analyzed here and the ζ and ζ/s decrease with decreasing of μ_B as we have noticed in the left panel of Fig. (36.1). At $\mu_B = 0$, the decreasing nature of our $\zeta/s(T)$ agrees with most of the earlier works as addressed elaborately in [4]. However, some investigations have reported it to increase with T and some time also have a peak structure near transition temperature [7].

Acknowledgements Authors thank UGC, DAE, DST Govt. of India for financial support during the period, when this work was carried out.

References

1. A. Bazavov et al., (HotQCD Collaboration), Equation of state in (2+1)-flavor QCD. Phys. Rev. D **90**, 094503 (2014)
2. S. Gavin, Transport coefficients in ultra-relativistic heavy-ion collisions. Nucl. Phys. A **435**, 826 (1985)
3. S. Ghosh, The nucleon thermal width due to pion-baryon loops and its contributions in Shear viscosity. Phys. Rev. C **90**, 025202 (2014)
4. S. Ghosh, S. Chatterjee, B. Mohanty, Bulk viscosity for pion and nucleon thermal fluctuation in the hadron resonance gas model. Phys. Rev. C **94**, 045208 (2016)

5. F. Karsch, K. Redlich, Probing freeze-out conditions in heavy ion collisions with moments of charge fluctuations. Phys. Lett. B **695**, 136 (2011)
6. G.P. Kadam, H. Mishra, Dissipative properties of hot and dense hadronic matter in an excluded-volume hadron resonance gas model. Phys. Rev. C **92**, 035203 (2015)
7. K. Saha, S. Upadhaya, S. Ghosh, A comparative study on two different approaches of bulk viscosity in the Polyakov Nambu Jona-Lasinio model. Mod. Phys. Lett. A **32**, 1750018 (2017)

Chapter 37
Higgs Searches via WW Decay Channel Using the CMS Detector

Ankita Mehta

37.1 Introduction

During the year 2012, the CMS [1] and ATLAS experiments at the LHC reported the discovery of a new boson with a mass near 125 GeV and having production and decay rates, spin, parity and coupling strengths consistent with those expected for the Standard Model (SM) Higgs boson within uncertainties [2, 3]. The SM Higgs boson is a neutral scalar particle predicted to arise from the Higgs field and is responsible for the electroweak symmetry breaking. At the LHC, the Higgs boson can be produced in several ways: gluon-gluon fusion, vector boson fusion, in association with a W/Z boson or a top quark pair.

This article reports the latest measurement on SM Higgs boson produced via gluon-gluon fusion mode and decaying to a pair of W bosons, at a center of mass energy of 13 TeV. The analyzed data corresponds to an integrated luminosity of 2.3 fb^{-1} collected during proton-proton collisions at the CMS detector [4]. The leptonic decay of W^+W^- is studied into an oppositely charged electron-muon ($e\mu$) pair with missing transverse energy (E_T^{Miss}) (coming from neutrinos), and in association with either zero or one jet. The major background contribution arises from the non-resonant WW production and top quark events ($t\bar{t}$ and single top quark), followed by other processes such as Drell–Yan, W+jets and electroweak productions.

Ankita Mehta on behalf of the CMS Collaboration.

A. Mehta (✉)
Panjab University, Chandigarh, India
e-mail: ankita.mehta@cern.ch

Md. Naimuddin (ed.), *XXII DAE High Energy Physics Symposium*, Springer Proceedings in Physics 203, https://doi.org/10.1007/978-3-319-73171-1_37

37.2 Analysis Strategy

The events are triggered using the single or double lepton triggers. Offline event selection criteria demands the presence of exactly one electron and one muon with opposite charges and a minimum p_T of 10 (13) GeV for the muon (electron) candidate. The two well identified and isolated leptons must originate from the primary vertex of the event and leading lepton should have a p_T greater than 20 GeV. The invariant mass of the two leptons (m_{ll}) is required to be greater than 12 GeV. Background processes involving multiple bosons and hence more than two leptons in the final state are suppressed using an additional lepton veto which rejects the events having three or more identified and isolated leptons with $p_T > 10$ GeV in the final state. To suppress the background arising from the Drell–Yan process, E_T^{Miss} of the event is required to be greater than 20 GeV. The di-lepton transverse momentum (p_T^{ll}) is required to be greater than 30 GeV to further reduce the contributions from Drell–Yan and the non-prompt lepton backgrounds. The contribution from top quark processes is reduced by requesting that no jets with $p_T > 20$ GeV are recognized to originate from the hadronization of a b quark.

Events are categorized according to different jet multiplicities, counting jets with $p_T > 30$ GeV. The zero jet (0-jet) category is dominated by the non-resonant WW background, while in the one jet (1-jet) category receives similar contributions from non-resonant WW and top quark events. To disentangle another important background, W+jets, where one jet is misidentified as a lepton, the 0-jet and 1-jet categories are further split according to the flavour of highest p_T lepton in the event: $e\mu$ and μe.

To extract the Higgs boson signal, the same strategy as in the Run-I analysis [5] is followed. Due to the presence of neutrinos, the Higgs boson invariant mass could not be reconstructed but the expected kinematics of the Higgs boson production and decay could be explored. Given the spin 0 nature of the SM Higgs, the two charged leptons are emitted close to each other. Also, the invariant mass of the two leptons in the signal is relatively small as compared to the one expected for a lepton pair arising from other background processes. The Higgs transverse mass $m_T^H = \sqrt{2p_T^{ll}E_T^{Miss}(1 - \cos\Delta\phi(ll, E_T^{Miss})}$, where $\Delta\phi(ll, E_T^{Miss})$ is the azimuthal angle between the di-lepton system and the E_T^{Miss}, could be used to disentangle signal from background events. An analysis based on bi-dimensional templates of m_{ll} versus m_T^H is performed to extract the Higgs signal. Background shapes and normalizations are estimated from the data, wherever possible. The non-prompt lepton background contribution is estimated from the data by measuring the lepton misidentification rates in a background enriched control sample of events, and extrapolating its contribution from this control region to the signal phase space. A low m_T^H phase space control region is defined to extract the normalization of Drell–Yan process, and events with at least one b-tagged jet for the top quark events.

37.3 Results

The final binned fit is performed using template histograms for signal and background processes obtained after all aforementioned selection criteria are applied. The signal and background templates, as well as the distribution observed in the data, are shown in Fig. 37.1 for the 0-jet and 1-jet, μe and eμ categories. Combining the four categories the observed (expected) significance is 0.7σ (2.0σ) for a SM Higgs boson with a mass of 125 GeV. The corresponding best fit signal strength, σ/σ_{SM}, which is the ratio of the measured H→WW signal yield to the expectation for the SM Higgs boson, is 0.3 ± 0.5.

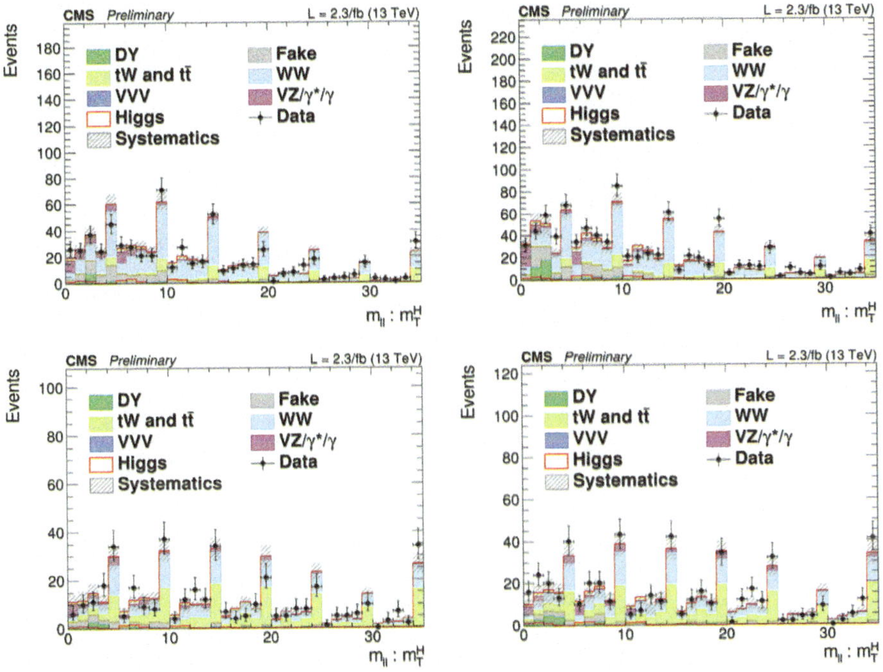

Fig. 37.1 Bi-dimensional distributions of the m_{ll} and m_T^H templates in the 0-jet (top) and 1-jet (bottom) and μe (left) and eμ (right) categories at the level of WW selection. The bi-dimensional templates ranges are $10 < m_{ll} < 110$ GeV and $0 < m_T^H < 200$ GeV with 5 bins in m_{ll} and 10 bins in m_T^H. The distributions are unrolled to one dimensional histograms such that identical values of m_T^H are in adjacent bins [4]

References

1. S. Chatrchyan et al., The CMS experiment at the CERN LHC. JINST **3**, S08004 (2008). https://doi.org/10.1088/1748-0221/3/08/S08004
2. S. Chatrchyan et al., Observation of a new boson at a mass of 125 GeV with the CMS experiment at the LHC. Phys. Lett. **B716**, 30–61 (2012). https://doi.org/10.1016/j.physletb.2012.08.021
3. G. Aad et al., Observation of a new particle in the search for the Standard Model Higgs boson with the ATLAS detector at the LHC. Phys. Lett. **B716**, 1–29 (2012). https://doi.org/10.1016/j.physletb.2012.08.020
4. CMS Collaboration, First results on Higgs to WW at $\sqrt{s} = 13$ TeV. CMS-PAS-HIG-15-003 (2016), https://cds.cern.ch/record/2161793
5. S. Chatrchyan et al., Measurement of Higgs boson production and properties in the WW decay channel with leptonic final states. JHEP **01**, 096 (2014). https://doi.org/10.1007/JHEP01(2014)096

Chapter 38
Digital System for Multi-parametric Analysis in Physics Application

Massimo Venaruzzo

38.1 Introduction

In recent years, the use of **waveform digitizers** for readout of radiation detectors has become popular in many different physics applications: the conventional analog electronics is going to be replaced by a full digital approach, where the detector output (with or without preamplifier, depending on the detector type) is directly connected to the digitizer input. This approach is especially beneficial in **multi-parametric** acquisition systems, where the analysis involves different quantities and parameters, such as energy, pulse shape and timing. In fact, thanks to **Digital Pulse Processing** (DPP) it is possible to apply dedicated algorithms on-line (typically in the FPGAs), which are designed to extract the information of interest from the raw waveform and allow the digitizer to implement in "one single box" the different functionalities of the old fashion TDC, QDC, Peak Sensing ADC, discriminator and other analog and logic modules [1]. Furthermore, the DPP allows the digital readout to be sustainable in terms of data throughput by eliminating the need to read out the full waveform, while retaining the capability to read out the full waveform for debugging and/or configuration purposes. This results in an **"all-in-one, multi-parametric digital DAQ for physics applications"**.

Massimo Venaruzzo on behalf of CAEN S.p.A.

M. Venaruzzo (✉)
CAEN S.p.A., via Vetraia 11, 55049 Viareggio (LU), Italy
email: m.venaruzzo@caen.it
URL: http://www.caen.it/

Md. Naimuddin (ed.), *XXII DAE High Energy Physics Symposium*, Springer Proceedings in Physics 203, https://doi.org/10.1007/978-3-319-73171-1_38

38.2 Overview of the Digital Pulse Processing Algorithms

CAEN has designed a complete family of digitizers that consists of several models differing in sampling frequency, resolution, number of channels, form factor, memory size and other parameters. In addition, CAEN developed also different digital pulse processing algorithms that allows the user to extract a set of significant information like energy, precise timing, PSD and so on. They are:

- **Pulse Height Analysis (DPP-PHA): Digital MCA**. The DPP-PHA firmware evaluates the input pulse energy implementing a digital trapezoidal filter [2] on the input pulse, which replaces the traditional analog chain of shaping amplifier and peak sensing ADC [3]. The digitizer is therefore directly connected to the charge sensitive preamplifier, with no need of additional devices [4].
- **Charge Integration and Pulse Shape Discrimination (DPP-PSD): Digital QDC and γ-n discrimination**. The DPP-PSD firmware is the digital version of the analog chain made of Charge Integrating ADC (QDC), Discriminator and Gate generator. In addition it implements a dual gated charge integrator, typically used for the separation of the fast and slow components of the signal (e.g. neutron-γ discrimination in liquid scintillators) [3, 5].
- **Timing Measurements: Digital CFD**. The DPP-PSD firmware implements also a fast timing filter, a digital Constant Fraction Discriminator (CFD), with a linear interpolation of the waveform at the output of the filter thus allowing a very good resolution in high precision timing measurements.
  If we consider an ideal signal without any electronic noise, the precision in finding the zero crossing depends on two factors: the Interpolation Error E_i (geometric error) due to deviation of the straight line from the actual pulse curve and the A/D Quantization Error E_q that affects the value of the two points used for the linear fit. The resolution of the fine time stamp is affected by the rising edge of the input signal, the sampling rate of the digitizer and the pulse amplitude compared to the LSB of the ADC (and so the number of bits of the ADC). Advanced calibration techniques for the E_i compensation has been implemented allowing, with our top level digitizer DT5730 (14bit - 500 MS/s), exceptional timing performance with rising edges as fast as 1 ns. Table 38.1 is a synthetic report of the timing tests performed in CAEN with the different signal sources (either generators or detectors).

38.3 Conclusions

Tests performed with a DT5730 (14 bit, 500 MS/s waveform digitizer) demonstrate that the digital CFD algorithm fully implemented on-line in the FPGA of the digitizer is capable of producing outstanding intrinsic resolution (2–10 ps RMS) when reading ideal signals coming from pulse generators with leading edge ranging from

Table 38.1 Timing resolution with different input pulses

Test	E_i corr	Rise time (ns)	Amplitude (mV)	ΔT Mean	ΔT sigma (ps)
Pulse generator	Enabled	1	450	11.21	9.5
Pulse generator	Enabled	5	450	11.23	10.1
Pulse generator	Enabled	20	450	11.27	1.9
BaF_2 to BaF_2	Disabled	1.3	130	11.15	538
BaF_2 to BaF_2	Enabled	1.3	130	11.04	107
BaF_2 to $LaBr_3$	Enabled	1.3/15	130/200	11.23	10.1

20 down to 1 ns and a considerable resolution with fast real detectors, featuring about 75 and 170 ps RMS with BaF_2 and $LaBr_3$ respectively (leading edge of 1.3 and 15 ns). Advanced calibration algorithm is applied to correct the interpolation error that dominates the timing resolution, especially in the cases where the rising edge of the signal is as fast as the sampling period or even faster. These tests prove that good time resolution can be achieved when one single point is taken from the leading edge of the signal, provided that the ADC has enough amplitude resolution (number of bits). In conclusion, a waveform digitizer with on-line Digital Pulse Processing is a high performance, "all-in-one" solution able to replace a multitude of conventional modules (CFD, TDC, QDC, peak ADC, Logic Units, etc.) in Multi-parametric acquisition systems for Nuclear Physics Applications.

References

1. W.R. Leo, *Techniques for Nuclear and Particle Physics Experiments* (Springer, Berlin, 1994)
2. V.T. Jordanov, G.F. Knoll, Digital Synthesis of pulse shapes in real time for high resolution radiation spectroscopy. NIM A **345**, 337 (1994)
3. G.F. Knoll, *Radiation Detection and Measurement* (Wiley, New York, 2010)
4. CAEN S.p.A., UM3182 - DPP-PHA and MC2Analyzer User Manual, http://www.caen.it/servlet/checkCaenManualFile?Id=12187
5. CAEN S.p.A., UM2580 - DPP-PSD User Manual, http://www.caen.it/servlet/checkCaenManualFile?Id=11970

Chapter 39
Search for Supersymmetry in Vector Boson Fusion Topology Using Proton–Proton Collision Data at the LHC

Priyanka Kumari, Amandeep Kaur Kalsi, Nitish Dhingra, Vipin Bhatnagar and J. B. Singh

39.1 Introduction

With the outstanding performance of the LHC machine, Standard Model (SM) discovered its long-sought "Higgs" particle in July 2012, which proved to be the biggest success for the SM as well as for the CMS [1] and ATLAS [2] experiments at LHC. But some mysteries like unification of forces, neutrino oscillations, matter-antimatter asymmetry and dark matter etc. remain unexplained. In order to solve such mysteries, various extensions of the SM have been developed by the physicists. One of them is known as Supersymmetry (SUSY) [3, 4] that associates every SM fermion with its "super-partner" boson and vice-versa. The SM and SUSY partners can be distinguished by a discrete quantum number called R-Parity. The R-parity conserving models [5] involve a weakly interacting particle known as the Lightest Supersymmetric Particle (LSP) which gives rise to transverse momentum imbalance (E_T^{miss}) in the detector and is considered to be a good dark matter candidate. SUSY searches have set limits on masses of gluinos and 1st/2nd generation squarks of the order of 1.7 TeV [6] while the limits on $\tilde{\chi}_1^{\pm}/\tilde{\chi}_2^0$ are relatively weaker in compressed-mass-spectra. In this paper, the search for electroweak SUSY in VBF topology is presented at $\sqrt{s} = 8$ TeV. Some prospects with Run2 data at $\sqrt{s} = 13$ TeV are also discussed.

Priyanka Kumari, Amandeep Kaur Kalsi, Nitish Dhingra, Vipin Bhatnagar and J.B. Singh on behalf of the CMS Collaboration.

P. Kumari (✉) · A. K. Kalsi · N. Dhingra · V. Bhatnagar · J. B. Singh
Panjab University, Chandigarh, India
e-mail: priyanka.kumari@cern.ch

N. Dhingra
G.H.G. Khalsa College, Ludhiana, India

Md. Naimuddin (ed.), *XXII DAE High Energy Physics Symposium*, Springer Proceedings in Physics 203, https://doi.org/10.1007/978-3-319-73171-1_39

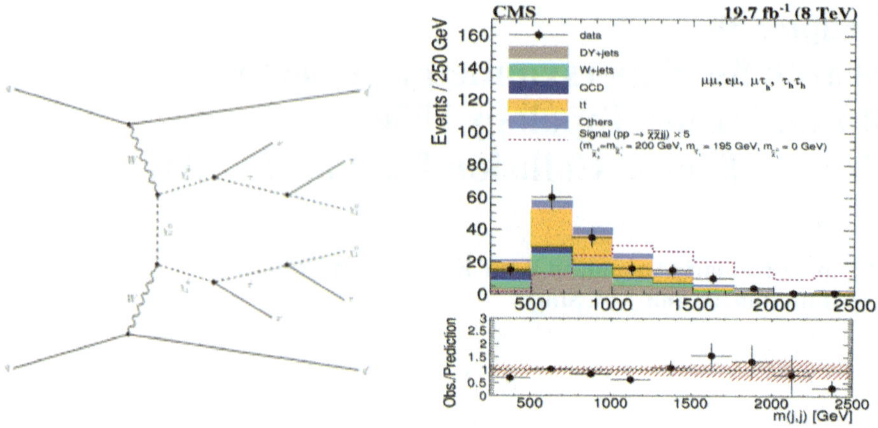

Fig. 39.1 Left: [7] Chargino–Chargino pair production by VBF processes, Right: [8] m_{jj} distribution in the signal region obtained by combining all final states

39.2 Analysis Strategy

The VBF processes are characterized by the presence of two forward jets in opposite hemispheres of the detector with large dijet invariant mass and large pseudorapidity gap. It provides a complementary tool to probe compressed-mass SUSY spectra and offers a unique handle for background suppression. Figure 39.1 (left) shows one of the representative Feynman diagram for pair-production of charginos in VBF processes where charginos decay to two τ-leptons through $\tilde{\tau}$s (same-sign or opposite-sign) and LSP. The SUSY search with 8 TeV data was performed in the 8 final states consisting of opposite-sign/same-sign lepton pairs in association with forward jets namely: $\mu\mu jj$, eμjj, $\mu\tau_h jj$ and $\tau_h\tau_h jj$. Event selection criteria is divided into 2 parts: Central selections and VBF selections. The central selections require the event to have two isolated leptons with p_T > 30/45 GeV (μ/τ), $|\eta| < 2.1$, $\Delta R(l_1, l_2)$ > 0.3 and E_T^{miss} > 75 GeV (>30 GeV only for $\tau_h\tau_h jj$). VBF selections requiring the presence of two VBF jets in opposite hemispheres ($\eta_1 * \eta_2 < 0$) with $p_T > 30/50$ GeV (Loose/Tight signal regions) and $|\eta| < 5.0$, reduces the background rate by a factor of 10^{-2}–10^{-4}. The b-jets with "combined secondary vertex loose" (CSVL) are vetoed to suppress $t\bar{t}$ background. The selected events are required to have a dijet candidate with m_{jj} > 250 GeV. Control regions are defined by some modifications in the nominal selection cuts to measure VBF efficiency and m_{jj} shapes from data in such a way that m_{jj} distribution remains unbiased. The data was found to be consistent with SM backgrounds predictions within uncertainty after combining all the final states as shown in Fig. 39.1 (right).

We are heading towards high luminosity of pp collision data at $\sqrt{s} = 13$ TeV, where focus is on single lepton final state namely: μjj, ejj and τjj. This search includes new invisible VBF trigger to get large signal efficiency. Signal selection

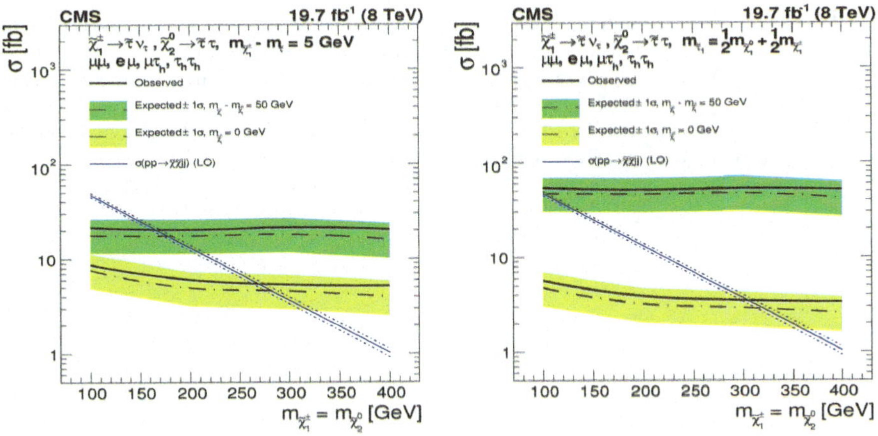

Fig. 39.2 Left: [8] Combined 95% CL upper limits on the cross section for scenario having $\tilde{\tau}$ closer to $m_{\tilde{\chi}_1^{\pm}/\tilde{\chi}_2^0}$ i.e. $\Delta m(\tilde{\chi}_1^{\pm}, \tilde{\tau}) = 5$ GeV for large mass gap scenarios where $m_{\tilde{\chi}_1^0} = 0$ GeV (yellow band) and the compressed mass scenario having a mass difference between chargino and LSP to be 50 GeV (green band), Right: [8] 95% CL on EWKino mass = 300 GeV for scenario having $m_{\tilde{\tau}} = 0.5m_{\tilde{\chi}_1^{\pm}} + 0.5m_{\tilde{\chi}_1^0}$

has been optimized for best signal significance and estimation of backgrounds using control regions is under progress to extend our reach substantially.

39.3 Results and Discussion

For the R-parity conserving MSSM models, results are simplified in four scenarios divided on the basis of masses of stau and LSP. The average mass assumption ($m_{\tilde{\tau}} = 0.5m_{\tilde{\chi}_1^{\pm}} + 0.5m_{\tilde{\chi}_1^0}$) with uncompressed mass spectra ($m_{\tilde{\chi}_1^0} = 0$) eliminates the possibility of $\tilde{\chi}_1^{\pm}/\tilde{\chi}_1^0$ with masses less than 300 GeV, and masses upto 170 GeV for compressed mass spectra ($m_{\tilde{\chi}_1^{\pm}} - m_{\tilde{\chi}_1^0} = 50$ GeV) with $\tilde{\tau}$ mass close to chargino mass i.e. $\Delta m(\tilde{\chi}_1^{\pm}, \tilde{\tau}) = 5$ GeV as shown in Fig. 39.2.

References

1. CMS Collaboration, The CMS experiment at the CERN LHC. JINST **3**, S08004 (2008)
2. ATLAS Collaboration, The ATLAS experiment at the CERN large hadron collider. JINST **3**, S08003 (2008)
3. S.P. Martin, A supersymmetry primer, arXiv:hep-ph/9709356v6
4. R. Barbieri, S. Ferrara, C.A. Savoy, Gauge models with spontaneously broken local supersymmetry. Phys. Lett. B **119**, 343 (1982)
5. G.R. Farrar, P. Fayet, Phys. Lett. B **76**, 575 (1978)

6. CMS Collaboration, Search for new physics in the multijet and missing transverse momentum final state in proton-proton collisions at $\sqrt{s} = 8$ TeV. JHEP **06**, 055 (2014)
7. B. Dutta et al., Vector Boson fusion processes as a probe of supersymmetric electroweak sectors at the LHC. Phys. Rev. D **87**, 035029 (2013)
8. CMS Collaboration, Search for supersymmetry in the vector-boson fusion topology in proton-proton collisions at $\sqrt{s} = 8$ TeV. JHEP **11**, 189 (2015)

Chapter 40
Test Beam Study of Gas Electron Multiplier (GEM) Detectors for the Upgrade of CMS Endcap Muon System

Ram Krishna Sharma, Md. Naimuddin, Brian Dorney, Jeremie Alexandre Merlin, Archana Sharma, Marek Michal Gruchala, Priyanka Kumari and Ankita Mehta

40.1 Introduction

LHC at CERN has two general purpose detectors. They are CMS [1] and ATLAS [2]. CMS collaboration is now preparing for the upgradation of the existing facilities of the detectors during Long Shutdown 2 (LS2, 2019–2020). This upgrade is motivated by the high luminosity that LHC will inject after LS2. It is expected that after LS2 the particle rate in the forward region of CMS detector will reach several tens of kHz/cm^2 and the total integrated charge will reach several C/cm^2. CMS collaboration approved the Gas Electron Multiplier (GEM) detector [3] for the forward muon system as these detectors have the excellent space and good time resolution along with high rate handling capability. Due to these excellent properties GEM detectors will improve the muon triggering and tracking capabilities by improving the muon momentum resolution for high pT muons and the timing resolution, as shown in Fig. 40.1.

Ram Krishna Sharma, Md. Naimuddin, Brian Dorney, Jeremie Alexandre Merlin, Archana Sharma, Marek Michal Gruchala, Priyanka Kumari and Ankita Mehta on behalf of the CMS-Muon Collaboration.

R. K. Sharma (✉) · Md. Naimuddin
University of Delhi, Delhi, India
e-mail: rasharma@cern.ch

B. Dorney · J. A. Merlin · A. Sharma
CERN, Geneva, Switzerland

M. M. Gruchala
Ghent University, Ghent, Belgium

P. Kumari · A. Mehta
Panjab University, Chandigarh, India

Md. Naimuddin (ed.), *XXII DAE High Energy Physics Symposium*, Springer Proceedings in Physics 203, https://doi.org/10.1007/978-3-319-73171-1_40

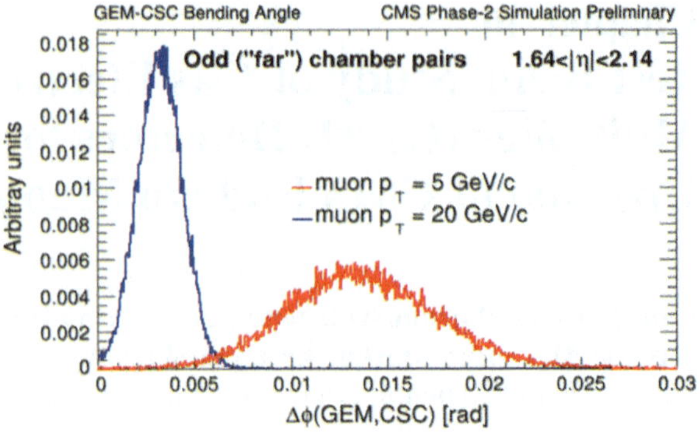

Fig. 40.1 Bending angle for GEM-CSC for soft and hard muons for odd numbered GEM chambers

40.2 GEM Detectors

The main part of a GEM detector is the GEM foil that act as an electron amplifier and an arrangement involving the sandwich of GEM foils between the two electrode is known as GEM detectors. GEM foil consists of thin metal-clad insulating foil with a high density of holes. After the application of suitable voltage between two sides, electrons released in the upper part of the detector will pass through the holes of foil and gets multiplied and transferred to the other side of the detector where it is collected or further amplified.

The prototype of GEM detector to be installed in the CMS detector is known as GE1/1 [4], where "G" denotes the GEM detector, "E" denotes the muon endcap station, the first "1" indicates the part of first muon station encountered by particles from the interaction point and the second "1" denotes the first ring of muon chamber going outward in radius from the beam line. The shape of GE1/1 chamber is trapezoidal with an active area of $990 \times (220–445)\,\text{mm}^2$. This size is fixed as per the constraint from the vacant high-η area in the CMS muon endcap. GE1/1 chamber consists of a Triple-GEM detector with gap configuration of 3/1/2/1 mm (drift/transfer 1/transfer 2/induction). The readout board is divided into eight η-partitions with 384 strips each oriented radially along the long side of the detector with a pitch varying from 0.6 mm (short side) to 1.2 mm (long side). Each partition is subdivided along the ϕ-coordinate into three readout sectors with 128 strips or channels each. In the test beam, we scanned several different sectors of the GE1/1 detectors. They are $(\text{i}\eta, \text{i}\phi) = (1,2), (5,2), (8,2)$. In Fig. 40.2, red and yellow colors show that which sectors of GE1/1s are exposed to the beam. The sector marked with red color was tested with gas $Ar/CO_2/CF_4$ (45/15/40), while the sectors marked with yellow was tested with gas Ar/CO_2 (70/30).

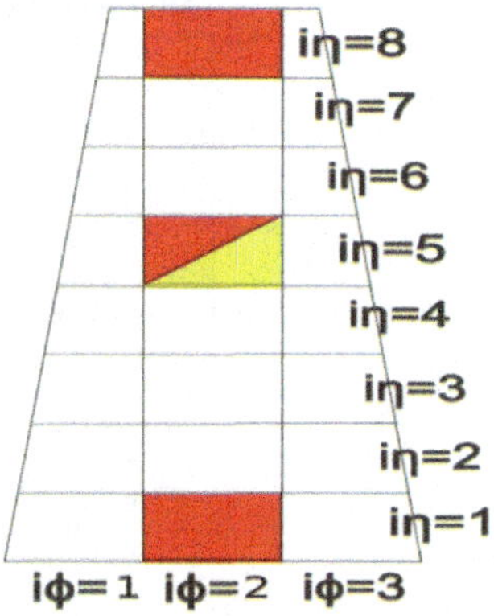

Fig. 40.2 Showing different ($i\eta$, $i\phi$) sector of GE1/1 prototype detectors

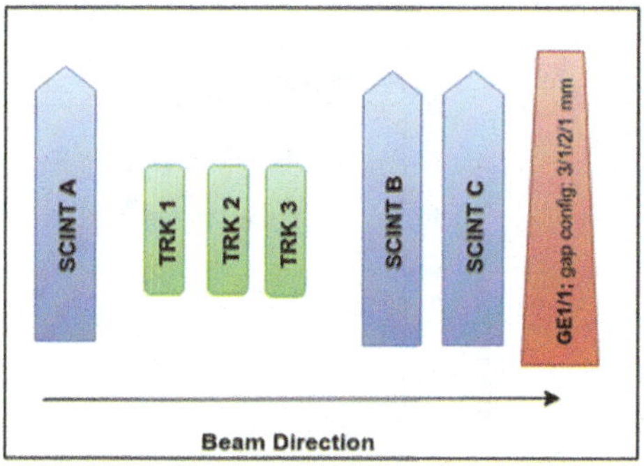

Fig. 40.3 Test beam setup

40.3 Experimental Setup

The GE1/1 detector was tested using muon beam having energy ~150 GeV. It was done in the CERN SPS test beam facility in 2014 [5]. The test beam setup is shown in Fig. 40.3. It consists of GE1/1 prototype detector, three trackers and three scintillators. The trackers are also a GEM detectors having active area of $10 \times 10\,\text{cm}^2$. It acts as the muon telescope which helps us to reconstruct the tracks and reduce backgrounds. And the trigger comes from the coincidence of the scintillators. The GE1/1 prototype was installed on a movable table for scanning the different sectors of the detector.

40.4 Result and Discussion

Efficiency and time resolution are important parameters for a gaseous detectors. Efficiency is defined as the ratio of events reconstructed by tracker+GEM to the number of events reconstructed by tracker only. Experimentally, the time resolution is calculated using the root mean square of the Gaussian distribution of the time taken by a particle to reach the detector from the scintillator. Further, we applied a

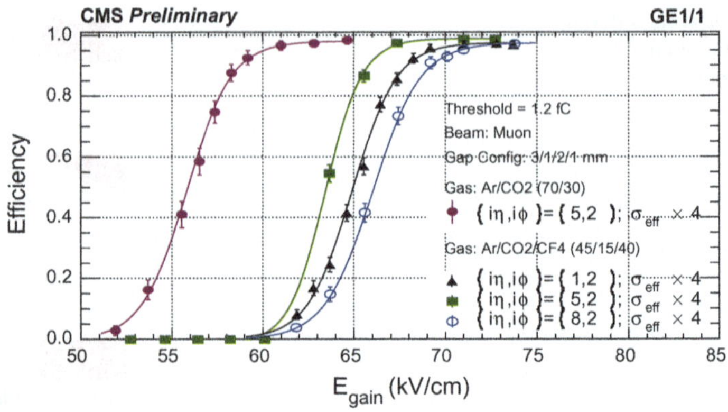

Fig. 40.4 Efficiency w.r.t. E_{gain} for different ($i\eta$, $i\phi$) sectors and for two different sectors

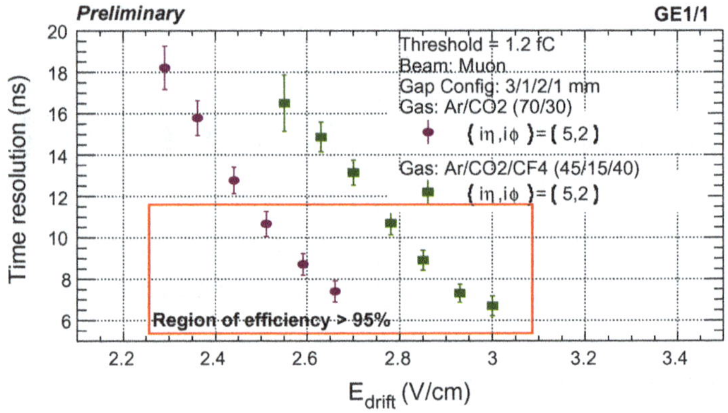

Fig. 40.5 Time-resolution w.r.t. E_{drift} for two different gases

fast 40 MHz clock pulse to cope up with the LHC bunch crossing. Thus, the detector's time response should be modelled as the Gaussian function convoluted with a square wave function having frequency 40MHz. To achieve the time resolution we fit the experimental data with the Gaussian function convoluted with the square wave and then using fit parameters we extracted the time resolution before convolution. Finally, we obtained the very good time resolution of 7ns and efficiency 98% with both gas mixture Ar/CO_2 (70/30) and $Ar/CO_2/CF_4$ (45/15/40), as shown in Figs. 40.4 and 40.5. At the end, by comparing the results of two different gas mixture we can conclude that GEM detectors can be operated without using CF_4, while maintaining its performance.

References

1. S. Chatrchyan, et al., [CMS Collaboration], The CMS experiment at the CERN LHC. JINST **3**, S08004 (2008)
2. G. Aad et al., ATLAS Collaboration, The ATLAS experiment at the CERN large hadron collider. JINST **3**, S08003 (2008)
3. F. Sauli, GEM: a new concept for electron amplification in gas detectors. Nucl. Instr. Meth. A **386**, 531 (1997)
4. CMS GEM Collaboration, CMS Technical Design Report for the muon endcap GEM upgrade, CMS-TDR-15-001-001
5. R. Sharma, et al., [CMS Collaboration], Charged particle detection performance of Gas Electron Multiplier (GEM) detectors for the upgrade of CMS endcap muon system at the CERN LHC. https://doi.org/10.1109/NSSMIC.2015.7581797

Chapter 41
Model Study of Two Particle Correlations with Identified Trigger Particles in p-Pb Collisions at LHC Energy

Debojit Sarkar, Subikash Choudhury and Subhasis Chattopadhyay

41.1 Introduction

A large baryon-to-meson enhancement at intermediate p_T has been observed in central heavy ion collisions both at the RHIC and the LHC energies. While at RHIC, the interpretation is given in terms of coalescence, at the LHC, data have also been explained in terms of radial flow. A similar baryon-to-meson enhancement is also observed in p-Pb collisions at $\sqrt{s_{NN}} = 5.02$ TeV as well [1].

In this work we have calculated the near-side jet-like yield associated with pion and proton triggers selected from the intermediate p_T where the inclusive proton to pion enhancement has been observed in p-Pb collisions at 5.02 TeV. Two models, EPOS 3 and AMPT string melting (SM) are used to study the effect of the hydrodynamical flow and the coalescence model of hadronization on this observable. In this analysis, the trigger particles are selected from the intermediate p_T range (1.5/2.0 $< p_T <$ 4.0 GeV/c) and associated particles have p_T in the range of $1.0 < p_T < 4.0$ GeV/c. The detailed description of the models, multiplicity class estimation, construction of the correlation function and the near-side jet-like yield determination have been discussed in [2, 3]. The bulk (soft triggered correlation) is estimated from large $|\Delta\eta|$ ($|\Delta\eta| \geq 1.1$) and subtracted from the near side jet peak ($|\Delta\eta| < 1.1$). After bulk subtraction, the event averaged near-side jet-like per trigger yield has been calculated by integrating the $\Delta\phi$ projection in the range $|\Delta\phi| < \pi/2$.

D. Sarkar (✉) · S. Choudhury · S. Chattopadhyay
Variable Energy Cyclotron Centre, HBNI, 1/AF-Bidhannagar, Kolkata 700064, India
e-mail: debojit03564@gmail.com

Md. Naimuddin (ed.), *XXII DAE High Energy Physics Symposium*, Springer Proceedings in Physics 203, https://doi.org/10.1007/978-3-319-73171-1_41

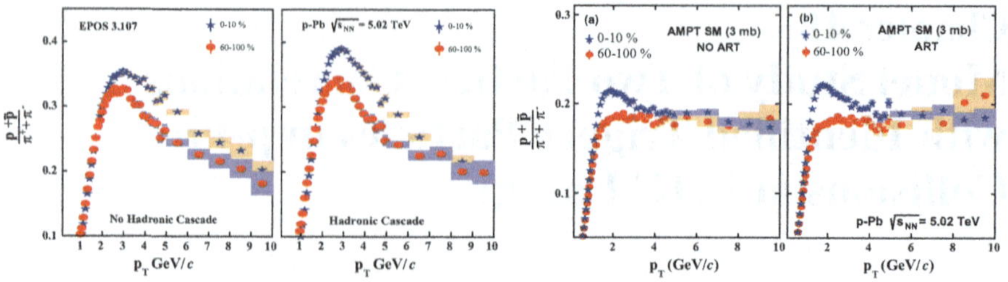

Fig. 41.1 Inclusive proton over pion ratio as obtained from EPOS 3.107 (left) and AMPT SM version (right) in 0–10% and 60–100% event classes of p-Pb collisions at 5.02 TeV

41.2 Results and Discussion

The inclusive proton to pion enhancement as obtained from EPOS 3 and AMPT are shown in Fig. 41.1. Hadrons pushed from lower to higher p_T by radial flow or generated by quark coalescence are expected not to exhibit short range jet-like correlation beyond the expected flow (ridge) like correlation. Thus the bulk/flow subtracted near-side jet peak is dominated by the hard triggered (jet) correlation. Since the correlation functions are normalized by both hard and soft triggers, the soft triggers without any jet-like correlated partners are expected to cause the trigger dilution effect in the per trigger jet-like yield. The proportion of soft triggers increase with multiplicity and the rate of increase has a species dependence as the soft processes like coalescence and/or radial flow favor proton production over the pion at intermediate p_T. So, the proton-triggered jet-like yield is expected to get diluted at a larger rate compared to the pion-triggered case with increase in multiplicity [2, 3].

In Figs. 41.2 and 41.3 the multiplicity evolution of the near-side jet-like yields associated with identified triggers are shown (Fig. 41.2). In case of AMPT the pion triggered yield increases with multiplicity whereas the proton triggered yield shows a slower rate of increase compared to pions - creating trigger dilution. Whereas, in EPOS 3, the pion triggered yield shows almost no variation with multiplicity but the proton triggered yield decreases gradually with increase in multiplicity- generating trigger dilution as shown in Fig. 41.3. The effect of the final state hadronic interactions has been found to be negligible in case of both AMPT and EPOS 3 [2, 3]. This suggests, although trigger dilution can originate from both coalescence and radial flow, the response of the individual pion and proton triggered yields, studied as a function of multiplicity, is sensitive to the underlying dynamics or the physics processes involved. In absence of jet quenching [2, 3], this observable can be used to probe the presence of collectivity in small collision systems and the hadronization at intermediate p_T as well.

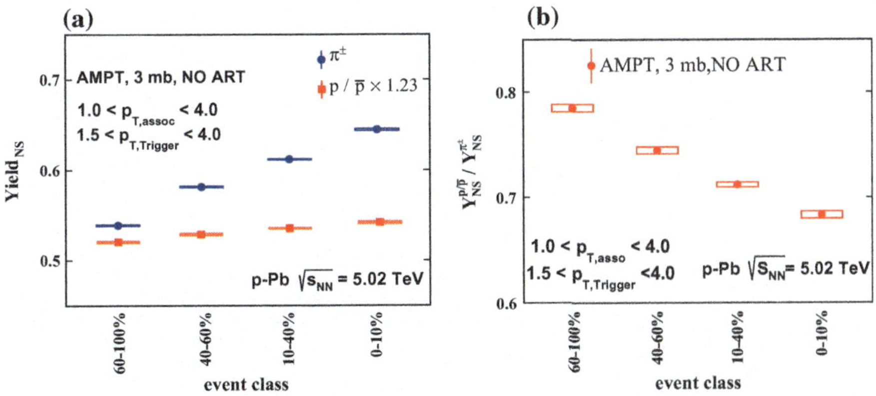

Fig. 41.2 **a** Multiplicity dependence of the near-side jet like per trigger yield associated with pion and proton triggers in p-Pb collisions at $\sqrt{s_{NN}} = 5.02$ TeV from AMPT SM without hadronic cascade (left). **b** Multiplicity dependence of the ratio of the near-side jet-like per trigger yield associated with proton and pion triggers (right)

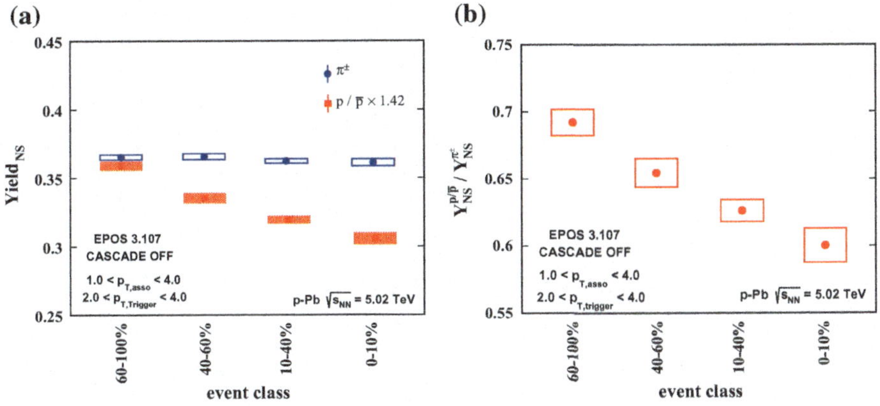

Fig. 41.3 **a** Multiplicity dependence of the near-side jet like per trigger yield associated with pion and proton triggers in p-Pb collisions at $\sqrt{s_{NN}} = 5.02$ TeV from EPOS 3 without hadronic cascade (left). **b** Multiplicity dependence of the ratio of the near-side jet-like per trigger yield associated with proton and pion triggers (right)

References

1. B. Ablev et al., ALICE Collaboration, Phys. Lett. B **728**, 2538 (2014)
2. D. Sarkar, S. Choudhury, S. Chattopadhyay, Phys. Lett. B **760**, 763768 (2016)
3. D. Sarkar, S. Choudhury, S. Chattopadhyay, Phys. Rev. C **94**, 044909 (2016)

Chapter 42
Searching for SUSY with Multijets and Missing Transverse Momentum

Aditee Rane

42.1 Introduction

Supersymmetry (SUSY) is a popular extension of the standard model (SM) of elementary particles and it proposes a superpartner to the every SM particle differing by spin-half. Depending on the SUSY model, the presence of new particles can be manifested in a variety of final states containing jets originating from light flavor quarks or gluons or bottom quarks, and an imbalance in total momentum in direction transverse to the beam direction (p_T^{miss}). A few representative models of production and decay of gluinos and top squarks are shown in Fig. 42.1.

This report presents an inclusive analysis to search for pair production of gluinos and squark-antisquark pairs using the data from proton-proton collisions at $\sqrt{s} = 13$ TeV collected by the CMS experiment at Large Hadron Collider (LHC) [1] in 2016 equivalent to an integrated luminosity of 12.9 fb^{-1}. Every gluino and squark decays directly or via a cascade into the quarks and lightest neutralino. The neutralino is assumed to be the lightest SUSY particle, neutral, weakly interacting and stable, hence is a source of true p_T^{miss} in the events. The quarks and gluons are measured as jets reconstructed from their hadronization productions and those originating from b-quarks are identified using their characteristic displaced vertices. To render this search sensitive to various final state topologies, the search regions are defined using the number of jets with $p_T > 30$ GeV and $|\eta| < 2.4$, and the number of b-tagged jets with $p_T > 30$ GeV and $|\eta| < 2.4$. In the following, these are denoted by N_{jet} and N_{b-jet} respectively. Each event is further categorized using scalar sum of p_T of jets, H_T, and magnitude of negative of vector sum of p_T of jets, H_T^{miss}.

Aditee Rane on behalf of the CMS collaboration.

A. Rane (✉)
Indian Institute of Science, Education and Research, Pune, India
e-mail: aditee.prabhakar.rane@cern.ch

Md. Naimuddin (ed.), *XXII DAE High Energy Physics Symposium*, Springer Proceedings in Physics 203, https://doi.org/10.1007/978-3-319-73171-1_42

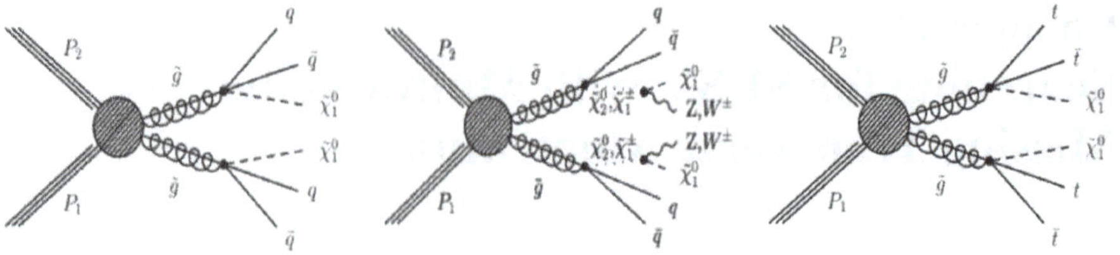

Fig. 42.1 Representative diagrams of the SUSY particle pair production: (left) pp $\rightarrow \widetilde{g}\,\widetilde{g}, \widetilde{g} \rightarrow q\overline{q} + \widetilde{\chi}_1^0$, (center) pp $\rightarrow \widetilde{g}\,\widetilde{g}, \widetilde{g} \rightarrow q\overline{q} + W/Z + \widetilde{\chi}_1^0$ via a $\widetilde{\chi}_1^{\pm}$ or $\widetilde{\chi}_2^0$, and (right) pp $\rightarrow \widetilde{g}\,\widetilde{g}, \widetilde{g} \rightarrow t\overline{t} + \widetilde{\chi}_1^0$ [2]

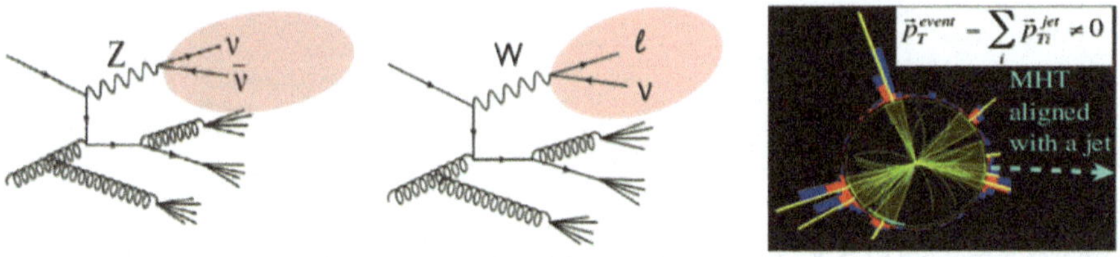

Fig. 42.2 Schematic sketch of SM backgrounds: (left) Z + jets, (center) W + jets, and (right) QCD multijet

42.2 The SM Backgrounds and Event Selection

Several SM processes can give rise to final states of jets and $p_{\mathrm{T}}^{\mathrm{miss}}$. To select the event likely to be SUSY like, we start with the events with $N_{\mathrm{jet}} \geq 3$, $H_{\mathrm{T}} > 300$ GeV, $H_{\mathrm{T}}^{\mathrm{miss}} > 300$ GeV. The Z+jets events in which Z boson decaying to a pair of neutrinos, is an irreducible background as shown in Fig. 42.2 (left). The events containing a W boson, direct production or from decay of top quarks, contribute to the backgrounds if it decays to a lepton (e, μ, or τ) and a neutrino (Fig. 42.2 center). This background is suppressed by vetoing the events which contain an isolated e, μ, or an isolated track. The QCD multijet events are a source of fake $p_{\mathrm{T}}^{\mathrm{miss}}$ resulting from jet mismeasurement or detector malfunctioning as well as true $p_{\mathrm{T}}^{\mathrm{miss}}$ from the semileptonic decay of b-hadrons. A characteristic feature of these events is that a high p_{T} jet is aligned with the direction of $H_{\mathrm{T}}^{\mathrm{miss}}$ in azimuth as shown in Fig. 42.2 (right). This background is effectively mitigated by rejecting the events with $\Delta\Phi$(jets, $H_{\mathrm{T}}^{\mathrm{miss}}$) $< 0.3, 0.3, 0.5, 0.5$ for four highest p_{T} jets.

The events surviving the above selections are further categorized in the bins of N_{jet}: [3–4], [5–6], [7–8], and $\geq$9. For separating the events containing more heavy flavors, we require $N_{\mathrm{b-jet}}$: 0, 1, 2, $\geq$3. Each [N_{jet},$N_{\mathrm{b-jet}}$] bin is further divided into 10 H_{T}–$H_{\mathrm{T}}^{\mathrm{miss}}$ bins as shown in Fig. 42.3.

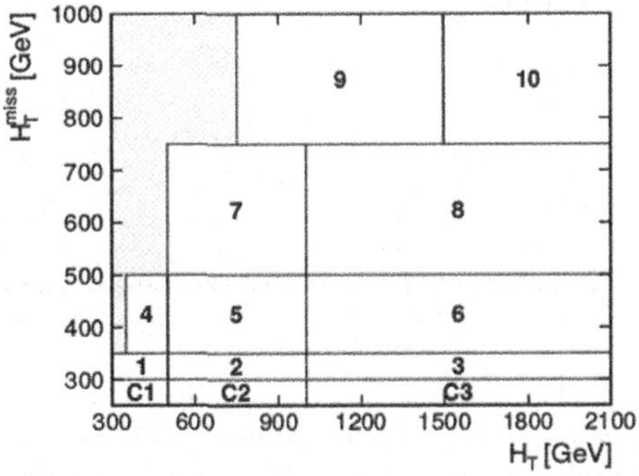

Fig. 42.3 Two-dimensional plane in H_{T} and $H_{\mathrm{T}}^{\mathrm{miss}}$ showing the signal bins and the QCD side-band bins [2]

42.3 Estimation of Backgrounds

Analysis uses data driven technique to estimate the remaining backgrounds in signal region. A method is developed to predict the background yield in signal region from corresponding yield in an orthogonal(signal free) region called control region. Method is validated using Monte Carlo (MC) samples and then applied on data control sample to get background prediction in signal region.

For hadronic tau background estimation $\mu + jets$ sample with exactly one well identified muon is used as a control sample. At the generator level, $\mu + jets$ and $\tau +$ jets have similar event kinematics but at detector level μ is reconstructed with good resolution while hadronic τ is reconstructed as a τ jet plus partial contribution to MET from τ neutrino. The response templates are derived using τ + jets MC events which give distribution of the ratio of p_{T} of reconstructed τ jet to p_{T} of gen τ in intervals of gen τ p_{T}. As reconstructed μ is equivalent to gen τ, we replace μ from μ + jets control sample event with response to mimic τ + jets event. The variables N_{jet}, H_{T}, $H_{\mathrm{T}}^{\mathrm{miss}}$ are recalculated after this replacement. Also we account for intrinsic b-mistag probability of τ jet obtained from τ + jets MC sample to redefine $N_{\mathrm{b-jet}}$. A bin defined by new [N_{jet}, $N_{\mathrm{b-jet}}$, H_{T}, $H_{\mathrm{T}}^{\mathrm{miss}}$] is filled with proper event weight. The Fig. 42.4 (left) shows validation of hadtau estimation method in 160 search bins. Direct count of hadtau events expected from MC in various bins is compared against those predicted from μ + jets control sample. The method closes well with expectation within 10% on average.

42.4 Results and Outlook

Similar to hadronic tau, all background teams validate their prediction method using MC. Then methods are applied on respective control samples from data to get background prediction in data signal region. Figure 42.4 (right) shows comparison of observed data against various background predictions.

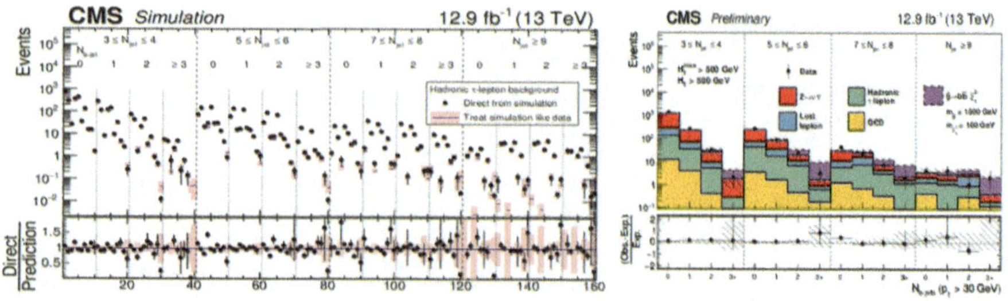

Fig. 42.4 (left) Hadronic τ background in 160 search bins as predicted directly from MC simulation (solid dots) and as predicted by data driven background estimation procedure (shaded regions). Simulation makes use of $t\bar{t}$, W+jets and single top MC event samples [2], (right) Observed number of events and pre-fit background predictions in all search bins obtained from data [2]

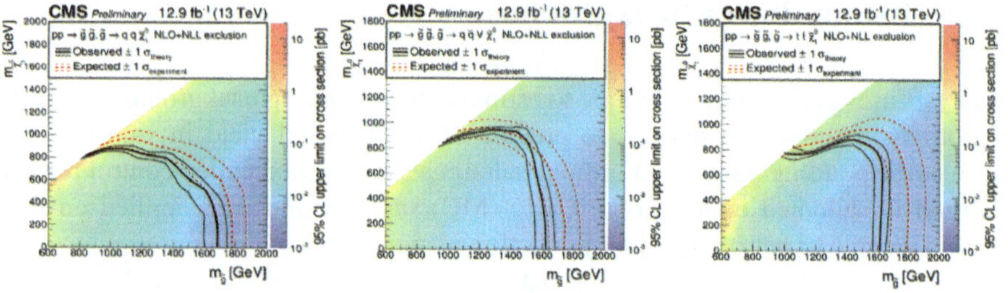

Fig. 42.5 Observed and Expected Upper Limit exclusion at 95% CL for T1qqqq (left), T5qqqqVV (center) and T1tttt (right) [2]

Data agrees with SM background prediction within uncertainties and no significant excess is observed in any of the search bins. Likelihood fit to the data based on SUSY signal strength, background yields and nuisance parameters associated with uncertainties is used to set limits on production cross section of various signal scenarios. Figure 42.5 shows exclusion plots for three different signal scenarios where gluino masses up to 1680, 1630 and 1610 GeV respectively are excluded to 95% confidence level (CL).

References

1. CMS Collaboration, JINST **3**, S08004 (2008)
2. CMS Collaboration, *Search for supersymmetry in the multijet and missing transverse momentum channel in pp collisions at 13 TeV*, CMS-PAS-SUS-16-014

Chapter 43
Modified TBM and Role of a Hidden $\mathbb{Z}_2$

Rome Samanta and Mainak Chakraborty

43.1 Introduction

In the minimally extended Standard Model (SM) with singlet RH neutrino fields N_{Ri}, $\mathcal{O}$(eV) neutrino masses are generated through Type-I seesaw mechanism. Relevant Lagrangian for the latter is written as

$$-\mathcal{L}^{\nu,N}_{mass} = \bar{\nu}_{L\alpha}(m_D)_{\alpha i}N_{Ri} + \frac{1}{2}\bar{N^C}_{iR}(M_R)_i\delta_{ij}N_{jR} + \text{h.c.} \tag{43.1}$$

with $N_i^C = C\bar{N}_i^T$. The effective light neutrino Majorana mass matrix M_ν which is then obtained by the standard seesaw formula $M_\nu = -m_D M_R^{-1} m_D^T$, can be put into a diagonal form as $U^T M_\nu U = M_\nu^d \equiv \text{diag}\,(\text{m}_1, \text{m}_2, \text{m}_3)$ with m_i assumed to be real. The effective low energy neutrino Majorana mass term that contains this M_ν comes out as

$$-\mathcal{L}^{\nu}_{mass} = \frac{1}{2}\bar{\nu^C_{L\alpha}}(M_\nu)_{\alpha\beta}\nu_{L\beta} + \text{h.c.} \tag{43.2}$$

Now in the basis where the charged lepton mass matrix M_ℓ is diagonal, U follows the standard parametrization [1]. Let us now have look at the latest 3σ ranges [2] for the relevant neutrino parameters obtained from oscillation data. solar: $\Delta m^2_{21} \equiv m_2^2 - m_1^2$: (7.02–8.09) $\times 10^{-5}$ eV^2, atmospheric: $|\Delta m^2_{31}| \equiv |m_3^2 - m_1^2|$: (2.32–2.59) $\times 10^{-3}$ eV^2, θ_{12}: 31.29°–35.91°, θ_{23}: 38.3°–53.3°, θ_{13}: 7.87°–9.11°. Finally, thanks to the

R. Samanta (✉)
Saha Institute of Nuclear Physics, HBNI, 1/AF Bidhannagar, Kolkata 700064, India
e-mail: rome.samanta@saha.ac.in

M. Chakraborty
Centre of Excellence in Theoretical and Mathematical Sciences, SOA University, Khandagiri Square, Bhubaneswar 751030, India
e-mail: mainak.chakraborty2@gmail.com

Md. Naimuddin (ed.), *XXII DAE High Energy Physics Symposium*, Springer Proceedings in Physics 203, https://doi.org/10.1007/978-3-319-73171-1_43

Planck for the observed upper bound on the sum of the light neutrino masses; $\Sigma_i m_i < 0.23$ eV.

In [3] it is argued that any horizontal symmetry of neutrino Majorana mass matrix M_ν is a residual $\mathbb{Z}_2 \times \mathbb{Z}_2$ flavour symmetry. The symmetry generators G_i obey the relation $U^\dagger G_i U = d_i$ with i = 2, 3, i.e. there are two independent G_i and hence d_i. We can choose these two independent d matrices as d_2 = diag $(-1, 1, -1)$ and d_3 = diag $(-1, -1, 1)$. Thus for a given U, one can calculate G_2 and G_3 corresponding to d_2 and d_3 respectively. In this work we focus particularly on the TBM mixing and calculate the corresponding G_i matrices; $G_{1,2}^{TBM}$ and $G_3^{\mu\tau}$. It can be justified theoretically as well as phenomenologically that G_1^{TBM} is the only symmetry which is viable one to exist as the unbroken $\mathbb{Z}_2$ generator in the Lagrangian. In the next section, we present an ephemeral discussion regarding the implementation of G_1^{TBM} and $G_3^{\mu\tau}$ on the neutrino fields.

43.2 Breaking of $\mathbb{Z}_2^{\mu\tau}$: Perturbation to the TBM Mass Matrices

Depending upon the residual symmetries on the neutrino fields and the phenomenological viability of the textures of the mass matrices, we discuss two cases.
Case 1. At the leading order G_1^{TBM} and $G_3^{\mu\tau}$ transform both the neutrino fields ν_L and N_R as $\nu_L \to G_i \nu_L$ and $N_R \to G_i N_R$. Now we choose a perturbation matrix $M_R^{G_1\epsilon}$ which violates $\mu\tau$ interchange in M_R but respect G_1^{TBM}. The leading order mass matrices and the perturbation matrix are of forms

$$m_D^0 = \begin{pmatrix} b-c-a & a & -a \\ a & b & c \\ -a & c & b \end{pmatrix}, \; M_R^0 = \begin{pmatrix} y & 0 & 0 \\ 0 & y & 0 \\ 0 & 0 & y \end{pmatrix}, \; M_R^{G_1\epsilon} = \begin{pmatrix} 0 & \epsilon & \epsilon' \\ \epsilon & \epsilon_4 & 0 \\ \epsilon' & 0 & \epsilon_6 \end{pmatrix} \quad (43.3)$$

where $\epsilon = \frac{1}{4}(3\epsilon_4 + \epsilon_6)$ and $\epsilon' = -\frac{1}{4}(3\epsilon_6 + \epsilon_4)$. Now the effective M_ν which is invariant under G_1^{TBM} is written as $M_{\nu 1}^{G_1^{TBM}} = -m_D^0 M_R^{-1} (m_D^0)^T$ with $M_R = M_R^0 + M_R^{G_1\epsilon}$. Since G_1^{TBM} invariance of the effective M_ν always fixes the first column of the mixing matrix to $(\sqrt{\frac{2}{3}}, -\sqrt{\frac{1}{6}}, \sqrt{\frac{1}{6}})^T$ up to some phases, a direct comparison of the latter with the U_{PMNS} matrix leads to a constraint relation between θ_{12} and θ_{13} as

$$\sin^2\theta_{12} = \frac{1}{3}(1 - 2\tan^2\theta_{13}). \quad (43.4)$$

Case 2. In this case, at the leading order, all the neutrino fields obey $G_3^{\mu\tau}$. However G_1^{TBM} of the over all M_ν is ensured only by the transformation $\nu_L \to G_1^{TBM}\nu_L$. Since the RH singlets are free from G_1^{TBM}, the perturbation matrix which is added with M_R is now arbitrary. Now the most general Dirac mass matrix m_D^0, the Majorana mass matrix M_R^0 and the perturbation matrix are of the forms

$$m_D^0 = \begin{pmatrix} a & \frac{1}{2}(b-c) & \frac{1}{2}(c-b) \\ a & b & c \\ -a & c & b \end{pmatrix}, \quad M_R^0 = \begin{pmatrix} x & 0 & 0 \\ 0 & y & 0 \\ 0 & 0 & y \end{pmatrix}, \quad M_R^{\epsilon} = \begin{pmatrix} 0 & 0 & 0 \\ 0 & \epsilon_4 & 0 \\ 0 & 0 & \epsilon_6 \end{pmatrix}. \tag{43.5}$$

Again the effective M_ν is calculated as $M_{\nu 2}^{G_1^{TBM}} = -m_D^0 M_R^{-1} (m_D^0)^T$ with $M_R = M_R^0 + M_R^{\epsilon}$. Besides reproducing the same relation as obtained in (43.4), another interesting point is realized that m_D^0 of (43.5) is of determinant zero due to the residual G_1^{TBM} symmetry; thus the $M_{\nu 2}^{G_1^{TBM}}$ matrix has one zero eigenvalue. For the remnant G_1^{TBM} symmetry, m_1 is of vanishing value.

43.3 Flavored Leptogenesis with Quasi Degenerate RH Neutrinos

Lepton number, CP violating and out of equilibrium decays of RH neutrinos create a lepton asymmetry [4]. A general expression for the CP asymmetry parameter ϵ_i^{α} for any RH mass spectrum is given by [5]

$$\frac{1}{4\pi v^2 \mathcal{H}_{ii}} \Big[\sum_{j \neq i} g(x_{ij}) \,\mathrm{Im}\, \mathcal{H}_{ij} (m_D)^{\dagger}_{i\alpha} (m_D)_{\alpha j} + \sum_{j \neq i} \frac{r \mathrm{Im}\, \mathcal{H}_{ji} (m_D)^{\dagger}_{i\alpha} (m_D)_{\alpha j}}{r^2 + \frac{\mathcal{H}_{jj}}{16\pi^2 v^4}} \Big]. \tag{43.6}$$

In (43.6), $r = (1 - x_{ij})$, $\mathcal{H} \equiv m_D{}^{\dagger} m_D$, $x_{ij} = M_j / M_i$ and $g(x_{ij})$ is given by

$$g(x_{ij}) = \frac{\sqrt{x_{ij}}(1 - x_{ij})}{(1 - x_{ij})^2 + \frac{\mathcal{H}_{jj}}{16\pi^2 v^4}} + f(x_{ij}). \tag{43.7}$$

The term proportional to $f(x_{ij})$ comes from the one loop vertex contribution while the remaining are from self energy diagram. Note that in the limit where the RH neutrinos are exactly degenerate, i.e., $x_{ij} = 1$, the self energy contribution vanishes and thus a nonzero value of CP asymmetry parameter ε_i^{α} is produced only through the vertex contribution. In our model RH neutrinos are quasi degenerate and thereby enhances the CP asymmetry parameter significantly through self energy contributions.

Another important issue is that the flavor effect [6] to the produced lepton asymmetry. In our upcoming work [7] we address this explicitly in detail, theoretically as well as numerically.

References

1. K.A Olive et al., [Particle Data Group], Chin. Phys. **C38**, 090001 (2014), R. Samanta, P. Roy, A. Ghosal, Eur. Phys. J. C **76**(12), 662 (2016)
2. M.C. Gonzalez-Garcia, M. Maltoni, T. Schwetz, Nucl. Phys. B **908**, 199 (2016)
3. C.S. Lam, Phys. Lett. B **656**, 193 (2007); Phys. Rev. Lett. **101**, 121602 (2008); Phys. Rev. D **78**, 073015 (2008)
4. M. Fukugita, T. Yanagida, Phys. Lett. B **174**, 45 (1986)
5. A. Pilaftsis, T.E.J. Underwood, Nucl. Phys. B **692**, 392 (2004)
6. A. Abada, S. Davidson, A. Ibarra, F.-X. Josse-Michaux, M. Losada, A. Riotto, JHEP **0609**, 010 (2006)
7. R. Samanta, M. Chakraborty [upcoming]

Chapter 44
Search for Excited Quark Resonances in the Photon+Jet Invariant Mass Spectrum at 13 TeV

Varun Sharma

44.1 Introduction

This document reports a search for compositeness of quarks using the data collected by the CMS experiment [1] at the CERN LHC. The standard model, despite its remarkable success, has not been able to address some of the key questions of the particle physics. The increase in collision energy at the LHC and improvement in detector technology make it possible to search for extensions of the standard model. Several of these extensions suggest quark substructure [2].

The study is focussed on the search for an excited state of a quark (q^*) in the photon + jet final state. The analysis looks for the signal process $qg \rightarrow q^* \rightarrow q\gamma$. The q^* signal model considered includes only the first generation of quarks ($q^* \in u^*, d^*$) with spin-$\frac{1}{2}$. In the model, it is assumed that the compositeness scale is equal to the mass of the excited quark, i.e., $\Lambda = \mathrm{M}_{q^*}$ and, also, below the center-of-mass energy of the collision. The major standard model backgrounds for this final state include $qg \rightarrow q\gamma$, $q\bar{q} \rightarrow g\gamma$, QCD multi-jet, and W/Z+γ.

The data set used in the analysis has been collected by the CMS experiment at $\sqrt{s} = 13\,\mathrm{TeV}$ in 2015 and corresponds to an integrated luminosity of $2.7\,\mathrm{fb}^{-1}$. A detailed description of the CMS detector, together with a definition of the coordinate system used and the relevant kinematic variables, can be found in [1].

Varun Sharma on behalf of the CMS Collaboration.

V. Sharma (✉)
Department of Physics and Astrophysics, University of Delhi, Delhi, India
e-mail: varun.sharma@cern.ch

Md. Naimuddin (ed.), *XXII DAE High Energy Physics Symposium*, Springer Proceedings in Physics 203, https://doi.org/10.1007/978-3-319-73171-1_44

44.2 Event Selection

We select events that pass the online trigger with a threshold of 165 GeV on the transverse momentum (p_T) of the photon. Photons in the event are required to have $p_T > 190$ GeV and pseudorapidity $|\eta| < 1.4442$. The photon with highest p_T (leading) amongst the ones, which satisfy the identification and isolation requirements [3], is used as the photon candidate in the event. The possible jet candidates are separated from the photon candidate by $\Delta R > 0.5$ (where, $\Delta R = \sqrt{\Delta\eta^2 + \Delta\phi^2}$) and satisfy stringent jet identification criteria [3]. They are also required to have a $p_T > 190$ GeV and $|\eta^{\text{jet}}| < 2.4$. If more than one photon or jet candidates exist in the event, the invariant mass for the $\gamma +$ jet system is calculated using the leading photon candidate and leading jet candidate.

Backgrounds arising from t-channel processes are suppressed by requiring $|\Delta\eta(\gamma, \text{jet})| < 1.8$. To retain a large acceptance for the signal, events with $|\Delta\phi(\gamma, \text{jet})| > 1.5$ are kept for selecting back-to-back photon and jet. To have a full kinematic acceptance, the invariant mass of the $\gamma +$ jet system is required to be $M_{\gamma,\text{jet}} > 695$ GeV. The signal selection efficiency goes from about 49% at $M_{q^*} = 1$ TeV to 58% at $M_{q^*} = 2$ TeV and 60% at $M_{q^*} = 5$ TeV.

To search for a bump in the $\gamma +$ jet spectrum, the non-resonant mass distribution is modeled with an empirical parameterization, $\frac{d\sigma}{dm} = \frac{p_0(1-m/\sqrt{s})^{p_1}}{(m/\sqrt{s})^{p_2+p_3\ln(m/\sqrt{s})}}$, where p_0, p_1, p_2, and p_3 are the four parameters used to describe the background shape. The invariant mass distribution for data and Monte Carlo (MC) predictions is shown in Fig. 44.1. The event with the highest invariant mass is observed at 3.29 TeV, whose three-dimensional view is presented in Fig. 44.2. Various sources of systematic uncertainties affecting signal are listed in Table 44.1. For the background shape uncertainty, the background parameters are marginalized with a flat prior.

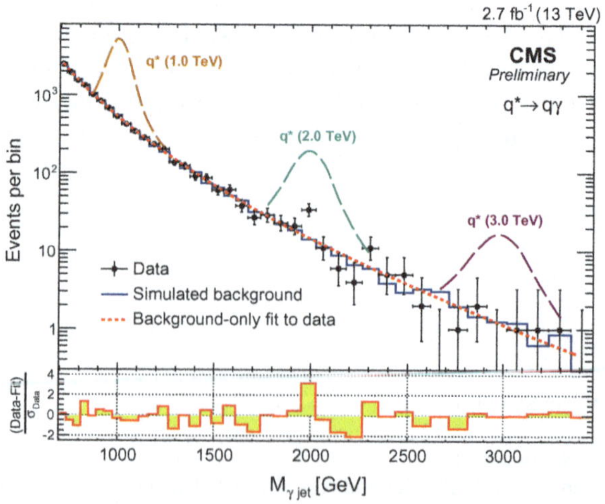

Fig. 44.1 Invariant mass distribution of $\gamma +$ jet in data and MC simulated events after final event selection. The dotted curve shows the result of the background-only fit to the data

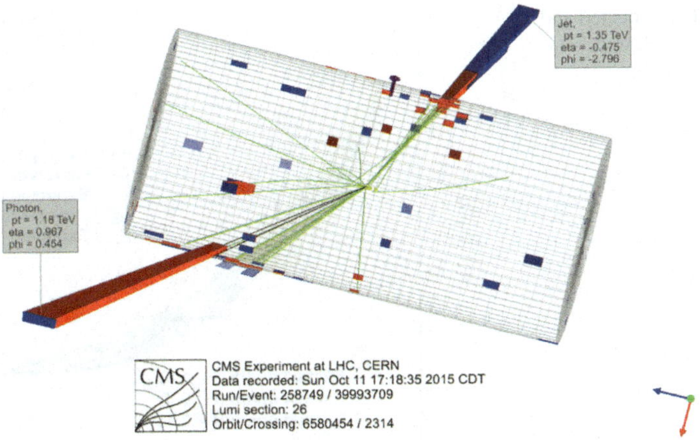

Fig. 44.2 Three-dimensional view of the highest invariant mass γ + jet candidate observed in data. The p_T, η, and ϕ values of the photon and jet are indicated

Table 44.1 Major sources of systematic uncertainty affecting the q* signal

Source	Uncertainty (%)
Jet energy resolution	10.0
Photon identification and isolation	2.0–4.0
Luminosity	2.7
Photon energy scale	1.0
Jet energy scale	0.5–0.8
Photon energy resolution	0.5

44.3 Results

A search for q^* in the γ + jet final state has been performed, using 2.7 fb^{-1} of pp collision data collected with the CMS experiment at $\sqrt{s} = 13$ TeV. A 95% confidence level (CL) upper limit is placed on $\sigma \times \mathcal{B}$ for q^* production using Bayesian formalism [4] as shown in Fig. 44.3.

Excited quark states with masses in range $1.0 < M_{q^*} < 4.37$ TeV are excluded at 95% CL for coupling multiplier $f = 1.0$ with 95% CL. Also, excited quarks with masses $1.0 < M_{q^*} < 3.64\,(1.36)$ TeV are excluded for $f = 0.5\,(0.1)$. Lastly, we present the excluded mass as a function of coupling strength in Fig. 44.4.

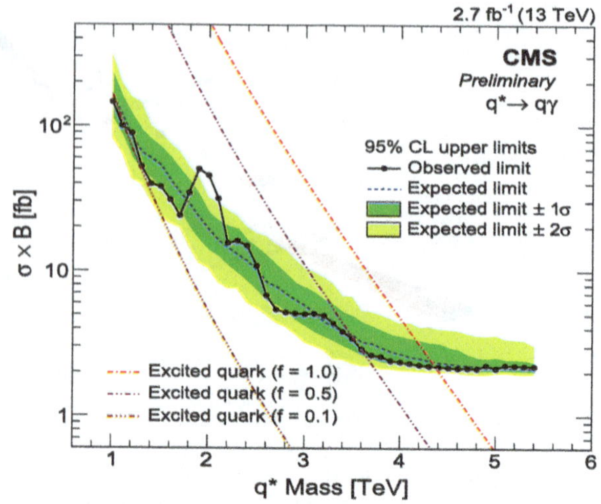

Fig. 44.3 The expected and observed 95% CL upper limits on $\sigma \times B$ for $q^* \rightarrow q\gamma$

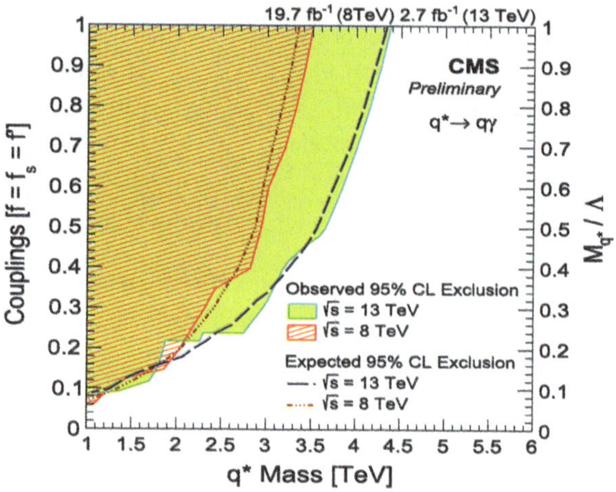

Fig. 44.4 The observed and expected excluded regions at 95% CL as a function of M_{q^*}

References

1. CMS Collaboration, The CMS experiment at the CERN LHC. J. Instrum. **3**, S08004 (2008)
2. U. Baur, I. Hinchliffe, D. Zeppenfeld, Excited quark production at hadron colliders. Int. J. Mod. Phys. A **2**, 1285 (1987)
3. CMS Collaboration, Search for excited quarks in the γ + jet final state in proton-proton collisions at $\sqrt{s} = 13$ TeV, CMS PAS EXO-16-015
4. Particle Data Group Collaboration, The review of particle physics (2017). Chin. Phys. C. **40**, 100001 (2016)

Chapter 45
Measurements of Higgs Boson Production and Properties in Di-photon Decay Channel Using Data Collected by CMS Detector at Center of Mass Energy of 13 TeV

Kuntal Mondal

45.1 Introduction

In the year 2012, during RunI at LHC, a Standard Model Higgs boson like particle was discovered by the CMS collaboration. Further studies are being made with the goal to measure the properties of the newly measured particle. Among the different decay channels, $H \rightarrow \gamma\gamma$ has a very small branching ratio (0.2%). Still the clear final state topology makes it the most sensitive channel for study of the Higgs boson. In the report we discuss the first measurements with data collected by CMS in the year 2016 during LHC RunII at center of mass energy of 13 TeV corresponding to integrated luminosity of 12.9 fb^{-1}.

45.2 Analysis Outline

The search for Higgs signal is based on looking for a fully reconstructed peak over a large continuously falling background. In order to maximise the signal sensitivity, the photon energy resolution needs to be optimised. The CMS electromagnetic calorimeter [1] response has been made uniform at single channel level, to correct for response variation with time and position. A multivariate approach has been used to correct per event per photon energy and estimate the energy resolution. A final energy scale and resolution correction is done by comparing data and simulation for $Z \rightarrow ee$ events.

Kuntal Mondal on behalf of CMS collaboration.
Presented at XXII DAE-BRNS High Energy Physics Symposium, December 12–16, 2016 at University of Delhi.

K. Mondal (✉)
Saha Institute of Nuclear Physics, Kolkata, India
e-mail: kuntal.mondal@cern.ch

Md. Naimuddin (ed.), *XXII DAE High Energy Physics Symposium*, Springer Proceedings in Physics 203, https://doi.org/10.1007/978-3-319-73171-1_45

45.2.1 Vertex and Photon Identification

The invariant mass for the two photons depend on the angle between them. So the correct vertex identification is done by a multivariate analysis technique (MVA) using kinematic correlations of the two photons, track momentum imbalance due to the photon pair, direction of tracks for e^+e^- in case of converted photons. For the analysis a vertex assignment is taken to be correct if its longitudinal displacement with respect to true primary vertex of the event, has much less effect on mass resolution compared to the impact of photon energy resolution. The average vertex assignment efficiency is measured to be 80%. The method has been validated for $Z \rightarrow \mu\mu$ events from data and simulation.

Events are first required to pass a diphoton trigger based on transverse energy of the photons, invariant mass of the photon pair, and combined cut on isolation variables and electromagnetic energy shower shape variables. Background for this process can be classified into two parts, irreducible diphoton production and reducible photon+jet and jet+jet production (where jets have been reconstructed as photons). To identify prompt and isolated photons MVA is used by utilizing kinematic properties, energy shower shape variables and isolation variables of photons. After selecting prompt photons, a diphoton MVA is used to select diphoton objects with signal-like kinematics and good mass resolution. The diphoton MVA uses per event mass resolution estimate, kinematic variables of photon pair, photon identification MVA output value and probability of assigning the correct vertex.

45.2.2 Event Categories

The selected events are further categorized in eight different production categories based on the additional particles present in the event along with the diphoton pair, in the following order. **Associated Higgs production with a $t\bar{t}$ pair (ttH)** is defined in two categories; when both W bosons coming from t, decay hadronically (labeled as *ttH Hadronic Tag*) and when atleast one W boson decays leptonically (labeled as *ttH Leptonic Tag*). **Vector boson fusion (VBF)**: MVA has been used to select events with two jets along with diphoton pair having characteristic kinematic properties of a VBF process. Two subcategories labeled as *VBF 0* and *1* have been defined in descending order of VBF MVA output value. **Gluon Gluon fusion (ggH)**: All remaining events correspond to ggH production mode and are further categorised as *Untagged 0, 1, 2, 3* in descending order of diphoton MVA output value.

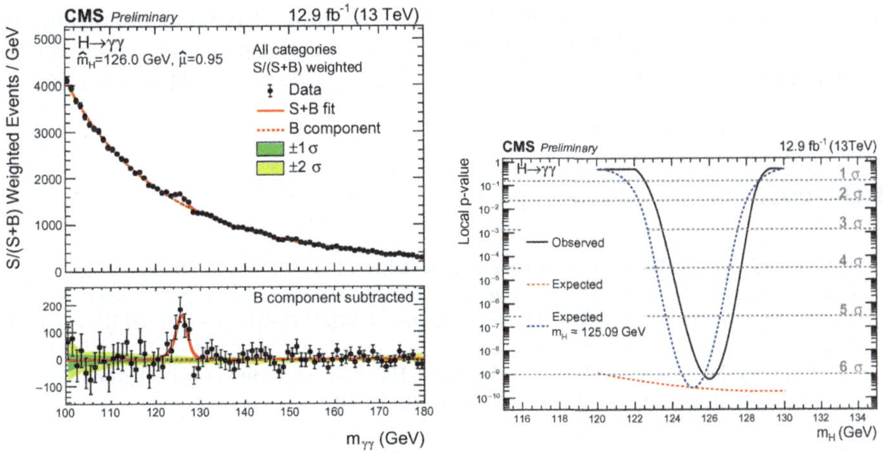

Fig. 45.1 Left: Diphoton mass distribution combined for all categories weighted to their respective signal over signal-plus-background ratio. Right: Local p-value observed (black line), expected local p-value for Higgs mass 125.09 GeV (blue dashed line), minimum expected local p-value as function of mass (red dashed line) [2]

45.2.3 Signal and Background Models

The signal model is extracted from simulation as a combination of several gaussian functions taking into account different correction and scale factors and is fitted for several mass points in the mass range of 120–130 GeV for each different category. The background models are completely data driven where a nuisance parameter is set to vary over a set of possible functional forms. This technique is discussed in detail in [3].

45.3 Results

Figure 45.1 shows the sum of events combining all categories weighted to their expected signal to signal-plus-background ratio. The maximum observed significance is 6.1σ with a best fit at mass 126.0 GeV, with a signal strength relative to standard model of $0.95 \pm 0.20 = 0.95 \pm 0.17$ (stat) $^{+0.10}_{-0.07}$ (syst) $^{+0.08}_{-0.05}$ (theo) [2].

45.4 Summary

This report summarizes the observation of the Higgs boson in diphoton decay channel with 12.9fb^{-1} data collected by CMS detector in $p-p$ collision at LHC at center of mass energy of 13 TeV in the year 2016. The signal has a maximum significance of 6.1σ at a mass of 126.0 GeV. The signal strength has also been measured individually for ttH, VBF, ggH production mode which are compatible with SM predictions.

References

1. S. Chatrchyan et al., The CMS experiment at the CERN LHC. JINST **3**, S08004 (2008)
2. *Updated measurements of Higgs boson production in the diphoton decay channel at* $\sqrt{s} = 13 TeV$ *in pp collisions at CMS*. Technical Report. CMS-PAS-HIG-16-020, CERN, Geneva, 2016
3. P.D. Dauncey, M. Kenzie, N. Wardle, G.J. Davies, Handling uncertainties in background shapes: the discrete profiling method. JINST **10**(04), P04015 (2015)

Chapter 46
Search for Dark Matter and Large Extra Dimensions in the Photon + MET Final State in pp Collisions at $\sqrt{s} = 13$ TeV

Ashim Roy

46.1 Introduction

A search in pp collisions at the Large Hadron Collider (LHC) for a final state consisting of a photon (γ) of large transverse momentum (p_T^γ) and missing transverse momentum (E_T^{miss}) is performed to investigate two extensions to the standard model (SM). We search for dark matter (DM), which is considered to be the dominant non-baryonic contribution to the matter density of the universe. However, its detection and identification in ground-based and spaceborne experiments remain elusive. At LHC, production of DM particles may be inferred from pp collisions with large E_T^{miss}, if the DM particles indirectly interact with the SM quarks and gluons via forces of the electroweak scale.

The second SM extension is the large extra dimension model which is motivated by the hierarchy problem, namely the large gap between the electroweak (M_{EW}) and Planck (M_{Pl}) scales. This model postulates that there exists n compactified extra dimensions, in which gravitons can propagate freely, and that the true scale of the gravitational interaction in this $4 + n$ dimension spacetime (M_D) is of the same order as M_{EW}. Production of such gravitons at the LHC will therefore also manifest itself as events with broad distribution in E_T^{miss}.

Events with large missing transverse momentum exist only if there are visible objects recoiling against the invisible particles. Among the many possibilities, a recoiling photon (γ) has the advantage of being identifiable with high efficiency and purity. The results of the search are interpreted in terms of the DM simplified models as proposed by the CMS-ATLAS Dark matter forum group and the ADD graviton production.

Ashim Roy on behalf of the CMS Collaboration.

A. Roy (✉)
Saha Institute of Nuclear Physics, HBNI, Kolkata 700064, India
e-mail: ashim.roy@saha.ac.in

Md. Naimuddin (ed.), *XXII DAE High Energy Physics Symposium*, Springer Proceedings in Physics 203, https://doi.org/10.1007/978-3-319-73171-1_46

46.2 Event Selection and Background Estimation

The analyzed data sample corresponds to an integrated luminosity of 12.9 fb^{-1} and is collected by CMS detector [1]. Events are required to have $E_T^{miss} > 170$ GeV and at least one photon with $p_T^{\gamma} > 175$ GeV in the central region ($|\eta| < 1.44$) of the detector and vetoed with well-identified electrons and muons. Events are rejected if the minimum azimuthal opening angle ($\Delta\phi$) between E_T^{miss} and up to four leading jets ($p_T > 30$ GeV and $|\eta| < 5$) is less than 0.5 rad. The candidate photon and E_T^{miss} must also be separated by more than 2 rad [2].

The major background from the $Z(\rightarrow \nu\overline{\nu}) + \gamma$ and $W(\rightarrow l\nu) + \gamma$ processes is estimated using simulated events with NNLO QCD and NLO EWK corrections and cross-checked with control data dominated by well-reconstructed $Z(\rightarrow l^+l^-) + \gamma$ and $W(\rightarrow l\nu) + \gamma$ events. Background from jets or electrons mis-identified as photons is estimated by measuring the mis-identification rates in control samples in data. Finally, non-collision backgrounds due to beam halo and ECAL spikes are estimated by fits to distributions of the azimuthal angle (ϕ) and the electromagnetic cluster seed time.

46.3 Results and Interpretation

A total of 400 events are observed in data, which is in agreement with the total expected background of 414.6 ± 38 events. Because no excess with respect to the SM prediction is observed, limits are set on the considered DM production models and ADD extra dimension scenarios.

The DM simplified models are designed to facilitate the comparison and translation of various DM search results by limiting the degrees of freedom of the DM production interaction. In the models considered in this analysis, Dirac DM particles couple to a vector or axial-vector mediator, which in turn has couplings to the SM quarks. Model points are identified by a set of four parameters: the DM mass m_{DM}, the mediator mass M_{med}, the universal mediator coupling to quarks g_q and the mediator coupling to DM g_{DM}. In this analysis, we fix the values of g_q and g_{DM} to 0.25 and 1 respectively [2], and scan the M_{med}–m_{DM} plane.

Figure 46.1 shows the cross section upper limits with respect to the corresponding theoretical cross section ($\mu = \sigma^{95\%}/\sigma_{Theory}$) for the vector and axial-vector mediator scenarios on the M_{med}–m_{DM} plane. The uncertainty on the expected upper limit includes the experimental uncertainties. The uncertainty in the theoretical cross section is translated to the uncertainty in the observed exclusion contour. For the simplified DM models considered, mediator masses of up to 760 GeV are excluded for small m_{DM} [2].

Figure 46.2 shows the upper limit and the theoretical calculation of ADD graviton production cross section for $n = 3$ as a function of M_D. Lower limits on M_D in different number of extra dimensions are compared to CMS Run1 results in Fig. 46.3.

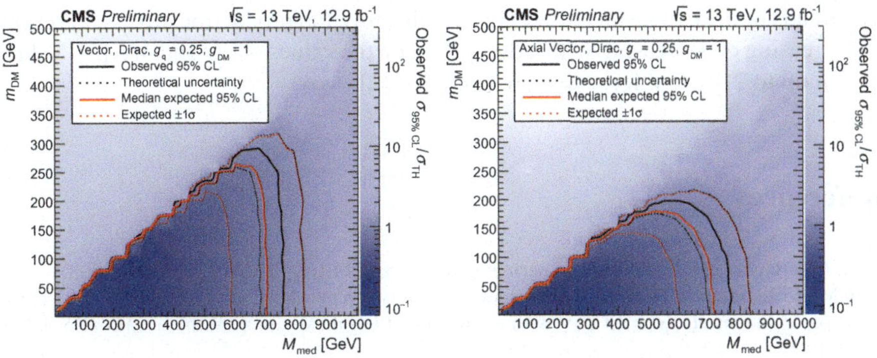

Fig. 46.1 95% CL upper limits on $\mu = \sigma/\sigma_{Theory}$ in the M_{med}–m_{DM} plane for vector and axial-vector mediator, assuming $g_q = 0.25$ and $g_{DM} = 1$. Expected and observed exclusion contours are overlaid, where mass points to the lower left of the curves are excluded [2]

Fig. 46.2 The 95% CL upper limits on the ADD graviton production cross section as a function of M_D for $n = 3$ [2]

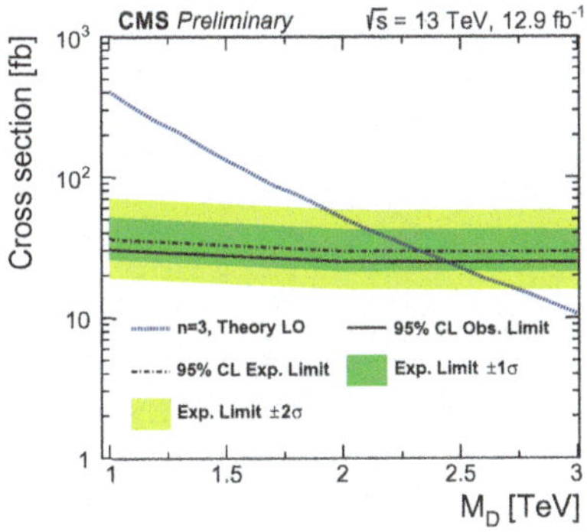

Fig. 46.3 Lower limit on M_D as a function of n [2]

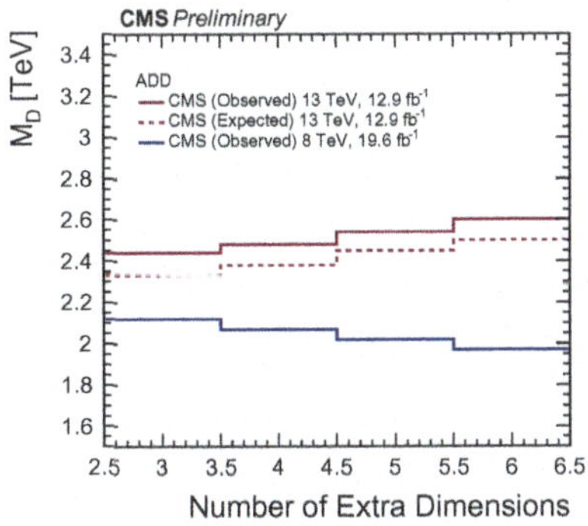

The trends of the two results differ because the graviton production cross section can be increasing or decreasing with n depending on the values of $\sqrt{s}$ and M_D [3]. Values of M_D up to 2.60 TeV for $n = 6$ are excluded [2].

References

1. S. Chatrchyan et al., The CMS experiment at the CERN LHC. JINST **3**, S08004 (2008)
2. CMS Collaboration, Search for dark matter and graviton produced in association with a photon in pp collisions at $\sqrt{s} = 13$ TeV with an integrated luminosity of 12.9 fb^{-1}. *CMS PAS*, EXO-16-039
3. Gian F. Giudice, Riccardo Rattazzi, James D. Wells, Quantum gravity and extra dimensions at high-energy colliders. Nucl. Phys. B **544**, 3–38 (1999)

Chapter 47
Prompt Muon Contribution at High Energies

Sharda Pandey, Satendra Kumar Chauhan, Jyotsna Singh and R. B. Singh

47.1 Introduction

Cosmic ray particles continuously interact with the atmospheric nuclei of the earth. It mainly consists of about 90% protons, 9% alpha particles and rest 1% heavy nuclei. The spectrum of the nature $\frac{dN}{dE} \approx E^{-\gamma}$, with spectral index γ. It is believed that some unknown astrophysical concepts are responsible for change in the spectral index of cosmic ray spectrum [1, 2]. Here, we have focused on creation of new heavy particles at knee energy and used pair-meter technique for the energy measurement of muon of better resolution.

47.2 Theoretical Calculation of Muon Flux

The transport equation for muon at underground includes only source term-

$$\frac{\partial \Phi_\mu(E,h)}{\partial h} = -\frac{\Phi_\mu(E,h)}{c\tau} + \sum_i G^i_\mu(E,h)(i \to \mu X) \tag{47.1}$$

where $\Phi_\mu(E,h)$ is vertical differential energy spectrum of muons at a depth h, and G^i_μ is the production function of muon in the atmosphere [3–6] (Table 47.1).

S. Pandey (✉) · S. K. Chauhan · J. Singh · R. B. Singh
Department of Physics, University of Lucknow, Lucknow 226007, India
e-mail: sharda2012536@gmail.com

S. K. Chauhan
e-mail: satenluck@gmail.com

J. Singh
e-mail: jyo2210@yahoo.co.in

Md. Naimuddin (ed.), *XXII DAE High Energy Physics Symposium*, Springer Proceedings in Physics 203, https://doi.org/10.1007/978-3-319-73171-1_47

Table 47.1 Table for the charmed hadrons

f	*Quark structure*	m_f	τ_f (s)	B_f (%)	E^c_f (GeV)
D^+, D^-	$c\bar{d}, \bar{c}d$	1.869	9.2×10^{-13}	18.0	4.3×10^7
$D^0, \bar{D}^0$	$c\bar{u}, \bar{c}u$	1.865	4.3×10^{-13}	7.0	9.2×10^7
λ_c^+	udc	2.281	2.3×10^{-13}	2.0	21.0×10^7
J/Ψ	$c\bar{c}$	3.097	8.0×10^{-21}	5.93	8.5×10^{15}

$$G^i_\mu(i \to \mu X) = \frac{B(i \to \mu X)}{c\tau_i}\left(1 - \frac{m^2_\mu}{m^2_i}\right)^{-1} \int_E^\infty dE_i \frac{\Phi_i(E_i)}{d_i} \tag{47.2}$$

where B is the branching ratio, τ_i is the mean life time of the particle and d_i is the interaction length; where ($i = \pi, k, D^\pm, D^0, \bar{D}^0, \lambda_c^+, J/\Psi$).

Solution has been obtained by using numerical analysis method. The differential vertical flux of Prompt muon is-

$$\frac{d\Phi_\mu}{dE_\mu} = \frac{\Phi_N(E_0)}{1 - Z_{pX}}\left[\left\{\frac{Z_{pD}(1-(r_D)^\gamma)}{1 + \frac{B_{D\mu}E_\mu}{\epsilon_D}}\right\} + \left\{\frac{Z_{p\lambda_c^+}(1-(r_{\lambda_c^+})^\gamma)}{1 + \frac{B_{\lambda_c^+\mu}E_\mu}{\epsilon_{\lambda_c^+}}}\right\} + \left\{\frac{Z_{pJ/\Psi}(1-(r_{J/\Psi})^\gamma)}{1 + \frac{B_{J/\Psi\mu}E_\mu}{\epsilon_{J/\Psi}}}\right\}\right] \tag{47.3}$$

where Φ_N is the primary spectrum of nucleon, Z_{pi} is the spectrum weight moment, γ is spectral index term varies from 2.7 to 3.1, ϵ_i is critical energy of charm particles, $r_i = \frac{m_\mu}{m_i}$; where ($i = D^+, D^-, D^0, \bar{D}^0, \lambda_c^+, J/\Psi$).

47.3 Production of Prompt Muons in Different Decay Channels

47.3.1 *Semi-leptonic Decay of D and J/ψ Meson*

Differential decay width of muon is-

$$\frac{d\Gamma}{dE_\mu} = \frac{G^2_F}{2(4\pi)^3 m_D}\int \frac{1}{4 \times 8}(E_i E_X E_l E_{\nu_l})\left[f_1(q^2)(P_i + P_X) - \frac{m^2_D}{q^2}\left(1 - \frac{m^2_X}{m^2_D}\right)\right]^2 |j_\mu|^2 . dE_X \tag{47.4}$$

where G_F is fermi constant, $f_1(q^2)$ is form factor, P is the momentum, j_μ is lepton current, q^2 is relative momentum transfer (Figs. 47.1, 47.2, 47.3, 47.4 and 47.5).

$$\Gamma = \frac{G^2_F}{(4\pi)^3} m^6_D (1 - 8r^2 + 8r^6 - r^8 - 24r^4 \ln r) \tag{47.5}$$

where $r = \frac{m_X}{m_i}$; where X is produced meson through the decay of charmed meson ($i = D^{\pm}, D^0, \bar{D^0}, J/\Psi$). From the theoretical calculation of decay width of charmed meson we get-

$$\Gamma(D^{\pm} \to K_s \mu \nu_\mu) = 4.898 \times 10^{11}\ \mathrm{s}^{-1}$$

$$\Gamma(J/\psi \to D_s \mu \nu_\mu) = 1.813 \times 10^{11}\ \mathrm{s}^{-1}.$$

47.3.2 Semi-leptonic Decay of λ_c^+ Baryon

Similarly we calculate the energy spectrum of semi-leptonic decay of $\lambda_c^+ \to \lambda_0 \mu \nu_\mu$. The differential decay rate of λ_c^+ is-

$$\frac{d\Gamma}{dE_\mu} = \frac{1}{(8\pi)^3 m_{\lambda_c^+} E_{\lambda_0}^2} E_{\lambda_0}^2 \int_{q_{max}^2}^{q_{min}^2} |\bar{M}|^2 dE \tag{47.6}$$

The matrix element is [7, 8]-

$$|\bar{M}| = \frac{G_F}{2\sqrt{2}} (\sqrt{E_{\lambda_c^+} E_{\lambda^0} E_l E_{\nu_l}}) u_{\bar{\lambda}^0} \gamma_\mu \left[\left(f_1(q^2) + \frac{m_{\lambda_0}}{m_{\lambda_c^+}} f_2(q^2) \right) - \gamma_5 \left(f_1(q^2) + \frac{m_{\lambda^0}}{m_{\lambda_c^+}} f_2(q^2) \right) \right.$$

$$\left. - \frac{i\sigma_{\mu\nu}}{(m_{\lambda_c^+})^2} (1 - \gamma^5) f_2(q^2) + v_\nu \left(\frac{1}{m_{\lambda_c^+}} f_1(q^2) - \gamma_5 \frac{1}{m_{\lambda_c^+}} f_2(q^2) \right) \right] u_{\lambda_c^+} j_\mu \tag{47.7}$$

where f_1, f_2 are the form factors, $u_{\lambda_c^+}$, $u_{\bar{\lambda}^0}$ are the spinor terms, and j_μ is leptonic current.

$$\Gamma = \frac{G_F^2}{6(4\pi)^3} m_{\lambda_c^+}^6 (1 + 6r - 8r^2 + 11r^3 + 8r - 5r^8 - 24r^3 lnr) \tag{47.8}$$

Through the theoretical calculation of decay width of λ_c^+ meson we get-

$$\Gamma(\lambda_c^+ \to \lambda_0 \mu \nu_\mu) = 3.266 \times 10^{11}\ \mathrm{s}^{-1}$$

47.4 Event Rates

The number of cascade events observed by the detector have been estimated for 1000 kton years by using pair-meter technique with the help of the given expression [4]-

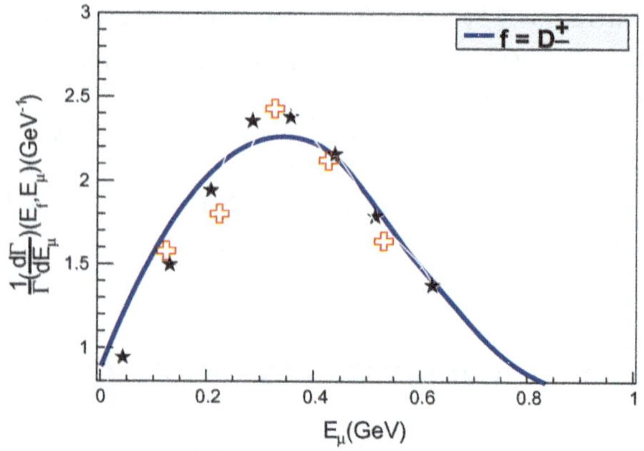

Fig. 47.1 Energy spectrum of muon from the semi-leptonic decay of $D^{\pm}$. The comparison has been made by bugev etal. (star) [3], and slac data (plus)

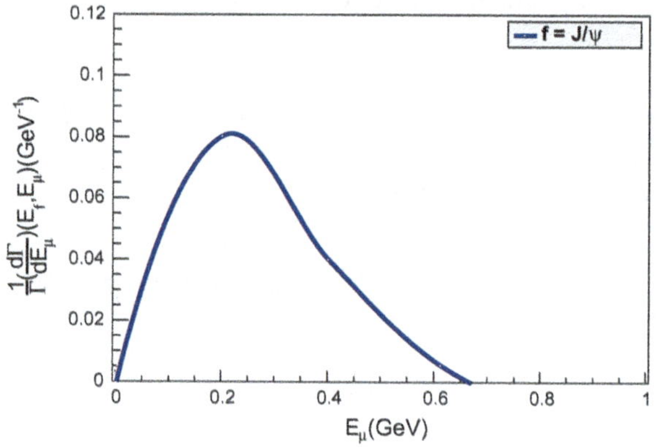

Fig. 47.2 Energy spectrum width of muon from the semi-leptonic decay of J/ψ

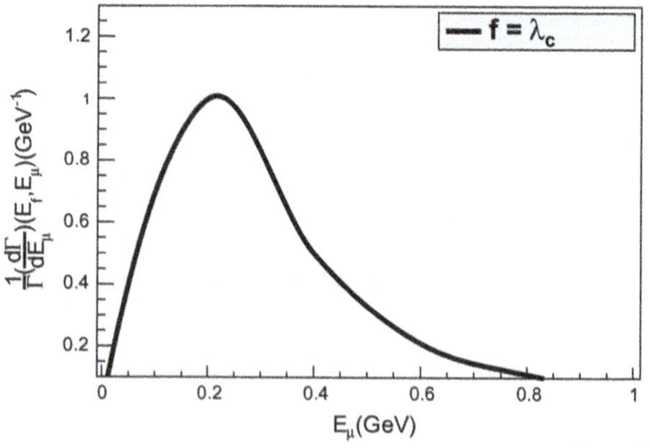

Fig. 47.3 Energy spectrum of muon from the semi-leptonic decay of λ_c^+

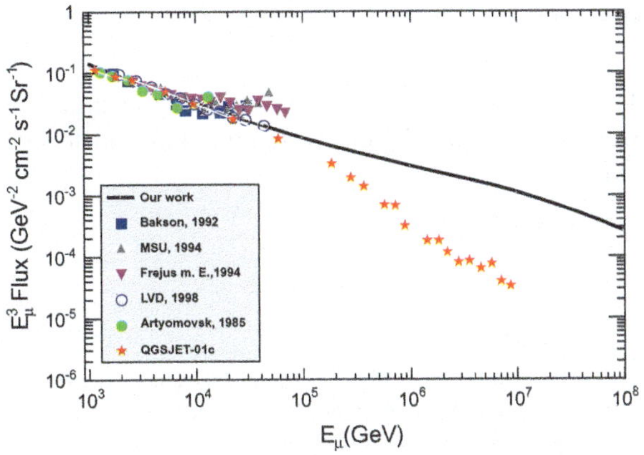

Fig. 47.4 Our theoretical muon flux has been compared with the known experimental data

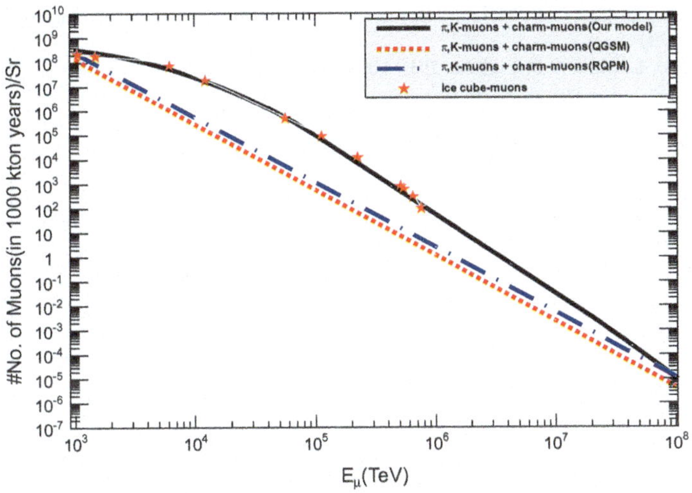

Fig. 47.5 Event rates of muon have been calculated for 1000 kton year [black], compared with the known theoretical [3] as well as experimental data (ice cube [orange])

$$Eventrate(n_\mu) = \int_{E_{th}}^{\infty} (\sigma_i(E, E_0) \times flux \times \rho \times A \times T \times t) dE \qquad (47.9)$$

where A = $4.87 \times 10^6\,\text{cm}^2$, $\rho = 7.87\,\text{g/cm}^2$, $T = 1500$ r.l.; T is the thickness of the target in radiation length (r.l.), $t = 10$ years.

47.5 Conclusions

This work has been attempted to study the vertical flux of high energy cosmic ray muons at sea level. Our theoretically calculated flux fairly matches with the known experimental data.

References

1. T.K. Gaisser, *Cosmic Rays and Particle Physics* (Cambridge University Press, Cambridge, 1990)
2. T. Stanev, *High Energy Cosmic Rays* (Springer-Praxis, Berlin, 2003)
3. E.V. Bugaev et al., Phys. Rev. D **58**, 054001 (1998)
4. R. Gandhi, P. Sukanta, 31 august 2006 arXiv:hep-ph/0512179
5. P. Lipari, Astropart. Phys. **1**, 195 (1993)
6. M. Thunman, G. Ingelman, P. Gondolo, Astropart. Phys. **5**, 309 (1995)
7. N. Cabibbo, G. Corbo, L. Maiani, Lepton spectrum in semileptonic charm decay. Nucl. Phys. B **155** (1979)
8. N. Cabibbo, L. Maiani, Phys. Lett. **79B**, 109 (1978)

Chapter 48
Performance of the CMS 2S p_T Module Prototype Using CBC2 Readout at Beam Tests

Suvankar Roy Chowdhury

48.1 Introduction

Beyond 2026, the planned upgrade of the LHC to the HL-LHC will result in an increase in instantaneous luminosity in p-p collisions reaching values up to $5-7.5 \times 10^{34}\,\mathrm{cm}^{-2}\,\mathrm{s}^{-1}$, leading to an average number of collisions per bunch crossing between 140 and 200. The current CMS tracker will be damaged by the radiation received during the Run 1 and Run 2 operations of the LHC and will not be able to run efficiently under the HL-LHC conditions. The CMS collaboration foresees to install a new radiation hard tracker which will be able to provide tracking information to the L1 trigger system to keep the Level-1 trigger rates within sustainable limits [1].

48.2 Future Tracker and the Concept of the p_T Module

The sketch of one quarter of the proposed CMS tracker is shown in Fig. 48.1a. The outer tracker will be equipped with p_T modules where each module will be made out of two silicon sensors separated by some distance (1.6–4 mm). The radial region between 200 and 600 mm will be equipped with modules with a macro-pixel sensor on one side and a strip sensor on the other side (**PS** p_T module) while the region above 600 mm will be populated with modules having strip sensors on both sides (**2S** p_T module). In the presence of a magnetic field of 3.8 T, the final state charged particles bend in a plane transverse to the direction of the beam. The radius of curvature of the track of a particle depends on the transverse momentum (p_T) of

Suvankar Roy Chowdhury on behalf of the CMS collaboration.

S. Roy Chowdhury (✉)
Saha Institute of Nuclear Physics, Kolkata, India
e-mail: suvankar.roy.chowdhury@cern.ch

Md. Naimuddin (ed.), *XXII DAE High Energy Physics Symposium*, Springer Proceedings in Physics 203, https://doi.org/10.1007/978-3-319-73171-1_48

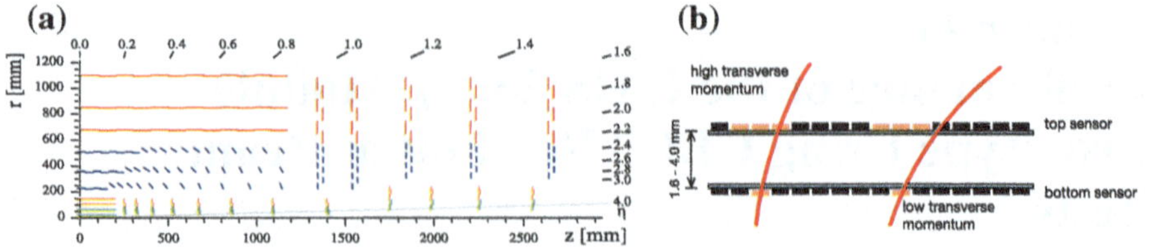

Fig. 48.1 The proposed CMS tracker layout in r-z view (**a**). The green and yellow lines correspond to inner pixel modules. The blue lines correspond to PS modules, while red lines correspond to 2S modules. Concept of a p_T module (**b**)

the particle. The p_T discrimination principle of a module is shown in Fig. 48.1b. As a charged particle passes through the module, it creates hits in the bottom and top sensors within the module. A hit in the bottom sensor is matched to coincident hits in the top sensor and if they are within a predefined window the two hits are combined to form a short track segment, a **stub**. The stubs are sent to the Level-1 trigger at bunch crossing frequency. The front end readout chips of the modules, capable of reading out both the bottom and top sensors, will provide the logic to discriminate between high and low p_T tracks. For the 2S p_T module, the readout chip will be the CMS Binary Chip (CBC) [2, 3]. In this proceeding, the performance of modules with version 2 of the CBC chips will be discussed.

48.3 Test Beam Setup

Two prototype mini 2S-modules were tested using a pion beam in the test beam facility at CERN. A 2S mini-module consists of two strip sensors, each sensor comprising 254 strips of 5 cm length. The module is read out by two CBC2 chips. One of the modules with a sensor spacing of 3.05 mm was irradiated to a fluence of $6 \times 10^{14} n_{eq}/cm^2$. The non-irradiated module has a sensor spacing of 2.75 mm. The schematic of the beam test setup is shown in Fig. 48.2. The detector under test (DUT) was placed within a telescope system [4] consisting of six layers of pixel detectors used for reconstructing the path of the incident particles.

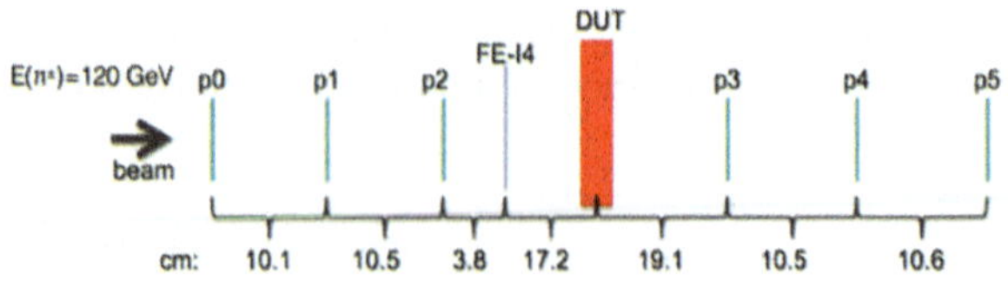

Fig. 48.2 Schematic drawing of the test beam setup; p0 to p6 refers to the telescope planes, DUT is the 2S prototype module under test

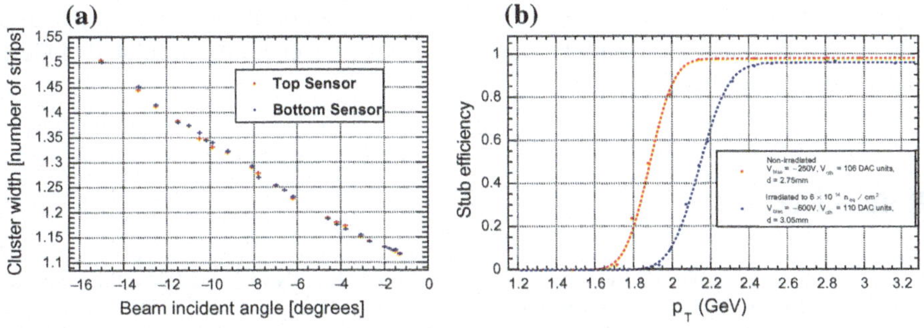

Fig. 48.3 Cluster width as a function of beam incident angle. (proposed CMS tracker layout). Stub efficiency for the irradiated (blue) and non-irradiated (red) module as a function of the particle p_T (proposed CMS tracker layout). A radius of 60 cm was used for the calculation of the p_T from the beam incident angle (Sect. 48.4)

48.4 Event Reconstruction and Results

The readout of the modules is binary and zero-suppressed. Only the number of fired strips having collected charged above a threshold is stored. The offline analysis of the data involves reconstruction of clusters (consecutive fired strips joined together) and stubs from the recorded hits in the event. To emulate the effect of the track bending inside a magnetic field, the DUT was rotated with respect to the incident beam. The variation of the mean cluster width (number of strips in a cluster) with the beam incident angle for the non-irradiated module is shown in Fig. 48.3a. As the incident angle of the particles increases, the charge deposited is shared by multiple strips and hence the cluster width increases. For the CMS field strength of $B = 3.8\,\text{T}$, the relationship between the beam incident angle (α) and the transverse momentum (p_T) of the traversing particle for a radial position of the module (R) is given by $p_T\ [\text{GeV}] \approx \frac{0.57 \cdot R[\text{m}]}{\sin(\alpha)}$. The stub efficiency, defined as the ratio of the number of events with stubs matched to a track to the number of events with a track, was measured for each incident angle. The stub efficiency as a function of p_T is shown in Fig. 48.3b. This turn-on curve is different for the two modules owing to different sensor spacing. The plot shows that the modules are efficient in selecting particle tracks above a $p_T \approx 2$ GeV with a resolution of 5.3%.

48.5 Summary

The performance of prototype 2S modules (both irradiated and non-irradiated) with the CBC2 readout chip has been studied. The cluster width increases with the incident angle as expected. The concept of p_T discrimination has been demonstrated with a p_T threshold of 2 GeV for both the irradiated and non-irradiated module. The efficiency above the p_T threshold is maintained at 95% even for the irradiated module under the beam test conditions.

References

1. S. Chatrchyan et al., The CMS experiment at the CERN LHC. JINST **3**, S08004 (2008). https://doi.org/10.1088/1748-0221/3/08/S08004
2. M. Raymond, D. Braga et al., The CMS binary chip for microstrip tracker readout at the SLH. JINST **7**, C01033 (2012). https://doi.org/10.1088/1748-0221/7/01/C01033
3. W. Ferguson, D. Braga et al., The CBC microstrip readout chip for CMS at the high luminosity LHC. JINST **7**, C08006 (2012). https://doi.org/10.1088/1748-0221/7/08/C08006
4. I. Rubinskiy, An EUDET/AIDA pixel beam telescope for detector development. Phys. Procedia **37**, 923 (2012). https://doi.org/10.1016/j.phpro.2012.02.434

Chapter 49
Quarkonium Production and Suppression with CMS Detector at LHC

Vineet Kumar

49.1 Introduction

The main goals of heavy-ion experiments is to validate the existence and study the properties of the quark-gluon plasma (QGP). The QGP is a state of deconfined quarks and gluons predicted by quantum chromodynamics (QCD) studies to exists at high temperatures and/or high energy density [1]. One of the most striking expected signatures of QGP formation is the suppression of quarkonium states. The yields of both of the charmonium (J/ψ, ψ(2S), χ_c, etc.) and the bottomonium (Υ(1S, 2S, 3S), χ_b, etc.) families are expected to suppress in heavy ion collisions. The suppression is predicted to occur above the critical temperature of the medium (T_c) and depends on the $Q\overline{Q}$ binding energy [2] alternatively the suppression can be understood in terms of quarkonium dissociation by collisions with gluons [3, 4]. Since the Large Hadron Collider (LHC) first performed Pb+Pb collisions at $\sqrt{s_{NN}} = 2.76$ TeV, a wealth of quarkonium results have become available [5, 6]. Some of these results are presented in this writeup and their significance is discussed. The latest results from the high statistics 5.02 TeV PbPb data collected in 2015 are also given in the writeup.

49.2 CMS Detector at LHC

The central feature of CMS is a superconducting solenoid, of 6 m internal diameter, providing a field of 3.8T. Within the field volume are the silicon pixel and strip tracker, the crystal electromagnetic calorimeter (ECAL) and the brass/scintillator hadron

---

Vineet Kumar (For CMS Collaboration).

V. Kumar (✉)
Nuclear Physics Division, Bhabha Atomic Research Center, Mumbai, India
e-mail: vineet.kumar@cern.ch

Md. Naimuddin (ed.), *XXII DAE High Energy Physics Symposium*, Springer Proceedings in Physics 203, https://doi.org/10.1007/978-3-319-73171-1_49

calorimeter (HCAL). Muons are detected in gas-ionization detectors embedded in the steel return yoke. The muons are measured in the pseudorapidity window $|\eta| \leq 2.4$, with detection planes made of three technologies: Drift Tubes, Cathode Strip Chambers, and Resistive Plate Chambers. A much more detailed description of CMS can be found in [7].

49.3 Quarkonium Measurement with CMS Detector at LHC

The production yields of $\Upsilon(1S)$, $\Upsilon(2S)$, and $\Upsilon(3S)$ quarkonium states are measured through their decays into muon pairs in the CMS detector, in PbPb and pp collisions at the centre-of-mass energy per nucleon pair of 2.76 and 5.02 TeV. Figure 49.1 shows Nuclear modification factors, R_{AA}, as a function of the Υ transverse momentum, p_t (left) and rapidity, $|y|$ (right) at $\sqrt{s_{NN}} = 2.76$ TeV [8]. The R_{AA} results show a suppression of a factor of ≈ 2 and 8 for $\Upsilon(1S)$ and $\Upsilon(2S)$ states, respectively. No pronounced dependence on the Υ meson kinematics is observed.

A useful variable to compare the strength of medium effects on the quarkonium ground state (1S) and excited state (nS) in PbPb collisions is the double ratio $((N_{nS}/N_{1S})_{PbPb}/(N_{nS}/N_{1S})_{pp})$ which is the ratio of the corresponding nuclear modification factors. Figure 49.2 (left) shows the double ratio of $\Upsilon(2S)$ as a function of collision centrality measured by CMS experiment at $\sqrt{s_{NN}} = 5.02$ TeV [9]. A large relative suppression of the $\Upsilon(2S)$ compared to the $\Upsilon(1S)$ in PbPb with respect to pp is observed. Figure 49.2 (right) shows the double ratio of $\psi(2S)$, in mid rapidity, as a function of collision centrality measured by CMS experiment at $\sqrt{s_{NN}} = 5.02$

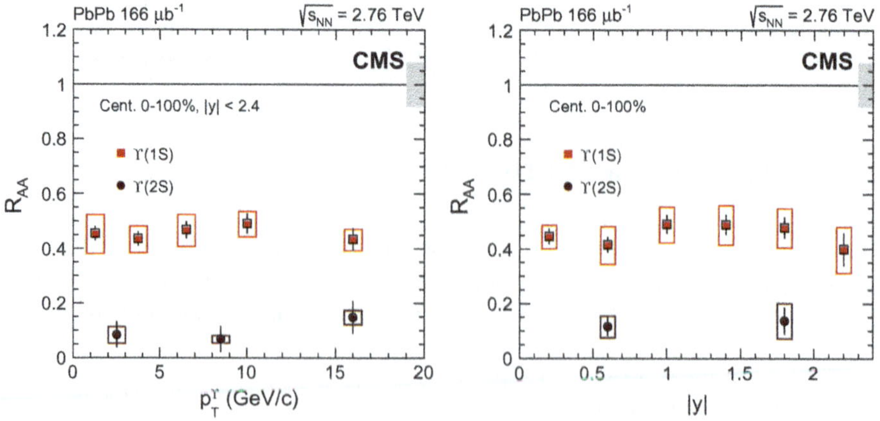

Fig. 49.1 The nuclear modification factor of $\Upsilon(1S)$ and $\Upsilon(2S)$ states in PbPb collisions at $\sqrt{s_{NN}} = 2.76$ TeV as a function of transverse momentum (left) and rapidity (right) measured by CMS experiment [8]

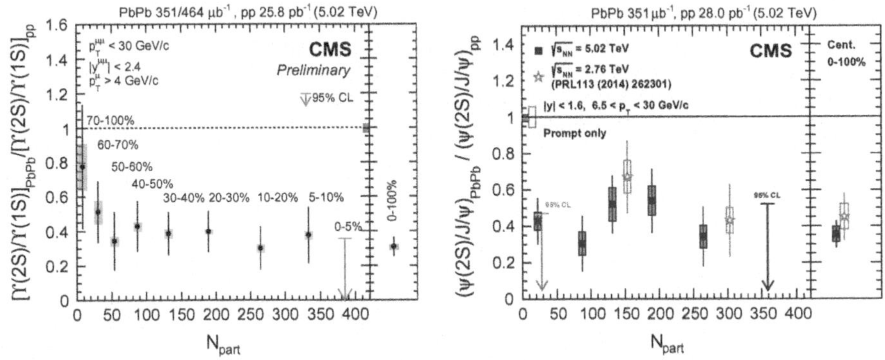

Fig. 49.2 The Double Ratio of $\Upsilon(2S)$ (left) and $\psi(2S)$ (right) as a function of collision centrality measured by CMS experiment at $\sqrt{s_{NN}} = 5.02$ TeV [9, 10]. The $\psi(2S)$ results are compared with an earlier measurement by CMS at $\sqrt{s_{NN}} = 2.76$ TeV [11]

TeV [10]. The double ratio is below unity in all the collision centralities studied, suggesting that the $\psi(2S)$ yield is more suppressed than the J/ψ yield in the kinematic range explored. No strong variations are observed with collision centrality. The double ratio shows reasonable agreement with the measurement made at $\sqrt{s_{NN}} = 2.76$ TeV in all the bins indicating no strong energy dependence on the relative suppression of charmonia states.

49.4 Summary

The $\Upsilon(1S)$ and $\Upsilon(2S)$ states are suppressed in PbPb relative to pp collisions at $\sqrt{s_{NN}} = 2.76$ TeV by factors of 2 and 8, respectively. The suppression does not show any strong kinematic dependence. The relative suppression of $\Upsilon(2S)$ is measured through the double ratios at $\sqrt{s_{NN}} = 5.02$. A large relative suppression of the $\Upsilon(2S)$ compared to the $\Upsilon(1S)$ in PbPb with respect to pp is observed. In the charmonia sector the excited state $\psi(2S)$ is found to be more suppressed than the ground state J/ψ at both centre of mass energies.

References

1. H. Satz, Nucl. Phys. A **862–863**, 4 (2011)
2. T. Matsui, H. Satz, Phys. Lett. B **178**, 416 (1986)
3. X.-M. Xu, D. Kharzeev, H. Satz, X.-N. Wang, Phys. Rev. C **53**, 3051 (1996)
4. V. Kumar, P. Shukla, R. Vogt, Phys. Rev. C **92**(2), 024908 (2015)
5. B. Muller, J. Schukraft, B. Wyslouch, Ann. Rev. Nucl. Part. Sci. **62**, 361 (2012)
6. P. Shukla, [CMS Collaboration], arXiv:1405.3810 [nucl-ex]

7. S. Chatrchyan et al., [CMS Collaboration], JINST **3**, S08004 (2008)
8. V. Khachatryan et al., [CMS Collaboration], arXiv:1611.01510 [nucl-ex]
9. CMS Collaboration, CMS-PAS-HIN-16-008
10. A.M. Sirunyan et al., [CMS Collaboration], arXiv:1611.01438 [nucl-ex]
11. V. Khachatryan et al., [CMS Collaboration], Phys. Rev. Lett. **113**, 262301 (2014)

Chapter 50
Development of a High Speed Data Acquisition System for the Detectors at High Luminosity LHC

Shuaib Ahmad Khan, Jubin Mitra and Tapan K Nayak

50.1 Introduction

During the long shutdown period of LHC at CERN in 2019 and 2020, the machine will be going for a luminosity upgrade, whereby the beam luminosities will increase by manifold. This is needed in order to acquire large amount of data to access exotic physics [1]. The major effect is a large increase in data volume, which needs to be handled and recorded properly. It leads to the need for a readout scheme that is high speed, error resilient, could perform in the harsh radiation environment, reconfigurable when required and flexible enough to adapt the experimental updates, less time to completion and at an affordable cost. In this article a suitable data acquisition scheme is proposed and the features are discussed. Implementation of such a scheme with the associated high speed links and the results are discussed in a separate section. This scheme also benefits the bio-medical instrumentation, radar and satellite applications.

50.2 Readout Architecture

In this proposed scheme, readout electronics for high energy physics experiments is divided into two parts. The first one is on-detector electronics, placed in radiation zone near to the detector and operates directly on the signals received from the particle detector. The other one is known as the data processing unit (DPU), placed in the counting room far from radiation zone as shown in Fig. 50.1.

The radiation tolerant 4.8-Gbps optical Gigabit Transceivers (GBT) links developed at CERN [2] has been used as an interace for digital transformation from the

S. A. Khan (✉) · J. Mitra · T. K Nayak
Variable Energy Cyclotron Centre, HBNI, Kolkata 64, India
e-mail: sakhan@cern.ch

Md. Naimuddin (ed.), *XXII DAE High Energy Physics Symposium*, Springer Proceedings in Physics 203, https://doi.org/10.1007/978-3-319-73171-1_50

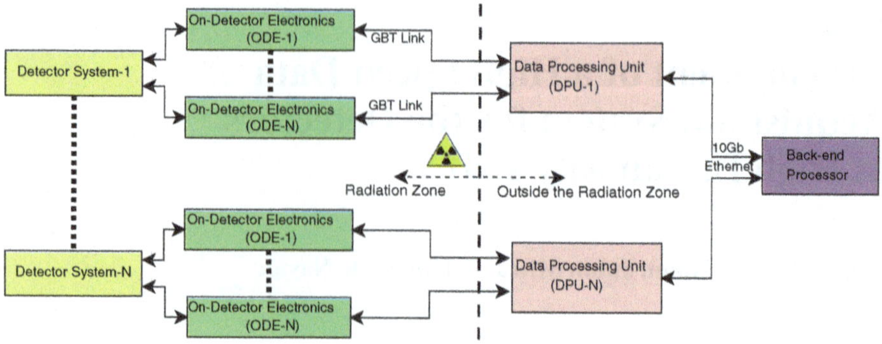

Fig. 50.1 Simplified readout architecture for high speed data acquisition system

detector electronics to the DPU. In the DPU, data are multiplexed, processed and formatted depending on the detector specifications and sent to back-end computing node using 10-Gb Ethernet protocol [3]. It increases the throughput and optimizes the system level cost. The application specific functionalities requires DPU to be implemented as custom designed electronics boards, with programmable functionality based on latest state-of-the art high processing power commercially available Field Programmable Gate Array (FPGA) technology. FPGAs are used as these are field reconfigurable with large resources, advantages like faster development time, no upfront non-recurring expenses, and convenient for prototyping designs before final ASIC production which perfectly supports an evolving design requirement.

50.3 Test Setup and Results

Two high speed links are used in this scheme; GBT-FPGA logic core [2] and the 10-Gb Ethernet protocol. Both are implemented on 28-nm Altera Stratix-V FPGA development board [4].

Testing and Performance Analysis of GBT-FPGA link: The GBT-FPGA Logic Core firmware is programmed in the Stratix-V FPGA. It mimics the GBT ASIC for receiving the GBT datagram. GBT Link is composed of a GBT transmitter (encodes, scrambles and crosses the clock domain), a GBT receiver (that aligns, decodes and descrambles the incoming data) and a Multi-Gigabit Transceiver (MGT) (serialises, transmits, receives and de-serialises the data) as shown in the Fig. 50.2a. The GBT protocol accepts data from the client interface in parallel data format and envelops it in GBT frame format to transmit in an optical link. A stable jitter free clock source was provided for proper operation of the protocol. Transceiver link signal integrity was validated by Altera transceiver tool kit. Quality of the GBT protocol operating at 4.8 Gbps is measured using a wide bandwidth high-end Lecroy serial data analyser.

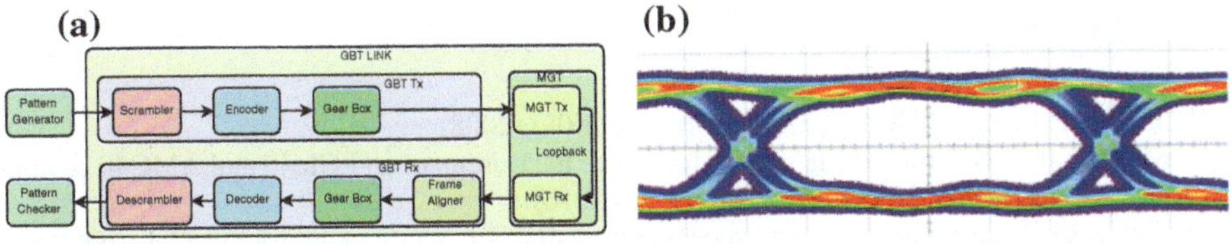

Fig. 50.2 **a** Internal architecture of GBT link **b** GBT link eye diagram at 4.8 Gbps

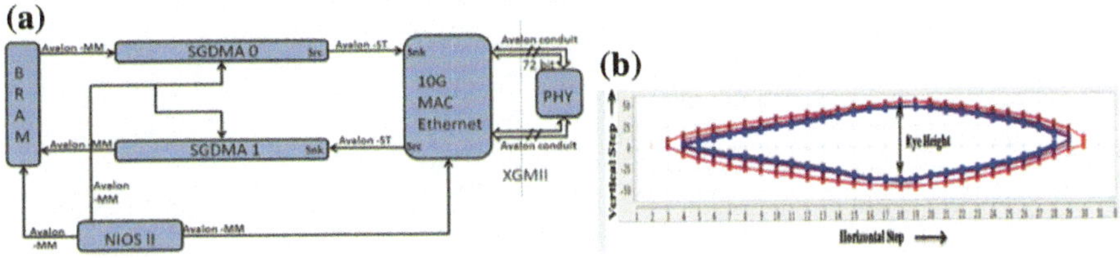

Fig. 50.3 **a** User logic for 10-Gb ethernet on FPGA **b** Eye-diagram on altera transceiver tool kit

Eye diagram is shown in Fig. 50.2b with BER of 5.525 × 10–12 and Eye width/height is 176.8ps/373 mV. This result is beneficial for further studies.

10-Gb Ethernet on FPGA testing and Analysis: The main purpose of this setup is to adapt a modular approach for the implementation of the 10-Gb Ethernet design in silicon, which is necessary for the rapid debugging of the constituent modules in case of faulty behaviour. It is a complete system on programmable chip. A NIOS-II 32 bit embedded processor [5] coordinates the design and provides overall system control. QSYS (Altera System Integration Tool) [5] was utilised with other peripheral components, as shown in Fig. 50.3a. The Scatter-Gather Direct Memory Access (SG-DMA) controller core links the transfers to non-contiguous memory and used for high-speed data transfer with minimal processor hold-up. It improves the overall system performance as compared to the DMA cores. This design includes an Ethernet Media Access controller (MAC) with 10G Base-R physical layer [5] and Avalon interface was used. The design was tested in simulation as well as on the Altera FPGA Stratix-V development board. A test data packet in the Ethernet frame format of the IEEE 802.3ae standard was developed for the simulation test. The design was successfully tested on FPGA in the internal transceiver loopback mode as shown in the Fig. 50.3b.

50.4 Conclusion

We have discussed an innovative readout architecture with optical links for high energy physics experiments. This design has been implemented on Altera Stratix-V FPGA development board. Testing and performance analysis in terms of signal integrity and functional tests are presented.

References

1. L. Evans, P. Bryant, LHC machine. J. Instrum. **3**(08) (2008)
2. J. Mitra et al., GBT link testing and performance measurement on PCIe40 and AMC40 custom design FPGA boards. JINST **11**(03), C03039 (2016)
3. S. Kolos, R. Hauser, The new inter process communication middle-ware for the atlas trigger and data acquisition system, Technical Report (2016)
4. Altera Corporations, *Stratix V Device Overview* (2016)
5. Altera Corporations, *Introduction to Intel FPGA IP Cores* (2017)

Chapter 51
Non-zero U_{e3} in the Presence of eV Scale Sterile Neutrino

S. Dev, Radha Raman Gautam and Desh Raj

51.1 Introduction

Recent anomalies at short baselines hint towards the existence of one or more sterile neutrinos at the eV scale or even higher. The evidence for $\overline{\nu}_\mu \rightarrow \overline{\nu}_e$ appearance in the LSND experiment [1], subsequently, confirmed by the MiniBooNE experiment [2] along with reactor $\overline{\nu}_e$ fluxes [3] is compatible with one or more extra sterile neutrinos at the eV scale. The hints for the presence of sterile neutrinos come, also, from Big Bang nucleosynthesis (BBN) and the structure of the universe. However, adding light sterile species would result in tension with the cosmological bounds. This conflict can be resolved if eV scale sterile neutrinos are partially thermalized before BBN era but equilibrate with active neutrinos at a later time. Harrison, Perkins and Scott first showed that experimentally obtained mixing matrix is close to the so-called tribimaximal (TBM) mixing [4]. In the present work, we attempt to generate deviations from the TBM mixing while keeping one of the columns of TBM fixed by incorporating sterile neutrinos to the known active neutrino spectrum. We impose the additional constraint of CP-conservation to simplify the analysis. The present work allows non-trivial mixing between active and sterile neutrinos.

S. Dev
Department of Physics School of Sciences, HNBG Central University, Srinagar 246174, Uttarakhand, India
e-mail: sdev@associates.iucaa.in

R. R. Gautam · D. Raj (✉)
Department of Physics, Himachal Pradesh University, Shimla 171005, India
e-mail: raj.physics88@gmail.com

D. Raj
Department of Applied Science and Humanities, J. N. Government Engineering College, Sundernagar 175018, India

Md. Naimuddin (ed.), *XXII DAE High Energy Physics Symposium*, Springer Proceedings in Physics 203, https://doi.org/10.1007/978-3-319-73171-1_51

51.2 Analysis

We consider an eV scale sterile neutrino in addition to the three known active neutrinos and set CP violating phases to be equal to zero. The neutrino mass matrix for the $3+1$ scheme is taken to be

$$M_\nu = \begin{pmatrix} \frac{1}{3}(2\bar{m}_1+\bar{m}_2) & \frac{1}{3}(\bar{m}_2-\bar{m}_1) & \frac{1}{3}(\bar{m}_2-\bar{m}_1) & e \\ \frac{1}{3}(\bar{m}_2-\bar{m}_1) & \frac{1}{6}(\bar{m}_1+2\bar{m}_2+3\bar{m}_3) & \frac{1}{6}(\bar{m}_1+2\bar{m}_2-3\bar{m}_3) & f \\ \frac{1}{3}(\bar{m}_2-\bar{m}_1) & \frac{1}{6}(\bar{m}_1+2\bar{m}_2-3\bar{m}_3) & \frac{1}{6}(\bar{m}_1+2\bar{m}_2+3\bar{m}_3) & g \\ e & f & g & \bar{m}_s \end{pmatrix} \quad (51.1)$$

where $\bar{m}_1, \bar{m}_2, \bar{m}_3$ are the mass eigenvalues of upper 3×3 active neutrino sector which is diagonalized by TBM matrix and the elements $(e, f, g, \bar{m}_s)$ belong to the sterile sector. For the $(3+1)$ neutrino mass matrix with vanishing CP violating phases, the following parametrization for the neutrino mixing matrix is used:

$$U_{4\times 4} = R(\theta_{34})R(\theta_{24})R(\theta_{14})R(\theta_{23})R(\theta_{13})R(\theta_{12}) \quad (51.2)$$

where $R(\theta_{ij})$ matrix describes rotation in the ij^{th} plane. In the present work, TM_1/TM_2 are 4×4 neutrino mixing matrices having the first/second column the same as that of TBM. The TM_1 mixing matrix, obtained by substituting $e=\frac{f+g}{2}$ in M_ν is, ruled out for the CP conserving case. The only viable case is TM_2 which can be obtained by substituting $e=-(f+g)$ in M_ν of (51.1). The modified mass matrix M_ν can be diagonalized as $M_{dig}=U_\nu^T M_\nu U_\nu$, where the mixing matrix $U_\nu = U_{TBM}\ R(\bar{\theta}_{34})\ R(\bar{\theta}_{14})\ R(\bar{\theta}_{13})$ takes the following form:

$$U_\nu = \begin{pmatrix} \sqrt{\frac{2}{3}}\bar{c}_{14}\bar{c}_{13} & \frac{1}{\sqrt{3}} & \sqrt{\frac{2}{3}}\bar{c}_{14}\bar{s}_{13} & \sqrt{\frac{2}{3}}\bar{s}_{14} \\ \frac{\bar{c}_{34}\bar{s}_{13}+\bar{c}_{13}\bar{s}_{14}\bar{s}_{34}}{\sqrt{2}} - \frac{\bar{c}_{14}\bar{c}_{13}}{\sqrt{6}} & \frac{1}{\sqrt{3}} & -\frac{3\bar{c}_{13}\bar{c}_{34}+\bar{s}_{13}\left(\sqrt{3}\bar{c}_{14}-3\bar{s}_{14}\bar{s}_{34}\right)}{3\sqrt{2}} & -\frac{\bar{s}_{14}}{\sqrt{6}} - \frac{\bar{c}_{14}\bar{s}_{34}}{\sqrt{2}} \\ -\frac{\bar{c}_{13}(\sqrt{3}\bar{c}_{14}+3\bar{s}_{14}\bar{s}_{34})+3\bar{c}_{34}\bar{s}_{13}}{3\sqrt{2}} & \frac{1}{\sqrt{3}} & \frac{3\bar{c}_{13}\bar{c}_{34}-\bar{s}_{13}\left(\sqrt{3}\bar{c}_{14}+3\bar{s}_{14}\bar{s}_{34}\right)}{3\sqrt{2}} & \frac{\bar{c}_{14}\bar{s}_{34}}{\sqrt{2}} - \frac{\bar{s}_{14}}{\sqrt{6}} \\ \bar{s}_{13}\bar{s}_{34} - \bar{c}_{13}\bar{c}_{34}\bar{s}_{14} & 0 & -\bar{c}_{34}\bar{s}_{14}\bar{s}_{13} - \bar{c}_{13}\bar{s}_{34} & \bar{c}_{14}\bar{c}_{34} \end{pmatrix} \quad (51.3)$$

with $\bar{c}_{ij}=\cos\bar{\theta}_{ij}, \bar{s}_{ij}=\sin\bar{\theta}_{ij}$ and rotation angles $\bar{\theta}_{14}, \bar{\theta}_{13}, \bar{\theta}_{34}$ depends on the mass matrix parameters $f, g, \bar{m}_1, \bar{m}_3, \bar{m}_s$. The neutrino masses are given by [5]

$$\begin{aligned} m_1 &= \bar{m}_1 - \frac{(\bar{m}_1-\bar{m}_3)\sin\bar{\theta}_{14}\sin\bar{\theta}_{13}\sec\bar{\theta}_{34}}{\sin\bar{\theta}_{14}\sin\bar{\theta}_{13}\cos\bar{\theta}_{34}+\cos\bar{\theta}_{13}\sin\bar{\theta}_{34}}, \\ m_2 &= \bar{m}_2, \\ m_3 &= \bar{m}_1 - \frac{(\bar{m}_1-\bar{m}_3)\sin\bar{\theta}_{14}\cos\bar{\theta}_{13}\sec\bar{\theta}_{34}}{\sin\bar{\theta}_{14}\cos\bar{\theta}_{13}\cos\bar{\theta}_{34}-\sin\bar{\theta}_{13}\sin\bar{\theta}_{34}}, \\ m_4 &= \bar{m}_1 + \frac{8(\bar{m}_1-\bar{m}_s)\cos^2\bar{\theta}_{14}\cos\bar{\theta}_{34}}{16\sin\bar{\theta}_{14}\cot(2\bar{\theta}_{13})\sin^3(\bar{\theta}_{34})+\cos(2\bar{\theta}_{14})(\cos(3\bar{\theta}_{34})-9\cos\bar{\theta}_{34})+6\sin(2\bar{\theta}_{34})\sin\bar{\theta}_{34})}. \end{aligned} \quad (51.4)$$

In our numerical analysis, we use the 3σ ranges of the neutrino oscillation parameters [6, 7]. We take the upper bound on the sum of the active neutrino masses $\Sigma m_\nu < 1$ eV and the free parameters $\bar{\theta}_{14}, \bar{\theta}_{13}, \bar{\theta}_{34}, \bar{m}_1$ and $\bar{m}_3$ are generated randomly. The six neutrino mixing angles and mass eigenvalues are calculated by using (51.3) and (51.4). The available experimental constraints on neutrino mass-squared differences and mixing matrix elements are used to restrict the unknown parameters. Allowed

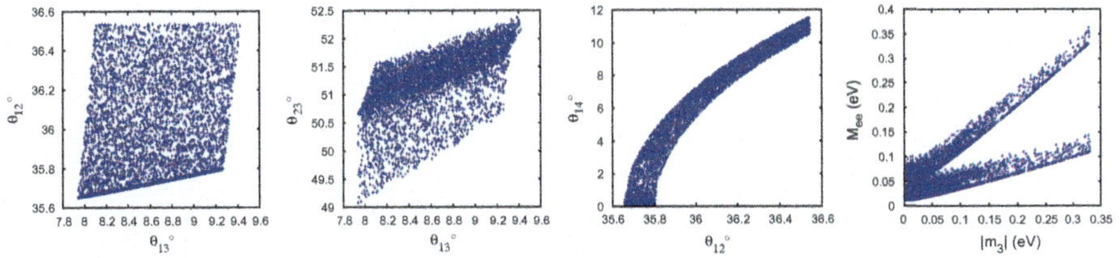

Fig. 51.1 Correlation plots between various neutrino parameters

ranges of $\bar{\theta}_{14}$, $\bar{\theta}_{13}$ and $\bar{\theta}_{34}$ come out to be (0–0.25), (0.17–0.21) and (0–0.35), respectively for both Normal (NO) and Inverted (IO) mass orderings. The ranges for other parameters $|\bar{m}_1|$, $|\bar{m}_2|$, $|\bar{m}_3|$ and $|\bar{m}_s|$ are 0–0.35 eV (0.045–0.4 eV), 0.008–0.35 eV (0.05–0.35 eV), 0–0.35 eV (0–0.42 eV) and 0.8–1.5 eV (0.8–1.5 eV) for NO (IO), respectively. The allowed range for the effective Majorana mass (M_{ee}) in our model for NO (IO) is 0–0.35 eV (0.015–0.4 eV) (Fig. 51.1).

51.3 Summary

In the present work, we have studied the phenomenological consequences of adding a light sterile neutrino to the active neutrinos. We examined the possibility of generating the necessary deviation from the TBM mixing by generating a non-zero U_{e3} from active-sterile mixing. The active-sterile neutrino mixing leads to the modification of the TBM pattern. It is found that a non-zero U_{e3} within its experimental range can be successfully generated and the atmospheric mixing angle (θ_{23}) is non-maximal in this model.

Acknowledgements R. R. G. acknowledges the financial support provided by Department of Science and Technology, Government of India under the Grant No. SB/FTP/PS-128/2013. S. D. gratefully acknowledges the financial support provided by Council for Scientific and Industrial Research, Govt. of India, New Delhi vide Grant No. 03(1333)/15/EMR-II.

References

1. L.S.N.D. Collaboration, A. Aguilar et al., Phys. Rev. D **64**, 112007 (2001)
2. MiniBooNE Collaboration, A.A. Aguilar-Arevalo et al., Phys. Rev. Lett. **105**, 181801 (2010)
3. G. Mention et al., Phys. Rev. D **83**, 073006 (2011)
4. P.F. Harrison, D.H. Perkins, W.G. Scott, Phys. Lett. B **530**, 167 (2002)
5. S. Dev, D. Raj, R.R. Gautam, Nucl. Phys. B **911**, 744–753 (2016)
6. J. Kopp, P.A.N. Machado, M. Maltoni, T. Schwetz, JHEP **1305**, 050 (2013)
7. D.V. Forero, M. Tortola, J.W.F. Valle, Phys. Rev. D **90**, 093006 (2014)

Chapter 52
Two Zeros in the Magic Neutrino Mass Matrix

Radha Raman Gautam and Sanjeev Kumar

52.1 Introduction

A non-vanishing value of θ_{13} [1] implies that the symmetries behind the tribimaximal (TBM) mixing [3] are not exact. However the mixing angles θ_{12} and θ_{23} are correctly predicted by the TBM mixing. We can view the TBM mixing as a combination of the $\mu - \tau$ symmetry [2] and the magic symmetry [4]. A non-zero θ_{13} implies that the neutrino mass matrix cannot be $\mu - \tau$ symmetric. Yet, it can still be magic. Such a mass matrix will give trimaximal (TM) mixing [5]. The mixing angles and CP-violating phase δ obey two sum-rules for the TM mixing. A more constraining approach [6] will be to combine the magic symmetry with texture zeros [7]. This approach is reviewed here in brief. The TM mixing matrix is

$$U_{TM} = \begin{pmatrix} \sqrt{\frac{2}{3}}\cos\theta & \frac{1}{\sqrt{3}} & \sqrt{\frac{2}{3}}\sin\theta \\ -\frac{\cos\theta}{\sqrt{6}} + \frac{e^{-i\phi}\sin\theta}{\sqrt{2}} & \frac{1}{\sqrt{3}} & -\frac{\sin\theta}{\sqrt{6}} - \frac{e^{-i\phi}\cos\theta}{\sqrt{2}} \\ -\frac{\cos\theta}{\sqrt{6}} - \frac{e^{-i\phi}\sin\theta}{\sqrt{2}} & \frac{1}{\sqrt{3}} & -\frac{\sin\theta}{\sqrt{6}} + \frac{e^{-i\phi}\cos\theta}{\sqrt{2}} \end{pmatrix}. \tag{52.1}$$

This mixing matrix has only two free parameters: θ and ϕ. It diagonalizes the magic neutrino mass matrix. So, we can parametrize a magic neutrino mass matrix as

$$M_{\text{magic}} = U^*_{\text{TM}} M_{\text{d}} U^\dagger_{\text{TM}}. \tag{52.2}$$

R. R. Gautam (✉)
Department of Physics, Himachal Pradesh University, Shimla 171005, India
e-mail: gautamrrg@gmail.com

S. Kumar
Department of Physics and Astrophysics, University of Delhi, Delhi 110007, India
e-mail: skverma@physics.du.ac.in

Md. Naimuddin (ed.), *XXII DAE High Energy Physics Symposium*, Springer Proceedings in Physics 203, https://doi.org/10.1007/978-3-319-73171-1_52

where $M_d = \text{diag}(m_1, m_2 e^{2i\alpha}, m_3 e^{2i\beta})$. Here, (m_1, m_2, m_3) are the neutrino masses and α and β are the Majorana phases.

52.2 Two Zeros in the Magic Neutrino Mass Matrix

There are only seven textures of the neutrino mass matrix with two zeros [7] in the diagonal charged lepton basis. The allowed patterns are shown in Table 52.1. Imposing these textures on the magic mass matrix M_{magic} [(52.2)], we find that only textures A_1 and A_2 are allowed experimentally.

The predictions for the texture A_1 can be obtained by solving the equations: $M_{\nu_{11}} = 0$ and $M_{\nu_{12}} = 0$. We obtain our predictions in terms of $r \equiv \frac{\Delta m^2_{21}}{\Delta m^2_{23}}$. The main prediction is $r = \tan^2\theta$. The three masses are

$$
\begin{aligned}
m_1^2 &= \Delta m_{21}^2 \frac{1 + 3r + 2\sqrt{3}\sqrt{r}\cos\phi}{x},\\
m_2^2 &= \frac{4\Delta m_{21}^2}{x},\\
m_3^2 &= \Delta m_{21}^2 \frac{3 + r - 2\sqrt{3}\sqrt{r}\cos\phi}{x}.
\end{aligned}
\tag{52.3}
$$

where $x \equiv 3 - 3r - 2\sqrt{3}\sqrt{r}\cos\phi$. The three mixing angles are

$$
\sin^2\theta_{12} = \frac{1+r}{3+r}, \quad \sin^2\theta_{23} = \frac{1}{2} + \frac{\sqrt{3}\sqrt{r}\cos\phi}{r+3}, \quad \sin^2\theta_{13} = \frac{2r}{3(r+1)}. \tag{52.4}
$$

The phenomenological predictions for A_2 can be obtained from the predictions of A_1 by the transformations $\theta_{23} \rightarrow \frac{\pi}{2} - \theta_{23}$ and $\delta = \pi - \delta$. For detailed phenomenology see [6].

Table 52.1 Allowed neutrino mass matrices with two texture zeros classified into three classes

Type	Constraining equations
A_1	$M_{ee} = 0, M_{e\mu} = 0$
A_2	$M_{ee} = 0, M_{e\tau} = 0$
B_1	$M_{e\tau} = 0, M_{\mu\mu} = 0$
B_2	$M_{e\mu} = 0, M_{\tau\tau} = 0$
B_3	$M_{e\mu} = 0, M_{\mu\mu} = 0$
B_4	$M_{e\tau} = 0, M_{\tau\tau} = 0$
C	$M_{\mu\mu} = 0, M_{\tau\tau} = 0$

Table 52.2 Transformation properties of various fields: D_{l_L} $(D_{e_L}, D_{\mu_L}, D_{\tau_L})^T$, l_R $(e_R, \mu_R, \tau_R)^T$, ν_{l_R}, ψ $(\psi_1, \psi_2, \psi_3)^T$, $\varphi(\varphi_1, \varphi_2, \varphi_3)^T$ and Δ

Fields	D_{l_L}	l_R	ν_R	ψ	φ	Δ
$SU(2)_L$	2	1	1	2	2	3
A_4	3	$1, 1', 1''$	1	3	3	$1''$
Z_2	1	1	−1	1	−1	1

The textures A_1 and A_2 can be realized using the non-Abelian discrete symmetry A_4. If various leptonic and scalar fields have the transformation properties as given in Table 52.2, then these transformations lead to neutrino mass matrices of type A_1 provided the various Higgs fields develop following vacuum expectation values: $\langle\psi\rangle_o = v_\psi(1, 1, 1)^T$ and $\langle\varphi\rangle_o = v_\varphi(0, -1, 1)^T$.

52.3 Conclusions

We investigate the phenomenological consequences of combining the magic symmetry with two texture zeros. We find that there are only two such textures that are compatible with the experimental data. We express all the three neutrino mixing angles, three neutrino masses and the three CP violating phases in terms of two known mass squared differences Δm^2_{21} and Δm^2_{23} and the phase ϕ. We also present a symmetry realization for these textures using the non-Abelian A_4 symmetry.

Acknowledgements R. R. G. acknowledges the financial support provided by the Department of Science and Technology, Government of India under the Grant No. SB/FTP/PS-128/2013.

References

1. K. Abe et al., [T2K Collaboration], Phys. Rev. Lett. **107**, 041801 (2011)
2. T. Fukuyama, H. Nishiura, arXiv:hep-ph/9702253
3. P.F. Harrison, D.H. Perkins, W.G. Scott, Phys. Lett. B **530**, 167 (2002)
4. C.S. Lam, Phys. Lett. B **640**, 260 (2006), arXiv:hep-ph/0606220
5. X.G. He, A. Zee, Phys. Lett. B **645**, 427 (2007), arXiv:hep-ph/0607163
6. R.R. Gautam, S. Kumar, Phys. Rev. D **94**(3), 036004 (2016), arXiv:1607.08328 [hep-ph]
7. P.H. Frampton, S.L. Glashow, D. Marfatia, Phys. Lett. B **536**, 79 (2002)

Chapter 53
Can We Measure θ_{23} Octant in 3 + 1 Scheme?

Sanjib Kumar Agarwalla, Sabya Sachi Chatterjee and Antonio Palazzo

Introduction: The resolution of octant[1] [1] of θ_{23} is one of the fundamental problems in neutrino oscillation. Long-baseline (LBL) experiments can resolve this octant ambiguity of θ_{23} with the help of $\nu_\mu \to \nu_e$ appearance channel, and the vital information coming from $\nu_\mu \to \nu_\mu$ disappearance channel also play an important role. Interestingly, at present, there are short-baseline anomalies which hint towards the existence of light eV-scale sterile neutrino [2]. Here, we expound in detail the capability of proposed LBL experiment DUNE to measure θ_{23} octant considering one light eV-scale sterile neutrino along with three active neutrinos.

Theoretical framework: In the 3 + 1 scheme, a new a mass eigenstate ν_4 appears on top of 3ν framework whose mixing is parametrized as

$$U = \tilde{R}_{34} R_{24} \tilde{R}_{14} R_{23} \tilde{R}_{13} R_{12} \,, \tag{53.1}$$

[1] According to the present 3ν best-fit [4], θ_{23} can have two solutions: one $< \pi/4$, labelled as lower octant (LO), and other $> \pi/4$, known as higher octant (HO).

S. S. Chatterjee—Speaker.

S. K. Agarwalla · S. S. Chatterjee (✉)
Institute of Physics, Sachivalaya Marg, Sainik School Post, Bhubaneswar 751005, India
e-mail: sabya@iopb.res.in

S. K. Agarwalla · S. S. Chatterjee
Homi Bhabha National Institute, Training School Complex,
Anushakti Nagar, Mumbai 400085, India
e-mail: sanjib@iopb.res.in

A. Palazzo
Dipartimento Interateneo di Fisica "Michelangelo Merlin",
Via Amendola 173, 70126 Bari, Italy
e-mail: palazzo@ba.infn.it

A. Palazzo
Istituto Nazionale di Fisica Nucleare, Sezione di Bari, Via Orabona 4, 70126 Bari, Italy

Md. Naimuddin (ed.), *XXII DAE High Energy Physics Symposium*, Springer Proceedings in Physics 203, https://doi.org/10.1007/978-3-319-73171-1_53

where R_{ij} ($\tilde{R}_{ij}$) is a real (complex) rotation in the (i, j) plane. The details of the parametrization of U can be seen in [3].

In [3], it was shown that the 4-flavor appearance probability can be approximately expressed as the sum of three terms $P_{\mu e}^{4\nu} \simeq P_0 + P_1 + P_2$, which in vacuum appears as

$$P_0 \simeq 4 s_{23}^2 s_{13}^2 \sin^2 \Delta \,, \tag{53.2}$$

$$P_1 \simeq 8 s_{13} s_{12} c_{12} s_{23} c_{23} (\alpha\Delta) \sin \Delta \cos(\Delta \pm \delta_{13}) \,, \tag{53.3}$$

$$P_2 \simeq 4 s_{14} s_{24} s_{13} s_{23} \sin \Delta \sin(\Delta \pm \delta_{13} \mp \delta_{14}) \,. \tag{53.4}$$

where $\Delta \equiv \Delta m_{31}^2 L/4E$ and $\alpha \equiv \Delta m_{21}^2/\Delta m_{31}^2$. In the double sign, upper (lower) sign corresponds to neutrinos (antineutrinos). The new interference term P_2 is governed by the interference between the atmospheric frequency and the large frequency related to the new mass eigenstate [3] which gets averaged out by the finite energy resolution of the detector. Recent global fits [4–6] suggest $s_{13} \sim s_{14} \sim s_{24} \sim 0.15\,(\sim \epsilon)$ and $\alpha = 0.03\,(\sim \epsilon^2)$ implying $P_0 \sim \epsilon^2$, $P_1 \sim \epsilon^3$, $P_2 \sim \epsilon^3$.

An experiment can measure the octant of θ_{23} even in the presence of unknown CP-phases, if there is a difference between the probabilities corresponding to the different octants, i.e.

$$\Delta P \equiv P_{\mu e}^{\rm HO}(\delta_{13}^{\rm HO}, \delta_{14}^{\rm HO}) - P_{\mu e}^{\rm LO}(\delta_{13}^{\rm LO}, \delta_{14}^{\rm LO}) \neq 0 \,, \tag{53.5}$$

where one of the two octants should be considered to generate data and the other octant should be used to simulate the theoretical model. From the expression of $P_{\mu e}^{4\nu}$, ΔP can be written as, $\Delta P = \Delta P_0 + \Delta P_1 + \Delta P_2$. Now for further discussions regarding the mathematical expressions of each term and their importance please see [7].

Results and discussion: Simulations for DUNE have been performed considering a total 248 kt.MW.yr of exposure, divided equally between ν and $\bar{\nu}$ mode.

Figure 53.1 depicts the discovery reach of θ_{23} octant in $[\sin^2\theta_{23}, \delta_{13}]$ (true) plane assuming NH as true choice. Left (right) panel shows the results for 3ν $(3+1)$ scheme. The following true values of the oscillation parameters have been considered in simulation: $\sin^2\theta_{12} = 0.304$, $\sin^2 2\theta_{13} = 0.085$, $\Delta m_{21}^2 = 7.5\times10^{-5}\,\text{eV}^2$, $\Delta m_{31}^2 = 2.475\times10^{-3}$ (-2.4×10^{-3}) eV^2 for NH (IH), $\sin^2\theta_{14} = \sin^2\theta_{24} = 0.025$, and $\sin^2\theta_{34} = 0$. In 3ν case, we have marginalized over mass hierarchy (MH), $\sin^2\theta_{23}$, and δ_{13} in test with the restriction that θ_{23} (test) lies in opposite octant with respect to true choice of the octant of θ_{23}. Finally we calculate minimum $\Delta\chi^2$ for each true data set. The solid blue, dashed magenta and dotted black curve correspond to the 2σ, 3σ and 4σ sensitivity of discrimination of one octant from another where the sensitivity has been defined as $\Delta\chi^2 = (4, 9, \text{and } 16)$ for $(2\sigma, 3\sigma, \text{and } 4\sigma)$ respectively. We notice that a minimum 2σ sensitivity can be achieved if $\sin^2\theta_{23}(\text{true}) \lesssim 0.47$ and $\sin^2\theta_{23}(\text{true}) \gtrsim 0.55$ irrespective of the choice of δ_{13} (true). In $3+1$ case, in addition, we marginalize over δ_{14} (true) and δ_{14} (test). In this case, we hardly have any octant sensitivity in the entire $[\sin^2\theta_{23}, \delta_{13}]$ (true) plane. This is because of the

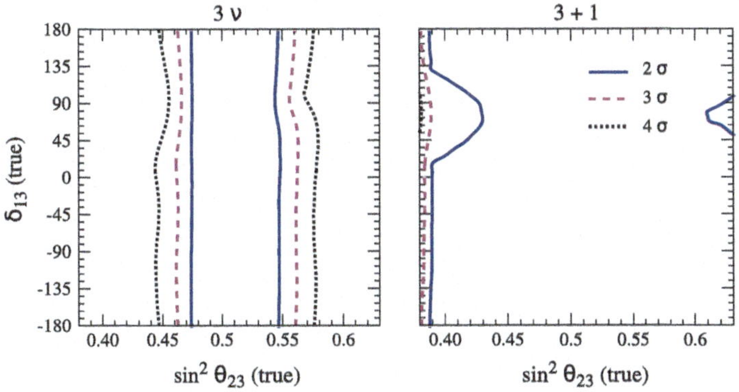

Fig. 53.1 Discovery potential for excluding the wrong octant in [$\sin^2\theta_{23}$, δ_{13}] (true) plane assuming NH as true choice. The left (right) panel corresponds to the 3ν (3 + 1) case. In 3-flavor scenario, we marginalize over (θ_{23}, δ_{13}) (test). In 3 + 1 case, in addition, we marginalize over δ_{14} (true) and δ_{14} (test) fixing $\theta_{14} = \theta_{24} = 9^0$ and $\theta_{34} = 0$. This figure has been taken from [7]

freedom of any choice of δ_{14} which comes through the new interference term P_2. For detailed discussion, see [7].

Conclusions: In this work, we have studied the impact of a light eV-scale sterile neutrino in measuring the octant of θ_{23} at DUNE. The sensitivity towards θ_{23} octant can be completely lost if there is active- sterile oscillations.

Acknowledgements S.S.C. would like to thank the organizers of XXII DAE-BRNS HEP Symposium 2016 for giving an opportunity to present this work. S.K.A. is supported by the DST/INSPIRE Research Grant [IFA-PH-12], Department of Science & Technology, India. A.P. is supported by the grant "Future In Research" *Beyond three neutrino families*, Fondo di Sviluppo e Coesione 2007–2013, APQ Ricerca Regione Puglia, Italy, Programma regionale a sostegno della specializzazione intelligente e della sostenibilità sociale ed ambientale. A.P. acknowledges partial support by the research project *TAsP* funded by the Instituto Nazionale di Fisica Nucleare (INFN).

References

1. G.L. Fogli, E. Lisi, Phys. Rev. D **54**, 3667 (1996)
2. K.N. Abazajian (UC, Irvine) et al.: *Light Sterile Neutrinos: A White Paper*, arXiv: 1204.5379
3. N. Klop, A. Palazzo, Phys. Rev. D **91**(7), 073017 (2015)
4. F. Capozzi, E. Lisi, A. Marrone, D. Montanino, A. Palazzo, Nucl. Phys. B **908**, 218 (2016)
5. C. Giunti, M. Laveder, Y.F. Li, H.W. Long, Phys. Rev. D **88**, 073008 (2013)
6. J. Kopp, P.A.N. Machado, M. Maltoni, T. Schwetz, JHEP **1305**, 050 (2013)
7. S.K. Agarwalla, S.S. Chatterjee, A. Palazzo, Phys. Rev. Lett. **118**(3), 031804 (2017)

Chapter 54
Right-Handed Currents in $B \to K^*\ell^+\ell^-$

Rusa Mandal

54.1 Introduction

The rare decay $B \to K^*\ell^+\ell^-$ is an important mode for indirect search of new physics (NP) due to the measurement of large number of observables in experiments. Using the most general parametric form of the amplitude in the standard model (SM), we probe NP in a theoretically clean approach and refer the reader to [1, 2] for detailed analysis.

54.2 New Physics Analysis

We start with the observables to be the longitudinal helicity fraction F_L and $F_\perp$, A_5, $A_{\rm FB}$, related to the LHCb measured observables S_3, S_5, $A_{\rm FB}^{\rm LHCb}$ [6] as $F_\perp = (1 - F_L + 2S_3)/2$, $A_5 = 3/4\, S_5$, $A_{\rm FB} = -A_{\rm FB}^{\rm LHCb}$. The observables are functions of transversity amplitudes and the most general parametric form in presence of RH currents, which includes all short-distance and long-distance effects, factorizable and non-factorizable contributions and resonance contributions is [1]

$$\mathcal{A}_\perp^{L,R} = \big((\widetilde{C}_9^\perp + C_9') \mp (C_{10} + C_{10}')\big)\mathcal{F}_\perp - \widetilde{\mathcal{G}}_\perp, \tag{54.1}$$

$$\mathcal{A}_{\|,0}^{L,R} = \big((\widetilde{C}_9^{\|,0} - C_9') \mp (C_{10} - C_{10}')\big)\mathcal{F}_{\|,0} - \widetilde{\mathcal{G}}_{\|,0} \tag{54.2}$$

R. Mandal—speaker.

R. Mandal (✉)
The Institute of Mathematical Sciences, Taramani, Chennai 600113, India
e-mail: rusam@imsc.res.in

Md. Naimuddin (ed.), *XXII DAE High Energy Physics Symposium*, Springer Proceedings in Physics 203, https://doi.org/10.1007/978-3-319-73171-1_54

Here C_9 and C_{10} are Wilson coefficients with $\widetilde{C}_9^{\lambda}$ being the redefined "effective" Wilson coefficient defined [1] as $\widetilde{C}_9^{\lambda} = C_9 + \Delta C_9^{(\text{fac})}(q^2) + \Delta C_9^{\lambda,(\text{non-fac})}(q^2)$, where $\Delta C_9^{(\text{fac})}(q^2)$, $\Delta C_9^{\lambda,(\text{non-fac})}(q^2)$ correspond to factorizable and soft gluon non-factorizable contributions, respectively. The C_9' and C_{10}' are the coefficients of RH current operators O_9' and O_{10}' and $\mathcal{F}_\lambda$ and $\widetilde{\mathcal{G}}_\lambda$ are the form factors for the decay mode. With the introduction of some notation: $r_\lambda = \text{Re}(\widetilde{\mathcal{G}}_\lambda)/\mathcal{F}_\lambda - \text{Re}(\widetilde{C}_9^{\lambda})$, $\xi = C_{10}'/C_{10}$, and $\xi' = C_9'/C_{10}$, we construct the following variables,

$$R_\perp = \left(\frac{r_\perp}{C_{10}} - \xi'\right) / (1+\xi)\,, \quad R_{\|,0} = \left(\frac{r_{\|,0}}{C_{10}} + \xi'\right) / (1-\xi)\,. \tag{54.3}$$

At low recoil energy of K^* meson, only three independent form factors describe the whole $B \to K^*\ell^+\ell^-$ decay and at leading order in $1/m_B$ expansion the relation among the form factors is given by [3, 4], $\widetilde{\mathcal{G}}_\|/\mathcal{F}_\| = \widetilde{\mathcal{G}}_\perp/\mathcal{F}_\perp = \widetilde{\mathcal{G}}_0/\mathcal{F}_0 = -\kappa\, 2m_b m_B C_7/q^2$, where $\kappa \approx 1$. Hence at the maximum point in q^2 i.e. the kinematic endpoint q^2_{max}, $r_0 = r_\| = r_\perp \equiv r$ (say). Therefore (54.3) implies that in the presence of RH currents $R_0 = R_\| \neq R_\perp$ is valid at $q^2 = q^2_{\text{max}}$ without any approximation and is unaltered by non-factorizable and resonance contributions [5]. To test the relation among R_λ's in light of LHCb data, first defining $\delta \equiv q^2_{\text{max}} - q^2$, we expand the observables F_L, $F_\perp$, A_{FB} and A_5 around q^2_{max} as,

$$\begin{aligned} F_L &= \frac{1}{3} + F_L^{(1)}\delta + F_L^{(2)}\delta^2 + F_L^{(3)}\delta^3, \quad F_\perp = F_\perp^{(1)}\delta + F_\perp^{(2)}\delta^2 + F_\perp^{(3)}\delta^3, \\ A_{\text{FB}} &= A_{\text{FB}}^{(1)}\delta^{\frac{1}{2}} + A_{\text{FB}}^{(2)}\delta^{\frac{3}{2}} + A_{\text{FB}}^{(3)}\delta^{\frac{5}{2}}, \quad A_5 = A_5^{(1)}\delta^{\frac{1}{2}} + A_5^{(2)}\delta^{\frac{3}{2}} + A_5^{(3)}\delta^{\frac{5}{2}}. \end{aligned} \tag{54.4}$$

The zeroth order coefficients of the observable expansions are assumed from the constraints arising from Lorentz invariance and decay kinematics derived in [5], whereas all the higher order coefficients are extracted by fitting the polynomials with 14 bin LHCb data [6] as shown in Fig. 54.1.

The limiting analytic expressions for R_λ at $q^2 = q^2_{\text{max}}$ are

$$R_\perp(q^2_{\text{max}}) = \frac{\omega_2 - \omega_1}{\omega_2\sqrt{\omega_1 - 1}}, \quad R_\|(q^2_{\text{max}}) = \frac{\sqrt{\omega_1 - 1}}{\omega_2 - 1} = R_0(q^2_{\text{max}}), \tag{54.5}$$

where ω_1, ω_2 contain coefficients which are extracted from data using (54.4) [2]. Thus the variables R_λ's can be estimated using data only and the allowed region is shown in gray bands in Fig. 54.2 left panel. A significant deviation is seen from a slope of 45° line (red line) which denotes $R_\perp = R_\| = R_0$ and hence hints toward the presence of RH currents without *using any estimate of hadronic contributions*. To quantify the RH couplings, we use (54.3) and the results are shown in last two panels of Fig. 54.2. The middle panel uses the SM estimate of parameter r/C_{10} [4] and the SM prediction for C_{10}'/C_{10} and C_9'/C_{10} (the origin) is at more than 5σ confidence level. We have performed another analysis where the input r/C_{10} is considered as nuisance parameter and the result is shown in the rightest panel of Fig. 54.2. It can

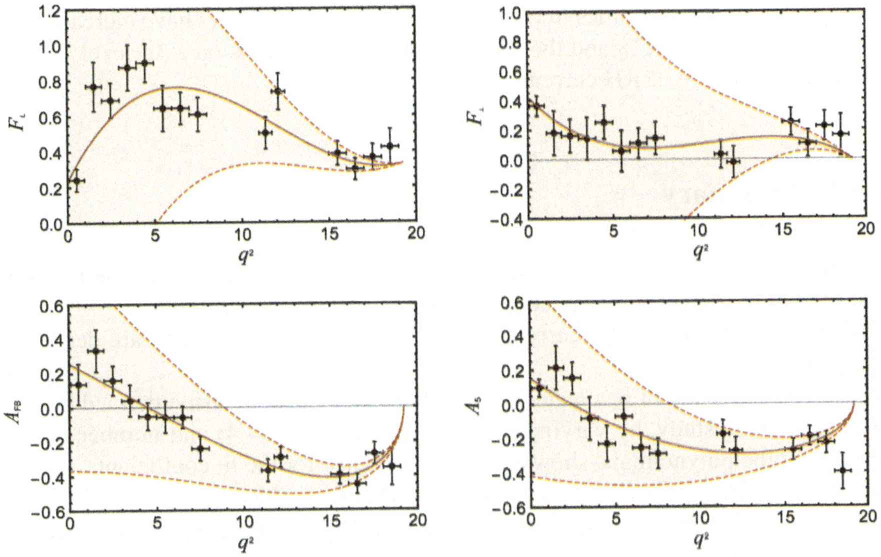

Fig. 54.1 An analytic fit to 14-bin LHCb data using Taylor expansion at $q^2_{\rm max}$ for the observables F_L, $F_\perp$, $A_{\rm FB}$ and A_5 are shown as the brown curves. The $\pm 1\sigma$ error bands are indicated by the light brown shaded regions, derived including correlation among all observables. The points with the black error bars are LHCb 14-bin measurements [6]

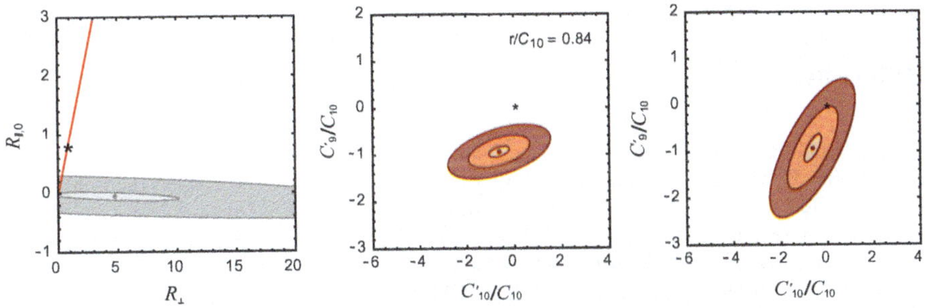

Fig. 54.2 (left panel) Allowed regions in $R_\perp - R_{\parallel,0}$ plane are shown in light and dark gray bands at 1σ and 5σ confidence level, respectively. The red straight line corresponds to the case $R_\perp = R_{\parallel,0}$ i.e. the absence of RH couplings. (middle panel) In C'_{10}/C_{10} - C'_9/C_{10} plane, the yellow, orange and red regions correspond to 1σ, 3σ and 5σ significance level, respectively where SM input for r/C_{10} [4] is used. The best fit values of C'_{10}/C_{10} and C'_9/C_{10}, with $\pm 1\sigma$ errors are -0.63 ± 0.43 and -0.92 ± 0.10, respectively. (right panel) Same color code as the middle panel figure. The input r/C_{10} is varied as a nuisance parameter and hence the obtained uncertainties in C'_{10}/C_{10} and C'_9/C_{10} are increased. The SM predictions for all the three plots are indicated by the stars. Strong evidence of RH current is pronounced from the plots

be seen that the uncertainties in C'_{10}/C_{10} and C'_9/C_{10} parameters have increased due to the variation of r/C_{10} and the SM prediction still remains on a 3σ level contour providing evidence of RH currents.

54.3 Summary

• A formalism has been developed to incorporate almost all possible effects within the SM and has no or minimal dependency on hadronic uncertainties.
• A strong evidence of RH currents is found where the conclusions are derived at endpoint limit.
• The detailed study of resonance effects strengthen the conclusion derived here.
• A systematic study, by varying the polynomial order (54.4) and number of bins used to fit the polynomials, shows good convergence for the fit coefficients.
• The finite width effect of K^* meson has also been considered.
• Need for more data to confirm the presence of the NP scenario presented here.

Acknowledgements We thank R. Sinha, T. E. Browder, A. K. Nayak and A. Karan for collaborating in this work and the organizers of XXII DAE-BRNS High Energy Physics Symposium-2016 for giving the opportunity to present our work in the conference.

References

1. R. Mandal, R. Sinha, D. Das, Phys. Rev. D **90**(9), 096006 (2014)
2. A. Karan, R. Mandal, A.K. Nayak, R. Sinha, T.E. Browder, arXiv:1603.04355 [hep-ph]
3. B. Grinstein, D. Prijol, Phys. Rev. D **70**, 114005 (2004)
4. C. Bobeth, G. Hiller, D. van Dyk, JHEP **1007**, 098 (2010)
5. G. Hiller, R. Zwicky, JHEP **1403**, 042 (2014)
6. R. Aaij et al., [LHCb Collaboration], JHEP **1602**, 104 (2016)

Chapter 55
Exclusion Limits on Minimal Anomaly Free U(1) Extensions of the Standard Model

Tanumoy Mandal, Andreas Ekstedt, Rikard Enberg, Gunnar Ingelman and Johan Löfgren

55.1 Introduction

Many new physics[2] models naturally predict the existence of heavy gauge bosons (Z') which are charge (electromagnetic) neutral and color singlet. Here, we consider a wide class of gauge anomaly free U(1) extensions of the Standard Model (SM) with some minimal assumptions. We derive exclusion limits using electroweak precision constraints and direct collider search data on the parameter space of a TeV-scale Z' which arise from the breaking of the extra U(1) symmetry by the complex scalar acquiring a vacuum expectation value (VEV).

In any renormalizable gauge theory, all sorts of gauge anomalies have to be canceled to preserve unitarity. Therefore, gauge anomaly cancellation imposes strong theoretical constraints on the models. The SM of particle physics is already an anomaly free theory. But any gauge extension of the SM introduces new conditions,

[1]Talk presented at the XXII DAE-BRNS High Energy Physics Symposium, Delhi University, December 12–16, 2016.

T. Mandal—Speaker.

T. Mandal (✉) · A. Ekstedt · R. Enberg · G. Ingelman · J. Löfgren
Department of Physics and Astronomy, Uppsala University, Box 516, 751 20 Uppsala, Sweden
e-mail: tanumoy.mandal@physics.uu.se

A. Ekstedt
e-mail: andreas.ekstedt@physics.uu.se

R. Enberg
e-mail: rikard.enberg@physics.uu.se

G. Ingelman
e-mail: gunnar.ingelman@physics.uu.se

J. Löfgren
e-mail: johan.lofgren@physics.uu.se

Md. Naimuddin (ed.), *XXII DAE High Energy Physics Symposium*, Springer Proceedings in Physics 203, https://doi.org/10.1007/978-3-319-73171-1_55

on the gauge charges of the particles present in the theory, that have to be satisfied in order to make the theory anomaly free. In this work, we take a minimalistic approach based on the following assumptions (i) the SM gauge group is extended by an additional U(1) gauge symmetry which is broken at the TeV-scale by a complex scalar acquiring a VEV (ii) three generations of SM singlet right-handed neutrinos which are charged under new U(1) group are introduced to cancel anomalies (iii) type-I seesaw mechanism is responsible for the generation of neutrino masses (iv) there are no new fermions other than the SM fermions that are charged under the SM gauge group (v) all gauge charges of fermions are generation independent and (vi) the SM electroweak symmetry breaking occurs as usual.

55.2 Anomaly Cancellation

To make a gauge theory involving chiral fermions free from anomalies, one should assign the gauge charges appropriately in such a way that respect all the anomaly cancellation conditions. Due to the presence of a new U(1), there are six types of gauge anomalies that arise from the anomalous triangle diagrams. This leads to six conditions relating the gauge charges of the fermions that have to fulfilled to ensure the anomaly free requirement of the gauge theory. If there is an extra non-anomalous $U(1)_z$ present, one can show that $U(1)_z$ charges of all the SM particles can be written in terms of two free $U(1)_z$ charges namely z_q (the $U(1)_z$ charges of the left-handed quark doublets) and z_u (the $U(1)_z$ charges of the right-handed up-type quarks). By requiring the right-handed neutrinos have both the Majorana and the Dirac mass terms from the seesaw mechanism, their $U(1)_z$ charges, z_k satisfies $z_k = 4z_q - z_u = 1/2$ relation. As discussed in [1], z_q and z_u charges are related by the relation $z_q = \kappa z_u$ for various types of U(1) extensions of the SM where κ is a constant. For instance, $\kappa = 1, -1, 1/4$ and 0 for the gauged $B - L$, SO(10)-GUT, Y-sequential and right-handed models, respectively. These models are briefly discussed in [1].

55.3 Exclusion Limits

In Fig. 55.1a, we show the lower limits on $M_{Z'}$ for various $U(1)_z$ models as described in [1]. These limits are obtained by comparing the 13 TeV ATLAS dilepton resonance search data [2]. To present our result model independently, we define κ-minimum, the value of κ for which $\sigma \times BR$ becomes minimum for a given $M_{Z'}$. Therefore, the parameter regions excluded for the κ-minimum model are also excluded for any other model that belongs to the κ framework. In Fig. 55.1b, we show the exclusion plots in the $M_{Z'} - g_z$ plane for the κ-minimum model using ATLAS dilepton [2] and dijet [3] data, T-parameter constraints and Z-width measurements.

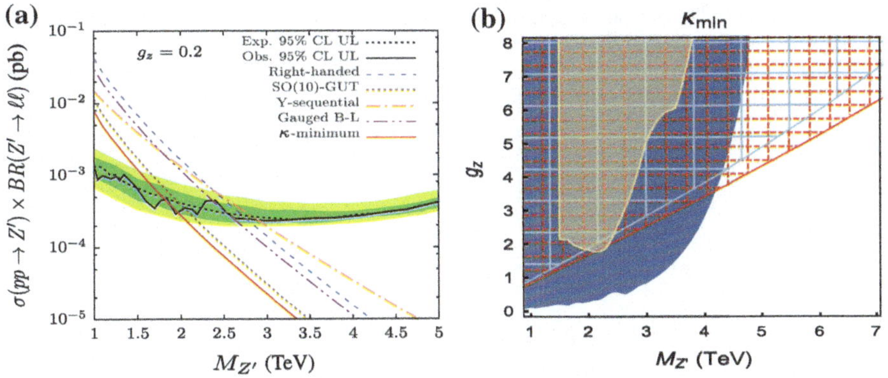

Fig. 55.1 (a) Lower limits on $M_{Z'}$ for various $\mathrm{U}(1)_z$ models as described in [1] obtained by comparing the observed upper limit on $\sigma \times BR$ for the 13 TeV ATLAS resonance search data in the dilepton channel. (b) Excluded parameter space in $M_{Z'} - g_z$ plane. The blue and beige regions are excluded by the 13 TeV ATLAS dilepton and dijet resonance search data, respectively. The electroweak precision data and Z-boson width measurements exclude regions shown by red dashed lines and blue solid lines, respectively

55.4 Summary and Conclusions

We derive exclusion limits on the parameter space of minimally extended anomaly free U(1) models using the LHC data and electroweak precision test constraints. We introduce a κ-framework where κ is the ratio of z_q and z_u as defined previously to present our results model independently. We find that a class of U(1) extensions of the SM can be characterized by the parameter κ when we require all the gauge anomalies are canceled. In this framework, the free parameters are the gauge coupling g_z of the new $\mathrm{U}(1)_z$, mass of the new gauge boson $M_{Z'}$ and the κ parameter. We find that the LHC dilepton data and electroweak precision data are mainly important in constraining the parameter space of a wide class of such U(1) models. Our derived exclusion limits are based on some minimal assumptions as stated in Sect. 55.1. It is always possible to alleviate these bounds by relaxing some of those assumptions. For example, by introducing new chiral fermions that are charged under $\mathrm{U}(1)_z$ or by considering generation dependent charges or perhaps by considering anomaly cancellation by the Green-Schwarz mechanism [4].

References

1. A. Ekstedt, R. Enberg, G. Ingelman, J. Lfgren, T. Mandal, JHEP **1611**, 071 (2016). https://doi.org/10.1007/JHEP11(2016)071, arXiv:1605.04855 [hep-ph]
2. M. Aaboud et al., [ATLAS Collaboration], Phys. Lett. B **761**, 372 (2016). https://doi.org/10.1016/j.physletb.2016.08.055, arXiv:1607.03669 [hep-ex]

3. G. Aad et al., [ATLAS Collaboration], Phys. Lett. B **754**, 302 (2016). https://doi.org/10.1016/j.physletb.2016.01.032, arXiv:1512.01530 [hep-ex]
4. P. Anastasopoulos, M. Bianchi, E. Dudas, E. Kiritsis, JHEP **0611**, 057 (2006). https://doi.org/10.1088/1126-6708/2006/11/057, arXiv:hep-th/0605225

Chapter 56
Identified Particle Production in U+U Collision at $\sqrt{s_{NN}} = 193$ GeV in STAR

Debadeepti Mishra

56.1 Introduction

The study of the bulk properties of the system formed in high-energy heavy-ion collisions shed light on the evolution of the system and the particle production mechanism. It is very interesting to study the effect of size, geometry and orientation of the two colliding nuclei on the finally measured bulk properties of the system. To address this question, data for U+U collisions at $\sqrt{s_{NN}} = 193$ GeV were taken by the STAR experiment at the Relativistic Heavy Ion Collider (RHIC). Uranium ($_{92}U^{238}$) nucleus is deformed in shape compared with Gold ($_{79}Au^{197}$) nucleus, which is almost spherical in shape. As a result, different orientations between the two colliding nuclei are possible in case of U+U collisions. Higher energy density and particle multiplicity are also expected in U+U collisions at $\sqrt{s_{NN}} = 193$ GeV than in Au + Au collisions at $\sqrt{s_{NN}} = 200$ GeV [1].

The minimum bias triggered events in the mid-rapidity region ($|y| < 0.1$) are selected for this analysis. The Time Projection Chamber (TPC) [2] and The Time Of Flight (TOF) [3] detectors in STAR are specifically used for particle identification. The raw yields of the identified particles (π^+, π^-, K^+, K^-, p, $\bar{p}$) are extracted using TPC and TOF by following different analysis techniques, which can be found in details in [4–6]. The yields of the produced particles as a function of transverse momentum p_T are corrected for energy loss, tracking efficiency, matching efficiency between TPC and TOF and for background contribution to the yields of pions and protons [5, 6]. The feed-down and muon contamination fractions from d+Au collisions at $\sqrt{s_{NN}} = 200$ GeV in STAR from [5] are used here for U+U collisions at $\sqrt{s_{NN}} = 193$ GeV.

D. Mishra (For the STAR Collaboration).

D. Mishra (✉)
National Institute of Science Education and Research, HBNI, Jatni 752050, Odisha, India
e-mail: debadeepti.m@niser.ac.in

Md. Naimuddin (ed.), *XXII DAE High Energy Physics Symposium*, Springer Proceedings in Physics 203, https://doi.org/10.1007/978-3-319-73171-1_56

56.2 Results and Discussions

Figure 56.1 shows the p_T spectra in U+U collisions at $\sqrt{s_{NN}} = 193$ GeV for π^+, K^+ and p. The statistical and systematic uncertainties have been added in quadrature. The spectra are shown for nine different centrality classes. The obtained spectra are fitted with functions to extract dN/dy and mean p_T for the unmeasured p_T regions. The fit functions are Bose–Einstein($A/[\exp(m_T/T_{BE}) - 1]$) for pions, m_T - exponential($A\exp(-m_T/T_{m_T})$) for kaons and double exponential ($A\exp(-p_T^2/T_1^2) + B\exp(-p_T^2/T_2^2)$) for (anti) protons. It is seen that, the yield decreases with p_T as well as from central to peripheral collisions. A hardening of the p_T spectra with increasing centrality and particle mass is also observed, which is due to stronger radial flow effects in central collisions than in peripheral collisions.

The left plot of Fig. 56.2 shows the mean p_T of π^+ and p as a function of number of participated nucleons N_{part} for U+U collisions at $\sqrt{s_{NN}} = 193$ GeV. The right two plots of Fig. 56.2 shows dN/dy of π^+, K^+, p and $\bar{p}$ normalized by ($0.5 \times N_{part}$) versus N_{part} for U+U collisions at $\sqrt{s_{NN}} = 193$ GeV. All the results are compared with Au+Au collisions at $\sqrt{s_{NN}} = 200$ GeV [5] as well as various center of mass energies from STAR [5, 6] are shown in order to observe the energy dependence. The data from U+U collisions at $\sqrt{s_{NN}} = 193$ GeV agree within error bars with the data points from Au+Au collisions at $\sqrt{s_{NN}} = 200$ GeV, which signifies that the evolution of the system is similar despite of their different geometric shapes. From the left plot of Fig. 56.2, we can also observe that the value of mean p_T increases from peripheral to central collisions indicating larger radial flow in case of central collisions. There is a slight increase of mean p_T with collision energy. The middle and the right panel of Fig. 56.2 shows dN/dy normalised by N_{part} increases slowly with N_{part} for all the particles in U+U collisions at $\sqrt{s_{NN}} = 193$ GeV, which suggests that particle production may also depend on the number of nucleon-nucleon binary collisions. Also, we can observe that particle yield increases for π^+, K^+ and $\bar{p}$, whereas it decreases for p with collision energy. This reflects the increase in baryon density due to baryon stopping at lower energies.

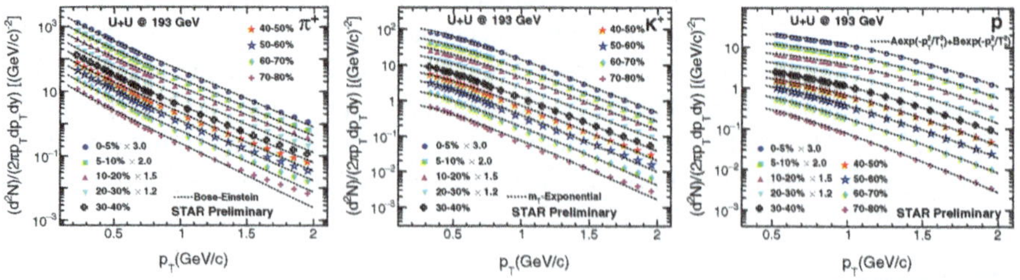

Fig. 56.1 Centrality dependence of identified particle spectra measured in mid-rapidity ($|y| < 0.1$) for U+U collisions at $\sqrt{s_{NN}} = 193$ GeV. Spectra are plotted at nine centrality bins and are scaled for different collision centralities for clarity. Curves represent Bose–Einstein function for pions, m_T-exponential for kaons, and double exponential for protons. The statistical and systematic uncertainties have been added in quadrature

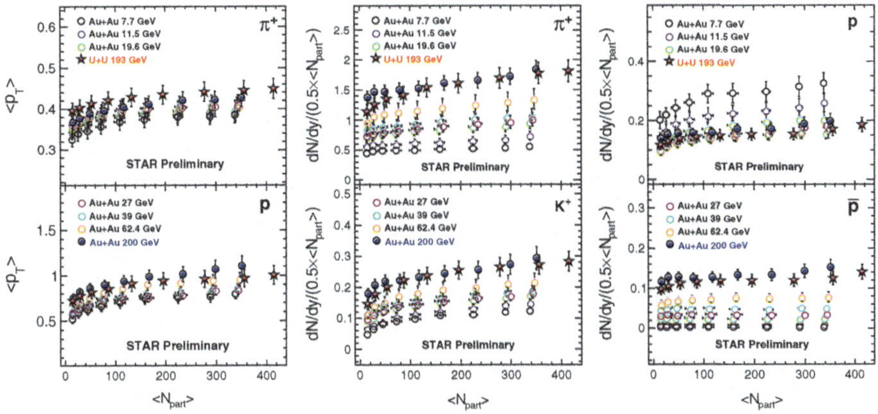

Fig. 56.2 (Left panel) Mean p_T of π^+ and p as a function of < $\mathrm{N_{part}}$ > for U+U collisions at $\sqrt{s_{NN}}$ = 193 GeV. (Middle and right panel) dN/dy of π^+, K^+, p and $\bar{p}$ scaled by 0.5× < $\mathrm{N_{part}}$ > as a function of < $\mathrm{N_{part}}$ > for U+U collisions at $\sqrt{s_{NN}}$ = 193 GeV. The results are shown compared with Au+Au collisions at $\sqrt{s_{NN}}$ = 200 GeV [5] along with other beam energies from STAR [5, 6]. The statistical and systematic uncertainties have been added in quadrature

56.3 Summary

Transverse momentum spectra, integrated particle yield and mean transverse momentum of the identified particles π^+, π^-, K^+, K^-, p and $\bar{p}$ are measured for U+U collisions at $\sqrt{s_{NN}}$ = 193 GeV. The results are compared with the published results from Au+Au collisions at $\sqrt{s_{NN}}$ = 200 GeV as well as with other beam energies from STAR.

The comparison of the observables in U+U and Au+Au collisions at similar energies seems to be almost insensitive to the initial size and geometry of the two colliding nuclei. It will be interesting to differentiate and select the specific orientation of the two colliding nuclei in U+U collision at $\sqrt{s_{NN}}$ = 193 GeV.

References

1. Md. Rihan Haque et al., Phys. Rev. C **85**, 034905 (2012)
2. M. Anderson et al., Nucl. Instrum. Meth. A **499**, 659 (2003)
3. W.J. Llope et al., Nucl. Instrum. Meth. A **522**, 252–273 (2004)
4. M. Shao et al., Nucl. Instr. Meth. A **558**, 419 (2006)
5. B.I. Abelev et al., (STAR Collaboration), Phys. Rev. C **79**, 034909 (2009)
6. L. Adamczyk et al., (STAR Collaboration), arXiv:1701.07065 [nucl-ex]

Chapter 57
Quark-Lepton Complementarity Model Based Predictions for θ_{23}^{PMNS} with Neutrino Mass Hierarchy

Gazal Sharma, Shankita Bhardwaj, B. C. Chauhan and Surender Verma

57.1 Introduction

The recent results from various neutrino oscillation experiments [1] in past several years have provided us a very strong sign of neutrinos being massive, the mixing of lepton flavours and their oscillations. After the successful investigation of θ_{13}^{PMNS} [2], there are certain issues that are not settled yet, out of which one is the problem of mass hierarchy in the neutrino mass spectrum and another is the quadrant of angle θ_{23}^{PMNS}, are the challenges that are to be settled.

The phenomenon of quark and lepton flavor mixing is described by a 3×3 unitary matrix called Cabibbo–Kobayashi–Maskawa (U_{CKM}) and Pontecorvo–Maki–Nakagawa–Sakata (U_{PMNS}) respectively. After investigating the global data fits of various experimental results, so far we know the values for the U_{PMNS} matrix which contains two large and a small mixing angles; i.e. the $\theta_{23}^{PMNS} \approx 45°$, the $\theta_{12}^{PMNS} \approx 34°$ and the $\theta_{13}^{PMNS} \approx 9°$. These results are observed along with the quark flavor mixing matrix (U_{CKM}), which is quite settled with three mixing angles that are small i.e. $\theta_{12}^{CKM} \approx 13°$, $\theta_{23}^{CKM} \approx 2.4°$ and $\theta_{13}^{CKM} \approx 0.2°$, which clearly indicates about a disparity-cum-complementarity between quark and lepton mixing angles. Since, the quarks and leptons are fundamental constituents of matter and Standard Model(SM),

---

G. Sharma · S. Bhardwaj · B. C. Chauhan (✉) · S. Verma
Department of Physics and Astronomical Science, School of Physical and Material Sciences, Central University of Himachal Pradesh (CUHP), Dharamshala, Kangra 176215, Himachal Pradesh, India
e-mail: chauhan@associate.iucaa.in

G. Sharma
e-mail: gazzal.sharma555@gmail.com

S. Bhardwaj
e-mail: shankita.bhardwaj982@gmail.com

S. Verma
e-mail: s_7verma@yahoo.co.in

Md. Naimuddin (ed.), *XXII DAE High Energy Physics Symposium*, Springer Proceedings in Physics 203, https://doi.org/10.1007/978-3-319-73171-1_57

the complementarity relation between these two families is seen as a consequence of a symmetry at some high energy scale. This complementarity termed as 'Quark-Lepton Complementarity'(QLC) has been studied by various authors [3].

The quark-lepton complementarity (QLC) relations hints about the depth of the structure that interrelates quarks and leptons. The disparity between the quark and lepton mixing angles has been expressed in terms of the QLC relations, which can be written as

$$\theta^l_{12} + \theta^q_{12} \simeq 45^\circ, \tag{57.1}$$

$$\theta^l_{23} + \theta^q_{23} \simeq 45^\circ, \tag{57.2}$$

$$\theta^l_{13} + \theta^q_{13} \simeq 0^\circ. \tag{57.3}$$

The above QLC relations indicate that, on the basis of certain flavor symmetry there could be a quark-lepton symmetry at some different energy scale.

Possible consequences of QLC have been investigated in the literature and in particular a simple correspondence between the U_{PMNS} and U_{CKM} matrices has been proposed and analyzed in terms of a correlation matrix [4]

$$V_c = U_{CKM} \cdot U_{PMNS}, \tag{57.4}$$

where V_c is the correlation matrix defined as a product of U_{PMNS} and U_{CKM}. In Sect. 57.2, we discuss in brief the theory of the QLC model along with the investigation of correlation matrix (V_c) and the methodology that we have followed to obtain the desired results. According to the model procedure, after using the most credible texture of the correlation matrix we derive the constraints on the θ_{23}^{PMNS} mixing angle for both normal and inverted neutrino mass hierarchies in the Sect. 57.3. Finally, in Sect. 57.4 we conclude and summarise our results.

57.2 The QLC Model and Theoretical Framework

The texture of V_c can be obtained under certain assumptions about the flavor structure of the theory [5]

$$V_c = U_{CKM} \cdot \psi \cdot U_{PMNS}, \tag{57.5}$$

where ψ is taken as diagonal matrix $\psi = diag(e^{(\iota\psi_i)})$ and the three phases ψ_i are assumed to free parameters as they are not constrained by any of the current experimental evidences.

This is more convenient to do because in Grand Unified Theories (GUTs) [6], once quarks and leptons are kept in the same representation of the underlying gauge group, one has to include an arbitrary but non-trivial phases between the quark and lepton mixing matrices in order to counter the phase mismatch. We take

$$U_{CKM} = \begin{bmatrix} 1-\lambda^2/2-\lambda^4/8 & \lambda & A\lambda^3(1+\lambda^2/2)(\bar{\rho}-\iota\bar{\eta}) \\ -\lambda+A^2\lambda^5(1/2-\bar{\rho}-\iota\bar{\eta}) & 1-\lambda^2/2-\lambda^4/8(1+4A^2) & A\lambda^2 \\ A\lambda^3(1-\bar{\rho}-\iota\bar{\eta}) & -A\lambda^2+A\lambda^4(1/2-\bar{\rho}-\iota\bar{\eta}) & 1-A^2\lambda^4/2 \end{bmatrix} + \mathcal{O}(\lambda^6)$$

and

$$U_{PMNS} = \begin{bmatrix} e^{\iota\phi_1}c_{12}c_{13} & e^{\iota\phi_2}c_{13}s_{12} & s_{13}e^{-\iota\phi} \\ e^{\iota\phi_1}(-c_{23}s_{12}-e^{\iota\phi}c_{12}s_{13}s_{23}) & e^{\iota\phi_2}(c_{12}c_{23}-e^{\iota\phi}s_{12}s_{13}s_{23}) & c_{13}s_{23} \\ e^{\iota\phi_1}(-e^{\iota\phi_1}c_{12}c_{23}c_{13}+s_{12}s_{23}) & e^{\iota\phi_2}(-e^{\iota\phi}c_{23}s_{12}s_{13}-c_{12}s_{23}) & c_{13}c_{23} \end{bmatrix}.$$

However, the values of quark (U_{CKM}) [1] and lepton (U_{PMNS}) [7] mixing parameters are at 1-σ range

$$\begin{aligned} \lambda &= 0.2255 \pm 0.0006, \\ A &= 0.818 \pm 0.015, \\ \bar{\rho} &= 0.124 \pm 0.024, \\ \bar{\eta} &= 0.354 \pm 0.015, \end{aligned} \tag{57.6}$$

$$\begin{aligned} \sin^2\theta_{13} &= 0.0218^{+0.0010}_{-0.0010}, \\ \sin^2\theta_{12} &= 0.304^{+0.013}_{-0.012}, \\ \sin^2\theta_{23} &= 0.452^{+0.052}_{-0.028}, \\ \phi &= (306^\circ)^{+39}_{-70}. \end{aligned} \tag{57.7}$$

For the unknown phases ϕ_1, ϕ_2 and the three ψ_i (57.5), we vary their values between the open interval $[0, 2\pi]$ in a flat distribution.

As per our model [5] procedure, in order to constrain the value of θ_{23}^{PMNS} we use the inverse equation obtained

$$U_{PMNS} = (U_{CKM} \cdot \Psi)^{-1} \cdot V_c. \tag{57.8}$$

We follow more familiar and democratic approach for the calculation of the correlation matrix i.e. it may take any form of texture as suggested by theoretical and experimental data from quark and lepton sectors. After doing Monte Carlo simulations we estimated the texture of the V_c matrix. We obtained predictions for U_{PMNS}^{23} for the two cases of neutrino mass hierarchies i.e. normal hierarchy(NH) and inverted hierarchy(IH).

57.3 Results

The PMNS matrix obtained in case of normal hierarchy is

$$U_{PMNS} = \begin{bmatrix} 0.68-0.87 & 0.40-0.67 & 0.02-0.35 \\ 0.20-0.59 & 0.40-0.70 & 0.59-0.74 \\ 0.38-0.43 & 0.56-0.61 & 0.65-0.72 \end{bmatrix}, \text{ where } V_C = \begin{bmatrix} 0.66-0.91 & 0.38-0.71 & 0.00-0.32 \\ 0.04-0.69 & 0.29-0.80 & 0.57-0.80 \\ 0.24-0.54 & 0.44-0.72 & 0.59-0.76 \end{bmatrix}.$$

The value thus obtained from above matrix is $39.73° - 48.47°$ having centre value $\theta_{23}^{PMNS} = 44.24°$.

In case of inverted hierarchy

$$U_{PMNS} = \begin{bmatrix} 0.67-0.86 & 0.40-0.66 & 0.02-0.36 \\ 0.18-0.57 & 0.37-0.67 & 0.62-0.77 \\ 0.40-0.44 & 0.58-0.63 & 0.61-0.68 \end{bmatrix}, \text{ where } V_C = \begin{bmatrix} 0.67-0.90 & 0.39-0.70 & 0.00-0.33 \\ 0.06-0.66 & 0.30-0.75 & 0.63-0.81 \\ 0.31-0.54 & 0.51-0.71 & 0.59-0.71 \end{bmatrix}.$$

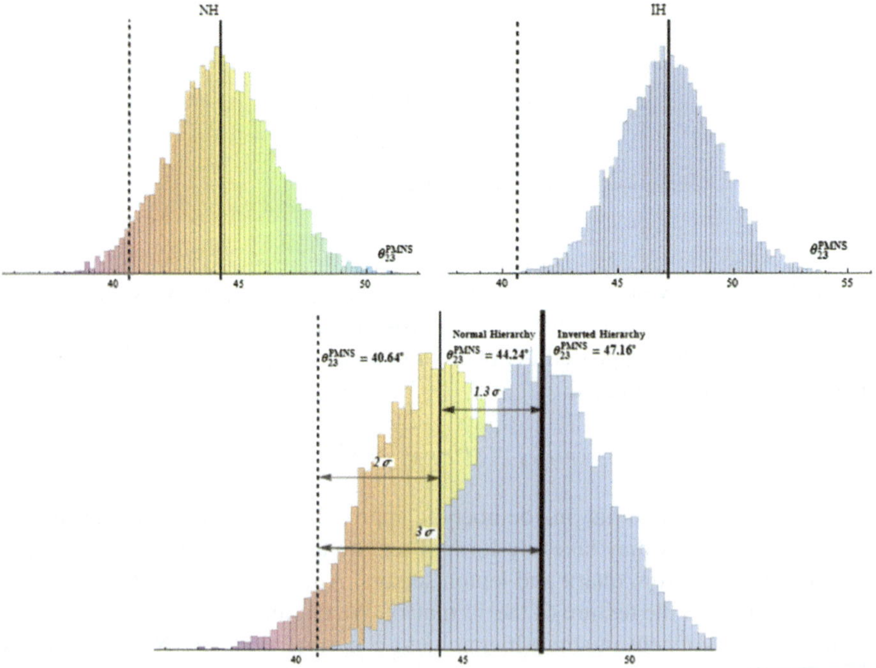

Fig. 57.1 Probability density distribution(PDF) plots of $\sin^2 \theta_{23}^{PMNS}$ including λ-terms upto 6th order for Normal and Inverted neutrino mass hierarchy and their comparison with value obtained in [8]

The value of θ_{23}^{PMNS} obtained for inverted hierarchy is 42.70°–52.38° with centre value as $\theta_{23}^{PMNS} = 47.16°$.

We have shown the probability density distribution of θ_{23}^{PMNS} for normal hierarchy (left) and inverted hierarchy (right) and their comparison with the peaked value of θ_{23}^{PMNS} when hierarchy is not considered [8] in Fig. 57.1. Here the dashed line is for θ_{23}^{PMNS} without considering hierarchy, the thin and thick solid lines are for the two cases of neutrino mass hierarchies i.e. NH and IH, respectively.

57.4 Conclusion

We use the QLC model, where the non trivial relation between the U_{PMNS} and U_{CKM} mixing matrix is taken as the phase mismatch between quark and leptons, via ψ the diagonal matrix. After following the model procedure the central values obtained for θ_{23}^{PMNS} are 44.24° and 47.16° for normal and inverted neutrino mass hierarchies respectively. It has been noticed that the precise values of θ_{23}^{PMNS} thus obtained for the two cases, NH and IH are about 2σ and 3σ away from our previously obtained result [8], which can give a strong hint for the hierarchy of neutrino masses.

As such, in future the better precision of θ_{23}^{PMNS} can give the strong hint about the neutrino mass hierarchy.

Acknowledgements B.C. Chauhan and all the other authors acknowledge the financial support provided by University Grants Commission(UGC), Government of India vide Grant No. UGC MRP-MAJOR-PHYS-2013-12281.

References

1. J. Beringer et al., (Particle Data Group), Phys. Rev. D **28**, 010001 (2012); A. Ceccucci, Z. Ligeti, Y. Sakai, PDG (2014). http://pdg.lbl.gov/2014/reviews/
2. F. An et al., [Daya Bay Collaboration], Phys. Rev. Lett. **108**, 171803 (2012); J. Ahn et al., [RENO Collaboration], Phys. Rev. Lett. **108**, 191802 (2012); K. Abe et al., [T2K Collaboration], Phys. Rev. Lett. **107**, 041801 (2011); P. Adamson et al., [MINOS Collaboration], Phys. Rev. Lett. **107**, 181802 (2011); Y. Abe et al., [Double Chooz Collaboration], Phys. Rev. D **86**, 052008 (2012)
3. H. Georgi, C. Jarlskog, Phys. Lett. B **86**, 297 (1979); H. Minakata, A.Y. Smirnov, Phys. Rev. D **70**, 073009 (2004); W. Rodejohann, Phys. Rev. D **69**, 033005 (2004); K.A. Hochmuth, W. Rodejohann, Phys. Rev. D **75**, 073001 (2007); S.K. Agarwalla, M.K. Parida, R. N. Mohapatra, G. Rajasekaran, Phys. Rev. D **75**, 033007 (2007)
4. Z.z. Xing, Phys. Lett. B **618**,141 (2005); A. Dighe, S. Goswami, P. Roy, Phys. Rev. D **73**, 071301 (2006); A.Y. Smirnov, arXiv:hep-ph/0604213; P.H. Frampton, S.T. Petcov, W. Rodejohann, Nucl. Phys. B **687**, 31 (2004); A. Datta, L. Everett, P. Ramond, Phys. Lett. B **620**, 42 (2005); J. Harada, Europhys. Lett. **75**, 248 (2006); J. Harada, Europhys. Lett. **103**, 21001 (2013); M. Picariello, B.C. Chauhan, J. Pulido, E. Torrente-Lujan, Mod. Phys. Lett. A **22**, 5860 (2008)

5. B.C. Chauhan, M. Picariello, J. Pulido, E. Torrente-Lujan, Eur. Phys. J. C **50**, 573 (2007)
6. G.G. Ross, *Grand Unified Theories* (Westview Press, Reading, MA, 1985)
7. F. Björkeroth et al., Testing constrained sequential dominance models of neutrinos, arXiv:1412.6996 [hep-ph] (2015)
8. G. Sharma, B.C. Chauhan, JHEP **1607**, 075 (2016)

Chapter 58
Probing CP Violation in Neutrino Oscillation Experiments and Leptonic Unitarity Quadrangle

Surender Verma, Shankita Bhardwaj, B. C. Chauhan and Gazal Sharma

58.1 Introduction

During the past decade, the neutrino physics has witnessed a great deal of theoretical and experimental advancement. Many new neutrino experiments have been planned to measure the *only* unknown parameter of the neutrino mixing matrix, i.e. Dirac-type CP violating phase. Dirac-type CP violating phase can be directly measured in the neutrino oscillation experiments. Also, in presence of the sterile neutrino(s) in addition to the standard three neutrino scenario, CP violation is, generally, expected to be violated due to the nontrivial complex phases in neutrino mixing matrix. In order to measure CP violation, in a rephasing invariant manner using Leptonic Unitarity Quadrangle (LUQ), one has to construct rephase invariants from 4×4 neutrino mixing matrix V given by $J^{ij}_{ff'} \equiv \Im\left(V_{fi}V_{f'j}V^*_{fj}V^*_{f'i}\right)$ [1], where $(i, j) = 0, 1, 2, 3$ and $(f, f') = s, e, \mu, \tau$. In the present work, we have considered the second approach to look for the prospects to measure CP violation in an rephasing invariant way in presence of one sterile neutrino [2].

S. Verma (✉) · S. Bhardwaj · B. C. Chauhan · G. Sharma
Department of Physics and Astronomical Science, Central University of Himachal Pradesh, Dharamshala 176215, India
e-mail: s_7verma@yahoo.co.in; sverma@cuhimachal.ac.in

S. Bhardwaj
e-mail: shankita.bhardwaj982@gmail.com

B. C. Chauhan
e-mail: chauhan@associate.iucaa.in

G. Sharma
e-mail: gazzal.sharma555@gmail.com

Md. Naimuddin (ed.), *XXII DAE High Energy Physics Symposium*, Springer Proceedings in Physics 203, https://doi.org/10.1007/978-3-319-73171-1_58

58.2 Relating LUQ Parameters with the Neutrino Mixing Matrix Elements

The unitarity of mixing matrix V ($V^\dagger V = VV^\dagger = 1$) provide twelve orthogonality relations which corresponds to twelve unitarity quadrangles in the complex plane. Now, let's consider a flavor state $|\nu_f\rangle$ that converts to $|\nu_{f'}\rangle$ after travelling a distance L km. The vacuum transition probability for this conversion is given by [3, 4]

$$\begin{aligned} P_{\nu_f \to \nu_{f'}} =& \delta_{ff'} - 4 \sum_{i<j} \left(\Re \left(V_{fi} V_{f'j} V^*_{fj} V^*_{f'i} \right) \sin^2 \left(X_{ij} \mp \lambda_{f'f;ij} \right) \right) \\ & + 2 \sum_{i<j} \left(J^{ij}_{ff'} \sin(2X_{ij} \mp \lambda_{f'f;ij}) \right) \end{aligned} \quad (58.1)$$

where upper (lower) sign in "$\mp$" sign correspond to neutrino (antineutrino) oscillations. Also, $X_{ij} \equiv 1.27 \Delta m^2_{ij} L/E$ with $\Delta m^2_{ij} \equiv m^2_j - m^2_i$, L is baseline length, and E is the neutrino beam energy. The flavor transition can be attributed to change in phase shift [4, 5] of the transition probability by phase angle $\lambda_{f'f:ij}$ defined as $\lambda_{f'f:ij} = arg \left(V_{f'i} V^*_{fi} V_{fj} V^*_{f'j} \right)$, with $\lambda_{f'f;ij} = -\lambda_{ff'ij} = -\lambda_{f'f;ji}$ and $X_{ij} = -X_{ji}$. The orthogonality relation $V_{f0} V^*_{f'0} + V_{f1} V^*_{f'1} + V_{f2} V^*_{f'2} + V_{f3} V^*_{f'3} = 0$ can be viewed as a quadrangle in complex plane, shown in Fig. 58.1.

The angles of the LUQ can be expressed in terms of elements of the 4×4 neutrino mixing matrix as

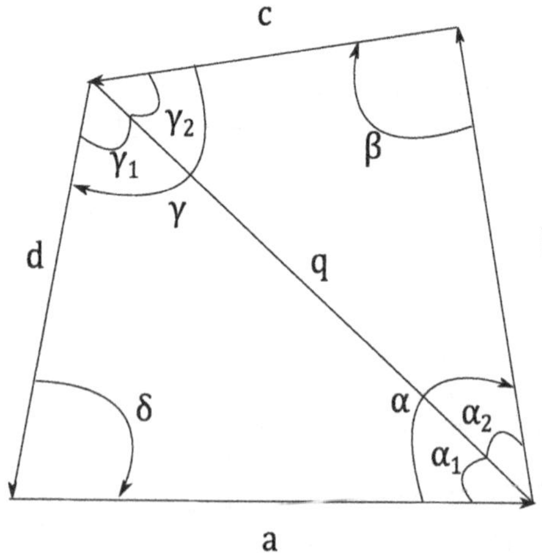

Fig. 58.1 The LUQ in the complex plane

$$\alpha = arg\left(-\frac{V_{f0}V^*_{f'0}}{V_{f1}V^*_{f'1}}\right), \beta = arg\left(-\frac{V_{f1}V^*_{f'1}}{V_{f2}V^*_{f'2}}\right),$$
$$\gamma = arg\left(-\frac{V_{f2}V^*_{f'2}}{V_{f3}V^*_{f'3}}\right), \delta = arg\left(-\frac{V_{f3}V^*_{f'3}}{V_{f0}V^*_{f'0}}\right). \tag{58.2}$$

From these relations we find that $\alpha = \pi - \lambda_{f'f;01}, \beta = \pi - \lambda_{f'f;12}, \gamma = \pi - \lambda_{f'f;23}, \delta = \pi - \lambda_{f'f;03}$. We write the oscillation probabilities (58.1) in terms of geometrical parameters of LUQ for neutrino and antineutrino as,

$$\begin{aligned} P = & \; a^2 + b^2 + c^2 + d^2 - 2ab\cos(2X_{01} \pm \alpha) - 2bc\cos(2X_{12} \pm \beta) \\ & -2cd\cos(2X_{23} \pm \gamma) - 2ad\cos(2X_{03} \pm \delta), \end{aligned} \tag{58.3}$$

58.3 CP Asymmetry in Terms of Independent Geometric Parameters of LUQ

In order to, uniquely, determine LUQ we choose two sides and three angles viz. $(b, c, \alpha, \beta, \gamma)$ as the five independent geometrical parameters. Then, all other parameters of the LUQ can be expressed in terms of these five independent parameters. From Fig. 58.1, we find that $\alpha + \beta + \gamma + \delta = 2\pi$, $\delta = 2\pi - \sigma$ and $a = -q\sin(\gamma - \gamma_2)\csc\sigma$, $d = -q\sin(\alpha - \alpha_2)\csc\sigma$, where, $q = \sqrt{b^2 + c^2 - 2bc\cos\beta}$, $\gamma_2 = \sin^{-1}\left(\frac{b\sin\beta}{q}\right)$, $\alpha_2 = \sin^{-1}\left(\frac{c\sin\beta}{q}\right)$ and $\sigma \equiv \alpha + \beta + \gamma$.

We write (58.3), in terms of five independent geometric parameters $(b, c, \alpha, \beta, \gamma)$. The oscillation probability can be written as

$$\begin{aligned} P = & \; q^2\sin^2(\gamma - \gamma_2)\csc^2\sigma + b^2 + c^2 + q^2\sin^2(\alpha - \alpha_2)\csc^2\sigma \\ & +2qb\sin(\gamma - \gamma_2)\csc\sigma\cos(2X_{01} \pm \alpha) \\ & -2bc\cos(2X_{12} \pm \beta) \\ & +2cq\sin(\alpha - \alpha_2)\csc\sigma\cos(2X_{23} \pm \gamma) \\ & -2q^2\sin(\gamma - \gamma_2)\sin(\alpha - \alpha_2)\csc^2\sigma\cos(2X_{03} \mp \sigma). \end{aligned} \tag{58.4}$$

Equation (58.4) provide the oscillation probability in terms of five independent geometric parameters of the LUQ. From current neutrino oscillation data [6] and considering the case of $E/L \sim \Delta m^2_{12}$, we find that X_{12} is $\mathcal{O}(1)$ and $X_{01}, X_{03} >> 1$. Thus, the oscillations induced by the oscillation frequencies X_{23}, X_{01} and X_{03} will be averaged out due to integration over the neutrino production region and energy resolution function. Thus, we can write oscillation probability for long baseline experiments as

$$P = q^2 \sin^2(\gamma - \gamma_2)\csc^2\sigma + b^2 + c^2 + q^2\sin^2(\alpha - \alpha_2)\csc^2\sigma \\ -2bc\cos(2X_{12} \pm \beta). \tag{58.5}$$

The CP asymmetry can be written as

$$\Delta P = 4bc\sin(2X_{12})\sin\beta. \tag{58.6}$$

In short baseline neutrino oscillation experiments, with neutrino energy $E = \mathcal{O}(1GeV)$ and $L = \mathcal{O}(1\,\text{km})$ ($E/L \sim 0.1eV^2$ such that $|X_{01}| \sim |X_{02}| \sim |X_{03}| \sim 1$), we can neglect the oscillations due to frequencies X_{12}, X_{23} as their contributions will be small. Under these approximations, (58.4) can be written as

$$P = q^2\sin^2(\gamma - \gamma_2)\csc^2\sigma + b^2 + c^2 + q^2\sin^2(\alpha - \alpha_2)\csc^2\sigma \\ +2qb\sin(\gamma - \gamma_2)\csc\sigma\cos(2X_{01} \pm \alpha) \\ -2q^2\sin(\gamma - \gamma_2)\sin(\alpha - \alpha_2)\csc^2\sigma\cos(2X_{03} \mp \sigma), \tag{58.7}$$

and CP asymmetry as

$$\Delta P = -4qb\sin(\gamma - \gamma_2)\csc\sigma\sin(2X_{01})\sin\alpha \\ -4q^2\sin(\gamma - \gamma_2)\sin(\alpha - \alpha_2)\csc\sigma\sin(2X_{03}). \tag{58.8}$$

The CP asymmetry is sensitive to all the five independent parameters viz. b, c, α, β, γ of LUQ. Thus, next generation short baseline experiments provide unique opportunity to construct leptonic unitarity quadrangle and to measure CP violation. Such opportunity to measure CP asymmetry will not be there if we try to directly measure CP phases under aforementioned approximations because the information on CP phases will be lost as new oscillations are, averaged out or small. We have, also,

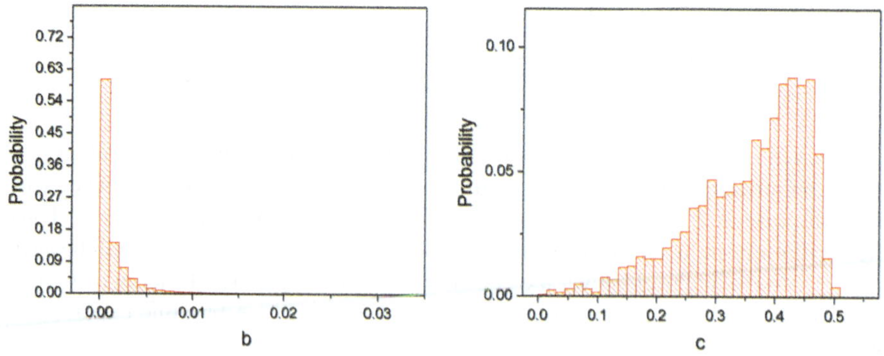

Fig. 58.2 Probability distribution of the geometric parameters of LUQ based on the current experimental data on neutrino masses and mixings [6]

obtained the probability distributions for two of the independent geometric parameters (sides b, c) of LUQ, as shown in Fig. 58.2. Similar plots can also be obtained for β and γ. These probability distributions for independent geometric parameters have been obtained using parametrisation as in [7] of the mixing matrix V. For other possible parametrisations of 4×4 mixing matrix, the relations obtained for the independent geometric parameters will remain identical. However, there may appear CP phase(s) depending upon a particular type of parametrisation considered.

58.4 Conclusions

In conclusion, we find that short baseline experiments have bright prospects to measure CP asymmetry as we can construct the LUQ and subsequently, measure CP violation because CP asymmetry depends on all the geometric parameters of LUQ.

Acknowledgements S. V. acknowledges the financial support provided by University Grants Commission (UGC)-Basic Science Research (BSR), Government of India vide Grant No. F.20-2(03)/2013(BSR). S. B. acknowledges the financial support provided by the Central University of Himachal Pradesh.

References

1. C. Jarlskog, Phys. Rev. Lett. **55**, 1039 (1985)
2. S. Verma, S. Bhardwaj, Nucl. Phys. B **907**, 249–257 (2016)
3. W. Guo, Z. Xing, Phys. Rev. D **65**, 073020 (2002)
4. S.M. Bilenky, S.T. Petcov, Rev. Mod. Phys. **59**, 671 (1987). and references therein
5. J. Beringer et al., (Particle Data Group), Phys. Rev. D **86**, 010001 (2012), and references therein
6. M. Gonzalez-Garcia, M. Maltoni, T. Schwetz, J. High Energy Phys. **052**, 1411 (2014)
7. H. Fritzsch, J. Plankl, Phys. Rev. D **35**, 1732 (1987)

Chapter 59
Weibull Approach to Study Multiplicity Moments in e^+e^- and $p\bar{p}$ Collisions

Ranjit Nayak, Ashutosh Kumar Pandey and Sadhana Dash

59.1 Introduction

Previous investigations based on normalized factorial moments have shown that intermittency is the general feature of particle productions which is related to the dynamical fluctuations in phase space. These fluctuations originate from random cascade processes involved in particle production [1] or branching processes in jet formation [2]. HERWIG [3] and Lund PS [4] model of QCD describe the charged particle production in e^+e^- annihilation. This showed that a sequential branching process plays a significant role in understanding the particle production mechanism based on general cascading processes.

Recently the Weibull model of characterizing charged particle multiplicity distributions in hadronic and e^+e^- collisions showed a good agreement with the measured data [5, 6]. It has two parameters namely λ and k. The former can be related to the mean multiplicity and the latter can be associated to the dynamics of particle production. As the source of intermittency or density fluctuations can be related to jet cascading or sequential branching processes, the signature of the same was seen in e^+e^- collisions. In this work, a complete analysis of higher order factorial moments and raw moments based on Weibull model of the charged particle multiplicity distribution is carried out. The calculated moments are compared to the measured moments obtained in e^+e^- annihilation recorded with the OPAL detector [7] at LEP energies and in $p\bar{p}$ collisions by UA5 and CMS experiment [8–10].

R. Nayak (✉) · A. K. Pandey · S. Dash
Indian Institute of Technology Bombay, Powai 400076, India
e-mail: ranjit@phy.iitb.ac.in

Md. Naimuddin (ed.), *XXII DAE High Energy Physics Symposium*, Springer Proceedings in Physics 203, https://doi.org/10.1007/978-3-319-73171-1_59

59.2 Normalized Raw and Factorial Moments

The probability distribution of a random variable "x" is given by

$$f(x;\lambda,k)=\frac{k}{\lambda}\left(\frac{x}{\lambda}\right)^{(k-1)}e^{-}\left(\frac{x}{\lambda}\right)^{k} \quad (59.1)$$

The nth raw moment of Weibull distribution is given by

$$m_n=\lambda^n\Gamma\left(1+\frac{n}{k}\right) \quad (59.2)$$

where $m_1=\lambda\Gamma\left(1+\frac{1}{k}\right)$ is the mean of Weibull distribution. The nth factorial moment of a variable "x" is defined as $f_n=\left\langle\frac{x!}{(x-n)!}\right\rangle$. The nth normalized raw "C_n" and factorial moments "F_n" are defined as

$$C_n=\frac{m_n}{m_1^n} \quad (59.3)$$

$$F_n=\frac{f_n}{m_1^n} \quad (59.4)$$

59.3 Results

The intermittency phenomena and genuine multi-particle correlations of charged particles are studied using the higher moments of Weibull distribution in various bins of rapidity, p_T and azimuthal angle. The model parameters, λ and k were extracted by using the expression for second order and third order normalized factorial moments. The top panel of Fig. 59.1 compares the calculated values of F_4 and F_5 to the measured data for various bins of rapidity. One can observe that the higher order factorial moments calculated from the Weibull model is able to nicely reproduce the data values within the error bars. As similar study was also carried out for the higher order moments in $p\bar{p}$ collisions using data sets of UA5 experiment (at 200, 540 GeV) and CMS experiment (at 0.9, 2.36 and 7 TeV). The comparison of the calculated moments with the measured data and their evolution with beam energy is shown in the bottom panel of Fig. 59.1.

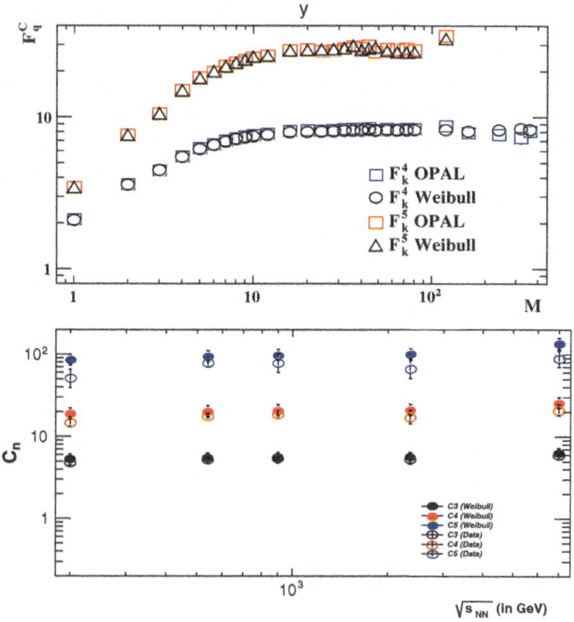

Fig. 59.1 Top: Factorial moments of order q = 4 and 5 as a function of M, where M is the number of bins of the 1-dimensional subspaces of the phase space of rapidity, in comparison with the predictions of the Weibull parametrization. The datas are taken from [7]. Bottom: Normalized raw moments plotted as a function of $\sqrt{SNN}$ for $p\bar{p}$ collisions. The datas are taken from [8–10]

59.4 Summary

The normalized higher order factorial moments and raw moments of the charged particle multiplicity distributions in e^+e^- and $p\bar{p}$ collisions have been studied using the Weibull model. The Weibull parametrization provides a nice description of the higher order factorial moments especially in smaller intermittency bins of the phase space of rapidity, azimuthal angle and transverse momentum for e^+e^- collisions. A good agreement is also obtained for higher order normalized raw and factorial moments in $p\bar{p}$ collisions. This study further establishes that the Weibull distribution is the optimal distribution to describe the charged particle multiplicity in e^+e^- and $p\bar{p}$ collisions compared to other available statistical models. [11, 12]

References

1. A. Bialas, K. Zalewski, Phys. Lett. B **228** (1989)
2. W. Ochs, J. Wosiek, preprint MPI-PAE/PTh 28/88
3. G. Marchesini et. al., Comput. Phys. Commun. **67**, 465 (1992)
4. M. Bengetsson, T. Sjostrand, Phys. Lett. B **185**, 435 (1987)
5. S. Dash, B.K. Nandi, P. Sett, Phys. Rev. D **93**, 114022 (2016)
6. S. Dash, B.K. Nandi, P. Sett, Phys. Rev. D **94**, 074044 (2016)
7. G. Abbiendi et al., OPAL Collaboration, Eur. Phys. J. C **11**, 239 (1999)

8. The CMS Collaboration, V. Khachatryan, A.M. Sirunyan et al., J. High Energy Phys. **2011** 79 (2011). https://doi.org/10.1007/JHEP01(2011)079
9. R.E. Ansorge et al., Z. Phys. C **43**, 357 (1989), CERN-EP-88-172
10. UA5 Collaboration. http://dx.doi.org/10.1016/0370-2693(85)91491-1
11. A. Bialas, R. Peschanski, Phys. Lett. B **207** (1988)
12. E.K.G. Sarkisyan, Phys. Lett. B **477**, 1–12 (2000)

Chapter 60
Study of Centrality and Beam Energy Dependence of $dN_{ch}/d\eta$ and $dE_T/d\eta$ at Midrapidity Using Two Component Approach

Ashutosh Kumar Pandey, Ranjit Nayak, Basanta. K. Nandi, Sadhana Dash and Priyanka Sett

60.1 Introduction

In heavy ion collisions, particle production is done by two processes: "hard" scattering processes, characterized by the production of high transverse momentum (p_T) particles, are expected to scale with the number of collisions, N_{coll} while, the production of low p_T particles (from soft processes) is expected to scale with the number of participants, N_{part} [1]. The energy and centrality dependence of the charged particle multiplicity density ($dN_{ch}/d\eta$) and the transverse energy density ($dE_T/d\eta$) with collision gives important information on the relative contribution of hard and soft processes to the particle production mechanism. In this work, we have studied the variation of $dN_{ch}/d\eta$ (and $dE_T/d\eta$) with centrality and beam energy using the two-component approach. The two component approach [1] parametrizes $dN_{ch}/d\eta$ by the following expression:

$$\frac{dN_{ch}}{d\eta} = n_{pp}\left((1-x)\frac{\langle N_{part}\rangle}{2} + x\langle N_{coll}\rangle\right) \tag{60.1}$$

Here n_{pp} represents the average charged-particle multiplicity in a single proton-proton collision and x refers to the fraction of "hard" partonic interactions. A similar approach was also used to explain the variation of $dE_T/d\eta$ dependence with collision centrality and energy. In the case of $dE_T/d\eta$, mean $E_{T_{pp}}$ was treated as a free parameter. We have used the x parameter values obtained from the $dN_{ch}/d\eta$ fit.

A. K. Pandey (✉) · R. Nayak · B. K. Nandi · S. Dash · P. Sett ·
Department of Physics, Indian Insititute of Technology Bombay, Powai, Mumbai, India
e-mail: ashutosh@phy.iitb.ac.in

Md. Naimuddin (ed.), *XXII DAE High Energy Physics Symposium*, Springer Proceedings in Physics 203, https://doi.org/10.1007/978-3-319-73171-1_60

60.2 Analysis

The data for $dN_{ch}/d\eta$ and $dE_T/d\eta$ for Au+Au collisions at mid-rapidity have been presented by PHENIX experiment for a large set of energies at RHIC [7]. The LHC data for $dN_{ch}/d\eta$ in Pb+Pb collisions at 2.76 and 5.02 TeV has been taken from ALICE experiment [6, 8] and the $dE_T/d\eta$ data for 2.76 TeV can be found in [9].

The values of n_{pp} were taken from p-p collisions for a particular energy. Since values of n_{pp} were not available for all energy, we have extrapolated $\sqrt{s}$ dependence of measured n_{pp} in pp collisions at different energies to obtain n_{pp}. The values of $\langle N_{part}\rangle$ are obtained from the Monte-Carlo Glauber model [10]. The values of $\langle N_{coll}\rangle$ can be obtained by a simple power law parameterization of $\langle N_{part}\rangle$ obtained from the Monte-Carlo Glauber model [10]. The dependence is given by, $N_{coll} = a \times {N_{part}}^{b}$.

60.3 Results: Charged Particle Multiplicity Densities

One can observe in Fig. 60.1 that at RHIC energies, the scaled multiplicity densities do not have a strong dependence on the $\langle N_{part}\rangle$ and have a flat behavior. However, the observed trend for LHC differs from RHIC. As a function of system size, the scaled multiplicity density shows an increasing trend, which one can expect from the large difference in collision energies.

A variable $R_{X/Y} = \frac{dN/d\eta}{\langle N_{part}\rangle/2}|_X / \frac{dN/d\eta}{\langle N_{part}\rangle/2}|_Y$, is used to denote the magnitude of the ratio of multiplicity densities at energy X and energy Y (where $X > Y$). The ratio is evaluated for similar $\langle N_{part}\rangle$ values for both the energies considered [2]. On careful observation, one can see that the trend does not follow strictly the energy difference only. For example, the values of $R_{2760/200}$ is consistently lower than $R_{200/19.6}$ although the collision energy difference is more for the former than the later. Therefore, one should also take into account an increase in the x factor along with an increase of energy.

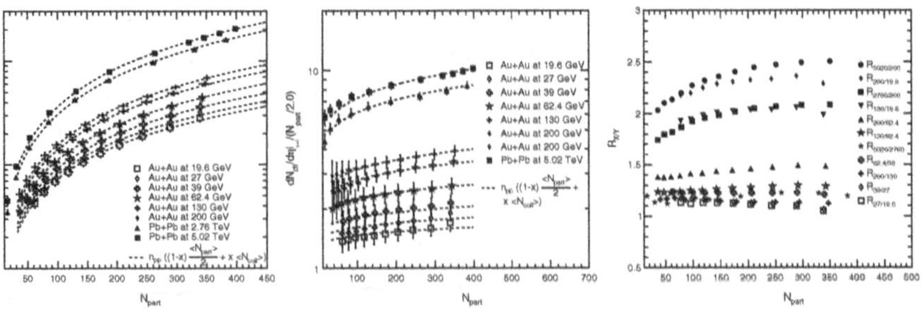

Fig. 60.1 $dN_{ch}/d\eta$ measured in heavy ion collisions. The dashed line represents the fits to (60.1) performed by χ^2 minimization. Most right plot shows, the variation of the double ratio of the charged particle multiplicity density (refer text) as a function of $\langle N_{part}\rangle$

60.4 Results: Transverse Energy Density

The results of the fit for $dE_T/d\eta$ are shown in Fig. 60.2. The model describes the data very nicely for all energies from RHIC to LHC. The extracted values of $E_{T_{pp}}$ is plotted as a function of $\sqrt{s}$ in Fig. 60.3. The scaled energy densities do not have any centrality dependence.

This behavior is also seen in the pair wise ratios of $dE_T/d\eta$ at two different set of energies as shown in Fig. 60.2. The magnitude of the ratios depends on the factor of difference in energy. One can conclude that the role of x is not significant in dictating the observed trend in case of transverse energy densities.

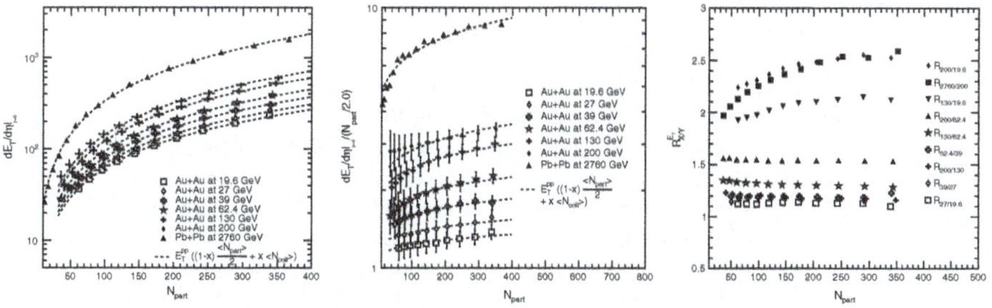

Fig. 60.2 $dE_T/d\eta$ measured in heavy ion collisions. The dashed line represents the fits to (60.1). Most right figure shows, the variation of the double ratio of the charged particle energy density (refer text) as a function of $\langle N_{part}\rangle$

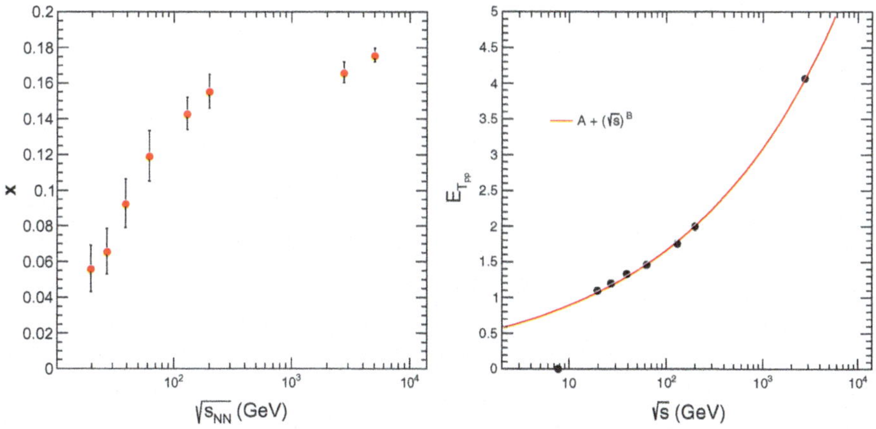

Fig. 60.3 Left, The variation of the parameters x. Right,The variation of $E_{T_{pp}}$ as a function of center of mass energy. This is parameterized by a power law shown by the solid line

60.5 Summary

The mid-rapidity $dN_{ch}/d\eta$ and $dE_T/d\eta$ were described nicely by the two-component Glauber approach for a broad range of energies. The value of x increases with an increase in beam energy (Fig. 60.3) but the increase is not very significant with respect to centrality. The pair wise ratios of the multiplicity densities at two different energies shows a scaling which depends on the interplay of energy difference and the contribution from the hard scattering component. The later plays no role in the observed order of ratios for the transverse energy densities.

References

1. D. Kharzeev, M. Nardi, Phys. Lett. B **507**, 121 (2001)
2. L. Zhou, G.S.F. Stephans, Phys. Rev. C **90**, 014902 (2014)
3. M.C. Abreu et al., (NA50 Collaboration), Phys. Rev. Lett. B **530**, 43 (2002)
4. M.M. Aggarwal et al., (WA98 Collaboration), Eur. Phys. J. C **18**, 651 (2001)
5. F. Antinori et al., (NA57 and WA97 Collaboration), J. Phys. G **27**, 391 (2001)
6. K. Aamodt et al., (ALICE Collaboration), Phys. Rev. Lett. **106**, 032301 (2011)
7. A. Adare et al., (PHENIX Collaboration), Phys. Rev. C **93**, 024901 (2016)
8. J. Adam et al., (ALICE Collaboration), Phys. Rev. Lett. **116**(22), 222302 (2016)
9. C. Loizides for ALICE Collaboration, J. Phys. G **38**, 124040 (2011)
10. M.M. Miller et al., Ann. Rev. Nucl. Part. Sci. **57**, 205–243 (2007)
11. B.B. Back et al., (PHOBOS Collaboration), Phys. Rev. C **70**, 021902 (2004)
12. J. Adam et al., (ALICE Collaboration), arXiv:1509.07541 [nucl-ex]
13. J. Adam et al., (ALICE Collaboration), Phys. Lett. B **753**, 319–329 (2016)

Chapter 61
Degenerate Quantum Vacua and Kerr Family of Black Holes

Sunita Singh, Supriya Kar, K. Priyabrat Pandey and Abhishek K. Singh

61.1 Introduction

Dirichlet (D) brane are describe as the non-perturbative objects in ten dimensional type II superstring theory. For a constant Neveu-Schwarz (NS) background in an open string theory, the U(1) gauge field turns out to be non-linear on a D-brane and has been shown to describe an open string metric [1]. In this article I will explore the dynamical aspect of NS two form. In the context, the Kalb-Ramond (KR) field dynamics are used to define a modified derivative in an effective gravity theory [2]. It has been shown to govern emergent curvatures on a gravitational pair of $(3\bar{3})$-brane. The Schwinger pair production mechanism is generalised to stringy pair production via the KR field. The non-perturbation consideration is important to describe Hawking radiation phenomenon at the black hole event horizon. The idea is used for the open strings used for the open strings pair production [3] by an electromagnetic field.

61.2 String-Brane Model

We consider the dynamics of KR field on a D_4-brane in presence of an background open-metric $G_{\mu\nu}^{(NS)}$ [1], sourced by constant NS two form, leads to the equation of motion:

S. Singh (✉)
Miranda House College, University of Delhi, Delhi, India
e-mail: informsunita@gmail.com

S. Kar
Department of Physics & Astrophysics, University of Delhi, Delhi, India

K. Priyabrat Pandey
Inderaprastha University, Delhi, India

A. K. Singh
Rajdhani College, University of Delhi, Delhi, India

Md. Naimuddin (ed.), *XXII DAE High Energy Physics Symposium*, Springer Proceedings in Physics 203, https://doi.org/10.1007/978-3-319-73171-1_61

$\partial_\lambda H^{\lambda\mu\nu} + \frac{1}{2} g^{\alpha\beta}\partial_\lambda\, g_{\alpha\beta}\, H^{\lambda\mu\nu} = 0$. A priori, identification of the gauge theoretic torsion H_3 with a connection modifies the covariant derivative to $\mathcal{D}_\mu$ [2]. This consideration leads to the absorption of KR field quanta and hence $\mathcal{D}_\mu$ describes a 'fat' brane or a string-brane model. It can be argued to describe a vacuum created pair of gravitational $(3\bar{3})$-brane underlying a geometric torsion curvature. This modification in the gauge connection results to the geometric torsion $\mathcal{H}_{\mu\nu\lambda} = \mathcal{D}_\mu B_{\nu\lambda} +$ cyclic in (μ, ν, λ). The $U(1)$ gauge invariance of $\mathcal{H}_3^2$ under NS field transformation incorporates a symmetric $f^q_{\mu\nu} = C\ \mathcal{H}_{\mu\alpha\beta}\bar{\mathcal{H}}_\nu^{\alpha\beta} = \tilde{C}\bar{\mathcal{F}}_{\mu\alpha}\bar{\mathcal{F}}^\alpha{}_\nu$ correction which in turn defines an emergent metric: $G_{\mu\nu} = G_{\mu\nu}^{(\text{NS})} \pm f_{\mu\nu}$. The generic curvature tensors are worked out using the commutator of the modified derivative operator [2]. Hence the dynamics of a geometric torsion, appear to modify the Riemannian curvature tensor to an effective curvature tensor $\tilde{\mathcal{K}}^\rho_{\mu\nu\lambda}$. In terms of the ir-reducible curvature tensor $\mathcal{K}$ the action governing D_3-brane dynamics is: $S^{\text{eff}}_{\text{D}_3} = \frac{1}{3\kappa^2}\int d^4x\sqrt{-G}\left(\mathcal{K} - \frac{3}{4}\bar{\mathcal{F}}_{\mu\nu}\mathcal{F}^{\mu\nu}\right)$.

61.3 Degenerate Kerr–Newman Black Hole

A priori, an effective D_3-brane may seen to be influenced by a generic torsion. Nevertheless, we argue that a geometric torsion completely decouples to yield a stable vacuum in the semi-classical and in the quantum regime. The Minkowski vacuum on a D_3-brane is defined with a flat metric $g_{\mu\nu}$ in Boyer-Lindquist coordinates as: $ds^2_{\text{flat}} = -dt^2 + \frac{\rho_a^2}{\Delta_a}dr^2 + \rho_a^2 d\theta^2 + \Delta_a \sin^2\theta d\phi^2$, where $\Delta_a = (r^2 + a^2)$ and $\rho_a^2 = (r^2 + a^2\cos^2\theta)$. The two form ansatz leading to a family of Kerr vacua in Einstein gravity in 5D is worked out [6]. Here gauge choice considered is: $A_t = -\frac{Qr}{\rho_a^2}$, $A_\phi = \frac{aQr\sin^2\theta}{\rho_a^2}$

$$B_{tr} = \left(\frac{2M}{\Delta_a}\right)^{1/2} \quad\text{and}\quad B_{r\theta} = \rho_a\left(\frac{a^2\sin^2\theta + 2M}{\Delta_a} - \frac{a^2\sin^2\theta - 2M}{\rho_a^2}\right)^{1/2} \tag{61.1}$$

where (M, Q) are arbitrary constants and are identified as mass and non-linear charge with a lower cutoff $(M, Q \geq a^2)$ in quantum gravity. The gauge ansatz freezes the local degrees of torsion on a D_3-brane within a vacuum created pair of brane/anti-brane. Hence the dynamics is solely contributed by an one form in four dimensions. In turns the nontrivial geometric field strength reduces in the gauge choice, *i.e.* $\mathcal{F}_2 \to \mathcal{F}_2^z$[6].

Mathematical engineering of the gauge ansatz in the effective metric leads to a quantum geometry with relatively wrong signs in the causal patches which can be corrected by introducing a projection matrix ([4, 6]). The metric so obtained in bane window $r^4 >> M^2$ and $r^8 >> Q^4$ yields:

$$\begin{aligned}
ds^2 &= \left(1 - \frac{2M}{\rho_a^2} \pm \frac{Q^2}{\rho_a^6}\left[\Delta_a - \frac{4r^2a^2\cos^2\theta}{\rho_a^2}\right]\right) dt_e^2 \\
&+ \left(1 - \frac{2M - a^2\sin^2\theta}{\rho_a^2} \pm \frac{Q^2}{\rho_a^2\Delta_a}\left[1 - \frac{4r^2a^2\cos^2\theta}{\rho_a^4}\right]\right)^{-1} dr^2 \\
&+ \left(1 + \frac{a^2Q^2}{\rho_a^6}\left[\sin^2\theta + \frac{4r^2a^2\cos^2\theta}{\rho_a^2}\right]\right)\rho_a^2\, d\theta^2 - \frac{2aQ^2\sqrt{\Delta_a}\sin\theta}{\rho_a^5}\, dt_e d\theta \\
&+ \left(1 + \frac{2M}{\rho_a^2}\left[1 + \frac{\Delta_a}{\rho_a^2}\right] + \frac{a^2\sin^2\theta}{\rho_a^2}\left[1 - \frac{\Delta_a}{\rho_a^2}\right] \mp \frac{4a^2Q^2r^2\cos^2\theta}{\rho_a^8}\right)\Delta_a\sin^2\theta d\phi^2 \\
&- \frac{2\sqrt{2M\Delta_a}}{\rho_a^2}\left(2M\left[1 + \frac{\Delta_a}{\rho_a^2}\right] + \left[1 - \frac{\Delta_a}{\rho_a^2}\right]a^2\sin^2\theta\right)^{1/2} \sin\theta\, dt_e d\phi.
\end{aligned} \tag{61.2}$$

The emergent quantum Kerr Newman(KN) geometries ensures $G_{tt}G_{rr} \neq 1$ which defines an ergo-sphere around the horizon(s). Importantly, for the black hole geometries, the horizon radius is $r_h = \sqrt{2M - a^2}$. Strikingly the event horizon in a charged quantum black hole is independent of the charge Q. Thus the physical quantity such as entropy remains unaffected by the charge. A thermodynamic analysis under $dU = TdS$, reveals a constant internal energy for a large class of black holes defined with different Q. It leads to a degeneracy in a quantum KN black hole. Now lets try to explore more on these quantum geometries in the low energy limit where lowest order terms $(2M/\rho_a^2)$ and (Q^2/ρ_a^4) become significant in the causal patches. Under the limit the KN quantum geometry with $+Q^2$ term (61.2) characterized by two horizons at

$$r \rightarrow: r_{\pm} = \sqrt{\frac{r_h^2 - a^2\cos^2\theta_{\pm}}{2}} \left(1 \pm \sqrt{1 + \frac{4a^2r_h^2\cos^2\theta_{\pm} - 4Q^2}{\left(r_h^2 - a^2\cos^2\theta_{\pm}\right)^2}}\right)^{1/2} .$$

It shows that in the limit, an event horizon develops a charge dependence and the geometry reduces to semi-classical KN blackhole. Interestingly, the degeneracy of quantum KN blackhole disappears and it may well be described by a Einstein–Maxwell theory.

The KN quantum geometry defined with $-Q^2$ term (61.2) in a low energy limit simplifies to a charged Kerr black hole and yields an event horizon:

$$r_e = \sqrt{\frac{r_h^2 - a^2\cos^2\theta_{+}}{2}} \left(1 + \sqrt{1 + \frac{4a^2r_h^2\cos^2\theta_{+} + 4Q^2}{\left(r_h^2 - a^2\cos^2\theta_{+}\right)^2}}\right)^{1/2} .$$

Unlike to a semi-classical KN black hole, a charged Kerr black hole is a priori described by two non-linear charges (M and Q) and a single horizon. It shows that the potential due to a two form and that due to an one form are in equivalence at the horizon of the charged Kerr black hole. Thus one of the non-linear charges may be absorbed by the other without any change in the characteristics of the black hole.

Conclusively, it implies that in the emergent curvature theory, we obtained degenerate quantum Kerr–Newman black holes which in the low energy limit reduces to semi-classical Kerr–Newman and classical Kerr black hole in 4D. Hence, the degeneracy disappears at the expense of the charge of the black hole.

References

1. N. Seiberg, E. Witten, String theory and noncommutative geometry. JHEP **09**, 032 (1999)
2. A.K. Singh, K.P. Pandey, S. Singh, S. Kar, Discrete torsion, de Sitter tunneling vacua and AdS brane. JHEP **05**, 033 (2013)
3. C. Bachas, M. Porrati, Pair creation of open strings in an electric field. Phys. Lett. **BB296**, 77 (1992)
4. K.P. Pandey, A.K. Singh, S. Singh, R. Kapoor, S. Kar, Quintessence and effective RN de Sitter brane geometries. Eur. Phys. J. **C74** (11), 3173 (2014)
5. S. Singh, K.P. Pandey, A.K. Singh, S. Kar, Quantum Kerr tunneling vacua on a $(D\bar{D})_4$-brane. Nucl. Phys. **B879**, 216 (2014)
6. S. Singh, K.P. Pandey, A.K. Singh, S. Kar, Quantum Kerr(Newman) degenerate stringy vacua in 4D on a non-BPS brane. Int. J. Mod. Phys. **A29**, 29 (2014)

Chapter 62
Measurement of Charged Jet Cross Sections and Jet Shapes in Proton–Proton Collisions at $\sqrt{s} = 2.76$ TeV with the ALICE Detector at LHC

Rathijit Biswas

62.1 Introduction

Jets are defined as the collimated stream of final state hadrons, produced from the fragmentation of hard scattered partons (quarks and gluons) in high energy collisions. Jets serve as tool to connect the final state hadrons with the primarily produced partons and thus they are probe to test perturbative quantum chromodynamics (pQCD). The study of jet shapes provides details of the parton to jet fragmentation with insight to jet transverse profile. Furthermore, the measurements in pp serves as a baseline reference for similar measurements in heavy-ion (AA) collisions, where a highly dense medium is created, to isolate the hot matter effects.

62.2 Data Analysis, Jet Reconstruction and Observables

For this analysis, 4.4 X 10^7 minimum bias events recorded by the ALICE detector at the Large Hadron Collider (LHC) during 2011 are used. Events with primary vertex within ± 10 cm along the beam axis from the geometrical interaction point are analysed. Charged tracks are reconstructed using the combined information from Time Projection Chamber (TPC) and Inner Tracking System (ITS) and tracks with

---

Rathijit Biswas for the ALICE Collaboration.

R. Biswas (✉)
Centre for Astroparticle Physics and Space Science, Bose Institute, Kolkata, India
e-mail: rathijit.biswas@cern.ch

Md. Naimuddin (ed.), *XXII DAE High Energy Physics Symposium*, Springer Proceedings in Physics 203, https://doi.org/10.1007/978-3-319-73171-1_62

$p_T^{track} > 150\,\text{MeV/c}$, $|\eta^{track}| < 0.9$ are selected for the analysis. Jets are reconstructed using sequential recombination anti-k_T [1] jet clustering algorithm from the FastJet package [2]. The differential charged jet cross sections are measured with R = 0.2, 0.3, 0.4 and 0.6 using the equation

$$\frac{d^2\sigma^{jet,ch}}{dp_T d\eta}(p_T^{jet,ch}) = \frac{1}{\mathcal{L}^{int}} \frac{\Delta N_{jets}}{\Delta p_T \Delta\eta}(p_T^{jet,ch}) \tag{62.1}$$

where $\mathcal{L}^{int}$ is the integrated luminosity (1.3 nb^{-1} in our case) and ΔN_{jets} the number of jets in the selected intervals of Δp_T and $\Delta\eta$. Jet shape observables such as average numbers of charged tracks in a leading jet ($\langle N_{ch}\rangle$), leading charged jet size (average radius containing 80% of the total jet p_T, $\langle R_{80}\rangle$) and radial distribution of p_T density about the jet axis (dp_T^{sum}/dr) are measured for R = 0.4.

62.3 Corrections and Systematic Uncertainties

The jet p_T spectra are corrected for instrumental effects by the Bayesian unfolding technique [3] using Monte Carlo (MC) events and full detector simulation with GEANT3 [4]. Jet shape observables are corrected using a bin-by-bin correction technique. All the observables are further corrected for the contribution of the underlying event (UE) from sources other than hard scattering. The UE is estimated following the technique used by ALICE [5] and subtracted event-by-event for jet p_T spectra and on a statistical basis for jet shape observables. The sources of systematic uncertainties include finite efficiency and momentum resolution, UE estimation and the dependence of correction factors on MC event generators.

62.4 Results

Figure 62.1 shows jet production cross sections for R = 0.2, 0.3, 0.4 and 0.6 (top), the $\langle N_{ch}\rangle$ distribution for R = 0.4 (bottom left) and the $\langle R_{80}\rangle$ distribution for R = 0.4 (bottom right) as a function of jet p_T. $\langle N_{ch}\rangle$ increases monotonically with increasing jet p_T as seen in earlier measurements. $\langle R_{80}\rangle$ is contained within half of the total radius for the lowest jet p_T, and it decreases for higher jet p_T. Different MC models such as HERWIG [6] and different tunes of PYTHIA6 [7] agree within 10% with the data whereas PHOJET [8] agrees within 15%. The results for $\langle N_{ch}\rangle$ and $\langle R_{80}\rangle$ at 2.76 TeV are also compared to earlier measurements at 7 TeV (Fig. 62.2, top left and

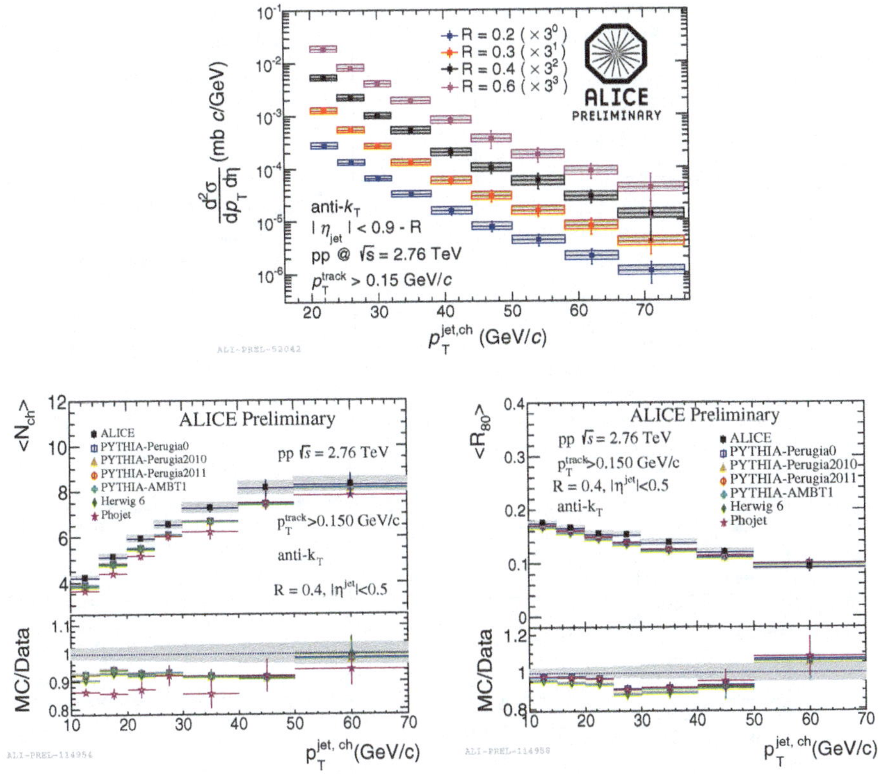

Fig. 62.1 Top: Charged jet production cross sections for R = 0.2, 0.3, 0.4 and 0.6. Bottom left: $\langle N_{ch} \rangle$ distributions for R = 0.4 with different MC generator predictions and their ratios. Bottom right: $\langle R_{80} \rangle$ distributions for R = 0.4 with different MC generator predictions and their ratios. The vertical error bars in data and MC stand for statistical errors. The boxes and the grey bands show the systematic uncertainties

right) and as such we do not see any $\sqrt{s}$ dependence within uncertainties. Figure 62.2 (bottom left) shows radial momentum distributions for four different jet p_T bins as a function of distance from the jet axis. $\langle dp_T{}^{sum}/dr \rangle$ is largest near the jet axis and it decreases towards the periphery. The measurements from various MC models are in good agreement with the data as shown in Fig. 62.2 (bottom right).

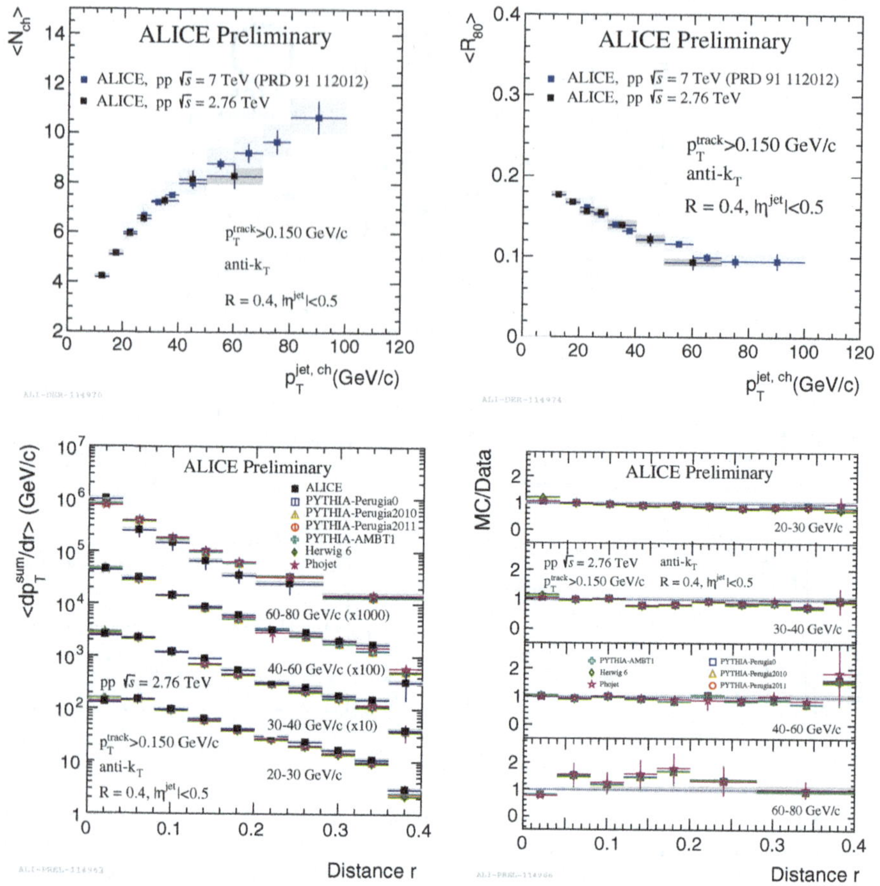

Fig. 62.2 ⟨N_{ch}⟩ (top left) and ⟨R_{80}⟩ (top right) at 2.76 TeV in comparison with that at 7 TeV. Bottom left: $p_T{}^{sum}$ distributions for different jet p_T bins for R = 0.4 with different MC generator predictions. Bottom right: Ratios MC/data. The vertical error bars in data and MC stand for statistical errors. The grey bands show the systematic uncertaintities

62.5 Conclusions

We reported measurements of charged jet cross section for R = 0.2, 0.3, 0.4 and 0.6 and jet shapes for R = 0.4 in pp collisions at 2.76 TeV with ALICE. ⟨N_{ch}⟩ increases with increasing jet p_T. Measurements of ⟨R_{80}⟩ and dp_T^{sum}/dr reveal that high p_T jets are more collimated. Various MC models well reproduce the data within uncertainties except ⟨N_{ch}⟩ which agrees within 15%. No $\sqrt{s}$ dependence is seen within the measured uncertainties between 2.76 and 7 TeV.

References

1. M. Cacciari, G.P. Salam, G. Soyez, JHEP **0804**, 063 (2008)
2. M. Cacciari, G.P. Salam, Phys. Lett. B **641**, 5761 (2006)
3. G. DAgostini, Nucl. Instrum. Meth. A **362**, 487498 (1995)
4. R. Brun, F. Carminati, S. Giani, CERN-W5013, CERN-W-5013, https://cds.cern.ch/record/1082634
5. ALICE Collaboration, JHEP **1207**, 116 (2012)
6. G. Corcella, I. Knowles, G. Marchesini, S. Moretti, K. Odagiri et al., JHEP **0101**, 010 (2001)
7. T. Sjostrand, S. Mrenna, P.Z. Skands, JHEP **0605**, 026 (2006)
8. S. Roesler, R. Engel, J. Ranft, arXiv:hep-ph/0012252

Chapter 63
Decay Properties of $J^{\pm}$ State of Quarkonia in Dirac Relativistic Formalism

Manan Shah, P. C. Vinodkumar, Tanvi Bhavsar and Bhavin Patel

63.1 Introduction

The recent experimental observations particularly in the quarkonia sector have generated renewed interest in the study of hadron spectroscopy [1–5]. The discovery of the η_c (2S), χ_{c_2} (2P) states and many high-precision experimental observations of various hadronic states [5] have necessitated reconsideration of the parameters involved in the previous studies. The prediction and discovery of the various excited states of the charmonium system encouraged the investigation of a variety of approaches to understand the systematic of charmonium spectra and decay properties, but new investigation of the Exotic meson, molecular and multiquark systems have created great interest in meson spectroscopy. Status of some of these states e.g. ψ(4160), X(4140), X(4274), ψ(4415), Z_b(10610) and Y_b(10890) are still controversial and there exist disparities related to their decay properties.

63.2 Theoretical Framework

The quark confining interaction of meson is considered to be produced by the non-perturbative multigluon mechanism and this mechanism is unfeasible to estimate theoretically from first principles of QCD. For the present study, we assume that the constituent quarks inside a meson are independently confined by an average potential

M. Shah (✉)
P. D. Patel Institute of Applied Sciences, CHARUSAT, Changa 388421, India
e-mail: mnshah09@gmail.com

P. C. Vinodkumar · T. Bhavsar
Department of Physics, Sardar Patel University, Vallabh Vidyanagar 388120, India

B. Patel
20 Tuxedo Crt, Scarborough, ON M1G 3S5, Canada

Md. Naimuddin (ed.), *XXII DAE High Energy Physics Symposium*, Springer Proceedings in Physics 203, https://doi.org/10.1007/978-3-319-73171-1_63

of the form [6, 7]

$$V(r) = \frac{1}{2}(1 + \gamma_0)(\lambda r^{0.1} + V_0) \tag{63.1}$$

where λ is the potential strength. In the stationary case, the spatial part of the quark wave functions $\psi(\mathbf{r})$ satisfies the Dirac equation given by

$$[\gamma^0 E_q - .\mathbf{P} - m_q - V(r)]\psi_q(\mathbf{r}) = 0. \tag{63.2}$$

The two component positive and negative energy solutions of the Dirac equation can be written as

$$\psi_A^{(+)}(\mathbf{r}) = N_{nlj}\begin{pmatrix} \frac{ig(r)}{r} \\ \frac{(\sigma.\hat{r})f(r)}{r} \end{pmatrix}\mathcal{Y}_{ljm}(\hat{r});\ \psi_B^{(-)}(\mathbf{r}) = N_{nlj}\begin{pmatrix} \frac{i(\sigma.\hat{r})f(r)}{r} \\ \frac{g(r)}{r} \end{pmatrix}(-1)^{j+m_j-l}\mathcal{Y}_{ljm}(\hat{r}) \tag{63.3}$$

Here $+/-$ represents the positive (quark) and negative (antiquark) energy states and N_{nlj} is the overall normalization constant. The optimised quark mass parameters m_b and m_c are 4.67, and 1.27 GeV respectively.

63.3 l^+l^-, $\gamma\gamma$ and gg Decay Widths

The right behavior of the wave function provides the correct predictions of the decay rates and important features of any successful model. The decay widths are computed using the expression given by [8, 9]

$$\begin{aligned}\Gamma(V \to l^+l^-) = \frac{8\alpha^2 e_Q^2}{\pi M^2}\Bigg|\int\Bigg[\left(\frac{2}{3} + \frac{m}{3E_p}\right) \times \hat{h}_{++}(p; S) + \left(\frac{2}{3} - \frac{m}{3E_p}\right)\hat{h}_{--}(p; S) \\ + \frac{2p}{3E_p}\hat{h}_{+-}(p; S) + \frac{\sqrt{2}}{3}\Bigg\{\left(1 - \frac{m}{E_p}\right) \times \hat{h}_{++}(p; D) - \left(1 - \frac{m}{E_p}\right)\hat{h}_{--}(p; D) \\ - \frac{2p}{E_p}\hat{h}_{+-}(p; D)\Bigg\}\Bigg] p^2 dp\Bigg|^2\end{aligned} \tag{63.4}$$

where $\hat{h}_{++}$, $\hat{h}_{+-}$ and $\hat{h}_{--}$ are energy amplitude. In the present model the expression for the vector decay constant f_V can be derived as

$$f_V = \left[\frac{3\Gamma(V \to l^+l^-)}{4\pi\alpha_{em}M_V}\right]^{\frac{1}{2}} \quad . \tag{63.5}$$

Table 63.1 Dileptonic and digamma decay widths (in keV)

	Dileptonic				Digamma				Digluon			
nS	[present]	[10]	[11]	Exp. [5]	[present]	[10]	[12]	Exp. [5]	[present]	[13]	[14]	Exp. [5]
1S	1.89	4.95	1.89	5.55	5.44	10.38	7.80	7.2 ± 2.0	39.33	31.24	39.79	26.7 ± 3.0
2S	1.34	1.69	1.04	2.35	3.68	3.38	3.50	7 ± 3.5	26.62	15.10	–	14 ± 0.7
3S	1.12	0.96	0.77	0.86	3.03	–	–	–	21.94	10.35	–	–
4S	0.99	0.65	0.65	0.58	2.67	–	–	–	19.35	–	–	–
5S	0.91	0.49	–	–	2.44	–	–	–	17.65	–	–	–
6S	0.85	0.39	–	–	2.27	–	–	–	16.44	–	–	–

Table 63.2 Vector decay constant of charmonia (in MeV)

nS	[present]	[15]	Exp. [5]
1S	466.93	393	416 ± 6
2S	428.22	293	304 ± 4
3S	409.36	258	187 ± 8

63.4 Result and Discussion

We have been able to predict the quarkonium s-wave masses of states which are in good agreement with the reported PDG values [5] and other theoretical models [10–14]. The predicted dileptons, digamma, digluon decay widths and decay constant are listed in Tables 63.1 and 63.2. The decay widths computed here would be useful for the identification of new mesonic resonances.

Acknowledgements One of the author (PCV) acknowledges DST-SERB, India for the financial support (Research Project No. SERB/F/8749/2015-16).

References

1. G. Bonvicini et al., (CLEO Collaboration), Phys. Rev. D **81**, 031104 (2010)
2. K.M. Ecklund et al., (CLEO Collaboration), Phys. Rev. D **78**, 091501 (2008)
3. B. Auger et al., (BABAR Collaboration), Phys. Rev. Lett. **103**, 161801 (2009)
4. A.K. Rai, B. Patel, P.C. Vinodkumar, Phys. Rev. C **78**, 055202 (2008)
5. C. Patrignani et al., (Particle Data Group), Chin. Phys. C **40**, 100001 (2016)
6. N. Barik, B.K. Dash, M. Das, Phys. Rev. D **31**, 1652 (1985); **26**, 2420 (1982)
7. M. Shah, B. Patel, P.C. Vinodkumar, Eur. Phys. J. C **76**, 36 (2016); Phys. Rev. D **90**, 014009 (2014)
8. I.T.O. Hitoshi, Prog. Theor. Phys. **63**, 2140 (1980)
9. Y. Abe, Prog. Theor. Phys. **60**, 639 (1978)
10. M. Shah, A. Parmar, P.C. Vinodkumar, Phys. Rev. D **86**, 034015 (2012)
11. S.F. Radford, W.W. Repko, Phys. Rev. D **75**, 074031 (2007)
12. G.A. Schuler, F.A. Berends, R. van Gulik, Nucl. Phys. B **523**, 423 (1998)
13. B. Patel, P.C. Vinodkumar, JOP G: Nucl. Part. Phys. **36**, 035003 (2009)
14. A. Parmar, B. Patel, P.C. Vinodkumar, Nucl. Phys. A **848**(3), 299–316 (2010)
15. O. Lakhina, E.S. Swanson, Phys. Rev. D **74**, 014012 (2006)

Chapter 64
Analyzing the Mesonic Spectrum Using the Method of Schottky Anomaly

Aritra Biswas

64.1 Introduction

The search for exotic hadrons has been rejuvenated in recent years with the discovery of the so called "pentaquark states" [1]. Along with the X, Y, Z states [2] discovered in the last decade by the different experimental groups [3] and the recent confirmation of the $\Lambda(1405)$ as a molecular state [4] (which had been proposed earlier [5]), this has rekindled the quest for a better theoretical understanding of these states. Various models have been proposed to this end over decades [6]. We propose a model independent analysis in order to identify states which differ in their underlying dynamics due to different interaction scales responsible for forming composite mesonic states using the method of Schottky anomaly.

64.2 C_V for an ideal system

Consider a two-level system with energy-gap Δ. The canonical partition function is given by $Z = 1 + e^{-\beta\Delta}$. Where $\beta = 1/k_B T$ is the inverse temperature.[1] The specific heat of the system at constant volume may be defined as

$$C_V = \beta^2 \left[\frac{1}{Z}\frac{\partial^2 Z}{\partial \beta^2} - \left(\frac{1}{Z}\frac{\partial Z}{\partial \beta} \right)^2 \right] = \beta^2[\langle E^2 \rangle - \langle E \rangle^2]. \tag{64.1}$$

[1]The temperature is introduced here simply as a mathematical parameter to define the partition function of the system and no assumption is made regarding the system being in a heat bath in equilibrium.

A. Biswas (✉)
The Institute of Mathematical Sciences (affiliated to HBNI), Chennai 600113, India
e-mail: aritrab@imsc.res.in

Md. Naimuddin (ed.), *XXII DAE High Energy Physics Symposium*, Springer Proceedings in Physics 203, https://doi.org/10.1007/978-3-319-73171-1_64

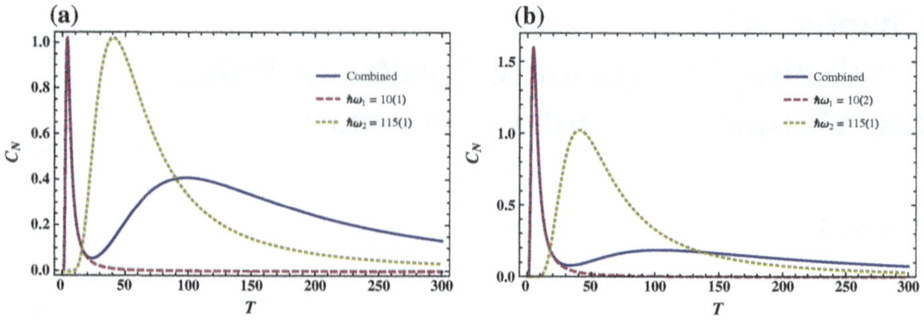

Fig. 64.1 Schematic illustration of the Schottky peaks in the ideal case when the data contains two scales. The cutoff in the principle quantum number n are indicated in brackets

Substituting for the partition function of the two level system we have

$$C_V = \beta^2 \frac{\Delta^2 e^{-\beta\Delta}}{(1 + e^{-\beta\Delta})^2}. \tag{64.2}$$

When C_V is plotted against $\beta\Delta$ the Schottky peak appears at a value $\beta\Delta \approx 2.4$. The location of the peak is a function of the energy gap in the system. A realistic spectra, in general, might have more than one scale in operation. In order to simulate the effect such a system we first consider the spectrum of a three-dimensional Harmonic oscillator. The partition function of the system is given by $Z_1 = \sum_{n=0}^{\infty} D(n)e^{-\beta\hbar\omega(n+3/2)}$ where the oscillator parameter ω defines the scale in the problem and $D(n)$ is the degeneracy of the level. We combine the spectra of two such systems with different values for the scale parameters (Fig. 64.1). The effect of combining is to normalize the specific heat with a single partition function given by $Z(\beta) = Z_1(\beta, \hbar\omega_1) + Z_2(\beta, \hbar\omega_2)$. The peaks appear when the spectrum has a cut off otherwise, it will saturate as a function of temperature. In the next section we give our results for the analysis of the current heavy mesonic spectra taken from [7]. For a detailed discussion the interested reader is referred to [8].

64.3 Analyzing the Experimental Spectra

- **The charmonium spectra**: When all the $c\bar{c}$ states are plotted together, the existence of two peaks at $T \approx 40$ MeV and $T \approx 190$ MeV are revealed, indicating the existence of two well defined scales, which might be identified with the "hyperfine" and the "confinement" scales respectively. When the states are grouped according to their individual J values $J = 0, 1, 2$, the hyperfine peak vanishes, and only the confinement peak is retained. However, when only the "exotic" states in the charmonium mass range are plotted, we again find a double-peaked structure, where the confinement peak shifts to 100 MeV along with a sharp peak below 10 MeV.

- **The bottomonium spectra** The two peaked structure with distinct hyperfine (22 MeV) and confinement (185 MeV) peaks are revealed for the case of the bottomonium spectrum also. The hyperfine peak disappears as expected when the states are grouped according to their J values and plotted again. The bottomonium exotics are however, scarce in number (three to be exact), with ill defined quantum numbers. It is hence premature to comment on these states. However, on plotting, they do retain the peak at 100 MeV, which is encouraging and may mean that the underlying mechanism for the charmonium and the bottomonium exotic states are the same.
- **The open charm spectra** Compared to those discussed previously, this case is more complicated due to the presence of the isospin (I) symmetry along with the J symmetry. However, on plotting all the open charm mesons together, the corresponding C_V vs T plot shows the two-peaked structure with the confinement peak at $T = 160$ MeV and the hyperfine peak at $T = 50$ MeV approximately. On plotting states with the same J and I together, the hyperfine peak vanishes and only the confinement peak is retained.
- **The open bottom spectra** The open bottom spectra are rather sparse as compared to the open charm scenario. Nevertheless, some features are already visible. The two peaked structure prevails on plotting all the states together. On grouping the states according to their J and I values, a well defined confinement peak is seen. However, due to the limited number of states, one should wait for the advent of more data in the future before drawing any conclusions for this sector.

64.4 Summary

I present a model independent analysis of the mesonic spectrum and show that it is possible to distinguish between the "exotic" and the regular states via to such an analysis, especially for the heavy meson sector. The light meson sector presents some difficulties, details of which can be found in [8].

References

1. R. Aaij et al., [LHCb Collaboration], Phys. Rev. Lett. **115**, 072001 (2015). https://doi.org/10.1103/PhysRevLett.115.072001, arXiv:1507.03414 [hep-ex]
2. S.L. Olsen, Front. Phys. (Beijing) **10**(2), 121 (2015). https://doi.org/10.1007/S11467-014-0449-6, arXiv:1411.7738 [hep-ex]
3. For more detailed experimental references for all the exotic states, see [2] and the references therein
4. J.M.M. Hall, W. Kamleh, D.B. Leinweber, B.J. Menadue, B.J. Owen, A.W. Thomas, R.D. Young, Phys. Rev. Lett. **114**(13), 132002 (2015). https://doi.org/10.1103/PhysRevLett.114.132002, arXiv:1411.3402 [hep-lat]
5. R.H. Dalitz, T.C. Wong, G. Rajasekaran, Phys. Rev. **153**, 1617 (1967). https://doi.org/10.1103/PhysRev.153.1617

6. S. Godfrey, S.L. Olsen, Ann. Rev. Nucl. Part. Sci. **58**, 51 (2008). https://doi.org/10.1146/annurev.nucl.58.110707.171145, arXiv:0801.3867 [hep-ph]
7. C. Patrignani et al., [Particle Data Group], Chin. Phys. C **40**(10), 100001 (2016). https://doi.org/10.1088/1674-1137/40/10/100001
8. A. Biswas, M.V.N. Murthy, N. Sinha, Phys. Rev. D **92**(11), 114012 (2015). https://doi.org/10.1103/PhysRevD.92.114012, arXiv:1509.06201 [hep-ph]

Chapter 65
Constraining Non-Standard Interactions of Neutrino Using ICAL Detector at INO

Amina Khatun, Sabya Sachi Chatterjee, Tarak Thakore and Sanjib Kumar Agarwalla

65.1 Introduction

The non-standard interactions (NSI's) of the neutrino are theoretically well motivated, and after the discovery of large θ_{13}, there is a huge interest in the community to look for these NSI's in the planned/proposed neutrino experiments. These NSI's can arise due to the presence of new heavy particles [1, 2], or, very light mediators [3], which have not been discovered yet. The proposed ICAL detector is designed to detect the atmospheric neutrinos having multi-GeV energy, and traversing vast ranges of baselines through the Earth matter. In the presence of flavor changing neutral current interactions, the MSW resonance of neutrino in the Earth matter gets modified and the ICAL detector can play an important role to explore these NSI's. The ICAL detector is capable enough to efficiently reconstruct the momentum of μ^- (μ^+) produced in the charge-current interaction of ν_μ ($\bar{\nu}_\mu$) with irons in the detector.[1] The hadrons are produced along with the $\mu^\pm$ in deep-inelastic scattering processes in multi-GeV energy, and they carry important information about the initial neutrino. The energy

[1]Detailed information about the ICAL detector can be found in [4].

A. Khatun (✉) · S. Sachi Chatterjee · S. Kumar Agarwalla
Institute of Physics, Sachivalaya Marg, Bhubaneswar 751005, India
e-mail: amina@iopb.res.in

S. Sachi Chatterjee
e-mail: sabya@iopb.res.in

S. Kumar Agarwalla
e-mail: sanjib@iopb.res.in

A. Khatun · S. Sachi Chatterjee · S. Kumar Agarwalla
Homi Bhabha National Institute, Training School Complex, Anushakti Nagar, Mumbai 400085, India

T. Thakore
Louisiana State University, Baton Rouge, LA 70803, USA
e-mail: tarakstar@gmail.com

Md. Naimuddin (ed.), *XXII DAE High Energy Physics Symposium*, Springer Proceedings in Physics 203, https://doi.org/10.1007/978-3-319-73171-1_65

of hadron ($E^{/}_{had} = E_{\nu} - E_{\mu}$) is calibrated using the number of hits in the detector due to hadron shower [5]. We study the future sensitivity of ICAL detector to place limit[2] on the NSI's using the observables E_{μ}, $\cos\theta_{\mu}$, and $E^{/}_{had}$ of each event.

65.2 Formalism of NSI's

We assume that the NSI's present only in the propagation through matter due to flavor changing neutral current interactions between neutrinos and matter fermions, $f = e, u, d$. The effective four fermion Lagrangian for such NSI can be written as:

$$L^{ff}_{\text{NSI}} = -2\sqrt{2}\, G_F\, \varepsilon^{Pf}_{\alpha\beta}\, (\bar{\nu}_\alpha \gamma^\rho L\, \nu_\beta)\, (\bar{f}\, \gamma^\rho\, P\, f), \tag{65.1}$$

where G_F is the Fermi constant. The superscript P represents the chiral projection operator $\{L, R = (1 \mp \gamma^5)/2\}$. The NSI relative coupling strength with respect to G_F is

$$\epsilon_{\alpha\beta} = \sum_{f=e,u,d} \frac{V_f}{V_e} \left(\varepsilon^{Lf}_{\alpha\beta} + \varepsilon^{Rf}_{\alpha\beta} \right), \tag{65.2}$$

where $V_f = \sqrt{2}\, G_F\, N_f$ with N_f as the density of matter fermion f in the Earth. The subscript α and β are used to indicate the neutrino flavor e, μ, and τ.

We explore $\varepsilon_{\mu\tau}$ keeping other NSI parameters zero in neutrino evolution equation in matter. Many attempts are made to constrain $\varepsilon_{\mu\tau}$ using the data of atmospheric neutrino experiments. Super-Kamiokande collaboration presented upper limit, $\epsilon_{\mu\tau} < 0.033$, at 90% C.L. in [7] by the combined analysis of SK I and SK II data. In the recent article [8], the strongest bound on $\varepsilon_{\mu\tau}$ is given using the one-year of high energy atmospheric neutrino data in IceCube experiment. The bound is $-6.0 \times 10^{-3} < \varepsilon_{\mu\tau} < 5.4 \times 10^{-3}$ @ 90% credible interval (C.I.).

We work in full 3 neutrino flavor paradigm using the numerically calculated oscillation probabilities with PREM profile of the Earth matter density. We generate the projected data at the ICAL detector using the oscillation parameters: $\sin^2\theta_{23} = 0.5$, $\sin^2 2\theta_{13} = 0.1$, $\sin^2 2\theta_{12} = 0.84$, $\Delta m^2_{21} = 7.5 \times 10^{-5}\,\text{eV}^2$, $\delta_{\text{CP}} = 0°$, and $\Delta m^2_{\text{eff}} = \pm 2.4 \times 10^{-3}\,\text{eV}^2$, where $+ (-)$ sign stands for normal (inverted) hierarchy. The effective mass square difference, Δm^2_{eff}, is related to atmospheric neutrino mass square difference Δm^2_{31} as expressed in [9]. While fitting, we marginalize over $\sin^2\theta_{23} \in [0.36, 0.66]$ and $\Delta m^2_{\text{eff}} \in \pm[2.1, 2.6] \times 10^{-3}\,\text{eV}^2$ keeping other oscillation parameters fixed at their true values. The binning scheme for the reconstructed E_μ, $\cos\theta_\mu$, and $E^{\prime}_{\text{had}}$ are given in Table 65.1.

[2]The sensitivity of ICAL to put the bound on NSI is presented in article [6] using reconstructed μ momentum (E_μ, $\cos\theta_\mu$).

Table 65.1 The binning scheme adopted for the reconstructed observable E_μ, $\cos\theta_\mu$, and E'_{had} for each muon polarity

Observable	Range	Bin width	Total bins
E_μ (GeV)	[1, 11]	1	10
	[11, 21]	5	2
$\cos\theta_\mu$	[−1.0, 0.0]	0.1	10
	[0.0, 1.0]	0.2	5
E'_{had} (GeV)	[0, 2]	1	2
	[2, 4]	2	1
	[4, 25]	21	1

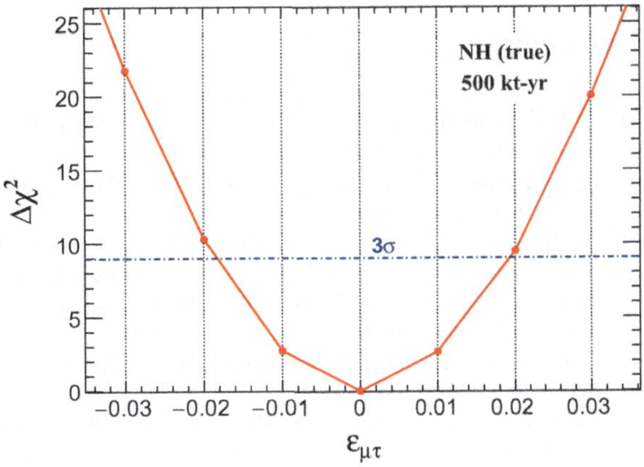

Fig. 65.1 Limit on $\varepsilon_{\mu\tau}$ using 500 kt-yr ICAL assuming normal hierarchy as true choice

65.2.1 Result

We produce the data with no new physics, and considering true values of oscillation parameters as mentioned above. In the fit, we introduce only the NSI parameter $\varepsilon_{\mu\tau}$ in the range $[-0.1, +0.1]$, and we marginalize over the five systematics uncertainties[3] and the oscillation parameters as mentioned before. We use the Poissonian χ^2 as expressed in [10].

Figure 65.1 shows the sensitivity of ICAL detector to constraint $\varepsilon_{\mu\tau}$ considering 10 years of run time assuming normal hierarchy (NH) as the true mass hierarchy. We can see from the figure that the ICAL detector can place at 3σ C.L. an upper limit on $|\varepsilon_{\mu\tau}|$ around 0.02.

---

[3]The details about these systematic uncertainties and their marginalization procedure using the pull method can be seen in [4, 10].

65.3 Final Remarks

The upcoming ICAL detector is capable to provide new informations on the neutrino oscillation parameters using atmospheric ν and $\bar{\nu}$ separately in the multi-GeV range, which traverse wide ranges of baselines. This provides an opportunity to explore whether the neutrinos feel NSI's while propagate through large distances in the Earth matter. In this paper, we set limit on the NSI parameter $\varepsilon_{\mu\tau}$ using reconstructed muon momentum and hadron energy of each event. We find that 500 kt-yr exposure of the ICAL detector can constrain the NSI parameter $|\varepsilon_{\mu\tau}| < 0.02$ at 3σ confidence level.

References

1. C. Biggio, M. Blennow, E. Fernandez-Martinez, JHEP **08**, 090 (2009). https://doi.org/10.1088/1126-6708/2009/08/090
2. O.G. Miranda, H. Nunokawa, New J. Phys. **17**(9), 095002 (2015). https://doi.org/10.1088/1367-2630/17/9/095002
3. Y. Farzan, I.M. Shoemaker, JHEP **07**, 033 (2016). https://doi.org/10.1007/JHEP07(2016)033
4. S. Ahmed, et al., Pramana **88**(5), 79 (2017). https://doi.org/10.1007/s12043-017-1373-4
5. M.M. Devi, A. Ghosh, D. Kaur, L.S. Mohan, S. Choubey, et al., JINST **8**, P11003 (2013). https://doi.org/10.1088/1748-0221/8/11/P11003
6. S. Choubey, A. Ghosh, T. Ohlsson, D. Tiwari, JHEP **12**, 126 (2015). https://doi.org/10.1007/JHEP12(2015)126
7. G. Mitsuka, et al., Phys.Rev. **D84**, 113008 (2011). https://doi.org/10.1103/PhysRevD.84.113008
8. J. Salvado, O. Mena, S. Palomares-Ruiz, N. Rius, JHEP **01**, 141 (2017). https://doi.org/10.1007/JHEP01(2017)141
9. H. Nunokawa, S.J. Parke, R. Zukanovich Funchal, Phys. Rev. **D72**, 013009 (2005). https://doi.org/10.1103/PhysRevD.72.013009
10. M.M. Devi, T. Thakore, S.K. Agarwalla, A. Dighe, JHEP **1410**, 189 (2014). https://doi.org/10.1007/JHEP10(2014)189

Chapter 66
Study of Trapping Probability in Proton Irradiated Silicon Pad Detectors Using Transient Current Technique Simulations

Geetika Jain, Chakresh Jain, Ranjeet Dalal, Ashutosh Bhardwaj and Kirti Ranjan

66.1 Introduction

In high energy physics experiments, prolonged exposure to radiation flux leads to radiation damage in the silicon (Si) detectors. This causes increase in leakage current, change in full depletion voltage and also charge carrier trapping, resulting in loss of charge collected (CC) and hindering particle detection operation [1]. The probability of a charge carrier to get trapped is given by the effective trapping probability, i.e., inverse of effective trapping time (τ_{eff}). Since τ_{eff} is not a directly measurable quantity, it can be estimated using an exponential term method described in [2]. In this method, a laser based characterization technique - the Transient Current Technique (TCT) is used to generate a time evolved carrier signal, also called the TCT signal. The integration of the TCT signal within a certain time window gives CC. Due to the charge trapping, CC decreases by an exponential trapping term, exp(-t/τ_{eff}), where t is the transient time, with increasing fluence. To compensate for the trapping, the TCT signal is corrected by introducing an exponential term, exp(t/τ_{r}), where τ_{r} is trapping time. The value of τ_{r} is varied to determine τ_{eff}, such that for $\tau_{\mathrm{r}} = \tau_{\mathrm{eff}}$, CC is independent and constant (after the full depletion voltage) as a function of bias voltage, for a particular fluence.

66.2 Simulated and Corrected TCT Signals

A similar procedure as described in the above section was carried out using TCAD device simulation tool. Silvaco [3] tool was used to generate TCT simulations at different voltages on proton irradiated silicon pad detectors [4]. The pad detectors were chosen to be p-on-n type with 15 kΩ·cm resistivity and 285 μm thickness.

G. Jain (✉) · C. Jain · R. Dalal · A. Bhardwaj · K. Ranjan
Centre for Detector and Related Software Technology, Department of Physics and Astrophysics, University of Delhi, New Delhi, India
e-mail: geetikajain.hep@gmail.com

Md. Naimuddin (ed.), *XXII DAE High Energy Physics Symposium*, Springer Proceedings in Physics 203, https://doi.org/10.1007/978-3-319-73171-1_66

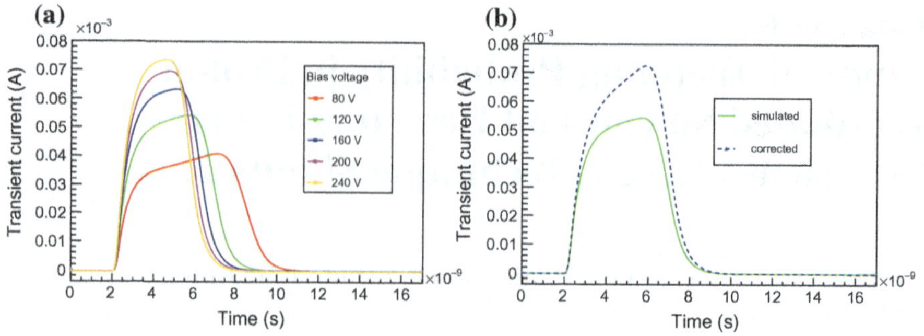

Fig. 66.1 TCT signal; **a** simulated at five different bias voltages, and **b** overlap of simulated and corrected for τ_r =20 ns, at 120 V; for fluence = $3 \times 10^{13} n_{eq} \cdot cm^{-2}$

The TCT signal at different bias voltages is shown in Fig. 66.1a as a function of time, for fluence value of $3 \times 10^{13} n_{eq} \cdot cm^{-2}$. Similar TCT signals are generated at different fluence values. Next, by choosing different values of τ_r, corrected TCT signals are generated. An example of the corrected TCT signal at 120 V is shown in Fig. 66.1b, for a value of τ_r equal to 20 ns.

66.3 Calculating Trapping Time Probability

CC is plotted for various τ_r as a function of bias voltage as shown in Fig. 66.2. It is observed that CC decreases for small τ_r and increases for large τ_r, with increasing bias voltage. Thus three cases can be thought of - I: if $\tau_r < \tau_{eff}$, this will cause incomplete correction of trapping; II: if $\tau_r > \tau_{eff}$, this will lead to more correction than what is required; and III: if $\tau_r = \tau_{eff}$, a complete and exact correction is accounted.

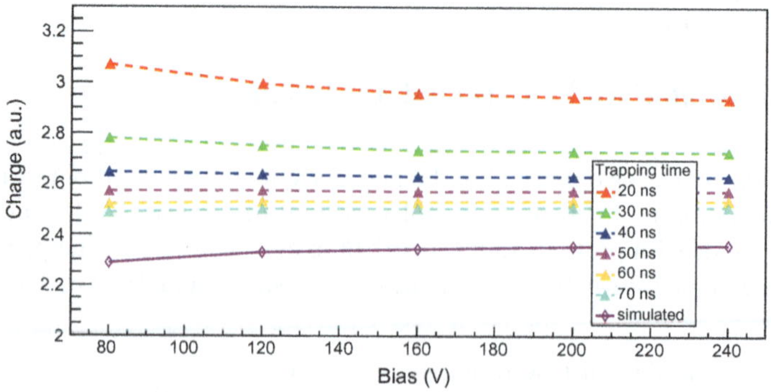

Fig. 66.2 The simulated and corrected charge versus voltage plot for different values of τ_r for fluence of $3 \times 10^{13} n_{eq} \cdot cm^{-2}$

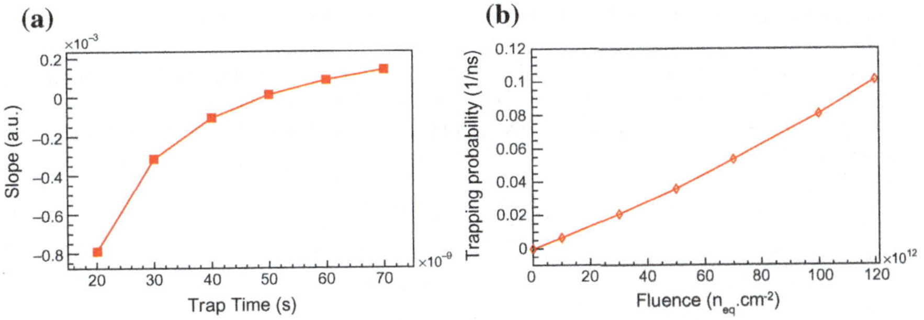

Fig. 66.3 **a** The slope of the linear fit to the corrected charge for the plot in Fig. 66.2 as a function of τ_r. **b** The effective trapping probability as a function of fluence

Hence, the slope of the linear fit for the plots is calculated. It is observed that the slope as a function of τ_r changes polarity (shown in Fig. 66.3a) from negative to positive, according to the three cases as mentioned above. The value of τ_r at which the curve intercepts the x-axis, i.e. when the slope goes to 0, is that value of τ_r for which CC is independent and constant as a function of bias voltage and hence equal to τ_{eff} for that fluence.

The above procedure is repeated for different values of the fluence to obtain τ_{eff}. Figure 66.3b shows the effective trapping probability plot as a function of fluence. Using a linear fit of the form $1/\tau_{eff} = \beta \cdot \phi$, the value of β is found to be equal to $6.66\times10^{-16} n_{eq}^{-1}\cdot cm^2\cdot ns^{-1}$. This is close to the value $4.2\times10^{-16} n_{eq}^{-1}\cdot cm^2\cdot ns^{-1}$ reported in [2].

66.4 Future Outlook

Further studies are envisaged at higher fluence, different detector depths, substrate resistivities, and also for p-type substrates. The effect of temperature on the effective trapping probabilities will also be investigated.

Acknowledgements Authors would like to thank University of Delhi R and D grant, DST, and UGC for research and financial support.

References

1. G. Lindstrom, M. Moll, E. Fretwurst, Radiation hardness of silicon detectors a challenge from high-energy physics. Nucl. Instrum. Methods Phys. Res. Sect. A **426**, 1–15 (1999). https://doi.org/10.1016/S0168-9002(98)01462-4
2. G. Kramberger et al., Determination of effective trapping times for electrons and holes in irradiated silicon. NIM A **476**, 645–651 (2002)

3. ATLAS Silvaco version 5.15.32.R *Users manual*, Silvaco webpage. (2009), http://www.silvaco.com
4. G. Jain, P. Kumari, K. Lalwani, R. Dalal, A. Bhardwaj, K. Ranjan, Development of transient current technique measurement setup at DU, in *Workshop on Contemporary Trends in High-Energy Physics and Experimentation*. Department of Physics, Panjab University, Chandigarh. 10–11 Mar 2014

Chapter 67
Charged Higgs Search in Bosonic Decays Using Jet Substructure at the LHC

Riley Patrick, Pankaj Sharma and Anthony G. Williams

67.1 Introduction

We study the prospects of a heavy charged Higgs at the Large Hadron Collider (LHC) in a type II two Higgs doublet model (2HDM-II) wherein the constraints coming from $b \to s\gamma$ decays dictate the charged Higgs boson mass $M_{H^\pm}$ be larger than 580 GeV [1]. We mainly focus on the bosonic decays of the charged Higgs, i.e., $H^\pm \to W^\pm h_i$. These decays become large and dominant in certain region of the parameter space. For the benchmark point, we take $M_{H^\pm} \sim M_H = 750$ GeV, $M_h = 125$ GeV and $M_A = 150$ GeV. For this hierarchical mass spectrum, interesting phenomenology occurs at the LHC.

Most importantly, because of large mass splitting between $H^\pm$ and A, the $W^\pm$ and A which are produced from $H^\pm$ decay, are highly boosted. These boosted bosons lead to very closely separated jets at the detector making them to resolve increasingly difficult. In such a scenario, jet substructure techniques prove to be quite useful. In this analysis we use boosted Higgs tagging algorithm to tag boosted Higgs arising from the decay of heavy charged Higgs.

R. Patrick · P. Sharma (✉) · A. G. Williams
Center of Excellence for Particle Physics at the Terascale, University of Adelaide, Adelaide, SA 5005, Australia
e-mail: pankaj.sharma@adelaide.edu.au

R. Patrick
e-mail: riley.patrick@adelaide.edu.au

A. G. Williams
e-mail: anthony.williams@adelaide.edu.au

Md. Naimuddin (ed.), *XXII DAE High Energy Physics Symposium*, Springer Proceedings in Physics 203, https://doi.org/10.1007/978-3-319-73171-1_67

67.2 Analysis

We use Cambridge/Aachen jet clustering algorithm with cone radius $R = 1.2$ to cluster the jets and call these jets to be "fat jets". Then we undo the last step of clustering and apply mass drop tagger to identify subjects. If two or more subjets are found, then these events are kept for further processing. After identifying two subjets, we apply b-tagging to each of them. If both the subjets are b-tagged, we refer them to be the Higgs jet. Subsequently, remaining hadronic constituents are clustered with anti-k_T algorithm and we call them to be "narrow jets".

Before proceeding with boosted decision tree (BDT) analysis, we identify two signal regions by acknowledging the fact that the W boson coming from $H^{\pm}$ tends to have large p_T from the one that come from top quark. So, we first reconstruct leptonic W boson from charged lepton and estimated neutrino momenta, and then apply the cut on its p_T. Events with W_{lep} with $p_T > 150\,\text{GeV}$ are attributed to signal region I (SRI) and vice versa to signal region (SRII).

In Fig. 67.1, we display the input kinematical variables in signal regions SRI (bottom) and SRII (top) which are passed through BDT algorithm to estimate the signal significance over background in each of the signal regions. In SRI, as the leptonic W boson emanates from the charged Higgs, the missing energy distribution tends to peak at higher values relative to background. Also, there is only one fat jet corresponding to pseudoscalar. We display the mass distribution of leading fat jet in Fig. 67.1 which peaks at the value of pseudoscalar mass. Finally charged Higgs is reconstructed from the momenta of leptonic W and reconstructed pseudoscalar. The

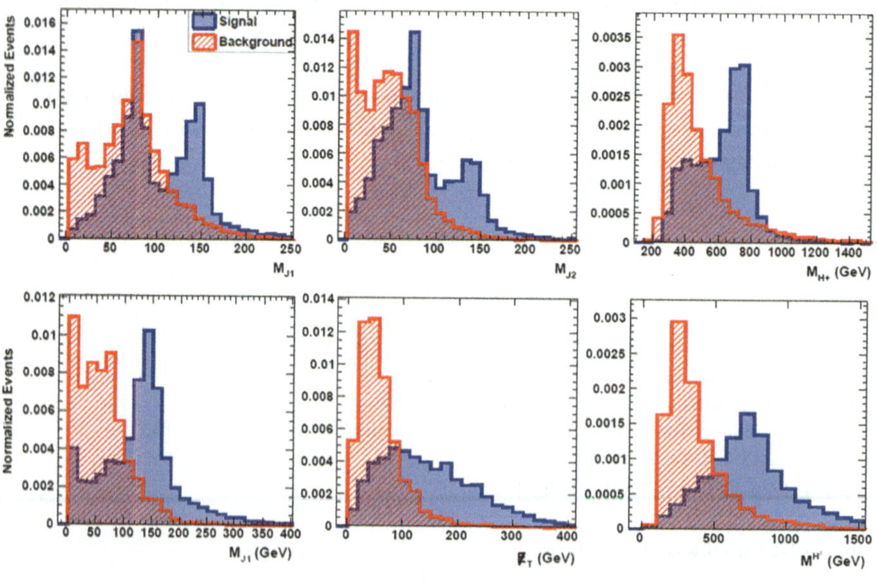

Fig. 67.1 Input kinematical variables for BDT analysis in SRI (bottom) and SRII (top)

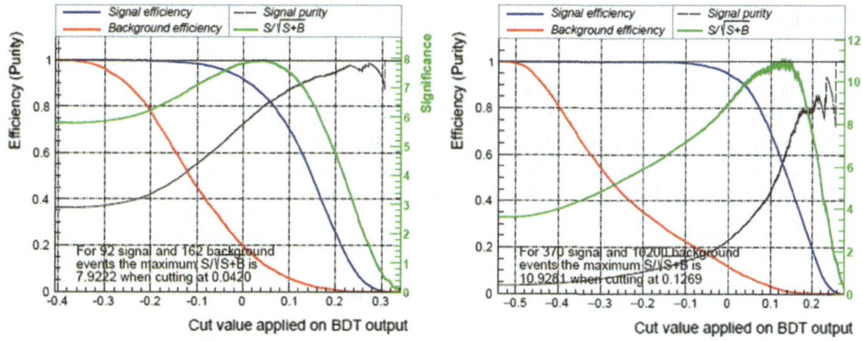

Fig. 67.2 Signal significance after BDT analysis in SRI (left) and SRII (right)

invariant mass peaks at the charged Higgs mass, thereby, showing the usefulness of jet substructure technique.

In the signal region SRII, as the charged Higgs decays hadronically into a hadronic W and a pseudoscalar, we expect to see two fat jets in these events. This peculiar feature is quite detrimental to backgrounds as very few background events have two fat jets with $p_T > 100\,\text{GeV}$. In the Fig. 67.1 top panel, we display the kinematical variables that posses highest separation capability between signal and background events. These variables have been passed through BDT to optimize the signal significance. The invariant mass distributions of first two leading jets show two peaks corresponding to a W boson and pseudoscalar. The charged Higgs is reconstructed from the momenta of reconstructed hadronic W and pseudoscalar and its distribution peaks at 750 GeV.

In Fig. 67.2, we display the signal significance as a function of cut over BDT response for SRI (left) and SRII (right). The signal significance in SRI with 100 fb^{-1} can be achieved to be at 8σ while for SRII, we find that with 1000 fb^{-1} of data, we can achieve around 10σ of significance which is a remarkable improvement over cut-based analysis.

67.3 Summary

To summarize, we study a heavy charged Higgs production at the LHC and focus on its bosonic decays which can be significant in certain regions of parameter space. The decay of such a $H^{\pm}$ leads to collimated final state jets that are harder to resolve. We apply jet substructure techniques to resolve them and reconstruct each object in an event. We then utilize BDT analysis to optimize the signal significance over the background.

References

1. M. Misiak et al., Phys. Rev. Lett. **114**, 221801 (2015)
2. J. Li, R. Patrick, P. Sharma, A.G. Williams, JHEP **1611**, 164 (2016)
3. R. Patrick, P. Sharma, A.G. Williams, Nucl. Phys. B **917**, 19 (2017)
4. S. Moretti, R. Santos, P. Sharma, Phys. Lett. B **760**, 697 (2016)
5. S. Moretti, R. Santos, P. Sharma, arXiv:1611.09082 [hep-ph]

Chapter 68
Light Nuclei Production in Heavy-Ion Collisions in STAR at RHIC BES Energies

Sabita Das

68.1 Introduction

The ultra-relativistic collisions of ions at high energy density provides the suitable conditions to produce light (anti-)nuclei. Light nuclei are expected to form at a later stage of the evolution of the system because of their low binding energies. The light (anti-)nuclei can be produced directly or via final state coalescence of produced nucleons and anti-nucleons or participant nucleons in such relativistic heavy-ion collisions [1, 2]. The probability of light nuclei formation is proportional to the product of the phase space densities of its constituent nucleons [1]. The invariant yields of light nuclei can be related to the primordial invariant yields of nucleons as,

$$E_A \frac{d^3 N_A}{d^3 p_A} = B_A (E_p \frac{d^3 N_p}{d^3 p_p})^Z (E_n \frac{d^3 N_n}{d^3 p_n})^{A-Z} \approx B_A (E_p \frac{d^3 N_p}{d^3 p_p})^A, \qquad (68.1)$$

where N_A, N_p, and N_n represent the yields of the particular nucleus, and of its constituent protons and neutrons, respectively. A and Z being the atomic mass number and atomic number, respectively. Here, protons and neutrons are assumed to be produced with identical momentum spectra and $p_A = A \times p_p$. B_A represents the coalescence parameter and it is related to the freeze-out correlation volume [2, 3]: $B_A \propto V_f^{1-A}$. The coalescence parameter, B_A, can be used to understand the space-time geometry of the system [4].

Sabita Das (for the STAR Collaboration).

S. Das (✉)
Institute of Particle Physics and Key Laboratory of Quark and Lepton Physics (MOE), Central China Normal University, Wuhan 430079, China
e-mail: sabita@rcf.rhic.bnl.gov

Md. Naimuddin (ed.), *XXII DAE High Energy Physics Symposium*, Springer Proceedings in Physics 203, https://doi.org/10.1007/978-3-319-73171-1_68

68.2 Results and Discussions

The results have been presented analysing the data sets collected in the year 2010 and 2011 by the Solenoidal Tracker at RHIC (STAR) detector [5] having a large uniform acceptance. The data sets include the first phase of Beam Energy Scan (BES) program from Au + Au collisions at $\sqrt{s_{NN}} = 7.7$, 11.5, 19.6, 27 and 39 GeV. In order to identify the light nuclei, the Time Projection Chamber (TPC) has been used. To extract the raw yield of light nuclei for a given p_T bin as shown in Fig. 68.1a, we have used a variable called Z which is in general defined as

$$Z_X = \log_e \frac{\langle dE/dx \rangle|_{\text{measured}}}{\langle dE/dx \rangle_X|_{\text{expected}}}, \quad (68.2)$$

where X is the particle type (d or $\bar{d}$, ^{3}He or $^3\overline{\text{He}}$), $\langle dE/dx \rangle|_{\text{measured}}$ is the measured energy loss of a track, $\langle dE/dx \rangle_X|_{\text{expected}}$ is the expected mean energy loss of particle X at a given momentum as obtained from modified Bethe-Bloch formula [6].

The raw yields are corrected on detector acceptance and tracking efficiency and also it includes momentum correction which are obtained from the embedding of Monte Carlo simulated data into the real data.

Figure 68.1b shows the p_T spectra for ^{3}He in Au + Au collisions at $\sqrt{s_{NN}} =$ 39 GeV for different centrality bins. Spectra are measured at mid-rapidity $|y| < 0.5$ and for four centrality bins. Errors of the p_T spectra are statistical only. The curves on the p_T spectra are Blast-wave model fits [7]. The average transverse momentum ($\langle p_T \rangle$) and integrated yields ($\frac{dN}{dy}$) have been calculated using the functional form for unmeasured region.

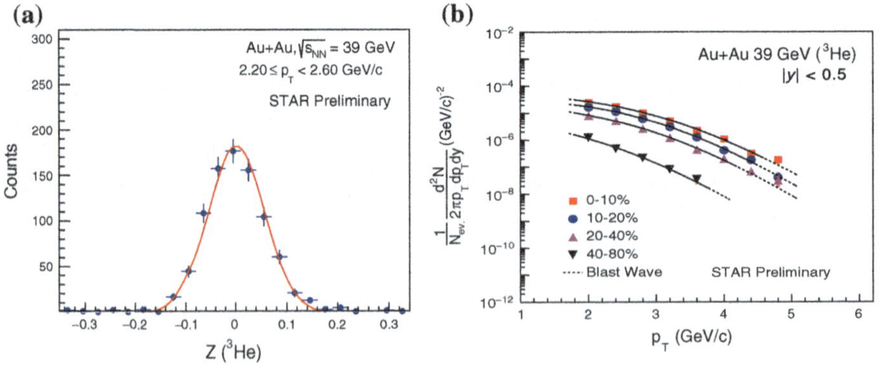

Fig. 68.1 **a** The Z distribution for ^{3}He in the TPC, at $2.2 < p_T < 2.6$ GeV/c and $|y| < 0.5$ in Au + Au collisions at $\sqrt{s_{NN}} = 39$ GeV. The curve represents the Gaussian fit. **b** Transverse momentum spectra for ^{3}He at mid-rapidity, $|y| < 0.5$ in Au + Au collisions at $\sqrt{s_{NN}} = 39$ GeV. Errors shown are statistical only. The curves represent the Blast-wave fits to the spectra

Table 68.1 The parameter (T) extracted from the exponential fit ($Texp(-x/p_0)$) to the $\frac{dN}{dy}$ of p, d and ^{3}He for 0–10% centrality. This can be compared with the chemical freeze-out parameters extracted using the $\frac{dN}{dy}$ of other light hadrons [8]

$\sqrt{s_{NN}}$ (GeV)	39	27	19.6	11.5	7.7
T (MeV)	165 ± 2	160 ± 2	161 ± 2	157 ± 2	149 ± 2

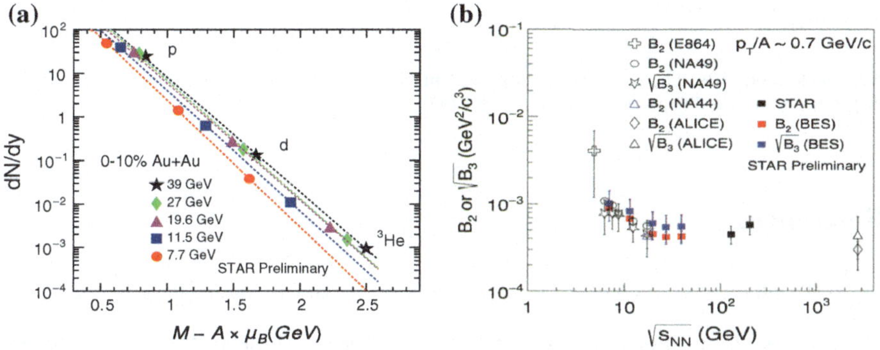

Fig. 68.2 **a** The $\frac{dN}{dy}$ of p, d and ^{3}He as a function of $M - A \times \mu_B$ for most central bins in Au + Au collisions at $\sqrt{s_{NN}} = 7.7, 11.5, 19.6, 27$, and 39 GeV. Errors shown for ^{3}He is statistical only and for p and d, errors are quadratic sum of statistical and systematic. The lines represent fit with an exponential function. **b** The coalescence parameters B_2 and $\sqrt{B_3}$ as a function of collision energy [9–11]

The integrated yield of ^{3}He have been compared with that of p and d. Figure 68.2a shows the $\frac{dN}{dy}$ of p, d and ^{3}He for 0–10% centrality in Au + Au collisions at RHIC BES energies, $\sqrt{s_{NN}} = 7.7, 11.5, 19.6, 27$, and 39 GeV. The lines represent the exponential fit to the yields and the extracted temperature is tabulated in Table 68.1. This shows an exponential decrease of $\frac{dN}{dy}$ with increase of mass of the particle. Similar behaviour have also been observed in lower and higher collision energies with different slopes [9–11].

Figure 68.2b shows the coalescence parameters B_2 and $\sqrt{B_3}$, which are derived by using the p_T spectra of p (feed-down corrected), d, and ^{3}He in (68.1), as a function of collision energy. B_A is found to be decreasing with increase of collision energies which corresponds to an increase in freeze-out volume. It is also observed that both B_2 and B_3 show strong centrality dependence in most central collisions. The smaller values of B_A in central collisions imply freeze-out volume at thermal freeze-out is larger than for peripheral collisions. This also indicates that the correlation length in nucleus coalescence grows with the system size.

68.3 Conclusion

We have reported the new results of light nuclei from Au + Au collisions at RHIC BES energies at $\sqrt{s_{NN}} = 7.7$, 11.5, 19.6, 27, and 39 GeV. The $\frac{dN}{dy}$ was to found decrease with the increases of mass of the particle. The extracted coalescence parameters B_2 and $\sqrt{B_3}$ were observed to have similar values. The smaller values of B_A for more central collisions imply an increasing source size with an increase in collision centrality.

Acknowledgements The work was supported in part by the MoST of China 973-Project No. 2015CB856901, NSFC under Grant No. 11575069.

References

1. H.H. Gutbrod et al., Phys. Rev. Lett. **37**, 667 (1976); H. Sato, K. Yazaki. Phys. Lett. B **98**, 153 (1981)
2. W.J. Llope et al., Phys. Rev. C **52**, 2004 (1995); R. Scheibl, U. Heinz, Phys. Rev. C **59**, 1585 (1999)
3. S.T. Butler, C.A. Pearson, Phys. Rev. **129**, 836 (1963); R. Scheibl, U. Heinz, Phys. Rev. C **59**, 1585 (1999)
4. A.Z. Mekjian, Phys. Rev. C **17**, 1051 (1978); S. Das Gupta, A.Z. Mekjian. Phys. Rep. **72**, 131 (1981)
5. K.H. Ackermann et al., Nucl. Instr. Meth. A **499**, 624 (2003)
6. H. Bichsel, Nucl. Instrum. Methods A **562**, 154 (2006)
7. E. Schnedermann et al., Phys. Rev. C **48**, 2462 (1993)
8. L. Adamczyk *et al.* (STAR Collaboration), arXiv:1701.07065
9. R. Arsenescu et al., NA52 collaboration. New J. Phys. **5**, 150 (2003)
10. J. Barrette et al., (E814 Collaboration), Phys. Rev. C **50**, 10771084 (1994)
11. T. Anticic et al., (NA49 Collaboration), Phys. Rev. C **94**, 044906 (2016); J. Adam et al. (ALICE Collaboration), Phys. Rev. C **93**, 2, 024917 (2016)

Chapter 69
Cosmological Pair Creation of Universe and Anti Universe at Big Bang

Abhishek K. Singh, K. Priyabrat Pandey, Sunita Singh and Supriya Kar

69.1 Introduction

In the article, we explore the generalized curvature in an effective D_4-brane to obtain a de Sitter Schwarzschild, Anti de Sitter Schwarzschild and a Topological de Sitter emergent black holes in a $U(1)$ gauge theory [1]. It is shown that the emergent geometries on an effective D_4-brane in a particular window identify themselves with the black hole geometries otherwise obtained in Einstein's gravity. The emergent gravity is sourced by a two form in a $U(1)$ gauge theory defined on a D_4-brane.

69.2 Effective Curvature on D_4-brane

We begin with a $U(1)$ gauge theory on a D_4-brane. In presence of the flat background metric $g_{\mu\nu}$, the A_μ-field dynamics is given by the $S = -\frac{1}{4C_1^2} \int d^5x \sqrt{-g}\, F^2$ where $C_1^2 = (4\pi^2 g_s)\alpha'^{1/2}$. Alternately, the Poincare dual to the electromagnetic field is described by a B_2-field which in turn describes a torsion. The two form gauge dynamics is given by the $S = -\frac{1}{12C_2^2} \int d^5x \sqrt{-g}\, H_{\mu\nu\lambda} H^{\mu\nu\lambda}$ where $C_2^2 = (8\pi^3 g_s)\alpha'^{3/2}$ and $H_{\mu\nu\lambda} = 3\nabla_{[\mu} B_{\nu\lambda]}$. In the context, an appropriate covariant derivative $\mathcal{D}_\mu$ on a brane may be constructed using the two form connections. The covariant derivative in a

A. K. Singh (✉)
Rajdhani College, University of Delhi, New Delhi, India

K. Priyabrat Pandey
Indraprastha University, Delhi, India

S. Singh
Miranda House College, University of Delhi, Delhi, India

S. Kar
Department of Physics and Astrophysics, University of Delhi, New Delhi, India

Md. Naimuddin (ed.), *XXII DAE High Energy Physics Symposium*, Springer Proceedings in Physics 203, https://doi.org/10.1007/978-3-319-73171-1_69

gauge theory becomes $D_\lambda B_{\mu\nu} = \nabla_\lambda B_{\mu\nu} - \Gamma_{\lambda\mu}{}^\rho B_{\rho\nu} + \Gamma_{\lambda\nu}{}^\rho B_{\rho\mu}$ where $-2\Gamma_{\mu\nu}{}^\rho = H_{\mu\nu}{}^\rho$. An iterative incorporation of B_2-corrections, to all orders, in the covariant derivative leads to an exact derivative. The covariant derivative modifies $H_{\mu\nu\lambda}$ to $\mathcal{H}_{\mu\nu\lambda} = H_{\mu\nu\lambda} + 3H_{[\mu\nu}{}^\alpha B^\beta{}_{\lambda]}\, g_{\alpha\beta} + ...$

Interestingly, a B_2-field dynamics in a first order formalism may be viewed as a torsion dynamics on an effective D_4-brane in a second order formalism leading to a fourth order generalized tensor $4\mathcal{K}_{\mu\nu\lambda}{}^\rho = 2\partial_\mu \mathcal{H}_{\nu\lambda}{}^\rho - 2\partial_\nu \mathcal{H}_{\mu\lambda}{}^\rho + \mathcal{H}_{\mu\lambda}{}^\sigma \mathcal{H}_{\nu\sigma}{}^\rho - \mathcal{H}_{\nu\lambda}{}^\sigma \mathcal{H}_{\mu\sigma}{}^\rho$. For a non-propagating torsion, $\mathcal{K}_{\mu\nu\lambda\rho} \to R_{\mu\nu\lambda\rho}$. In a second order formalism the $\mathcal{H}_3$ dynamics on an effective D_4-brane may be approximated by

$$S^{\text{eff}}_{\text{D}_4} = \frac{1}{3C_2^2} \int d^5x \sqrt{-\tilde{G}}\; \mathcal{K}^{(5)} , \tag{69.1}$$

where $\tilde{G} = \det \tilde{G}_{\mu\nu}$. Generically the emergent metric takes a form $\tilde{G}_{\mu\nu} = \left(\tilde{g}_{\mu\nu} + C\, \bar{\mathcal{H}}_{\mu\lambda\rho} \mathcal{H}^{\lambda\rho}{}_\nu\right)$, where $\tilde{g}_{\mu\nu} = \left(g_{\mu\nu} - B_{\mu\lambda} B^\lambda{}_\nu\right)$ and $B_{\mu\nu}$ signify the background fluctuations.

69.3 Tunneling Emergent de Sitter Geometries

The two form ansatz which solve the equation of motions are given by $B_{t\psi} = B_{r\psi} = \frac{b}{(2\pi\alpha')^{1/2}}$, $B_{\theta\psi} = \frac{\tilde{p}^3}{(2\pi\alpha')^{3/2}} \sin^2\psi \cot\theta$and $B_{\psi\phi} = \frac{p^3}{(2\pi\alpha')^{3/2}} \sin^2\psi \cos\theta$. A geometric torsion, sourced by a two form in a gauge theory, is worked out to yield $\mathcal{H}_{\theta\phi}{}^\psi = (2\pi\alpha')^{-1} \frac{p^3}{r^2} (\sin^2\psi \sin\theta)$, $\mathcal{H}_{\theta\phi}{}^t = -\mathcal{H}_{\theta\phi}{}^r = (2\pi\alpha')^{-3/2} \frac{bP^3}{r^2}(\sin^2\psi \sin\theta)$. In the brane window given by $\frac{P^6}{r^6} << \frac{r^2}{b^2} << 1$ and using Weyl scaling $\tilde{G}_{\mu\nu} = \frac{b^2}{r^2}\tilde{G}'_{\mu\nu}$ a large conformal factor due to the allowed range of r enforces a quantum geometry. The effective D_4-brane geometries in the quantum regime becomes

$$\begin{aligned} ds^2 = &- \left(-\left(1 - \frac{r^2}{b^2}\right) \mp \frac{p^6}{r^6}\right) dt^2 + \left(\left(1 - \frac{r^2}{b^2}\right) \mp \frac{p^6}{r^6}\right)^{-1} dr^2 + \frac{2r}{b}(dt + dr)\, r d\psi \\ &+ \left(1 \pm \frac{p^6}{r^6}\right)\left(2dtdr - \frac{2r}{b}(dt + dr)\, r d\psi + \frac{r^4}{b^2} d\Omega_3^2\right) . \end{aligned} \tag{69.2}$$

69.4 Projection Matrix: Sewing up of Quantum Patches

The projection matrix given by:

$$M = \frac{1}{2} \begin{pmatrix} \tilde{G}'_{tt}(S) & \tilde{G}'_{rr}(S) \\ \tilde{G}'_{rr}(T) & \tilde{G}'_{tt}(T) \end{pmatrix} . \tag{69.3}$$

where ; $\tilde{G}'_{tt}(S) = \left(1 - \frac{r^2}{b^2} - \frac{p^6}{r^6}\right)$, $\tilde{G}'_{rr}(S) = \left(1 - \frac{r^2}{b^2} - \frac{p^6}{r^6}\right)^{-1}$, $\tilde{G}'_{tt}(T) = \left(1 - \frac{r^2}{b^2} + \frac{p^6}{r^6}\right)$ and $\tilde{G}'_{rr}(T) = \left(1 - \frac{r^2}{b^2} + \frac{p^6}{r^6}\right)^{-1}$ yields present brane geometry but action of M^{-1} gives Schwarzschild de Sitter solution for first sign and Topological de Sitter solution in the

$$ds^2 = -\left(1 - \frac{r^2}{b^2} \mp \frac{P^6}{r^6}\right) dt^2 + \left(1 - \frac{r^2}{b^2} \mp \frac{P^6}{r^6}\right)^{-1} dr^2 \pm 2\left(1 - \frac{P^6}{r^6}\right) dtdr$$
$$\pm \frac{2P^6}{br^4} (dt + dr)\, d\psi + \left(\frac{r^2}{b^2} \mp \frac{P^6}{b^2 r^4}\right) r^2 d\Omega_3^2 . \quad (69.4)$$

One may interpret the action of M^{-1} as rejoining of casually disconnected quantum space time patches.

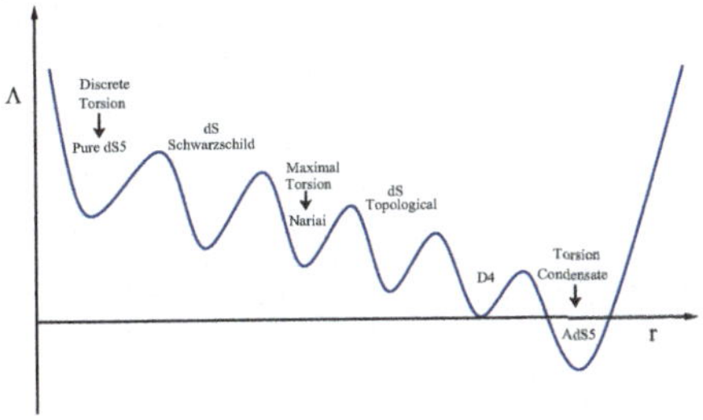

69.5 Tunneling

The pure de Sitter geometry represents degenerate matrix which may be interpreted as Big Bang singularity. It absorbs discrete quanta of Kalb Ramond and stabilizes into pair creation of Schwarzschild de Sitter (SDS) geometry and under various dynamical growth of parameters we observe SDS tunnels to Narai limit, and after that observer is in temporal phase. Further growth of parameter "p" leads to Topological de Sitter solution.

69.6 Conclusion

The effective space time underlying torsion dynamics has been proposed in this model. We expect at high energy the torsion to play dynamical role and at low energy the torsion freezes dynamical degree to generate well understood Einstein classical gravity solutions.

Reference

1. A.K. Singh, K.P. Pandey, S. Singh, S. Kar, JHEP **05**, 033 (2013)

Chapter 70
Effects of Leptonic Non-unitarity on Charged Lepton Flavor Violation, Leptogenesis and Lightest Neutrino Mass

Gayatri Ghosh and Kalpana Bora

70.1 Introduction

The cosmological constraints of the sum of the ν masses bound is $\sum_i m(\nu_i) < 0.23$ eV from CMB, Planck 2015 data (CMB15+ LRG+ lensing + H_0) [1]. We note that the lepton mixing matrix U has a big mixing and we know almost nothing about the phases. The discoveries of neutrino mass and leptonic mixing have come from the observation of neutrino flavor change, $\nu_\alpha \to \nu_\beta$. CP violation interchanges every particle in a process by its antiparticle. In the usual unitarity scenario, the three active neutrinos, the flavor eigen states ν_e, ν_μ, ν_τ are connected to the mass eigen states ν_1, ν_2, ν_3 via $\nu_\alpha = N_{\alpha i}\nu_i$, where $N^\dagger N = 1$. Here N is the generalised ν mixing matrix which could be both unitary and nonunitary. In the See-Saw picture, we assume that, just as there are 3 light neutrinos ν_1, ν_2, ν_3, there are 3 heavy right handed neutrinos M_1, M_2, M_3, where $M_R \sim 10^{9-14}$ GeV,$M_R \sim M_1$, M_2, M_3 which were there in the Hot Big Bang. The M_R decays modes are:

$$M \to l^- + H^+, M \to l^+ + H^-, M \to \nu + H^0, M \to \bar{\nu} + \bar{H^0} \tag{70.1}$$

where, l^- are e^-, μ^-, τ^- and H^+, H^-, H^0 are the Higgs field. CP violation effects in the M_R decays, may result from phases in the decay coupling constants. This leads to unequal numbers of leptons ($l^-$and ν) and antileptons ($l^+$and $\bar{\nu}$) in the Universe.

$$\Gamma(M \to l^+ + H^-) \neq \Gamma(M \to l^- + H^+) \tag{70.2}$$

The region of temperatures belonging to $10^9 < T/GeV < 10^{12}$ and $T/GeV < 10^9$ are respectively denoted as two and three flavor regimes of leptogenesis. The paper is organized as follows. In Sect. 70.2, we show the effect of low energy

G. Ghosh (✉) · K. Bora
Physics Department, Gauhati University, Guwahati 781014, Assam, India
e-mail: gayatrighsh@gmail.com

Md. Naimuddin (ed.), *XXII DAE High Energy Physics Symposium*, Springer Proceedings in Physics 203, https://doi.org/10.1007/978-3-319-73171-1_70

Table 70.1 Our calculated constraints on non-unitarity parameter $\eta_{\tau e}, \eta_{\tau\mu}, \eta_{\mu e}$ from latest cLFV decays

Branching ratios on cLFV decays	Calculated bounds on $\|\eta\|_{\alpha\beta}$
$BR(\mu \to e + \gamma) = 4.2 \times 10^{-13}$	$\|\eta_{\mu e}\| = 6.64733013 \times 10^{-6}$
$BR(\tau \to \mu + \gamma) = 4.4 \times 10^{-8}$	$\|\eta_{\tau\mu}\| = 5.11766 \times 10^{-3}$
$BR(\tau \to e + \gamma) = 3.3 \times 10^{-8}$	$\|\eta_{\tau e}\| = 7.021 \times 10^{-3}$

phenomonology of non unitarity on charged lepton flavor violating decays in type I seesaw theories and present the values of various parameters used in our analysis for the generation of baryon asymmetry of the Universe through the mechanism of leptogenesis. Section 70.3 contains our calculations and results. In this paper we calculate values of lightest neutrino mass, which is important as its value is not known precisely yet.

70.2 Low Energy Phenomonology of Non-unitarity and Leptogenesis

One interesting feature of non-unitarity of the PMNS matrix lies in Lepton Flavor Violation (LFV). Ee calculate latest updated bounds on $|\eta_{\tau e}|, |\eta_{\mu e}|, |\eta_{\mu e}|, |\eta_{\tau\mu}|$. The calculations are summarised in Table 70.1.

Returning to leptogenesis, the baryon asymmetry should lie in the interval, $5.8 \times 10^{-10} < Y_B < 6.6 \times 10^{-10}$ [2]. For the Normally ordered light ν masses, we have

$$M_R^{diag} = \text{diag}(M_1, M_2, M_3) = M_1 \text{diag}\left(1, \frac{M_2}{M_1}, \frac{M_3}{M_1}\right) = M_1 \text{diag}\left(1, \frac{m_1}{m_2}, \frac{m_1}{m_3}\right) \tag{70.3}$$

With $m_1 \in [10^{-6} eV, 10^{-1} eV], and, m_2^2 - m_1^2 = 7.60 \times 10^{-5} eV^2, m_3^2 - m_1^2 = 2.48 \times 10^{-3} eV^2$ as is evident from the ν oscillation data [3], m_1 being the lightest of three ν masses. For flavored leptogenesis regime, we take $M_1 \sim 10^{10}$ GeV. Next we do the parameter scan for flavored leptogenesis of a minimal seesaw model satisfying the Planck data on baryon to photon ratio of the universe.

70.3 Calculations and Results

We show in Fig. 70.1a that in the two flavor regime of leptogenesis, normal hierarchical structure of neutrino masses, non-unitarity texture of PMNS matrix gives rise to correct baryon asymmetry of the Universe, $5.8 \times 10^{-10} < Y_B < 6.6 \times 10^{-10}$, if the lightest ν mass lies around 0.023–0.03 eV, 0.034–0.038 eV and also around

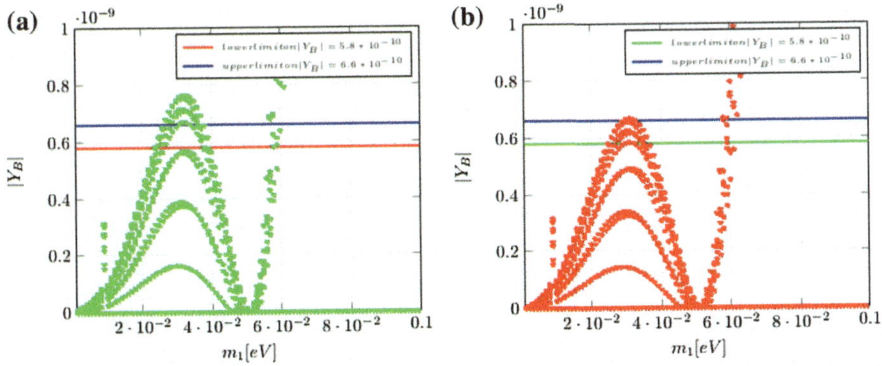

Fig. 70.1 Figure 70.1**a** Scatter plot of the lightest neutrino mass m_1 against the baryon asymmetry of the Universe with normal hierarchy, non-unitarity case in two flavored leptogenesis regime. Figure 70.1**b** Variation of the lightest neutrino mass m_1 against the baryon asymmetry of the Universe with normal hierarchy, unitarity in two flavor regime of leptogenesis

0.058–0.06 eV. If $M_1 \sim 10^{10}$ GeV, then 2 flavored leptogenesis is favored in the light of baryon asymmetry. We show in Fig. 70.1b, the scatter plot of the lightest neutrino mass m_1 against the baryon asymmetry of the Universe with normal hierarchy, unitarity of U_{PMNS} in two flavor regime. For Y_B to be in the region, $5.8 \times 10^{-10} < \eta_B < 6.6 \times 10^{-10}$, m_1 lies between 0.023–0.037 eV and 0.06 eV. For Y_B to be in the order 10^{-10}, m_1 is mostly concentrated in the region 0.01–0.065 eV. Detailed results, including results on one flavored, two flavored and three flavored leptogenesis will be presented elsewhere.

References

1. A.J. Cuesta (Barcelona, IEEC), V. Niro (Madrid, Autonoma U. Madrid, IFT), L. Verde (Barcelona, IEEC Harvard-Smithsonian Ctr. Astrophys. ICREA, Barcelona, Oslo U.), Phys. Dark Univ. **13** (2016), arXiv: 1511.05983
2. B.D. Fields, P. Molarto, S. Sarkar, Big Bang Nucleosynthesis, in Review of PDG-2014 (Astrophysical Constants and Parameters)
3. L.M. Cebola, D.E. Costa, R.G. Felipe, Eur. Phys. J. C **76**(3),156 (2016), arXiv:1601.06150

Chapter 71
A Density of States for QGP Fireball Formation in Heavy Ion Collisions Incorporating Hydrodynamical Features in the Model

Agam K. Jha, R. Ramanathan, K. K. Gupta and S. S. Singh

More data on the behaviour of Quark–Gluon Plasma (QGP) and its thermodynamics from ongoing Ultra Relativistic Heavy Ion Collisions (URHIC) experiments especially at LHC are expected. Albeit the main theoretical results from Lattice Gauge Computations [1] can be tested and counter checked with data, there is enough opportunity to test semi-phenomenological models of QGP. The thermodynamics of the system can easily be analysed with the parametrisation proposed in the paper. Ever since the suggestion that the QGP can be treated as a gas in thermodynamic quasi-equilibrium with its hadron gas environment, there have been a number of proposals for modelling the gas and the density of states of quarks and gluons in them in the literature [3]. In natural units ($\hbar = c = 1$), the relativistic momentum 'k' of the quarks and gluons is the argument of the density of states $\rho(k)$, where the total number of states '$N_{q,g}$' is given by

$$N_{q,g} = \lambda_{q,g} \int_0^\infty \rho(k)dk \,, \tag{71.1}$$

where the subscripts q, g refer to quarks and gluons and $\lambda_{q,g}$ refers to the QCD multiplicites for quarks and gluons. we parametrise $\rho(k)$ as

$$\rho(k) = \frac{\alpha}{k} + \beta k + \delta k^2 \,, \tag{71.2}$$

A. K. Jha (✉)
Department of Physics, Kirori Mal College, University of Delhi, Delhi 110007, India
e-mail: agamjha_2001@yahoo.co.in

R. Ramanathan · S. S. Singh
Department of Physics and Astrophysics, University of Delhi, Delhi 110007, India

K. K. Gupta
Department of Physics, Ramjas College, University of Delhi, Delhi 110007, India

Md. Naimuddin (ed.), *XXII DAE High Energy Physics Symposium*, Springer Proceedings in Physics 203, https://doi.org/10.1007/978-3-319-73171-1_71

where α, β and δ are scalar parameters obviously dimensional. In the expression, the first term is generated by hydrodynamical flow. Assuming only the relativistic shear viscous drag as the source of this term, the relativistic Navier–Stokes equation

$$\pi^{\mu\nu} = \eta \nabla^{\nu} u^{\mu} , \tag{71.3}$$

where $\pi^{\mu\nu}$ is the shear tensor in spacetime, u^{μ} is the shear velocity of hydrodynamical flow and η is the shear viscosity coefficient [5]. A comparison of the terms shows that to leading order the following approximate relation holds:

$$\frac{\eta}{4\pi} \sim \alpha , \tag{71.4}$$

The Free-energy of the system under pionic medium [3] is made up of the following pieces, namely the free energies F_q of the u,d,s quarks, the free energy F_g of gluons and the free energy F_π of the pionic hadrons.

The total free energy F by

$$F = \sum_{i=u,d,s} F_{q_i} + F_g + F_\pi \tag{71.5}$$

for the system of QGP in quasi-equilibrium with the pionic medium.

As in the literature [3], we have the following expression for the free energy pieces,

$$F_{q_i,g} = \mp T \lambda_{q_i,g} \int dk \rho(k) \ln(1 \pm e^{-(\sqrt{m_i^2+k^2})/T}) , \tag{71.6}$$

and

$$F_\pi = (3T/2\pi^2)\nu \int_0^\infty k^2 dk \ln(1 - e^{-\sqrt{m_\pi^2+k^2}/T}) , \tag{71.7}$$

where we take the multiplicities as usual for quark $\lambda_q = 6$ and for gluon $\lambda_g = 8$ and the masses $m_u = m_d = 0$ and $m_s = 150\,\text{MeV}$ and ν is the QGP volume which we take $\sim 1\ fm^3$ as in previous paper [3].

With these ingredients the entropy as usual is $S = -(\frac{\partial F}{\partial T})_V$, and the specific heat $C_V = -(\frac{\partial^2 F}{\partial^2 T})_V$. The velocity of sound in the system is $C_s^2 = \frac{S}{C_V}$.

On the basis of above equations and constructed three simultaneous equations with data input [1], the numerical values are $\alpha = 5.7 \times 10^4$, $\beta = -0.74\,\text{MeV}^{-2}$, $\delta = 1.3 \times 10^{-2}\,\text{MeV}^{-3}$ and $k_m \equiv 140\,\text{MeV}$ [4].

Our model only gives results for the quasi static equilibrium states of the QGP with some hydrodynamical features. Therefore our numerical values at different temperatures may be interpreted as static snapshots of a basically dynamical process. Therefore if we estimate the entropy density at different temperatures say at an initial value of $T = 0.35\,\text{GeV}$ and at a final freezeout temperature $T = 0.17\,\text{GeV}$ it turns out that the entropy density difference $s \sim 10^7$ in natural units. So that the experimentally verified rate $\frac{\eta}{s} \sim \frac{4\pi\alpha}{10^7} \sim 10^{-1}$ in order of magnitude. This agrees with the hydrodynamical model estimate of Luzum and Romatschke for this very important ratio [6].

References

1. P. Petreczky, Nucl. Phys. A **830** (2009)
2. F. Karsch, et.al., Nucl. Phys. B (proc. Suppl.) **94**, 411 (2001)
3. R. Balian et al., Ann. Phys. (N. Y.), **64**, 401 (1970); G.Neergaard et. al., Phys. Rev. D **62**, 034005 (2000); R. Ramanathan, Y.K. Mathur, K.K. Gupta, A.K. Jha Phys. Rev. C **70**, 027903 (2004); R. Ramanathan, K.K. Gupta, A.K. Jha, S.S. Singh. Pramana **68**, 757 (2007)
4. R. Ramanathan, A.K. Jha, S.S. Singh (2011), arXiv:1103.5049v1 [hep-ph]
5. R. Baier et. al., JHEP04 **100** (2008)
6. M. Luzum, P. Romatschke, Phys. Rev. C **78**, 034915 (2008)

Chapter 72
Study of $B_s^0 \to \phi\phi \to$ KKKK with the CMS Phase II Detector

Rajarshi Bhattacharya, Suchandra Dutta and Subir Sarkar

72.1 Introduction

In the Standard Model (SM) the decay $B_s^0 \to \phi\phi$ (with $\phi \to K^+K^-$) proceeds predominantly via a $b \to s\bar{s}s$ penguin amplitude (cf. Fig. 72.1). The first evidence for this decay was obtained by the CDF experiment at the Tevatron [1]. To date, the most precise determination of the phase $\phi_s = -0.17 \pm 0.15_{\text{stat}} \pm 0.03_{\text{syst}}$ was performed by the LHCb experiment, based on approximately 4000 B_s^0 candidates [2, 3]. In the SM, ϕ_s in the decay $B_s^0 \to \phi\phi$ is expected to be very small: QCD factorization calculations provide an upper limit of $\phi_s < 0.02$ [4–6].

The present analysis is a study of the CMS sensitivity for the overall process $B_s^0 \to \phi\phi \to$ KKKK at the LHC Phase II [7]. In particular, the goal of the analysis is to investigate how well the decay $B_s^0 \to \phi\phi$ can be triggered with the Level-1 (L1) tracks, whether the L1 trigger rate is manageable, and what overall signal event yield can be achieved. This is the first purely hadronic B decay investigated in the CMS experiment.

Phase II CMS Tracker

The present tracking system of CMS will be replaced with new detectors to cope with the High Luminosity LHC (HL-LHC) conditions where 140–200 collisions per bunch crossing, also known as pileup (PU) will take place [8]. The outer part of

On Behalf of CMS Collaboration, HBNI.

R. Bhattacharya (✉) · S. Dutta · S. Sarkar
Saha Institute of Nuclear Physics, Kolkata, India
e-mail: rajarshi.bhattacharya@cern.ch

S. Dutta
e-mail: suchandra.dutta@cern.ch

S. Sarkar
e-mail: subir.sarkar@cern.ch

Md. Naimuddin (ed.), *XXII DAE High Energy Physics Symposium*, Springer Proceedings in Physics 203, https://doi.org/10.1007/978-3-319-73171-1_72

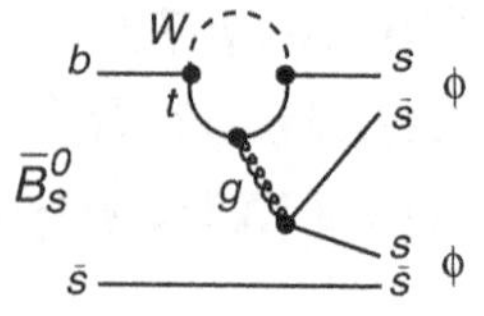

Fig. 72.1 Feynman diagram of the dominant process contributing to the decay $B_s^0 \rightarrow \phi\phi$

Table 72.1 Efficiency and rate for loose, medium and tight baselines, for different Pileup compositions. Uncertainty is statistical only

Baseline	Efficiency (%)		Rate (kHz)		
	L1	Offline	<PU>= 70	<PU>= 140	<PU>= 200
Loose	41.6 ± 1.2	64.6 ± 1.4	6.3 ± 1.5	27.9 ± 1.7	61.8 ± 5.2
Medium	36.6 ± 1.1	57.7 ± 1.3	2.5 ± 0.9	13.3 ± 1.2	29.6 ± 3.6
Tight	31.1 ± 1.0	57.6 ± 1.3	1.4 ± 0.7	5.1 ± 0.7	12.2 ± 2.3

the tracking system will be equipped with detectors capable of identifying high p_T tracks (>2 GeV) which will enable CMS to use tracking information at the L1 trigger system at an input rate of 40 MHz and significantly improve the performance of the L1 trigger decision to cope with the huge interaction rate [9].

Analysis

The signal events are generated with Pythia6 [10] and EvtGen at $\sqrt{s} = 14$ TeV with kaons having $p_T > 1.95$ GeV. At the analysis level the kaons are further required to have $p_T > 2$ GeV. On each signal event an average of 140 PU events are superimposed. MinimumBias events with $<PU> = 70$, 140, 200 compositions are used to study event rate.

ϕ candidates in an event are first reconstructed from all pairs of oppositely charged kaon tracks coming from the same vertex. Then, B_s^0 candidates are formed from a pair of reconstructed ϕ candidates compatible with the same vertex. The lowest-p_T kaon p_T lies very close to the threshold of the Level-1 tracking which is the major cause for loss of signal efficiency. Signal efficiency is studied both at the L1 and offline. The offline study involving tracks reconstructed with much higher precision is performed to check if one should expect any further significant reduction in efficiency after L1. Event rate is studied only at L1.

Three different baselines for event selection, namely loose, medium and tight have been used to find the optimum working point acceptable by CMS at L1. Figure 72.2 (Left) shows the reconstructed B_s^0 mass with the loose baseline selection. More than one B_s^0 candidates within the selected mass window can contribute to the same event.

Table 72.1 summarizes the efficiency and rate for all the selection baselines. The same results have also been presented in Fig. 72.2 (Right).

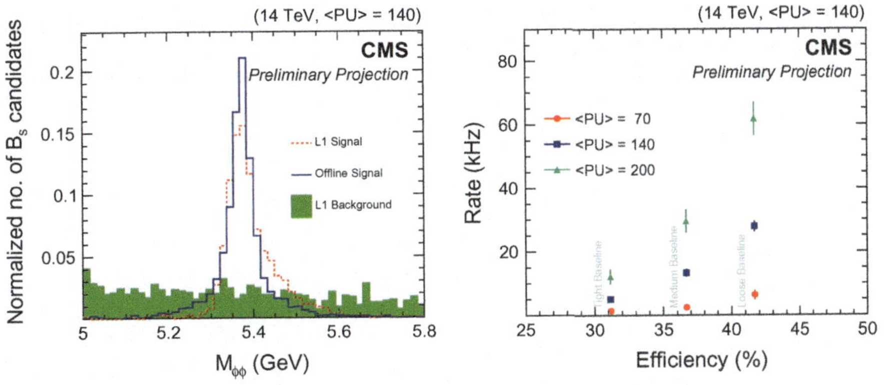

Fig. 72.2 (Left) ϕ-pair invariant mass [7]. Plots are normalized to unit area. The blue and red lines refer to offline and L1 signals respectively. The L1 background is shown as green filled area. (Right) Efficiency and rate for different baseline selection and for different pileup scenarios [7]. Uncertainty is statistical only

Summary

$B_s^0 \to \phi\phi \to$ KKKK is an important benchmark channel to study the capabilities of the CMS detector in low-p_T fully-hadronic final states. Signal efficiency and L1 trigger rate have been studied for several selection baselines. It has been found that the offline efficiency for the events triggered at L1 is high.

Acknowledgements The authors gratefully acknowledge Fabrizio Palla, Mario Galanti and Urs Langenegger for their invaluable help.

References

1. CDF Collaboration, First evidence for $B_s^0 \to \phi\phi$ decay and measurements of branching ratio and A_{CP} for $B^+ \to \phi K^+$. Phys. Rev. Lett. **95**, 031801 (2005)
2. LHCb Collaboration, Measurement of CP violation in $B_s^0 \to \phi\phi$ decays. Phys. Rev.**D90**(5), 052011 (2014)
3. LHCb Collaboration, Measurement of the $B_s^0 \to \phi\phi$ branching fraction and search for the decay $B^0 \to \phi\phi$. JHEP **10**, 053 (2015)
4. M. Bartsch, G. Buchalla, C. Kraus, $B \to V(L)V(L)$ Decays at Next-to-Leading Order in QCD, arXiv:0810.0249
5. M. Beneke, J. Rohrer, D. Yang, Branching fractions, polarisation and asymmetries of $B \to VV$ decays. Nucl. Phys. B **774**, 64–101 (2007)
6. H.-Y. Cheng, C.-K. Chua, QCD factorization for charmless hadronic B_S decays revisited. Phys. Rev. D **80**, 114026 (2009)
7. CMS Collaboration, Updates on Projections of Physics Reach with the Upgraded CMS Detector for High Luminosity LHC, CMS Performance Note CMS DP-2016/064 (2016)
8. CMS Collaboration, Technical Proposal for the Phase-II Upgrade of the CMS Detector, CMS Technical Proposal 2015/010 (2015)

9. M. d'Alfonso et.al, Use of tracking in the CMS L1 trigger for the phase-2 upgrade, CMS Detector Note CMS DN-2014/002 (2014)
10. T. Sjöstrand, S. Mrenna, P. Skands, PYTHIA 6.4 physics and manual. JHEP **05**, 026 (2006)

Chapter 73
Study of the Rare Decays $B^*_{s,d} \to \mu^+\mu^-$

Suchismita Sahoo and Rukmani Mohanta

73.1 Introduction

In recent times, the rare B meson decays mediated by flavour changing neutral current (FCNC) $b \to s, d$ transitions have been providing crucial information in our search for new physics (NP) beyond the standard model (SM). The LHCb Collaboration has reported several anomalies in semileptonic B decays at the level of few standard deviations. The rare $B_{s,d} \to \mu^+\mu^-$ processes are highly suppressed in the SM as they proceed through one loop and box diagrams. These decays also encounter additional helicity suppression. Their theoretical branching fractions are given by [1]

$$\begin{aligned} \text{BR}(B_s \to \mu^+\mu^-)|_{\text{SM}} &= (3.65 \pm 0.23) \times 10^{-9}, \\ \text{BR}(B_d \to \mu^+\mu^-)|_{\text{SM}} &= (1.06 \pm 0.09) \times 10^{-10}\,. \end{aligned} \tag{73.1}$$

In this work, we investigate the $B^*_{s,d} \to \mu^+\mu^-$ processes in the context of leptoquark (LQ) and Z' model. These processes don't suffer from helicity suppression like $B_{s,d} \to \mu^+\mu^-$ processes and the only non-perturbative quantity involved is the decay constant of $B^*_{s,d}$ mesons, which can be precisely calculated from lattice. LQs are color triplet bosons that couples to both quarks and leptons. The baryon and lepton number violating LQs have mass near the grand unification scale to avoid rapid proton decay. Thus we consider the baryon and lepton number conserving LQs, which do not induce proton decay and could be light enough to be accessible in accelerator searches. On the other hand, Z' is a color singlet boson, which could be naturally derived from the extension of electroweak symmetry of the SM by adding additional $U(1)'$ gauge symmetry. The family non-universal Z' model [2] is the sim-

S. Sahoo (✉) · R. Mohanta
School of Physics, University of Hyderabad, Hyderabad 500046, India
e-mail: suchismita@uohyd.ac.in

R. Mohanta
e-mail: rukmani98@gmail.com

Md. Naimuddin (ed.), *XXII DAE High Energy Physics Symposium*, Springer Proceedings in Physics 203, https://doi.org/10.1007/978-3-319-73171-1_73

plest one to explore the inconsistency between the observed experimental data and the corresponding SM predicted values in some of the observables associated with $b \to sl^+l^-$ decays. The popular πK puzzle in the hadronic $B \to \pi K$ decays [3] and other anomalies associated with $b \to s\mu^+\mu^-$ transitions observed at LHCb could be explained in the Z' model.

The paper is organised as follows. In Sect. 73.2, we discuss the effective Hamiltonian of $b \to s, d$ transitions in the SM. We then compute the $B^*_{s,d} \to \mu^+\mu^-$ processes in the SM as well as LQ and Z' model. Section 73.3 contains the conclusion.

73.2 $B^*_{s,d} \to \mu^+\mu^-$ processes

In the SM, the effective Hamiltonian of rare processes involving the quark-level transitions $b \to ql^+l^-$, where $q = d, s$ is given by [4]

$$\mathcal{H}_{\text{eff}} = -\frac{G_F}{\sqrt{2}}\left[\lambda_t^{(q)}\mathcal{H}_{\text{eff}}^{(t)} + \lambda_u^{(q)}\mathcal{H}_{\text{eff}}^{(u)}\right] + h.c., \tag{73.2}$$

where

$$\begin{aligned} \mathcal{H}_{\text{eff}}^{(u)} &= C_1(\mathcal{O}_1^c - \mathcal{O}_1^u) + C_2(\mathcal{O}_2^c - \mathcal{O}_2^u), \\ \mathcal{H}_{\text{eff}}^{(t)} &= C_1\mathcal{O}_1^c + C_2\mathcal{O}_2^c + \sum_{i=3}^{10} C_i\mathcal{O}_i, \end{aligned} \tag{73.3}$$

G_F is the Fermi constant, $\lambda_k^{(q)} = V_{kb}V_{kq}^*$ and C_i's are the Wilson coefficients.

The decay widths of $B_q^* \to \mu^+\mu^-$ processes are given by [5]

$$\begin{aligned} \Gamma(B_q^* \to \mu^+\mu^-) = \frac{G_F^2\alpha^2}{96\pi^3}|V_{tb}V_{tq}^*|^2 f_{B_q^*}^2 m_{B_q^*}^2 \sqrt{m_{B_q^*}^2 - 4m_l^2} \times \\ \left[\left|C_9^{\text{eff}} + 2\frac{m_b}{m_{B_q^*}}C_7^{\text{eff}}\right|^2 + \left|C_{10}\right|^2\right], \end{aligned} \tag{73.4}$$

where α is the fine structure constant, $f_{B_q^*}$ and $m_{B_q^*}$ are the decay constant and mass of B_q^* meson respectively and m_l is the mass of lepton. These processes are sensitive to the $C_{7,9}^{\text{eff}}$ Wilson coefficients, i.e., $\mathcal{O}_7$ and $\mathcal{O}_9$ operators, whereas the contributions from these operators vanish in the case of $B_{s,d} \to \mu^+\mu^-$ processes. In order to calculate the branching fractions, we need to know the total decay widths of $B^*_{s,d}$ vector bosons. However, these are neither measured nor precisely known theoretically except for the decay widths of $B^*_{s,d} \to B_{s,d}\gamma$ decays. Now using the particle masses from [6] and taking the decay widths of radiative $B^*_{s,d}$ bosons from [7], the branching fractions are found to be

Table 73.1 Predicted branching fractions of $B^*_{s,d} \to l^+l^-$ decays in the LQ and Z' model

Decay processes	Values in $Y = 1/6$ LQ model	Values in $Y = 7/6$ LQ model	Values in Z' model
$B_s^* \to \mu^+\mu^-$	$(1.7–3.19) \times 10^{-11}$	$(1.7–1.93) \times 10^{-11}$	$(1.7–2.2) \times 10^{-11}$
$B_d^* \to \mu^+\mu^-$	$(2.38–8.99) \times 10^{-13}$	$(2.47–5.4) \times 10^{-13}$	$(1.67–2.23) \times 10^{-13}$

$$\mathrm{BR}(B_s^* \to \mu^+\mu^-)|_{\mathrm{SM}} = (1.7 \pm 0.2)\left(\frac{0.07\ \mathrm{KeV}}{\Gamma^{\mathrm{tot}}_{B_s^*}}\right) \times 10^{-11},$$

$$\mathrm{BR}(B_d^* \to \mu^+\mu^-)|_{\mathrm{SM}} = (1.86 \pm 0.21)\left(\frac{0.2\ \mathrm{KeV}}{\Gamma^{\mathrm{tot}}_{B_d^*}}\right) \times 10^{-13}\,. \qquad (73.5)$$

These predicted branching fractions are sizable, about two order lower than the branching fractions of the corresponding pseudoscalar mesons decay.

The effective Hamiltonian in (73.2) can be modified in the both the LQ and Z' model. We consider $S_1(3, 2, 7/6)$ and $S_2(3, 2, 1/6)$ scalar LQ multiplets, which are invariant under the SM $SU(3)_C \times SU(2)_L \times U(1)_Y$ gauge group. These will contribute new $C_{9,10}^{(\prime)LQ}$ coefficients to the SM. In the NP model, the Wilson coefficients in (73.4) will be replaced by $C_{9,10} \to C_{9,10} + C_{9,10}^{(\prime)NP}$. Now comparing the branching fraction of $B_{s,d} \to \mu^+\mu^-$ with the 1σ uncertainty of experimental data, the constraints on the LQ couplings (λ) for $M_{LQ} = 1$ TeV are found to be [8]

$$0 \le \frac{|\lambda^{32}\lambda^{22*}|}{M_S^2} \le 5 \times 10^{-3}, \quad 1.5 \times 10^{-3} \le \frac{|\lambda^{32}\lambda^{12*}|}{M_S^2} \le 3.9 \times 10^{-3}\,. \quad (73.6)$$

Using the constrained couplings, we predict the branching fractions of $B^*_{s,d} \to \mu^+\mu^-$ in both $S_{1,2}$ LQ model and the corresponding values are listed in Table 73.1.

Similar to the LQ model, the presence of Z' boson will also provide $C_{9,10}^{Z'}$ new coefficients. Varying the mass difference of $B^0_{s,d} - \bar{B}^0_{s,d}$ mixing within the 2σ allowed range of experimental value, the constraints on ρ_q^L parameters are [4]

$$0 \le \rho_s^L \le 0.5 \times 10^{-3} \quad 1 \times 10^{-4} \le \rho_d^L \le 1.25 \times 10^{-4}, \qquad (73.7)$$

where $\rho_q^L = \frac{g_2 M_Z}{g_1 M_Z'} B_{qb}^L$ and B_{qb}^L are the left handed FCNC $b_L - q_L - Z'$ couplings [9]. We consider the coupling of Z' with the leptons as SM like. Using the above constrained couplings, the predicted branching fractions are given in Table 73.1. The detailed calculation of these processes in the SM as well as in both the LQ and Z' model can be found in [5] and [4], respectively.

73.3 Conclusions

We have studied the rare leptonic decays of $B^*_{s,d}$ bosons in both the scalar LQ and the family nonuniversal Z' model. We constrain the NP parameters using the measured branching fractions of $B_{s,d} \to \mu^+\mu^-$ processes and the $B^0_{s,d} - \bar{B}^0_{s,d}$ mixing data. We then estimated the branching fractions of $B^*_{s,d} \to \mu^+\mu^-$ processes, which are found to be sizable and within the reach of LHC experiments.

We thank Science and Engineering Research Board (SERB), Government of India for financial support through the grant No. SB/S2/HEP-017/2013.

References

1. C. Bobeth, M. Gorbahn, T. Hermann, M. Misiak, E. Stamou, M. Steinhauser, Phys. Rev. Lett. **112**, 101801 (2014)
2. P. Langacker, M. Plumacher, Phys. Rev. D **62**, 013006 (2000)
3. Q. Chang, X.Q. Li, Y.D. Yang, JHEP **0905**, 056 (2009), arXiv:0903.0275; V. Barger, C.W. Chiang, P. Langacker, H.S. Lee, Phys. Lett. B **598**, 218 (2004)
4. S. Sahoo, R. Mohanta, J. Phys. G **44**, 035001 (2017), arXiv:1612.02543
5. B. Grinstein, J.M. Camalich, Phys. Rev. Lett. **116**, 141801 (2016)
6. K.A. Olive et al., Particle data group. Chin. Phys. C **38**, 090001 (2014)
7. A. Khodjamirian, T. Mannel, A.A. Petrov, JHEP **11**, 142 (2015)
8. S. Sahoo, R. Mohanta, Phys. Rev. D **91**, 094019 (2015), arXiv:1501.05193
9. V. Barger, C.W. Chiang, J. Jiang, P. Langacker, Phys. Lett. B **596**, 229 (2004)

Chapter 74
Heavy-Flavour Measurements in p-Pb Collisions with ALICE at the LHC

Jitendra Kumar

74.1 Introduction

Heavy quarks (charm and beauty) due to their large masses are predominantly produced in hard-scattering processes in the initial phase of hadronic collisions. Therefore, they are excellent probes to study the properties of the Quark–Gluon Plasma created in relativistic heavy-ion collisions. The measurement of their production in p-Pb collisions is important to disentangle the hot nuclear matter effects present in heavy-ion collisions from cold nuclear matter (CNM) effects, such as transverse momentum broadening, nuclear modification of the parton distribution functions, initial-state multiple scatterings and energy loss. These effects can be investigated by measuring the nuclear modification factor R_{pPb}, defined as the ratio of particle cross section $\mathrm{d}\sigma/\mathrm{d}p_{\mathrm{T}}$ measured in p-Pb collisions to that measured in pp collisions scaled by the atomic mass number of Pb nuclei. In the absence of CNM effects R_{pPb} is expected to be unity. The R_{pPb} of D mesons and leptons from charm and beauty hadron decays at central and forward rapidities was studied in p-Pb collisions at $\sqrt{s_{\mathrm{NN}}} = 5.02$ TeV with ALICE.

74.2 Analysis Details

Prompt D mesons and their charge conjugates are reconstructed via their hadronic decay channels: $\mathrm{D}^0 \rightarrow \mathrm{K}^+\pi^-$, $\mathrm{D}^+ \rightarrow \mathrm{K}^-\pi^+\pi^+$ and $\mathrm{D}^{*+} \rightarrow \mathrm{D}^0\pi^+$ [1]. The extraction of the signal is based on an invariant mass analysis of reconstructed decay vertices

Jitendra Kumar for the ALICE collaboration.

J. Kumar (✉)
Indian Institute of Technology Bombay, Mumbai, India
e-mail: jitendra.kumar@cern.ch

Md. Naimuddin (ed.), *XXII DAE High Energy Physics Symposium*, Springer Proceedings in Physics 203, https://doi.org/10.1007/978-3-319-73171-1_74

displaced from the primary vertex by few hundred microns. The necessary spatial resolution on the track position is guaranteed by the Inner Tracking System (ITS) and the Time Projection Chamber (TPC) covering a pseudorapidity region $|\eta| < 0.8$. Particle identification (PID) of the decay particle species is also exploiting using the measurement of specific energy loss ($\mathrm{d}E/\mathrm{d}x$) in the TPC and of the time of flight with the Time-Of-Flight (TOF) detector. Kaons and pions are identified up to $p_{\mathrm{T}} = 2$ GeV/c. The electrons from heavy-flavour (HF) hadron decays are identified using ITS, TPC and TOF detectors in the range $0.5 < p_{\mathrm{T}} < 6\,\mathrm{GeV}/c$ and using the TPC and the Electromagnetic Calorimeter (EMCal) for $p_{\mathrm{T}} > 6$ GeV/c [2]. The background from π^0 and η Dalitz decay and from photon conversions is subtracted via the invariant mass method, and the hadron contamination decays is statistically subtracted [2]. The muons from heavy-flavour hadron decays are measured with the muon spectrometer in pseudorapidity range, $2.5 < y_{\mathrm{lab}} < 4$ [3] (additional information in [4]). The background from π and K decays is subtracted using a data-tuned Monte Carlo cocktail.

74.3 Results

The R_{pPb} of prompt D mesons (D^0, D^+ , and D^{*+} average) is found compatible with unity, as shown in Fig. 74.1 (left plot), and described by models which include CNM effects [5]. The comparison to the nuclear modification factor in Pb–Pb collisions, R_{AA}, is reported in Fig. 74.1 (right plot) and highlights a strong suppression for $p_{\mathrm{T}} > 3\,\mathrm{GeV}/c$ in central (0–10%) and semi-central Pb–Pb collisions (30–50%) [6]. This comparison allows to conclude that the suppression observed in Pb–Pb collisions is due to final-state effects induced by the interaction of heavy quarks with the

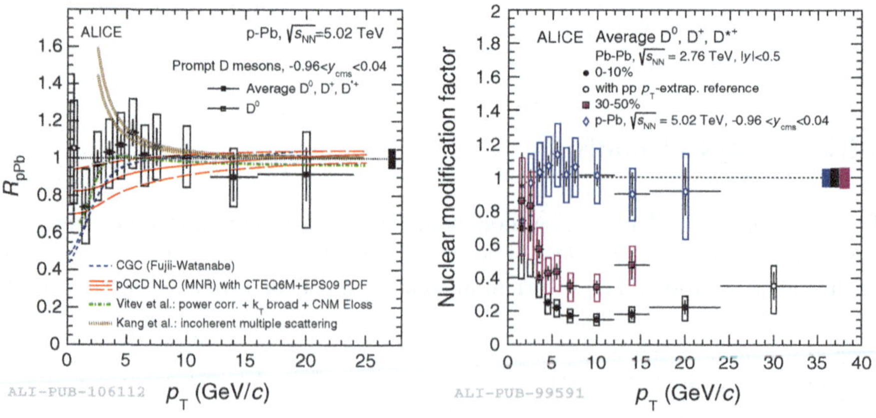

Fig. 74.1 R_{pPb} of D mesons (D^0, D^+, and D^{*+} average) compared with models, including CNM effects (left) and comparison with the R_{AA} for central and semi-central Pb–Pb collisions at $\sqrt{s_{\mathrm{NN}}} = 2.76\,\mathrm{TeV}$ (right) [5]

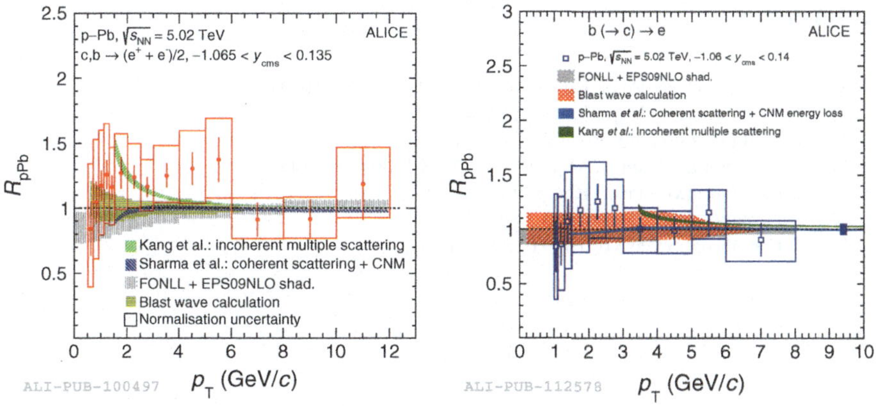

Fig. 74.2 R_{pPb} of inclusive electrons from HF-hadron decays (left plot, [2]) and beauty-hadron decay electrons (right plot, [7]) along with comparison to the models

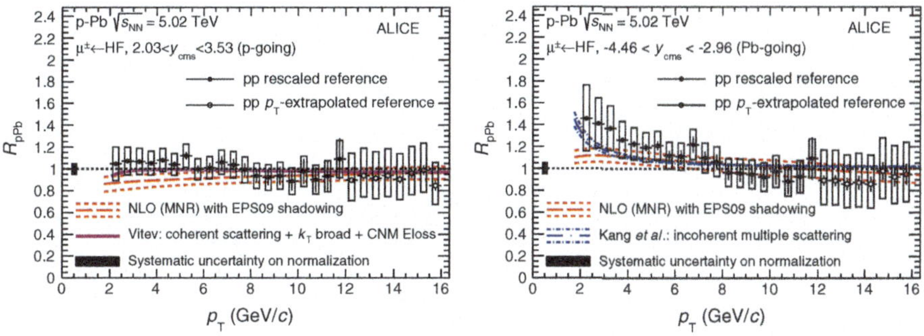

Fig. 74.3 R_{pPb} of muons from heavy-flavour hadron decays and comparison with models. Left (right): forward (backward) rapidity, $2.03 < y_{cms} < 3.53$ ($-4.46 < y_{cms} < -2.96$) [3]

QGP produced in these collisions. The R_{pPb} of HF-hadron decay electrons shown in Fig. 74.2 (left plot) is consistent with unity and also described by various models considering CNM effects [2]. The impact parameter distributions of beauty decay electrons is expected to be broader than that of charm decay electrons due to the larger separation between the primary and decay vertices. Therefore, one can separate the contributions of charm and beauty production. The R_{pPb} of beauty decay electrons is shown in the right panel of Fig. 74.2 [7]. The results are similar and consistent with unity within the uncertainties.

Figure 74.3 shows the R_{pPb} of heavy-flavour hadron decay muons, which is also consistent with unity at both forward ($2.03 < y_{cms} < 3.53$, left panel) and backward ($-4.46 < y_{cms} < -2.96$, right panel) rapidities. However an enhancement is observed above unity at backward rapidity for $2 < p_T < 4$ GeV/c [3]. The results in both rapidity ranges are described within uncertainties by model calculations that include CNM effects.

References

1. ALICE Collaboration, Phys. Rev. Lett. **113** (23), 232301 (2014)
2. ALICE Collaboration, Phys. Lett. B **754**, 81 (2016)
3. ALICE Collaboration, arXiv:1702.01479 [nucl-ex]
4. ALICE Collaboration, Int. J. Mod. Phys. **A29**, 1430044 (2014)
5. ALICE Collaboration, Phys. Rev. C **94**, 054908 (2016)
6. ALICE Collaboration, JHEP **1603**, 081 (2016)
7. ALICE Collaboration, arXiv:1609.03898 [nucl-ex]

Chapter 75
New Tensor Interaction as the Source of the Observed CP Asymmetry in $\tau \rightarrow K_S \pi \nu_\tau$

Lobsang Dhargyal

Introduction

The study of CP violation in tau decays has always been of much interest for beyond the Standard Model studies in the past two decades. In SM, the only source of CP violation is the one phase in the Kobayashi Maskawa (KM) matrix. While the Kobayashi Maskawa ansatz for CP violation within the Standard Model in the quark sector has been clearly verified by the plethora of data from the B factories, this is unable to account for the observed baryon asymmetry of the Universe. Hence, one needs to look for other sources of CP violation, including searches in the leptonic sector. Apart from the CP phases that may arise in the neutrino mixing matrix, the decays of the tau lepton may allow us to explore nonstandard CP-violating interactions. Various experimental groups have been involved in exploring CP violation in tau decays in the last decade or more. In 2002, the CLEO collaboration, and more recently the Belle Collaboration, studied the angular distribution of the decay products in $\tau \rightarrow K_s \pi \nu_\tau$ in search of CP violation; however, neither study revealed any CP asymmetry. The BABAR collaboration for the first time reported a sign anomaly in the integrated decay rate asymmetry $A_{cp}(\tau \rightarrow K_s \pi \nu_\tau)$ of

$$A_{cp}^{Exp} = (-0.36 \pm 0.23 \pm 0.11)\%. \tag{75.1}$$

However for $\tau^\pm \rightarrow K_s^0 \pi^\pm \nu_\tau \rightarrow [\pi\pi]_K^0 \pi^\pm \nu_\tau$, the predicted SM integrated decay rate asymmetry is

$$A_{cp}^{SM} = (0.33 \pm 0.01)\%. \tag{75.2}$$

Comparing the rate asymmetries for decays to neutral kaons of the taus with that of D mesons, Grossman and Nir have pointed out that since $\tau^+(\tau^-)$ decays initially to a $K^0(\bar{K}^0)$ whereas $D^+(D^-)$ decays initially to $\bar{K}^0(K^0)$, the time-integrated

L. Dhargyal (✉)
Institute of Mathematical Sciences, Chennai, India
e-mail: dhargyal@imsc.res.in

Md. Naimuddin (ed.), *XXII DAE High Energy Physics Symposium*, Springer Proceedings in Physics 203, https://doi.org/10.1007/978-3-319-73171-1_75

decay-rate CP asymmetry (arising from oscillations of the neutral kaons) of τ decays must have a sign opposite to that of D decays. The observation of a CP asymmetry in τ decays to K_s having the same sign as that in D decays, and moreover of the same magnitude but opposite in sign to the SM expectation, implies that this asymmetry cannot be accounted for by the CP violation in $K^0 - \bar{K}^0$ mixing.

1. Naively one may expect that the simplest way to account for the observed anomaly would be to introduce a direct CP violation via a new CP violating charged scalar exchange. However, it turns out that the charged scalar type of exchange may contribute in the angular distributions, but its mixing with SM term in the integrated decay rate goes to zero.
2. Now the next candidate of NP would be a new CP violating charged vector exchange, but CP violation from vector type NP will be observable only if both vector current and axial vector currents contributes to the same final states [2]. Since in two pseudo scalar meson final states only vector current can contribute due to parity conservation of strong interaction, vector type of NP can contribute in general to CP violation in three or more pseudo scalar meson final states but not in two pseudo scalar meson final states such as $K_s\pi$.
3. Now the only possibility left is tensor type of NP.

In this presentation, we will give the results from the materials contianed in the [1, 2].

Results

To a good approximation we can take $A^k_{pc}A^{\tau}_{cp} \approx 0$, then the (29) and (30) of [2] becomes:

$$A_{cp}(\tau \to K_s\pi\nu_\tau) = A^K_{cp} + A^{\tau}_{cp} \tag{75.3}$$

and

$$Br(\tau \to K_s\pi\nu_\tau) = \frac{(\Gamma^{\tau^+} + \Gamma^{\tau^-})}{2}\tau_\tau = (\Gamma_{SM} + \Gamma_T)\tau_\tau, \tag{75.4}$$

where A^k_{cp} is the known SM CPV from the $K - \bar{K}^0$ mixing, Γ_{SM} is the SM decay rate corresponding to fitted form factors from Belle fits, Γ_T is the decay rate due to purely tensor term and τ_τ is the life time of τ lepton. From (75.9), (10) and using F^a_T from (75.7) the best fitted value of the complex parameter $Im(C^{\tau}_T)$ to the two data points gives at $\chi^2 \approx 4.5$:

$$Im(C^{\tau}_T) = -0.071, \tag{75.5}$$

which gives

$$Br(\tau \to K_0\pi\nu_\tau)^{(Th)} = 2Br(\tau \to K_s\pi\nu_\tau)^{(Th)} = (0.756 \pm 0.085)\% \tag{75.6}$$

and

$$A^{\tau(Th)}_{cp} = (-0.703 \pm 0.54)\% \tag{75.7}$$

whereas the experimental values of these observables are given as

$$A_{cp}^{(Exp-SM)} = A_{cp}^{\tau(Exp)} - (A_{cp}^{k})^{SM} = (-0.69 \pm 0.26)\%, \tag{75.8}$$

and

$$Br(\tau \to K_0 \pi \nu_\tau)^{(Exp)} = 2Br(\tau \to K_s \pi \nu_\tau)^{(Exp)} = (0.84 \pm 0.04)\%. \tag{75.9}$$

Comparing (13), (14) and (12), (15) we see that the theoretical predicted values fit with the experimental values within 1σ.

Conclusion

Babar collaboration has reported an intriguing opposite sign in the integrated decay rate asymmetry $A_{cp}(\tau \to K_s \pi \nu_\tau)$ than that of SM prediction from the known $K^0 - \bar{K}^0$ mixing. Babar's result deviate from the SM prediction by about 2.7σ. In this work we have presented an improved analysis of our previous work on the contributions coming from tensorial current to this observable. Assuming the real part of the NP coupling is negligible compare to its imaginary part, the best fitted value of the parameter $Im(C_T^\tau)$ to the two data points $A_{cp}(\tau \to K_s \pi \nu_\tau)$ and $B_r(\tau \to K_0 \pi \nu_\tau)$ is given by $Im(C_T^\tau) = -0.071$ which gives $A_{cp}^{Th} = (-0.703 \pm 0.51) \times 10^{-2}$ compare to the experimental minus SM value of $A_{cp}^{(Exp-SM)} = (A_{cp}^{Exp} - A_{cp}^{SM}) = (-0.69 \pm 0.26) \times 10^{-2}$. And similarly we have $Br(\tau \to K_0 \pi \nu_\tau)^{(Th)} = (0.756 \pm 0.084) \times 10^{-2}$ comapre to the $Br(\tau \to K_0 \pi \nu_\tau)^{(Exp)} = (0.84 \pm 0.04) \times 10^{-2}$. As we can see the theoretical predictions fit with the experimental results within 1σ for both observables.

Acknowledgements Author did part of this work with Nita Sinha and H. Zeen Devi, Institute of Mathematical Sciences.

References

1. H.Z. Devi, L. Dhargyal, N. Sinha. Phys. Rev. D **90**, 013016 (2014)
2. L. Dhargyal, arXiv:1605.00629 [hep-ph] & references there in

Chapter 76
Phase Portraits of Higher Dimensional FRW Cosmology in $R^p exp(\lambda R)$ Gravity Filled with Non-perfect Fluid

Sebika K. Banik, Debika K. Banik and Kalyan Bhuyan

76.1 Introduction

The discovery of accelerated expansion of the universe is one of the most revolutionary advancements in modern cosmology. In order to study this cosmic acceleration, the gravitational theory gets modified in $f(R)$ gravity theory [1, 2]. Several recent observations also suggest that the universe is homogeneous and isotropic on large scale. However, some small anisotropies are needed in the early universe to explain the structure formation in the universe. Higher dimensional FRW model [3] with the different scale factors in the extra dimensions can provide interesting dynamics in comparison to normal four dimensional model. In this paper we extend our work given in [4] by analysing the phase space for higher dimensional FRW model in $R^p exp(\lambda R)$ gravity theory.

76.2 Dynamical Analysis

For $(1 + 3 + D)$ dimensional spacetime with three normal spatial dimensions and D $(I, J = 4, 5, \ldots, (D + 3))$ extra spatial dimensions, given by [3]

$$ds^2 = -dt^2 + A^2(t)\delta_{ij}dx^i dx^j + B^2(t)\delta_{IJ}dX^I dX^J, \tag{76.1}$$

S. K. Banik (✉) · D. K. Banik · K. Bhuyan
Department of Physics, Dibrugarh University, Dibrugarh 786004, Assam, India
e-mail: sebikabanik063@gmail.com

D. K. Banik
e-mail: debikabanik05@gmail.com

K. Bhuyan
e-mail: kalyanbhuyan@gmail.com

Md. Naimuddin (ed.), *XXII DAE High Energy Physics Symposium*, Springer Proceedings in Physics 203, https://doi.org/10.1007/978-3-319-73171-1_76

and considering the energy momentum tensor of a general imperfect fluid [3], the Friedmann's Field equation takes the form

$$\frac{(n-1)}{2n}\Theta^2 - \sigma^2 + \Theta\frac{\dot{f}'}{f'} - \frac{1}{2f}(Rf' - f) - \frac{\rho}{f'} = 0 \quad (76.2)$$

Let us now introduce the set of variables suggested by the above equations

$$\Sigma = \sqrt{\frac{2n}{n-1}}\frac{\sigma}{\Theta},\ x = \frac{2n}{n-1}\frac{\dot{f}'}{f'\Theta},\ y = \frac{n}{n-1}\frac{R}{\Theta^2},$$
$$z = \frac{n}{n-1}\frac{f}{f'\Theta^2},\ \Omega = \frac{2n}{n-1}\frac{\rho}{f'\Theta^2}. \quad (76.3)$$

Differentiating (76.3) with respect to time $\tau = lna$ and putting $f(R) = R^p exp(\lambda R)$, we have the system

$$\frac{d\Sigma}{d\tau} = \frac{(n-1)}{2}\Sigma[-2y + z - \Omega],$$
$$\frac{dy}{d\tau} = \frac{(n-1)}{2}y\left[\frac{2(n+1)}{n-1} + 2\Sigma^2 - 2y + \frac{(-1+\Sigma^2+y-z+\Omega)yz}{y^2 - pz^2}\right],$$
$$\frac{dz}{d\tau} = \frac{(n-1)}{2}z\left[\frac{3n+1}{n-1} + \Sigma^2 - 3y + z - \Omega + \frac{(-1+\Sigma^2+y-z+\Omega)y^2}{y^2 - pz^2}\right],$$
$$\frac{d\Omega}{d\tau} = \frac{(n-1)}{2}\Omega\left[\frac{1+3(1-2w_A)+D(1-2w_B)}{(n-1)}\right.$$
$$\left. - 2\sqrt{\frac{2D}{(n-1)}}\Sigma(w_B - w_A) + \Sigma^2 - 3y + z - \Omega\right],$$
$$1 - \Sigma^2 + x - y + z - \Omega = 0. \quad (76.4)$$

Setting (76.4) equal to zero we have obtained the following fixed points:
$A : (\Sigma, y, z, \Omega) = (0, 0, 0, 0)$, $B : (\Sigma, y, z, \Omega) = \left(0, \frac{n+1}{n-1}, 0, 0\right)$,
$C : (\Sigma, y, z, \Omega) = \left(0, \frac{n+1}{n-1}, \frac{2}{n-1}, 0\right)$, $D : (\Sigma, y, z, \Omega) = \left(0, 0, -\frac{3n+1}{n-1}, 0\right)$,
$E : (\Sigma, y, z, \Omega) = \left(0, \frac{p(2np-3n+2p-1)}{(n-1)(p-1)(2p-1)}, \frac{(2np-3n+2p-1)}{(n-1)(p-1)(2p-1)}, 0\right)$,
$F : (\Sigma, y, z, \Omega) = \left(0, 0, 0, \frac{1+3(1-2w_A)+D(1-2w_B)}{n-1}\right)$,
$H : (\Sigma, y, z, \Omega) = \left(0, \frac{(n+1)p-3(1+w_A)-D(1+w_B)}{(n-1)p}, \frac{(n+1)p-3(1+w_A)-D(1+w_B)}{(n-1)p^2}, -\frac{\left(3(1+w_A)+D(1+w_B)\right)(2p^2-3p+1)+2p^2-(n+1)p}{(n-1)p^2}\right)$, $L : (\Sigma, y, z, \Omega) = (\Sigma_*, 0, 0, 0)$.

76.3 Conclusion

In this analysis, besides obtaining the de Sitter solutions B and C, we also have found the fixed point E, which shows power law accelerated expansion for $p < \frac{(1-\sqrt{n})}{2}$ and $\frac{(1+\sqrt{n})}{2} < p < \frac{n+1}{2}$. In these intervals for p, the line of anisotropic fixed points L is repulsive in nature and the points B and E are attractors, which means that the universe starts in an anisotropic state, isotropize and smoothly approach to a phase of accelerated expansion at late times. In Figs. 76.1, 76.2, 76.3 and 76.4 we have ploted the phase portraits of the system in the intervals $p = -1.5$ and $p = 2$. There is another important point H. It behaves as a saddle when the standard matter is dust or radiation, for all values of p. Thus, in this case we have a transient matter (dust or radiation) dominated phase which is a saddle followed by a stable accelerating universe.

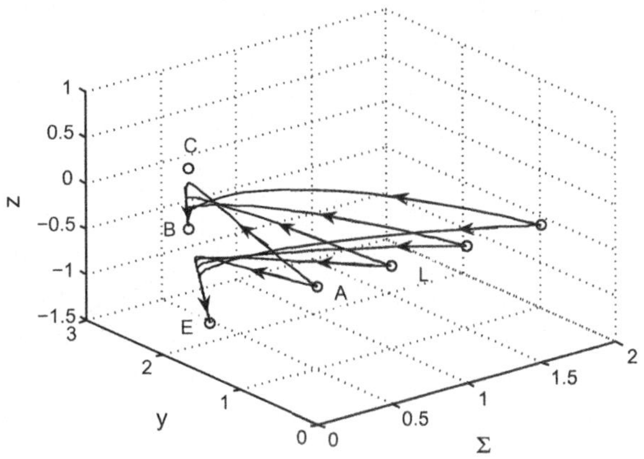

Fig. 76.1 Phase portrait for $n = 4, D = 1, \Omega = 0, p = -1.5$

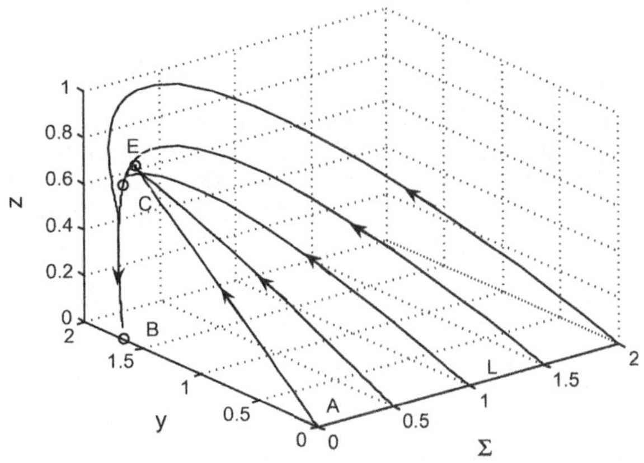

Fig. 76.2 Phase portrait for $n = 4, D = 1, \Omega = 0, p = 2$

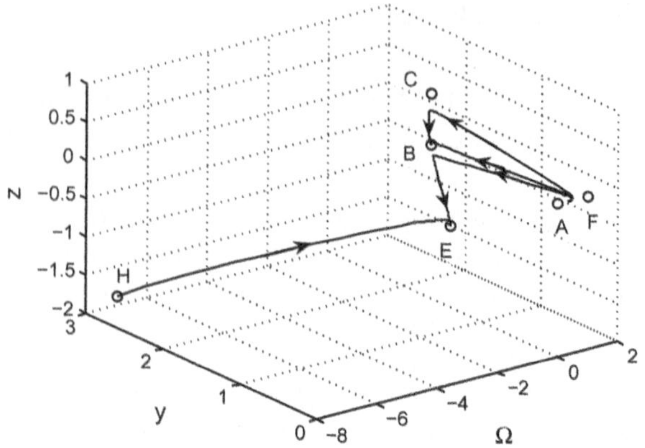

Fig. 76.3 Phase portrait for $n = 4, D = 1, w_A = 1/3, w_B = 0, p = -1.5$

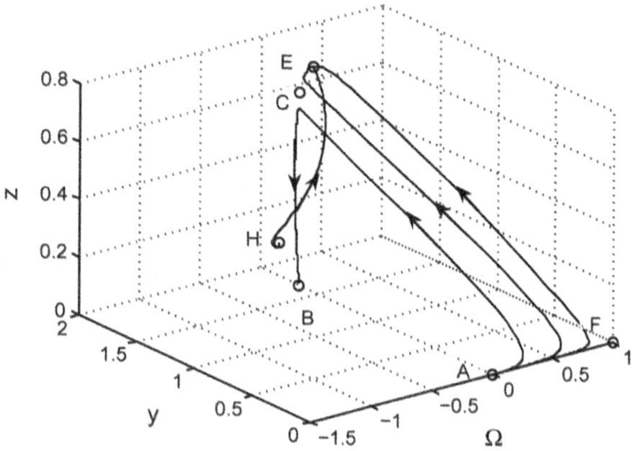

Fig. 76.4 Phase portrait for $n = 4, D = 1, w_A = 1/3, w_B = 0, p = 2$

References

1. T.P. Sotiriou, V. Faraoni, $f(R)$ theories of gravity. Rev. Mod. Phy. **82**, 451–497 (2010)
2. A.D. Felice, S. Tsujikawa, $f(R)$ Theories. Living Rev. Rel. **13**, 3–161 (2010)
3. I. Pahwa, H. Nandan, U.D. Goswami, Shear dynamics in higher dimensional FLRW cosmology. Astrophys. Space Sci. **360**, 58 (2015). https://doi.org/10.1007/s10509-015-2580-4
4. S.K. Banik, K. Bhuyan, Dynamics of higher-dimensional FRW cosmology in $R^p exp(\lambda R)$ gravity. Pramana J. Phys. **88**, 26 (2017). https://doi.org/10.1007/s12043-016-1335-2

Chapter 77
Optimal Renormalization and the Extraction of Strange Quark Mass from Semi-leptonic τ-Decay

Balasubramanian Ananthanarayan and Diganta Das

77.1 Introduction

The relevant quantities to extract strange quark mass m_s from semi-leptonic τ-decay are the polarization functions, the data on which were made available some years ago by ALEPH [1] and OPAL [2] collaborations. One of the important theoretical objective in these extractions is to account for the renormalization group (RG) running of all the parameters that enter the evolutions. There have been several schemes, for example *fixed-order perturbation theory* (FOPT) [3], *contour-improved perturbation theory* (CIPT) [4] and *method of effective charges* (MEC) [5] to account for these runnings. We follow the optimal renormalization group analysis [6, 7] in [8] that uses the RG constraints of a given perturbation series to obtain a closed form sum all the RG-accessible logarithms which substantially reduce the RG scale (μ) dependence. The RG-accessible logarithms are defined as all the leading and the next-to-leading logarithms that can be accessed through RG equation. We call this scheme *renormalization group summed perturbation theory* (RGSPT).

After a brief summary of the formalism of semi-leptonic τ-decay in Sect. 77.2, we describe the derivations of the closed form summation in Sect. 77.3. The extraction of m_s is described in Sect. 77.4 and we conclude in Sect. 77.5.

D. Das—Speaker

B. Ananthanarayan
Centre for High Energy Physics, Indian Institute of Science,
Bangalore 560 012, India
e-mail: anant@chep.iisc.ernet.in

D. Das (✉)
Physical Research Laboratory, Navrangpura 380 009, Ahmedabad, India
e-mail: diganta@prl.res.in

Md. Naimuddin (ed.), *XXII DAE High Energy Physics Symposium*, Springer Proceedings in Physics 203, https://doi.org/10.1007/978-3-319-73171-1_77

77.2 Formalism

The experimentally measurable quantities relevant for the extraction of m_s from Cabibbo suppressed semi-leptonic τ-decay are the moments of polarization functions. These can be written as a contour integral of the polarization functions with suitable weight functions in the complex q^2 plane running counter clockwise along the circle $|q^2| = M_\tau^2$

$$R_\tau^{kl} = \frac{6i}{2\pi}\oint \rho_{k,l}(q^2)\frac{m_s^2}{q^2}\Pi^{mq}\frac{dq^2}{M_\tau^2} - \frac{6i}{2\pi}\oint\left[\frac{m_s^2}{q^2}D^{mg}\int \rho_{k,l}(q^2)\frac{dq^2}{M_\tau^2}\right]\frac{dq^2}{M_\tau^2}. \tag{77.1}$$

Here $\rho_{k,l}$ are the weight functions, Π^{mq} is the τ-decay polarization function and D^{mg} is the Adler function related to the polarization function Π^{mg} and defined as $D^{mg} = -Q^2\frac{d\Pi^{mg}}{dQ^2}$, where $Q^2 = -q^2$. The functions Π^{mq} [9, 10] and D^{mg} [11] are known to order α_s^3 in the perturbation theory, and they have in addition to Q^2 dependence, μ dependence through terms like $\alpha_s^n(\mu^2)\ln^{n-k}(\mu^2/Q^2)$, at each order n in the perturbation expansion. Both Π^{mq} and D^{mg} satisfy homogeneous RG equations. In the next section we show how the RG equations of these functions can be used to obtain closed form sums of all the RG accessible logarithms which reduce μ dependence.

77.3 Closed Form Sum RG-Accessible Logarithms

We demonstrate the derivations of RGSPT series with D^{mg}. We write down the un-summed and RGSPT series as $D^{mg} = \sum_{n=0}^{\infty}\sum_{k=0}^{n} d_{n,k}^{mg} a_s^n L^k$ and $D_{\text{RGSPT}}^{mg} = \sum_{n=0}^{\infty} a_s^n \Pi_n^{mg}[a_s L]$, respectively. Here the intermediate quantities are defined as $D_k^{mg}[a_s L] = \sum_{n=k}^{\infty} d_{n,n-k}^{mg}(a_s L)^{n-k}$, d^{mg} are the series coefficients that can be extracted from the expression of D^{mg} given in the appendix of [8], $L = \ln(\mu^2/Q^2)$, and $a_s = \alpha_s(\mu)/\pi$. We substitute the un-summed expression of D^{mg} in the homogeneous RG equation that it satisfies and collect the coefficients of $a_s^n L^{n-1-k}$ which leads to a recursion relation in terms of the series coefficients d^{mg}. The recursion relation is then multiplied by $(a_s L)^{n-1-k}$ and is summed from $n = 1 + k$ to infinity, which following the definitions of the intermediate quantities $D_k^{mg}[a_s L]$ results in differential equations for them. These differential equations are solved with suitable boundary conditions and the resultant solutions are the closed form expressions of $D_k^{mg}[a_s L]$. The closed form expressions of $D_k^{mg}[a_s L]$ are given in [8].

77.4 Extraction of m_s from ALEPH and OPAL Data

The extraction of m_s is possible by the measurements of the $SU(3)$ flavor breaking terms $\delta R_\tau^{kl} = N_c S_{EW}\Big(- R^{kl} - 4\pi^2 \frac{m_s^2(M_\tau^2)}{M_\tau} \frac{\langle \bar{s}s \rangle}{M_\tau^3} f_{kl}\Big)$ where the second term within the bracket is the contribution of the condensate [12]. The vales of S_{EW}, f_{kl} and $\langle \bar{s}s \rangle$ are collected in [8]. The flavor breaking term δR_τ^{kl} for different moments (k, l) have been measured by ALEPH [1] and OPAL [2] collaborations. For our determination, we have used the moments $(0, 0)$, $(1, 0)$, $(2, 0)$ of ALEPH and $(2, 0)$, $(3, 0)$ and $(4, 0)$ of OPAL. The determinations for individual moments for both ALEPH and OPAL are tabulated in [8] and are compared with the determinations in FOPT, CIPT and MEC schemes. We find that RGSPT is consistent with other schemes. By doing a weighted average of the individual determinations we obtain the following final values of m_s at the τ mass scale $m_s(M_\tau^2) = 110.18 \pm 9.67\,\text{MeV}$, $m_s(M_\tau^2) = 76.90 \pm 8.03\,\text{MeV}$ for ALEPH and OPAL data, respectively. Here the errors are dominated by experimental uncertainties. Evolving these determinations to 2 GeV using [13] we get

$$m_s(2\text{GeV}) = 106.70 \pm 9.36\,\text{MeV}, \quad m_s(2\text{GeV}) = 74.47 \pm 7.77\,\text{MeV}, \qquad (77.2)$$

for ALEPH and OPAL data, respectively. Within errors our determinations are comparable with that in lattice QCD.

77.5 Summary and Conclusion

We have applied for the first time the optimal renormalization group analysis to τ-decay polarization functions in the context of extraction of strange quark mass. Our determinations are consistent with other schemes and show that the strange quark mass extracted from τ-decay is insensitive to the choice of renormalization schemes. Therefore, in the future when the experimental data improve, precise determinations of m_s can be made.

References

1. R. Barate [ALEPH Collaboration], Z. Phys. C 76, 15 (1997); R. Barate et al., ALEPH Collaboration. Eur. Phys. J. C **4**, 409 (1998)
2. G. Abbiendi [OPAL Collaboration], Eur. Phys. J. C 35, 437 (2004); K. Ackerstaff et al., OPAL Collaboration. Eur. Phys. J. C **7**, 571 (1999)
3. M. Jamin, JHEP **0509**, 058 (2005)
4. A.A. Pivovarov, Sov. J. Nucl. Phys. **53**, 467 (1991); [Nuovo Cim. A **105**, 641 (1992)] [Yad. Fiz. **53**, 746 (1991)]
5. S. Groote, J.G. Korner, A.A. Pivovarov, Phys. Lett. B **407**, 66 (1997). Mod. Phys. Lett. A **13**, 637 (1998)

6. C.J. Maxwell, Nucl. Phys. Proc. Suppl. **86**, 74 (2000); C.J. Maxwell, A. Mirjalili, Nucl. Phys. B **577**, 209 (2000)
7. M.R. Ahmady, F.A. Chishtie, V. Elias, A.H. Fariborz, N. Fattahi, D.G.C. McKeon, T.N. Sherry, T.G. Steele, Phys. Rev. D **66**, 014010 (2002)
8. B. Ananthanarayan, D. Das, Phys. Rev. D **94**(11), 116014 (2016)
9. K.G. Chetyrkin, A. Kwiatkowski, Z. Phys. C **59**, 525 (1993)
10. P.A. Baikov, K.G. Chetyrkin, J.H. Kuhn, Phys. Rev. Lett. **95**, 012003 (2005)
11. P.A. Baikov, K.G. Chetyrkin, J.H. Kuhn, Phys. Rev. D **67**, 074026 (2003)
12. E. Braaten, S. Narison, A. Pich, Nucl. Phys. B **373**, 581 (1992)
13. K.G. Chetyrkin, J.H. Kuhn, M. Steinhauser, Comput. Phys. Commun. **133**, 43 (2000)

Chapter 78
Extraction of the Strong Coupling Constant from the Measurement of Inclusive Multijet Event Cross Sections in pp Collisions at Center of Mass Energy of 8 TeV

Anterpreet Kaur

78.1 Introduction

Quantum Chromodynamics (QCD) is the theory which describes the strong interactions between the partons, the (anti-)quarks and gluons. During the collision of hadrons such as protons (p), the scattered partons hadronize into highly collimated bunches of particles called jets. These jets preserve energy and momentum of the initial partons. In the context of perturbative QCD, the cross section of a high transverse momentum (p_{T}) scattering can be expressed as a sum of terms with increasing powers of the strong coupling constant, α_S, convoluted with the parton momentum distribution functions (PDFs) of the proton. The lowest-order α_S^2 and higher-order α_S^3 terms represent the production of two-parton and multi-parton final states respectively.

The investigation of the inclusive multijet event cross sections $\sigma_{\mathrm{i-jet}}$ for the process pp→ i jets + X, as a function of jet p_{T} and rapidity y, is proportional to α_S^i. Hence, it provides essential information about the PDFs and $\alpha_S(M_Z)$. The cross section ratios, $R_{\mathrm{mn}} = \frac{\sigma_{\mathrm{m-jet}}}{\sigma_{\mathrm{n-jet}}}$ (m > n), where numerous theoretical and experimental uncertainties cancel, provide an ideal tool to determine $\alpha_S(M_Z)$.

A measurement of inclusive 2- and 3-jet event cross sections is presented [1, 2] using an event sample collected by the Compact Muon Solenoid (CMS) experiment [3] during 2012 at the LHC and corresponding to an integrated luminosity of 19.7 fb^{-1} of pp collisions at a centre-of-mass energy of 8 TeV.

---

on behalf of the CMS collaboration.

A. Kaur (✉)
Panjab University, Chandigarh, India
e-mail: anterpreet.kaur@cern.ch

Md. Naimuddin (ed.), *XXII DAE High Energy Physics Symposium*, Springer Proceedings in Physics 203, https://doi.org/10.1007/978-3-319-73171-1_78

78.2 Results and Discussion

The inclusive 2-jet and 3-jet event cross sections are measured as a function of the average p_T of the two leading jets, referred to as $H_{T,2}/2$ here. After unfolding for detector effects, these are compared to the NLOJET++ predictions using the CT10 PDF set, corrected for non-perturbative (NP) and also for electroweak (EWK) effects in the 2-jet case. The comparison is also made to predictions from MADGRAPH5 + PYTHIA6 with tune Z2*, corrected for EWK effects in the 2-jet case. The data are in agreement with the NLO predictions over the whole range of $H_{T,2}/2$ from 0.3 TeV up to 2.0 (2-jet) and 1.68 TeV (3-jet) respectively. In Fig. 78.1 [1], the top plot presents the cross section ratio, R_{32} obtained from unfolded data in comparison to that from NLO pQCD. The error bars correspond to the total experimental uncertainty. The NLO predictions using the CT10 NLO PDF set corrected with NP corrections are shown for a series of values assumed for $\alpha_S(M_Z)$ (dashed lines) together with the central prediction (solid line) where $\alpha_S(M_Z) = 0.118$. The assumption on $\alpha_S(M_Z)$is varied in steps of 0.001 in the range of 0.112–0.127. The value of $\alpha_S(M_Z)$ is determined from the measured inclusive 2-jet and 3-jet event cross sections and their ratio, by minimizing the χ^2 between the experimental measurement and the theoretical predictions. The fit procedure follows closely the one used in [4, 5].

Using the MSTW2008 PDF set, which dates from before the LHC start, the $\alpha_S(M_Z)$ from R_{32} is finally determined to

$$\begin{aligned}\alpha_S(M_Z) &= 0.1150 \pm 0.0010\,(\text{exp}) \pm 0.0013\,(\text{PDF}) \pm 0.0015\,(\text{NP})\,^{+0.0050}_{-0.0000}\,(\text{scale})\\ &= 0.1150 \pm 0.0023\,(\text{all except scale})\,^{+0.0050}_{-0.0000}\,(\text{scale})\,.\end{aligned}$$

From the extracted $\alpha_S(M_Z)$ value for each range in $H_{T,2}/2$, the $\alpha_S(Q)$ values are evolved, with total uncertainty, to the respective cross-section averaged scale $\langle Q \rangle$. The evolution is performed for five flavours at 2-loop order with the RUNDEC program. The obtained $\alpha_S(Q)$ points are illustrated in bottom plot of Fig. 78.1 [1] together with the world average [9] and results from other experiments [4–8].

78.3 Summary

A measurement of the inclusive 2-jet (3-jet) event cross sections has been presented in a range of $0.3 < H_{T,2}/2 < 2.0$ TeV ($0.3 < H_{T,2}/2 < 1.68$ TeV) for the average p_T of the two leading jets at central rapidity of $|y| < 2.5$. The data are found to be well described by calculations at NLO in pQCD complemented with NP corrections that are important at low $H_{T,2}/2$. NLO QCD provides an adequate description of R_{32} in the accessible range of $H_{T,2}/2$. By contrast, LO tree-level MC predictions exhibit significant deviations. Based on the observed agreement, the $\alpha_S(M_Z)$ is determined in a fit to the R_{32} measurement to $\alpha_S(M_Z) = 0.1150^{+0.0055}_{-0.0023}$, using the MSTW2008 PDF

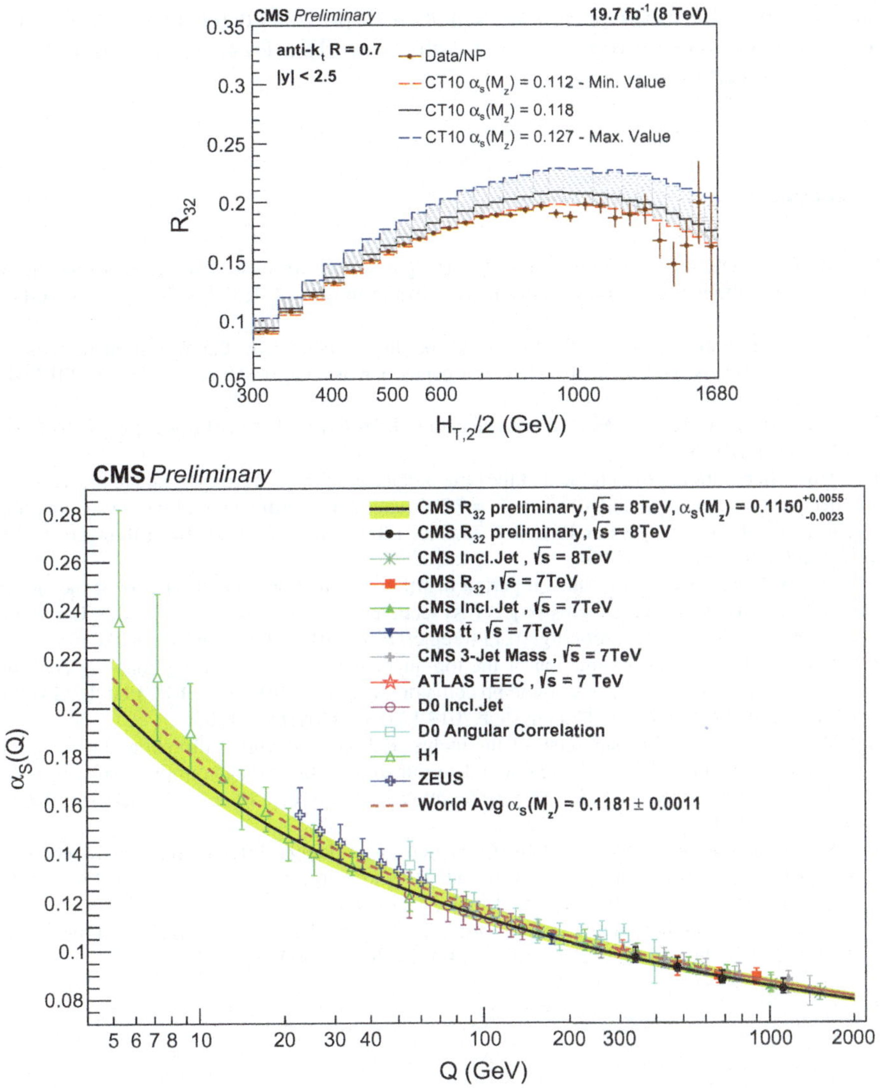

Fig. 78.1 Top: Cross section ratio R_{32} as a function of $H_{T,2}/2$ calculated from data (solid circles) in comparison to that from NLO pQCD (lines). Bottom: The running $\alpha_S(Q)$ as a function of the scale Q is shown as obtained by using the MSTW2008 NLO PDF set

set. The result for $\alpha_S(M_Z)$ is in agreement with previous determinations obtained by the CMS collaborations [4–8] and with the world average value of $\alpha_S(M_Z) = 0.1181 \pm 0.0011$ derived in [9].

References

1. CMS Collaboration, Determination of the strong coupling constant from the measurement of inclusive multijet event cross sections in pp collisions at $\sqrt{s} = 8$ TeV, CMS-PAS-SMP-16-008 (2017)
2. CMS Collaboration, Extraction of the strong coupling constant from the measurement of inclusive multijet cross-sections in pp collisions at center of mass energy of 8 TeV, CMS-CR-2017-047 (2017)
3. CMS Collaboration, The CMS Experiment at the CERN LHC, *JINST* **3**. https://doi.org/10.1088/1748-0221/3/08/S08004
4. CMS Collaboration, Measurement of the ratio of the inclusive 3-jet cross section to the inclusive 2-jet cross section in *pp* collisions at $\sqrt{s} = 7$ TeV and first determination of the strong coupling constant in the TeV range, Eur. Phys. J. C **73**, 2604 (2013), https://doi.org/10.1140/epjc/s10052-013-2604-6, arXiv:1304.7498
5. CMS Collaboration, Constraints on parton distribution functions and extraction of the strong coupling constant from the inclusive jet cross section in pp collisions at $\sqrt{s} = 7$ TeV. Eur. Phys. J. C **75**, 288 (2015), https://doi.org/10.1140/epjc/s10052-015-3499-1, arXiv:1410.6765
6. CMS Collaboration Determination of the top-quark pole mass and strong coupling constant from the $t\bar{t}$ production cross section in pp collisions at $\sqrt{s} = 7$ TeV TeV. Phys. Lett. B **728**, 496 (2014), https://doi.org/10.1016/j.physletb.2013.12.009, arXiv:1307.1907
7. CMS Collaboration, Measurement of the inclusive 3-jet production differential cross section in proton-proton collisions at 7 TeV and determination of the strong coupling constant in the TeV range, Eur. Phys. J. C **75**, 186 (2015), https://doi.org/10.1140/epjc/s10052-015-3376-y, arXiv:1412.1633
8. CMS Collaboration, Measurement and QCD analysis of double-differential inclusive jet cross-sections in pp collisions at $\sqrt{s} = 8$ TeV and ratios to 2.76 and 7 TeV. JHEP **03**, 156 (2017), https://doi.org/10.1007/JHEP03(2017)156, arXiv:1609.05331
9. C. Patrignani and others (Particle Data Group), Review of Particle Physics, Chin. Phys. C **40**, 100001 (2016), https://doi.org/10.1088/1674-1137/40/10/100001

Chapter 79
Coalescing Versus Merging of Energy Levels in One-Dimensional Potentials

Zafar Ahmed, Sachin Kumar, Achint Kumar and Mohammad Irfan

In one dimensional quantum mechanics there is one to one correspondence between eigenvalues and eigenstates, there is an absence of degeneracy. So when a parameter of the Hamiltonian is varied slowly, curves can not cross but they can come quite close and then diverge from each other (Avoided Crossing). Crossings and avoided crossings of levels is commonly observed in the spectra of two or three dimensional systems. Mostly, in one dimensional systems if a parameter of the potential is varied slowly, the eigenvalues increase or decrease monotonically. For particle in an infinitely deep well of width a, $E_n(= \frac{n^2\pi^2\hbar^2}{2\mu a^2})$ decrease as function of a. For harmonic oscillator potential, $E_n = (n + 1/2)\hbar\omega$ increase linearly as function of the frequency parameter ω.

Two levels coming very close may either display merging of two levels or their avoided crossing. The former is well known to occur in symmetric double-well potentials wherein the sub-barrier doublets of energy levels merge [1, 2] into the levels of the independent wells when the inter-well distance is increased slowly. On the other hand AC is observed rather rarely in one dimensional systems. Recently

Z. Ahmed
Nuclear Physics Division, Bhabha Atomic Research Centre,
Trombay Mumbai-85, India
e-mail: zahmed@barc.gov.in

S. Kumar (✉)
Theoretical Physics Section, Bhabha Atomic Research Centre,
Trombay Mumbai-85, India
e-mail: sachinv@barc.gov.in

A. Kumar
Department of Physics, Birla Institute of Technology and Science,
Pilani, Goa 403726, India
e-mail: achint1994@gmail.com

M. Irfan
Department of Physics, Indian Institute of Science Education and Research,
Bhopal 462066, India
e-mail: mohai@iiserb.ac.in

Md. Naimuddin (ed.), *XXII DAE High Energy Physics Symposium*, Springer Proceedings in Physics 203, https://doi.org/10.1007/978-3-319-73171-1_79

[2, 3] it has been shown that in double-well potential if the width or depth of the potential is varied slowly very interesting level crossings can be observed. Notably the double-well becomes asymmetric.

For one-dimensional non-Hermitian Hamiltonions, it is known that two complex eigenvalues may become real at one special value of the parameter($\lambda = \lambda_c$) of the potential after this point these two eigenvalues may again be complex. Such special values of the parameter are called Exceptional Point (EP) [4]. More interestingly, when a potential PT-symmetric (invariant under the joint action of Parity: $x \to -x$ and Time-reversal ($i \to -i$), the two discrete eigenvalues make a transition from real to complex-conjugate or *vice versa*. For instance, in the complex PT-symmetric potential: $V(x) = -V_1 \text{sech}^2 x + i|V_2| \text{sech} x \tanh x$, $|V_2| = V_1 + 1/4 = V_c$ $(2\mu = 1 = \hbar^2)$ is the EP of this potential when $|V_2| \leq V_c$ eigenvalues are real, otherwise these are complex conjugate pairs [5]. In both merging and coalescing two eigen values meet when a parameter λ is varied. In coalescing $dE/d\lambda = \infty$ critically at $\lambda = \lambda_c$. The merging of two levels is not a critical phenomenon. It is actually an asymptotic phenomenon where $\Delta E = |E_1 - E_2| \to 0$ as $\lambda \to \infty$. Here in our model below, $\lambda = b$, which is being varied. In the following, we study the evolution of real part of eigenvalues ($g = 0$ and $g \neq 0$) of the double-well potential when the distance between wells is increased slowly.

The solution of Schrödinger equation for the square double-well potential (Fig. 79.1) is given as, $\psi(x < -a) = Ae^{px} + Be^{-px}$, $\psi(-a \leq x \leq -b) = Ce^{iqx} + De^{-iqx}$, $\psi(-b < x < b) = Fe^{px} + Ge^{-px}$, $\psi(b \leq x \leq a) = He^{irx} + Ke^{-irx}$ and $\psi(x > a) = Le^{px} + Me^{-px}$, where,

$$p = \sqrt{-2\mu E}/\hbar, \quad q = \sqrt{2\mu(E + u + ig)}/\hbar, \quad r = \sqrt{2\mu(E + u - ig)}/\hbar. \tag{79.1}$$

$$\begin{aligned}
Ae^{-pa} + Be^{pa} &= Ce^{-iqa} + De^{iqa}, & Ape^{-pa} - Bpe^{pa} &= iCqe^{-iqa} - iDqe^{iqa} \\
Ce^{-iqb} + De^{iqb} &= Fe^{-pb} + Ge^{pb}, & iCqe^{-iqb} - iDqe^{iqb} &= Fpe^{-pb} - Gpe^{pb} \\
Fe^{pb} + Ge^{-pb} &= He^{irb} + Ke^{-irb}, & Fpe^{pb} - Gpe^{-pb} &= iHre^{irb} - iKre^{-irb} \\
He^{ira} + Ke^{-ira} &= Le^{pa} + Me^{-pa}, & iHre^{ira} - iKre^{-ira} &= Lpe^{pa} - Mpe^{-pa}
\end{aligned} \tag{79.2}$$

In matrix notation, these (79.2) can be re-written as,

$$M_1 \begin{pmatrix} A \\ B \end{pmatrix} = M_2 \begin{pmatrix} C \\ D \end{pmatrix}, \quad M_3 \begin{pmatrix} C \\ D \end{pmatrix} = M_4 \begin{pmatrix} F \\ G \end{pmatrix} \tag{79.3}$$
$$M_5 \begin{pmatrix} F \\ G \end{pmatrix} = M_6 \begin{pmatrix} H \\ K \end{pmatrix}, \quad M_7 \begin{pmatrix} H \\ K \end{pmatrix} = M_8 \begin{pmatrix} L \\ M \end{pmatrix}$$

$$\begin{pmatrix} A \\ B \end{pmatrix} = M_1^{-1} M_2 M_3^{-1} M_4 M_5^{-1} M_6 M_7^{-1} M_8 \begin{pmatrix} L \\ M \end{pmatrix} = \begin{pmatrix} m_{11}(E) \; m_{12}(E) \\ m_{21}(E) \; m_{22}(E) \end{pmatrix} \begin{pmatrix} L \\ M \end{pmatrix} \tag{79.4}$$

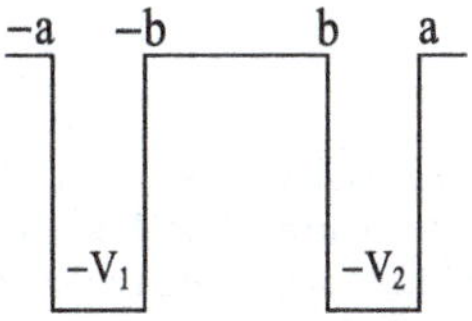

Fig. 79.1 Depiction of square double-well potential perturbed by imaginary ($g \neq 0$) PT-symmetric potential. Here, $V_1 = u + ig$, $V_2 = u - ig$

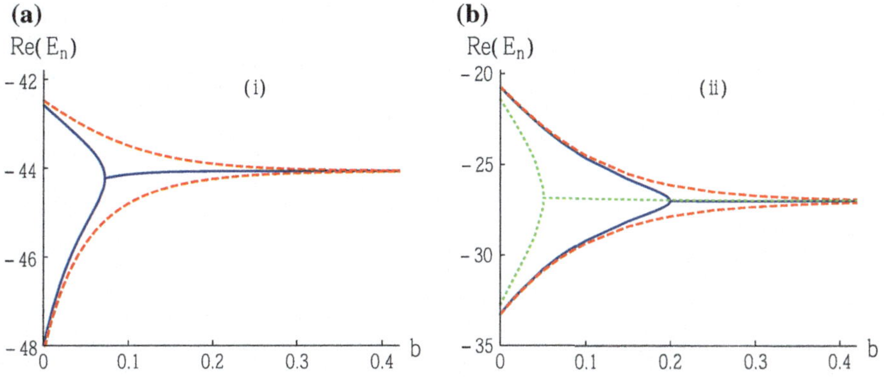

Fig. 79.2 Real part of E_n of the levels of the Square double-well (Fig. 79.1) when b is varied: (i) $V_1 = 50 = V_2$ (dashed line), (ii) $V_1 = 50 + i$, $V_2 = 50 - i$ (solid line), (iii) $V_1 = 50 + 5i$, $V_2 = 50 - 5i$ (dotted line). Excepting dashed lines, others represent coalescing of eigenvalues in complex PT-symmetric case. Dashed lines present merging of levels in Hermitian case. **a** E_0, E_1 for $g = 0, 1$ **b** E_2, E_3 for $g = 0, 1$ and E_0, E_1 for $g = 5$. The equation $m_{22}(E) = 0$ has been used here

For bound states, we demand $B = 0 = L$. From (79.4), we have $B = L\ m_{21}(E) + M\ m_{22}(E)$, where $M \neq 0$. Finally we get, $m_{22}(E) = 0$, gives the eigenvalues of bound states of double-well in Fig. 79.1.

In Fig. 79.2, we present the variation of eigenvalues when the distance between wells in Fig. 79.1 is increased slowly. The solid lines present the coalescing of two levels when the total potential is mildly complex PT-symmetric (g is small). See the dashed lines for Hermitian double-well ($g = 0$) representing merging of two levels. Dotted lines arise when non-Hermiticity parameter becomes large, then the coalesced levels are not contained between the merging levels (dashed lines). The same scenario has been shown for other interesting potentials in [6].

Lastly, we conclude that the model discussed here brings the spectral phenomena of coalescing and merging of energy levels closer, however we know that they occur in two different domains: Hermitian and complex PT-symmetric.

References

1. E. Merzbacher, *Quantum Mechanics* (Wiley, New York, 1970), pp. 128–39
2. Z. Ahmed, S. Kumar, M. Sharma, V. Sharma, Eur. J. Phys. **37**, 045406 (2016)
3. Z. Ahmed, S. Pavaskar, D. Sharma L. Prakash, arXiv: 1508.00661 [quant-ph]
4. T. Kato, *Perturbation Theory of Linear Operators* (Springer, New York, 1980)
5. Z. Ahmed, Phys. Lett. A **282**, 343 (2001); **287**, 295 (2001)
6. Z. Ahmed, S. Kumar, A. Kumar, M. Irfan, arXiv: 1702.08335 [quant-ph]

Chapter 80
Differential Cross Section Measurement of the Drell-Yan Process at 13 TeV proton-proton Collisions with the CMS Detector

Ridhi Chawla

80.1 Introduction

Drell-Yan (DY) process [1] in hadron-hadron collisions is described in the Standard Model (SM) by s-channel γ^*/Z exchange. Theoretical calculations of the DY differential cross section are well established up to the next-to-next-to-leading order (NNLO). Comparison between various theoretical predictions and experimental measurements provide tests of perturbative quantum chromodynamics (QCD) and effective input for constraints on parton distribution functions (PDFs). In addition, the study of DY lepton-pair production is important for various physics analysis at the Large Hadron Collider (LHC) as a background source for $t\bar{t}$ and diboson measurements, as well as for searches of new physics beyond the SM.

A measurement of DY differential cross section $d\sigma/dm$ where m is the invariant mass of the dilepton pair is presented [2] in dimuon channel in the mass range $15 < \mathrm{m} < 3000\,\mathrm{GeV}$ using an integrated luminosity of $2.8\,\mathrm{fb}^{-1}$ of proton-proton collision data collected using the Compact Muon Solenoid (CMS) detector [3] at the LHC at $\sqrt{s} = 13\,\mathrm{TeV}$. The measured differential cross section is obtained using the following formula:

$$\sigma = \frac{N_u}{A \cdot \epsilon \cdot \rho \cdot L_{int}} \tag{80.1}$$

where N_u denotes the signal yield obtained after subtracting the backgrounds and applying an unfolding technique [4] to the background subtracted data in order to correct for bin-to-bin migration effect due to detector resolution and final-state Quantum electrodynamics radiation (FSR) effect. A and ϵ denote the acceptance and efficiency for signal events and are obtained from MC simulation. In addition, ρ is the scale

On behalf of the CMS collaboration.

R. Chawla (✉)
Panjab University, Chandigarh, India
e-mail: ridhi.chawla@cern.ch

Md. Naimuddin (ed.), *XXII DAE High Energy Physics Symposium*, Springer Proceedings in Physics 203, https://doi.org/10.1007/978-3-319-73171-1_80

factor which accounts for the difference in the efficiency between data and MC. L_{int} is the integrated luminosity corresponding to 2015 dataset.

The events are collected using an isolated single muon trigger with $p_T > 20$ GeV. The muons are required to pass the standard CMS muon identification and quality control criteria. The leading muon p_T in the event is required to have $p_T > 22$ GeV and subleading muon $p_T > 10$ GeV. The muons should be within the acceptance of the muon system ($|\eta| < 2.4$). The two muons are required to have opposite charge and in case of more than one muon pair in an event, the one with smallest χ^2 for the dimuon vertex is selected. At least one of the two muons selected in each event should match the HLT trigger object. The $d\sigma/dm$ measurement is performed in 43 dilepton invariant mass bins. The edge of mass bins is identical to the previous measurement [5] performed at $\sqrt{s} = 8$ TeV. The highest mass bin is extended to 3000 GeV considering the highest mass event observed in the dataset to be 2.3 TeV. The systematic uncertainties from different sources [2] are obtained in each mass bin. The dominant uncertainty vary depending on the mass range from 15 to 3000 GeV.

80.2 Results and Discussion

The DY differential cross section in the full phase space is measured after applying all the corrections discussed in the previous section. The cross section is presented as a function of dimuon invariant mass in the range of 15–3000 GeV. The results are compared to the NNLO theoretical prediction in Fig. 80.1 (left) which are calculated using FEWZ 3.1 [6] with NNPDF3.0 [7] and NLO EW correction as well as MADGRAPH5_AMC@NLO [8] predictions with NNPDF3.0 (NLO). The ratio between

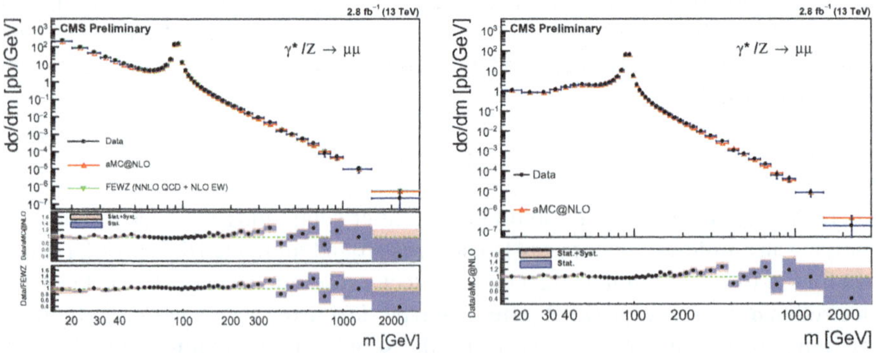

Fig. 80.1 The DY differential cross section measurement as a function of dimuon invariant mass is shown for the full phase space (left) and the fiducial phase space (right) [2]. Overlaid on the left panel are the NNLO theoretical prediction of FEWZ (red) and the NLO prediction of MADGRAPH5_AMC@NLO (green) with FSR correction, while on the right the NLO theoretical prediction using MADGRAPH5_AMC@NLO (red) is shown at particle level

data and theoretical prediction is shown in the middle and bottom plots, where red colour denotes total uncertainty which is the combination of statistical, systematical, theoretical and luminosity uncertainties. The band with purple colour denotes statistical uncertainty only.

In addition to the fully corrected DY differential cross section measurement, the fiducial cross section within the detector acceptance and without FSR correction is produced. Figure 80.1 (right) shows the results compared to the NLO prediction by MadGraph5_aMC@NLO. The ratio between data and theoretical prediction is shown in the bottom plot, where red colour denotes the total uncertainty and the band with purple colour denotes statistical uncertainty only.

80.3 Summary

A measurement of the DY differential cross section $d\sigma/dm$ is presented in the dimuon channel in the mass range $15 < m < 3000\,\text{GeV}$. The measurement is performed with $2.8\,\text{fb}^{-1}$ of proton-proton collision data collected using the CMS detector at the LHC at a center-of-mass energy of $\sqrt{s} = 13\,\text{TeV}$. The results are corrected for detector resolution effect resulting in event migration between mass bins, efficiency which considers the difference between data and MC simulation, acceptance to take into account the coverage of CMS detector and FSR effects dominant below the Z peak region. The results are in good agreement with the SM theoretical predictions at NNLO prediction calculated with FEWZ and NLO prediction calculated with MadGraph5_aMC@NLO.

References

1. S.D. Drell, T.M. Yan, Massive lepton pair production in hadron-hadron collisions at high-energies. Phys. Rev. Lett. **25**, 316 (1970). https://doi.org/10.1103/PhysRevLett.25.316
2. CMS Collaboration, Measurement of the differential Drell-Yan cross section in proton-proton collisions at $\sqrt{s}$ = 13 TeV, CMS-PAS-SMP-16-009 (2016)
3. CMS Collaboration, The CMS Experiment at the CERN LHC. *JINST* **3**, S08004 (2008). https://doi.org/10.1088/1748-0221/3/08/S08004
4. G. D'Agostini, A Multidimensional unfolding method based on Bayes' theorem. Nucl. Instrum. Meth. A **362**, 487 (1995). https://doi.org/10.1016/0168-9002(95)00274-X
5. CMS Collaboration, Measurement of differential and double-differential Drell-Yan cross sections in proton-proton collisions at $\sqrt{s}$ = 8 TeV. Eur. Phys. J. C **75**, 147 (2015). https://doi.org/10.1140/epjc/s10052-015-3364-2, arXiv:1412.1115
6. Y. Li, F. Petriello, Combining QCD and electroweak corrections to dilepton production in FEWZ. Phys. Rev. D **86**, 094034 (2012). https://doi.org/10.1103/PhysRevD.86.094034, arXiv:1208.5967
7. R.D. Ball et al., A first unbiased global NLO determination of parton distributions and their uncertainties. Nucl. Phys. B **838**, 136 (2010). https://doi.org/10.1016/j.nuclphysb.2010.05.008, arXiv:1002.4407

8. J. Alwall et al., The automated computation of tree-level and next-to-leading order differential cross sections, and their matching to parton shower simulations. JHEP **07**, 079 (2014). https://doi.org/10.1007/JHEP07(2014)079, arXiv:1405.0301

Chapter 81
Galactic Cosmic Energy Spectrum Based Simulation of Total Equivalent Dose in Human Phantom

Kajal Garg and Sonali Bhatnagar

81.1 Introduction

Astronauts in space are exposed to cosmic rays of very high intensity and frequency which may affect their central nervous system and cause cancer [1]. Earth is protected from this space radiation by it's magnetic field so once the astronauts leave the orbit, they are exposed to constant shower of various radioactive and high energy particles. ALTEA (Anomalous Long Term Effect in Astronauts) experiment [2] which has been developed by ASI (Italian space agency) is also working on the cosmic radiation which pass through crew-members brain, measuring their brain activity and visual perception. Space radiation is of three types which are identified by their origin; GCR which is produced by the acceleration of solar plasma by strong electromotive force on solar surface, SEP which are produced by acceleration across the transition shock boundary of propagating coronal mass ejection and last is trapped radiation which is produced near the magnetic field lines of earth [3]. SEP has always been the primary concern for operation outside the earths protective magnetic shield but after space flights began it was realized that GCR are much more harmful than SEP when astronaut leave earths protecting shield. Energy of SEP ranges from 10 MeV to 100 MeV and GCR ranges from 100s of MeV to 10 GeV. In GCR proton is 85%, alpha is 12% and 3% are other light particle. The salient feature of these radiations is that a significant number of these particles have high charge which affects the means by which energy is transferred to tissues hence increasing the cancer probability by 5 times as compare to the person on earth. Mars mission showed that the maximum

K. Garg · S. Bhatnagar (✉)
Dayalbagh Educational Institute, Agra, India
e-mail: deisonali.bhatnagar@gmail.com

K. Garg
e-mail: kajalgarg.dei@gmail.com

Md. Naimuddin (ed.), *XXII DAE High Energy Physics Symposium*, Springer Proceedings in Physics 203, https://doi.org/10.1007/978-3-319-73171-1_81

exposure from GCR is 2 Gy in which 50 % is from high energy charge particles and is 1 Gy from SEP. Equivalent Dose is given by

$$H = QD \tag{81.1}$$

here Q = Quality factor, D = absorbed Dose, Q = 10 (for proton).

$$D = \frac{\int_0^\infty f(E)dE}{\rho V} \tag{81.2}$$

here ρ is detector density, V is the volume and f(E) is energy distribution function.

81.2 Methodology and Simulation

Hadronic binary model has been used for the simulation [4] because it has hadronic interaction above 10 MeV. In this study, the intra nuclear cascade model generates the final state for hadronic elastic scattering. The propagation of the incident particle through the nucleus and the production of secondaries is modeled by two particle collision cascading series. Secondaries are created during the decay of resonances formed during the collisions. This model is valid within the energy range of 40 MeV to 10 GeV. The simulation in geant4 is divided into two parts: first initialization and second event processing. In the initialization the geometries (shape of the object defined) of the phantom (human considered as a square box), vehicle, SPE and GCR shielding (75 cm thick) are designed (Fig. 81.1). Particle interaction processes and their interaction mechanisms are defined according to the model. For visualization purpose, HepRep driver is used. Secondly the event processing defines how many events, how the decay process should take place (at how much distance it should decay) and finally how all these processes should be saved in desired format. The data is analysed in ROOT which is an open source software developed by CERN. It saves the output in histogram and also has an interface to analyze data for zooming, fitting, filtration, marking data points to differentiate between two plots etc. This simulation has two shielding, one for SEP and other for GCR. By keeping the SEP shielding material fixed as water, we have observed the effect of GCR shielding with three materials Aluminium, Polystyrene and polyethylene on astronauts.

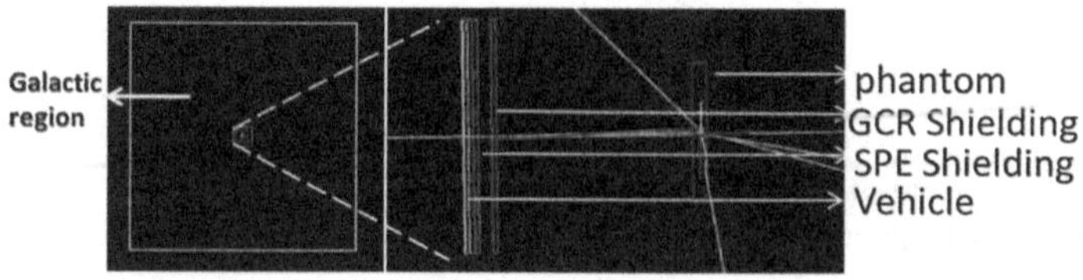

Fig. 81.1 Setup in simulation

81.3 Result and Discussion

Each event has 1000 runs. The primary proton obeying GCR energy spectrum defined in [4] has been incident on the vehicle in the galactic atmosphere. The primary and secondaries reaching to the phantom after passing through vehicle and shielding is given in Table 81.1. The data for materials considered is shown in Table 81.1 with graph of few are in Fig. 81.2 is showing the energy spectrum for polystyrene, for rest of the material value is given in Table 81.1. Secondary particles are produced from the nuclear reaction that happen in shielding, this includes neutron, proton, mesons, gamma rays etc. When there is no shielding, intensity of the secondaries reaching phantom is very less.

Table 81.1 Energy, number of particle deposited in phantom and equivalent dose in different materials

Results		Aluminum	Polyethylene	Polystyrene	No shielding
Primary reaching to phantom	Energy (MeV)	762 ± 27	774 ± 28	789 ± 29	970 ± 31
	Intensity	184 ± 14	197 ± 16	227 ± 19	999 ± 32
Secondaries reaching to phantom	Energy (MeV)	$5-6$	$5-6$	$5-6$	$5-6$
	Intensity	1572 ± 32	1955 ± 31	1902 ± 30	14 ± 27
Equivalent dose (calculated)	(sv)	140	117	107	217 ± 26
Expected dose [1]	(sv)	130	113	100	220 ± 23

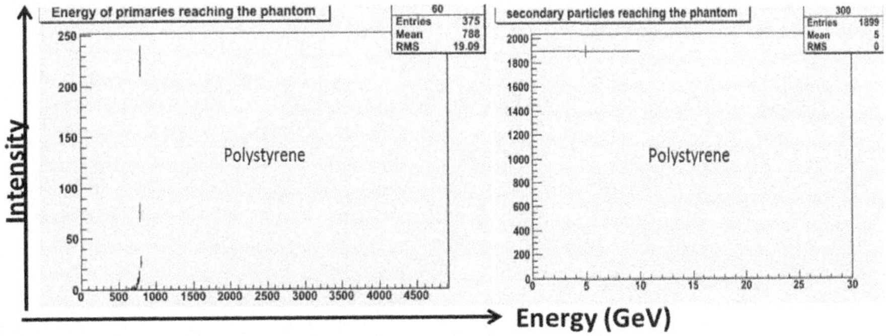

Fig. 81.2 (Left) Energy spectrum of primaries reaching to phantom, (Right) energy spectrum of secondaries reaching to phantom

81.4 Conclusion

Aluminium structure attenuate radiation effects over most of the range of depth used in human vehicles and Poly materials are material with high hydrogen (H) which is efficient to break GCR particles into small fragments and low atomic number (Z) which produce more secondary particles. From this study Aluminium (Al) is found to be a poor shielding material as far as dose equivalent is concerned. Polyethylene, Polystyrene are examined as potentially useful material and demonstrates important advantages as an alternative to Al. It is observed that equivalent dose is minimum in Polystyrene as compared to the other material.

References

1. F.A. Cucinotta et al., Space radiation risks to the central nervous system. Life Sci. Space Res. **2**, 54–69 (2014)
2. L. Narici et al., ALTEA: anomalous long term effects in astronauts. A probe on the influence of cosmic radiation and microgravity non central nervous system during long flights. Adv. Space Res. **31**(1), 141–146 (2003)
3. T. Koi, D.H. Wright, G. Folger, Validation of hadronic models, in *Geant4 Hadronic Shower Simulations Workshop, Batavia* (2006)
4. A.J. Tylka et al., CREME96: a revision of the cosmic ray effects on micro-electronics code. IEEE Trans. Nucl. Sci **44**(6) (1997)

Chapter 82
Light Flavor Hadron Production as a Function of Multiplicity in pp Collisions at $\sqrt{s} = 7\,\text{TeV}$ Measured with ALICE

Kishora Nayak

82.1 Introduction

In recent measurements, several collective features similar to Pb–Pb collisions have been observed in high multiplicity pp and p–Pb collisions. Two-particle correlation studies indicate that there are flow-like effects in high-multiplicity pp and p–Pb collisions systems analogous to Pb–Pb collisions [1–4]. Particle ratio (e.g. Λ/K_s^0) measurements in p–Pb collisions show similar features to Pb–Pb collisions [5]. In order to understand the origin of these intriguing features, a comprehensive study of identified particle production in pp collisions is needed.

The ALICE detector has excellent particle identification capabilities. This enables us to study identified particle production over a wide range of transverse momentum at mid-rapidity ($|y| < 0.5$). Particle identification is performed by using different detectors such as the Inner Tracking System (ITS), Time Projection Chamber (TPC), Time Of Flight (TOF) and High Momentum Particle IDentification (HMPID). This analysis has been performed with events having low pile-up collected in 2010 using a minimum bias trigger and requiring at least one charged particle within $|\eta| < 1$ (INEL > 0) in pp collisions at $\sqrt{s} = 7\,\text{TeV}$. The event multiplicity is estimated by using the V0 detector. The whole data sample has been divided into ten event multiplicity classes from highest (I) to lowest (X).

Kishora Nayak for the ALICE Collaboration.

K. Nayak (✉)
National Institute of Science Education and Research, HBNI, Jatni 752050, India
e-mail: kishora.nayak@cern.ch

Md. Naimuddin (ed.), *XXII DAE High Energy Physics Symposium*, Springer Proceedings in Physics 203, https://doi.org/10.1007/978-3-319-73171-1_82

82.2 Results

The transverse momentum spectra of different particles (π, K, p, K^{*0}, K^0_s, Λ, Ξ, Ω) have been measured as function of event multiplicity in pp collisions at $\sqrt{s} =$ 7 TeV [6]. A comparison of the p_T-differential Λ/K^0_s ratio for the pp, p–Pb and Pb–Pb collision systems is shown in Fig. 82.1. The ratio is qualitatively similar for all three colliding systems and changes in a similar way across multiplicity classes. It has a maximum value around $p_T \sim$ 2–3 GeV/c. The magnitude of increase in the ratio from low to high event multiplicities is larger in Pb–Pb followed by p–Pb and then pp collisions. However, it is worth noting that the absolute value of $\langle dN_{ch}/d\eta\rangle_{|\eta|<0.5}$ for the highest multiplicity class is much larger in Pb–Pb in comparison to p–Pb and pp collisions. A similar behaviour for the p_T-differential ratio of p/π is also observed, with depletion at low-p_T and enhancement at intermediate p_T in all three collisions systems [5].

To obtain the p_T-integrated yield, the spectra are fitted with a Lévy–Tsallis function and then extrapolated to the full p_T range. The integrated yield ratios of baryons to pions, normalized to their corresponding yield ratio of in the 0–100% multiplicity event class, are shown in Fig. 82.2 as a function of multiplicity in pp and p–Pb collisions. There is a multiplicity dependent increase in the normalized yield for particles having a net strangeness content (Λ, Ξ, Ω). The p_T-integrated particle ratios show a continuous evolution from pp to p–Pb collisions as a function of event multiplicity. The rate of increase is found to be proportional to the number of constituent strange quarks. In Fig. 82.2, commonly used MC models fail to describe all the observed features of high multiplicity pp collisions [6].

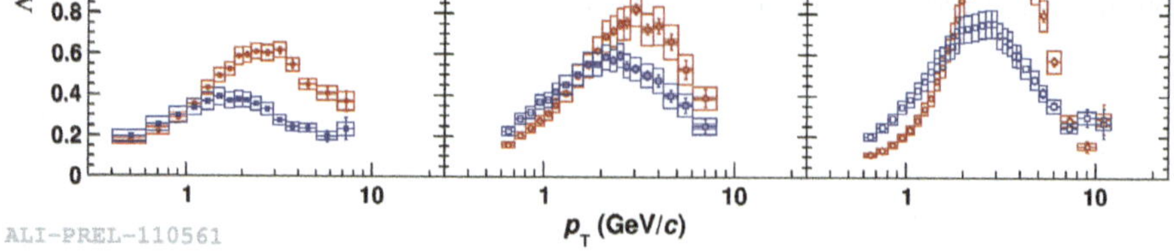

Fig. 82.1 p_T dependent Λ/K^0_s ratio for different V0 multiplicity classes in pp (left), p–Pb (middle), Pb–Pb (right) collisions at $\sqrt{s_{NN}}$ = 7, 5.02 and 2.76 TeV

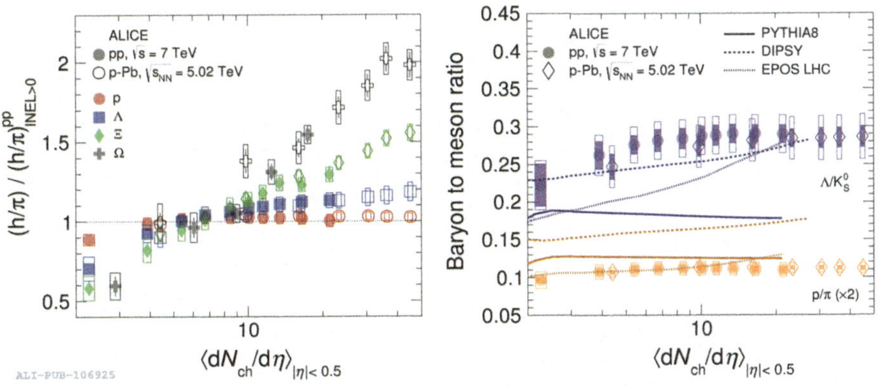

Fig. 82.2 Ratios of baryons (p, Λ, Ξ, Ω) to pions, normalized to the corresponding ratio in the 0–100% multiplicity class (Left). Baryon to meson ratio as a function of average charged particle multiplicity in pp and p–Pb collisions (Right)

82.3 Conclusion

The production of identified light flavor hadrons as a function of event multiplicity in pp collisions at $\sqrt{s} = 7\,\text{TeV}$ has been presented by the ALICE Collaboration. In p_T-dependent baryon to meson ratios (Λ/K^0_s, p/π), a qualitatively similar evolution of the baryon enhancement at intermediate p_T has been observed in pp, p–Pb and Pb–Pb collisions. The p_T-integrated yield ratios to pions show an increase in the relative production of strange particles as function of multiplicity in both pp and p–Pb collisions. This relative increase is more prominent for particles with higher strangeness content. There is no difference observed in the production of the particles with same number of constituent strange quarks, indicating the enhancement is more related to the strangeness content rather than particle masses.

References

1. V. Khachatryan et al. (CMS Collaboration), JHEP **1009**, 091 (2010)
2. G. Aad et al. (ATLAS Collaboration), Phys. Rev. Lett. **116**, 172301 (2016)
3. B. Abelev et al. (ALICE Collaboration), Phys. Lett. B **719**, 29–41 (2013)
4. G. Aad et al. (ATLAS Collaboration), Phys. Rev. Lett. **110**, 182302 (2013)
5. B. Abelev et al. (ALICE Collaboration), Phys. Lett. B **728**, 25 (2014)
6. Adam J et al. (ALICE Collaboration), arXiv:1606.07424 (2016)

Chapter 83
Different Freezeout Scenarios in Large and Small Systems

Sandeep Chatterjee, Ajay Kumar Dash and Bedangadas Mohanty

83.1 Introduction

The phase transition predicted by Quantum Chromodynamics (QCD) from ordinary matter to a deconfined Quark-Gluon Plasma (QGP) is being studied in high-energy heavy-ion (Pb+Pb) collisions at the Large Hadron Collider (LHC) by producing a hot and dense strongly interacting medium called the fireball. The expansion of the fireball causes dilution of the energy density of the system. The medium mean free path elongates and finally drives the fireball out of equilibrium. The identified hadron spectra and integrated yield provide insight into the particle production mechanism and interaction in the hadronic and QGP phases. The analysis of hadron yields allows an access to the properties of the chemical freezeout (CFO) surface where the hadronic inelastic collisions ceased [1–4]. Measurements in smaller collision systems such as proton-proton (p+p) and proton-lead (p+Pb) constitute a fundamental reference for the interpretation of the heavy-ion results. Here we have analysed the mid-rapidity data for hadron yields in Pb+Pb [5], p+Pb [6, 7] and p+p [8, 9] collisions at $\sqrt{s_{\rm NN}} = 2.76$, 5.02 and 7 TeV respectively across all the available centralities. At these energies, the data reveals good particle-antiparticle symmetry which allow us to take the hadron chemical potentials to be zero. Thus the remaining parameters to be extracted from thermal fits are only the fireball volume, temperature and strangeness suppression factor at the time of freezeout.

S. Chatterjee · A. K. Dash (✉) · B. Mohanty
School of Physical Sciences, NISER, HBNI, Jatni 752050, India
e-mail: ajayd@niser.ac.in

Md. Naimuddin (ed.), *XXII DAE High Energy Physics Symposium*, Springer Proceedings in Physics 203, https://doi.org/10.1007/978-3-319-73171-1_83

83.2 The Model and Results

The particle multiplicity N_i of the ith hadron in the thermal model is given by

$$N_i = \frac{g_i V}{2\pi^2} \sum_{k=1}^{\infty} (\pm 1)^{k+1} \frac{m_i^2 T}{k} K_2\left(\frac{k m_i}{T}\right) \gamma_S^{k|S_i|}. \tag{83.1}$$

Here, V and T refer to the fireball volume and temperature respectively. g_i, m_i and $|S_i|$ are the degeneracy factor, mass and strangeness content of the ith hadron respectively. We have omitted the chemical potential dependence in (83.1) as they have been taken to be zero throughout this work. The total yield $N_i^{\rm tot}$ is obtained by including the resonance decay contribution to the above yield ($N_i^{\rm tot} = N_i + \sum_j \text{B.R.}_{ij} \times N_j$, where B.R._{ij} is the branching ratio for the jth hadron to ith hadron). We have used three different freezeout scheme for this study. In 1CFO freezeout scheme, complete thermal and chemical equilibrium for all hadrons is assumed and γ_S is fixed at unity. In the 1CFO+γ_S scheme, non-equilibrium production of strangeness is considered and the γ_S is treated as a fitting parameter. In the 2CFO scheme we stay within complete equilibrium setup. However, we allow for different freezeout thermal parameters for the non-strange and strange hadrons.

The goodness of fit in terms of χ^2/N_{df} for the different freezeout schemes across all available centralities for Pb+Pb, p+Pb and p+p collisions is shown in the top pannel of Fig. 83.1. In all schemes the fit quality deteriorates as one goes from Pb+Pb to p+Pb and p+p collisions. The 2CFO scheme describes the Pb+Pb data quite well but fails to explain the p+Pb and p+p data though the number of free parameters is double to that of 1CFO. From Fig. 83.1 (top pannel), it is clear that the better fits in the 2CFO scheme in heavy ion collisions (HICs) is not due to mere doubling of parameters- 2CFO would then always provide the best fit across all systems. The fact that there is a very clear system size dependence over the choice of freezeout scheme and that 2CFO works best only in the expected regime and fail miserably where it is expected to provides a strong evidence that the better performance of 2CFO in HICs is of physics origin. For smaller systems, 2CFO clearly performs worse than the other scenarios. Thus, our study strengthens the claim of 2CFO as a viable candidate of freezeout in HICs [10]. Depending on the best fit, the temperature, volume and the strangeness suppression factor γ_S are shown in Fig. 83.1 as a function of $\langle \text{dN}_{ch}/\text{d}\eta \rangle$ for all the three systems. The extracted fireball volume varies over three orders of magnitudes whereas the temperature varies from 145 to 160 MeV. In Pb+Pb the strange hadrons freeze earlier at 160 MeV whereas the non-strange hadrons freezes at 145 MeV. The γ_S increases from lower to higher multiplicity in p+p and p+Pb and saturates in Pb+Pb. A smooth transition of the freezeout parameters are observed among all the three colliding systems.

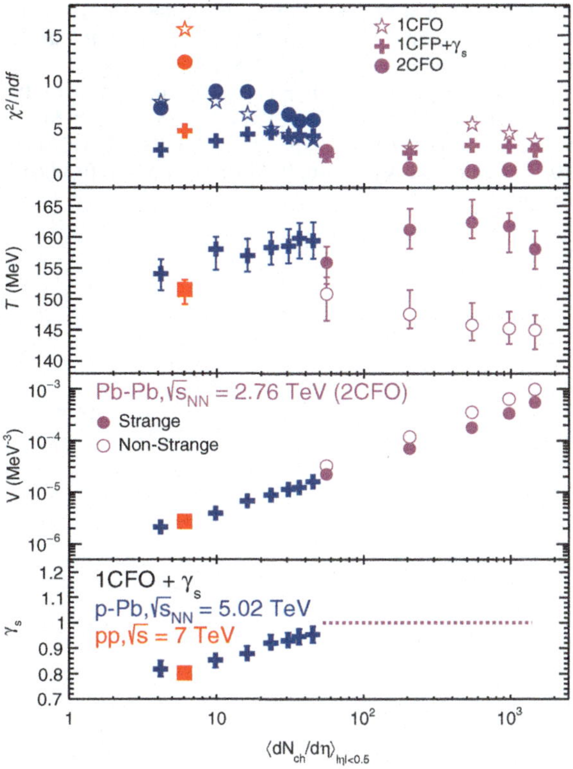

Fig. 83.1 Freezeout parameters (temperature T, volume V and strangeness non-equilibrium factor γ_S) as a function of average charged particle multiplicity in three different freezeout schemes for the three colliding systems Pb+Pb [5], p+Pb [6, 7] and p+p [8, 9]

83.3 Summary

We have studied the system size dependence in the CFO conditions in p+p, p+Pb and Pb+Pb collisions at the LHC. The late stage hadronic interactions in Pb+Pb are found to cause a sequential freezeout with early freeze-out of strangeness as compared to the sudden and unified freezeout in p+p and p+Pb due to domination of expansion over interaction. Our study confirms that the analysis of the hadron yields within thermal models is sensitive to the physics of the CFO and the χ^2/N_{df} is a good measure to discriminate between the different CFO scenarios.

References

1. P. Braun-Munzinger et al., Phys. Lett. B **365**, 1 (1996)
2. G.D. Yen, M.I. Gorenstein, Phys. Rev. C **59**, 2788 (1999)
3. F. Becattini et al., Phys. Rev. C **64**, 024901 (2001)
4. A. Andronic, P. Braun-Munzinger, J. Stachel, Nucl. Phys. A **772**, 167–199 (2006)

5. B. Abelev et al. (ALICE), Phys. Rev. C **88**, 044910 (2013); Phys. Rev. C **91**, 024609 (2015); Phys. Rev. Lett. **111**, 222301 (2013); Phys. Lett. B **728**, 216–227 (2014)
6. B. Abelev et al., Phys. Lett. B **728**, 25–38 (2014)
7. J. Adam et al. (ALICE), Eur. Phys. J. C **76**, 245 (2016); Phys. Lett. B **758**, 389–401 (2016)
8. J. Adam et al. (ALICE), Eur. Phys. J. C **75**, 226 (2015)
9. B. Abelev et al. (ALICE), Eur. Phys. J. C **72**, 2183 (2012); Phys. Lett. B **712**, 309–318 (2012)
10. S. Chatterjee, A.K. Dash, B. Mohanty, arXiv:1608.00643

Chapter 84
Study of Z' Mediated FCNC Effects on $B_s \to \phi\mu^+\mu^-$ Semileptonic Decay

Barilang Mawlong and Soram Robertson Singh

84.1 Introduction

The rare decay mode $B_s \to \phi\mu^+\mu^-$ proceeds through the quark level transition $b \to s\mu^+\mu^-$. The LHCb measurement [1] of the differential branching fraction in the low q^2 region found a deviation of more than 3σ below SM predictions for this mode. We therefore analyse this observable within the 331 model [2] to investigate whether the Z'-mediated FCNC arising from this model has any effect on it or not. We also examine the 331 Z' effects on the forward-backward asymmetry and the lepton polarisation asymmetries. Since the Z' has not yet been discovered, its exact mass is unknown. Many extensions of the SM predict this mass to be in the range 0.6–5 TeV. LHC bounds on $M_{Z'}$ are 3.4 and 2.8 TeV [3] for Z' originating from sequential standard model (SSM) and E_6 grand unification models. In our work, we explore the effects of the 331 Z' for masses ranging from 1–5 TeV.

84.2 Standard Model Analysis

The effective Hamiltonian describing the quark level transition $b \to sl^+l^-$ is given by

$$\mathcal{H}_{\text{eff}} = -4\frac{G_F}{\sqrt{2}} V_{tb} V_{ts}^* \sum_{i=1}^{10} C_i(\mu) O_i(\mu), \qquad (84.1)$$

B. Mawlong (✉) · S. R. Singh
School of Physics, University of Hyderabad, Hyderabad 500046, India
e-mail: barilang05@gmail.com

S. R. Singh
e-mail: robsoram@gmail.com

Md. Naimuddin (ed.), *XXII DAE High Energy Physics Symposium*, Springer Proceedings in Physics 203, https://doi.org/10.1007/978-3-319-73171-1_84

The differential decay width for $B_s \to \phi\mu^+\mu^-$ is given by

$$\frac{d\Gamma}{d\hat{s}} = \frac{G_F^2\,\alpha^2\,m_B^5}{2^{10}\pi^5}\left|V_{ts}^*V_{tb}\right|^2\,\hat{u}(\hat{s})\Delta \tag{84.2}$$

where the form of $\hat{u}(\hat{s})$ and Δ can be found in [4].

In addition, we also find the normalised forward-backward asymmetry as defined in [4] as well as the longitudinal polarisation asymmetry defined in [5].

84.3 331 Model Analysis

The 331 model based on the gauge group $SU_C(3) \otimes SU_L(3) \otimes U_X(1)$ predicts several new particles which includes the Z' boson. The electric charge is defined as a linear combination of the diagonal generators of the group

$$Q = T_3 + \beta T_8 + XI \tag{84.3}$$

where β is a key parameter that gives rise to different variants of the 331 model. We follow [6] where we consider $\beta = \pm\frac{1}{\sqrt{3}}$ and $\beta = \pm\frac{2}{\sqrt{3}}$ values only. The dominant Z' contributions affect only C_9 and C_{10}. The new contributions are

$$\sin^2\theta_W C_9' = -\frac{1}{g_{\text{SM}}^2 M_{Z'}^2}\frac{\Delta_L^{sb}(Z')\Delta_V^{\mu\bar{\mu}}(Z')}{V_{ts}^*V_{tb}}, \tag{84.4}$$

$$\sin^2\theta_W C_{10}' = -\frac{1}{g_{\text{SM}}^2 M_{Z'}^2}\frac{\Delta_L^{sb}(Z')\Delta_A^{\mu\bar{\mu}}(Z')}{V_{ts}^*V_{tb}}. \tag{84.5}$$

where $\Delta_L^{sb}(Z')$, $\Delta_V^{\mu\bar{\mu}}(Z')$, $\Delta_A^{\mu\bar{\mu}}(Z')$ are the relevant couplings of Z' to quarks and leptons. The quark couplings $\Delta_L^{sb}(Z')$ are parametrized as $\Delta_L^{sb}(Z') = -s_{23}e^{-i\delta_{23}}$ and are constrained by flavor observables like ΔM_s, CP asymmetry $S_{\psi\phi}$ and also by the branching ratio of $B_s \to \mu^+\mu^-$ [7]. The values of the lepton couplings are taken from [6]. All other input parameters are taken from [8].

84.4 Results and Discussion

The variation of differential branching fraction, forward-backward asymmetry and lepton longitudinal polarisation asymmetry with respect to the dilepton invariant mass q^2 for the range $0 < q^2 < 6\,\text{GeV}^2$ is shown in Fig. 84.1. The bands we have drawn correspond to uncertainties in the parameter space of the new Z' couplings. The allowed parameter space has been obtained by considering constraints as mentioned in Sect. 84.3.

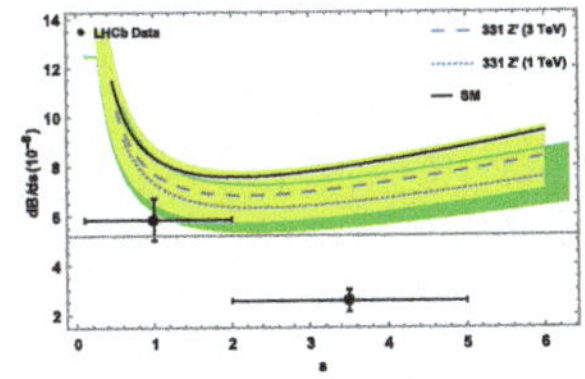

(a) dB/ds

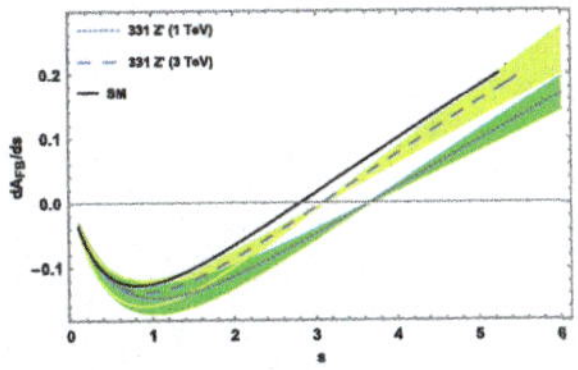

(b) dA_{FB}/ds

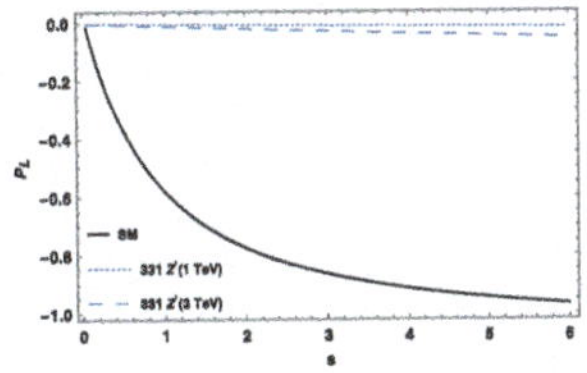

(c) P_L

Fig. 84.1 The differential branching fraction, forward-backward asymmetry and lepton longitudinal polarisation asymmetry for $\beta = -1/\sqrt{3}$. The green and yellow bands correspond to the 331 Z' contributions for $M'_Z = 1$ and 3 TeV, respectively

It is found that only $\beta = -1/\sqrt{3}$ and $\beta = -2/\sqrt{3}$ values have significant effects on the observables considered. The (s_{23}, δ_{23}) pair central values generated for these β values are (0.01, 5.3) and (0.03, 5.4) for two representative values of M'_Z, i.e., $M'_Z = 1$ TeV and $M'_Z = 3$ TeV, respectively. We have shown only the $\beta = -1/\sqrt{3}$ results here. It can be seen from the figures that the differential branching ratio, dA_{FB}/ds and P_L are sensitive to the new parameters of the 331 model. There is a reasonable deviation from SM expectation for differential branching ratio. However, we find that the model cannot satisfactorily explain the discrepancy between theory and data. For P_L, the deviation is quite significant. The forward-backward asymmetry also shows deviation from SM expectation. The zero crossing shifts to a new position. To conclude, since $B_s \to \phi\mu^+\mu^-$ is a rare decay mode, it can provide sensitive probe for new physics (NP) like a new Z' boson. If NP is discovered in the future, we can understand the tension between theory and data for this mode and others in a far better way.

Acknowledgements S. R. Singh would like to acknowledge DST Inspire for financial support.

References

1. R. Aaij et al. (LHCb Collaboration), JHEP **09**, 179 (2015)
2. F. Pisano, V. Pleitez, Phys. Rev. D **46**, 410 (1992); P.H. Frampton, Phys. Rev. Lett. **69**, 2889 (1992)
3. ATLAS-CONF-2015-070 (Dec 2015); CMS-EXO-15-005 (Sep 2016)
4. A. Ali, P. Ball, L.T. Handoko, G. Hiller, Phys. Rev. D **61**, 074024 (2000)
5. G. Erkol, G. Turan, J.W. Wagenaar, Eur. Phys. J. C **41**, 189 (2005)
6. A.J. Buras, F.D. Fazio, J. Girrbach, JHEP **02**, 112 (2014); A.J. buras, F. De Fazio, J. Girrbach, M.V. Carlucci, JHEP **1302**, 023 (2013); A.J. Buras, F. De Fazio, J. Girrbach, JHEP **02**, 116 (2013)
7. Y. Amhis et al. (HFAG), arXiv:1612.07233; R. Aaij et al. (LHCb collaboration), Phys. Lett. B **762**, 253 (2016); CMS Collaboration and LHCb Collaboration, Nature **522**, 68 (2015)
8. C. Patrignani et al. (Particle Data Group), Chin. Phys. C **40**, 100001 (2016)

Chapter 85
Timing and Induced Charge Profile of Large Size RPC Detector for INO-ICAL Experiment

Ankit Gaur, Aman Phogat, Moh. Rafik, Ashok Kumar and Md. Naimuddin

85.1 Introduction

The India-based Neutrino Observatory (INO) [1] is an approved project aimed at building a world-class underground laboratory to study fundamental issues related to the atmospheric neutrinos. The INO will be located in the Bodi West Hills, Theni District of Tamil Nadu. The primary detector proposed to be built at INO is the Iron Calorimeter (ICAL) detector. The ICAL detector is a magnetized sampling calorimeter with a mass of about 51 Kton. The complete detector will have a modular structure, having the dimensions of $48 \times 16 \times 14.4\,\text{m}^3$. The detector will utilize magnetised iron plates as the target material and about 28000 Resistive Plate Chambers (RPCs) as sensitive elements. The spatial localization of the muons produced from the muon neutrino induced charge current events will be done by the two dimensional orthogonal readout of the RPCs and the layer number, while the directionality information revealed via the timing information provided by the sensitive detectors. In this paper, we present the timing and charge response of the $1\,\text{m} \times 1\,\text{m}$ size single gap glass based Resistive Plate Chamber under the different gas compositions.

85.2 Resistive Plate Chambers

Resistive Plate Chambers [2] are the particle detectors that combine good spatial and timing resolution and are extensively used in the experiments that require fast triggering. Various studies concerning choice of appropriate electrodes, choosing of suitable gas mixture and performance study of these detectors [3–7] have been already done. In our present study we have fabricated $1\,\text{m} \times 1\,\text{m}$ size RPC, using

A. Gaur (✉) · A. Phogat · Moh. Rafik · A. Kumar · Md. Naimuddin
Department of Physics & Astrophysics, University of Delhi, Delhi, India
e-mail: ankitphysics09@gmail.com

Md. Naimuddin (ed.), *XXII DAE High Energy Physics Symposium*, Springer Proceedings in Physics 203, https://doi.org/10.1007/978-3-319-73171-1_85

Saint Gobain glass as an electrodes material. The fabricated RPC was tested for their timing and charge response under two gas mixtures. In this study the following two gas mixtures have been used.

1. Freon (95.0%), Isobutane (5.0%).
2. Freon (95.0%), Isobutane (4.5%), Sulfur Hexaflouride (0.5%).

85.3 Experimental Setup

The performances of the fabricated RPC was tested using the cosmic ray test setup shown in Fig. 85.1. In order to determine timing response of the RPC, a device named Time to Digital Converter (TDC) has been used.

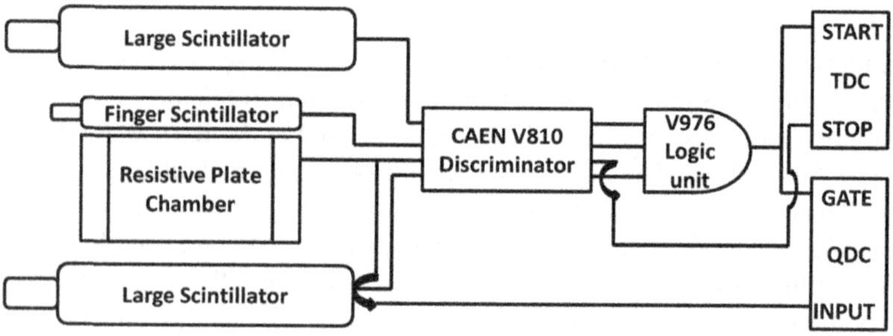

Fig. 85.1 Schematic diagram of the experimental set-up used for the timing and charge response measurements

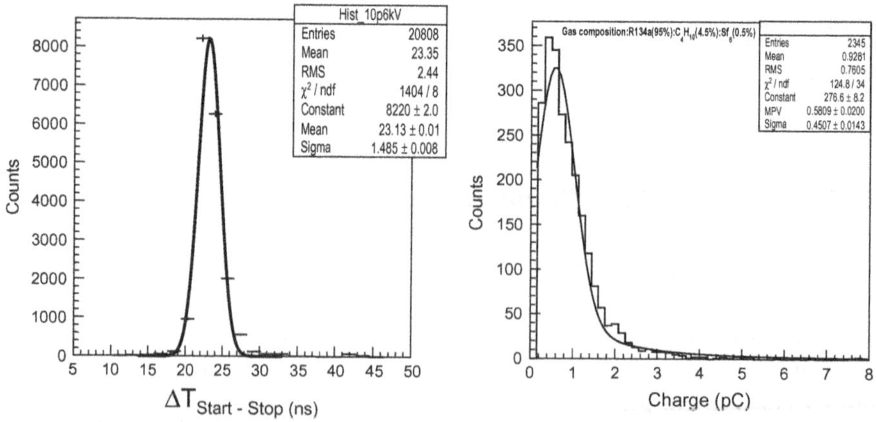

Fig. 85.2 Time distribution(left) and charge spectra (right) of the Saint Gobain RPC detector at a bias voltage of 10.6 kV, under SF_6 concentration of 0.5%

The output pulses of the different scintillators and the RPC are transferred into a CAEN V814 discriminator for the digital conversion. Triple coincidence of the scintillators pulses (Digital) have been done using a CAEN V976 logic unit. The output of the logic unit utilize as a START pulse and the discriminated output pulse of the RPC used as a STOP pulse for a CAEN V775 TDC. A CAEN V965 Charge to Digital Converter (QDC) is used to measure the charge content of the pulses. The output of the CAEN V976 logic unit is taken as a GATE pulse and the analog pulse of the RPC used as a input for the CAEN V965 QDC. Figure 85.2 shows the timing response and charge spectra of the Saint Gobain RPC at applied bias voltage of 10.6 kV for the gas mixture Freon (95.0%), Isobutane (4.5%), Sulfur Hexafluoride (0.5%) respectively.

85.4 Results

We have investigated the timing and charge response of 1 m × 1 m glass based single gap Resistive Plate Chamber under different gas mixtures. Addition of the SF_6 gas in the mixture results in reduction of charge and fluctuations, hence leads to stable operation of the detector. The time resolution of the fabricated RPC is found in between 1.96 and 1.71 ns for the gas mixture Freon (95.0%), Isobutane (5.0%) and 2.05 - 1.48 ns for the gas mixture Freon (95.0%), Isobutane (4.5%), Sulfur Hexafluoride (0.5%) respectively. Charge spectra shows peak charge collection between 0.4 and 2 pC on average under both the gas mixtures.

References

1. S. Atthar et al., Technical report of INO, INO/2006/01, June 2006, http://www.ino.tifr.res.in/ino/OpenReports/INOReport.pdf
2. R. Santonico, R. Cardarelli, Development of resistive plate counters. Nucl. Instr. Meth. A **187**, 377 (1981)
3. A. Kumar et al., Study of RPC bakelite electrodes and detector performance for INO-ICAL, 2014 JINST 9 C10042
4. Md. Naimuddin et al., Characterisation of glass electrodes and RPC detectors for INO-ICAL experiment, 2014 JINST 9 C10039
5. D. Kaur et al., Characterization of 3 mm glass electrodes and development of RPC detectors for INO-ICAL experiment. Nucl. Instr. Meth. A **774**, 74 (2015)
6. A. Kumar et al., RPC detector characteristics and performance for INO-ICAL experiment. J. Instr. **11**, C03034 (2016)
7. A. Gaur et al., Performance study of glass RPC detectors for INO-ICAL experiment. Nucl. Instr. Meth. Phys. Res. A **845**, 363–366 (2017)

Chapter 86
Explaining $R(D^{(*)})$, R_K and $(g-2)_\mu$ in a E_6 Motivated Left-Right Model

Diganta Das, Chandan Hati, Girish Kumar and Namit Mahajan

Introduction: The recently reported measurements of semileptonic B decays $B \to D^{(*)}\tau\bar{\nu}$ and $B \to K\ell^+\ell^-$ by the LHCb collaboration [1, 2] and B factories [3] show significant deviations from the Standard Model (SM) predictions. To reduce theoretical uncertainties, one defines observables $R(D^{(*)})$ and R_K as ratios of branching ratios:

$$R(D^{(*)}) = \frac{\text{BR}(B \to D^{(*)}\tau\bar{\nu}_\tau)}{\text{BR}(B \to D^{(*)}\ell\bar{\nu}_\ell)}, \quad R_K = \frac{\text{BR}(B \to K\mu^+\mu^-)}{\text{BR}(B \to Ke^+e^-)}. \tag{86.1}$$

The current experimental world averages [4] for observable $R(D^{(*)})$ are $R(D) = 0.397 \pm 0.040 \pm 0.028$ and $R(D^*) = 0.316 \pm 0.016 \pm 0.010$ corresponding to 1.9σ and 3.3σ deviations from the SM values $R(D)^{\text{SM}} = 0.300 \pm 0.008$ [5] and $R(D^*)^{\text{SM}} = 0.252 \pm 0.003$ [6], respectively. The measured value of $R_K = 0.745^{+0.090}_{-0.074} \pm 0.036$ [2] in the dilepton invariant mass bin $1\,\text{GeV}^2 < q^2 < 6\,\text{GeV}^2$ corresponds to a 2.6σ departure from the SM prediction, $R_K^{\text{SM}} = 1.0003 \pm 0.0001$ [7].

Another interesting deviation from the SM has been seen in the measurement of anomalous magnetic moment of the muon $(g-2)_\mu$. The measured value of $(g-2)_\mu$ by E821-BNL experiment [8] differs from the SM value by $\Delta a_\mu = a_\mu^{\text{exp}} - a_\mu^{\text{SM}} = (2.9 \pm 0.9) \times 10^{-9}$ [9] corresponding to about 3σ discrepancy.

In this article, we show that these anomalies can be explained within the framework of neutral left-right symmetric model (NLRSM) gauge group $SU(3)_C \times SU(2)_L \times SU(2)_N \times U(1)_Y$ [10] motivated by one of the low energy subgroups of E_6 which naturally contains leptoquarks. Our primary interest in this paper is to study the effects of interactions involving leptoquarks and their supersymmetric partners on concerned flavor processes [11].

Superpotential of the model: The superpotential governing the interactions among the fields of NLRSM is given by

D. Das · C. Hati · G. Kumar (✉) · N. Mahajan
Physical Research Laboratory, Ahmedabad 380009, India
e-mail: girishk@prl.res.in

Md. Naimuddin (ed.), *XXII DAE High Energy Physics Symposium*, Springer Proceedings in Physics 203, https://doi.org/10.1007/978-3-319-73171-1_86

$$W = \lambda^1 \left(\nu_e N_L^c N_E^c + e E^c N_L^c + \nu_E N_E^c n + E E^c n\right) + \lambda^2 \left(d^c N_L^c h + h h^c n\right)$$
$$+ \lambda^3 u^c e^c h + \lambda^4 \left(u u^c N_E^c + u^c d E^c\right) + \lambda^5 \left(\nu_e e^c E + e e^c \nu_E\right)$$
$$+ \lambda^6 \left(u d^c E + d d^c \nu_E + u h^c e + d h^c \nu_e\right), \qquad (86.2)$$

where we have suppressed the family indices of superpotential couplings λ for simplicity. Here h is a leptoquark with charge $-1/3$ and ν_E, E, E^c, N_E^c, n and N^c are new exotic fields in this model.

New physics explanation of anomalies: From the superpotential (86.2) it follows that there are two new Feynman diagrams, involving tree-level exchange of the scalar leptoquark ($\tilde{h}^*$) and slepton ($\tilde{E}$), which contribute to decays $B \to D^{(*)}\tau\bar{\nu}$. The corresponding new physics (NP) effective Hamiltonian is given by

$$\mathcal{H}_{\text{eff}} = -\sum_{i,k=1}^{3} V_{2i} \left(\frac{\lambda^5_{33k}\lambda^{6*}_{i3k}}{m^2_{\tilde{E}^k}} \bar{c}_L b_R \; \bar{\tau}_R \nu_L + \frac{\lambda^6_{33k}\lambda^{6*}_{i3k}}{2m^2_{\tilde{h}^{k*}}} \bar{c}_L \gamma^\mu b_L \; \bar{\tau}_L \gamma_\mu \nu_L \right), \quad (86.3)$$

where V_{ij} are the CKM matrix elements and $m_{\tilde{h}}(m_{\tilde{E}})$ is the mass of scalar leptoquark $\tilde{h}^{k*}$ (slepton $\tilde{E}^k$). Under one operator dominance hypothesis we find that the NP Wilson coefficient of the vector operator $(\bar{c}_L\gamma^\mu b_L)$ $(\bar{\tau}_L\gamma_\mu\nu_L)$ is capable of explaining both $R(D)$ and $R(D^*)$ simultaneously, while that of the scalar operator $(\bar{c}_L b_R)$ $(\bar{\tau}_R\nu_L)$ cannot. The NP couplings (λ^6_{33k} and λ^6_{23k}) corresponding to vector operator also contribute to other flavor processes, e.g., $B \to \tau\nu$, $D_s^+ \to \tau\nu$, $D^+ \to \tau\nu$ and $D^0 - \bar{D}^0$. We use the experimental information available on these processes to constrain these NP couplings. We find that taking $m_{\tilde{h}^{k*}} = 750\,\text{GeV}$ (larger values are also allowed), one can explain both $R(D)$ and $R(D^*)$ experimental data quite well with values of λ^6_{33k} and λ^6_{23k} being less than unity consistent with the constraints from the flavor sector.

On the other hand, in NLRSM the leptoquark can induce new contribution to R_K through one loop Feynman diagrams involving the scalar leptoquark and its supersymmetric partner as virtual particles in the loop. We find that contributions of γ- and Z- penguin diagrams vanish while the box diagrams give the following contribution to NP Wilson coefficients of operator $(\bar{s}_L\gamma_\mu b_L)(\bar{\ell}_L\gamma^\mu\ell_L)$:

$$C^\mu_{LL} = \frac{\lambda^6_{32k}\lambda^{6*}_{32k}}{8\pi\alpha_e}\left(\frac{m_t}{m_{\tilde{h}_j}}\right)^2 - \frac{\lambda^6_{3jk}\lambda^{6*}_{2jl}\lambda^6_{i2k}\lambda^{6*}_{i2l}}{32\sqrt{2}\,G_F\,V_{tb}V^*_{ts}\pi\alpha_e}\left[\frac{1}{m^2_{\tilde{h}}} + \frac{1}{m^2_{\tilde{h}}}\, g\left(\frac{m^2_{\tilde{u}_i}}{m^2_{\tilde{h}}}, 1, \frac{m^2_{\tilde{\nu}_j}}{m^2_{\tilde{h}}}\right)\right], \qquad (86.4)$$

where $g(x, y, z) = x^2 \log x/[(x-1)(x-y)(x-z)] + (\text{cycl. perm.})$ is the loop function. The superpotential couplings appearing in the expression of C^μ_{LL} can be constrained from the present experimental measurements of $B - \bar{B}$ mixing, $Z \to \mu^+\mu^-$, $t \to b\mu\bar{\nu}_\mu$, $D_s \to \mu\bar{\nu}_\mu$ etc.. The global fit to $b \to s\ell\ell$ data suggests that R_K can be explained with $-1.5 \lesssim C^{\mu}_{LL} \lesssim -0.7$ and $-1.9 \lesssim C^{\mu}_{LL} - C^{\mu}_{LR} \lesssim 0$ [12]. Taking $m_{\tilde{h}} \sim 750\,\text{GeV}$ and $m_h \sim 600\,\text{GeV}$, we find that one can easily obtain the standard solution $C^{\mu}_{LL} = -1$ and $C^{\mu}_{LR} = 0$ to explain R_K while satisfying the constraints from other processes as discussed above.

In NLRSM, one can also address the anomalous magnetic moment of the muon through one loop vertex diagrams induced by λ^5 term in the superpotential (86.2). The details can be found in our original work [11].

Conclusion: We have studied the NP contribution to semileptonic B decays $B \to D^{(*)}\tau\bar{\nu}$ and $B \to K\ell^+\ell^-$, and anomalous magnetic moment of the muon arising in NLRSM and shown that discrepancies observed in the measurement of $R(D^{(*)})$, R_K and $(g-2)_\mu$ can be explained in NLRSM with the required size of NP couplings being compatible with the present experimental constraints from B and D (semi)leptonic decays and neutral B and D meson mixings.

References

1. R. Aaij et al. [LHCb Collaboration], Phys. Rev. Lett. **115**(11), 111803 (2015); Addendum: Phys. Rev. Lett. **115**(15), 159901 (2015). https://doi.org/10.1103/PhysRevLett.115.159901, https://doi.org/10.1103/PhysRevLett.115.111803
2. R. Aaij et al. [LHCb Collaboration], Phys. Rev. Lett. **113**, 151601 (2014). https://doi.org/10.1103/PhysRevLett.113.151601
3. J.P. Lees et al. [BaBar Collaboration], Phys. Rev. D **88**(7), 072012 (2013). https://doi.org/10.1103/PhysRevD.88.072012; M. Huschle et al. [Belle Collaboration], Phys. Rev. D **92**(7), 072014 (2015). https://doi.org/10.1103/PhysRevD.92.072014; A. Abdesselam et al. [Belle Collaboration]. arXiv:1603.06711 [hep-ex]
4. Heavy Flavour Averaging Group, http://www.slac.stanford.edu/xorg/hfag/semi/winter16/winter16_dtaunu.html
5. H. Na et al. [HPQCD Collaboration], Phys. Rev. D **92**(5), 054510 (2015); Erratum: Phys. Rev. D **93**(11), 119906 (2016). https://doi.org/10.1103/PhysRevD.93.119906, https://doi.org/10.1103/PhysRevD.92.054510
6. S. Fajfer, J.F. Kamenik, I. Nisandzic, Phys. Rev. D **85**, 094025 (2012). https://doi.org/10.1103/PhysRevD.85.094025
7. C. Bobeth, G. Hiller, G. Piranishvili, JHEP **0712**, 040 (2007). https://doi.org/10.1088/1126-6708/2007/12/040
8. G.W. Bennett et al. [Muon g-2 Collaboration], Phys. Rev. D **73**, 072003 (2006). https://doi.org/10.1103/PhysRevD.73.072003
9. F. Jegerlehner, A. Nyffeler, Phys. Rep. **477**, 1 (2009). https://doi.org/10.1016/j.physrep.2009.04.003
10. D. London, J.L. Rosner, Phys. Rev. D **34**, 1530 (1986). https://doi.org/10.1103/PhysRevD.34.1530
11. D. Das, C. Hati, G. Kumar, N. Mahajan, Phys. Rev. D **94**, 055034 (2016). https://doi.org/10.1103/PhysRevD.94.055034
12. G. Hiller, M. Schmaltz, Phys. Rev. D **90**, 054014 (2014). https://doi.org/10.1103/PhysRevD.90.054014

Chapter 87
SU(6) Grand Unification of 3-3-1 Model

Frank F. Deppisch, Chandan Hati, Sudhanwa Patra, Utpal Sarkar and José W. F. Valle

Introduction: The $SU(3)_c \times SU(3)_L \times U(1)_X$ [1] gauge extension of the Standard Model (SM) is one of the most popular candidates for new physics which can be verified at the next generation accelerators [2, 3]. This model can also shed light on our understanding of non-vanishing neutrino masses. The $SU(3)_c \times SU(3)_L \times U(1)_X$ model originally proposed by Singer, Valle and Schechter (SVS) [2] is not free of anomaly for each generation of fermions. However, once all three generations of fermions are included in the calculation the model becomes anomaly free. Consequently, multiplicities of different multiplets of the $SU(3)_c \times SU(3)_L \times U(1)_X$ group appear with differently and it becomes a challenging task to find a usual grand unified theory to embed such a theory. In [4] we have studied the unification of such a theory in a minimal SU(6) gauge theory which can in turn emerge from an E(6) GUT. Interestingly, the SVS 3-3-1 theory can readily be rearranged into an generation wise anomaly free structure which can be then readily be embedded in a minimal anomaly free combination of SU(6) representations. We will refer to this new variant as the

F. F. Deppisch
Department of Physics and Astronomy, University College London,
London WC1E 6BT, UK

C. Hati (✉)
Theoretical Physics, Physical Research Laboratory, Ahmedabad 380009, India
e-mail: hati.official@gmail.com

S. Patra
Center of Excellence in Theoretical and Mathematical Sciences,
Siksha 'O' Anusandhan University, Bhubaneswar 751030, India

U. Sarkar
Department of Physics, Indian Institute of Technology Kharagpur,
Kharagpur 721302, India

J. W. F. Valle
AHEP Group, Institut de Fisica Corpuscular – C.S.I.C./Universitat de Valencia,
Parc Cientific de Paterna, C/ Catedràtico Josè Beltràn, 2,
46980 Paterna (Valencia), Spain

Md. Naimuddin (ed.), *XXII DAE High Energy Physics Symposium*, Springer Proceedings in Physics 203, https://doi.org/10.1007/978-3-319-73171-1_87

sequential 3-3-1 model. This variant of the 3-3-1 model is particularly intriguing since this SU(6) embedding of this variant does not need any new bulk exotic fields to explain the chiral families. In this sense it is a truly minimal unification in the same spirit as the minimal SU(5) unification [5].

The sequential $SU(3)_c \times SU(3)_L \times U(1)_X$ **Model**: In this model the assignment of the fields are done in such a way that the anomalies cancel separately for each of the generation. The field assignments given by

$$\begin{aligned} Q_{aL} &= \left(u_{aL}, d_{aL}, D_{aL}\right)^T \equiv [3, 3, 0],\ u_{aR} \equiv [3, 1, 2/3],\ d_{aR} \equiv [3, 1, -1/3], \\ D_{aR} &\equiv [3, 1, -1/3], \psi_{aL} = \left(e^-_{aL}\ \nu_{aL}\ N^1_{aL}\right)^T \equiv [1, 3^*, -1/3], \\ \xi_{aL} &= \left(E^-_{aL}, N^2_{aL}, N^3_{aL}\right)^T \equiv [1, 3^*, -1/3],\ \chi_{aL} = \left(N^4_{aL}, E^+_{aL}, e^+_{aL}\right) \equiv [1, 3^*, 2/3]. \end{aligned} \tag{87.1}$$

We assume the Higgs sector to be the same as the SVS 3-3-1 model for symmetry breaking and generation the charged fermion masses. The Yukawa Lagrangian for the quark sector is given by

$$\mathcal{L}_{\text{quarks}} = y_{u_a}\overline{Q_{aL}}u_{aR}\phi_0^* + y^i_{d_a}\overline{Q_{aL}}d_{aR}\phi_i^* + y^i_{D_a}\overline{Q_{aL}}D_{aR}\phi_i^* + \text{h.c.} \quad , \tag{87.2}$$

where $a = 1, 2, 3$, $i = 1, 2$ (neglecting any flavour mixing). The relevant Yukawa interactions in the leptonic sector can be written as

$$\mathcal{L}_{\text{leptons}} = \epsilon_{\alpha\beta\gamma}\left[\psi^T_{\alpha L}C^{-1}\left(y_1\xi_{\beta L}\phi_{0\gamma} + y^i_2\chi_{\beta L}\phi_{i\gamma}\right) + \xi^T_{\alpha L}C^{-1}y^i_3\chi_{\beta L}\phi_{i\gamma}\right] + \text{h.c.} \quad , \tag{87.3}$$

where α, β, γ are the $SU(3)_L$ tensor indices ensuring Dirac mass terms are anti-symmetric. Here C denotes charge conjugation matrix, and $i = 1, 2$. Post symmetry breaking the 5×5 neutrino mass matrix for each generation can be diagonalized to obtain two quasi-Dirac heavy neutrinos with mass around $SU(3)_c \times SU(3)_L \times U(1)_X$ symmetry breaking scale, two Dirac neutrinos with the electroweak symmetry breaking scale mass, and a light Majorana neutrino.

SU(6) Grand Unification: It is straightforward to check that each generation of the fermionic multiplets of the sequential 3-3-1 model given in (87.1) can be fitted exactly in the anomaly free combination of SU(6) representations given by: $\bar{6} + \bar{6}' + 15$, where $\bar{6}$ contains $d^c_L \equiv [3, 1, -1/3]$ and $\psi_L \equiv [1, 3^*, -1/3]$; $\bar{6}'$ contains $D^c_L \equiv [3, 1, -1/3]$ and $\xi_L \equiv [1, 3^*, -1/3]$; and 15 contains $u^c_L \equiv [3^*, 1, -2/3]$, $\chi_L \equiv [1, 3^*, 2/3]$ and $Q_L \equiv [3, 3, 0]$. Now the E(6) fundamental representation 27 decomposes under the maximal $SU(2) \otimes SU(6)$ subgroup as $27 = [2, \bar{6}] + [1, 15]$. Three 27s of E(6) which contain three sets of multiplets $\bar{6} + \bar{6}' + 15$ can accommodate the three generations of the fermionic fields of the sequential 3-3-1 model. However, it turns out that with the minimal content of the sequential 3-3-1 model one cannot obtain a successful unification. Interestingly,

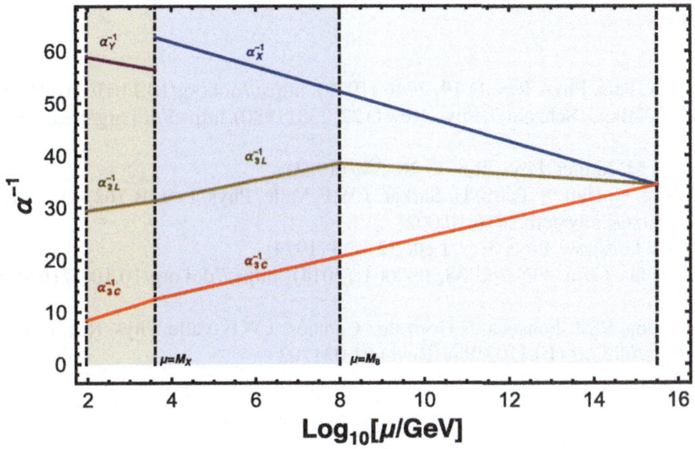

Fig. 87.1 The RG running of the gauge couplings for the sequential 3-3-1 Model with sequential fermionic octets with $SU(3)_c \times SU(3)_L \times U(1)_X$ symmetry breaking scale $M_X = 4000\,\text{GeV}$ and octet mass $M_8 = 10^8\,\text{GeV}$, demonstrating successful gauge unification at the scale $M_U = 10^{15.5}\,\text{GeV}$ with $n_Y = \sqrt{5/3}$ and $n_X = 2/\sqrt{3}$

if three generations of the fermionic octets are added then one can have a successful gauge coupling unification.

The one-loop beta-coefficients for the energy scale between the electroweak symmetry breaking and the $SU(3)_c \times SU(3)_L \times U(1)_X$ symmetry breaking (M_Z to M_X) are given by $b_{2L} = -19/6$, $b_Y = 41/10$, $b_{3C} = -7$. The one-loop beta-coefficients for the energy scale between the $SU(3)_c \times SU(3)_L \times U(1)_X$ symmetry breaking scale and the octet mass (M_X to M_8) are given by $b_{3L} = -9/2$, $b_X = 13/2$, $b_{3C}^{331} = -5$. Finally, the one-loop beta-coefficients for the energy scale between the octet mass to the unification scale (M_8 to M_U) are given by $b_{3L}^8 = 2n - 9/2$, where n is the number of generations of the fermionic octets ($\Omega \equiv [1, 8^*, 0]$), $b_X^8 = 13/2$, $b_{3C}^8 = -5$. Figure 87.1 shows the gauge coupling running of the sequential $SU(3)_c \times SU(3)_L \times U(1)_X$ model and fermionic octets with $SU(3)_c \times SU(3)_L \times U(1)_X$ symmetry breaking scale $M_X = 4000\,\text{GeV}$ and octet mass $M_8 = 10^8\,\text{GeV}$, showing a successful gauge unification at the scale $M_U = 10^{15.5}\,\text{GeV}$ for $n_Y = \sqrt{5/3}$ and $n_X = 2/\sqrt{3}$. Taking $M_U = 10^{15.5}\,\text{GeV}$ and $\alpha_{\text{GUT}}^{-1} \sim 35$ in sequential 3-3-1 model, the proton decay lifetime for the mode $p \to e^+\pi^0$ is roughly $\sim 10^{34}$ yrs and is consistent with the limits from current experimental [6].

Concluding Remarks: We have discussed a minimal SU(6) grand unification of the sequential variant of the $SU(3)_c \times SU(3)_L \times U(1)_X$ model which allows for a TeV scale $SU(3)_c \times SU(3)_L \times U(1)_X$ breaking as well as neutrino masses induced via seesaw. The unification of the gauge couplings can be connected to the presence of three generation of leptonic octets with mass between the 3-3-1 symmetry breaking scale and the unification scale. The presence of the octet scale can have interesting implications for radiative origin of neutrino masses [7].

References

1. H. Georgi, A. Pais, Phys. Rev. D **19**, 2746 (1979). https://doi.org/10.1103/PhysRevD.19.2746
2. M. Singer, J. Valle, J. Schechter, Phys. Rev. D **22**, 738 (1980). https://doi.org/10.1103/PhysRevD.22.738
3. J.W.F. Valle, M. Singer, Phys. Rev. D **28**, 540 (1983)
4. F.F. Deppisch, C. Hati, S. Patra, U. Sarkar, J.W.F. Valle, Phys. Lett. B **762**, 432 (2016). https://doi.org/10.1016/j.physletb.2016.10.002
5. H. Georgi, S. Glashow, Phys. Rev. Lett. **32**, 438 (1974)
6. K. Olive, et al., Chin. Phys. C **38**, 090001 (2014). https://doi.org/10.1088/1674-1137/38/9/090001
7. S.M. Boucenna, R.M. Fonseca, F. Gonzalez-Canales, J.W.F. Valle, Phys. Rev. D **91**(3), 031702 (2015). https://doi.org/10.1103/PhysRevD.91.031702

Chapter 88
Search for Majorana Neutrino via B_c Decay Modes

Sanjoy Mandal

88.1 Introduction

The discovery of neutrino oscillation confirms neutrino mass. The seesaw mechanism [1] provides a natural explanation of the smallness of neutrino mass. The simplest realization of the seesaw, the so-called type-I seesaw, requires the existence of a set of heavy electroweak singlet (sterile) lepton number violating (LNV) right handed Majorana fermions, N_{bR} (b $= 1, 2, 3 \ldots$ n).

We extend the SM to include n right-handed SM singlets along with the three generation of left-handed SM SU(2) doublets [2]. In this model, flavor eigenstates $\nu_{\ell L}$ can be written in terms of the mass eigenstates as, $\nu_{\ell L} = \sum_{m=1}^{3} U_{\ell m}\nu_{mL} + \sum_{m'=4}^{3+n} V_{\ell m'} N^{c}_{m'L}$, with $UU^{\dagger} + VV^{\dagger} = 1$. In this work, we consider only one heavy neutrino N, with heavy-light mixing parameters $V_{\ell N}$ (the mixing parameters of the fourth mass eigenstate N with the SM flavor neutrinos ν_{ℓ}). We take a phenomenological approach regarding the mass and mixing elements of the heavy neutrino, taking them to be free parameters, constrained only by experimental observations.

88.2 Evaluation of the $B_c^- \to \overline{B}_s^0 \ell_1^- \ell_2^- \pi^+$, $J/\psi \ell_1^- \ell_2^- \pi^+$ Decays

For the four-body decay $B_c^-(p) \to \overline{B}_s^0(k_1)\ell_1(k_2)\ell_2(k_3)\pi^+(k_4)$, where $\ell_1, \ell_2 = e, \mu$, only s-channel diagrams shown in Fig. 88.1 contribute. Hence, the Majorana neutrino N that induces this LNV process can appear as an intermediate on mass shell state, leading to an enhancement of the decay rate. The decay amplitude for the processes depicted in Fig. 88.1 can be expressed as,

S. Mandal (✉)
The Institute of Mathematical Sciences, C.I.T Campus,
Tharamani, Chennai 600113, India
e-mail: smandal@imsc.res.in

Md. Naimuddin (ed.), *XXII DAE High Energy Physics Symposium*, Springer Proceedings in Physics 203, https://doi.org/10.1007/978-3-319-73171-1_88

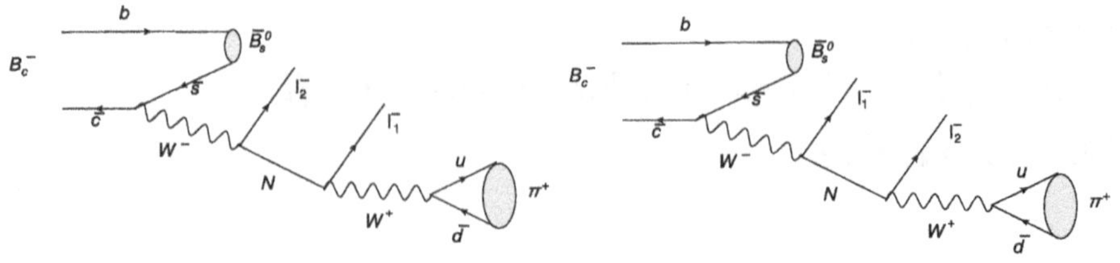

Fig. 88.1 Feynman diagrams for the decay $B_c^- \to \overline{B}_s^0 \ell_1^- \ell_2^- \pi^+$

$$\mathcal{M} = \frac{G_F^2 V_{cs} V_{ud} V_{\ell_1 N}^* V_{\ell_2 N} f_\pi}{(p - k_1 - k_2)^2 - m_N^2 + i m_N \Gamma_N} \left(F_+(q^2)(p + k_1)^\mu + F_-(q^2)(p - k_1)^\mu\right)$$
$$\bar{u}(k_3)\gamma_\beta \gamma_\mu (1 + \gamma_5)\, v(k_2) k_4^\beta + (k_2 \leftrightarrow k_3, \ell_1 \leftrightarrow \ell_2)\ .$$

The form factors for $B_c^- \to \overline{B}_s^0$ have been calculated in the framework of 3-point QCD sum rule in [3]. Γ_N is the total decay width of the heavy neutrino N, obtained by summing over all accessible final states. The total decay width Γ_N is given by [4]:

$$\Gamma_N = a_e(m_N) \mid V_{eN} \mid^2 + a_\mu(m_N) \mid V_{\mu N} \mid^2 + a_\tau(m_N) \mid V_{\tau N} \mid^2\ , \tag{88.1}$$

where, a_e, a_μ and a_τ are functions of the Majorana neutrino mass and hence will differ from mode to mode. To calculate the four-body phase space required for evaluating the decay rate $\Gamma(B_c^-(p) \to \overline{B}_s^0(k_1)\ell_1(k_2)\ell_2(k_3)\pi^+(k_4)) = \frac{1}{2m}\int d_4(ps) \mid \mathcal{M} \mid^2$, the final particles can be partitioned into two subsystems X_{12} and X_{34}, each of which subsequently decays into a two-body state.

$$d_4(ps) = d_2\left(ps\ B_c^- \to X_{12}X_{34}\right) d_2\left(ps\ X_{12} \to k_1 k_2\right) d_2\left(ps\ X_{34} \to k_3 k_4\right) dM_{12}^2 dM_{34}^2,$$

where $X_{12} = (k_1 + k_2)$, $X_{34} = (k_3 + k_4)$, $X_{12}^2 = M_{12}^2$ and $X_{34}^2 = M_{34}^2$, $p^2 = m^2$ and $k_i^2 = m_i^2$. m, m_1, m_2, m_3, and m_4 are the masses of B_c^-, $\overline{B}_s^0$, ℓ_1, ℓ_2 and π^+ respectively. For detailed discussion and explicit form of the above four momenta, four body phase space see [4].

88.2.1 Bounds on Mixing Angles Using Upper Limits on the Branching Ratios for $B_c^- \to \overline{B}_s^0 \ell_1^- \ell_2^- \pi^+$, $J/\psi \ell_1^- \ell_2^- \pi^+$ Decays

The LNV branching ratios in general can be written as: $\mathcal{B}\left(B_c^- \to \overline{B}_s^0 \ell_1^- \ell_2^- \pi^+\right) = G_{\ell_1 \ell_2}(m_N) \frac{|V_{\ell_1 N}|^2 |V_{\ell_2 N}|^2}{\Gamma_N}$, where, $G_{\ell_1 \ell_2}$ are functions of the Majorana mass and depend on the explicit matrix element and phase space for each of the processes. Now

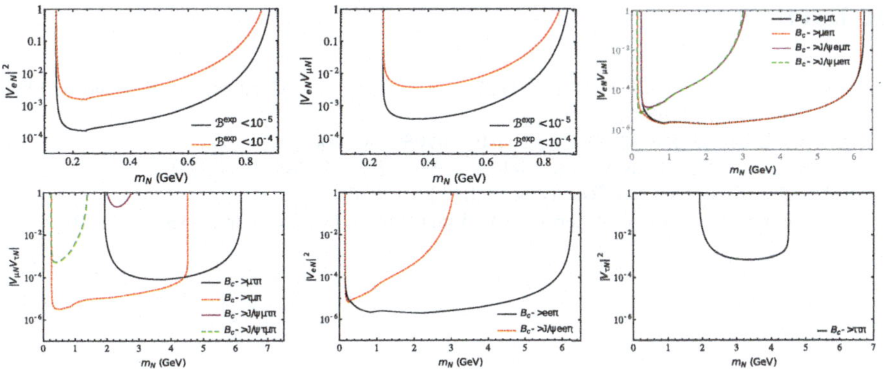

Fig. 88.2 Exclusion curves for mixing elements from the possible upper limits for different decay modes. The first two figure are for $\overline{B}_s^0 \ell_1^- \ell_2^- \pi^+$

define, $F_{\ell_1\ell_2} \equiv \frac{\mathcal{B}^{exp}\left(B_c^- \to \overline{B}_s^0 \ell_1^- \ell_2^- \pi^+\right)}{G_{\ell_1\ell_2}(m_N)}$. The upper limits on the $\mathcal{B}^{exp}$ can be very simply translated into the upper limits on, $\mid V_{eN} \mid^2$, $\mid V_{\mu N} \mid^2$, $\mid V_{eN} V_{\mu N} \mid$ under the assumption, $\mid V_{eN} \mid \sim \mid V_{\mu N} \mid \sim \mid V_{\tau N} \mid$ in Γ_N. This result in the constraints, $\mid V_{eN} \mid^2 < F_{ee}\left(a_e + a_\mu + a_\tau\right)$; $\mid V_{\mu N} \mid^2 < F_{\mu\mu}\left(a_e + a_\mu + a_\tau\right)$; $\mid V_{eN} V_{\mu N} \mid < F_{e\mu}/F_{\mu e}$ $\left(a_e + a_\mu + a_\tau\right)$. According to [5] at the LHC with $\sqrt{s} = 14\,\text{TeV}$, the expected no. of B_c events can be $\sim\mathcal{O}\left(10^{10}\right)$ per year. It may be possible to set upper limit on The BR of $10^{-9} - 10^{-7}$. However since the final state B_c have to reconstructed, we can only set limit $\sim\mathcal{O}\left(10^{-5} - 10^{-4}\right)$. On the other hand for the mode $J/\psi \ell_1^- \ell_2^- \pi^+$, final state J/ψ reconstruction results in suppression factor of $\sim\mathcal{O}(10^{-2})$, so tighter upper limits on the BR $\sim\mathcal{O}(10^{-7})$ may be achievable. For the case of $B_c^- \to \pi^+ \ell_1^- \ell_2^-$ we can even set limit on BR $\sim\mathcal{O}(10^{-9})$. We have shown the exclusion curves for mixing elements in Fig. 88.2. These limits are mostly tighter or comparable than those obtained from other heavy meson and tau three body decay modes in earlier studies [2].

88.3 Conclusions

We propose several B_c decay modes for Majorana neutrino searches. We find tighter constraints on $\mid V_{eN} \mid^2$, $\mid V_{\mu N} \mid^2$, $\mid V_{eN} V_{\mu N} \mid$ in a large range of the heavy neutrino mass. Exclusion regions for $\mid V_{eN} V_{\tau N} \mid$, $\mid V_{\mu N} V_{\tau N} \mid$ can also be obtained for masses larger than those accessible in tau decays. Also, upper limits for $\mid V_{\tau N} \mid^2$ can be obtained in the mass range (0.3–5.0) GeV, where it is so far unconstrained.

References

1. J. Schechter, J.W.F. Valle, Phys. Rev. D **22**, 2227 (1980); R.N. Mohapatra, G. Senjanovic, Phys. Rev. Lett. **44**, 912 (1980)
2. A. Atre, T. Han, S. Pascoli, B. Zhang, JHEP **0905**, 030 (2009)
3. V.V. Kiselev, A.E. Kovalsky, A.K. Likhoded, Nucl. Phys. B **585**, 353 (2000)
4. S. Mandal, N. Sinha, Phys. Rev. D **94**(3), 033001 (2016)
5. C.H. Chang, C. Driouichi, P. Eerola, X.G. Wu, Comput. Phys. Commun. **159**, 192 (2004)

Chapter 89
Impact of Active-Sterile Neutrino Mixing on Physics Potential of Long Baseline Neutrino Oscillation Experiments

C. Soumya and Rukmani Mohanta

89.1 Introduction

The discovery of neutrino oscillation in Super-Kamiokande [1], SNO [2], and KamLAND [3] experiments can be considered as a first evidence in favour of new physics beyond the Standard Model (SM) of particle physics. Moreover, the three flavor neutrino oscillation framework is very successful in explaining the observed oscillation data, which involves six parameters (three mixing angles: $\theta_{12}, \theta_{13}, \theta_{23}$, one phase: δ_{CP}, and two mass squared differences: $\Delta m^2_{21}, \Delta m^2_{31}$) and considered as a standard picture of flavour transitions of neutrinos. At this point of time, the investigation of neutrino oscillation enters to an era of precision measurement with the measurement of the reactor mixing angle $\sin^2 2\theta_{13}$ by both reactor (Daya Bay [4], RENO [5], and Double Chooze [6]) and accelerator (T2K [7]) experiments, which opens up a way to determine the current unknowns in 3-flavor framework such as neutrino Mass Hierarchy (MH), CP-violating phase, and the octant of atmospheric mixing angle. The current and future neutrino oscillation experiments are intend to determine these unknowns.

Though the 3 flavor framework could accommodate the remarkable experimental results from solar, atmospheric, reactor, and long baseline experiments, it can not explain the experimental results from short baseline experiments. They are collectively known as short baseline anomalies (LSND anomaly, MiniBooNE anomaly, reactor and gallium anomaly). All these anomalies point towards the existence of one or more eV-scale neutrino states. In addition to this, there are theoretical models, which are introduced in the literature to explain observed neutrino mass, predicts existence of eV-scale sterile neutrino, for instance the neutrino mass model in wrapped

C. Soumya (✉) · R. Mohanta
University of Hyderabad, Hyderabad 500046, India
e-mail: soumyac20@gmail.com

R. Mohanta
e-mail: rukmani98@gmail.com

Md. Naimuddin (ed.), *XXII DAE High Energy Physics Symposium*, Springer Proceedings in Physics 203, https://doi.org/10.1007/978-3-319-73171-1_89

extra-dimension predicts existence of eV-scale sterile neutrino [8]. However, the invisible decay width of Z boson at LEP experiment shows that the effective number of neutrinos participate in weak interaction is three. This implies that the new neutrino states do not participate in weak interaction and such neutrinos are known as sterile neutrinos. Though the sterile neutrinos do not have any SM interactions, they can mix with the three active neutrinos. These active-sterile mixing can affect the sensitivity of long baseline experiment. In this regard, in this paper, we investigate the effect of active-sterile neutrino mixing on the determination of unknowns at NOνA experiment in presence of one sterile neutrino.

89.1.1 3+1 Framework

In the presence of one sterile neutrino, the 4×4 neutrino mixing matrix can be parametrized with six mixing angles and three CP-violating phases and a convenient form of it is given by,

$$U_{PMNS} = V_{34} V_{24} R_{14} R_{23} V_{13} R_{12},$$

where V_{ij} and R_{ij} are complex and real rotation matrix with rotation in (i, j) plane whose 2×2 sub-matrix is given by,

$$V_{ij} = \begin{bmatrix} \cos\theta_{ij} & \sin\theta_{ij} e^{-\delta_{ij}} \\ -\sin\theta_{ij} e^{\delta_{ij}} & \cos\theta_{ij} \end{bmatrix}, \quad R_{ij} = \begin{bmatrix} \cos\theta_{ij} & \sin\theta_{ij} \\ -\sin\theta_{ij} & \cos\theta_{ij} \end{bmatrix}.$$

The constraints on the active-sterile neutrino mixing angles are given in Table 89.1, which are obtained from $\nu_e/\bar{\nu}_e$ disappearance and appearance searches, and $\nu_\mu/\bar{\nu}_\mu$ disappearance searches using various neutrino oscillation experiments.

89.1.2 Simulation Details

The long baseline neutrino oscillation experiment NOνA, which uses NuMI beam with a power of 700 kW, has a 14 kt far detector situated almost 810 km away from the Fermilab. We use GLoBES [11, 12] package to simulate this experiment and the experimental specifications are explicitly given in [13].

Table 89.1 The constraints on mixing angles at 95 % C.L [9]

U_{PMNS} element	Mixing angles
$\lvert U_{e4}\rvert^2 \in [0, 0.1]$	$\theta_{14} \in [0°, 13°]$
$\lvert U_{\mu4}\rvert^2 \in [0, 0.03]$	$\theta_{24} \in [0°, 7°]$
$\lvert U_{\tau4}\rvert^2 \in [0, 0.3]$	$\theta_{34} \in [0°, 26°]$

Table 89.2 The oscillation parameters used in the analysis [10]

3-flavor parameters	4-flavor parameters
$\sin^2\theta_{12} = 0.32$	$\theta_{14} = 13°$
$\sin^2 2\theta_{13} = 0.085$	$\theta_{24} = 7°$
$\sin^2\theta_{23} = 0.41$	$\theta_{34} = 26°$
$\Delta m^2_{atm} = 2.4 \times 10^{-3}\,\text{eV}^2$	$\Delta m^2_{41} = 1\,\text{eV}^2$
$\Delta m^2_{21} = 7.6 \times 10^{-5}\,\text{eV}^2$	$\delta_{24} = 0°$
$\delta_{CP} = -90°$	$\delta_{34} = 0°$.

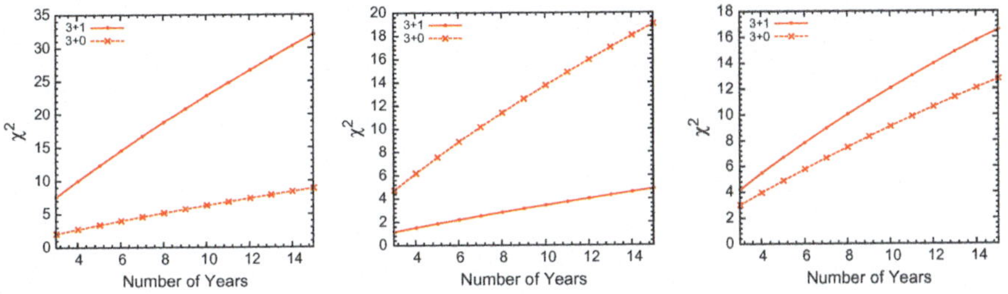

Fig. 89.1 The CPV (left panel), octant (middle panel), and MH (right panel) sensitivities of NOνA. These results correspond to lower octant ($\sin^2\theta_{23}^{true} = 0.41$)

89.1.3 Results and Conclusions

In this section, we present the sensitivity of NOνA to discriminate the CP-conserving points, wrong octant, and wrong mass hierarchy in 3-flavor framework and in presence of sterile neutrino. Recent experimental results hint towards NH with $\delta_{CP} = -\pi/2$. Therefore, we assume the true values of oscillation parameters as given in Table 89.2. The sensitivity as a function of number of years of run is given in Fig. 89.1. From the figure, we can see that there is a significant change in the sensitivities of NOνA in presence of one sterile neutrino.

References

1. Y. Fukuda et al., SK Collaboration, Phys. Rev. Lett. **82**, 2644 (1999)
2. Q.R. Ahmad et al., SNO Collaboration, Phys. Rev. Lett. **89**, 011301 (2002)
3. K. Eguchi et al., KamLAND Collaboration, Phys. Rev. Lett. **90**, 021802 (2003)
4. F.P. An et al., Daya Bay Collaboration, Phys. Rev. Lett. **115**, 111802 (2015)
5. S.B. Kim, arXiv:1412.2199 (hep-ex)
6. T. Abrahao et al., Double Chooz Collaboration, arXiv:1510.08937 (hep-ex)
7. K. Abe et al., T2K Collaboration, Phys. Rev. D **91**, 072010 (2015)
8. C.S. Fong et al., Phys. Lett. B **704**, 171–178 (2011)
9. J. Kopp et al., JHEP **1305**, 050 (2013)

10. M.C. Gonzalez-Garcia et.al., Nucl. Phys. B **908** (2016)
11. P. Huber et al., Comput. Phys. Commun. **167**, 195 (2005)
12. P. Huber et al., Comput. Phys. Commun. **177**, 432 (2007)
13. C. Soumya, R. Mohanta, Eur. Phys J. C **76**, 302 (2016)

Chapter 90
Measurement of Z + Jets Differential Cross Section at $\sqrt{s} = 8$ and 13 TeV with the CMS Detector

Suman B. Beri, Vipin Bhatnagar and Sandeep Kaur

90.1 Introduction

The large centre-of-mass energy at the Large Hadron Collider (LHC) allows production of events with high jet transverse momenta and high jet multiplicities associated with an electroweak bosons. The data corresponds to an integrated luminosity of $2.25\,\text{fb}^{-1}$ at 13 TeV proton-proton (pp) collision, recorded by the Compact Muon Solenoid (CMS) [1] detector during the Run II of the LHC. The differential cross sections are measured as a function of the inclusive and exclusive jet multiplicities and jet kinematical variables including the jet transverse momentum (p_{T}), the absolute jet rapidity ($|y|$) and the jet transverse momenta scalar sum (H_{T}). The measurements are compared to multileg NLO predictions using the FxFx jet merging scheme.

90.2 Analysis Strategy

The Z+jets signal process is generated with MG5_aMC interfaced with Pythia8, using the FxFx merging scheme. The $t\bar{t}$ and single top backgrounds are generated using POWHEG interfaced to Pythia8. Background MC samples corresponding to double vector boson electroweak production are generated with aMC@NLO (WZ), POWHEG (WW), both interfaced to Pythia8 and with Pythia8 alone (ZZ). The final state particles in Z+jets decays are identified and reconstructed with the particle-flow (PF) algorithm. The reconstruction of muons is described in detail in [2]. The trigger

Suman B. Beri, Vipin Bhatnagar, Sandeep Kaur (on behalf of the CMS Collaboration).

S. B. Beri · V. Bhatnagar · S. Kaur (✉)
Panjab University, Chandigarh, India
e-mail: sandeep.kaur@cern.ch

Md. Naimuddin (ed.), *XXII DAE High Energy Physics Symposium*, Springer Proceedings in Physics 203, https://doi.org/10.1007/978-3-319-73171-1_90

selects events with two isolated muons with transverse momenta of at least 8 and 17 GeV. After reconstruction, muons with $p_T > 20$ GeV and $|\eta| < 2.4$ are selected. We require that the two muons with highest transverse momenta to be oppositely charged and form a pair with an invariant mass within a window of 91 ± 20 GeV. Jets are reconstructed by clustering the PF candidates with the anti-k_T algorithm, with a size parameter of 0.4. Jets are required to have $p_T > 30$ GeV, $|y| < 2.4$ and to have a spatial separation from muon candidates of $\Delta R = \sqrt{\Delta\eta^2 + \Delta\phi^2} > 0.4$. Jets are further required to satisfy the loose identification criteria.

The Z decay in $\tau^+\tau^-$ is considered as background and is subtracted during the unfolding procedure. The background contamination is below the percent level for the inclusive selection, but increases with the number of jets mainly due to $t\bar{t}$ production, reaching 15% for a jet multiplicity of 4 and above. The fiducial cross sections are obtained by subtracting the simulated backgrounds from the data distributions and correcting the background-subtracted data distributions back to the particle level using an unfolding procedure. This is performed using the iterative Bayesian method [3]. A response matrix is computed using the simulated Z+jets events.

In this analysis, the systematic uncertainties are considered from various sources. The dominant source of systematic uncertainty is the jet energy scale uncertainty, which is up to 7% for a jet multiplicity of 1. The uncertainty on the cross section of the largest background contribution from top pairs, is estimated to be 10%. A systematic uncertainty associated with the generator used to build the unfolding response matrix is assessed by weighting the simulation to agree with the data in each distribution and constructing a reweighted response matrix to unfold the data. The difference between the unfolded results obtained using the weighted response matrix and the nominal results is taken as the systematic uncertainty associated with the unfolding response matrix. The uncertainty on the integrated luminosity is respectively 2.6% for 8 TeV dataset and 4.6% for 13 TeV dataset.

90.3 Results and Summary

The measured cross sections for 8 and 13 TeV are shown in Fig. 90.1. For 8 TeV, measurements are compared with the three theoretical prediction [4]. For 13 TeV, measurements are compared with the prediction of MG5_AMC@NLO. Good agreement between reconstructed data and theoretical predictions is observed up to four jets. The predictions generally describe the measured cross sections within the uncertainties.

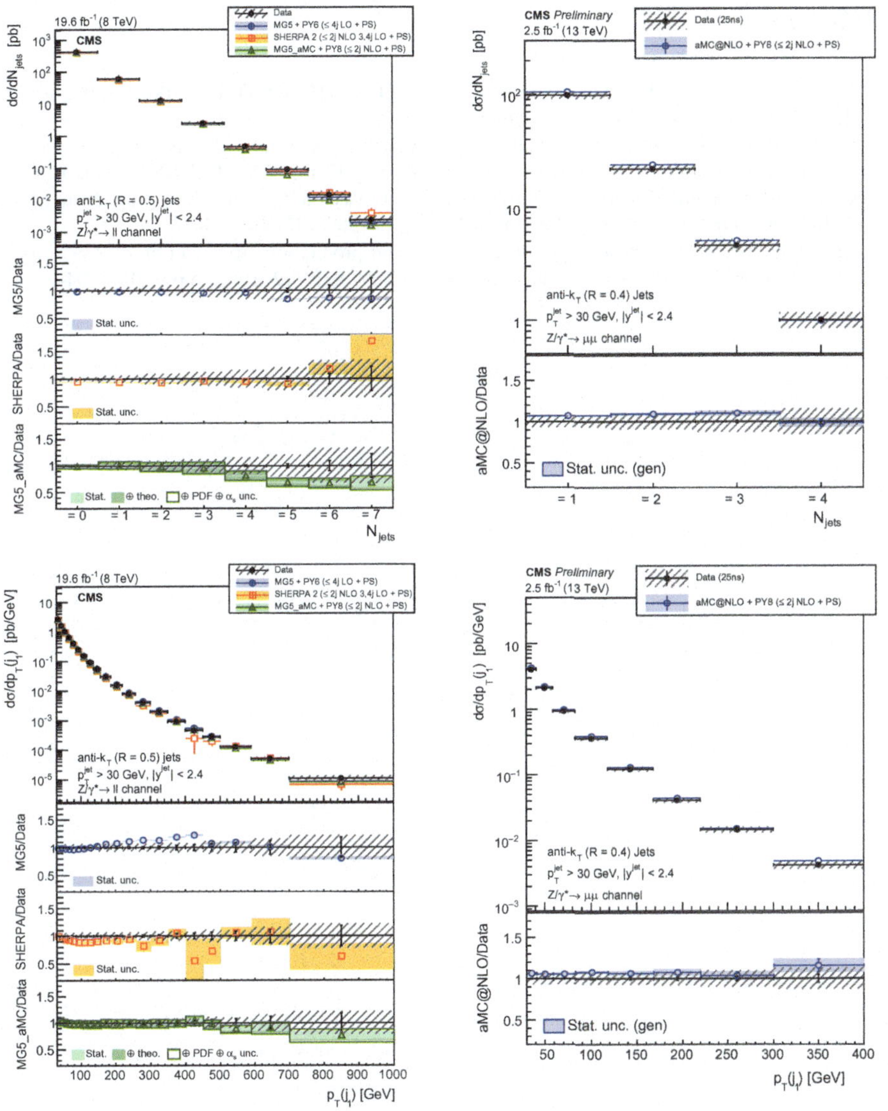

Fig. 90.1 Measured cross section as a function of the jets exclusive multiplicity at 8 TeV (left) [2] and 13 TeV (right) [4]. In the bottom plots, cross section is measured as a function of the leading jet p_T at 8 TeV (left) [2] and 13 TeV (right) [4]

References

1. CMS Collaboration, The CMS experiment at the CERN LHC. JINST **3**, S08004 (2008). https://doi.org/10.1088/1748-0221/3/08/S08004
2. CMS Collaboration, Measurement of the differential cross section of Z boson production in association with jets in proton-proton collisions at $\sqrt{s} = 13$ TeV, CMS PAS SMP-15-010 (2015)
3. G. DAgostini, A multidimensional unfolding method based on Bayes theorem. Nucl. Instr. Meth. A **362**, 487–498 (1995). https://doi.org/10.1016/0168-9002(95)00274-X
4. CMS Collaboration, Measurement of the differential production cross section of Z bosons in association with jets in pp collisions at sqrt(s) = 8 TeV, JHEP (2016), arXiv:1611.03844

Chapter 91
Detection of Low Mass Vector Mesons in the Muon Detector of CBM Experiment

Ekata Nandy and Subhasis Chattopadhyay

91.1 Introduction

FAIR is an upcoming facility that will conduct 4 different experiments like APPA, CBM, NUSTAR, PANDA. Among them, CBM aims to study the exotic state of nuclear matter at very high net baryon density. In CBM, heavy ions will be collided in the energy range 4–40 AGeV. This energy region will probe the phase diagram of QCD matter at moderate temperature and very high net baryon density. Dimuon measurement is a significant part of CBM research program. As muons are weakly interacting particles, they will not be affected by the final state effect of strongly interacting medium produced in heavy ion collision and thus carry undistorted information of the fireball. Here our focus is on dimuon measurement enabled by muon detection system (MUCH) of the CBM experimental set up. The entire dimuon spectrum can be broadly divided into three regions– the low mass region ($M_{\mu^+\mu^-} < M_\phi$), the intermediate mass region ($M_\phi < M_{\mu^+\mu^-} < M_{J/\psi}$), and the high mass region ($>M_{J/\psi}$). Dimuons in the low mass range originate from the decay of LMVMs. Detection of such low momentum muons at a very high interaction rate in the FAIR energy range is highly challenging due to very high background contamination. In this article, we will report the feasibility of LMVMs detection via their dimuon decay channel by the muon detection system at CBM.

91.2 Muon Detection Setup at CBM

The muon chamber [1] in CBM is a conical shaped setup with detector angular acceptance 5.7° to 25°. MUCH is placed after magnet. Inside magnet, there is Silicon

E. Nandy (✉) · S. Chattopadhyay
Variable Energy Cyclotron Centre, 1/AF, Bidhan Nagar, Kolkata 700064, India
e-mail: ekatanandy@gmail.com

Md. Naimuddin (ed.), *XXII DAE High Energy Physics Symposium*, Springer Proceedings in Physics 203, https://doi.org/10.1007/978-3-319-73171-1_91

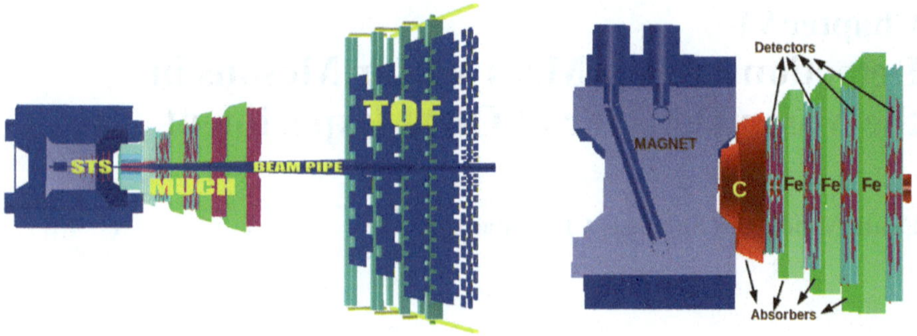

Fig. 91.1 CBM muon detection setup

Tracking Stations (STS) for tracking and momentum determination & after MUCH there is Time of Flight (TOF). MUCH consists of sliced absorbers & detectors are placed in between them. Detection of LMVMs at 8 AGeV require 4 absorbers & 4 stations as shown in Fig. 91.1.

Unlike other high energy physics experiments where single thick absorber is used in muon detection system, the CBM-MUCH has sliced absorber system that ensures detection of low momentum muons. If instead a single thick absorber is used, most of the low momentum tracks will stop inside the absorber. The first absorber layer is made of Graphite (C) of thickness 60 cm & rest are made of Iron (Fe) of thicknesses 20, 20, 30 cm respectively. Three detector chambers, called station, with a inter-spacing of 10 cm, are placed in between two absorbers. The Gas Electron Multiplier (GEM) is used as an active detector component in all stations in the simulation. All these dimensions and thicknesses are optimized to get best possible efficiency and S/B for LMVMs.

91.3 Simulation for $\mu^{\pm}$ Reconstruction and Identification

The simulation is done using CBMROOT framework with GEANT3 transport code. For particle generation PLUTO [2] and URQMD [3] event generators are used. PLUTO generates signal LMVM particles and decay them into dimuons. Whereas, URQMD is used for background particle generation. After transporting these particles through CBM setup, track reconstruction is done in STS based on cellular automaton method and then extrapolated in MUCH using Kalman Filter technique. Finally these global tracks (tracks that pass through all detectors in the present setup) are used for selection of muon candidates using the following selection criteria: STS hits ≥ 7, MUCH hits ≥ 11 and $\chi^2_{Vertex} \leq 2.0$, $\chi^2_{MUCH} \leq 1.3$. Usage of these cuts reduces the background significantly. But for rejection of high momentum background tracks, mainly the punched-through hadrons, a Time of Flight (TOF) mass cut of $m^2 <= 0.05\,\text{GeV}/c^2$ is used.

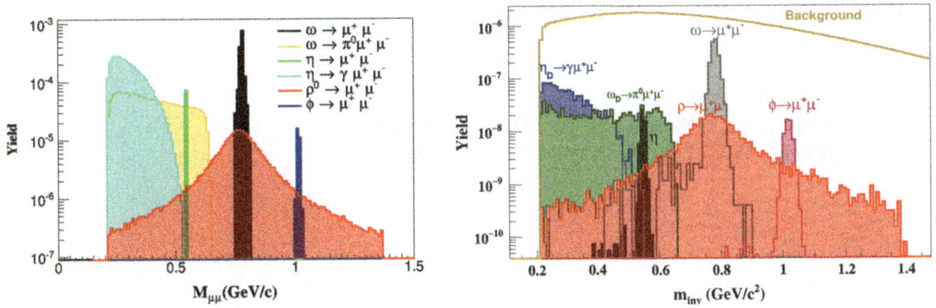

Fig. 91.2 Hadronic cocktail distribution from input PLUTO (left) and after full pair reconstruction (right)

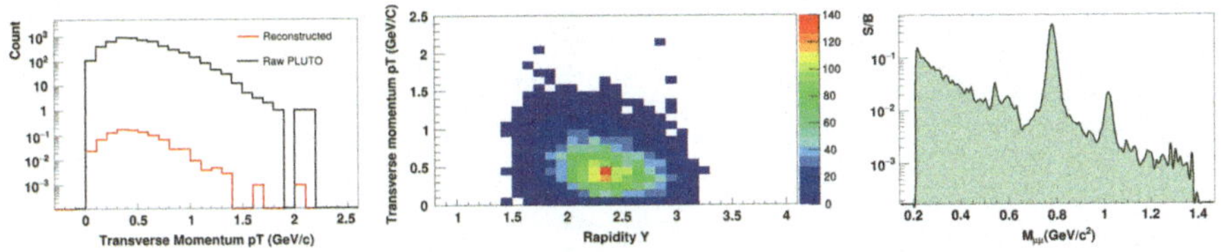

Fig. 91.3 Transverse momentum (p_T) distribution of ρ^0 (left). Y-p_T acceptance of the same (middle). Signal to background (S/B) ratio (Right)

Table 91.1 Table for efficiency and S/B for different LMVMs for latest geometries

	ρ^0	ω	ϕ	η	η^D	ω^D
Efficiency (%)	1.03	1.01	1.53	0.56	0.23	0.37
S/B	0.005	0.29	0.005	0.004	0.092	0.004

91.3.1 Simulation Results and Discussions

Simulation has been performed with all latest geometries (STS, MAGNET, MUCH, TOF) at 8 AGeV central Au+Au collision. Figure 91.2 shows the invariant mass distribution of hadronic cocktails including LMVMs at the generator level i.e. directly from PLUTO (left) and at reconstructed level (right). By combining μ^+ and μ^-, we get invariant mass of mother particles. Reconstructed background has also been shown in Fig. 91.2 (right).

Figure 91.3 (left) shows the p_T-spectra of ρ-meson from raw PLUTO & at the reconstructed level. It shows that yields are reduced at the reconstructed level which is because of finite acceptance of detector and absorption in the absorber. Figure 91.3 (middle) shows Y-p_T coverage of the same at the reconstructed level. It shows that the detector coverage is in the forward rapidity. Finally Fig. 91.3 (right) shows the S/B of hadronic cocktails as a function of dimuon invariant mass. The values of efficiency and S/B of hadronic cocktails tabulated in Table 91.1 which suggests that MUCH can detect LMVM with reasonable efficiency and S/B.

References

1. MUCH, Technical design report
2. I. Froehlichet et al., J. Phys. Conf. Ser. **219**, 032039 (2010)
3. S.A. Bass et al., Prog. Part. Nucl. Phys. **41**, 225 (1998)

Chapter 92
Type Ia Supernovae: Non-Gaussianity and Direction Dependence in Union2 Catalogue

Meghendra Singh, Shashikant Gupta, Amit Sharma, Satendra Sharma and Anshu Gupta

92.1 Introduction

Cosmological Principle (hereafter CP) states that the Universe is homogeneous and isotropic on large scales [1]. Along with the observations of Supernovae type Ia (hereafter SNe Ia) it led to establish that the two third of the constituents of the Universe is dark energy and one third is dark matter. This model of the universe is known as the standard model of cosmology or ΛCDM cosmology. After release of first year Wilkinson Microwave Anisotropy Probe (WMAP) data [2–4] reported a direction dependence in cosmic microwave background (CMB). Using extreme value statistics, [5–7] show that the two supernova data sets, [8] Gold data release in 2004 (hereafter GD04) and [9] Gold data release in 2007 (hereafter GD07), do show some evidence for direction dependence in SNe Ia data. [10] have also shown a preferred axis using the Union2 catalogue. This dependence could arise due to deviations from CP. However, it is also possible that CP is correct, but there may be issues with the quality of data. Along with the direction dependence, non-Gaussianity is also reported in the SNe Ia data [5, 7].

M. Singh (✉)
Dr. A.P.J. Abdul Kalam Technical University,
Lucknow 226021, Uttar Pradesh, India
e-mail: meghendrasingh_db@yahoo.co.in

S. Gupta · A. Gupta
GD Goenka University, Gurugram, Haryana, India

A. Sharma
Amity University Gurugram, Gurugram, Haryana, India

S. Sharma
Yobe State University, Damaturu, Nigeria

Md. Naimuddin (ed.), *XXII DAE High Energy Physics Symposium*, Springer Proceedings in Physics 203, https://doi.org/10.1007/978-3-319-73171-1_92

In this paper our main task is to look for direction dependent systematic effects and non-Gaussianity in the latest SNe Ia data Union2. This paper is arranged as follows. In Sect. 92.2, we introduce the method we have used. In Sect. 92.3 we provide our results with conclusions.

92.2 Data and Methods

The latest SNe Ia data is accessible from [11] and is termed as Union2 data. The observables in the data are the redshift (z), distance modulus (μ) and positions. Below we discuss the methodology in brief, the detailed discussion is available in [5, 7]. First we calculate the bestfit parameters for ΛCDM cosmology using maximum likelihood method. If the ith observed distance modulus is μ_i^o and the corresponding error is σ_i then, we define $\chi_i = (\mu_i^o - \mu_i^{bf}(z))/\sigma_i$. Where $\mu_i^{bf}(z))$ is calculated using bestfit parameters. We consider two subsets of data defined by two hemispheres labeled by the direction vector $\hat{n}$, containing N_{north} and N_{south} SNe, where the total number of SNe, $N = N_{\text{north}} + N_{\text{south}}$, and define the quantity

$$\Delta\chi_{\hat{n}} = \frac{1}{\sqrt{N}} \left(\sum_{i=1}^{N_{\text{north}}} \chi_i - \sum_{j=1}^{N_{\text{south}}} \chi_j \right) . \tag{92.1}$$

Clearly $\langle \Delta\chi_{\hat{n}} \rangle = 0$ and $\langle (\Delta\chi_{\hat{n}})^2 \rangle = 1$. From the Central Limit Theorem [12] it follows that for $N \gg 1$, the quantity $\Delta\chi_{\hat{n}}$ follows a Gaussian distribution with a zero mean and unit variance. As we maximize this quantity by varying the direction $\hat{n}$ across the sky to obtain the maximum absolute difference

$$\Delta_\chi = \max\{|\Delta\chi_{\hat{n}}|\} . \tag{92.2}$$

Extreme value theory shows that the distribution of Δ_χ is described by a two parameter distribution known as Gumbel distribution given by [13]

$$P(\Delta) = \frac{1}{s} \exp\left[-\frac{\Delta - m}{s} \right] \exp\left[-\exp\left(-\frac{\Delta - m}{s} \right) \right] , \tag{92.3}$$

The Δ_χ statistic has an advantage over Δ_{χ^2} used in [5, 7] that the position parameter m and the shape parameters s can be determined analytically. If N_d is the number of direction over which $\Delta\chi_{\hat{n}}$ is maximize and $N_d \gg 1$, then

$$m = \sqrt{2 \log N_d - \log\log N_d - \log 4\pi} \quad ; s = \frac{1}{m} . \tag{92.4}$$

where we have additionally assumed that the number of SNe $N \gg 1$, since the distribution for χ_is tends to Gaussian only in this limit.

92.3 Results and Conclusions

First we obtain the bestfit parameters $\Omega_{\rm M}$ and H_0 for the Union2 data. Table 92.1 presents a comparison of these values with the other SNe Ia data [5–7] and Planck results [14]. It is clear from Table 92.1 that Union2 favors slightly smaller matter density ($\Omega_{\rm M}$) and higher expansion rate (H_0).

Now we calculate Δ_χ for the Union2 data as defined in (92.2). The numerical value of Δ_χ is 4.3 for Union2 data. The bootstrap and theoretical distributions are plotted in Fig. 92.2. The Δ_χ for Union2 is more than 1σ away from the mode of bootstrap distribution indicating slight direction dependence in the data.

The bootstrap distribution is generated by assuming that errors are Gaussian and hence χ_is follow standard normal distribution. If we draw a sample of size N from standard normal distribution and calculate its maximum, this procedure is similar to calculation of Δ_χ. The distribution of above maxima thus should be compared to bootstrap distribution. It is represented as theoretical distribution in Fig. 92.2. If the errors in data are drawn from Gaussian distribution than the two distribution should match. However a specific bias in Δ_{χ^2} and Δ_χ statics is discussed in [5–7]. There it has been shown that theoretical χ_is are unbounded while the bootstrap χ_is are bounded. Thus bootstrap distribution should lay slightly on the left of the theoretical distribution, this is shown through simulation in Fig. 92.1. In contrast to the above expectation Fig. 92.2 shows that the bootstrap distribution match quite well with the theoretical distribution, indicating slight non-Gaussian features in the errors.

In addition the analytic distribution using (92.3) with parameters defined in (92.4) is plotted in Fig. 92.2. The shape of analytic distribution is quite different. The mismatch could arise if all the directions are not independent. Thus number of actually independent directions is less than N_d and hence actual value of m is smaller than calculated in (92.4). This also makes the spread larger in bootstrap and theoretical distributions compared to the analytic distribution.

Table 92.1 The model parameters for various data set are tabulated here

Data/Mission	$\Omega_{\rm M}$	H_0
Union2	0.27	70.0
GD04	0.30	64.5
GD07	0.33	63.0
Planck	0.30	67.0

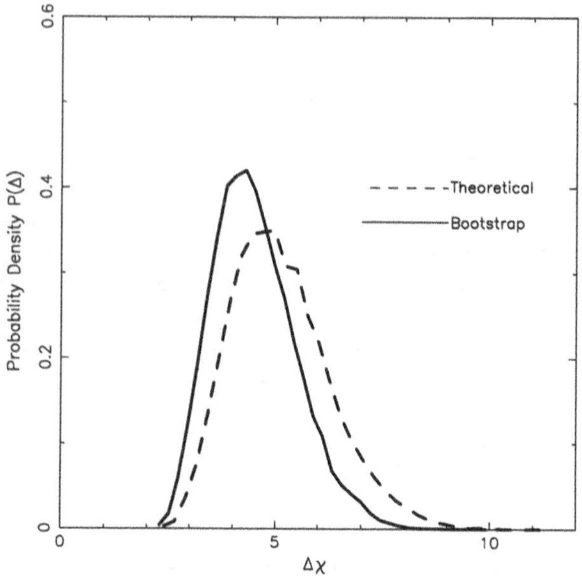

Fig. 92.1 Comparison of theoretical and bootstrap probability distributions for simulated data. The data comprises 557 SNe,whose positions on the sky were generated randomly

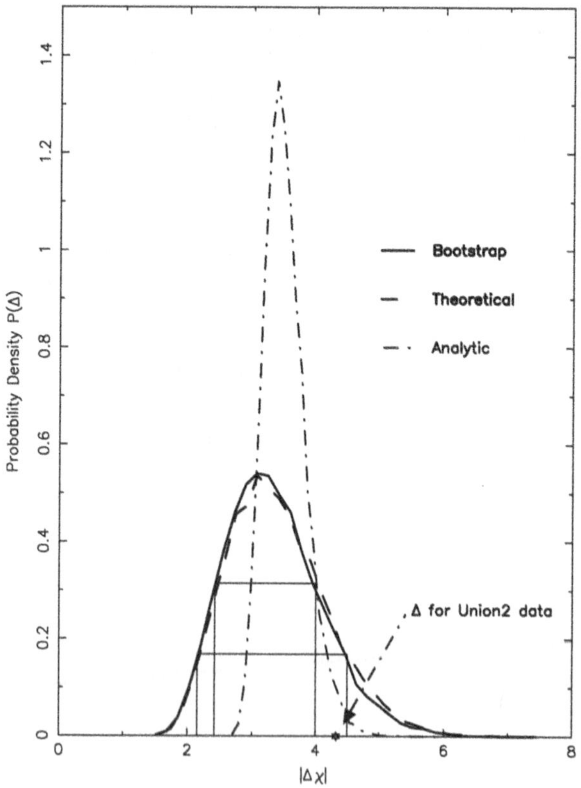

Fig. 92.2 The theoretical, analytic and the bootstrap probability distributions for the Δ_χ statistic for Union2. The analytic distribution uses the limiting values of the shape and position parameters for the Gumbel distribution

Acknowledgements Meghendra Singh thanks DMRC for support, Shashikant Gupta thanks Tarun Deep Saini for discussion.

References

1. P.J.E. Peebles, *Princeton Series in Physics* (Princeton University Press, Princeton, 1993), p. c1993
2. C.L. Bennett, M. Halpern, G. Hinshaw et al., ApJS **148**(1) (2003)
3. H.K. Eriksen et al., Astrophys. J. **605**, 14 (2004)
4. F.K. Hansen, A.J. Banday et al., Mon. Not. R. Astron. Soc. **354**, 641 (2004b)
5. S. Gupta, T.D. Saini, Mon. Not. R. Astron. Soc. **407**, 651 (2010)
6. S. Gupta, T.D. Saini, T. Laskar, Mon. Not. R. Astron. Soc. **388**, 242 (2008)
7. S. Gupta, M. Singh, Mon. Not. R. Astron. Soc. **440**, 3257 (2014)
8. A.G. Riess et al., Astrophys. J. **607**, 665 (2004)
9. A.G. Riess et al., Astrophys. J. **659**, 98 (2007)
10. I. Antoniou, L. Perivolaropoulos, J. Cosmol. Astropart. Phys. **12**, 12 (2010)
11. R. Amanullah et al., Astrophys. J. **716**, 712 (2010)
12. M. Kendall, A. Stuart, *1977*, 4th edn. (Griffin, London, 1977)
13. L. Haan, A. Ferreira, *Extreme Value Theory: An Introduction* (Springer, Berlin, 2006)
14. P.A.R. Ade et al., Planck 2015 results-XIII. Cosmological parameters. Astron. Astrophys. **594**, A13 (2016)

Chapter 93
Probing Non-holomorphic MSSM via Precision Constraints, Dark Matter and LHC Data

Utpal Chattopadhyay and Abhishek Dey

93.1 Non-holomorphic MSSM

The Higgs Boson @ 125 GeV requires large radiative corrections demanding heavier stops, or effectively a large trilinear stop mixing soft parameter. A large region of pMSSM parameter space associated with large $\tan\beta$ is ruled out via $\mathrm{Br}(B \to X_s + \gamma)$. On the other hand, in a single component dark matter (DM) setup, a well motivated higgsino DM of the right relic density leads to enhanced Electro-Weak Fine Tuning (EWFT). Moreover, $(g-2)_\mu$ limits require either a light lighter chargino or light smuons, hardly allowed by LHC data in generic scenarios. With identical particle content as that of MSSM Non-holomorphic MSSM (NHSSM) can successfully cover these issues by including NH soft terms.

In general, MSSM can be extended with different SUSY breaking terms that are classified into "soft", "may be soft" and "hard" [1]. Choosing the middle option, the NH terms read as follows [2, 3]

$$-\mathcal{L}'_{soft} \supset \tilde{Q} \cdot H_d^c A'_t \tilde{U} + \tilde{Q} \cdot H_u^c A'_b \tilde{D} + \tilde{L} \cdot H_u^c A'_\tau \tilde{E} + \mu' \tilde{H}_u \cdot \tilde{H}_d + h.c. \quad (93.1)$$

The above trilinear soft terms modify the left-right mixing of sfermion matrices, $(A_t - \mu\cot\beta) \to [A_t - (\mu + A'_t)\cot\beta]$ whereas the bilinear term affects the higgsino content of the electroweakino sector.

U. Chattopadhyay
Department of Theoretical Physics, Indian Association for the Cultivation of Science, 2A & B Raja S. C. Mullick Road, Jadavpur, Kolkata 700032, India
e-mail: tpuc@iacs.res.in

A. Dey (✉)
Maulana Azad College, Government of West Bengal, 8 Rafi Ahmed Kidwai Road, Kolkata 700013, India
e-mail: dey.abhishek111@gmail.com

Md. Naimuddin (ed.), *XXII DAE High Energy Physics Symposium*, Springer Proceedings in Physics 203, https://doi.org/10.1007/978-3-319-73171-1_93

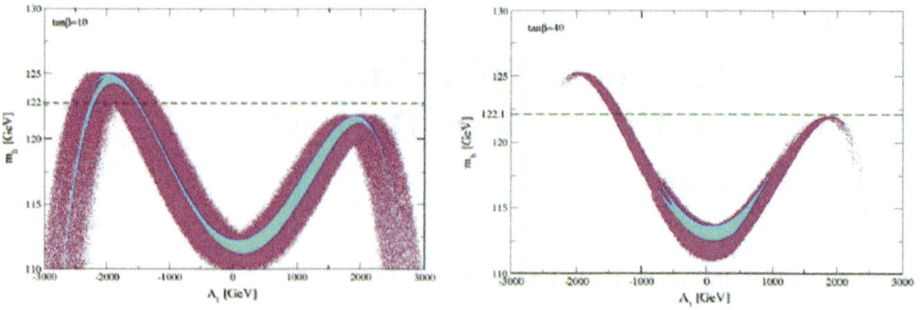

Fig. 93.1 Scatter plot of m_h against A_t for $\tan\beta = 10$ (left) and $\tan\beta = 40$ (right)

Effect of NH parameters on m_h, $\mathrm{Br}(B \to X_s + \gamma)$ **and Higgsino Dark Matter**: Figure 93.1 shows the variation of m_h with A_t where the scan is given over A_t, A'_t, μ and μ' [4]. The magenta and cyan regions correspond to NHSSM and MSSM respectively. We find that for low $\tan\beta$, there is a significant change (1–2 GeV) in m_h due to the NH parameters [4]. The effect is reduced for large $\tan\beta$ since the contribution of A'_t in the stop L-R mixing becomes small. A large value of A_t in MSSM that is required for $m_h \sim 125\,\mathrm{GeV}$ can not accommodate the $\mathrm{Br}(B \to X_s + \gamma)$ limits when $\tan\beta$ becomes large thus discarding a lot of parameter space [4]. Appropriate μ' and A'_t in NHSSM can bring back the above parameter space while accommodating both the $\mathrm{Br}(B \to X_s + \gamma)$ and the Higgs mass data.

Electro-weak Fine Tuning and Higgsino Dark Matter: The neutral scalar potential of NHSSM at the tree level is identical to that of MSSM since the NH terms are not associated with any neutral colorless scalar. The minimization of same leads to the following well known results [5].

$$\frac{m_Z^2}{2} = \frac{m_{H_d}^2 - m_{H_u}^2 \tan^2\beta}{\tan^2\beta - 1} - |\mu|^2 \quad \text{and} \quad \sin 2\beta = \frac{2b}{m_{H_d}^2 + m_{H_u}^2 + 2|\mu|^2}. \tag{93.2}$$

Terms involving SUSY breaking parameters m_{H_d}, m_{H_u} and SUSY preserving parameter μ undergo fine cancellation to obtain $m_Z^2/2$ where m_Z refers to the mass of Z-boson. The degree of cancellation broadly indicates a measure of EWFT and it is obviously lowered for smaller μ. Specifically, EWFT is obtained as $\Delta_{Total} = \sqrt{\sum_i \Delta_{p_i}^2}$ where Δ_{p_i} is given by $\Delta_{p_i} = \left|\frac{\partial \ln m_Z^2(p_i)}{\partial \ln p_i}\right|$, with $p_i \equiv \{\mu^2, b, m_{H_u}, m_{H_d}\}$[4]. Unlike MSSM, here we can reduce EWFT to a very low value even close to zero while still satisfying the lighter chargino bound via a limit on $|\mu - \mu'|$. In turn, this easily accommodates a higgsino DM (~1 TeV) with a very small EWFT.

$(g-2)_\mu$ **in NHSSM**: The NH trilinear parameter (A'_μ) can significantly affect the SUSY contributions to $(g-2)_\mu$, namely a_μ^{SUSY} when the latter is dominated by the $\tilde{\chi}_1^0 - \tilde{\mu}$ loop. The L-R mixing, controlled by $A'_\mu \tan\beta$ can result into large a_μ^{SUSY}

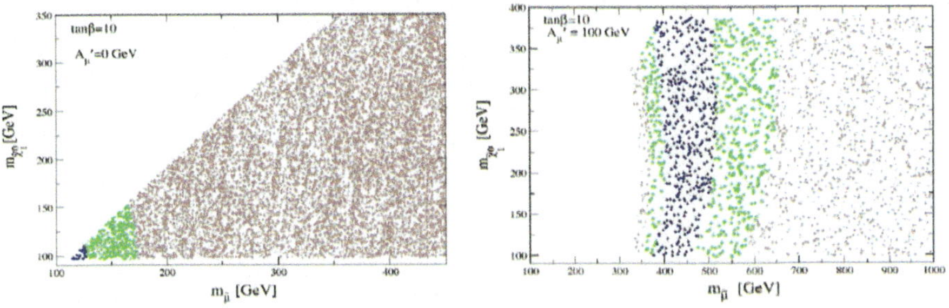

Fig. 93.2 Scatter plot of $m_{\tilde{\chi}_1^0}$ against $m_{\tilde{\mu}_1}$ for $\tan\beta = 10$

even for a small A'_μ[4]. Figure 93.2 below identifies regions satisfying the $a_\mu^{\rm SUSY}$ data within 1σ, 2σ and 3σ limits, shown in blue, green and brown respectively. It is evident that when $A'_\mu = 100\,\text{GeV}$, $a_\mu^{\rm SUSY}$ is obtained within 1σ for significantly heavier smuon masses compared to that of MSSM. Thus LHC data on sparticle masses can easily be accommodated even for a small $\tan\beta$.

93.2 Conclusion

NHSSM has unique features in accommodating various phenomenological data. It allows to have a higgsino DM with low electroweak fine-tuning, while accommodating the Higgs mass data with a relatively smaller A_t. $\text{Br}(B \to X_s + \gamma)$ is respected even for a large $\tan\beta$, whereas the SUSY contributions to $(g-2)_\mu$ can be very large even for a small $\tan\beta$ and quite a large value of smuon masses. The framework only needs relevant soft terms and does not involve any more additional sparticles aside from MSSM.

References

1. S.P. Martin, Phys. Rev. D **61**, 035004 (2000). https://doi.org/10.1103/PhysRevD.61.035004 [hep-ph/9907550]
2. A. Sabanci, A. Hayreter, L. Solmaz, Phys. Lett. B **661**, 154 (2008). https://doi.org/10.1016/j.physletb.2008.01.071, arXiv:0801.2029 [hep-ph]
3. C.S. Ün, H. Tanyildizi, S. Kerman, L. Solmaz, Phys. Rev. D **91**(10), 105033 (2015). https://doi.org/10.1103/PhysRevD.91.105033, arXiv:1412.1440 [hep-ph]
4. U. Chattopadhyay, A. Dey, JHEP **1610**, 027 (2016). https://doi.org/10.1007/JHEP10(2016)027, arXiv:1604.06367 [hep-ph]
5. M. Drees, P. Roy, R.M. Godbole, *Theory and Phenomenology of Sparticles* (World Scientific, Singapore, 2005)

Chapter 94
Simulation of Response of Detector Materials to Muon Induced Neutrons for DINO Experiment

K. K Meghna and Bedangadas Mohanty

94.1 Introduction

Dark matter at INO (DINO) is an upcoming dark matter experiment in the proposed India-based Neutrino Observatory (INO) [1]. DINO will be looking for the signature of Weakly Interacting Massive Particles (WIMPs) using cryogenic scintillator detectors. WIMPs are one of the most favoured dark matter candidates. WIMPs are expected to have mass in the range $\sim$10 GeV to a few TeV and very weak interaction cross section [2]. They are expected to interact via elastic scattering with the detector material to produce nuclear recoils. This energy can be detected via phonon and light signals. The first phase of the experiment is expected to be done in Jaduguda UCIL mine. The laboratory will be located at a depth of 555 m from the surface. This overburden will help in reducing low energy cosmic ray muon background. The Jaduguda rock contains 8 ppm of U and 16 ppm of Th [3]. The average density of the rock is about 2.9 g/cc.

Rare event search experiments like those of dark matter need high sensitivity and very low background. Inorganic scintillators, CsI and $Gd_3Ga_3Al_2O_{12}$ (GGAG) have been considered as possible detector materials. The light yield of CsI(Tl) is about 54 photons/keV and that of GGAG is about 45 photons/keV [4, 5]. The background radiation can be either cosmogenic or radiogenic. Cosmogenic backgrounds, especially low energy cosmic rays will be reduced significantly in the underground laboratories. High energy cosmic ray muons, muon induced neutrons, natural and induced radio activity will act as potential sources of background. Neutrons interact with the detector in a similar way as that of WIMPs in terms of recoil signal. Therefore accurate estimation of neutron flux at the site is very crucial. In underground laboratories, neutrons can be generated mainly by spontaneous fission of U, by U/Th (α, n) reactions with rock and in the interaction of muons with rock and

K. K. Meghna (✉) · B. Mohanty
School of Physical Sciences, National Institute of Science Education and Research, HBNI, Jatni 752050, Odisha, India
e-mail: meghna@niser.ac.in

Md. Naimuddin (ed.), *XXII DAE High Energy Physics Symposium*, Springer Proceedings in Physics 203, https://doi.org/10.1007/978-3-319-73171-1_94

Table 94.1 The composition of Jaduguda rock

Material	Conc (%)	Material	Conc (%)
SiO_2	66.45	K_2O	2.61
Al_2O_3	18.20	TiO_2	0.59
Fe_2O_3	0.26	P_2O_5	0.18
FeO	4.61	MnO	0.03
CaO	1.82	U_3O_8	0.005
MgO	1.39	Mo	0.002
Na_2O	1.60	H_2O	0.25

shielding materials. Here we study the neutrons generated in muon interaction with rock. The Jaduguda rock composition, obtained from rock sample analysis, is given in Table 94.1. This was used for simulation.

94.2 Description of GEANT4 Simulation and Results

A Monte Carlo simulation code using GEANT4 [6] was used for defining detector geometry as well as production and propagation of particles. The reference physics list "Shielding" was used for defining physics processes. Secondary particle production cuts were set to 0.7 mm for gammas and e^+/e^-.

10^8 muons, generated according to Gaisser's parameterization for cosmic ray muons at sea level [7], are propagated from random positions from a plane of dimension (0.5 m × 0.5 m) (ABCD in Fig. 94.1) and neutrons coming out of the other side of the rock are recorded. The simulation is repeated for different rock thickness. The input muon energy distribution is given in the Fig. 94.2 (left). The number of neutrons produced in the interaction increases as the thickness increases for a particular muon energy. But since lower energy neutrons get absorbed in the rock volume, only a fraction of neutron will be transmitted through rock. The ratio of number of neutrons transmitted to the number of neutrons produced (Transmission ratio) is found for different thicknesses and is given as a function of thickness in Fig. 94.2 (right).

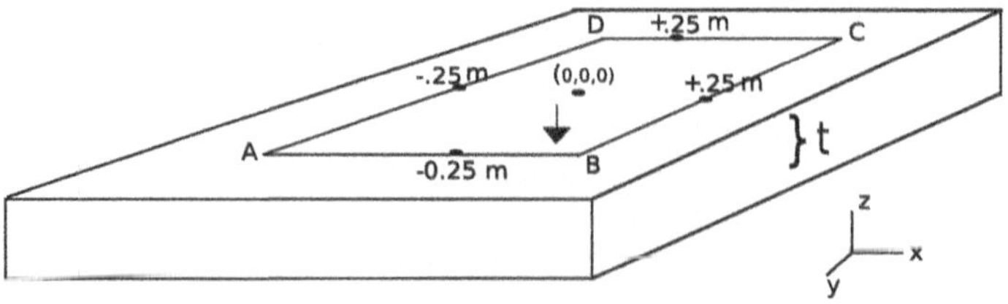

Fig. 94.1 The geometry used for studying different rock thickness

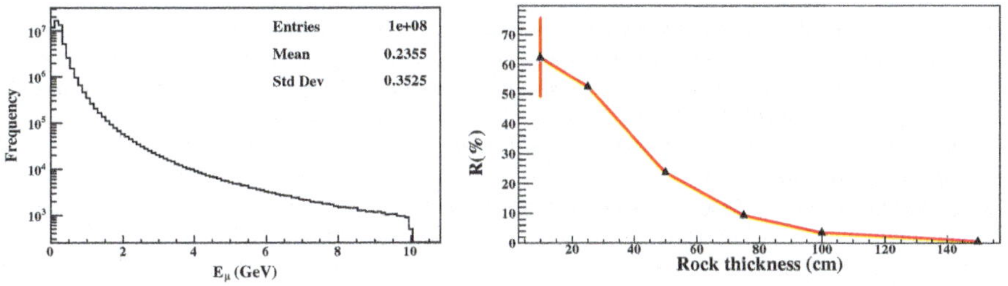

Fig. 94.2 Energy distribution of muon in the range 0.1–10 GeV obtained from Gaisser's parameterization (left). Transmission ratio of neutrons as a function of rock thickness (right)

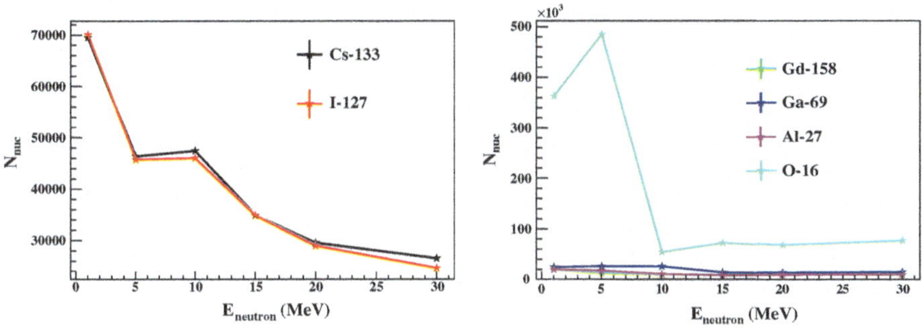

Fig. 94.3 Number of recoil nuclei as a function of neutron energy for CsI (left) and GGAG (right)

94.2.1 Response of Detector Materials to Neutrons

In order to study the response of detector materials to neutrons, a target volume, which is a cube of side 1 cm made of CsI or GGAG is considered. 10^6 fixed energy neutrons with energies 1, 5, 10, 15, 20 and 30 MeV incident on one face of the crystal cube are propagated through the target volume of the crystal. The production cut for secondary particles in GEANT4 was set to zero for recording low energy recoil nucleus. It was found that there are more interactions in GGAG compared to CsI, which gives more number of recoil nuclei in GGAG. The number of recoil nuclei reduces as the neutron energy increases for both crystals. The absorption of neutrons is more in GGAG because of the presence of Gadolinium. Number of recoil nuclei as a function of neutron energy for individual nuclei are given in Fig. 94.3. As neutron energy increases the number of nuclei recoiled decreases.

References

1. S. Ahmed et al., Physics potential of the ICAL detector at the India-based Neutrino Observatory (INO), ICAL Collaboration (2015), arXiv:1505.07380 [physics.ins-det]

2. K.A. Olive et al., (Particle Data Group), Chin. Phys. C **38**, 090001 (2014)
3. Jaduguda rock analysis, Atomic Minerals Directorate for Exploration and Research
4. Ling Han et al., *Proceedings of SPIE–the International Society for Optical Engineering, 12 Sep 2014*, 9214:92140D (2014)
5. M. Tyagi et al., J. Phys. D Appl. Phys. **46**, 475302 (2013)
6. http://geant4.cern.ch/2018
7. K. Nakamura et al., (PDG), JP G **37**, 075021 (2010)

Chapter 95
Searching the Inert Scalars Through the Dijet Plus Missing Transverse Energy Final States at High Luminosity LHC

P. Poulose, Shibananda Sahoo and K. Sridhar

It is well-known that the Standard Higgs mechanism with the observed physical scalar at LHC is too minimal, and the possibility of EWSB with more than one Higgs field is not ruled out yet. The inert version of two Higgs doublet model is one such possibility with one of the doublet fields not having any direct (at the level of the Lagrangian) interaction with the SM fermions, is a viable candidate in this direction. This is possible by demanding the Lagrangian is invariant under Z_2 symmetry. Under this symmetry, one of the doublets is odd, while all other fields are even. Such an Inert Doublet Model (IDM) [1] has quite distinct Higgs phenomenology, compared to that of MSSM or the usual 2HDM scenarios. Our main aim in this article is to explore the additional scalars of this model via dijet plus MET at LHC. Denoting the SM scalar doublet as Φ_1, additional scalar doublet as Φ_2 which is odd under a discrete Z_2 symmetry, the scalar potential respecting gauge invariance same as SM is given by

$$V(\Phi_1, \Phi_2) = \mu_1^2|\Phi_1|^2 + \mu_2^2|\Phi_2|^2 + \frac{\lambda_1}{2}|\Phi_1|^4 + \frac{\lambda_2}{2}|\Phi_2|^4 + \lambda_3|\Phi_1|^2|\Phi_2|^2 + \lambda_4|\Phi_1^\dagger\Phi_2|^2 + \left\{\frac{\lambda_5}{2}(\Phi_1^\dagger\Phi_2)^2 + H.c.\right\}, \quad (95.1)$$

One can write these scalar doublets after EWSB in the unitary gauge,

---

P. . Poulose · S. Sahoo (✉)
Department of Physics, Indian Institute of Technology Guwahati,
Guwahati 781039, Assam, India
e-mail: shibasipu@gmail.com

P. . Poulose
e-mail: poulose@iitg.ernet.in

K. Sridhar
Department of Theoretical Physics, Tata Institute of Fundamental Research,
Homi Bhabha Road, Mumbai 400005, India
e-mail: sridhar@theory.tifr.res.in

Md. Naimuddin (ed.), *XXII DAE High Energy Physics Symposium*, Springer Proceedings in Physics 203, https://doi.org/10.1007/978-3-319-73171-1_95

$$\Phi_1 = \begin{pmatrix} 0 \\ \frac{v+h}{\sqrt{2}} \end{pmatrix}, \Phi_2 = \begin{pmatrix} H^+ \\ \frac{H+iA}{\sqrt{2}} \end{pmatrix} \tag{95.2}$$

where $v = 246\,\text{GeV}$ is the vacuum expectation value of Φ_1. Apart from the SM-like Higgs h, this model has additional scalars i.e. a neutral scalar, H, a neutral pseudoscalar, A, and two charged Higgs bosons $H^{\pm}$. Here H is the candidate of dark matter. We can express the masses of these physical scalars in terms of parameters of the potential and v as

$$\begin{aligned}
m_h^2 &= \lambda_1 v^2, \qquad m_{H^+}^2 = \mu_2^2 + \frac{1}{2}\lambda_3 v^2, \\
m_H^2 &= \mu_2^2 + \frac{1}{2}(\lambda_3 + \lambda_4 + \lambda_5)v^2 = m_{H^\pm}^2 + \frac{1}{2}(\lambda_4 + \lambda_5)\, v^2, \\
m_A^2 &= \mu_2^2 + \frac{1}{2}(\lambda_3 + \lambda_4 - \lambda_5)v^2 = m_{H^\pm}^2 + \frac{1}{2}(\lambda_4 - \lambda_5)\, v^2
\end{aligned} \tag{95.3}$$

For our study we have taken $\{m_{H^+},\ m_A,\ m_H,\ m_h,\ \lambda_l,\ \lambda_2\ \}$ as our free parameter set. We have selected few benchmark points[1] which are given below considering all the dark matter and collider constraints.

$$\begin{aligned}
&\text{BP1}: m_{H^+} = 80, m_A = 75.4, m_H = 65, m_h = 125.1, \lambda_l = 0.006, \lambda_2 = 0.1 \\
&\text{BP2}: m_{H^+} = 150, m_A = 138.6, m_H = 65, m_h = 125.1, \lambda_l = 0.009, \lambda_2 = 0.1 \\
&\text{BP3}: m_{H^+} = 200, m_A = 189.5, m_H = 65, m_h = 125.1, \lambda_l = 0.009, \lambda_2 = 0.1 \\
&\text{BP4}: m_{H^+} = 300, m_A = 289.3, m_H = 65, m_h = 125.1, \lambda_l = 0.009, \lambda_2 = 0.1 \\
&\text{BP5}: m_{H^+} = 400, m_A = 397.6, m_H = 65, m_h = 125.1, \lambda_l = 0.009, \lambda_2 = 0.1 \\
&\text{BP6}: m_{H^+} = 500, m_A = 494.0, m_H = 65, m_h = 125.1, \lambda_l = 0.009, \lambda_2 = 0.1
\end{aligned}$$

The parton level cross sections are given in Table 95.1 for the processes which contribute to the dijet plus MET final states. We have used MADGRAPH5 for event generation with the IDM UFO files available in FeynRules wiki page. PYTHIA6 inside MADGRAPH5 is being used for showering and hadronization purpose. After studying the $p_T(jets)$ of the signal, we have employed the basic cuts of $p_T(j_1) > 80\,\text{GeV}$, and $p_T(j_2) > 50\,\text{GeV}$, $|\eta_{jets}| < 5.0$, and a jet separation of $0.4 < \Delta R_{j_1 j_2} < 2.0$, at the generation level, without loosing the signal events much. After this, the effective fiducial cross sections of the signal reduces to 1.3, 14, 16, 9.8, 4.8 and 2.5 fb, for the cases of BP1, BP2, BP3, BP4, BP5 and BP6, respectively. For signal, 50000 events have been generated in all cases. The background cross sections are reduced to 12.94 and 0.405 pb for $jj + \nu\bar{\nu}$ and Wjj with W decaying to leptons, respectively. 1300000 and 100000 events have been generated for these two backgrounds respectively, which provides sufficient statistics at $100\,\text{fb}^{-1}$ luminosity. Event analysis is done with MADANALYSIS5. These results are summarized in Table 95.1. One can find the detailed analysis in [3] (Table 95.2).

[1] We have taken BP's from A. Ilnicka and T. Robens [2].

Table 95.1 Parton level cross section (in fb) at $\sqrt{s} = 13\,\text{TeV}$ for specific benchmark points

Benchmark points	$2j + HH$ cross sections in fb (different channels)				$2j + \nu\bar{\nu}HH$ cross sections in fb (different channels)		(A)+(B)
	(i) $H^{\pm}H$	(ii): HA	Total cascade (i) + (ii)	(A):$pp \to 2j + HH$	$H^{\pm}A$ (all inclusive)	(B):$pp \to 2j + \nu\bar{\nu}HH$ (all inclusive)	
BP1	1.2	0.008	1.2	1.9	0.2	27.5	29.4
BP2	134.5	36.3	170.8	184.4	9.2	17.5	201.9
BP3	53.1	29.0	82.1	86.7	2.9	6.5	93.2
BP4	15.3	7.8	23.1	27.5	0.6	1.6	29.1
BP5	5.5	2.5	8.0	13.9	0.2	0.5	14.4
BP6	2.3	1.1	3.4	10.5	0.005	0.2	10.7

Table 95.2 Selection criteria used to optimise the S/B ratio. Corresponding signal significances are quoted at a 13 TeV LHC

Cuts employed	B	BP's	S	$\frac{S}{B}$ %	$\frac{S}{\sqrt{S+B}}$		$\frac{S}{\sqrt{B+(0.01\times B)^2+(0.1\times S)^2}}$	
					1 /ab	3 /ab	1 /ab	3 /ab
$N(j)=2$, $N(b)=0$, $N(l)=0$, $MET >$ 260 GeV, $p_T(j_1) > 110$, $p_T(j_2) > 90$ $75 \leq M_{j_1 j_2} \leq 90$ GeV, $\Delta R_{j_1 j_2} < 1.8$	8500	BP2	90	1.05	0.97	1.68	0.72	0.89
		BP3	198	2.32	2.12	3.68	1.56	1.94
		BP4	168	1.97	1.80	3.13	1.33	1.65
		BP5	120	1.41	1.29	2.24	0.95	1.19
		BP6	70	0.82	0.76	1.31	0.56	0.70

To conclude, we have considered the possibility to probe IDM through $2j + MET$ signal at the LHC with high luminosity. We have taken a few benchmark points with m_{H^+} ranging from 80 to 500 GeV. We found that the best scenarios are the cases with m_{H^+} around 200–400 GeV, which could be explored via the final states considered in this report at the LHC with about 3000 fb^{-1} integrated luminosity with a signal significance of about 2 for $m_{H^+} = 200$ GeV. The low mass scenarios with $m_{H^+} = 80$ GeV is also hard to probe via the final state considered in this analysis. For low mass scenarios, jets coming from the decay of inert scalars are too soft and also difficult to separate from the QCD background.

References

1. N.G. Deshpande, E. Ma, Phys. Rev. D **18**, 2574 (1978). https://doi.org/10.1103/PhysRevD.18.2574
2. A. Ilnicka, M. Krawczyk, T. Robens, Phys. Rev. D **93**(5), 055026 (2016). https://doi.org/10.1103/PhysRevD.93.055026, arXiv:1508.01671 [hep-ph]
3. P. Poulose, S. Sahoo, K. Sridhar, Phys. Lett. B **765**, 300 (2017). https://doi.org/10.1016/j.physletb.2016.12.022, arXiv:1604.03045 [hep-ph], TIFR-TH-16-13

Chapter 96
Measurement of D-Meson Production in pp Collisions with ALICE at the LHC

Ankita Sharma

96.1 Introduction

ALICE is the dedicated heavy-ion experiment at the Large Hadron Collider (LHC). Its main purpose is to investigate the properties of the deconfined state of strongly interacting matter produced in heavy-ion collisions, the Quark Gluon Plasma (QGP). The measurement of charm and beauty hadron production in Pb-Pb collisions is one of the most important goal of the ALICE physics program. Charm quarks are produced in hard scattering in the early stages of high energy nucleus-nucleus collisions and, due to their long life time, they are expected to be a powerful tool to investigate nuclear effects on heavy-flavour production. The measurements of charm and beauty production in proton-nucleus collisions are crucial to understand the cold nuclear matter effects on heavy-flavour production. These measurements in proton-proton collisions provide a baseline to compare with the results obtained from nuclear collisions and gives a chance to test the validity of perturbative Quantum Chromo Dynamic (pQCD) based models in a new energy domain.

96.2 D-Meson Production Cross Section

D mesons were reconstructed in their hadronic decay channels by means of the invariant mass analysis [1]. The measurement of the production cross section of the charmed D mesons was done in pp collisons at $\sqrt{s} = 8\,\text{TeV}$, where D-mesons and their charge conjugates; $\text{D}^0(\rightarrow K^-{+}\pi^+)$, $\text{D}^+(\rightarrow K^-{+}\pi^+{+}\pi^+)$, and

---

Ankita Sharma, for the ALICE Collaboration.

A. Sharma (✉)
University of Jammu, Jammu, India
e-mail: ankita.sharma@cern.ch

Md. Naimuddin (ed.), *XXII DAE High Energy Physics Symposium*, Springer Proceedings in Physics 203, https://doi.org/10.1007/978-3-319-73171-1_96

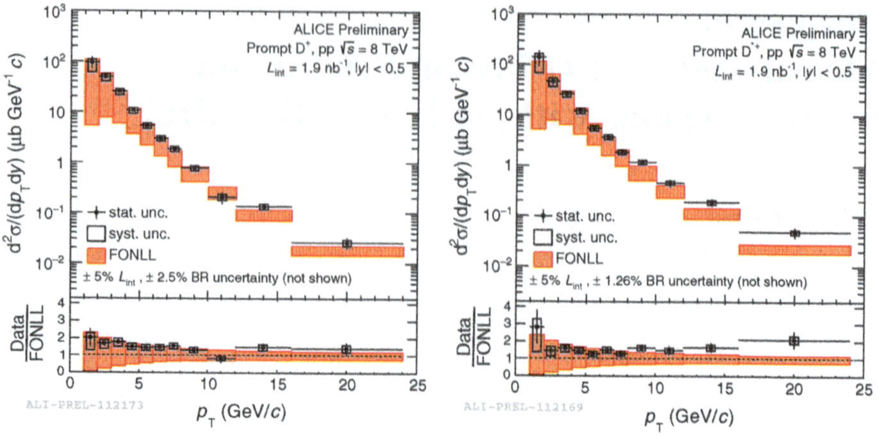

Fig. 96.1 p_{T}-differential cross sections for D^{+} and D^{*+} in pp collisons at $\sqrt{s} = 8\,\mathrm{TeV}$ compared with FONLL theoretical predictions. Bottom: the ratio of the measured cross section and the central FONLL calculations

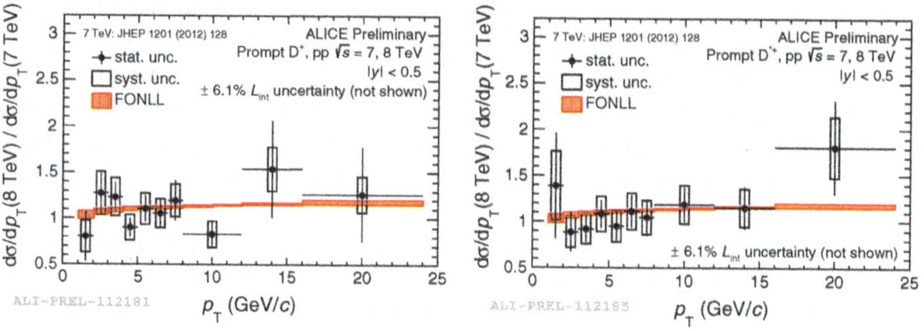

Fig. 96.2 Ratio of p_{T}-differential cross sections for D^{+} and D^{*+} at $\sqrt{s} = 7$ and $8\,\mathrm{TeV}$

$\mathrm{D}^{*+}(\rightarrow \mathrm{D}^{0}+\pi^{+})$, were reconstructed in the transverse momentum range $1 < p_{\mathrm{T}} < 24\,\mathrm{GeV}/c$, covering the rapidity interval $|y| < 0.5$. Figure 96.1 shows the prompt D-meson differential cross sections as a function of transverse momentum. The measurements are compared with the FONLL calculations [2]. It can be noted that, while fully compatible, FONLL predictions are on average lower than the measured cross sections. Therefore, FONLL tends to underestimate the charm production in pp at $\sqrt{s} = 8\,\mathrm{TeV}$, as was already noted at lower energies [1].

These measurements were also compared with the same measurements at $\sqrt{s} = 7\,\mathrm{TeV}$ shown in Fig. 96.2. From these figures it is clear that within the statistical fluctuations the 8 and 7 TeV results are compatible, also the ratio of the cross sections at the two energies is compatible with FONLL.

96.3 D-Meson Yield as a Function of Charged Particle Multiplicity

The measurement of heavy-flavour production in pp collisions as a function of the charged-particle multiplicity could provide an insight into the processes occurring in the collision at the partonic level and the interplay between the hard and soft mechanisms in particle production. The analysis results are presented as a function of the relative charged-particle multiplicity at central rapidity measured in pp collisions at $\sqrt{s} = 7\,\text{TeV}$ with at least one charged particle in $|\eta| < 1.0$. Figure 96.3 shows the relative yields of the D^0, D^+ and D^{*+} as a function of the relative charged-particle multiplicity in various p_T bins [3]. The charm hadron yield increases with the charged particle multiplicity at central rapidity with a faster than linear increase at large multiplicities. The enhancement is qualitatively described by models including multi-parton interactions, like PYTHIA [4], EPOS [5] and the percolation model [6].

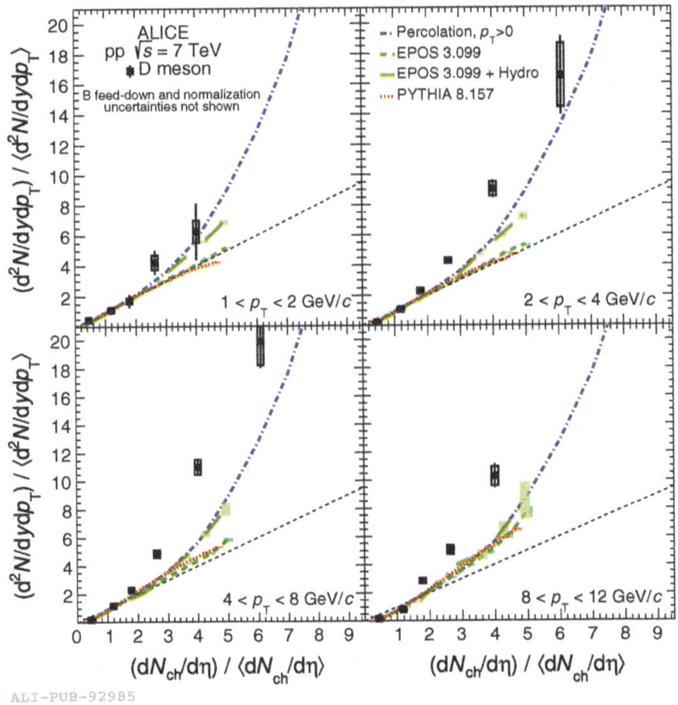

Fig. 96.3 Average D-meson relative yields as a function of the relative charged-particle multiplicity at central rapidity in different p_T intervals [3]

References

1. ALICE Collaboration, JHEP **1201**, 128 (2012)
2. M. Cacciari, S. Frixione, N. Houdaeu, M.L. Mangano, P.Nason, G. Ridolfi, JHEP **0103**, 006 (2001)
3. ALICE Collaboration, JHEP **09**, 148 (2012)
4. T. Sjostrand, S. Mrenna, P.Z. Skands, Comput. Phys. Commun. **178**, 852–867 (2008)
5. E.G. Ferreiro, C. Pajares, Phys. Rev. C **86**, 034903 (2012)
6. H. Drescher, M. Hladik, S. Ostapchenko, T. Pierog, K. Werner, Phys. Rep. **350** (2001)

Chapter 97
Event-by-Event Charge Separation in Au+Au Collisions at $\sqrt{s_{NN}} = 200\,\text{GeV}$ with the STAR Detector at RHIC

Anjali Attri

97.1 Introduction

The relativistic heavy-ion collisions provide a unique environment to investigate the particle interactions at very high temperature and extremely energy density. The distinguished features related to a non-central heavy-ion collision are the large orbital angular momentum and a very strong magnetic field ($B \sim 10^{15}$ T) produced by fast moving ions. The interaction of the magnetic field and the deconfined state created in these collisions might result in the phenomena of Chiral Magnetic Effect (CME) [1–3]. The CME is expected to cause the separation of positively and negatively charged particles along the axis of the magnetic field and perpendicular to the reaction plane.

The multi-particle correlator $\langle \cos(\phi_\alpha + \phi_\beta - 2\Psi_{\mathrm{RP}}) \rangle$ has been proposed by Voloshin [4] for the measurement of this effect, where ϕ_α, ϕ_β denote the azimuthal angles of the particles α, β and Ψ_{RP} is the reaction plane angle. Since the reaction plane angle can not be determined experimentally, one can use the three-particle correlator $\langle \cos(\phi_a + \phi_b - 2\phi_c) \rangle$, here ϕ_a, ϕ_b and ϕ_c are the azimuthal directions of particles a, b and c, respectively.

97.2 Sliding Dumbbell Method

The STAR experiment at RHIC and the ALICE experiment at LHC have made efforts to investigate the charge separation by measuring the three particle correlator [5]. We have studied the event-by-event charge separation using Sliding Dumbbell Method

Anjali Attri (for the STAR Collaboration).

A. Attri (✉)
Panjab University, Chandigarh 160014, India
e-mail: anjali@rcf.rhic.bnl.gov

Md. Naimuddin (ed.), *XXII DAE High Energy Physics Symposium*, Springer Proceedings in Physics 203, https://doi.org/10.1007/978-3-319-73171-1_97

(SDM) similar to the Sliding Window Method [6] used for the study of neutral-charged fluctuations in Pb+Pb collisions at 158 AGeV at SPS [7]. In this method we evaluate the quantity Db_{+-}, which is defined as:

$$Db_{+-} = \frac{N_{+}^{L}}{(N_{+}^{L} + N_{-}^{L})} + \frac{N_{-}^{R}}{(N_{+}^{R} + N_{-}^{R})} . \tag{97.1}$$

N_{+}^{L} and N_{-}^{L} are the numbers of positively and negatively charged particles on the left side of the dumbbell respectively. N_{+}^{R} and N_{-}^{R} are the numbers of positively and negatively charged particles on the right side of the dumbbell respectively.

The whole azimuthal plane is scanned by sliding the $\Delta\phi = 90°$ dumbbell in the steps of $\delta\phi = 1°$ and calculating the quantity Db_{+-} for each $\Delta\phi$ region to extract the maximum value of Db_{+-} in each event. Db_{+-} has also been obtained in each event, for the dumbbell placed in the azimuthal plane at randomly chosen ϕ.

97.3 Analysis Details

The data used in this analysis were recorded in the 2004 RHIC run. The data set contains about one million minimum-bias-trigger events. The TPC tracks in the pseudo-rapidity region $|\eta| < 1.0$ and transverse momentum range $0.15 < p_T < 2.0$ GeV/c are used. Standard STAR track quality cuts, $N_{Hits} > 14$ and $N_{HitsFit}/N_{HitsPoss} > 0.52$ have been applied.

97.4 Results and Discussions

The quantity Db_{+-}, has been evaluated for azimuthal dumbbells of different sizes using the Sliding Dumbbell Method and random dumbbell method. Figure 97.1 shows the Db_{+-} distributions for 30–80% central events obtained using SDM and random method for the 90^0 dumbbell. The dotted curve is for random method and solid curve is for SDM. The Db_{+-} distribution for the random method peaks at ~1 whereas, for the SDM it is seen to be shifted towards 2. This shift indicates that the SDM indeed quantifies the maximal charge separation in an event.

Analyzing the tail events of Db_{+-} distribution obtained using SDM, some interesting events showing back to back charge separation are found. Figure 97.2 shows the scatter plots ($x = 200 * \sin\theta \cos\phi$, $y = 200 * \sin\theta \sin\phi$) of the hit distribution of positively and negatively charged particles in the azimuthal plane for four such events. Here, hits of positively charged particles are indicated by red circles and those of negatively charged particles are indicated by blue triangles. The arcs drawn opposite to each other are to make the charge separation to be clearly visible. It is observed in Fig. 97.2 that red circles (positive particles) are more in number as compared to the blue triangles (negative particles) in one arc and vice-versa. Note the results reported here are from only 1 million events. More analysis with full statistics are underway.

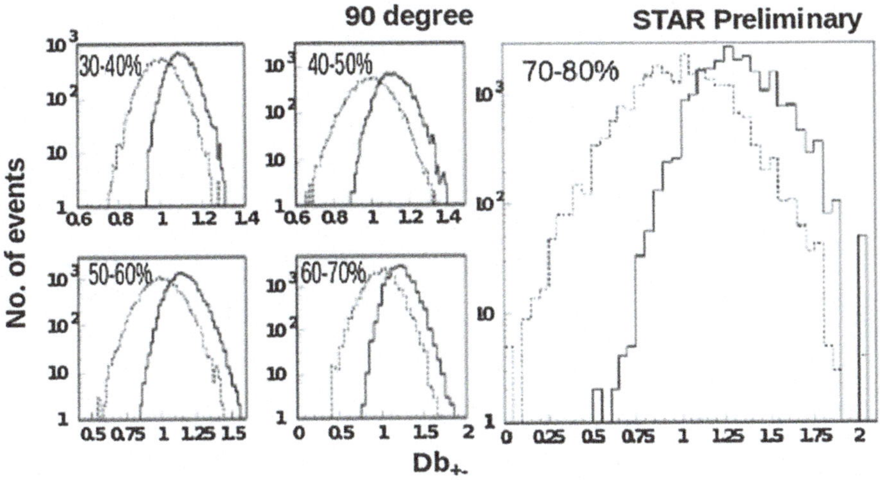

Fig. 97.1 Db_{+-} distributions obtained for random dumbbell (dotted lines) and sliding dumbbell (solid lines) in the 30–80 % centrality intervals

Fig. 97.2 Scatter plots of positively charged particles (red circles) and negatively charged particles (blue triangles) in the azimuthal plane for four events with different collision centrality (top left 50–60%, top right 70–80%, bottom left 60–70% and bottom right 50–60%), belonging to higher Db_{+-} values

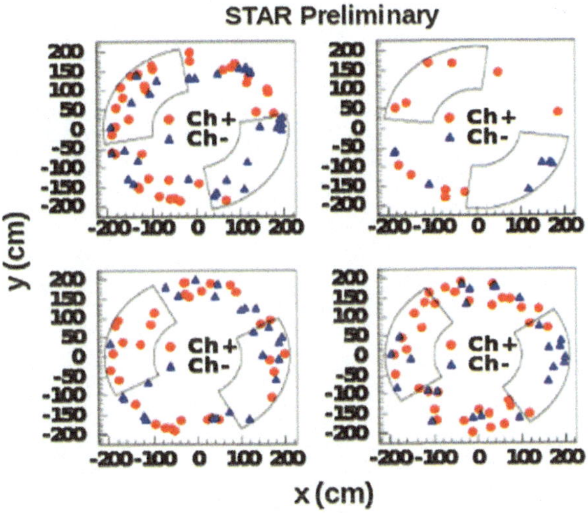

References

1. D.E. Kharzeev et al., Prog. Part. Nucl. Phys. **88**, 1–28 (2016)
2. I. Selyuzhenkov (2009), arXiv: 0910.0464
3. A. Bzdak, V. Koch, J. Liao, Phys. Rev. C **83**, 014905 (2011)
4. S.A. Voloshin, Phys. Rev. C **70**, 057901 (2004)
5. B.I. Abelev et al., (STAR Collaboration), Phys. Rev. C **81**, 054908 (2010); B. Abelev et al., (ALICE Collaboration), Phys. Rev. Lett. **110**, 012301 (2013)
6. M.M. Aggarwal, G. Sood, Y.P. Viyogi, Phys. Lett. B **638**, 39 (2006)
7. M.M. Aggarwal et al., (WA98 Collaboration), Phys. Lett. B **701**, 300 (2011)

Chapter 98
The INO-ICAL Sensitivity for the Separate Measurement of Neutrinos/Anti-neutrinos Parameters

Daljeet Kaur, Zubair Ahmad Dar, Sanjeev Kumar and Md. Naimuddin

98.1 Introduction

The India-based Neutrino Observatory (INO)[1], an approved project, is a neutrino experiment to study the atmospheric neutrino properties with a Magnetized Iron Calorimeter (ICAL) detector. Atmospheric ν_μ or $\bar{\nu}_\mu$ on interaction with the iron target produces muons and hadrons through a Charge-Current (CC) interaction process. The ICAL has a distinguished feature of charge identification through an applied magnetic field of strength 1 Tesla which will help to distinguish between atmospheric neutrinos and anti-neutrinos. The energies and direction of these interacting ν_μ or $\bar{\nu}_\mu$ can be determined from the reconstructed energies and direction of muons and hadrons. This paper presents the ability of ICAL detector to measure oscillation parameters separately from ν and $\bar{\nu}$ events. We will also show for the first time the ICAL potential to find out any difference in the atmospheric mass square differences described for neutrino and anti-neutrino i.e. $|\Delta m^2_{32}| - |\Delta \bar{m}^2_{32}|$.

98.2 Analysis Procedure

The atmospheric neutrino oscillation parameters can be extracted by a χ^2 analysis. In the present analysis, we use the same data set and procedure as mentioned in [2]. Here, all the events with detector resolutions and efficiencies are binned into

D. Kaur (✉)
S.G.T.B. Khalsa College, University of Delhi, Delhi, India
e-mail: daljeet.kaur97@gmail.com

Z. A. Dar
Aligarh Muslim University, Aligarh, Uttar Pradesh, India

S. Kumar · Md. Naimuddin
Department of Physics and Astrophysics, University of Delhi, Delhi, India

Md. Naimuddin (ed.), *XXII DAE High Energy Physics Symposium*, Springer Proceedings in Physics 203, https://doi.org/10.1007/978-3-319-73171-1_98

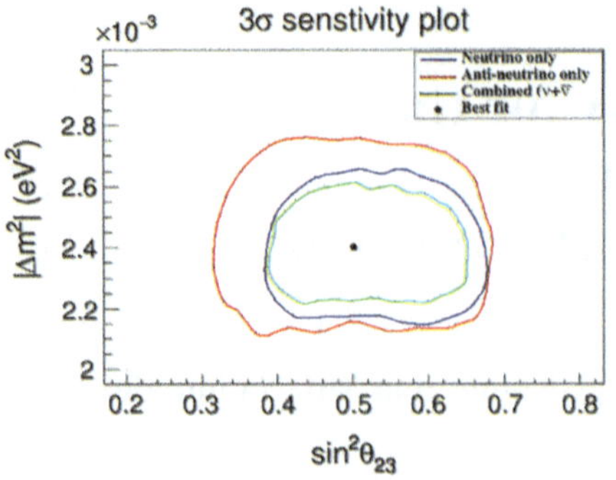

Analysis	$\sin^2(\theta_{23})$	Δm^2_{32} (eV^2)
ν_μ events	27.10%	10.41%
$\bar{\nu}_\mu$ events	38.0%	13.4%
Combined($\nu + \bar{\nu}$)	25.0%	8.3%

Fig. 98.1 ICAL sensitivity in Δm^2_{32} and $\sin^2(\theta_{23})$ plane at 3σ confidence level (Left) and corresponding precision table (Right)

neutrino energy, muon energy and muon direction with optimized bin width. The above mentioned procedure is followed for both ν_μ and $\bar{\nu}_\mu$ events separately. A "pulled" χ^2 has been calculated using systematic uncertainties as outlined in [2]. A separate measurement of $\chi^2(\nu_\mu)$ and $\chi^2(\bar{\nu}_\mu)$ has been obtained and then these two are added to get total χ^2_{total} for the combined results.

In this paper, the neutrino and anti-neutrino oscillations are considered independently over a two dimensional oscillation space such that they have identical oscillation parameters ($|\Delta m^2_{32}| = |\Delta \bar{m}^2_{32}|$=2.4 × 10^{-3}, $\sin^2\theta_{23} = \sin^2\bar{\theta}_{23}$ =0.5). We kept other oscillation parameters fixed at δ_{CP}= 0 degree, $\sin^2(\theta_{13})$= 0.03, Δm^2_{21}=7.6 × 10^{-5} and $\sin^2(2\theta_{12}) = 0.86$ as they do not show any impact on the analysis results. The χ^2_ν (or $\chi^2_{\bar{\nu}}$) is then minimized with respect to ($|\Delta m^2_{32}|$, $\sin^2\theta_{23}$) (or ($|\Delta \bar{m}^2_{32}|$, $\sin^2\bar{\theta}_{23}$)) and the five nuisance parameters.

Figure 98.1 shows the resulting contours plots at 3σ confidence level obtained for the ($|\Delta m^2_{32}|$, $\sin^2\theta_{32}$) or ($|\Delta \bar{m}^2_{32}|$, $\sin^2\bar{\theta}_{23}$) planes. These results are also compared with combined results of neutrino and anti-neutrinos events.

98.2.1 Variation in True Values

We performed analysis in which we allow ν_μ and $\bar{\nu}_\mu$ to have different mass square differences i.e. true values of $|\Delta m^2_{32}| \neq |\Delta \bar{m}^2_{32}|$ and rest all the oscillation parameters are same. A χ^2 is minimized for the observed $|\Delta m^2_{32}|$=$|\Delta \bar{m}^2_{32}|$ and a χ^2 plot as a function of difference of the true values ($|\Delta m^2_{32}| - |\Delta \bar{m}^2_{32}|$) have been obtained as shown in Fig. 98.2a.

In Fig. 98.2a, X-axis corresponds to the difference of the true values ($|\Delta m^2_{32}| - |\Delta \bar{m}^2_{32}|$) while Y-axis corresponds to the χ^2 values only for those events, where $|\Delta m^2_{32}|$=$|\Delta \bar{m}^2_{32}|$. For a particular $(|\Delta m^2_{32}| - |\Delta \bar{m}^2_{32}|)_{True}$, there may have several χ^2 values which indicates the points where $|\Delta m^2_{32}|$=$|\Delta \bar{m}^2_{32}|$. These values represent the null hypothesis ($|\Delta m^2_{32}|$=$|\Delta \bar{m}^2_{32}|$) while the lowest χ^2 value for a specific

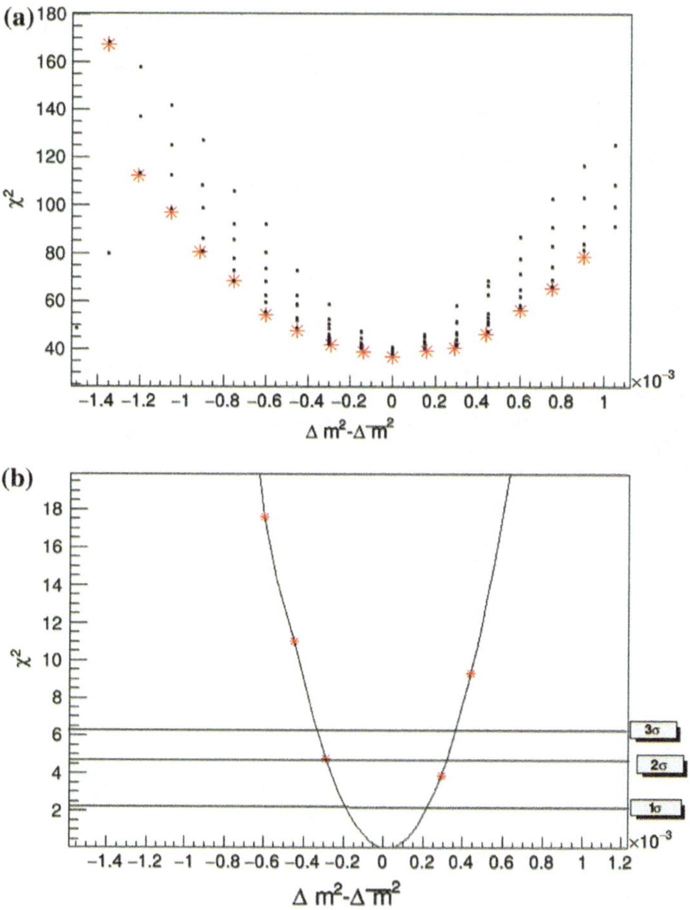

Fig. 98.2 **a**χ^2 Vs $(|\Delta m^2_{32}| - |\Delta \bar{m}^2_{32}|)_{True}(eV^2)$ and **b** $\Delta\chi^2$ Vs $(|\Delta m^2_{32}| - |\Delta \bar{m}^2_{32}|)_{True}(eV^2)$

difference shows the ICAL sensitivity for that particular difference. It is clear form the Fig. 98.2a that $\chi^2_{min} = 37$. Finally, the ICAL sensitivity for the differences of $(|\Delta m^2_{32}| - |\Delta \bar{m}^2_{32}|)_{True}$ at different confidence interval has been shown in Fig. 98.2b.

98.3 Results and Conclusion

Assuming identical oscillation parameters, the ICAL can measure the atmospheric neutrino parameters with a better precision as compared to anti-neutrino oscillation parameters as shown in Fig. 98.1(Right). As expected, the combined ν_μ+$\bar{\nu}_\mu$ events show a better sensitivity with a precision of 8.7% for $|\Delta m^2_{32}|$ and 25.0% for $\sin^2\theta_{23}$ due to larger events in χ^2. As shown in the Fig. 98.2b, we conclude that ICAL discovery potential for the non-zero $|\Delta m_{32}^2| = |\Delta \bar{m}_{32}^2|$ is more than 3σ level if the difference of true values of $|\Delta m_{32}^2| - |\Delta \bar{m}_{32}^2| >= +0.35$ or $|\Delta m_{32}^2| = |\Delta \bar{m}_{32}^2| <= -0.35$.

Acknowledgements We would like to thank the INO physics analysis group for their important comments and suggestions. We also thank CSIR & DST for supporting this work.

References

1. The ICAL Collaboration, Physics potential of the ICAL detector at the India-based Neutrino Observatory (INO) (2015), arXiv:1505.07380v1 [physics.ins-det]
2. D. Kaur et al., INO-ICAL detector sensitivity for the neutrino oscillation parameters. Euro. Phys. J. C **75**, 156 (2015)

Chapter 99
CP Sensitive Observable Exploring Tau Lepton Pairs from Higgs at the LHC

Partha Konar, Pankaj Sharma and Abhaya Kumar Swain

99.1 Introduction

Enormous effort has been poured into studying the different properties of Higgs boson, after its recent discovery. Any deviation in the measured properties of the Higgs from the Standard Model (SM) value can indicate the existence of the physics beyond (BSM). The current LHC data still allows significant deviation from the SM predictions in terms of the couplings and mass of the Higgs, which gives enough freedom to study many new physics scenarios in this sector. In this direction, scrutinising the Higgs coupling to the third generation fermions is very important because of the large Yukawa coupling. One property is the CP in Higgs sector which can shed light on the CP violation (CPV). It is one of the three Sakharov's conditions which plays crucial role in explaining the observed baryon asymmetry of the universe. Experimentally it was first observed in K and B mesons systems and studied extensively. The SM has the CPV information in the Cabibbo–Kobayashi–Maskawa mixing matrix in terms of phase but it is not sufficient to explain the observed baryon asymmetry. The SM Higgs sector is CP conserving as the couplings of Higgs to all SM particles are CP even but some BSM, like two-Higgs doublet and minimal supersymmetric standard model contains additional bosons which leads to CP odd couplings.

Although the current study excludes a pure pseudoscalar boson with 95% CL [1], the CP admixture of both scalar and pseudoscalar component is still allowed. Thus, it is very essential to determine the CP composition of the Higgs which in a way

P. Konar · A. K. Swain (✉)
Physical Research Laboratory (PRL), Ahmedabad 380009, Gujarat, India
e-mail: abhaya@prl.res.in

P. Konar
e-mail: konar@prl.res.in

P. Sharma
Department of Physics, ARC Center of Excellence for Particle Physics at the Terascale, University of Adelaide, Adelaide, SA 5005, Australia
e-mail: pankaj.sharma@adelaide.edu.au

Md. Naimuddin (ed.), *XXII DAE High Energy Physics Symposium*, Springer Proceedings in Physics 203, https://doi.org/10.1007/978-3-319-73171-1_99

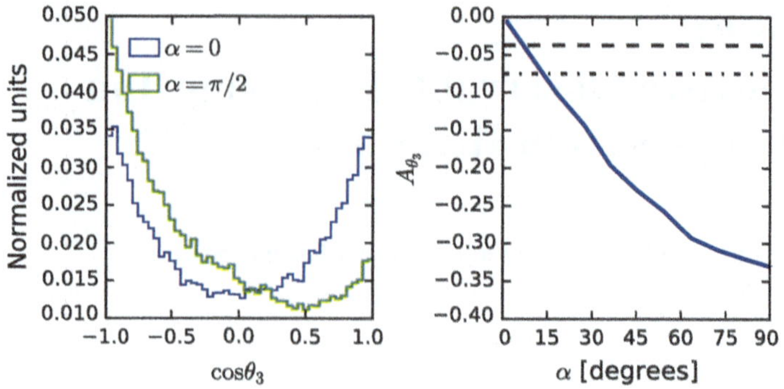

Fig. 99.1 The distrbution of the $\cos\theta_3$ with two values of the CP phase $\alpha = 0, \pi/2$ are shown. We also have constructed their asymmetry and shown in the right plot

tells the CPV couplings to other SM particles. In this analysis, we investigate the CP violating coupling of Higgs in the leptonic sector mainly the tau lepton pair production from the Higgs owing to its relatively larger Yukawa coupling. The tau lepton further decays hadronically leading to two invisible neutrinos in the final state. Although our analysis is applicable to all hadronic decay channels of the tau, we take two charged pions and two neutrinos in the final state as an example. There are several studies in the literature which focus on the measuring CPV nature of τ Yukawa coupling at the LHC [2–7]. We construct observable based on the CP and T transformation properties which are momentum correlations of the visible decay products and tau momenta.

99.2 CP Sensitive Observable

In this analysis, we consider Higgs to be a CP admixture, does not have definite CP transformation. The Yukawa term for this Higgs boson is, $\mathcal{L} \supset -m_\tau \bar{\tau}\tau - \frac{y_\tau}{\sqrt{2}} H\bar{\tau}(\cos\alpha + i\gamma_5 \sin\alpha)\tau$. Where τ and H are the physical fields, respectively. With y_τ is the $\tau-$Yukawa coupling and α is the degree of the mixing of the scalar and pseudoscalar component of the Higgs boson. For the SM, α is zero; $\alpha = \pi/2$ corresponds to pure psedoscalar and, $\alpha = \pi/4$ is for maximally CP violating case. In this study, we choose the y_τ to its SM value, and vary α to realise the effect of this in the expectation value of the CP observable. We define the CP observable as shown in the Table 99.1 with the ZMF corresponds to the Higgs rest frame and prime frame refers to a frame where part of the momenta are in the Higgs rest frame and other part is in tau rest frame which are presented in h and τ superscript, respectively. The distribution of $\cos\theta_3 = \hat{P}.\hat{Q}$ are shown in the Fig. 99.1 with P and Q are the first and second terms in the vector triple product of each observable. We also define asymmetries corresponding to each observables and details of which can be found in the [7]. The asymmetry for this observable is as large as 35% and with 300 fb^{-1} integrated luminosity of the LHC data we can constrain the CP phase up to 15°.

Table 99.1 CP sensitive variable for $h \to \tau^+\tau^- \to \pi^+\pi^-\nu\bar{\nu}$ at the LHC and have the definite CP and time reversal transformation. Frames are described in text

Observables	Frame
$\mathcal{O}_1 = (\mathbf{p}_{\tau^-} - \mathbf{p}_{\tau^+}).(\mathbf{p}_{\pi^-} \times \mathbf{p}_{\pi^+})$	ZMF
$\mathcal{O}_2 = (\mathbf{p}_{\tau^-} - \mathbf{p}_{\tau^+})^h.(\mathbf{p}_{\pi^-} \times \mathbf{p}_{\pi^+})^\tau$	Prime
$\mathcal{O}_3 = (\mathbf{p}_{\pi^-} - \mathbf{p}_{\pi^+})^\tau.(\mathbf{p}_{\pi^-} \times \mathbf{p}_{\pi^+})^h$	Prime
$\mathcal{O}_4 = (\mathbf{p}_{\pi^-} - \mathbf{p}_{\pi^+})^h.(\mathbf{p}_{\pi^-} \times \mathbf{p}_{\pi^+})^\tau$	Prime

99.3 Results and Conclusion

The study of CP violation is essential in understanding the observed baryon asymmetry of the universe. Higgs decays abundantly to tau leptons, because of the large Yukawa coupling, gives opportunities to study CP properties in this channel at the present run of LHC. We construct CP sensitive triple product momentum correlations using visible and tau momenta and also studied their asymmetries. We found that the asymmetries can be as large as 35%. The reconstruction efficiency of $\tau^\pm$ pair events and statistical sensitivity is also taken into account in studying the asymmetries. We found that the CP phase can be constrained to 15° at the LHC with 300 fb^{-1} of integrated luminosity.

References

1. G. Aad et al., Evidence for the spin-0 nature of the Higgs boson using ATLAS data. Phys. Lett. B **726**, 120–144 (2013)
2. S. Berge, W. Bernreuther, S. Kirchner, Determination of the Higgs CP-mixing angle in the tau decay channels at the LHC including the Drell-Yan background. Eur. Phys. J. C **74**(11), 3164 (2014)
3. R. Harnik, A. Martin, T. Okui, R. Primulando, Yu. Felix, Measuring CP violation in $h \to \tau^+\tau^-$ at colliders. Phys. Rev. D **88**(7), 076009 (2013)
4. M.J. Dolan, P. Harris, M. Jankowiak, M. Spannowsky, Constraining CP-violating Higgs sectors at the LHC using gluon fusion. Phys. Rev. D **90**, 073008 (2014)
5. A. Askew, P. Jaiswal, T. Okui, H.B. Prosper, N. Sato, Prospect for measuring the CP phase in the $h\tau\tau$ coupling at the LHC. Phys. Rev. D **91**(7), 075014 (2015)
6. A. Hayreter, X.-G. He, G. Valencia, CP violation in $h \to \tau\tau$ and LFV $h \to \mu\tau$. Phys. Lett. B **760**, 175–177 (2016)
7. P. Konar, P. Sharma, A.K. Swain. Exploring CP phase in τ-lepton Yukawa coupling in Higgs decays at the LHC (2016)

Chapter 100
Sensitivities of INO–ICAL to 2–3 Oscillation Parameters by the Extension of Observed Muon Energy Range and by Constraining the $\nu_\mu - \bar{\nu}_\mu$ Flux Ratio

Lakshmi S. Mohan and D. Indumathi

100.1 Introduction

In this study we show how the extension of the observed energy range of the final state muon in a charged current (CC) ν_μ or $\bar{\nu}_\mu$ interaction to 0.5–25 GeV and the addition of a new systematic pull in the form of a constraint on the $\nu_\mu - \bar{\nu}_\mu$ flux ratio improves the sensitivity to 2–3 oscillation parameters of a 50 kTon magnetised atmospheric neutrino detector [1], ICAL [2] at INO.

100.2 χ^2 Analysis

The details of true parameters and their 3σ ranges used to generate events, how the 3-flavour oscillations in matter are applied, the application of detector smearing and resolutions and the binning of events in several observed bins are explained in detail in [1]. Here a new systematic uncertainty in the form of $\nu_\mu - \bar{\nu}_\mu$ flux ratio is included. Then, χ^2 for ICAL is:

L. S. Mohan (✉)
Physical Research Laboratory, Ahmedabad 380009, India
e-mail: slakshmi@prl.res.in

D. Indumathi
The Institute of Mathematical Sciences, Chennai 600113, India

D. Indumathi
Homi Bhabha National Institute, Mumbai 400094, India

Md. Naimuddin (ed.), *XXII DAE High Energy Physics Symposium*, Springer Proceedings in Physics 203, https://doi.org/10.1007/978-3-319-73171-1_100

$$\chi^2_{11} = \min_{\xi_l^{\pm},\xi_6} \sum_{i=1}^{N_{E_\mu^{obs}}} \sum_{j=1}^{N_{\cos\theta_\mu^{obs}}} \left(\sum_{k=1}^{N_{E'^{obs}_{had}}} \right) 2\left[\left(T^+_{ij(k)} - D^+_{ij(k)}\right) - D^+_{ij(k)} \ln\left(\frac{T^+_{ij(k)}}{D^+_{ij(k)}}\right)\right] +$$
$$2\left[\left(T^-_{ij(k)} - D^-_{ij(k)}\right) - D^-_{ij(k)} \ln\left(\frac{T^-_{ij(k)}}{D^-_{ij(k)}}\right)\right] + \sum_{l^+=1}^{5} \xi^2_{l^+} + \sum_{l^-=1}^{5} \xi^2_{l^-} + \xi_6^2 \,, \quad (100.1)$$

where i, j, k are the muon energy, direction and hadron energy indices, and the number of theory events in each bin, with systematic errors, are given by: $T^{\pm}_{ij(k)} = T^{0\pm}_{ij(k)}\left(1 + \sum_{l^+=1}^{5} \pi^{l^+}_{ij(k)} \xi_{l^+} \pm \pi_6 \xi_6\right)$; $T^{0\pm}_{ij(k)}$ being the number of events without systematic errors and $D^{\pm}_{ij(k)}$ the number of "data" events per bin. All the previous analyses by the ICAL collaboration [3] have used only 10 pulls $\xi_l^{\pm}$, corresponding to the same five systematic uncertainties, l = 1,. . .,5, corresponding to ν and $\bar{\nu}$ contributions, where $\pi_1, \ldots, 5$ are 20, 10, 5, 5 and 5% for flux normalisation, cross-section, tilt, zenith angle and overall normalisation respectively. The 11th pull is denoted by ξ_6 and is defined such that it accounts for the uncertainty in the flux ratio:

$$\frac{N^+}{N^-} \simeq \frac{T^{0+}}{T^{0-}} \frac{(1 + \pi_6 \xi_6)}{(1 - \pi_6 \xi_6)} \simeq \frac{T^{0+}}{T^{0-}} (1 + 2\pi_6 \xi_6) \,. \quad (100.2)$$

Here $2\pi_6$ corresponds to the 1σ error (when $\xi_6 = 1$); hence the 1σ error on the ratio is 5%. A prior of 8% at 1σ is added to $\sin^2 2\theta_{13}$ so that $\chi^2_{ICAL} = \chi^2_{11} + \chi^2_{prior}$.

100.3 Results

From Fig. 100.1a, b it can be seen that the extension of energy range improves the precision on both $\sin^2\theta_{23}$ and $|\Delta m^2_{32}|$ compared to 1–11 GeV; the addition of the 11th pull improves that on $\sin^2\theta_{23}$, leaving that on $|\Delta m^2_{32}|$ unaltered. This is because, a deviation in the value of $\sin^2\theta_{23}$ from the true (input) value changes the number of ν and $\bar{\nu}$ events in opposite directions - these are now constrained by the 11th pull on the flux ratio and hence the inclusion of this pull tightens the limits on $\sin^2\theta_{23}$. The best relative 1σ precisions on $\sin^2\theta_{23}$ and $|\Delta m^2_{32}|$ are 8.9% (9.25%) and 2.57% (2.48%) respectively for NH (IH). The hierarchy sensitivity improves with the extension of energy range, but remains the same when the 11th pull is applied when NH is the true hierarchy as shown in Fig. 100.2a. If IH is the true hierarchy, addition of the 11th pull and extension of E_μ^{obs} both increase the hierarchy sensitivity as shown Fig. 100.2b. The best mass hierarchy sensitivity for 500 kTon year exposure of ICAL is $\chi^2_{MH-ICAL} = 8.5$ (9.82) for true NH (IH). More details of the analysis presented here are available in [1].

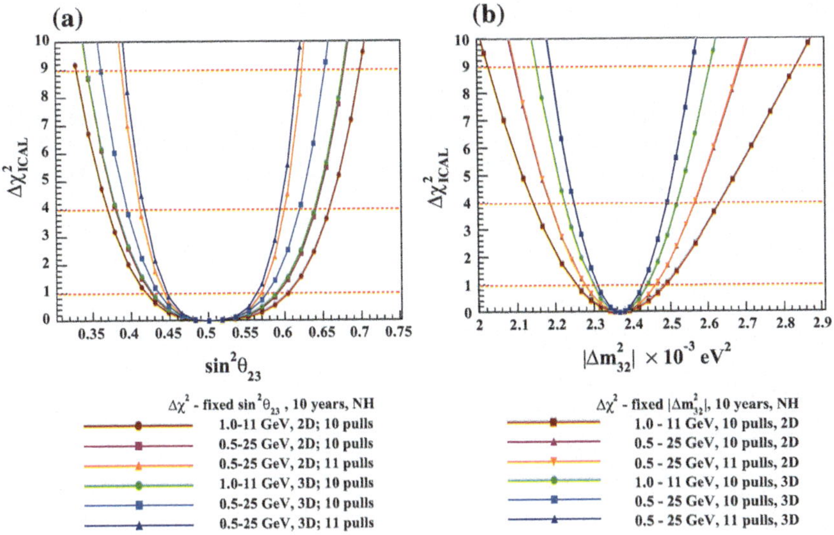

Fig. 100.1 Precision on (left) $\sin^2\theta_{23}$ and (right) $|\Delta m^2_{32}|$

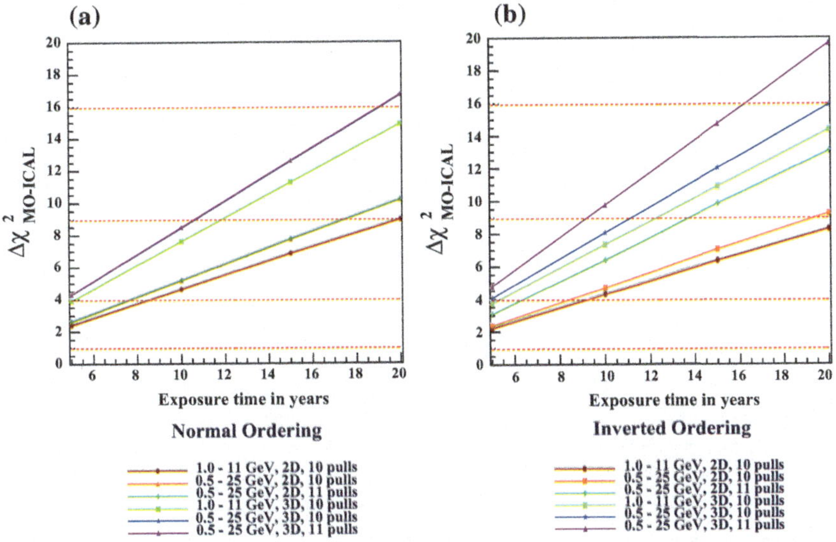

Fig. 100.2 Mass hierarchy sensitivity of ICAL for **a** NH and **b** IH as the true hierarchy

Acknowledgements Prof. Nita Sinha, IMSc, Dr. Meghna K.K, former INO student, INO internal referees, DAE India and DST India and the excellent computing facility of IMSc, Chennai, and the organisers of the DAE-HEP symposium.

References

1. S. Lakshmi, D. Mohan, Indumathi. Eur. Phys. J. C **77**, 54 (2017)
2. A. Kumar, A. M. Vinod Kumar, A. Jash et al., Invited review: Physics potential of the ICAL detector at the India-based Neutrino Observatory (INO). Pramana J. Phys. **88**, 79 (2017).
3. M.M. Devi et al., JHEP **10**, 189 (2014); T. Thakore et al., JHEP **5**, 058 (2013); A. Ghosh et al., JHEP **4**, 009 (2013)

Chapter 101
Diphoton Emission from Equilibrium Quark-Gluon Plasma

S. Somorendro Singh

101.1 Introduction

Many theorists and experimentalists try to specify the nature of QCD feature through the lattice calculation and the present ongoing experiments of heavy-ion physics. The idea enhances the attraction of more scientists towards the study of heavy-ion experiments carried out around the globe. So there are a number of ultra-relativistic research like heavy-ion collision at BNL and large hadron collider at CERN. Now these experiments have come out as a basic platform in the study of QGP. Besides these, there are experiments like NICA at Dubna and FAIR at Darmtart. These experiments are specially focused on compress baryon matter. It is already predicted that the existence of matter is for a few microseconds after the big-bang and this short period of existence has made the system a difficult task in the detection of strongly interacting matter.

Indirect possibilities are now considered as a signal for the detection of this matter through strangeness enhancement [1], J/ψ suppression [2] and radiation of dileptons and photons [3, 4] etc. One indirect probes of dileptons and diphotons are assumed to be the best signal for the formation of QGP and their production is involved through the electromagnetic interaction. In this short paper we consider the quark mass as a finite value and it removes the infrared (IR) divergence produced in the diphoton productions [5]. We use it as effective mass generated due to thermal interactions of quarks and anti-quarks and quarks and gluons. It is defined as [6]:

$$m_q^2(T, p) = \frac{N_c^2 - 1}{8N_c} T^2 G^2(p) \quad (101.1)$$

S. Somorendro Singh (✉)
University of Delhi, Delhi 110007, India
e-mail: sssingh@physics.du.ac.in

Md. Naimuddin (ed.), *XXII DAE High Energy Physics Symposium*, Springer Proceedings in Physics 203, https://doi.org/10.1007/978-3-319-73171-1_101

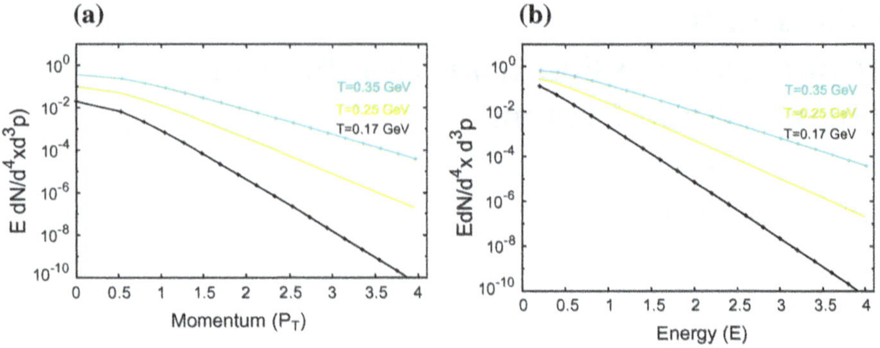

Fig. 101.1 **a** The diphoton emission rate, $\log(E\frac{dN}{d^4xd^3p})$ versus transverse momentum at initial temperatures $T_0 = 0.17, 0.25, 0.35\,\text{GeV}$. **b** The diphoton emission rate, $\log(E\frac{dN}{d^4xd^3p})$ versus photon energy at initial temperatures $T_0 = 0.17, 0.25, 0.35\,\text{GeV}$

Its value gives the minimum finite value on the basis of critical temperature. $G^2(p) = 4\pi\alpha_s(p)[1 + \frac{\alpha_s(p)a_1}{4\pi}]$ known as the QCD strong coupling factor where, α_s which depend on the momentum cut off [7]. a_1 is one loop correction factor chosen on the number of quark degrees of freedom. We focus to study the diphoton radiations at the different temperatures for the quark flavor 3. So we extend to calculate photon production at transition $T = 0.17\,\text{GeV}$ to hot phase of temperature $T = 0.35$ (0.40) GeV considering one loop correction factor in the stable QGP droplet formation. Then we show the results of diphoton production with these correction in the coupling.

We organize the paper as: In Sect. 101.2, we present the diphoton production. In Sect. 101.3, we show the results and conclusion.

101.2 Diphoton Production from QGP

Diphoton production from high energy collision of QGP has become very interesting theoretical problem since many years. It was a little understood as the system was just before the process of thermalization. Now the system is considered as a thermalized state after big-bang process and the system takes a longer time compared to time scale associated with the diphoton productions. On these factors, there are a number of theoretical research works for productions of photon from quark-gluon plasma. Most works are considered as quark antiquark annihilation process and Compton process. The one loop calculation of compton process plus annihilation is given as by the $q(\bar{q})g \rightarrow q(\bar{q})\gamma$ and $q\bar{q} \rightarrow g\gamma$.

Then the formulation of diphoton production with one loop order of Compton plus annihilation is given by the expression [8, 9]

$$E\frac{dN}{d^4xd^3p} = \frac{1}{3\pi^2}\alpha^2T^2e^{-E/T}\left[\ln\left(\frac{4ET}{2m_q^2}\right) - C_E\right] \tag{101.2}$$

where $C_E = 1.415$.

101.3 Results and Discussion

In the results it presents numerical results for diphoton production from high energy collision of quark-gluon plasma through Compton and annihilation process at finite temperature with the loop correction factor in the coupling value. Diphoton emission is performed with a suitable choice of quark flavors $n_f = 3$.

In Fig. 101.1a we show diphoton emission rate at the different initial temperatures with the change of transverse momentum. We plot the figure of the production rate with the transverse momentum and photon energy at temperatures $0.17 \leq T \leq 0.35$ GeV. The emission rates are found to be increasing with temperature of the QGP fireball. The increase in the emission rate is highly effected by temperature of the system, and it seems to be large near creation of quark matter, that is considered to exist at very hot temperature. So as we keep on increasing the temperature the production of diphoton is increasing at the lower transverse momentum and suppression is obtained at the higher transverse momentum.

In Fig. 101.1b we again show diphoton production of Compton with annihilation process with photon energy. The figure shows the production rate at different temperatures and showed for the spectra with changed in photon energy. At the larger finite temperature $T = 0.35$ GeV the result showed a significant production and at the lower temperature $T = 0.17$ GeV the results are highly reduced. So we summarize that total diphoton spectra with the model of quark mass through this one loop correction in the coupling value has a sizable production rate of diphoton in this process. Lastly it indicates that the model gives significant contributions in the production of diphoton spectra. This implies that the consideration of one loop correction in the coupling value has important role in the photon measurements of the high energy heavy ion collisions.

Acknowledgements I am very much thankful to Dr. Naimuddin, faculty, University of Delhi, Delhi-7, for his support in publishing the manuscript in this reputed journal series. I thank University of Delhi, Delhi-11007 for providing the R and D research grant, 2016 as financial support.

References

1. J. Rafelski, B. Müller, Phys. Rev. Lett. **48**, 1066 (1982)
2. T. Matsui, H. Satz, Phys. Lett. B **178**, 416 (1986)
3. E. Shuryak, Phys. Rep. **61**, 71 (1980)
4. K. Kajantie, J. Kapusta, L. McLerran, A. Mekjian, Phys. Rev. D **34**, 2746 (1986)

5. B. Braaten, R.D. Pisarski, Nucl. Phys. B **337**, 569 (1990)
6. Y. Kumar, S.S. Singh, Can. J. Phys. **90**, 955 (2012)
7. S.S. Singh, R. Ramanathan, Prog. Theo. Expt. Phys. **102D** (2014)
8. R. Baier, H. Nakkagawa, A. Niegawa, K. Redlich, Phys. Rev. D **45**, 4323 (1992)
9. S.S. Singh, Y. Kumar, Int. J. Mod. Phys. A **29**, 1450110 (2014)

Chapter 102
Vacuum Stability with Leptoquark

Rusa Mandal

102.1 Introduction

Leptoquark, a proposed particle, can decay to a lepton and a quark at tree level has drawn attention because of explaining various lepton non universal anomalies seen in rare B meson decays [1]. In this note we consider an extension of the standard model (SM) with a scalar leptoquark having $(\mathbf{3}, \mathbf{1}, -1/3)$ charge under SM gauge group $SU(3)_C \times SU(2)_L \times U(1)_Y$ and briefly discuss the effect of the leptoquark at electroweak (EW) vacuum stability, Higgs mass hierarchy and refer the reader to [2] for detailed description. Introduction of the leptoquark adds additional terms in SM Lagrangian which is given by,

$$\mathcal{L}_\phi = (D_\mu\phi)^\dagger D^\mu\phi - m_\phi^2|\phi|^2 - g_{h\phi}|\Phi|^2|\phi|^2 + \bar{Q}^c Y^{\mathrm{L}} i\tau_2 L\phi^* + \bar{u}_R^c Y^{\mathrm{R}} e_R \phi^* + h.c. \tag{102.1}$$

Here Φ, Q and L are the usual SM Higgs, quark and lepton $SU(2)_L$ doublet and u_R^c and e_R are the right-handed singlet up type quark and charged lepton, respectively. General Y^{L} and Y^{R} matrices with off-diagonal terms lead to lepton-quark flavor as well as generation violating couplings. The off diagonal couplings and first generation diagonal couplings are strongly constrained by various rare meson decay modes [4] and hence for most of our analysis, we will assume them to be vanishing as our results are unaffected by their small values. We introduce a notation for diagonal terms in Y^{L} and Y^{R} matrices after performing the rotations for moving to mass basis as,

$$Y^{\mathrm{L,R}} \to \mathrm{dia}\left(Y_{11}^{\mathrm{L,R}}, Y_{22}^{\mathrm{L,R}}, Y_{33}^{\mathrm{L,R}}\right). \tag{102.2}$$

Rusa Mandal Speaker.

R. Mandal (✉)
The Institute of Mathematical Sciences, Taramani, Chennai 600113, India
e-mail: rusam@imsc.res.in

Md. Naimuddin (ed.), *XXII DAE High Energy Physics Symposium*, Springer Proceedings in Physics 203, https://doi.org/10.1007/978-3-319-73171-1_102

102.2 Stability Analysis

In this section we discuss the stability of EW vacuum where at large Higgs field (h) values, the RG improved effective scalar potential can be approximated as $V^{\text{eff}} \simeq \lambda^{\text{eff}}(h,\mu)/8h^4$, where the one-loop expression for effective Higgs-self coupling λ^{eff} is given by,

$$\lambda^{\text{eff}}(h,\mu) \simeq e^{4\Gamma(\mu)}\Bigg\{\lambda(\mu) + \frac{1}{8\pi^2}\sum_{i=W,Z,h,G,t} N_i\kappa_i^2(\mu)\left[\ln\frac{\kappa_i(\mu)e^{2\Gamma(\mu)}h_c^2}{\mu^2} - C_i\right] + \frac{1}{8\pi^2}\frac{3g_{h\phi}^2(\mu)}{2}\left[\ln\frac{g_{h\phi}(\mu)e^{2\Gamma(\mu)}h_c^2}{2\mu^2} - \frac{3}{2}\right]\Bigg\}. \quad (102.3)$$

The detailed expression for each term is described in [2]. We note that the first line of (102.3) depicts contributions from the SM particles and the terms in second line arise due to the presence of the leptoquark. Depending upon the running of λ^{eff}, the EW vacuum can be categorized in three different regions; stable, metastable and unstable vacuum and the results are shown in Fig. 102.1. We use two loop beta functions for the running of all couplings present in the theory. The EW vacuum remains (meta)stable up to the Planck scale for $g_{h\phi}(M_Z) > 0.3$.

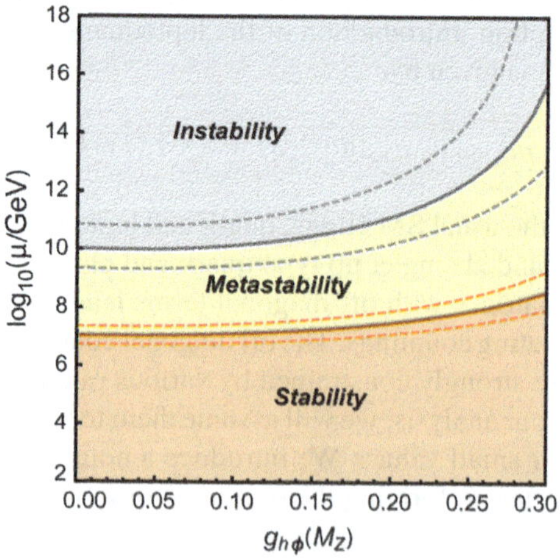

Fig. 102.1 The stability, metastability [3] and instability regions for EW vacuum are shown as a function leptoquark-Higgs coupling constant $g_{h\phi}$ value at the EW scale. The light brown region (stability) below the brown solid curve assures the effective Higgs self-coupling $\lambda^{\text{eff}}(\mu) > 0$. The yellow region (metastability) denotes $\lambda^{\text{eff}}(\mu) > \frac{-0.065}{1-0.01\ln(v/\mu)}$ and the gray region (instability) corresponds to $\lambda^{\text{eff}}(\mu) < \frac{-0.065}{1-0.01\ln(v/\mu)}$. The dashed lines show the uncertainty due to top quark mass $M_t = 173.3 \pm 0.87$ GeV in the corresponding cases. The plot is obtained for $\alpha_s = 0.1185$ and NP coupling $Y_{33}^{\text{L, R}} = 0.4$. For $g_{h\phi} > 0.3$ the vacuum is (meta)stable up to the Planck scale

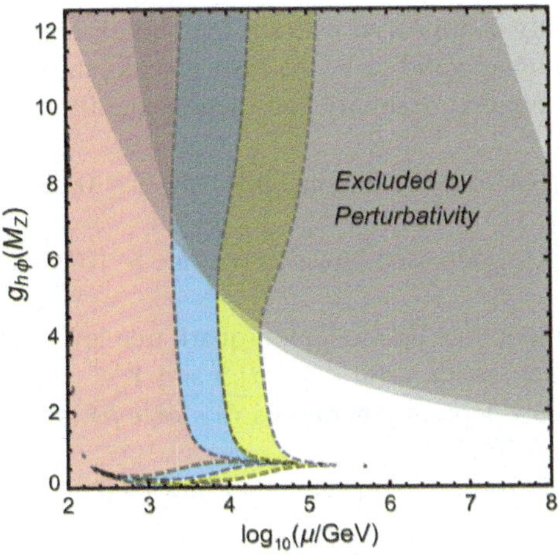

Fig. 102.2 The allowed region in $g_{h\phi} - \log_{10}(\mu/\mathrm{GeV})$ plane for different values of the fine-tune parameter '$1/f$'. The pink, blue and yellow regions denote no fine-tuning, up to 10% fine-tuning and up to 100% fine-tuning required for the model to remain valid up to the corresponding cut-off scale, respectively. The gray shaded region is excluded for perturbativity of the leptoquark-Higgs coupling $g_{h\phi}$ and Higgs self-coupling λ. The plot is obtained for $Y_{33}^{\mathrm{L}} = 0.5$ and $Y_{33}^{\mathrm{R}} = 0.5$ values at the EW scale

102.3 Fine-Tuning

The Higgs mass receives quadratically divergent contributions proportional to the cut-off scale (Λ) of the theory. The leptoquark modifies the Higgs mass correction at one-loop via the leptoquark-Higgs coupling $g_{h\phi}$ as,

$$\delta M_h^2 = \frac{\Lambda^2}{16\pi^2}\left(3\lambda + \frac{9}{4}g^2 + \frac{3}{4}g'^2 - 6y_t^2 + 3g_{h\phi}\right), \tag{102.4}$$

where λ, g_3, g, g' and y_t are SM Higgs self-coupling, three SM renormalized gauge couplings and top quark Yukawa coupling constant, respectively.

Defining a fine-tune parameter $1/f \equiv |\delta M_h^2|/M_h^2$, in Fig. 102.2 we highlight various different regions in $g_{h\phi} - \log_{10}(\mu/\mathrm{GeV})$ plane allowed by the corresponding fine tuning in this leptoquark model. It can be seen that even with $\sim$100% fine-tuning the validity scale of the theory remains within $\sim 10^6$ GeV.

102.4 Summary

- The leptoquark is motivated to explain some anomalies in rare B decays and muon $g-2$.

- EW vacuum stability is improved with the leptoquark-Higgs coupling.
- For the model to remain valid up to the Planck scale, the combined bound on the parameter space from perturbativity and vacuum stability is obtained as

$$0.3 \leq g_{h\phi}(M_Z) \leq 0.65 \quad \text{and} \quad Y_{ii}^{\text{L, R}}(M_Z) \leq 0.65; \quad i \in \{2, 3\}.$$

- Hierarchy problem is addressed however regions $> 10^6$ GeV is not accessible by huge fine tuning.
- Interesting phenomenology due to lepton-quark tree level coupling–
  For a TeV mass leptoquark, with $g_{h\phi} = 0.5$ and $Y_{ii}^{\text{L, R}} = 0.5$, branching fraction to the first two quark-lepton generations ($u\,e$ or $c\,\mu$) is $\sim$25% and to the third generation ($t\tau$) is $\sim$23%.
- Possibility to extend further to answer other unsolved questions of the SM.

Acknowledgements We thank P. Bandyopadhyay for collaborating in this work and the organizers of XXII DAE-BRNS High Energy Physics Symposium-2016 for giving the opportunity to present our work in the conference.

References

1. M. Bauer, M. Neubert, Phys. Rev. Lett. **116**(14), 141802 (2016)
2. P. Bandyopadhyay, R. Mandal, Phys. Rev. D **95**(3), 035007 (2017)
3. G. Isidori, G. Ridolfi, A. Strumia, Nucl. Phys. B **609**, 387 (2001)
4. S. Davidson, D.C. Bailey, B.A. Campbell, Z. Phys. C **61**, 613 (1994)

Chapter 103
Search for Z' Resonances Decaying to Tau Pairs in pp Collisions Using CMS Detector at the LHC

Amandeep Kaur Kalsi, J. B. Singh and Vipin Bhatnagar

103.1 Introduction

The Standard Model (SM) of particle physics has been explored successfully by various experimental observations at the LHC. But still, it is not the complete theory of nature. Many models have been proposed to explain physics beyond the SM which predict the existence of new heavy gauge bosons W' and Z' by including an additional U(1) group to the SM gauge group [1]. The universality of couplings is not necessary for these new gauge bosons. There exist models that incorporate generation-dependent couplings which results in Z' bosons preferentially decaying to the third generational fermions. Such models motivate a search for Z' resonances that decay to a pair of τ leptons such as topcolor-assisted technicolor (TAT) models [2–4]. A widely used benchmark model in searches for Z' bosons is the sequential standard model (SSM), which predicts a neutral spin-1 Z' boson (Z'_{SSM}) with the same couplings to quarks and leptons as that of the SM Z boson. Z'_{SSM} decaying to a pair of electrons and muons have been excluded below 3.4 [5] and 3.2 [6] TeV respectively while the exclusion limit is 2.0 TeV in tau lepton pair final state [7–9]. In this paper, searches for Z' resonance decaying to tau pairs in SSM and TAT models have been presented on a data sample of pp collisions corresponding to an integrated luminosity of 2.2 fb^{-1} at $\sqrt{s}$ = 13 TeV recorded with the CMS detector [10] at the LHC.

Amandeep Kaur Kalsi, J. B. Singh, Vipin Bhatnagar—On behalf of CMS Collaboration.

A. K. Kalsi (✉) · J. B. Singh · V. Bhatnagar
Panjab University, Chandigarh, India
e-mail: amandeep.kaur.kalsi@cern.ch

Md. Naimuddin (ed.), *XXII DAE High Energy Physics Symposium*, Springer Proceedings in Physics 203, https://doi.org/10.1007/978-3-319-73171-1_103

103.2 Analysis Strategy

The analysis has been performed in four final states: $\tau_h\tau_h$, $\tau_e\tau_h$, $\tau_\mu\tau_h$ and $\tau_e\tau_\mu$ depending upon the hadronic or leptonic decay of tau leptons. The signal events consist of high p_T, well-reconstructed and isolated τ leptons. The muons (electrons) are required to have $p_T > 30$ (35) GeV and τ_h's are required to have $p_T > 20$ and 60 GeV in $\tau_l\tau_h$ (l = e, μ) and $\tau_h\tau_h$ final states. Both τ_l and τ_h candidates are required to have $|\eta| < 2.1$. The $\tau\tau$ pairs are required to be oppositely charged and should be back-to-back in ϕ *i.e.* $\cos\Delta\phi(\tau_1, \tau_2) < -0.95$. The events are required to have missing transverse energy $E_T^{miss} > 30$ GeV. The CDF-ζ variable [11] defined as $\zeta = p_\zeta - 3.1 \cdot p_\zeta^{vis.}$ is required to be greater than -50 GeV. Here, $p_\zeta^{vis.} = (\vec{p}_T^{\tau_1} + \vec{p}_T^{\tau_2}) \cdot \hat{\zeta}$ and $p_\zeta = (\vec{p}_T^{\tau_1} + \vec{p}_T^{\tau_2} + \vec{p}_T^{miss}) \cdot \hat{\zeta}$ represents two projection variables obtained using the visible decay products of tau leptons and $\vec{p}_T^{miss}$, and $\hat{\zeta}$ is a unit vector along the bisector between p_T directions of the two τ leptons. The contribution of events having b-tagged jets is rejected. The mass reconstructed from the energies and momenta of visible tau decay products and E_T^{miss}, defined as $m(\tau_1, \tau_2, \vec{p}_T^{miss}) = \sqrt{(E_{\tau_1} + E_{\tau_2} + E_T^{miss})^2 - (\vec{p}_{\tau_1} + \vec{p}_{\tau_2} + \vec{p}_T^{miss})^2}$ is used to distinguish the signal from background processes. Broad enhancement in $m(\tau_1, \tau_2, \vec{p}_T^{miss})$ distribution is an indication for the signal region, consistent with physics beyond the SM. The background contributions in the signal region has been estimated using data-driven control regions, wherever possible.

103.3 Results

The $m(\tau_1, \tau_2, \vec{p}_T^{miss})$ distribution in four final states is used as a discriminating variable. The observed mass spectra do not reveal evidence for new particles decaying to $\tau\tau$ pairs as shown in Fig. 103.1(left) [12]. Therefore, the upper limits have been placed at 95% confidence level (CL) on the product of the signal cross section and the branching fraction as a function of Z' mass using modified frequentist approach for a Z' decaying to $\tau\tau$ pairs. Figure 103.1(right) [12] represents the combined observed and expected limits including the theoretical predictions from SSM and TAT models. Z'_{SSM} and Z'_{TAT} have been excluded having masses less than 2.1 TeV (1.9 TeV expected) and 1.7 TeV (1.5 TeV expected) respectively, at 95% CL combining the four final states [12]. These exclusion limits on Z'_{SSM} at $\sqrt{s}$ = 13 TeV have already extended the previous limits [7–9] in tau pairs final state. The exclusion limit obtained on Z' resonances in TAT model is the most stringent limit to date.

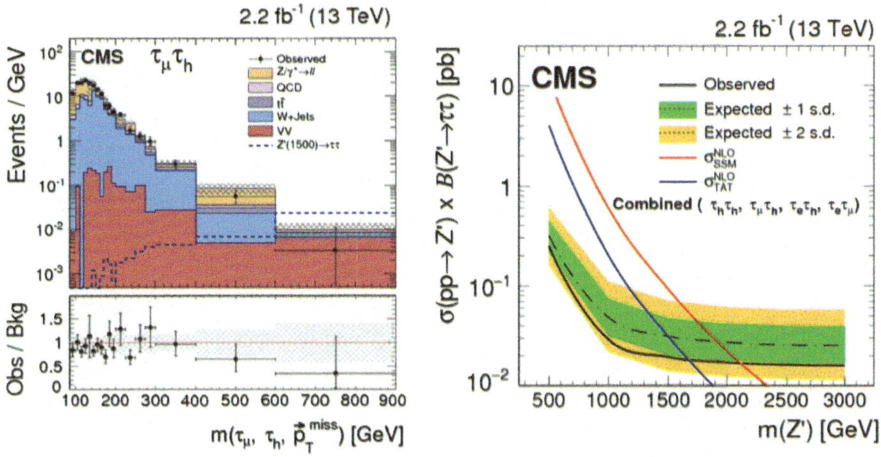

Fig. 103.1 Observed $m(\tau_1, \tau_2, \overrightarrow{p}_T^{miss})$ distribution in $\tau_\mu \tau_h$ final state (left). The dashed blue line shows the distribution expected for a Z'_{SSM} with mass 1.5 TeV. The lower panel represents the ratio of the observed events to the total predicted background. The shaded bands show the total uncertainty in the background prediction. The observed 95% CL upper limits on the product of the cross section and branching fraction into τ lepton pairs as a function of Z' mass, for the combined four channels are shown in right plot [12]. The expected and observed limits are shown with dashed and solid black lines. The red and blue curves represent the predictions of the NLO theory cross sections in SSM and TAT models, respectively

References

1. P. Langacker, The physics of heavy Z' gauge bosons. Rev. Mod. Phys. **81**, 1199–1228 (2009)
2. K.R. Lynch, S. Mrenna, M. Narain, E.H. Simmons, Finding Z' bosons coupled preferentially to the third family at CERN LEP and the Fermilab Tevatron. Phys. Rev. D **63**, 035006 (2001)
3. C.T. Hill, Topcolor assisted technicolor. Phys. Lett. B **345**, 483–489 (1995)
4. K.D. Lane, A new model of topcolor assisted technicolor. Phys. Lett. B **433**, 96–101 (1998)
5. ATLAS Collaboration, Search for high-mass new phenomena in the dilepton final state using proton-proton collisions at $\sqrt{s} = 13$ TeV with the ATLAS detector. Phys. Lett. B **761**, 372–392 (2016)
6. CMS Collaboration, Search for narrow resonances in dilepton mass spectra in proton-proton collisions at $\sqrt{s} = 13$ TeV and combination with 8 TeV data. Phys. Lett. B (2016) (Submitted)
7. CMS Collaboration, Search for high-mass resonances decaying into τ-lepton pairs in pp collisions at $\sqrt{s} = 7$ TeV. Phys. Lett. B **716**(1), 82–102 (2012)
8. ATLAS Collaboration, A search for high-mass resonances decaying to $\tau^+\tau^-$ in pp collisions at $\sqrt{s} = 7$ TeV with the ATLAS detector. Phys. Lett. B **719**, 242–260 (2013)
9. ATLAS Collaboration, A search for high-mass resonances decaying to $\tau^+\tau^-$ in pp collisions at $\sqrt{s} = 8$ TeV with the ATLAS detector. JHEP **07**, 157 (2015)
10. CMS Collaboration, The CMS experiment at the CERN LHC. JINST **3**, S08004 (2008)
11. CDF Collaboration, Search for neutral higgs bosons of the Minimal Supersymmetric Standard Model decaying to τ pairs in $p\overline{p}$ collisions at $\sqrt{s} = 1.96$TeV. Phys. Rev. Lett. **96**, 011802 (2006)
12. CMS Collaboration, Search for heavy resonances decaying to tau lepton pairs in proton-proton collisions at $\sqrt{s} = 13$ TeV. JHEP **1702**, 048 (2017). https://doi.org/10.1007/JHEP02(2017)048, arXiv:1611.06594 [hep-ex]

Chapter 104
Bounds on Neutrino Decay Lifetime with ICAL Detector

Chandan Gupta, Sandhya Choubey, Srubabati Goswami, S. M. Lakshmi and Tarak Thakore

104.1 Introduction

Neutrino oscillation experiments have confirmed neutrinos are massive. However in literature the possibility of the mass states being unstable have been discussed [1, 2]. India based Neutrino Observatory (INO) is a proposed experiment with a 50 kton magnetized Iron CALorimeter (ICAL) for observing atmospheric neutrinos. Besides having excellent potential to settle the mass hierarchy problem and precision studies, it is also helpful in studying non-standard physics [3]. Here we have investigated the capability of the ICAL detector for probing neutrino decay along with oscillations. We assume normal hierarchy (NH) to be the true hierarchy in nature and the highest neutrino mass eigenstate (ν_3) to decay invisibly [4].

In our analysis, we have studied the charge current (CC) interactions of ν_μ and $\bar{\nu}_\mu$ with the target material iron in ICAL. As atmospheric neutrino flux comprises of both $\nu_e(\bar{\nu}_e)$ and $\nu_\mu(\bar{\nu}_\mu)$, the final signal will get contribution from both $\nu_e \rightarrow \nu_\mu$ and $\nu_\mu \rightarrow \nu_\mu$. Figure 104.1 shows how these oscillation probabilities get modified in the presence of decay.

C. Gupta (✉) · S. Goswami · S. M. Lakshmi
Physical Research Laboratory, Ahmedabad 380009, India
e-mail: ph10c009@gmail.com

C. Gupta · S. Choubey
Homi Bhabha National Institute, Mumbai 400094, India

S. Choubey
Harish-Chandra Research Institute, Allahabad 211019, India

T. Thakore
Louisiana State University, Baton Rouge, LA 70803, USA

Md. Naimuddin (ed.), *XXII DAE High Energy Physics Symposium*, Springer Proceedings in Physics 203, https://doi.org/10.1007/978-3-319-73171-1_104

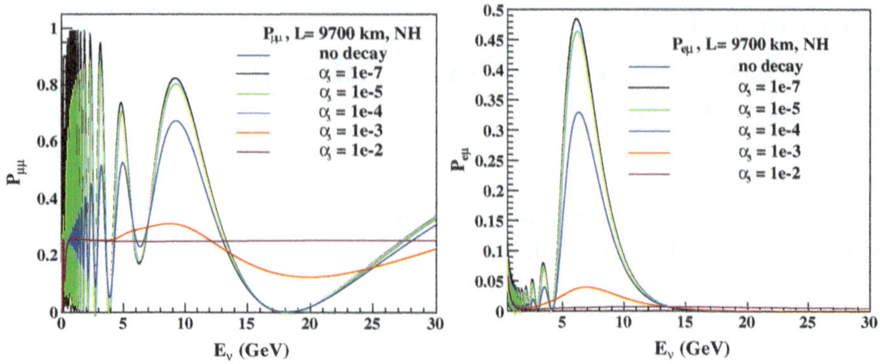

Fig. 104.1 $P_{\mu\mu}$ (Left) and $P_{e\mu}$ (Right) in matter with different values of decay parameter α_3 (eV^2) for the baseline L = 9700 km with NH as the true hierarchy. [$\alpha_3 = 0$ corresponds no decay]

104.2 Results

The α_3 sensitivity of ICAL with 500 kton year exposure has been studied assuming normal hierarchy (NH) to be the true hierarchy. Data is simulated for oscillation only, which is fitted with a theory having both oscillation and decay. The analysis procedure is discussed in [5, 6]. Figure 104.2 shows that CC interaction analysis of ICAL puts an upper bound on $\alpha_3 < 4.36 \times 10^{-6}\,eV^2$ at 90% CL (marginalized case), which is two orders of magnitude improvement over MINOS (charge + neutral current analysis) bound [4].

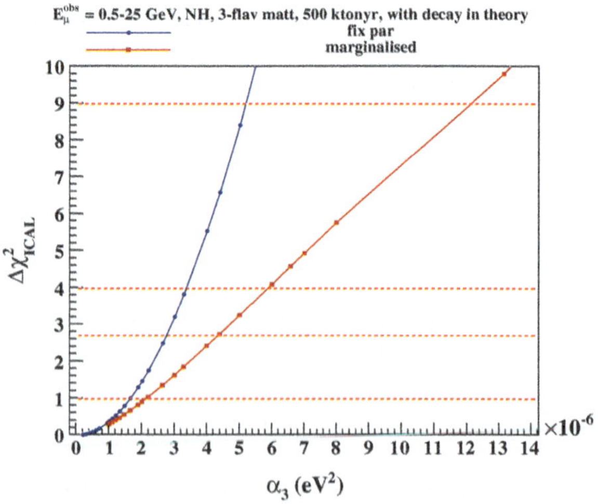

Fig. 104.2 Bound on the allowed value of α_3 with 500 kton year exposure of ICAL with or without marginalization over parameters (summarized in Table 104.1)

Table 104.1 True parameter values and their marginalized ranges

Parameter	True value	Marginalization range
θ_{13}	8.5°	[7.80°, 9.11°]
$\sin^2\theta_{23}$	0.5	[0.39, 0.64]
Δm^2_{32}	2.3663×10^{-3} eV^2	[2.3, 2.6]$\times 10^{-3}$ eV^2 (NH)
$\sin^2_{12}$	0.304	Not marginalised
Δm^2_{21}	7.6×10^{-5} eV^2	Not marginalised
δ_{CP}	0°	Not marginalised

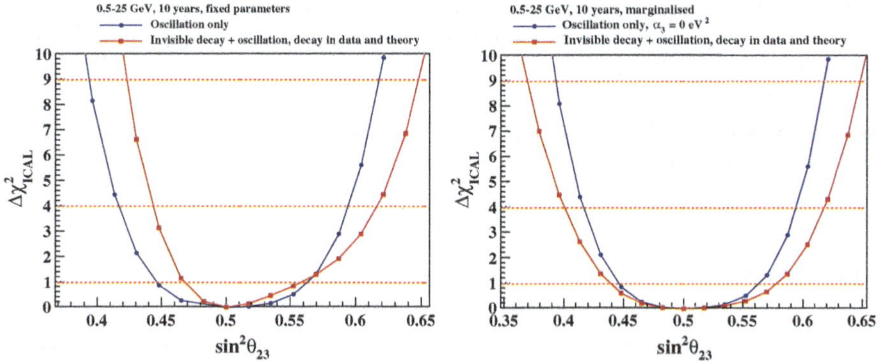

Fig. 104.3 Precision on $\sin^2\theta_{23}$ in the presence and absence of invisible decay for fixed parameter case (Left) marginalized case (Right)

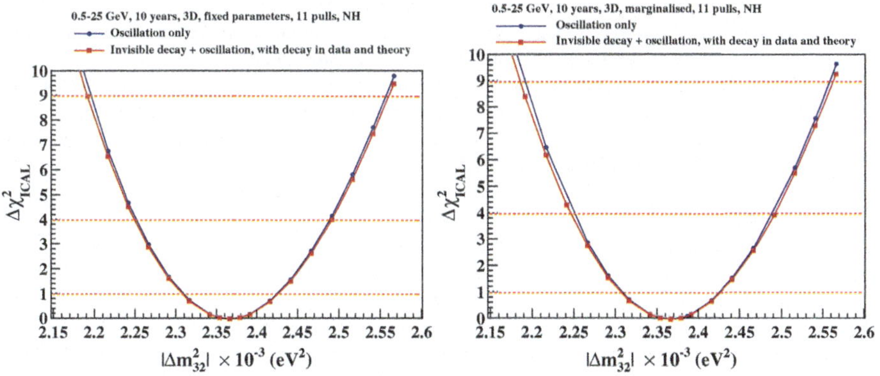

Fig. 104.4 Precision on $|\Delta m^2_{32}|$ in the presence and absence of invisible decay for fixed parameter case (Left) marginalized case (Right)

Next, we have studied the effect of invisible decay on precision measurements where both data and theory assume neutrino decay with oscillations. $\alpha_3 = 1 \times 10^{-5}\,eV^2$ is taken for fixed parameter study while it is varied over $[0, 2.35 \times 10^{-4}]\,eV^2$ for marginalized case.

The effect on the precision measurement of $\sin^2\theta_{23}$ in the presence of decay is shown in Fig. 104.3. It is evident that for fixed parameters, the relative 1σ precision on $\sin^2\theta_{23}$ in presence of decay ($\sim$8.9%) is almost similar to only oscillation case ($\sim$8.6%) but the overall shift happens towards higher octant as in [4] for MINOS. While marginalization worsens the sensitivity of α_3 ($\sim$10.85%) as compared to only oscillation case ($\sim$8.87%). On the other hand Fig. 104.4 shows that neutrino decay does not affect the precision measurement of $|\Delta m^2_{32}|$. It is due to the fact that the presence of decay modifies the amplitude of oscillation and not the frequency which is governed by $|\Delta m^2_{32}|$.

Acknowledgements LSM acknowledges Prof. D. Indumathi IMSc, Chennai for the analysis code and the computing facility of IMSc Chennai.

References

1. V. Barger, J.G. Learned, S. Pakvasa, T.J. Weiler, Neutrino decay as an explanation of atmospheric neutrino observations. Phys. Rev. Lett. **82**, 2640 (1999)
2. S. Choubey, S. Goswami, Is neutrino decay really ruled out as a solution to the atmospheric neutrino problem from Super-Kamiokande data? Astropart. Phys. **14**, 67–78 (2000)
3. S. Ahmed et al., Physics Potential of the ICAL detector at the India-based Neutrino Observatory (INO), *INO/ICAL/PHY/NOTE/2015-01* (2015), arXiv:1505.07380
4. R.A. Gomes, A.L.G. Gomes, O.L.G. Peres, Constraints on neutrino decay lifetime using long-baseline charged and neutral current data. Phys. Lett. B **740**, 345 (2015)
5. L.S. Mohan, D. Indumathi, Pinning down neutrino oscillation parameters in the 2–3 sector with a magnetised atmospheric neutrino detector: a new study. Eur. Phys. J. C **77**(54) (2017)
6. M.M. Devi et al., Enhancing sensitivity to neutrino parameters at INO combining muon and hadron information. JHEP **10**, 189 (2014)

Chapter 105
Measurement of Azimuthal Correlations Between D Mesons and Any Charged Particle in pp Collisions at $\sqrt{s} = 7$ TeV with ALICE

Sonia Rajput

105.1 Introduction

The goal of the ALICE experiment at the LHC is to study nuclear matter at extreme conditions of high temperature and high density at which quarks are no more confined into nucleons, giving rise to a new state of matter known as Quark Gluon Plasma (QGP) [1]. Due to their large masses, heavy quarks (charm and beauty), are produced primarily in the early stages of heavy-ion collision, in hard partonic scattering processes, and experience the full evolution of the system propagating through the medium produced in such collisions. Therefore, they are an effective probe to study the medium and provide information on its evolution. The analysis of angular correlations between heavy-flavour particles and charged particles serves as a tool to characterize the heavy quark fragmentation process and its possible modification in the medium, formed in heavy-ion collision. To this purpose, the measurement in pp collisions is necessary to obtain a reference for the studies in heavy-ion collisions. The measurements in p-Pb collisions and their comparison with the pp measurements can give insight on how heavy-quark production and hadronization process is affected by the cold nuclear matter effects in p-Pb collisions.

The ALICE apparatus has excellent capabilities for heavy flavour measurements. A detailed description of the ALICE detector and its performance can be found in [2]. D mesons and their charge conjugates are reconstructed in hadronic decay channels and selected exploiting the typical displaced-like topology of the decay vertices, PID (particle identification) and reconstruction quality cuts on the daughter tracks [3].

Sonia Rajput, for the ALICE Collaboration.

S. Rajput (✉)
University of Jammu, Jammu, India
e-mail: sonia.rajput@cern.ch

Md. Naimuddin (ed.), *XXII DAE High Energy Physics Symposium*, Springer Proceedings in Physics 203, https://doi.org/10.1007/978-3-319-73171-1_105

105.2 D Meson-Charged Particle Angular Correlation

Two-dimensional angular correlations of D mesons (D^0, D^+, D^{*+}) with charged particles are evaluated in pp collisions at $\sqrt{s} = 7$ TeV for different ranges of D-meson and associated charged particle p_T. The contribution of background candidates is removed using the correlation distribution from the candidates in the sidebands of the D-meson invariant mass distribution. To account for the limited detector acceptance and detector spatial inhomogeneities, a correction factor is applied using the event mixing technique.

Corrections for reconstruction and selection efficiency of D mesons and charged particles are also applied. PYTHIA simulations [4] were used to produce template distributions of angular correlations between D mesons from B-meson decays and charged particles. These distributions, normalized to the expected amount of feed-down contribution evaluated from FONLL calculations [5] and considering the reconstruction efficiency of feed-down D mesons, are subtracted from the inclusive correlation distributions obtained from data. The fully corrected 2D correlations were

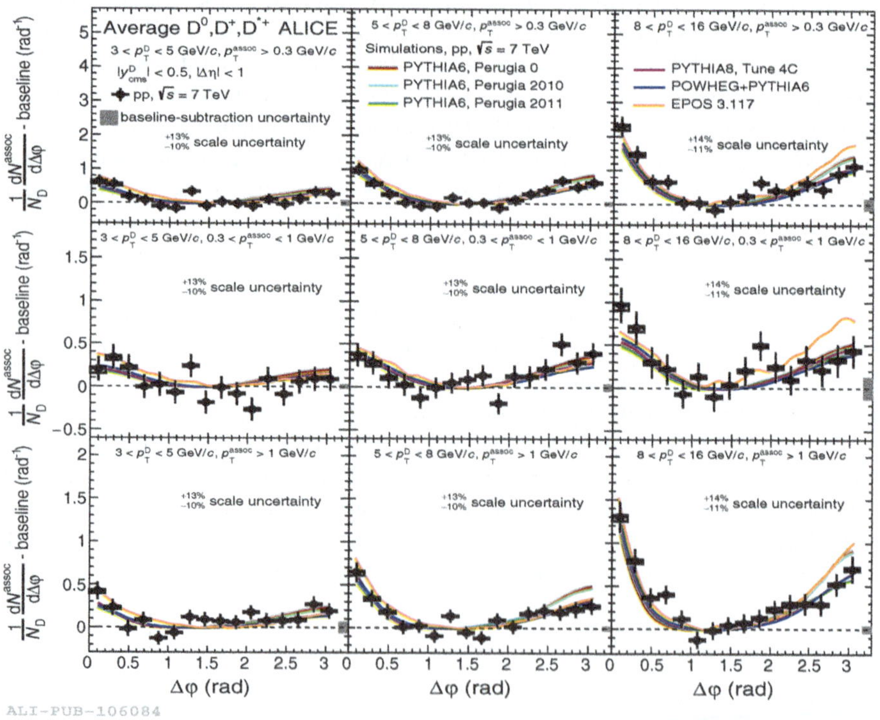

Fig. 105.1 Comparison of the baseline-subtracted D meson-charged particle azimuthal correlation distributions in pp collisions and predictions from different models for different ranges of D-meson and the associated charged particle p_T [6]

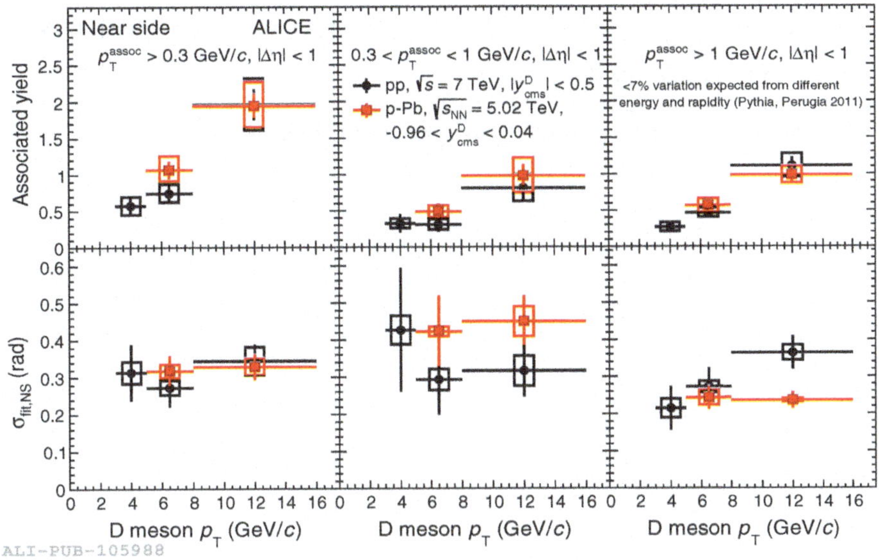

Fig. 105.2 Comparison of near-side associated yields (top) and near-side widths (bottom) extracted in pp and p-Pb collisions as a function of D-meson p_T [6]. The near-side yields are extracted by integrating the near-side Gaussian component of the fit function

projected onto the $\Delta\phi$ axis, producing azimuthal correlation distributions normalized to the number of trigger D mesons. A weighted average of the results for the three D-meson species was then performed to reduce the statistical uncertainty. A fit with a function composed of a constant term and two Gaussian functions modelling the near-side ($\Delta\phi \sim 0$) and away-side correlation peaks was done to estimate the near-side peak associated yield and peak width and the height of the baseline.

The azimuthal correlation distributions in pp collisions are described well within uncertainties by expectations of various event generators (PYTHIA [4], POWHEG [7] and EPOS [8]) in all kinematic ranges, as shown in Fig. 105.1. Figure 105.2 shows the compatibility of the near-side yields measured in pp collisions at $\sqrt{s} = 7$ TeV and p-Pb collisions at $\sqrt{s_{NN}} = 5.02$ TeV indicating that within the current uncertainties, no significant effect of cold nuclear matter effects in p-Pb collisions is observed from the data.

References

1. S. Sarkar et al., *The Physics of the Quark-Gluon Plasma* (Springer, Berlin, 2010)
2. ALICE Collaboration, Int. J. Mod. Phys. A **29**, 1430044 (2014)
3. ALICE Collaboration, JHEP **01**, 128 (2012)
4. T. Sjostrand, S. Mrenna, P.Z. Skands, JHEP **05**, 26 (2006)

5. M. Cacciari, S. Frixione, N. Houdaeu, M.L. Mangano, P. Nason, G. Ridolfi, JHEP **1210**, 137 (2012)
6. ALICE Collaboration, arXiv:1605.06963
7. S. Frixione, P. Nason, C. Oleari, JHEP **11**, 070 (2007)
8. K. Werner, B. Guiot, I. Karpenko, T. Pierog, Phys. Rev. C **89**(6), 064903 (2014)

Chapter 106
Gas Proportion Studies on the Operation of Resistive Plate Chambers

K. Raveendrababu and P. K. Behera

106.1 Introduction

The India-based Neutrino Observatory (INO) has proposed to build a 50 kton magnetized Iron Calorimeter (ICAL) to study neutrino oscillations. Glass resistive plate chambers (RPCs) of $2 \times 2\,m^2$ size are going to be the active elements for the ICAL detector. The gas mixture of freon ($C_2H_2F_4$), iso-butane (C_4H_{10}) and SF_6 will be used to operate the RPCs in the avalanche mode [1]. Systematic characterization studies were undertaken on the performance of RPCs for various gas proportions and the results are summarized in this paper.

106.2 Experimental Arrangement

RPCs of $30 \times 30\,cm^2$ size were built using 3 mm thick float glass plates from Saint-Gobain glass manufacturer. A cosmic ray muon telescope was set up with three plastic scintillator counters. The dimensions of scintillator counters in length $\times$ width $\times$ thickness are $30 \times 2 \times 1\,cm^3$ (top), $30 \times 3 \times 1\,cm^3$ (middle), and $30 \times 5 \times 1\,cm^3$ (bottom). The RPCs were stacked between top and middle scintillator counters. The telescope window that is defined by a 2 cm wide finger paddle was centered on the 2.8 cm wide central strip of the RPC. The detailed experimental arrangement of RPCs, scintillator counters and electronic circuit are shown in Fig. 106.1.

K. Raveendrababu (✉) · P. K. Behera
Physics Department, Indian Institute of Technology Madras, Chennai 600036, Tamil Nadu, India
e-mail: ravi2ramana@gmail.com

K. Raveendrababu
Physical Science Division, Homi Bhabha National Institute, Anushaktinagar, Mumbai 400094, India

Md. Naimuddin (ed.), *XXII DAE High Energy Physics Symposium*, Springer Proceedings in Physics 203, https://doi.org/10.1007/978-3-319-73171-1_106

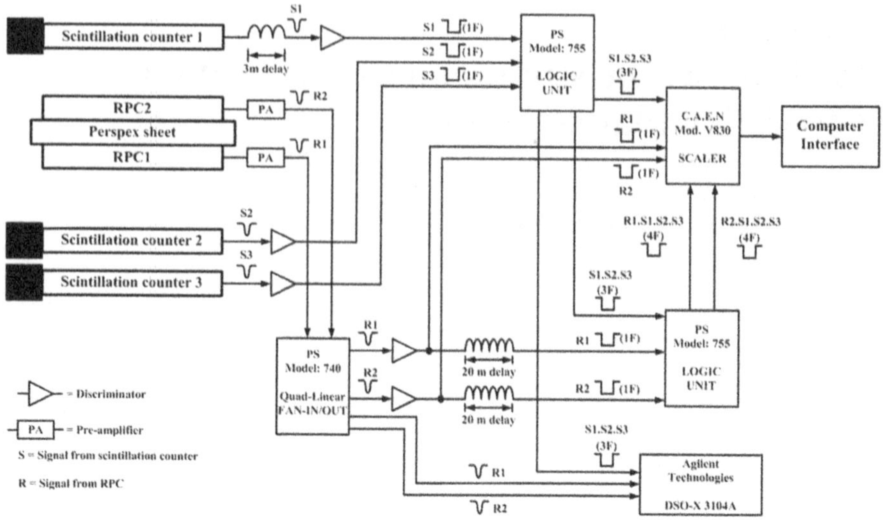

Fig. 106.1 Schematic of the experimental arrangement

106.3 Test Results

Using the experimental arrangement shown in Fig. 106.1, the following gas proportion studies were performed on the RPCs.

106.3.1 Efficiencies of the RPCs

A gas mixture of $C_2H_2F_4/C_4H_{10}/SF_6 = 95.2/4.5/0.3$ [2] was flown through the RPCs (RPC1 and RPC2) with a total flow rate of 10 SCCM and operated in the avalanche mode. The efficiencies of RPCs were measured using the cosmic-ray muons. The knee of efficiency plateau of RPCs starts at 11.8 kV and showed greater than 95% efficiency [3].

106.3.2 Efficiency of RPC as a Function of Iso-C_4H_{10}

The fraction of SF_6 was kept at 0.3% and various fractions of C_4H_{10} was added to the RPC1 gas mixture. The RPC1 was operated at 12.2 kV and measured its efficiency for various fractions of C_4H_{10}. These results are shown in Fig. 106.2. There was no observation of significant variation in the RPC1 efficiency.

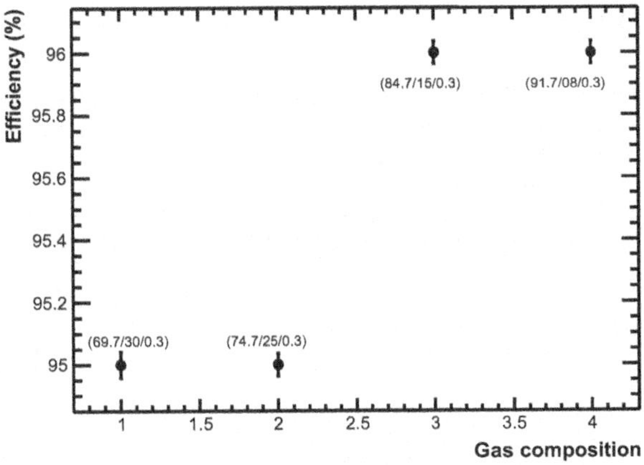

Fig. 106.2 Efficiency of RPC1 at 12.2 kV as a function of gas composition. The gas mixture is in the proportions of $C_2H_2F_4/C_4H_{10}/SF_6$. Here, the fraction of SF_6 was kept constant at 0.3%

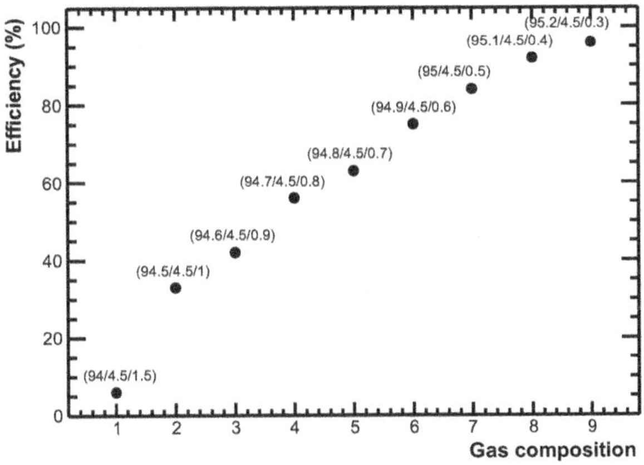

Fig. 106.3 Efficiency of RPC1 at 12.2 kV as a function of gas composition. The gas mixture is in the proportions of $C_2H_2F_4/C_4H_{10}/SF_6$. Here, the fraction of C_4H_{10} was kept constant at 5.4%

106.3.3 Efficiency of RPC as a Function of SF_6

The fraction of C_4H_{10} was kept at 5.4% and that of SF_6 was increased gradually from 0.3 to 1.5%. It was observed a gradual deterioration in the RPC1 efficiency with the increase of SF_6 fraction as shown in Fig. 106.3. It is due to the large electron affinity of SF_6.

106.4 Conclusions

RPCs of 30 × 30 cm^2 size were built using Saint-Gobain glass plates and operated them in the avalanche mode. They showed greater than 95% efficiency on the plateau region. Larger fractions of iso-butane have not shown significant effect on the efficiency of RPC. Larger fractions of SF_6 severely deteriorated the RPC efficiency.

References

1. ICAL collaboration, S. Ahmed et al., Physics potential of the ICAL detector at the India-based Neutrino Observatory (INO), arXiv:1505.07380
2. M. Salim et al., Experimental and numerical studies on the effect of SF_6 in a glass RPC. JINST **7**, P11019 (2012)
3. K. Raveendrababu et al., Effect of electrical properties of glass electrodes on the performance of RPC detectors for the INO-ICAL experiment. JINST **11**, P08024 (2016)

Chapter 107
Variation of Fine Structure Constant and Electron Properties

S. Ghosh

107.1 Introduction

In the first half of the twentieth century Milne [1] proposed the possibility of time variation of the fundamental physical constants. Following that the variation of fundamental physical constants was doubted and predicted by Dirac [2]. When Milne and Dirac were actually thinking of the time variation of G, then Teller [3] considered the fine structure constant to be varying in time. Gamow in 1967 [4] recommended fine structure constant as varying one whereas G would behave as constant. Then in 1972 Dyson [5] proposed that the variation can be introspected from the renormalization of quantum electrodynamics.

With some cosmological consequences the theory of time varying fine structure constant was first introduced by Bekenstein [6]. This provides even a generalization of the Maxwell's equations. But the effects of the gravitational field equations were overlooked. Sandvik, Magueijo and Barrow presented a modified Bekenstein model with the inclusion of the the gravitational fields [7]. This modified model is better known as BBSM (Bekenstein–Barrow–Sandvik–Magueijo) model [8]. Webb et al. [9] continued to fine-tune it.

Additional dimensions of space are necessary for unified theorem. Moving forward towards additional dimensions may lead to the change of the physical constants [7, 10]. From recent observation of quasar spectra at medium red shift the first evidence came out which hints about probable change in fine structure constant with cosmological time, which introduce α as varying (growing) constant with $\frac{\Delta\alpha}{\alpha} = -0.72 \pm 0.18 \times 10^{-5}$ for $z \approx 0.5 - 3.5$ [8]. At the present scenario it is very essential to know the exact reasons of this variation of α. As well as we need to observe the effect of this variation on different particle properties and the calculations if at all.

S. Ghosh (✉)
PG Department of Physics, Vijaya College, R. V. Road, Bangalore 560004, India
e-mail: gsovan@gmail.com

Md. Naimuddin (ed.), *XXII DAE High Energy Physics Symposium*, Springer Proceedings in Physics 203, https://doi.org/10.1007/978-3-319-73171-1_107

107.2 Time Varying Charge and the Speed of Light

Fine structure constant can be written using the charge, the speed of light and Planck's constant as $\alpha = \frac{e^2}{\hbar c}$, where e, $\hbar$ and c are known as constants. But the variation of the fine structure raises questions about the constancy of the charge and the speed of light. The variation of the charge is assumed by Bekenstein [6]. The variation of the speed of light is hinted by Albrecht and Magueijo [11, 12].

According to Bekenstein [6], the charge of electron is composed of two integral parts, out of which one plays the role of constant and the other is a variable and that is a function of space and time both. Hence the charge of any particle can be expressed as $e = e_0\varepsilon(x^\nu)$, where e_0 is constant and $\varepsilon(x^\nu)$ is the variable [6].

Magnetic moment of electron with first order correction (Schwinger correction) [13] can be written with Bekenstein's proposal as $\mu = \frac{e_0\varepsilon\hbar}{2mc}\left(1 + \frac{\alpha}{2\pi}\right)$. Hence we are left with the conditions of variations of the fine structure constant either along with variation of speed of light or without the variation of speed of light. Without considering the variation of speed of light time variation of α comes out to be

$$\frac{\dot{\alpha}}{\alpha} = 2c\frac{\dot{\varepsilon}}{\varepsilon}. \tag{107.1}$$

Corresponding variation in magnetic moment is then expressed as

$$\frac{\dot{\mu}}{\mu} = \frac{\dot{\varepsilon}}{\varepsilon} + \frac{\dot{\alpha}}{2\pi\left(1 + \frac{\alpha}{2\pi}\right)} \quad . \tag{107.2}$$

Considering varying speed of light, the time variation of the fine structure constant becomes

$$\frac{\dot{\alpha}}{\alpha} = 2c\frac{\dot{\varepsilon}}{\varepsilon} - \dot{c}. \tag{107.3}$$

This leads to the time variation in the of the magnetic moment of the electron as

$$\frac{\dot{\mu}}{\mu} = \frac{\dot{\varepsilon}}{\varepsilon} - \varepsilon\frac{\dot{c}}{c} + \frac{\dot{\alpha}}{2\pi}\left(\frac{1}{1 + \frac{\alpha}{2\pi}}\right). \tag{107.4}$$

Equations (107.1) and (107.3) are related with the time variation of α whereas (107.2) and (107.4) are related with time variation of magnetic moment of the electron. Equation (107.3) varies from (107.1) only due to the time variation of the speed of light.

Magnetic moment of the electron with recent experimental and theoretical corrections is expressed in different orders of fine structure constant [14–17]. If we introduce $\alpha = e^{yt}$, $\varepsilon = e^{xt}$, $c = e^{wt}$ and $\mu = e^{ut}$ in the above four equations, then the parameters α, ε, c, μ can be expanded in higher orders. Hence magnetic moment of the electron can be expressed using higher order corrections of fine structure (as a function of time). Thus the introduction of the new parameters can assist us to match our calculation with the updated expression of μ. Thus the temporal variation of fine structure can be correlated with the corrected expression of the magnetic moment of electron, which is the basic goal of this work.

107.3 Conclusion

The variation of the magnetic moment of the electron caused by the variation of the fine structure constant is expressed here. As magnetic moment can provide us different valuable information about the particle, we expect to have better realization about particle properties by above equations.

Acknowledgements The author is supported with partial financial assistance from Vijaya College, R. V. Road, Bangalore. For some useful technical support author conveys the sincere gratitude to Srinath N, M Y Anand and Nagaraja B S. Author is grateful to his wife Moumita Ghosh for an untiring inspiration and support without which this work was not possible.

References

1. E.A. Milne, Proc. R. Soc. A **158**, 324 (1937)
2. P.A.M. Dirac, Nat. Lett. **139**, 323 (1937)
3. E. Teller, Phys. Rev. **73**, 801 (1948)
4. G. Gamow, Phys. Rev. Lett. **19**, 759 (1967)
5. F.J. Dyson, Phys. Rev. Lett. **19**, 1291 (1967)
6. J.D. Bekenstein, Phys. Rev. D **1527**, 25 (1982)
7. J.D. Barrow, Ann. Phys. (Berlin) **19**, 202 (2010)
8. H.B. Sandvik, J.D. Barrow, J. Magueijo, Phys. Rev. Lett. **88**, 031302 (2002)
9. J.K. Webb et al., Phys. Rev. Lett. **884**, 82 (1999)
10. V.A. Dzuba, V.V. Flambaum, J.K. Webb, Phys. Rev. Lett. **82**, 888 (1999)
11. A. Albrecht, J. Magueijo, Phys. Rev. D **59**, 043516 (1999)
12. J.D. Barrow, Phys. Rev. D **59**, 043515 (1999)
13. J. Schwinger, Phys. Rev. **73**, 416 (1948)
14. D. Hanneke, S. Fogwell, G. Gabrielse, Phys. Rev. Lett. **100**, 120801 (2008)
15. B. Odom, D. Hanneke, B. D'Usro, G. Gabrielse, Phys. Rev. Lett. **97**, 030801 (2006)
16. A. Czarnecki, Nature **442**, 516 (2006)
17. S. Ghosh, A. Choudhury, J.K. Sarma, Apeiron **19**, 247 (2012)

Chapter 108
Intersecting D3-Branes at Finite Temperature

Varun Sethi, Sudipto Paul Chowdhury and Swarnendu Sarkar

108.1 Introduction

In [1], a holographic model for strong coupling BCS superconductivity is proposed. It is based on Sakai–Sugimoto model [2] in the sense that there are two probe D8-branes in the background of N_c D4 branes (where N_c is number of colour degrees of freedom and is large). It is shown that at finite baryon density the two D8-branes separate with an opening angle in the bulk. Such a configuration is known to have tachyons in the spectrum [3]. This instability is proposed to be dual of the BCS instability in superconductors. For a more detailed overview of this model, see [5]. Here, we consider a toy model with intersecting D3 branes in flat background. This is the continuation of the work initiated in [4] where two intersecting D1 branes are considered. The action for this configuration, in the Yang–Mills approximation, is obtained by dimensionally reducing 10-dimensional $\mathcal{N} = 1$ supersymmetric Yang–Mills theory to 4-dimensions. The resulting $\mathcal{N} = 4$ SYM theory has the field content (bosons:) Φ_I, $\tilde{\Phi}_I$ (scalars), $I = 1, 2, 3$, A_μ (gauge bosons), $\mu = 0, \ldots, 3$, and (fermions:) λ_i, $i = 1, \ldots, 4$. (Gauge indices are suppressed throughout.) The thermal stability of this configuration is analysed by computing one-loop correction to 2-point tachyon amplitude at finite temperature in Yang–Mills approximation. Here, the temperature is introduced by compactification of the Euclidean time.

V. Sethi (✉) · S. Sarkar
Department of Physics and Astrophysics, University of Delhi, Delhi 110007, India
e-mail: vsethi@physics.du.ac.in

S. Sarkar
e-mail: ssarkar@physics.du.ac.in

S. P. Chowdhury
Institute of Physics, Sachivalaya Marg, Bhubaneswar 751005, India
e-mail: sudipto@iopb.res.in

Md. Naimuddin (ed.), *XXII DAE High Energy Physics Symposium*, Springer Proceedings in Physics 203, https://doi.org/10.1007/978-3-319-73171-1_108

108.2 Spectrum

In the framework of SU(2) Yang–Mills theory, the quadratic part of the bosonic action can be written compactly as

$$S_b = \int d^4z \left[\frac{1}{2}\xi^T \mathcal{O}_B \xi + \frac{1}{2}\xi'^T \mathcal{O}'_B \xi' + \mathcal{L}(A^3_\mu, \Phi_I, \tilde{\Phi}_J)\right] \tag{108.1}$$

where

$$\xi = \begin{pmatrix} \Phi^1_1 \\ A^2_1 \\ A^2_2 \\ A^2_3 \end{pmatrix} ; \zeta = \begin{pmatrix} \Phi^1_1 \\ A^2_1 \end{pmatrix} \quad ; \quad \xi' = \begin{pmatrix} \Phi^2_1 \\ A^1_1 \\ A^1_2 \\ A^1_3 \end{pmatrix} ; \zeta' = \begin{pmatrix} \Phi^2_1 \\ A^1_1 \end{pmatrix} \tag{108.2}$$

and $\mathcal{O}_B$ is the matrix of the differential operators acting on the above multiplets. Further, $A^a_0 = 0$ is the temporal gauge choice and the intersecting D-branes configuration is captured by fixing the background value of Φ^3_1 equal to qx. The eigenfunctions of the operator $\mathcal{O}_B$ can be written in terms of the Hermite polynomials. With these as the basis, the doublet ζ is expanded with $C(m, n, \mathbf{k})$ & $\tilde{A}^2_1$ as the Fourier coefficients. The latter can be clubbed along with the Fourier coefficients $\tilde{A}^2_{2,3}$ in expansion of $A^2_{2,3}$, while former is tachyonic for $n = 0$ and for all values of m and $\mathbf{k}$. This can also be seen from the eigenvalues of $\mathcal{O}_B$, which are $(2n-1)q$, n being a whole number. Similar analysis holds for ζ', which is expanded with the Fourier coefficients $C'(m, n, \mathbf{k})$ & $\tilde{A}^1_1$. The last term of (108.1) gives decoupled eigenvalue equations for scalar fields with eigenvalues $(2n+1)q$. The fields $\Phi^{1,2}_I$ and $\tilde{\Phi}^{1,2}_I$ $(I = 2, 3)$, are expanded in the basis of harmonic oscillator wavefunctions with the Fourier coefficients $\Phi^{1,2}_{2,3}(m, n, \mathbf{k})$. Lastly, all the fields with gauge index 3 can be Fourier expanded as plane waves as these do not couple with the background.

The quadratic part of the fermionic terms in the action is

$$\mathcal{L}' = -\frac{i}{2}\left[\bar{\chi}_1 \mathcal{O}_F \chi_1 + \bar{\chi}_2 \mathcal{O}_F \chi_2 + \bar{\chi}_3 \tilde{\mathcal{O}}_F \chi_3 + \bar{\chi}_4 \tilde{\mathcal{O}}_F \chi_4 + \bar{\lambda}^3_k \gamma^\mu \partial_\mu \lambda^3_k\right] \tag{108.3}$$

where

$$\chi_1 = \begin{pmatrix} \lambda^1_1 \\ \lambda^2_4 \end{pmatrix} \quad \chi_2 = \begin{pmatrix} \lambda^1_2 \\ \lambda^2_3 \end{pmatrix} \quad \chi_3 = \begin{pmatrix} \lambda^1_3 \\ \lambda^2_2 \end{pmatrix} \quad \chi_4 = \begin{pmatrix} \lambda^1_4 \\ \lambda^2_1 \end{pmatrix}. \tag{108.4}$$

With eigenfunctions of $\mathcal{O}_F$ and $\tilde{\mathcal{O}}_F$ as basis, the Fourier expansion of χ_i is written in terms of the fields $\theta_i(\omega, n, \mathbf{k})$ & $\theta^*_i(\omega, n, \mathbf{k})$, with the corresponding eigenvalues being $-i\sqrt{2nq}$. In addition, there are $\lambda^3_l(\omega, k_x, \mathbf{k})$ fields in the Fourier expansion of λ^3_l.

Using the expansions discussed above, actions (108.1) and (108.3) can be rewritten in terms of the Fourier coefficients. These can be used to read off the propagators (refer Appendix C in [6]).

108.3 One-Loop Two-Point Tachyon Amplitude

As seen in Sect. 108.1, there are tachyonic modes in the bosonic tree-level spectrum. Next, we would like to compute one-loop correction to the tachyon propagator at finite temperature. For this purpose, we need Feynman diagrams with tachyons on external legs. Using the interaction terms in the action and the mode expansions for various fields, all required vertices can be read off. These are listed in appendix C in [6]. The Quantum corrections to the tachyon propagator have contributions coming from both bosons and fermions in the loop. Figure 108.1 illustrates both type of contributions.

Infrared divergence. One-loop corrections to tachyon amplitude also have massless modes in the loop. These modes are the fields with the gauge index $a = 3$, namely $\Phi_J^3/\tilde{\Phi}_J^3 (J = 1, 2, 3)$ and $A_i^3 (i = 1, 2, 3)$. The reason is that the spectrum of these fields is not gapped since these do not couple with the background. To handle IR divergence, we first find the one-loop correction to the propagators for these fields (requisite vertices are shown in appendix E in [6]) and then use the corrected propagators to evaluate the one-loop two-point tachyon amplitude.

Ultraviolet behaviour. Though supersymmetry is broken by the background, the supersymmetry breaking scale is finite. Consequently, it is expected that the UV behaviour remains same as the underlying $\mathcal{N} = 4$ SYM theory. This has been shown analytically in the large n limit, where the amplitudes with bosons and fermions in the loop are equal in magnitude and opposite in sign.

Since the amplitude is finite in both UV and IR, we can compute the temperature at which the effective mass-squared of tachyon becomes zero. This critical value of temperature, at which tachyon ceases to exist, comes out to be $T_c/\sqrt{q} = 0.29$ [6]. This behaviour signals a phase transition.

Fig. 108.1 Feynman diagrams having 4 and 3-point vertices

Acknowledgements V. S. acknowledges CSIR, India, for support through SRF grant 09/045(1355)/2014-EMR-I.

References

1. S. Kalyana Rama, S. Sarkar, B. Sathiapalan, N. Sircar, Strong coupling BCS superconductivity and holography. Nucl. Phys. B **852**, 634 (2011). https://doi.org/10.1016/j.nuclphysb.2011.07.011, arXiv:1104.2843 [hep-th]
2. T. Sakai, S. Sugimoto, Low energy hadron physics in holographic QCD. Prog. Theor. Phys. **113**, 843 (2005). https://doi.org/10.1143/PTP.113.843, arXiv:hep-th/0412141
3. K. Hashimoto, S. Nagaoka, Recombination of intersecting D-branes by local tachyon condensation. JHEP **0306**, 034 (2003). https://doi.org/10.1088/1126-6708/2003/06/034, arXiv:hep-th/0303204
4. S.P. Chowdhury, S. Sarkar, B. Sathiapalan, BCS instability and finite temperature corrections to Tachyon Mass in intersecting D1-branes. JHEP **1409**, 063 (2014). https://doi.org/10.1007/JHEP09(2014)063, arXiv:1403.0389 [hep-th]
5. S. Sarkar, S.P. Chowdhury, S.K. Rama, B. Sathiapalan, N. Sircar, A Holographic description of BCS instability at finite baryon density. Proceedings report in this volume (Chapter 143)
6. V. Sethi, S.P. Chowdhury, S. Sarkar, Finite temperature corrections to Tachyon Mass in intersecting D-branes. JHEP **1704**, 109 (2017). https://doi.org/10.1007/JHEP04(2017)109, arXiv:1610.07140 [hep-th]

Chapter 109
Jet Identification with Zest

Ankita Budhraja and Ambar Jain

109.1 Introduction

The standard model heavy particles, namely, W and Z bosons, Higgs boson and the top quark, dominantly decay to light quarks and gluons which in turn shower into jets of hadrons. Since jets are invariably produced at LHC, heavy SM particles decaying to jets are often faked by jets from light quarks and gluons (collectively parton jets). One might expect that the jet mass may cut out undesired gluon jets largely, however, as it happens to be the case, gluon jet cross-section has a large and long tail that is worsened by the underlying event and the pile-up. A great deal of effort [1] is underway in jet identification by constructing observables that can reject the parton jets and thus reduce the background with a goal to improve the signal rate to miss-tag rate ratio. The purpose of this work is to improve upon such strategies by studying a new observable *zest* similar to transverse-zeal introduced in context of jet quenching studies [2].

109.2 Zest of a Jet

For a jet composed of hadrons, zest [3] is defined as

$$\zeta = \frac{-1}{\log\left(\sum_{i\in\text{Jet}} e^{-P_T/|\mathbf{p}_{\perp i}|}\right)}, \tag{109.1}$$

where $P_T = \sum_{i\in\text{Jet}} |\mathbf{p}_{\perp i}|$ and $\mathbf{p}_{\perp i}$ is the transverse momentum of the ith particle in the jet with respect to the jet axis. Here we have assumed that a jet has been

A. Budhraja (✉) · A. Jain
Indian Institute of Science Education and Research, Bhopal 462066, Madhya Pradesh, India
e-mail: ankitab@iiserb.ac.in

Md. Naimuddin (ed.), *XXII DAE High Energy Physics Symposium*, Springer Proceedings in Physics 203, https://doi.org/10.1007/978-3-319-73171-1_109

constructed using a suitable jet algorithm with quintessential grooming [4]. Zest has some interesting properties: (1) ζ is invariant under boosts along the jet axis, (2) it is mostly sensitive to particles with large transverse momenta and insensitive to the soft particles in the jet, (3) it is largely insensitive to the global color flow of partons. Due to its little sensitivity to the soft particles in the jet, zest distributions are stable against inclusion or exclusion of a few soft particles [3]. We further expect it to be stable against jet reconstruction methods, and jet grooming techniques. Owing to its salient properties, ζ-distribution of gluons are narrow, independent of jet energy and have little overlap with similar distribution of most other particles. ζ is highly non-linear and infrared unsafe due to which it cannot be calculated in perturbation theory, however it can be computed using Monte Carlo based event generators. Zest provides a perspective into the jet substructure that may not be accessible with infrared safe observables. Zest is hadronization model dependent but we expect zest based discrimination is not [3].

109.3 Results

All jets are simulated using Pythia 8 [5] by inserting particles with energy 500 GeV along the z-axis and allowing them to decay to hadronic modes only. For gluon jets, one energetic gluon of energy 500 GeV and off-shellness same as the heavy particle mass was inserted along the z-axis, while a soft on-shell gluon of energy 1 MeV with opposite color was inserted along the negative z-axis to shower a color singlet event. The gluon jet was identified by constructing the thrust axis [6] and taking all particles in the forward hemisphere. The resulting jets were accepted if their masses were within 5 GeV of the heavy particle. Likewise, for top quarks, a top anti-top pair of opposite color were inserted back to back for simulation, whereafter jets were constructed by dividing the event into two hemispheres perpendicular to the thrust axis. Zest distributions for gluons and heavy particles are shown in Fig. 109.1a. We note that the gluon zest distribution peaks around $\zeta \sim 0.1$ while heavy particle jets are dominantly distributed at higher values of ζ. The near jet-mass independence of gluon zest distribution makes it very useful for vetoing gluon jets. In Fig. 109.1b, we have presented the relative operating characteristic curves (dashed lines) for zest based filter, if it alone was used for heavy particle jet identification. We note that, zest provides good signal statistics after high background rejection, for example, nearly 90% of signal stays in the accepted sample after a zest cut to remove 90% of gluon jets.

Discrimination for the originating particle can be further improved with multivariate analysis. We introduce a simple infrared safe observable *boost-invariant broadening* (*bib* for brevity), defined by $b = \frac{1}{m_{\text{Jet}}} \sum_{i \in \text{Jet}} |\mathbf{p}_{\perp i}|$. Bib is similar to jet broadening defined in [7], however it is invariant under boosts along the jet direction. In Fig. 109.2a–c, we show scatter plot of simulated jets in zest-bib plane. Gluons occupy small zest and smaller bib region, while heavy particles occupy a large zest and large bib region, thus providing a strong statistical discrimination between gluon

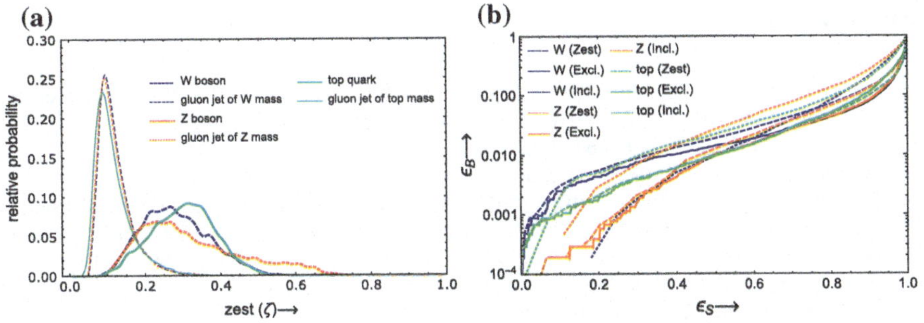

Fig. 109.1 Zest distribution for W, Z and top quark along with corresponding gluon jets

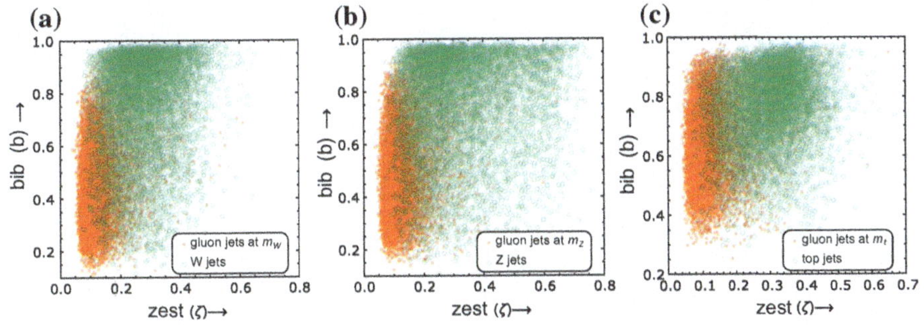

Fig. 109.2 **a–c** Scatter plots: each point represents a jet

jets and heavy particle jets. In Fig. 109.1b, we present three types of relative operating characteristic (ROC) curves (a) zest-based cuts only – represented by dashed lines (b) cuts based on inclusion zones constructed by drawing contours of equal heights on the heavy particle scatter plot – represented by dotted lines, and (c) cuts based on exclusion zones constructed by drawing contours of equal height on the gluon scatter plot – represented by solid lines. We note that for high signal rate, exclusion zone statistics provide slightly better discrimination than only-zest-based cuts for weak bosons, however they provide no significant improvement for top quark. On the other hand, inclusion zone statistics is more efficient when high gluon rejection is required and small signal rate is acceptable.

Although results presented here are preliminary, due to salient properties of zest, we expect that these results will remain mostly unaffected. A complete study will be performed in a future work.

AJ thanks SERB, DST for providing support through *Ramanujan Fellowship*.

References

1. Kasieczka et. al., JHEP **1506**, 203 (2015); Thaler et. al., JHEP **1103**, 015 (2011); Plehn et. al., Phys. Rev. Lett. **104**, 111801 (2010); Larkoski et. al., JHEP **1605**, 117 (2016)
2. R. Gavai et al., arXiv:1509.04671 [hep-ph]
3. A. Budhraja, A. Jain, Under preparation (2017)
4. Ellis et al., Phys. Rev. D **81**, 094023 (2010); Krohn et al., JHEP **1002**, 084 (2010); Dasgupta et. al., Eur. Phys. J. C **73**(11), 2623 (2013); Dasgupta et al., JHEP **1508**, 079 (2015); Larkoski et al., JHEP **1405**, 146 (2014); Butterworth et al., Phys. Rev. Lett. **100**, 242001 (2008)
5. T. Sjostrand et al., Comput. Phys. Commun. **178**, 852–867 (2008)
6. E. Farhi, Phys. Rev. Lett. **39**, 1587 (1977)
7. S. Catani et. al., Phys. Lett. B **295**, 269 (1992)

Chapter 110
Trimaximal TM_1 and TM_2 Mixings as Perturbation of Tri-Bimaximal Mixing

Kanwaljeet S. Channey and Sanjeev Kumar

110.1 Introduction

Reconstruction of the neutrino mass matrix from the current experimental neutrino data leads to many possible structures. After the measurement of non-zero θ_{13} [1], many such structures were ruled out. For example, the neutrino mass matrix (M_{TBM}) corresponding to Tri-Bimaximal mixing [2] is incompatible with the current experimental data. However it can be made compatible with the experiments by adding perturbations that break the underlying symmetry of TBM mixing. Such perturbations need not break the symmetry completely and may preserve the first or second column of TBM mixing matrix. In the present work, we perturb the mass matrix (M_{TBM}) by adding an extra complex term (M') in such a way that either first or second column of TBM mass matrix remains unchanged. Such mixing schemes are called Trimaximal Mixing [3] (TM_1 and TM_2) schemes of first and second kind. We consider only those real TBM mass matrices for which the smallest eigen value is zero.

We can write general form of Tri-Bimaximal mass matrix and the corresponding mixing matrix as:

$$U_{TBM} = \begin{pmatrix} \underbrace{\begin{matrix} \sqrt{\frac{2}{3}} \\ -\frac{1}{\sqrt{6}} \\ -\frac{1}{\sqrt{6}} \end{matrix}}_{TM_1} & \underbrace{\begin{matrix} \frac{1}{\sqrt{3}} \\ \frac{1}{\sqrt{3}} \\ \frac{1}{\sqrt{3}} \end{matrix}}_{TM_2} & \underbrace{\begin{matrix} 0 \\ \frac{1}{\sqrt{2}} \\ -\frac{1}{\sqrt{2}} \end{matrix}}_{\mu\tau} \end{pmatrix}; \quad M_{TBM} = \begin{pmatrix} 4a+b & b-2a & b-2a \\ b-2a & a+b+c & a+b-c \\ b-2a & a+b-c & a+b+c \end{pmatrix}. \tag{110.1}$$

K. S. Channey (✉) · S. Kumar
University of Delhi, Delhi 110007, India
e-mail: kjschanney@yahoo.co.in

Md. Naimuddin (ed.), *XXII DAE High Energy Physics Symposium*, Springer Proceedings in Physics 203, https://doi.org/10.1007/978-3-319-73171-1_110

Similarly the mixing and mass matrices corresponding to the TM_1 and TM_2 magic symmetries are given by

$$U_{TM_1} = \begin{pmatrix} \sqrt{\frac{2}{3}} & \frac{\cos\theta}{\sqrt{3}} & \frac{\sin\theta}{\sqrt{3}} \\ -\frac{1}{\sqrt{6}} & \frac{\sqrt{\frac{2}{3}}\cos\theta - e^{i\phi}\sin\theta}{\sqrt{2}} & \frac{e^{i\phi}\cos\theta + \sqrt{\frac{2}{3}}\sin\theta}{\sqrt{2}} \\ -\frac{1}{\sqrt{6}} & \frac{\sqrt{\frac{2}{3}}\cos\theta + e^{i\phi}\sin\theta}{\sqrt{2}} & \frac{\sqrt{\frac{2}{3}}\sin\theta - e^{i\phi}\cos\theta}{\sqrt{2}} \end{pmatrix}; M_{TM_1} = \begin{pmatrix} 2a & 2b & 2c \\ 2b & b-c+d & 2a+2b-d \\ 2c & 2a+2b-d & -3b+3c+d \end{pmatrix} \tag{110.2}$$

and

$$U_{TM_2} = \begin{pmatrix} \sqrt{\frac{2}{3}}\cos\theta & \frac{1}{\sqrt{3}} & \sqrt{\frac{2}{3}}\sin\theta \\ \frac{e^{i\phi}\sin\theta - \frac{\cos\theta}{\sqrt{3}}}{\sqrt{2}} & \frac{1}{\sqrt{3}} & \frac{-e^{i\phi}\cos\theta - \frac{\sin\theta}{\sqrt{3}}}{\sqrt{2}} \\ \frac{-\frac{\cos\theta}{\sqrt{3}} - e^{i\phi}\sin\theta}{\sqrt{2}} & \frac{1}{\sqrt{3}} & \frac{e^{i\phi}\cos\theta - \frac{\sin\theta}{\sqrt{3}}}{\sqrt{2}} \end{pmatrix}; M_{TM_2} = \begin{pmatrix} a & b & c \\ b & d & a+c-d \\ c & a+c-d & b-c+d \end{pmatrix}. \tag{110.3}$$

The neutrino mass matrix M can be diagonalized by neutrino mixing matrix U by using the following relations

$$U^T M U = M^{Diag}, \tag{110.4}$$

where M^{Diag} is the diagonal matrix of the form diag(m_1, $m_2 e^{2\iota\alpha}$, $m_3 e^{2\iota\beta}$) with α, β as the Majorana phases. The mixing angles can be calculated from U using the relations

$$\sin^2\theta_{12} = \frac{|U_{12}|^2}{1-|U_{13}|^2}; \qquad \sin^2\theta_{23} = \frac{|U_{23}|^2}{1-|U_{13}|^2}, \tag{110.5}$$

$$\sin^2_{13} = |U_{13}|^2; \qquad J = Im(U_{11}U_{12}^*U_{21}^*U_{22}). \tag{110.6}$$

We can use (110.4), (110.5) and (110.6) to find the mixing angles and CP violating phase δ in terms of θ and ϕ.

We can break the neutrino mass matrices corresponding to either TM_1 or TM_2 mixing in terms of $\mu - \tau$ breaking and $\mu - \tau$ preserving terms as follows

$$M_{TM_{(1,2)}} = M_{TBM} + M'_{TM_{(1,2)}}, \tag{110.7}$$

where M_{TBM} is given by (110.1). $M'_{TM_{(1,2)}}$ breaks the $\mu - \tau$ exchange symmetry while preserving the corresponding $TM_{(1,2)}$ symmetry.

110.2 Calculations

In the present work, we propose some simple textures for M_{TM_1} and M_{TM_2} by making minimal assumptions on (110.7). We take M_{TBM} to be real with vanishing lowest eigen value. Using these assumptions in (110.1) and

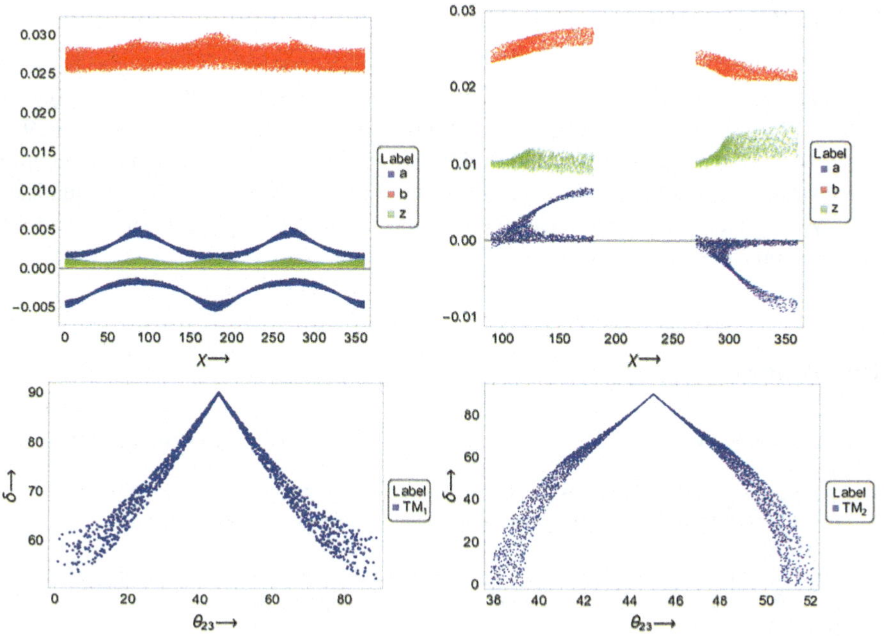

Fig. 110.1 Upper graphs show variation of the neutrino mass matrix parameters with χ for TM_1 (left) and TM_2 (right). Lower graphs show variation of CP phase δ with mixing angle θ_{23}

$$M'_{TM_1} = \begin{pmatrix} 0 & \frac{1}{4}e^{\iota\chi}z & -\frac{1}{4}e^{\iota\chi}z \\ \frac{1}{4}e^{\iota\chi}z & e^{\iota\chi}z & -\frac{1}{2}e^{\iota\chi}z \\ -\frac{1}{4}e^{\iota\chi}z & -\frac{1}{2}e^{\iota\chi}z & 0 \end{pmatrix}; \; M'_{TM_2} = \begin{pmatrix} 0 & 0 & ze^{\iota\chi} \\ 0 & ze^{\iota\chi} & 0 \\ ze^{\iota\chi} & 0 & 0 \end{pmatrix}, \tag{110.8}$$

we can get $M_{TM_{(1,2)}}$ from (110.7). Using $M_{TM_{(1,2)}}$, $U_{TM_{(1,2)}}$ and (110.4), we can find θ and ϕ for TM_1 and TM_2 respectively as

$$\cot\phi = \csc\chi\left(-\frac{z}{2b} - \cos\chi\right) \; ; \quad \sin^2 2\theta = \frac{\sqrt{3}z\sqrt{4b^2 + 4bz\cos\chi + z^2}}{2b(2b + z\cos\chi)}, \tag{110.9}$$

$$\cot\phi = \frac{(2a+3b)\csc\chi(\cos\chi + z)}{2a - 3b} \; ; \; \tan^2\theta = \frac{3z\left(9a^2 + 12ab\cos 2\chi + 2z(3a+2b)\cos\chi + 4b^2 + z^2\right)}{2\left(-9a^2 + 4b^2 + 4bz\cos\chi + z^2\right)^2}. \tag{110.10}$$

From these values of θ and ϕ, we can calculate all the mixing angles and CP violating phase using (110.5) and (110.6) (Fig. 110.1).

110.3 Conclusions

The parameters a, b and z are constrained in the ranges $0.001 \leq |a| \leq 0.005$, $0.025 \leq b \leq 0.030$ and $0.0003 \leq z \leq 0.0015$ for T_1 and in the ranges $-0.009 \leq a \leq 0.007$, $0.021 \leq b \leq 0.028$ and $0.009 \leq z \leq 0.015$ for T_2. The phase χ remains unconstrained for TM_1 while it is constrained in the second and fourth quadrant for TM_2. χ will further be constrained by the future measurements of θ_{23} and CP violating phase δ.

References

1. K. Abe, Others, Phys. Rev. Lett. **107**, 41801 (2011). https://doi.org/10.1103/PhysRevLett.107.041801
2. P.F. Harrison, D.H. Perkins, W.G. Scott, Phys. Lett. B **530**(1–4), 167 (2002). https://doi.org/10.1016/S0370-2693(02)01336-9, http://www.sciencedirect.com/science/article/pii/S0370269302013369
3. C.H. Albright, W. Rodejohann, Eur. Phys. J. C **62**(3), 599 (2009). https://doi.org/10.1140/epjc/s10052-009-1074-3, arXiv:0812.0436, http://www.springerlink.com/index/10.1140/epjc/s10052-009-1074-3

Chapter 111
Study of Hyperon Polarization at T2K, MicroBooNE and MINERνA

F. Akbar, M. Rafi Alam, M. Sajjad Athar, S. Chauhan and S. K. Singh

At the energies below the threshold of associated particle production, the hyperons can be produced by antineutrinos through the QE processes

$$\bar{\nu}_\mu + p \rightarrow \mu^+ + \Lambda^0(\Sigma^0); \qquad \bar{\nu}_\mu + n \rightarrow \mu^+ + \Sigma^-. \tag{111.1}$$

The production of these hyperons is however Cabibbo suppressed as compared to the Δ production, but could be significant at the lower energies due to the threshold effects. It has been shown that in the nuclear targets the pions produced by the hyperons are comparable to the pions produced through the Δ excitation in the lower energy region of $E_{\bar{\nu}_l} \sim 500$–$700\,\text{MeV}$ [1]. The asymmetry in the angular distribution of pions will give information about the hyperon polarization extending the scope of the weak interactions in $|\Delta S| = 1$ sector. We have in this work calculated, the differential cross section and the polarization of hyperons produced in the reaction (111.1). The details of the calculations are given in [2]. The cross section $d\sigma$ and the four vector polarization ξ^τ of the final hyperon in the reaction (111.1) are given by [3]

$$d\sigma = \frac{1}{(2\pi)^2}\frac{1}{4E_{\bar{\nu}_\mu} m_N}\delta^4(k + p - k' - p')\frac{d^3k'}{2E_{k'}}\frac{d^3p'}{2E_{p'}}\sum\overline{\sum}|\mathcal{M}|^2, \tag{111.2}$$

$$\xi^\tau = \left(g^{\tau\sigma} - \frac{p'^\tau p'^\sigma}{m_Y^2}\right)\frac{\mathcal{L}^{\alpha\beta}\mathrm{Tr}\left[\gamma_\sigma\gamma_5\Lambda(p')J_\alpha\Lambda(p)\tilde{J}_\beta\right]}{\mathcal{L}^{\alpha\beta}\mathrm{Tr}\left[\Lambda(p')J_\alpha\Lambda(p)\tilde{J}_\beta\right]}. \tag{111.3}$$

where $\mathcal{M} = \frac{G_F}{\sqrt{2}}\sin\theta_c\, l^\mu J_\mu$ with

F. Akbar (✉) · M. R. Alam · M. S. Athar · S. Chauhan · S. K. Singh
Department of Physics, Aligarh Muslim University, Aligarh 202002, India
e-mail: faiza.akbar.amu@gmail.com

Md. Naimuddin (ed.), *XXII DAE High Energy Physics Symposium*, Springer Proceedings in Physics 203, https://doi.org/10.1007/978-3-319-73171-1_111

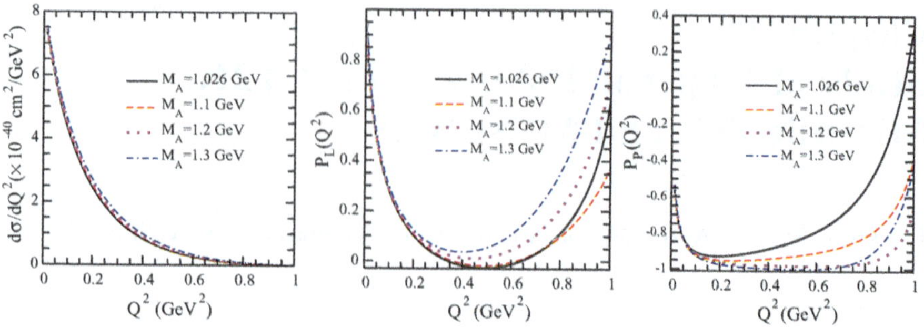

Fig. 111.1 $\frac{d\sigma}{dQ^2}$, $P_L(Q^2)$ and $P_P(Q^2)$ versus Q^2 for the process $\bar{\nu}_\mu p \rightarrow \mu^+ \Lambda$ at $E_{\bar{\nu}_\mu} = 1\,\text{GeV}$ for different values of M_A viz. 1.026 GeV (solid line), 1.1 GeV (dashed), 1.2 GeV (dotted line) and 1.3 GeV (double dashed-dotted line) with $m_\mu = 0$

$$l^\mu = \bar{u}(k')\gamma^\mu(1+\gamma_5)u(k)$$

$$J_\mu = \left[\gamma_\mu f_1^{NY}(Q^2) + i\sigma_{\mu\nu}\frac{q^\nu}{m_N+m_Y}f_2^{NY}(Q^2) - \gamma_\mu\gamma_5 g_1^{NY}(Q^2) - \frac{q_\mu}{m_N+m_Y}g_3^{NY}(Q^2)\gamma_5\right] \tag{111.4}$$

$\Lambda(p) = \frac{\not{p}+M}{2M}$, $\Lambda(p') = \frac{\not{p}'+M_\Lambda}{2M_\Lambda}$ and $\mathcal{L}^{\mu\nu} = \Sigma l^\mu l^{\nu\dagger}$. $f_{1,2}^{NY}(Q^2)$ and $g_{1,3}^{NY}(Q^2)$ are the strangeness changing transition form factors which are determined in terms of the nucleon form factors using SU(3) symmetry [4]. $f_1^{NY}(Q^2)$ and $f_2^{NY}(Q^2)$ are determined in terms of the electromagnetic form factors of the nucleon for which the parameterization of Bradford et al. [5] has been used. The axial vector form factor $g_1^{NY}(Q^2)$ is given in terms of the axial form factor $g_A^{n\rightarrow p}(Q^2)$ corresponding to n–p transitions, for which a dipole form is taken with $M_A = 1.026\,\text{GeV}$. The pseudoscalar form factor $g_3^{NY}(Q^2)$ is obtained in terms of $g_1^{NY}(Q^2)$ assuming PCAC and Goldberger–Treiman (GT) relation extended to strangeness sector as discussed by Marshak et al. [6] and Nambu [7]. Using (111.2) and (111.3), the expressions for differential cross section $\frac{d\sigma}{dQ^2}$ and longitudinal ($P_L(Q^2)$) and perpendicular ($P_P(Q^2)$) components of the hyperon polarization are calculated to be

$$\frac{d\sigma}{dQ^2} = \frac{G_F^2\sin^2\theta_c}{8\pi m_N E_{\bar{\nu}_\mu}^2}\mathcal{N}(Q^2, E_{\bar{\nu}_\mu}) \tag{111.5}$$

$$\frac{d\sigma}{dQ^2}P_L(Q^2) = \frac{G_F^2\sin^2\theta_c}{8\pi|\mathbf{q}|E_{p'}m_N E_{\bar{\nu}_\mu}^2}\left[\left(E_{\bar{\nu}_\mu}^2 - E_\mu^2 + m_\mu^2\right)m_Y\mathcal{A}(Q^2, E_{\bar{\nu}_\mu}) + |\mathbf{q}|^2\mathcal{B}(Q^2, E_{\bar{\nu}_\mu})\right] \tag{111.6}$$

$$\frac{d\sigma}{dQ^2}P_P(Q^2) = -\frac{G_F^2\sin^2\theta_c}{4\pi}\frac{|\mathbf{k}'|}{|\mathbf{q}|}\frac{\mathcal{A}(Q^2, E_{\bar{\nu}_\mu})\sin\theta}{m_N E_{\bar{\nu}_\mu}}, \tag{111.7}$$

where in the lab frame $E_{p'} = \sqrt{|\mathbf{q}^2| + m_Y^2}$ and θ is the scattering angle and $\mathcal{N}(Q^2, E_{\bar{\nu}_\mu})$, $\mathcal{A}(Q^2, E_{\bar{\nu}_\mu})$ and $\mathcal{B}(Q^2, E_{\bar{\nu}_\mu})$ are given in [2].

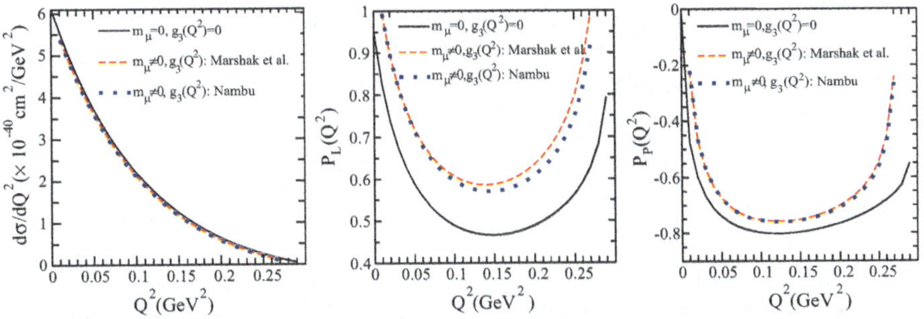

Fig. 111.2 $\frac{d\sigma}{dQ^2}$, $P_L(Q^2)$ and $P_P(Q^2)$ versus Q^2 ($M_A = 1.026\,\text{GeV}$) for the process $\bar{\nu}_\mu p \to \mu^+ \Lambda$ at $E_{\bar{\nu}_\mu} = 0.5\,\text{GeV}$ using $f_1^{p\Lambda}(Q^2)$, $f_2^{p\Lambda}(Q^2)$, $g_1^{p\Lambda}(Q^2)$ with $m_\mu = 0$ (solid line), $m_\mu \neq 0$ and $g_3^{p\Lambda} \neq 0$ from Marshak et al. [6] (dashed line) and $m_\mu \neq 0$ and $g_3^{p\Lambda} \neq 0$ from Nambu [7] (dotted line)

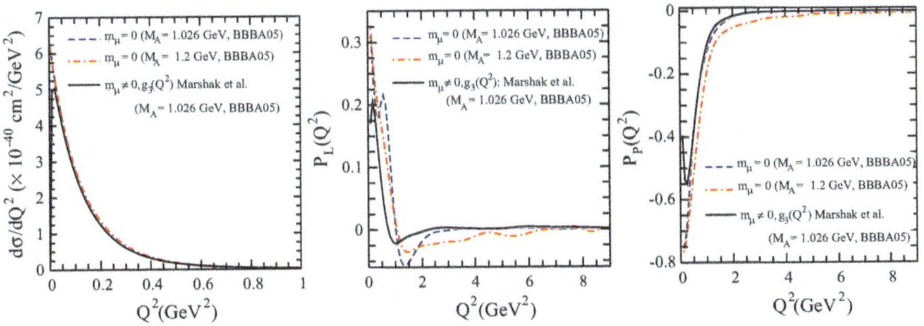

Fig. 111.3 $\frac{d\sigma}{dQ^2}$, $P_L(Q^2)$ and $P_P(Q^2)$ versus Q^2 for the process $\bar{\nu}_\mu p \to \mu^+ \Lambda$(^{12}C target) averaged over T2K spectrum, using $f_1^{p\Lambda}(Q^2)$, $f_2^{p\Lambda}(Q^2)$, $g_1^{p\Lambda}(Q^2)$ with $m_\mu = 0$ and $M_A = 1.026\,\text{GeV}$ (dashed line), $m_\mu = 0$ and $M_A = 1.2\,\text{GeV}$ (dashed–dotted line) and $m_\mu \neq 0$, $M_A = 1.026\,\text{GeV}$ with $g_3^{p\Lambda}(Q^2)$ from Marshak et al. [6] (solid line)

In Fig. 111.1, we have studied the sensitivity of $\frac{d\sigma}{dQ^2}$, $P_L(Q^2)$ and $P_P(Q^2)$ on M_A by varying the value of M_A from 1.026 to 1.3 GeV for $\bar{\nu}_\mu p \to \mu^+ \Lambda$ process at $E_{\bar{\nu}_\mu} = 1\,\text{GeV}$. We observe that in the case of $\frac{d\sigma}{dQ^2}$, the sensitivity on M_A is very low, whereas $P_L(Q^2)$ and $P_P(Q^2)$ are strongly sensitive to the variation of M_A. Hence, we may be able to measure M_A using the polarization observables independent of the cross section. To study the effect of $g_3(Q^2)$ on $\frac{d\sigma}{dQ^2}$, $P_L(Q^2)$ and $P_P(Q^2)$ in $|\Delta S| = 1$ sector, we have used two different parameterizations for $g_3(Q^2)$, given by Marshak et al. [6] and Nambu [7], and the results are presented in Fig. 111.2 for the process $\bar{\nu}_\mu p \to \mu^+ \Lambda$ at $E_{\bar{\nu}_\mu} = 0.5\,\text{GeV}$. We find that $P_L(Q^2)$ and $P_P(Q^2)$ are sensitive to $g_3(Q^2)$ in the region of low antineutrino energies.

In Fig. 111.3, we present flux averaged $\langle\frac{d\sigma}{dQ^2}\rangle$, $\langle P_L(Q^2)\rangle$ and $\langle P_P(Q^2)\rangle$ for $\bar{\nu}_\mu p \to \mu^+ \Lambda$ process corresponding to the T2K antineutrino flux for ^{12}C target. It may be observed from the figure that polarization measurements in the T2K experiment will enable us to independently determine the value of axial vector, pseudoscalar form factor and it may be possible to test the hypothesis of PCAC in the strangeness sector.

References

1. M.R. Alam, S. Chauhan, M.S. Athar, S.K. Singh, Phys. Rev. D **88**, 077301 (2013)
2. F. Akbar, M. Rafi, Alam, M. Sajjad Athar, S. K. Singh. Phys. Rev. D **94**, 114031 (2016)
3. S.M. Bilenky, E. Christova, J. Phys. G **40**, 075004 (2013)
4. N. Cabibbo, E.C. Swallow, R. Winston, Ann. Rev. Nucl. Part. Sci. **53**, 39 (2003)
5. R. Bradford et al., Nucl. Phys. Proc. Suppl. **159**, 127 (2006)
6. R.E. Marshak, Riazuddin, C.P. Ryan, *Theory of Weak Interactions in Particle Physics* (Wiley-Interscience, 1969)
7. Y. Nambu, Phys. Rev. Lett. **4**, 380 (1960)

Chapter 112
Sequential Quarkonium Production via Recombination in Heavy-Ion Collisions

Captain R. Singh, S. Ganesh and M. Mishra

112.1 Introduction

The production of heavy quarkonia (bottomonium and charmonium) is modified in the heavy-ion collisions as compared with elementary nuclear collisions. This modification in the production of quarkonia is due to the hot and dense medium formed in heavy ion collisions named as Quark-Gluon Plasma (QGP) which is expected not to be produced in the elementary collisions. In such a thermalized medium, quarks and gluons can cause the dissociation of heavy mesons present in the medium. Likewise, bottomonium and charmonium present in the medium can be destroyed, but the constituent quarks (q) anti-quarks $(\bar{q})$ will likely stay spatially correlated due to small mean free paths in this system. In the due course of time, they can recombine and form color neutral states. As charm quark and aniquarks have less mass than bottom quark-anti quarks, they can be present in the medium in sufficient numbers from the initial time of the QGP. These are called un-correlated $c\bar{c}$ pairs. These correlated as well as uncorrelated $c\bar{c}$ pair can recombine to form charmonium states (e.g., $J/\psi,\ \chi_c,\ \psi^{'}$). So, charmonium regeneration can happen in two ways, first one is through correlated $c\bar{c}$ and another is by un-correlated $c\bar{c}$ pair. But because of the very heavy mass, the number of uncorrelated $b\bar{b}$ pair would be too small. Thus, the possibility of the recombination of un-correlated $b\bar{b}$ pairs in the medium is negligible. That is why regeneration of bottomonium (normally due to un-correlated pairs) in QGP has been neglected so far. But regeneration of bottomonia due to correlated $b\bar{b}$ pair may be present in the medium which can play a significant role in explaining the bottomonium nuclear modification factor, R_{AA}. Quarkonium suppression in a thermal QCD medium created in heavy ion collisions is a complex interplay of various physical processes, e.g., Gluonic dissociation, Collisional damping, Color screening and Cold Nuclear Effects [1–3]. Here we plan to describe the production

C. R. Singh (✉) · S. Ganesh · M. Mishra
Department of Physics, Birla Institute of Technology and Science, Pilani 333031, India
e-mail: captainriturajsingh@gmail.com

Md. Naimuddin (ed.), *XXII DAE High Energy Physics Symposium*, Springer Proceedings in Physics 203, https://doi.org/10.1007/978-3-319-73171-1_112

of charmonium and bottomonium via recombination and attempt to explain their observed nuclear modification factor, R_{AA}. We have developed a phenomenological model and we named it 'Unified Model of quarkonia suppression (UMQS)'. In this model, we included hot matter effect (such as color screening, gluonic dissociation, collisions damping), cold nuclear effect (CNM) along with regeneration effect in a unified way to calculate the quarkonium suppression. Using this model, we analyze the centrality and transverse momentum, p_T dependent recombination factor. Finally, we calculated the quarkonia survival probability and compared to the available R_{AA} obtained from LHC at $\sqrt{s_{NN}} = 2.76$ TeV.

112.2 Model Formalism

The production of quark (q), anti-quark ($\bar{q}$) and their bound states is governed by a rate equation given by Thews et al., [4]. We will describe here the rate equations in the context of $\Upsilon(1S)$. We will utilize the same for J/ψ states [5].

$$\frac{dN_\Upsilon}{d\tau} = \Gamma_F N_b \, N_{\bar{b}} \, [V(\tau)]^{-1} - \Gamma_D N_\Upsilon \tag{112.1}$$

The first term in the RHS of (112.1) is formation term and second is the dissociation term of bottomonia. We have modified the solution of the transport equation by including the shadowing factor in the initially produced bottomonia, $N_\Upsilon(\tau_0, b)$. So, the modified analytical solution of (112.1) as follows [5]:

$$N_\Upsilon(\tau_f, b) = \epsilon(\tau_f) \left[N^i_\Upsilon(\tau_0, b) + N^2_{b\bar{b}} \int_{\tau_0}^{\tau_f} \Gamma_F [V(\tau)\epsilon(\tau)]^{-1} d\tau \right] \tag{112.2}$$

where, $$N^i_\Upsilon(\tau_0, b) = N_\Upsilon(\tau_0, b) * S_{sh}$$

The term, $\tau_0 = 0.5 \ fm$ is the initial thermalization time of QGP and $\tau_f = 6.5 \ fm$ is the life time of the QGP at the LHC $\sqrt{s_{NN}} = 2.76$ TeV. $N^i_\Upsilon(\tau_0, b)$ is the initially suppressed production due to shadowing effect, S_{sh} [3] and $N_\Upsilon(\tau_f, b)$ is the finally survived number of Υ meson. $\epsilon(\tau_f)$ and $\epsilon(\tau)$ are the suppression factors. They include gluonic dissociation and collisional damping suppression mechanisms [6]. Here Γ_F is the recombination factor obtained using the detailed balance [5]. Other variables used in (112.2), have been discussed in details in referred article [5].

We have calculated the initial $\Upsilon(N_\Upsilon(\tau_0, b))$ production using the production cross section, $\sigma^\Upsilon_{NN} = 75 \ nb$ [7]. For calculating the $N_{b\bar{b}}$, we have used the $b\bar{b}$ production cross section $\sigma^{b\bar{b}}_{NN} = 43 \ \mu b$ [7].

To refine the estimate of bottomonia suppression in QGP medium, we have introduced the effective temperature in our UMQS model calculation.

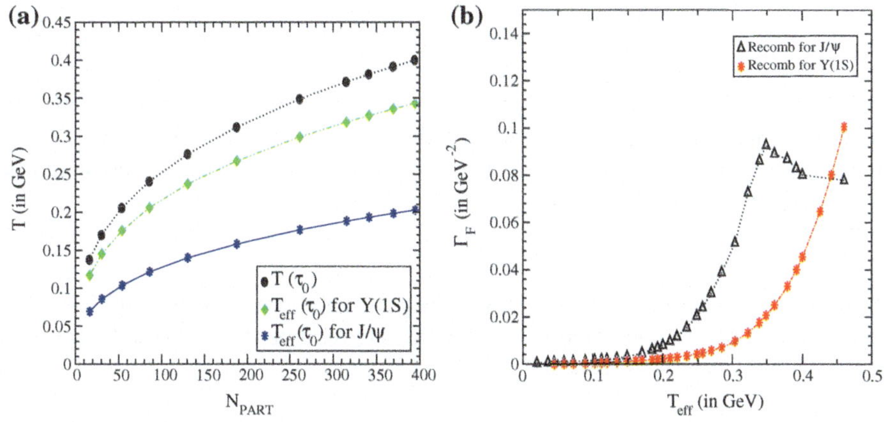

Fig. 112.1 **a** Represents the medium temperature and effective temperature of $\Upsilon(1S)$ and J/ψ in medium as the function of centrality. **b** Shows the variation of Γ_F with T_{eff} for $\Upsilon(1S)$ and J/ψ in medium

112.2.1 Effective Temperature, T_{eff}

It has been assumed here that the quarkonium states do not experience the same temperature as the medium because of their heavy mass scale and formation time. So the modeling of quarkonia temperature becomes necessary and it should play an important role to explain the quarkonia production in QGP medium. The effective temperature is given as [8]:

$$T_{eff}(v_{rel}, \theta) = \frac{T(b, \tau)\sqrt{1 - v_{rel}^2}}{1 - v_{rel}\cos\theta}, \tag{112.3}$$

where, v_{rel} and θ are the relative velocity and scattering angle between the quarkonia and medium. $T(b, \tau)$ is centrality and time dependent medium temperature [5]. Here, we have used Quasi Particle Model (QPM) equation of state for medium evolution.

112.2.2 Net Suppression

We have modeled the net production of bottomonia and charmonia by incorporating the suppression mechanisms along with recombination process. So, very first we put together gluonic dissociation, collisional damping, shadowing and regeneration and then calculated the survival probability, S_g^{Υ} given as:

$$S_g^{\Upsilon} = \frac{N_{\Upsilon}(\tau_f, b)}{N_{\Upsilon}(\tau_0, b)} \tag{112.4}$$

In UMQS model, we assumed that color screening would act independently on quarkonia. So, we calculated color screening survival probability, S_c^{Υ} using the formalism developed by Mishra et al. [9]. The net survival probability for $\Upsilon(1S)$ in QGP medium is defined as:

$$S_p^{\Upsilon} = S_g^{\Upsilon} S_c^{\Upsilon} \tag{112.5}$$

We have used same formalism to calculate the charmonium survival probability, $S_p^{J/\psi}$, and results of charmonium and bottomonium suppression are shown and discussed in Result and Discussion section.

112.3 Results and Discussions

As shown in Fig. 112.1a, the quarkonium states $\Upsilon(1S)$ and J/ψ do not experience the same temperature as the medium due to their heavy mass scale. The $\Upsilon(1S)$ effective temperature is found more than J/ψ at each centrality. It is because of high dissociation temperature, T_d of $\Upsilon(1S)$. Figure 112.1b, shows possibility of J/ψ recombination is more than Υ. For J/ψ un-correlated $c\bar{c}$ can recombine and/or spatially correlated $c\bar{c}$ pair can make transition from color octet to color singlet states. But near J/ψ dissociation temperature ($T_d = 335\,\text{MeV}$ [9]), the resultant J/ψ production decreases. The high dissociation temperature ($T_d = 668\,\text{MeV}$ [6]) of $\Upsilon(1S)$ leads the explanation of continuously increasing recombination factor for $\Upsilon(1S)$ at higher temperature values. This difference in dissociation temperature of J/ψ and $\Upsilon(1S)$, the recombination factor of J/ψ decreases near the temperature 340–400 MeV while $\Upsilon(1S)$ is continuously increasing with temperature as shown in Fig. 112.1b. This recombination of J/ψ and $\Upsilon(1S)$ plays a significant role to explain the their respective nuclear modification factor, R_{AA} at LHC $\sqrt{s_{NN}} = 2.76\,\text{TeV}$ [10, 11] as shown below:

In Fig. 112.2a Γ_F is increasing with p_T because at low p_T $q\bar{q}$ pairs go through multi-scattering with medium constituents which restrict them to recombine. But with increasing p_T multi-scattering of $q\bar{q}$ pairs with medium decreases, so Γ_F increases with p_T. In Fig. 112.2b we have found that recombination for bottomonia is negligible at lower centrality bins but in most central bins it plays a key role to explain the suppression data. It may be because of quadratic dependence of $N_{b\bar{b}}$, produced in sufficient numbers at most central bins than lower centrality bins. Figure 112.3a, b follow the same explanation as Fig. 112.2b.

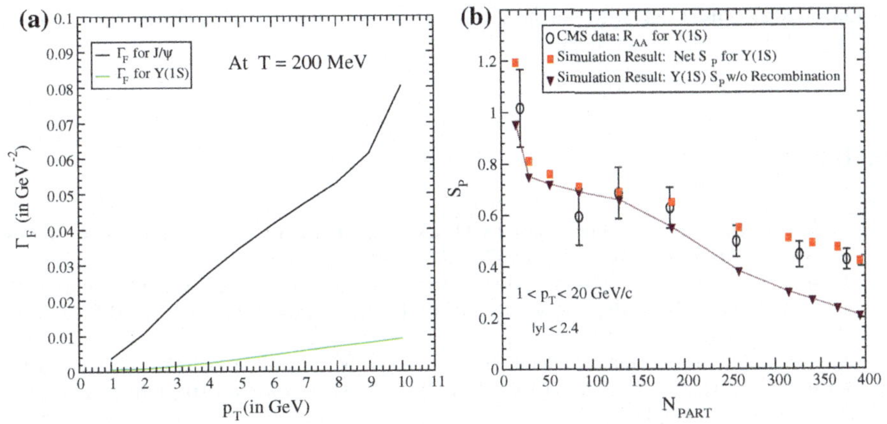

Fig. 112.2 **a** Γ_F for $\Upsilon(1S)$ and J/ψ as the function of their respective transverse momentum, p_T **b** $\Upsilon(1S)$ net survival probability, S_p^{Υ}, as the function of centrality

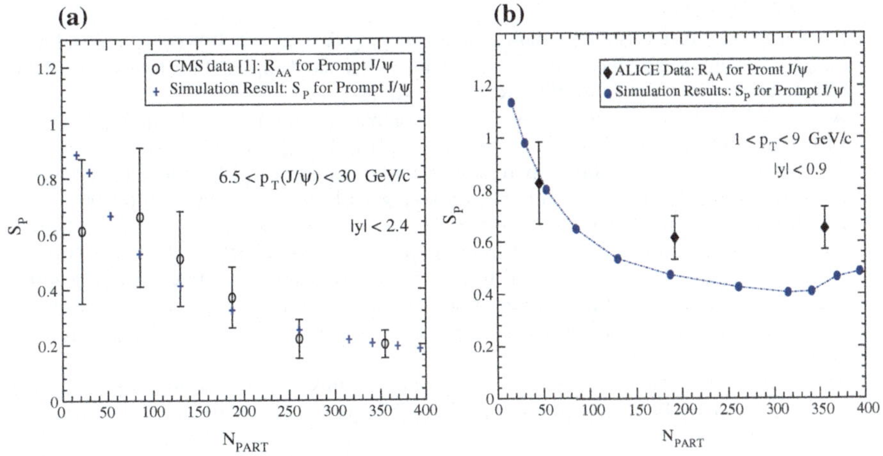

Fig. 112.3 **a** Net $S_p^{J/\psi}$ as the function centrality at high p_T. **b** Net $S_p^{J/\psi}$ as the function centrality at low p_T

112.4 Conclusion

The UMQS model developed to explain the quarkonia suppression at mid rapidity in heavy ion collisions at LHC $\sqrt{s_{NN}} = 2.76$ TeV. The model explains the Υ and J/ψ experimental data reasonably well. At low p_T regime the calculated J/ψ survival probability shows some variation with J/ψ nuclear modification factor, R_{AA} at ALICE, but qualitatively it is following the trend. So, the J/ψ suppression in heavy ion collisions at low p_T regime needs more study.

Acknowledgements Captain R. Singh acknowledges the financial support from the Dept. of Physics, BITS -Pilani, Pilani Campus.

References

1. T. Matsui, H. Satz, J/ψ suppression by Quark-Gluon plasma formation. Phys. Lett. B **178**, 416 (1986). https://doi.org/10.1016/0370-2693(86)91404-8
2. F. Nendzig, G. Wolschin, Υ suppression in Pb-Pb collisions at energies available at the CERN Large Hadron Collider. Phys. Rev. C **87**, 024911 (2013). https://doi.org/10.1103/PhysRevC.87.024911
3. R. Vogt, Cold nuclear matter effects on J/ψ and Υ production at energies available at the CERN Large Hadron Collider (LHC). Phys. Rev. C **81**, 044903 (2010). https://doi.org/10.1103/PhysRevC.81.044903
4. R.L. Thews et al., Enhanced J/ψ production in deconfined quark matter. Phys. Rev. C **63**, 054905 (2001). https://doi.org/10.1103/PhysRevC.63.054905
5. C.R. Singh et al., Unified description of charmonium suppression in a quark-gluon plasma medium at RHIC and LHC energies. Phy. Rev. C **92**, 034916 (2015). https://doi.org/10.1103/PhysRevC.92.034916
6. S. Ganesh, M. Mishra, Bottomonium suppression at $\sqrt{s_{NN}} = 2.76$ TeV using a model based on color screening and gluonic dissociation with collisional damping. Phy. Rev. C **88**, 044908 (2013). https://doi.org/10.1103/PhysRevC.88.044908
7. A. Emerick et al., Bottomonia in the quark-gluon plasma and their production at RHIC and LHC. Eur. Phys. J. A **48**, 72 (2012). https://doi.org/10.1140/epja/i2012-12072-y
8. G. Wolschin et al., Bottomium suppression in Pb-Pb collisions at LHC energies. J. Phys. G: Nucl. Part. Phys. **41**, 095003 (2014). https://doi.org/10.1088/0954-3899/41/9/095003
9. P.K. Srivastava et al., Color screening scenario for quarkonia suppression in a quasiparticle model compared with data obtained from experiments at the CERN SPS, BNL RHIC, and CERN LHC. Phys. Rev. C **87**, 034903 (2013). https://doi.org/10.1103/PhysRevC.87.034903
10. The CMS Collaboration, Suppression of non-prompt J/ψ, prompt J/ψ, and Υ in Pb-Pb collisions at $\sqrt{s_{NN}} = 2.76$ TeV, J. High Energy Phys. **05**, 063 (2012). https://doi.org/10.1007/JHEP05(2012)063
11. The CMS Collaboration, Suppression and azimuthal anisotropy of prompt and nonprompt J/ψ production in Pb-Pb collisions at $\sqrt{s_{NN}} = 2.76$ TeV, arXiv:1610.00613v1 [nucl-ex] 03 Oct (2016)

Chapter 113
Studies on Reconstruction Algorithms for Material Imaging with MST

Sridhar Tripathy, Abhik Jash, Nayana Majumdar, Supratik Mukhopadhyay, Sandip Sarkar and Satyajit Saha

113.1 Introduction

Cosmic ray muons are the secondary products of interactions between highly energetic cosmic rays and the nuclei of atmospheric particles. Their flux at sea level is around 10,000/m^2/min (160 Hz), that changes with the zenith angle θ as $\cos^2\theta$ [1]. Their energy varies from 10 MeV to 10 GeV with most probable value around 1 GeV [2]. Monitoring the cosmic-muon tracks using radiation detectors and using reconstruction algorithm, the imaging of high-Z material is feasible.

113.2 Muon Interaction with Matter

Muons are leptons and they undergo two types of electromagnetic interactions, which are the ionization of atoms and Coulomb scattering with charge particles or nuclei. Muons undergo multiple small deflections which is termed as Multiple Coulomb Scattering, the amount of scattering depends upon the atomic properties. The distribution of scattering angles can be approximated to a Gaussian distribution [3], with mean at zero and standard deviation given by

$$\sigma = \frac{13.6}{\beta c p}\sqrt{\frac{L}{X_0}}\left(1 + 0.038 ln\left(\frac{L}{X_0}\right)\right) \quad X_0 = \frac{716.4 g/cm^2}{\rho}\frac{A}{Z(Z+1)ln(\frac{287}{\sqrt{Z}})} \tag{113.1}$$

S. Tripathy (✉) · A. Jash · N. Majumdar · S. Mukhopadhyay · S. Sarkar · S. Saha
Saha Institute of Nuclear Physics, 1/AF Bidhannagar, Kolkata 700064, India
e-mail: sridhar.tripathy@saha.ac.in

A. Jash
HBNI, Training School Complex, Anushakti Nagar, Mumbai 400094, India

Md. Naimuddin (ed.), *XXII DAE High Energy Physics Symposium*, Springer Proceedings in Physics 203, https://doi.org/10.1007/978-3-319-73171-1_113

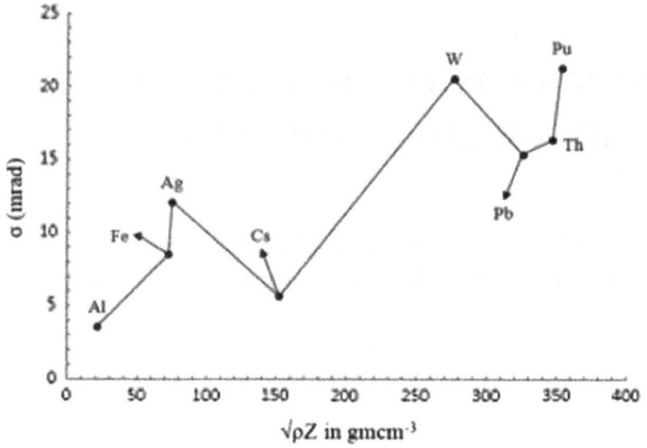

Fig. 113.1 The plot of σ versus $\sqrt{\rho}Z$ for different materials

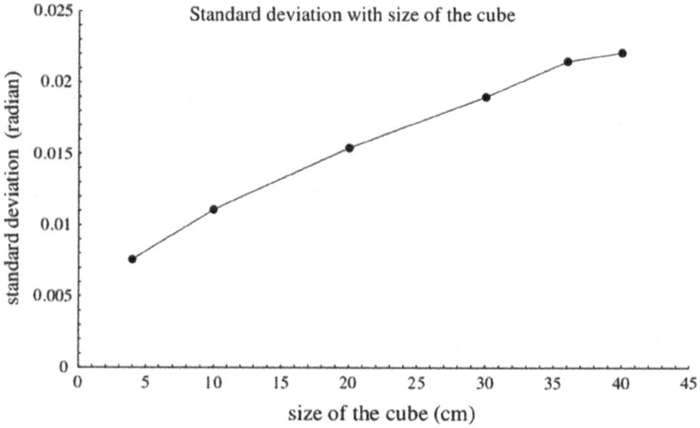

Fig. 113.2 The variation of standard deviation with size of the cube

where L is the length of the medium traversed, β is the ratio of speed of the muon to speed of light, p is the momentum of the muon, X_0 is the radiation length, Z is the atomic number and ρ is the density.

In GEANT4 [4] model, muons of energy 1 GeV were incident from a point source on cubes having side 20 cm made of materials of widely varying ρ and Z. Detector planes of 30 $\times$ 30 cm dimensions were placed both above and below the target. Incoming and outgoing muon tracks were created based on the hit information. Hence found scattering angles were plotted and fitted with Gaussian distribution. The variation of standard deviations σ with ρ, Z and the side of the cube is shown in Figs. 113.1 and 113.2 respectively.

113.3 Comparison: POCA and GEANT4 Estimated Points

The POCA [5] algorithm is a geometrical algorithm which considers a single scattering per event and no multiple scattering [6]. The algorithm first intakes the hit points of all 4 detector planes; i.e. $A(x_1, y_1, z_1)$, $B(x_2, y_2, z_2)$, $C(x_3, y_3, z_3)$, $D(x_4, y_4, z_4)$. For a particular event, it calculates $\mathbf{u} = B - A$ and $\mathbf{v} = D - C$. Then it finds the straight lines joining A, B and C, D following the equations $P(s) = A + s\mathbf{u}$; $Q(t) = C + t\mathbf{v}$. From which $W(s,t) = P(s) - Q(t)$; and $w_0 = A - C$ are found. To find the closest point, two linear equations given in (113.2) are solved. The point hence found (s_c, t_c) is the POCA.

$$(\mathbf{u}.\mathbf{u})s_c - (\mathbf{u}.\mathbf{v})t_c = -\mathbf{u}.w_0 \quad (\mathbf{v}.\mathbf{u})s_c - (\mathbf{v}.\mathbf{v})t_c = -\mathbf{v}.w_0 \tag{113.2}$$

For 13022 events, hits on the detector planes were used to find the tracks and then the approximation points were generated from the POCA algorithm. For the same events in GEANT4, the scattering points inside the material volume were located. For more than one scattering in an event, the geometrical average was taken to find out a center of scattering according to the formula in (113.3) where (x_n, y_n, z_n) are the scattering point coordinates inside the material cube. The spatial distance between the points were found between POCA and $G(x, y, z)$ by (113.3) where Δx, Δy, Δz represents the difference of POCA estimated point and simulated point. The so found distances were plotted in a histogram which has a mean around 2.32 cm. This is shown in Fig. 113.3.

$$G(x, y, z) = \left(\frac{\Sigma x_n}{n}, \frac{\Sigma y_n}{n}, \frac{\Sigma z_n}{n}\right) \quad d = \sqrt{(\Delta x)^2 + (\Delta y)^2 + (\Delta z)^2} \tag{113.3}$$

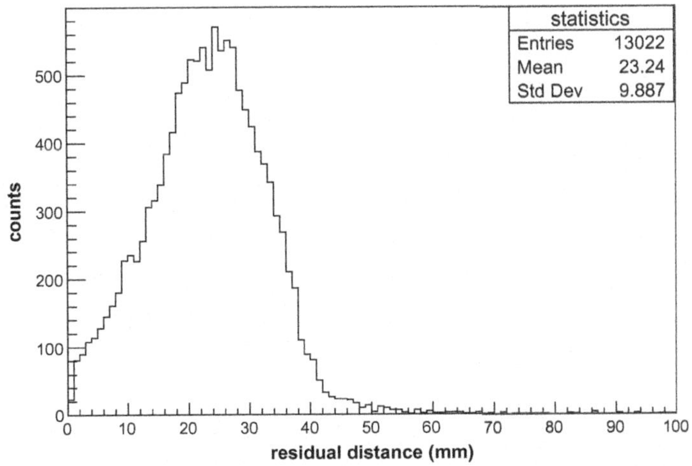

Fig. 113.3 Residual plot of POCA and GEANT4 estimated point

113.4 Conclusions and Future Goals:

The residual plot shows that within a cubical volume of lead with a side of 20 cm, the POCA algorithm has a mean inaccuracy of **2.32 cm** in comparison to GEANT4 in the present calculation. This work describes the POCA algorithm applied to a case where only coordinates of the muon hits on the detector plane are known, and the volume of the objects are not discretized to voxels. The methods like Expectation Maximization likelihood method which discritizes the object volume and finds the track by forming a weighted matrix may lead to better estimation of the muon track. Moreover, it is planned to set-up Bakelite plate Resistive Plate Chamber (RPC) detector based muon-tracking system to discriminate high Z materials. The corresponding data acquisition system is being constructed. Triggering scintillators will be used to select the muon events passing through the tracking system and data will be stored and we need fast and high storage clusters for analyzing the data.

Acknowledgements The author is grateful to DST Inspire, Govt. of India for the financial support. The help and support of the members of Applied Nuclear Physics Division in this project has been enormous.

References

1. K. Borozdin et al., Nature **422**, 277 (2003)
2. T. Sato et al., J. Nucl. Sci. Technol. **50**, 913–923 (2013)
3. H.A. Bethe et al., Phys. Rev. **89**, 1256 (1953)
4. S. Agostinelli et al., Nucl. Instrum. Method A **506**, 250 (2003)
5. D. Sunday, (http://geometryalgorithms.com/Archive/algorithm0106/algorithm0106.htm)
6. L. Scrutz, Ph.D. thesis COSMIC RAY MUON RADIOGRAPHY

Chapter 114
Open Charm Hadroproduction as a Probe of Gluon Sivers Function

Rohini M. Godbole, Abhiram Kaushik and Anuradha Misra

Transverse single spin asymmetry (TSSA) for an inclusive process $A^{\uparrow} + B \rightarrow C + X$ is defined as,

$$A_N = \frac{d\sigma^{\uparrow} - d\sigma^{\downarrow}}{d\sigma^{\uparrow} + d\sigma^{\downarrow}} \quad (114.1)$$

where $d\sigma^{\uparrow}$ ($d\sigma^{\downarrow}$) is the cross-section for scattering of a transversely polarised hadron A off an unpolarised hadron B, with the polarisation being upwards (downwards) w.r.t plane of production of C.

In this work, we treat TSSAs in the framework of a generalized parton model (GPM) where the parton distribution functions of QCD are generalized to include the transverse momenta of the partons. The Sivers function is one such transverse momentum dependent distribution parametrising the azimuthal anisotropy (in k_T-space) of partons inside the hadron:

$$f_{a/p^{\uparrow}}(x, \mathbf{k}_{\perp}) = f_{a/p}(x, \mathbf{k}_{\perp}) + \Delta^N f_{a/p^{\uparrow}}(x, k_{\perp}) \mathbf{S}.(\mathbf{p} \times \mathbf{k}_{\perp}) \quad (114.2)$$

Recently the first indirect estimates of the GSF have been obtained by D'Alesio et al. [1], by fitting to midrapidity data on $pp^{\uparrow} \rightarrow \pi^0 + X$ measured by the PHENIX collaboration and using QSFs extracted earlier from SIDIS data. We will refer to these as the DMP fits. It is necessary to identify processes where the predictions of these fits can be compared with data in a straightforward manner.

R. M. Godbole · A. Kaushik (✉)
Centre for High Energy Physics, Indian Institute of Science, Banglore 560012, India
e-mail: abhiramk@iisc.ac.in

R. M. Godbole
e-mail: rohini@iisc.ac.in

A. Misra
Department of Physics, University of Mumbai, Mumbai 400098, India
e-mail: misra@physics.mu.ac.in

Md. Naimuddin (ed.), *XXII DAE High Energy Physics Symposium*, Springer Proceedings in Physics 203, https://doi.org/10.1007/978-3-319-73171-1_114

Open charm production, being highly gluon dominated, would be ideally suited for this. We therefore look at the process $pp^\uparrow \to D + X$. TSSA in this process was first studied by Anselmino et al. [2] with two extreme cases for the GSF - zero and maximal, and asymmetries of up to 25% were predicted with the maximal GSF. Here we present estimates of TSSA in this process using the DMP fits. We also look at the effect of TMD evolution on asymmetry predictions by considering the case of a saturated GSF evolved with and without TMD evolution.

The GPM expression for the numerator of the asymmetry is [3],

$$
\begin{aligned}
d\sigma^\uparrow - d\sigma^\downarrow &= \frac{E_D\, d\sigma^{p^\uparrow p \to DX}}{d^3 \boldsymbol{p}_D} - \frac{E_D\, d\sigma^{p^\downarrow p \to DX}}{d^3 \boldsymbol{p}_D} \qquad (114.3)\\
&= \int dx_a\, dx_b\, dz\, d^2\boldsymbol{k}_{\perp a}\, d^2\boldsymbol{k}_{\perp b}\, d^3\boldsymbol{k}_D\, \delta(\boldsymbol{k}_D \cdot \hat{\boldsymbol{p}}_c)\, \delta(\hat{s} + \hat{t} + \hat{u} - 2m_Q^2)\, \mathcal{C}(x_a, x_b, z, \boldsymbol{k}_D)\\
&\times \sum_{i=q,g} \left[\Delta^N f_{i/p^\uparrow}(x_a, \boldsymbol{k}_{\perp a})\, \hat{f}_{\bar{i}/p}(x_b, \boldsymbol{k}_{\perp b})\, \frac{d\hat{\sigma}^{i\bar{i} \to Q\bar{Q}}}{d}\hat{t}(x_a, x_b, \boldsymbol{k}_{\perp a}, \boldsymbol{k}_{\perp b}, \boldsymbol{k}_D)\, \hat{D}_{D/Q}(z, \boldsymbol{k}_D) \right]
\end{aligned}
$$

where $\Delta^N f_{g/p^\uparrow}(x, \boldsymbol{k}_\perp) \equiv \Delta^N f_{g/p^\uparrow}(x, k_\perp)\mathbf{S}.(\mathbf{p} \times \mathbf{k}_\perp)$ is the GSF and all other symbols have their usual meaning.

The unpolarized TMD PDF is parametrized with a gausian $\mathbf{k}_\perp$ dependence:

$$
f_{i/p}(x, \mathbf{k}_\perp) = f_{i/p}(x) \frac{e^{-k_\perp^2/\langle k_\perp^2 \rangle}}{\pi \langle k_\perp^2 \rangle} \qquad (114.4)
$$

where $f_{i/p}(x)$ is the standard PDF for parton i. $\langle k_\perp^2 \rangle$=0.25 GeV^2. The Sivers function is parametrized as:

$$
\Delta^N f_{i/p^\uparrow}(x, k_\perp) = 2\mathcal{N}_i(x)\, \sqrt{2e}\, \frac{k_\perp}{M'}\, e^{-k_\perp^2/M'^2}\, f_{i/p}(x, \mathbf{k}_\perp) \qquad (114.5)
$$

$$
\mathcal{N}_i(x) = N_i x^{\alpha_i} (1-x)^{\beta_i}\, \frac{(\alpha_i + \beta_i)^{(\alpha_i + \beta_i)}}{\alpha_i^{\alpha_i}\, \beta_i^{\beta_i}}, \qquad (114.6)
$$

where M', N_i, α_i and β_i are obtained from fits to data.

In Fig. 114.1, we present our estimates of asymmetry in $pp^\uparrow \to D + X$ at $\sqrt{s} = 200$ GeV using DMP fits [1], whose parameters are given in Table 114.1. In Fig. 114.2, we have presented comparison of predictions using DGLAP evolved and TMD evolved maximal GSF.

We find that the DMP-SIDIS1 fit of the GSF gives sizeable asymmetries in the regions $0.1 \lesssim x_F \lesssim 0.7$ with $P_T = 1.5$ GeV and $1.0 \lesssim P_T \lesssim 3.0$ with pseudorapidity $\eta = 3.8$. The DMP-SIDIS2 fit gives much smaller asymmetry in these kinematic ranges. Nevertheless, it is still non-negligible and dominates over the quark contributions. The inclusion of TMD evolution causes overall asymmetry predictions to diminsh. The peak asymmetry prediction obtained with a maximal gluon Sivers

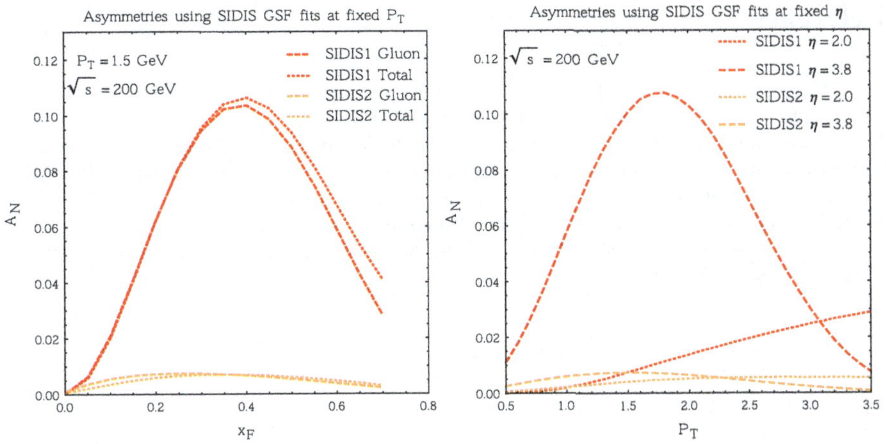

Fig. 114.1 A_N predictions for $pp^{\uparrow} \to D + X$ using DMP fits [1]. x_F-distribution in left and p_T-distribution in right

Table 114.1 DMP fits of the GSF

DMP-SIDIS 1	$N_g = 0.65$	$\alpha_g = 2.8$	$\beta_g = 2.8$	$\rho = 0.687$	$\langle k_\perp^2 \rangle = 0.25$ GeV^2
DMP-SIDIS 2	$N_g = 0.05$	$\alpha_g = 0.8$	$\beta_g = 1.4$	$\rho = 0.576$	

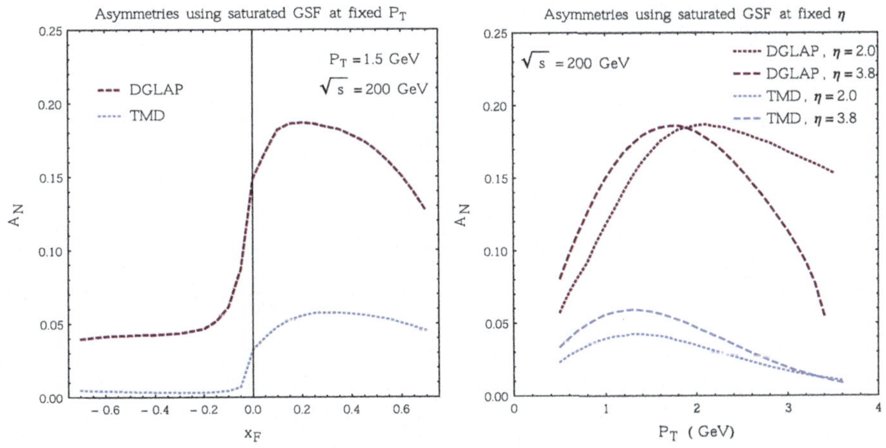

Fig. 114.2 Comparing predictions for A_N in $pp^{\uparrow} \to D + X$, obtained with DGLAP and TMD evolved densities. GSF taken to be maximal in both cases

function and TMD evolved densities goes down to less than a third of the peak value obtained with DGLAP evolved densities.

References

1. U. D'Alesio, F. Murgia, C. Pisano, Towards a first estimate of the gluon Sivers function from A_N data in pp collisions at RHIC. JHEP **09**, 119 (2015)
2. M. Anselmino, M. Boglione, U. D'Alesio, E. Leader, F. Murgia, Accessing Sivers gluon distribution via transverse single spin asymmetries in $p^{\uparrow} p \to D + X$ processes at RHIC. Phys. Rev. D **70**, 074025 (2004)
3. R.M. Godbole, A. Kaushik, A. Misra, Transverse single spin asymmetry in $p + p^{\uparrow} \to D + X$. Phys. Rev. D **94**, 114022 (2016) https://doi.org/10.1103/PhysRevD.94.114022, arXiv:1606.01818 [hep-ph]

Chapter 115
Effect of Initial Fluctuations on Quarkonia Suppression

Partha Bagchi, Nirupam Dutta and Ajit M. Srivastava

Introduction

Intensive investigation is going on in the field of quarkonia suppression in relativistic heavy-ion collisions. Matsui and Satz [1] first proposed J/ψ (bound state of charm and anti-charm) suppression as a signal for Quark Gluon Plasma (QGP) formation in Heavy Ion Collisions (HIC). The basic picture of this suppression is that due to the presence of deconfined medium, potential between quark (q) and anti-quark ($\bar{q}$) gets Debye screened. As a result the bound state swells in its size. If the Debye screening length of the medium is less than the radius of the state, then $q\bar{q}$ may not form bound state, leading to melting of the initial state. Due to this melting, the yield of the initial quarkonia state will be *suppressed*. Experimentally the phenomena was observed at CERN NA50 Experiment [2]. And J/ψ suppression pattern measured in Pb-Pb collisions at the CERN-SPS was used as evidence for deconfinement of quarks and gluons. However, there are other factors too that can lead to the suppression of J/ψ because of which it has not been possible to use J/ψ suppression as a clean signal for QGP.

The above picture of evolution of quarkonia in medium is taken as either static picture or *adiabatic* evolution of quantum states under Time-dependent perturbation. That means the quarkonia evolves as a quantum states corresponding to a Hamiltonian which *slowly* changes with time. But the reality for HIC is far away from this scenario. The fireball produced in HIC evolves rapidly. Starting from the time of collision it takes only few fm time for hadronisation. That means if quarkonia is described in potential model then $q\bar{q}$ potential is no-doubt time dependent. It has been shown that

P. Bagchi (✉) · N. Dutta
Variable Energy Cyclotron Centre, Kolkata, India
e-mail: p.bagchi@vecc.gov.in

N. Dutta
e-mail: niripamdu@gmail.com

A. M. Srivastava
Institute of Physics, Bhubaneswar, India
e-mail: ajit@iopb.res.in

Md. Naimuddin (ed.), *XXII DAE High Energy Physics Symposium*, Springer Proceedings in Physics 203, https://doi.org/10.1007/978-3-319-73171-1_115

the evolution of quarkonia is highly non-adiabatic during thermalization [3] and also during cooling [4] stage of evolution of fireball.

Initial Energy Density Fluctuation and Evolution of Quarkonia

The initial energy density of QGP is not homogeneous for the case of HIC. Though the pre thermalisation dynamics is not completely understood yet and this is initial condition for the post thermalisation dynamics, still there are few model calculation which shows that the initial energy density is highly fluctuating in space [5–8] and as time evolves, it also fluctuates in time. Since 4th power of the local temperature (T) is proportional to the local energy density of medium, the temperature of the medium is also fluctuating.

The Debye screened potential between quark and anti-quarks in QGP can be written as [9],

$$V(r) = -\frac{\alpha}{r} exp(-\omega_D r) + \frac{\sigma}{\omega_D}(1 - exp(-\omega_D r)) \tag{115.1}$$

where $\alpha_s = \frac{3}{4}\alpha$ is the strong coupling constant, σ is the string tension and Debye mass $\omega_D = T\sqrt{6\pi\alpha_s}$. That means the potential between quark and anti-quark in medium is also temperature dependent. If we consider a quarkonia is moving in QGP from one point to other point, then it *might* see huge change in temperature. As a result the potential (Hamiltonian) will also change accordingly. Hence the evolution of quarkonia should not be taken as adiabatic evolution. One should solve full time dependent Schrödinger equation to find the population of different resonance of charmonia and bottomonia.

Results and Discussions

We have used Glauber initial condition [10] to produce fluctuating initial condition for hydrodynamic evolution of the fireball created in HIC. We have generated temperature profile (a quarkonia feels) with time as quarkonia with different velocity (v) moves in QGP. And we found that the temperature changes with time as quarkonia moves from one point to another point. The change in temperature depends on several factors like velocity of quarkonia, initial position, direction of motion etc. In Fig. 115.1 (left) we have shown the profile of the temperature when a quarkonia moves along $+x$ axis (perpendicular to the beam direction) and started from $-4fm$ from the center of the medium.

One can see (Fig. 115.1 left) that the temperature, a quarkonia feels, changes rapidly. Accordingly the Hamiltonian will also change rapidly. As a result, the evolution of quarkonia can not be taken as adiabatic. Dissociation probability is calculated (using 1st order perturbation theory) for different velocity of J/ψ (Fig. 115.1 right) by choosing different temperature profile corresponding to different velocity (Fig. 115.1 left) with same initial position. To show velocity dependence of dissociation probability we varied v from $0.1c$ to c. But in reality the velocity may not be close to the speed of light. Again the J/ψ wave function remains non relativistic as it is calculated in J/ψ rest frame. As velocity increases, initially the dissociation

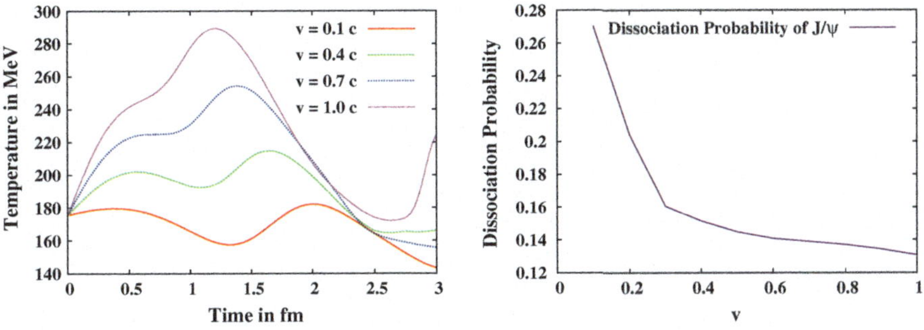

Fig. 115.1 Left: Temperature profile for different velocity of quarkonia. Right: Dissociation probability as a function of velocity

probability decreases fast then it almost saturates. That means J/ψ with high $p_\perp$ will survive more. This is distinguishable feature of our mechanism of quarkonia dissociation.

Acknowledgements We are very thankful to Arpan Das and Shreyansh S. Dave for useful discussions.

References

1. T. Matsui, H. Satz, Phys. Lett. B **178**, 416 (1986)
2. M. Abreu et al., NA50 Collaboration. Phys. Lett. **B477**; 28 (2000); H. Satz, J. Phys. **G 32**, R25 (2006)
3. P. Bagchi, A.M. Srivastava, Mod. Phys. Lett. A **30**(32), 1550162 (2015)
4. N. Dutta, N. Borghini, Mod. Phys. Lett. A **30**(37), 1550205 (2015)
5. S. Plumari, G.L. Guardo, F. Scardina, V. Greco, Phys. Rev. C **92**(5), 054902 (2015)
6. V.Y. Naboka et al., Phys. Rev. C **91**(1), 014906 (2015)
7. K. Paech, A. Dumitru, Phys. Lett. B **623**, 200 (2005)
8. B. Schenke, P. Tribedy, R. Venugopalan, Phys. Rev. C **89**(6), 064908 (2014)
9. H. Satz, J. Phys. Conf. Ser. **455**, 012045 (2013)
10. W. Broniowski, P. Bozek, M. Rybczynski, Phys. Rev. C **76**, 054905 (2007)

Chapter 116
Identified Hadron Production in Proton-Proton Collisions at $\sqrt{s} = 13\,\text{TeV}$ with ALICE at the LHC

Sourav Kundu

116.1 Introduction

After a shutdown period of two years, the Large Hadron Collider (LHC) has resumed its physics program in June 2015, with proton-proton (pp) collisions at $\sqrt{s} = 13$ TeV, which is the highest centre-of-mass energy ever reached in a collider experiment. Soft particle production in pp collisions cannot be fully described by perturbative Quantum Chromodynamics (pQCD) and hence pQCD inspired MC event generators need to be tuned based on experimental data. The measurement of identified particle production in pp collisions gives input for event generators to model soft parton interactions and the hadronization processes. Results from pp collisions are also used as a baseline to understand the medium effects present in heavy-ion collisions. ALICE measurements [1] from lower energies in pp collisions show that model predictions give a qualitative description of p/π and K/π[1] ratios as a function of p_{T}, but fail to describe it quantitatively.

Stable particles like $\pi^{\pm}$, $K^{\pm}$, p and $\bar{p}$ are identified in the rapidity region $|y| < 0.5$ with various ALICE sub-detectors: the Inner Tracking System (ITS), the Time Projection Chamber (TPC), the Time Of Flight detector (TOF) and the High Momentum Particle Identification Detector (HMPID), by using different PID techniques [1]. Resonances (K^{*0}, ϕ) and multi-strange baryons ($\Omega^{\pm}$, $\Xi^{\pm}$) are reconstructed through their hadronic decay daughters using invariant mass technique. Full description of the analysis technique can be found in [2, 3].

[1] As particles and anti-particles are produced in approximately equal abundance at LHC energies, the notations p/π and K/π refer to $(p + \bar{p})/(\pi^+ + \pi^-)$ and $(K^+ + K^-)/(\pi^+ + \pi^-)$ respectively.

Sourav Kundu for the ALICE Collaboration.

S. Kundu (✉)
National Institute of Science Education and Research, HBNI, Jatni 752050, India
e-mail: sourav.kundu@cern.ch

Md. Naimuddin (ed.), *XXII DAE High Energy Physics Symposium*, Springer Proceedings in Physics 203, https://doi.org/10.1007/978-3-319-73171-1_116

116.2 Results

The p/π and K/π ratios measured in $|y| < 0.5$ as a function of p_T are shown in Fig. 116.1 for $\sqrt{s} = 0.9, 7$, and 13 TeV. These ratios are also compared with PYTHIA 8 Monash-2013 [4]. The p_T dependent K/π ratio at $\sqrt{s} = 13$ TeV is consistent with the two lower energy measurements, but the maximum in the p/π ratio shows a slight shift towards higher p_T with increasing $\sqrt{s}$. Figure 116.2 shows the K^{*0}/K (left) and the ϕ/K (right) ratios as a function of $\sqrt{s}$. Results from pp collisions at $\sqrt{s} = 13$ TeV are compared with other lower energy measurements [2, 5–7]. No significant energy dependence is observed for these ratios in pp collisions. ϕ/K ratios in heavy-ion collisions are consistent with the measurements from pp collisions. However, K^{*0}/K ratios in heavy-ion collisions are found to be lower with respect to pp collisions. This decrease in K^{*0}/K ratios can be understood as the re-scattering of K^{*0} decay daughters in the hadronic phase [7]. The p_T integrated K/π and p/π ratios are observed to be independent of collision energies for $\sqrt{s} > 0.9$ TeV.

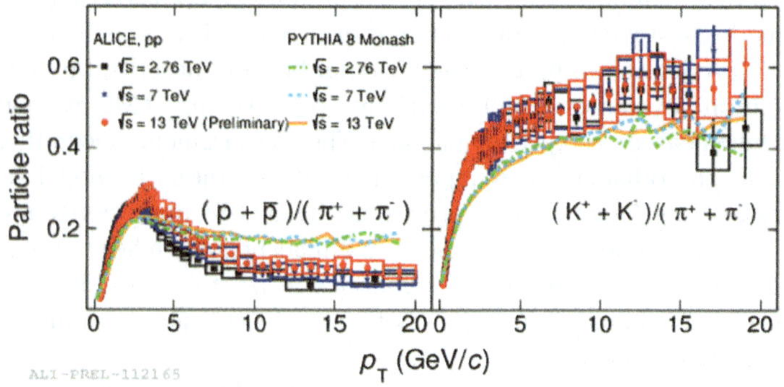

Fig. 116.1 Left panel shows the p/π ratios and right panel shows the K/π ratios as a function of p_T for three different collision energies. The lines show PYTHIA 8 Tune Monash-2013 predictions

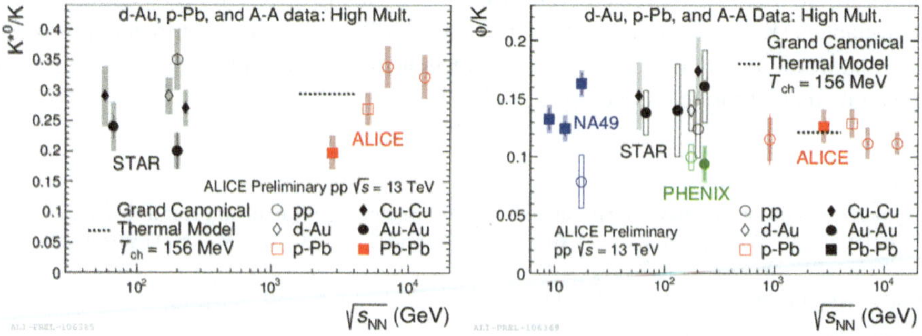

Fig. 116.2 p_T integrated K^{*0}/K (left) and ϕ/K (right) ratios as a function of $\sqrt{s}$. The results are compared with measurements at lower energies

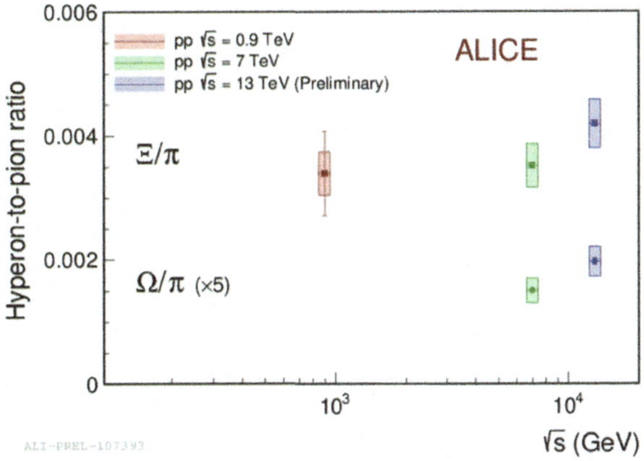

Fig. 116.3 p_{T} integrated hyperon-to-pion ratios as a function of $\sqrt{s}$

The p_{T} integrated hyperon to pion ratios as a function of $\sqrt{s}$ are shown in Fig. 116.3. A hint of increase is observed for hyperon to pion ratios at $\sqrt{s} = 13\,\text{TeV}$ with respect to lower energy data. This enhancement in multi-strange baryon production may be related to the increase of strange and multi-strange particle production as a function of multiplicity in pp collisions [8], as the charged particle multiplicity density increases with the collision energy.

116.3 Conclusions

The p_{T} dependent p/π and K/π ratios do not show any significant energy dependence at LHC energies and PYTHIA 8 describes the shape of these ratio qualitatively but not quantitatively. The p_{T} integrated K/π, p/π, K^{*0}/K and ϕ/K ratios are energy independent for $\sqrt{s} > 0.9$ TeV, however Ω/π and Ξ/π ratios are showing a hint of an increase with $\sqrt{s}$. In the future, multiplicity dependent studies in pp collisions at 13 TeV will help to disentangle the effect of collision energy and event multiplicity in the production of identified particles in pp collisions.

References

1. ALICE Collaboration, Eur Phys. J. C **75**, 226 (2015)
2. ALICE Collaboration, Eur. Phys. J. C **72**, 2183 (2012)
3. ALICE Collaboration, Phys. Lett. B **712**, 309–318 (2012)
4. P.Z. Skands, S. Carrazza, J. Rojo, Eur. Phys. J. C **74** (2014)
5. ALICE Collaboration, Eur. Phys. J. C **71**, 1594 (2011)
6. ALICE Collaboration, Eur. Phys. J. C **76**, 245 (2016)
7. ALICE Collaboration, Phys. Rev. C **91**, 024609 (2015)
8. ALICE Collaboration, Nat. Phys. **13**, 535 (2017)

Chapter 117
Constraints on Leptophilic Light Dark Matter from Internal Heat Flux of Earth

Bhavesh Chauhan and Subhendra Mohanty

117.1 Introduction

There is evidence for gravitational interaction of Dark Matter (DM) from rotation curves, and lensing measurements [1, 2]. However, the particle nature of DM (i.e. mass, spin, and interactions) remains a mystery. The search for recoil from scattering of DM particles in terrestrial experiments has only given null results thus far. In these experiments, the ionization of electrons is attributed to background radioactivity, and only non-ionizing scatterings are considered. As a result, the bounds on leptophilic DM which scatter with the electrons are weak [3].

We investigate the gravitational capture of leptophilic DM, and its subsequent annihilation/decay in Earth. DM in our local neighbourhood can scatter off electrons of the Earth, lose energy, and be gravitationally captured by an Earth. After accumulating inside, these DM particles can annihilate/decay into SM states, thus heating up the Earth. We estimate the heat produced from this process and compare with the experimentally measure value of 44 TW [4]. Measurement of geo-neutrinos indicates that 20–25 TW is contributed by the nuclear reactions [5] while the origin of remaining heat output is uncertain. In this work, we have taken a conservative estimate of 25 TW as the upper limit of heating by DM annihilations.

This article is organised as follows. In Sect. 117.2 we evaluate the capture rate of DM by Earth, and the heat produced by subsequent annihilations/decays. In Sect. 117.3 we discuss the results and compare with bounds from XENON10 experiment.

B. Chauhan (✉) · S. Mohanty
Physical Research Laboratory, Ahmedabad, India
e-mail: bhavesh@prl.res.in

B. Chauhan
Indian Institute of Technology, Gandhinagar, India

Md. Naimuddin (ed.), *XXII DAE High Energy Physics Symposium*, Springer Proceedings in Physics 203, https://doi.org/10.1007/978-3-319-73171-1_117

117.2 Heat Flux from Dark Matter

We assume that the galactic DM is thermal and follows Maxwellian distribution characterised by $\bar{v} = 300\,\mathrm{km/s}$. The local dark matter density is assumed to be $0.4\,\mathrm{GeV/cm^3}$ [6]. Following [7], the rate capture of DM particles by Earth is given as,

$$\mathcal{F} = \xi_\eta \sqrt{\pi} \left(\frac{3}{2}\right)^{3/2} n_\chi \frac{w^4}{\bar{v}^3} R_E^2 g(\sigma_{\chi e}) \int_0^{x_{max}} dx \int_0^{1.42x+1.72} \exp\left(\frac{-3w^2}{2\bar{v}^2}x\right) dy \tag{117.1}$$

this expression was derived using the conditions that in order for the incident DM particle to be gravitationally captured, it must first intersect the Earth such that it travels at least a distance R_E inside. It is also necessary that such a particle must lose sufficient energy via scatterings with electrons of Earth such that it is scattered inside the Hill sphere of Earth. Such constraints appear as limits of integration in the above expression.

The mass of DM particle (m_χ) and the scattering cross section with electrons appear in x_{max} and $g(\sigma_{\chi e})$ [7]. With reasonable assumptions, it can be seen that for both annihilations and decays, the heat produced by Earth is given by,

$$\mathcal{H} = \mathcal{F} \times m_\chi \tag{117.2}$$

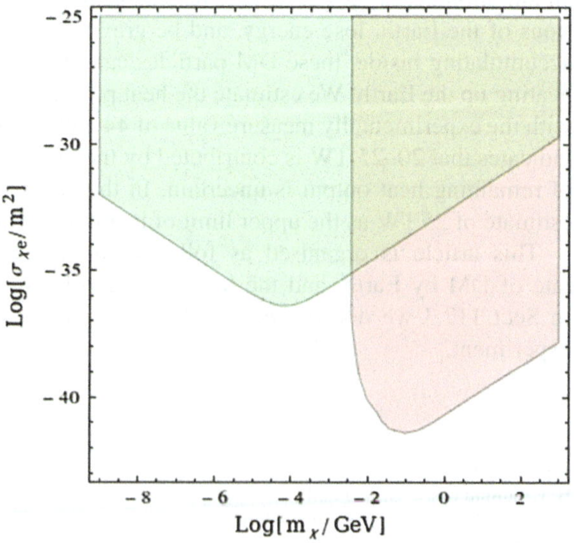

Fig. 117.1 The region shown in green produces more than 25 TW heat flux and is excluded. The result from XENON10 is shown in red

Given the mass of DM and scattering cross section with electron, the heat flux can be estimated. We exclude the parameter space that gives $\mathcal{H} \geq 25\,\text{TW}$ in Fig. 117.1. Similar analysis has been done for the Moon and Jupiter, however, the best and most reliable constraints are generated by Earth [7].

117.3 Results

Using the heat flux of Earth, we have excluded parameter space for low mass DM which has not been ruled out yet by direct detection experiments. In Fig. 117.1 we show the latest bounds on sub-GeV dark-matter from [3], as well as the new bounds from this analysis. We put novel limits on the scattering cross section of leptophilic DM and electrons in the mass range 1 eV–3 MeV. This result will be useful for testing already established and future models of leptophilic dark matter with mass in sub-MeV domain.

References

1. Katherine Garrett, Gintaras Duda, Dark matter: a primer. Adv. Astron. **2011**, 968283 (2011)
2. S. Profumo, Astrophysical probes of dark matter, in *Proceedings of the Theoretical Advanced Study Institute in Elementary Particle Physics: Searching for New Physics at Small and Large Scales (TASI 2012)* (2013), pp. 143–189
3. Rouven Essig, Aaron Manalaysay, Jeremy Mardon, Peter Sorensen, Tomer Volansky, First direct detection limits on sub-GeV dark matter from XENON10. Phys. Rev. Lett. **109**, 021301 (2012)
4. Henry N. Pollack, Suzanne J. Hurter, Jeffrey R. Johnson, Heat flow from the earth's interior: analysis of the global data set. Rev. Geophys. **31**(3), 267–280 (1993)
5. A. Gando et al., Partial radiogenic heat model for earth revealed by geoneutrino measurements. Nature Geo. **4**, 647–651 (2011)
6. Riccardo Catena, Piero Ullio, A novel determination of the local dark matter density. JCAP **1008**, 004 (2010)
7. Bhavesh Chauhan, Subhendra Mohanty, Constraints on leptophilic light dark matter from internal heat flux of Earth. Phys. Rev. D **94**(3), 035024 (2016)

Chapter 118
Ion Backflow Studies Related to a Proposed Micromegas Based TPC for the International Linear Collider

Deb Sankar Bhattacharya, Purba Bhattacharya, Supratik Mukhopadhyay, Nayana Majumdar, Sandip Sarkar, Paul Colas, David Attié, Serguei Ganjour, Aparajita Bhattacharya and Sudeb Bhattacharya

118.1 Introduction

The IBF is a secondary effect, occurring even in the MPGDs, that limits the performance of a TPC in a high-rate environment [1, 2]. The ions produced during charge multiplication partially flow back to the drift volume of the TPC. IBF is estimated as the ratio of the average number of backflowing ions (N_b) to the average number of ions produced (N_t) during amplification. Thus, the ion backflow fraction, $IBF = \frac{N_b}{N_t}$. Micromegas is a kind of MPGD which has stable high gain, good space and energy resolutions and intrinsically low ion feedback. It is an efficient candidate for the ILD-TPC. Previously [3] we have reported the basics of IBF measurement in MM. Here we report few advancements in our study so as to measure IBF more accurately in the context of a TPC.

D. S. Bhattacharya (✉)
Institute of Physics, Bhubaneswar 751005, India
e-mail: dsb.physics@gmail.com

P. Bhattacharya
Weizmann Institute of Science, Herzl St. 234, 7610001 Rehovot, Israel

S. Mukhopadhyay · N. Majumdar · S. Sarkar · A. Bhattacharya · S. Bhattacharya
Saha Institute of Nuclear Physics, Kolkata 700064, India

P. Colas · D. Attié · S. Ganjour
CEA Saclay, 91191 Gif Sur Yvette, France

A. Bhattacharya
Department of Physics, Jadavpur University, Kolkata 700032, India

Md. Naimuddin (ed.), *XXII DAE High Energy Physics Symposium*, Springer Proceedings in Physics 203, https://doi.org/10.1007/978-3-319-73171-1_118

118.2 Experimental and Numerical Studies

At the SINP test bench, we have installed a small drift chamber (Fig. 118.1a) setup that consists of a Micromegas readout, two drift meshes and a Wilson seal system to mount the Fe_{55} source. The details of the setup may be found in [3]. We have measured ion currents from the lower drift mesh (I_C) and the Micromesh (I_M). The sum, $I_C + I_M$ gives the average ion produced during avalanche and I_C is the number of ions that flow back to the drift volume. Hence experimentally we define IBF as $IBF = \frac{I_C}{I_C+I_M}$. We used the Danfysik and CAEN AH401D pico-ammeter to measure I_M and I_C respectively. While measuring currents, the respective electrode is grounded and the applied high voltages of other electrodes are accordingly changed such that the field configuration remains the same. The reason of using two drift meshes at same potential is to adopt a TPC-like environment with the small drift chamber. When there is only one drift mesh, the ions produced within the region, above the drift mesh will also be collected on the drift mesh (Fig. 118.1a). Therefore, considering the case of a TPC-cathode, the ion current on the drift mesh will be higher than what is expected. To block those ions we used another drift mesh on top of the existing drift mesh. For further advancement, we have then applied and varied an electric field, E_Z (direction of Z is shown in Fig. 118.1a) between the two drift meshes in order to improve the measurement.

Results: We have used standard Micromegas of amplification gap of 128 μm, and pitch of 63 μm. IBF in two different gases- Ar95%:$CF_4$3%:Iso2% (T2K gas for ILD-TPC), and in Ar95%:Iso5% have been measured. Figure 118.1b shows that IBF in T2K gas is smaller and the reason is that T2K has smaller diffusion coefficient. A set of different fields have been applied between the drift meshes. As shown in Fig. 118.2a, IBF is smaller when the field is positive, that is, upper drift mesh is in negative potential. The positive field helps the upper drift mesh to collect more of

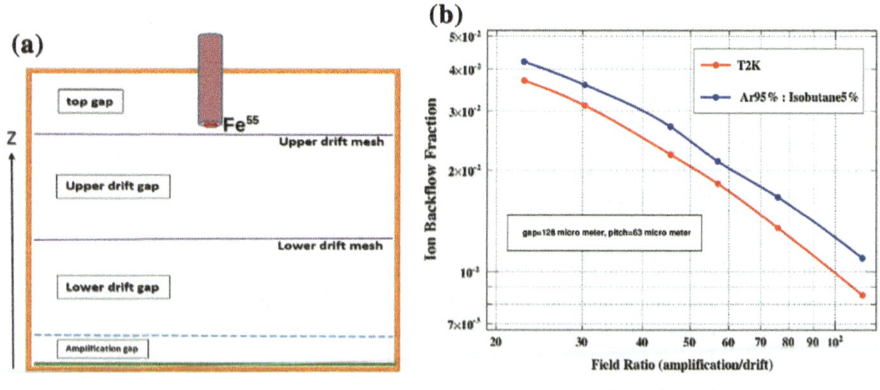

Fig. 118.1 **a** Schematic of the Drift chamber **b** IFB for two different gas mixtures

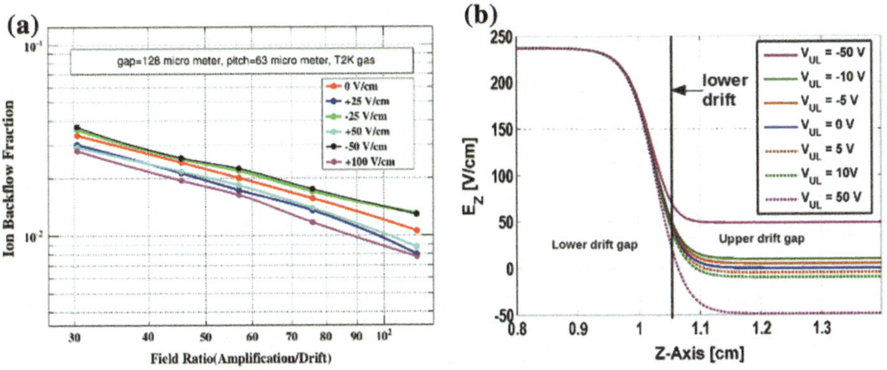

Fig. 118.2 **a** IBF with different fields between the drift meshes. **b** Simulated field values for the same configurations

those ions that are produced in lower and upper drift gaps (Fig. 118.1a). This also gives proper I_C measurement from lower drift mesh, like a TPC.

Numerical Study: Along with the experimental study, we have also pursued numerical study to comprehend the field orientation between the two drift meshes. In Fig. 118.2b, the filed variation is showed. When the upper drift mesh is negative with respect to the lower one (e.g., $V_{UL} = -50\,V$), the variation of field in lower drift gap is relatively smaller than other configurations. This ensures uniform drift field for the setup. A very high positive field, on the either hand, is not expected in the upper drift gap since that may cause to start ions collection from lower drift gap as well (Fig. 118.1a).

118.3 Conclusion and Acknowledgement

The IBF in a Micromegas is measured to be intrinsically low (below 1%). Effect of different gas mixtures is studied. We applied different fields between the two drift meshes. A positive field (from lower to upper drift mesh) is helpful to get proper ion current measurement from the lower drift and accurate IBF estimation. We Acknowledge IFCPAR/CEFIPRA (project 4304–1) for partial funding.

References

1. T. Behnke et al. (eds.), The international linear collider technical design report - Volume 4: Detectors (2013). arXiv:1306.6329v1 [physics.ins-det]
2. Y. Giomataris et al., MICROMEGAS: a high-granularity position-sensitive gaseous detector for high particle-flux environments. https://doi.org/10.1016/0168-9002(96)00175-1
3. P. Bhattacharya et al., J. Instrum. **10**, P09017 (2015)

Chapter 119
Time Resolution and Characteristic Studies of MWPC Detectors with Argon Based Gas Mixtures

Rajendra Nath Patra, R. N. Singaraju, S. Biswas, T. K. Nayak and Y. P. Viyogi

119.1 Introduction

The development of MWPC by Charpak [1] revolutionized the field of experimental nuclear and high energy physics. It found immense applications in medical imaging also [2, 3]. In VECC, Kolkata we have developed many small scale prototypes of MWPC detectors for the characteristic studies in terms of its efficiency, gain and time resolution with various gas mixtures.

119.2 Experimental Details and the Test Results

The MWPC detector contains an anode wire plane in between two conducting planes named as drift plane and the readout plane. The distance of the anode plane is 3 mm from both the drift and readout planes. Anode wire diameter is 20 μm and wire spacing is 2.8 mm. More details about the detector are given in [4]. The detector is tested using Ar/CO_2 gas mixture in 70:30 and 90:10 ratios in flow mode at laboratory temperature and pressure.

R. N. Patra (✉) · R. N. Singaraju · T. K. Nayak · Y. P. Viyogi
Variable Energy Cyclotron Centre, HBNI, Kolkata 700064, India
e-mail: rajendrapatra07@gmail.com

S. Biswas
Bose Institute, 93/1 APC Road, Kolkata 700009, India

T. K. Nayak
CERN, 1211 Geneva 23, CH, Switzerland

Md. Naimuddin (ed.), *XXII DAE High Energy Physics Symposium*, Springer Proceedings in Physics 203, https://doi.org/10.1007/978-3-319-73171-1_119

119.2.1 Efficiency and Gain

The efficiency of the detector is measured with ^{106}Ru β-source using a 3-fold trigger setup made up with three scintillator detectors - two crossed scintillators above and one scintillator below the MWPC under study. The efficiency as a function of applied high voltage (HV) is shown in Fig. 119.1 for both gas mixtures. In the plateau region the efficiency is ∼94% in both cases.

The detector is also tested using ^{55}Fe 5.9 keV X-ray source to measure the gas gain. The anode energy spectrum of ^{55}Fe at 1900 V in Ar/CO_2 (70:30) gas is given in Fig. 119.2a. The anode signal induces an opposite polarity signal in the readout. The induced charge is shared by the readout and the drift plane equally. We measured the charge sharing fraction. This is shown in Fig. 119.2b. The charge fraction is found to be ∼48 and ∼47% for the two cases.

The gain of the detector is calculated from the mean ADC of the ^{55}Fe spectrum with a Gaussian fitting is shown in Fig. 119.3a as a function of HV. We find that in

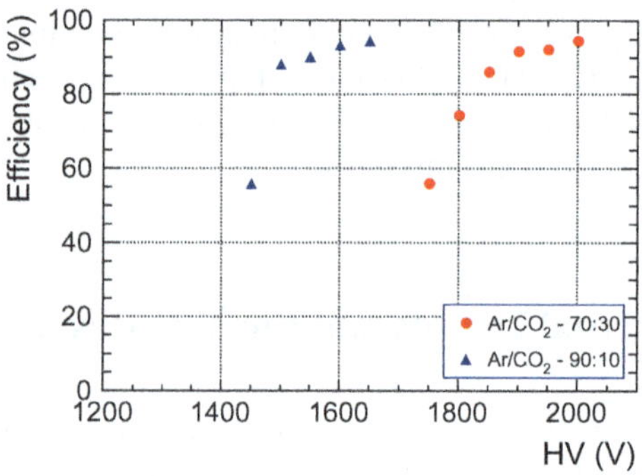

Fig. 119.1 Efficiency as a function of HV in Ar/CO_2 70:30 and 90:10 gas mixtures

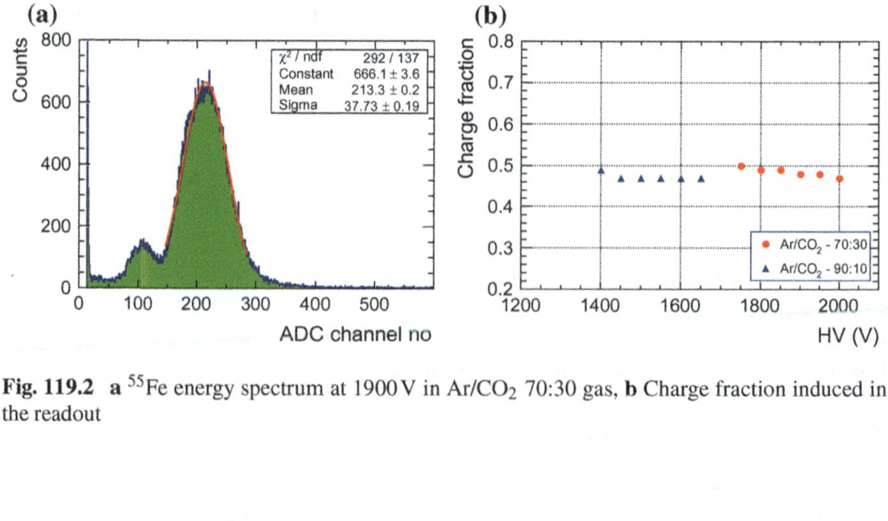

Fig. 119.2 **a** ^{55}Fe energy spectrum at 1900 V in Ar/CO_2 70:30 gas, **b** Charge fraction induced in the readout

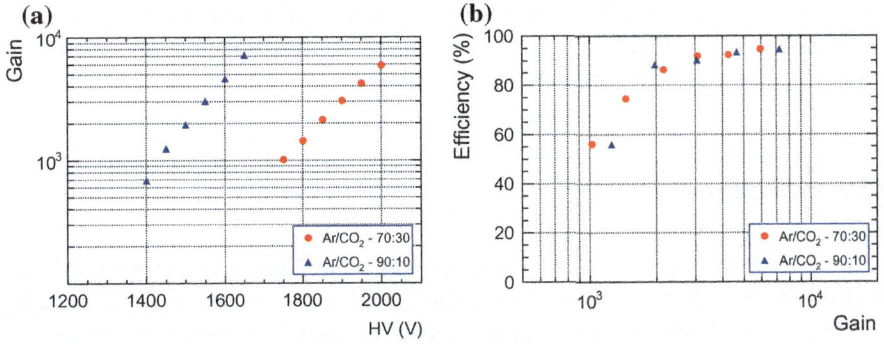

Fig. 119.3 **a** Gain variation as a function of HV in Ar/CO_2 70:30 and 90:10 gas, **b** Efficiency as a function of gain

case of Ar/CO_2 90:10 gas has higher gain at low operating voltage compared to the Ar/CO_2 70:30 because of larger fraction of Argon content.

Efficiency of the detector should depend on the gain only, providing all other setup condition same. This is demonstrated in Fig. 119.3b where efficiency is plotted versus gain for the given HV.

119.2.2 Time Resolution

Time resolution of a detector depends on spatial distribution of the primary charge cluster, electric field, gas and pressure. In this study time resolution was measured with both gas mixtures. It was measured using the 3-fold scintillator signal as a start signal of a ORTEC TAC and the fast logic signal of the detector as stop signal.

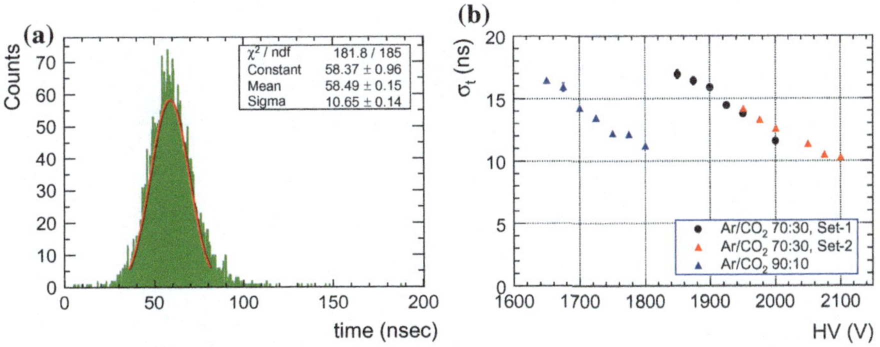

Fig. 119.4 **a** Time spectrum at 2075 V in Ar/CO_2 70:30 gas, **b** σ_t variation as a function of HV

Fig. 119.4a shows a typical time spectrum at 2075 V with Ar/CO_2 70:30 gas mixture, using which we obtained $\sigma_t \sim 10$ ns. The time resolution (σ_t) as a function of HV for the two gas mixtures is plotted in Fig. 119.4b.

119.3 Summary

A number of small prototype MWPC detectors were fabricated and operated with Ar/CO_2 70:30 and 90:10 gas mixtures. The efficiency, charge fraction, gain, timing resolution have been measured using radioactive sources. A comparative study of gain and efficiency has also been performed. Gas gain approaching 10^4 is achieved for both the gas mixtures. Time resolution of the detector is also measured with different gas mixtures and σ_t value of $\sim$10 ns is obtained.

Acknowledgements RNP acknowledges the receipt of UGC-NET fellowship for this work and SB acknowledges the support of DST-SERB Ramanujan Fellowship. YPV thanks INSA for the Senior Scientist position.

References

1. G. Charpak et al., Nucl Instrum. Methods **62**, 262 (1968)
2. W.T. Chu et al., Physics and Engineering in Medical Imaging, *Proceedings of the SPIE* (1982), vol. 221, p. 03772. https://doi.org/10.1117/12.934519
3. Jeffrey L. Lacy et al., J Nucl. Med. **25**, 1003–1012 (1984)
4. R.N. Patra et al., Proc. DAE-BRNS Symp. Nucl. Phys. **61**, 1052 (2016)

Chapter 120
BRST Qantization on Torus Knot

Vipul Kumar Pandey and Bhabani Prasad Mandal

120.1 Introduction

The celebrated BRST transformation [1] characterized by an infinitesimal, anti-commuting, global parameter are extremely powerful tools in describing various renormalizable field theoretic models. Thus it is important to study BRST formulation for all models. To the best of our knowledge there were no BRST formulation for a particle which is constrained to move on the surface of torus knot. Knot theory [2, 3], based on mathematical concepts has found immense applications in various branch of frontier physics. Knot invariants in physical systems were introduced long ago and has got considerable impact during last one and half decades [2–6], especially when interpreted as Wilson loop observable in Chern-Simon(CS) theory [5].

In this paper we discuss for the first time the BRST/anti-BRST formulation for a particle on the torus knot [7]. We further construct the nilpotent charges which generates these transformations by using the constraints of the system.

120.2 Particle on a Torus Knot

Lagrangian for a particle on surface of torus knot is given by [4, 6]

$$L = \frac{1}{2}ma^2 \frac{\dot{\eta}^2 + \dot{\theta}^2 + \sinh\eta^2\dot{\phi}^2}{(\cosh\eta - \cos\theta)^2} - \lambda(p\theta + q\phi) \tag{120.1}$$

V. K. Pandey (✉) · B. P. Mandal
Banaras Hindu University, Varanasi 221005, India
e-mail: vipulvaranasi@gmail.com

B. P. Mandal
e-mail: bhabani.mandal@gmail.com

Md. Naimuddin (ed.), *XXII DAE High Energy Physics Symposium*, Springer Proceedings in Physics 203, https://doi.org/10.1007/978-3-319-73171-1_120

where (η, θ, ϕ) are the toroidal coordinates for toric geometry, p and q are co-prime integers, and λ is Lagrange multiplier. The constraint that forces the particle to move in knot is imposed as $(p\theta + q\phi) \approx 0$. The Hamiltonian for the particle on torus knot corresponding to this Lagrangian is then written as

$$H = \frac{(\cosh\eta - \cos\theta)^2}{ma^2}\left[p_\eta^2 + p_\theta^2 + \frac{p_\phi^2}{\sinh\eta^2}\right] + \lambda(p\theta + q\phi) \qquad (120.2)$$

This model is analogous to the rigid rotor model discussed in [8].

To quantize the system and realize the physics, we have to diagonalize the system in presence of the constraints $\Phi = (p\theta + q\phi) \approx 0$. To execute this, we will discard the [8] terms proportional to p_θ^2 and p_ϕ^2 as they don't commute with the constraint equation. The effective Hamiltonian for this system keeping the essential physical content of the theory intact is then written as

$$H_c = \frac{(\cosh\eta - \cos\theta)^2}{ma^2}p_\eta^2 + \lambda(p\theta + q\phi) \qquad (120.3)$$

120.3 Particle on Torus Knot as Gauge Theory

The constraint which doesn't commute with unphysical operators is to be identified with the generator of gauge transformations. In this case the gauge transformation is an unitary operator of the form [8]

$$U_f = e^{-if(t)\Phi} \qquad (120.4)$$

where $f(t)$ is an arbitrary c number and function of t. Under this transformation only p_θ and p_ϕ transform non-trivially. First order Lagrangian for torus knot is written as

$$L_c = p_\eta\dot{\eta} + p_\theta\dot{\theta} + p_\phi\dot{\phi} - \frac{(\cosh\eta - \cos\theta)^2}{ma^2}p_\eta^2 - \lambda(p\theta + q\phi) \qquad (120.5)$$

Here we have enlarged the dynamical degrees of freedom to include the Lagrange multiplier λ and its canonically conjugate momentum p_λ. This Lagrangian is invariant under following time dependent gauge transformation [8]

$$\delta\lambda = \dot{f}(t), \quad \delta p_\theta = pf(t), \quad \delta p_\phi = qf(t), \quad \delta p_\eta = \delta\eta = \delta\theta = \delta\phi = \delta p_\lambda = 0 \qquad (120.6)$$

as under this gauge transformation the Lagrangian corresponding canonical Hamiltonian H_c changes only by a total time derivative, $\frac{d}{dt}[f(t)(p\theta + q\phi)]$.

Hence the action is invariant under gauge transformation in (120.6).

120.4 Gauge Fixing and BRST Transformation

To construct the BRST and anti-BRST symmetry for the particle on the surface of a torus knot we consider the gauge invariant Lagrangian of the torus knot obtained in last section and replace the gauge transformation in (120.6) by introducing new anti-commuting variables c and $\bar{c}$ and a commuting variable b such that the BRST transformations

$$\begin{aligned} \delta\lambda = -\dot{c}, \quad \delta p_\theta = pc, \quad \delta p_\phi = qc, \quad \delta\bar{c} = -p_\lambda = b \\ \delta p_\eta = \delta\eta = \delta\theta = \delta\phi = \delta p_\lambda = \delta c = \delta b = 0 \end{aligned} \tag{120.7}$$

are nilpotent in nature. The gauge fixed effective Lagrangian which is invariant under this kind of BRST transformation is then constructed as

$$\begin{aligned} L_{eff} &= p_\eta\dot{\eta} + p_\theta\dot{\theta} + p_\phi\dot{\phi} - \frac{(\cosh\eta - \cos\theta)^2}{ma^2} p_\eta^2 - \lambda(p\theta + q\phi) - b\left(\dot{\lambda} - \frac{p_\theta}{2p} - \frac{p_\phi}{2q}\right) \\ &- \frac{b^2}{2} + \dot{\bar{c}}\dot{c} - \bar{c}c \end{aligned} \tag{120.8}$$

The BRST charge which generates the transformation in (120.7) is written as

$$Q_b = -ic(p\theta + q\phi) - i\dot{c}p_\lambda. \tag{120.9}$$

The Lagrangian (120.8) is also invariant under anti-BRST transformation in which role of c and $-\bar{c}$ are interchanged.

The nilpotent charge for the anti-BRST symmetry is written as

$$Q_{ab} = i\bar{c}(p\theta + q\phi) + i\dot{\bar{c}}p_\lambda \tag{120.10}$$

It is straight forward to check that these charges are nilpotent i.e. $Q_b^2 = 0 = Q_{ab}^2$ and satisfy

$$\{Q_b, \ Q_{ab}\} = 0 \tag{120.11}$$

Notes and Comments. We for the first time construct BRST/anti-BRST transformations for particle moving on torus knot. The nilpotent BRST/anti-BRST charges are constructed which produce such nilpotent transformations. Our formulation will be helpful in characterizing various field theoretic models defined on torus manifold.

References

1. C. Becchi, A. Rouet, R. Stora, Ann. Phys. **98**(2), 287321 (1976), I.V. Tyutin, Gauge invariance in field theory and statistical physics in operator formalism. Lebedev Phys. Inst. prepr. **39**, (1975)
2. L. Faddeev, A.J. Niemi, Nature **387**, 58 (1997)
3. P.M. Morse, H. Feshbach, *Methods of Theoretical Physics* (McGraw Hill Publication, 1953)
4. V.V. Sreedhar, Ann. Phys. **359**, 20 (2015)
5. A. Gorsky, A. Milekhin, Condensates and instanton - torus knot duality. Hidden Physics at UV scale, arXiv:1412.8455
6. P. Das, S. Ghosh, Particle on a torus knot: a Hamiltonian analysis, arXiv:1508.06411
7. V. K. Pandey, B. P. Mandal, BRST Symmetry for Torus Knot, arxiv:1702.03212
8. D. Nemeschansky, C. Preitschopf, M. Weinstein, Ann. Phys. **183**, 226268 (1988)

Chapter 121
Single and Double Differential Drell-Yan Cross Section Measurements Using the CMS Detector

Genius Walia

121.1 Introduction

The Drell-Yan (DY) [1] event is the best studied physics process at the Large Hadron Collider (LHC) due to very clean final state of a pair of isolated and opposite sign leptons. The total and differential cross sections, for the DY process have been calculated up to the next-to-next-to-leading order (NNLO) in the perturbation theory. At the LHC, the measured cross section as a function of the accessible dilepton invariant mass varies over about 9 orders of magnitude and it matches the prediction within few percent. A thorough understanding of the origin of the transverse momentum (p_T) of the vector bosons is essential for future high precision measurement of W-boson mass. In recent years, a new variable ϕ^* [2] has been proposed which utilizes the angular correlation among the leptons as a measure to probe p_T. The first measurement of the differential cross section $d\sigma/d\phi^*$ for the DY in the Compact Muon Solenoid (CMS) experiment at $\sqrt{s} = 8\,\text{TeV}$ is presented [3]. The rapidity dependence of this distribution is measured in terms of a double-differential cross section $d^2\sigma/d\phi^* d|y|$ and is reported here.

121.2 Analysis Strategy

The work presented in this paper is based on single muon data collected by the CMS [4] experiment during the 2012 run corresponding to the total integrated luminosity of $19.\,\text{fb}^{-1}$. The simulation samples are used to model the signal (DY) and various

---

On behalf of CMS Collaboration.

G. Walia (✉)
Panjab University, Chandigarh, India
e-mail: genius.walia@cern.ch

Md. Naimuddin (ed.), *XXII DAE High Energy Physics Symposium*, Springer Proceedings in Physics 203, https://doi.org/10.1007/978-3-319-73171-1_121

background processes. The hard-scatter part of signal and some of the dominant background processes have been generated using matrix element (ME) calculations: either MADGRAPH [5] at leading order(LO) or POWHEG [6] at next-to-leading order (NLO).

The online di-muon event selection is done using inclusive single-muon trigger requiring $p_T > 24\,\mathrm{GeV}$ and muon pseudo-rapidity ($|\eta|$) < 2.1. For muon reconstruction, Particle Flow Algorithm is used, which relies on an optimised combination of information from the various elements of the CMS detector. Reconstructed muon candidates are then required to satisfy CMS recommended identification and isolation selection criteria. The oppositely charged dimuon events with invariant mass in the range $60 < M_{\mu\mu} < 120\,\mathrm{GeV}$ are retained, in which one muon, consistent with the trigger, satisfies $p_T > 30\,\mathrm{GeV}$ and $|\eta| < 2.1$, while the other has $p_T > 20\,\mathrm{GeV}$ and $|\eta| < 2.4$. For data simulation comparison, various correction factors are applied on both the data and simulated samples.

The event selections described above leads to a very pure Z-boson events, with less than 0.1% background contamination. The inclusive production of $t\bar{t}$ and di-boson production are the major background processes and are estimated from the simulation. After the removal of estimated background using bin-by-bin subtraction, the data is unfolded from the reconstructed distribution to the true produced ϕ^* distribution. The variation of reconstructed ϕ^* as function of generated ϕ^* is of the order of 1% for 1-dimensional case. For the two dimensional unfolding, 216 bins has been defined corresponding to 36 ϕ^* bins and 6 $|y|$ bins and the bin-migration is of the order of 6%.

121.3 Results and Discussion

The measured cross sections are compared with three different theoretical predictions: (i) ResBos [7], (ii) POWHEG interfaced with PYTHIA6 and (iii) MADGRAPH. The total cross sections for the theoretical predictions, except for ResBos, are scaled to the value calculated using Fully Exclusive W, Z Production (FEWZ) simulation code. The uncertainty of the data distribution is dominated by the fully correlated uncertainty on the integrated luminosity (2.6%) of the data sample. The uncertainty on the MADGRAPH distribution is dominated by the Parton Distribution Function (PDF) uncertainty of the FEWZ calculated on the cross-section. For POWHEG the uncertainty is dominated by the PDF uncertainty calculated by varying the CT10 PDF.

The absolute and normalised cross sections measured in the di-electron and di-muon channels are combined using the best linear unbiased estimator (BLUE) [8]. None of the predictions matches the measurements perfectly for the entire range of ϕ^* covered in this analysis. Figure 121.1 shows MADGRAPH provides best description with a disagreement of at most 4% over the entire range. ResBos successfully describes the data at low ϕ^* and has a disagreement up to 9% for $\phi^* > 0.1$. POWHEG

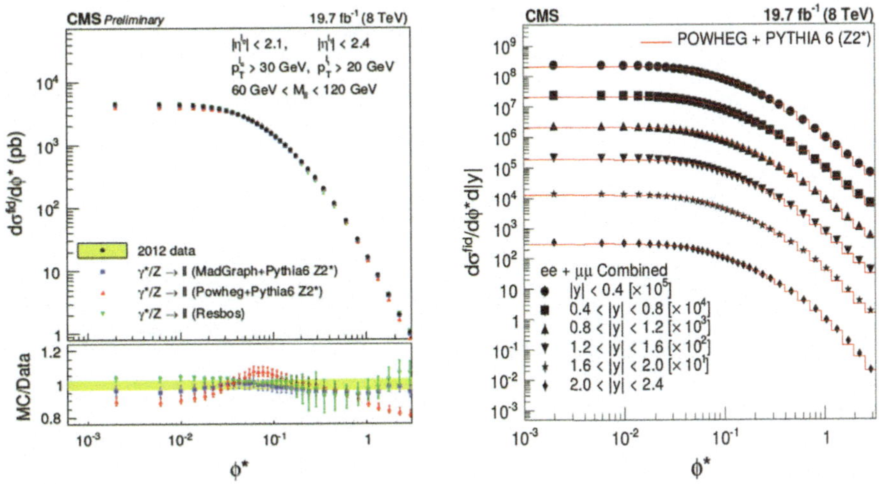

Fig. 121.1 The measured single (left) and double (right) differential cross sections after BLUE combination [3]

interfaced with PYTHIA6 provides the least accurate prediction, with a disagreement up to 8% for $\phi^* < 0.1$ and up to 18% for $\phi^* > 0.1$.

References

1. S.D. Drell, T.M. Yan, Massive lepton-pair production in Hadron-Hadron collisions at high energies. Phys. Rev. Lett. **25**, 316 (1970)
2. A. Banfi et al., Predictions for Drell-Yan $\phi_\eta{}^*$ and Q_T observables at the LHC. Phys. Lett. B **701** (2011)
3. CMS Collaboration, Measurements of ϕ^* differential cross sections for DrellYan events in pp collisions at $\sqrt{s}$ = 8 TeV. CMS-PAS-SMP-15-002 (2016)
4. C.M.S. Collaboration, The CMS experiment at the CERN LHC. JINST **3**, S08004 (2008)
5. J. Alwall et al., MadGraph 5: going beyond. JHEP **1106**, 128 (2011)
6. P. Nason, A new method for combining NLO QCD with shower Monte Carlo 321 algorithms. JHEP **11**, 040 (2004)
7. G. Ladinsky, C. Yuan, The Nonperturbative regime in QCD resummation for gauge boson production at hadron colliders. Phys.Rev. **D50** (1994)
8. A. Valassi, Combining correlated measurements of several different physical quantities. Nucl. Instrum. Methods **A500**, 391405 (2003)

Chapter 122
Study of the Associated Production of Higgs Boson with a Single Top Quark

Pallabi Das

122.1 Introduction

According to the Standard Model (SM), the Higgs to fermion coupling (y_f) is proportional to the fermion mass. Top being the heaviest fermion, top to Higgs coupling (y_t), is expected to be the largest and is of $\mathcal{O}(1)$. At the LHC, associated production of Higgs with top quark pair $pp \rightarrow t\bar{t}H$ is used for measuring y_t. However, y_t can also be estimated with the $pp \rightarrow tHq$ process [2], representative Feynman diagrams are shown in Fig. 122.1.

The magnitude of tHq production cross section is driven by the interference of two diagrams where the Higgs couples to either a W boson or a top quark. In SM, these two amplitudes interfere destructively; as a result, the cross section is very low: ~ 71 fb in pp collisions at 13 TeV center of mass energy. If there are effects beyond SM, due to anomalous coupling of Higgs, it can lead to a large enhancement of the cross section. Hence, in addition to measuring the y_t, this process is also sensitive to the relative signs of coupling of Higgs to W boson (WWH) and top quark (ttH).

122.2 Event Selection

Criteria for the selection of signal events have been derived from the study of simulated samples. In the multi-leptonic final state of the signal, the major contribution is from Higgs decays through W boson pairs ($H \rightarrow WW$). We consider the case where one of the W decays leptonically and the other hadronically. For the top decay

Pallabi Das on behalf of the CMS Collaboration.

P. Das (✉)
Tata Institute of Fundamental Research, Mumbai, India
e-mail: pallabi.das@cern.ch

Md. Naimuddin (ed.), *XXII DAE High Energy Physics Symposium*, Springer Proceedings in Physics 203, https://doi.org/10.1007/978-3-319-73171-1_122

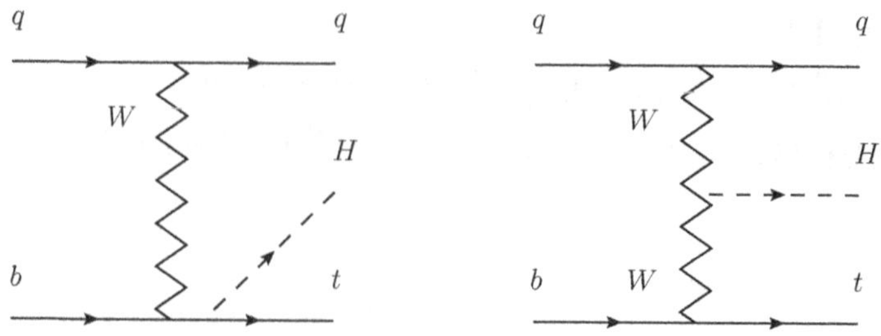

Fig. 122.1 Leading-order diagrams of $pp \rightarrow tHq$ process

($t \rightarrow Wb$), leptonic decay of W has been considered ($W \rightarrow l\nu$), leading to two leptons in the final state. We choose the same sign pair of leptons since it has lesser background as compared to the opposite signed pairs. For $\mu^{\pm}\mu^{\pm}$ final state, the p_T of the two same sign leading muons satisfying the stringent criteria based on reconstruction and isolation, is required to be above 20 GeV within a pseudorapidity region $|\eta| < 2.4$. Any event having a third electron/muon which also passes the stringent criteria for selection with $p_T > 20$ GeV is rejected. Also, there should not be any hadronically decaying tau in the selected event.

Jets are required to have p_T greater than 25 GeV within $|\eta| < 4.5$. They may be identified as originating from a b quark based on a set of special criteria within $|\eta| < 2.4$. The event is required to contain at least one b tagged jet from the top decay. The light quark in the tHq final state is expected to be in the forward rapidity region. The jet with the highest $|\eta|$ which fails the b tag requirement is tagged as the forward jet. As one W decays hadronically, we require at least three jets in the event.

122.3 Signal Discrimination

The simulated tHq signal is discriminated from the main backgrounds, $t\bar{t}$ and $t\bar{t}H$ production processes, using a boosted decision tree (BDT) classifier. The classifier performs a multi-variate analysis based on event characteristics and the output value determines whether the event is signal or background like. From the BDT output, we have obtained moderate separation between signal and background like events as can be seen in Fig. 122.2.

The set of discriminating variables given as inputs to the classifier are shown in Fig. 122.3.

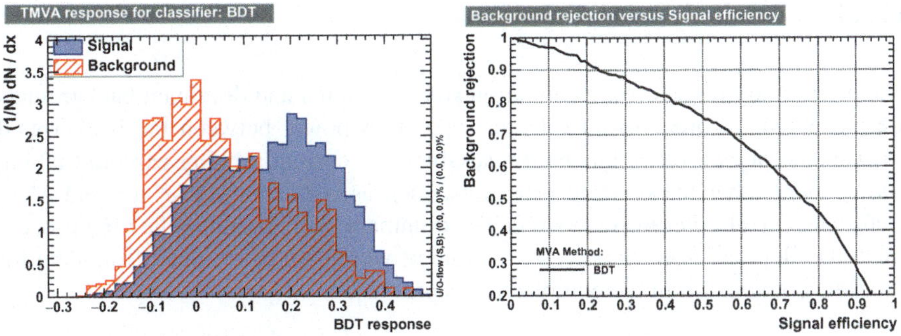

Fig. 122.2 The BDT MVA output

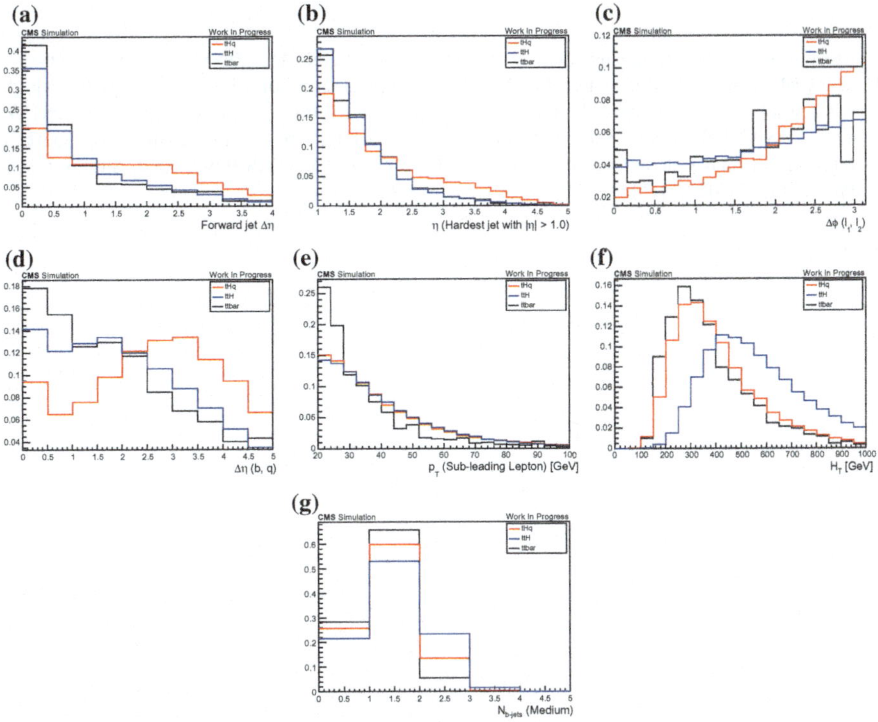

Fig. 122.3 Classifer inputs: **a** η gap between the forward jet and the nearest jet or lepton, **b** maximum $|\eta|$ of light quark jets with $|\eta| > 1.0$, **c** ϕ difference between the two leading muons, **d** η gap between the best b tagged jet and the forward jet, **e** p_T of the sub-leading muon, **f** scalar sum of p_T of all jets, **g** number of b jets in the event

122.4 Conclusions and Outlook

The preliminary results from simulation study of signal and dominant backgrounds are presented. In order to have better separation power between the background process and signal, other kinematic variables are currently being explored taking into account other background processes, such as $t\bar{t}Z$, $t\bar{t}W$ etc. $e^{\pm}\mu^{\pm}$ and $e^{\pm}e^{\pm}$ final states are also being considered. The dominant background to the tHq process are "fake" leptons from $t\bar{t}$ process, the rate of which are determined from data for data/MC comparison in the completed analysis [3].

References

1. The CMS Collaboration, The CMS experiment at the CERN LHC. JINST3:S08004 (2008)
2. The CMS Collaboration, Search for the associated production of a Higgs boson with a single top quark in proton-proton collisions at $\sqrt{s} = 8$ TeV. JHEP **06**, 177 (2016), arXiv:1509.08159
3. The CMS Collaboration, Search for production of a Higgs boson and a single top quark in multilepton final states in proton collisions at $\sqrt{s} = 13$ TeV. CMS Phys. Anal. Summ. CMS-PAS-HIG-17-005 (2017)

Chapter 123
Search for Model Independent Di-Higgs Production in the bb$\tau_h\tau_h$ channel at CMS

Ram Krishna Dewanjee

123.1 Introduction

While the mass of the Standard Model (SM) higgs boson (H) is now known, it is still not sufficient to fully unravel the nature of electroweak symmetry breaking. Measurement of Higgs self-coupling allows to probe higher order terms in the Higgs potential which in SM is given by:

$$V = \frac{1}{2}m_H^2\Phi_H^2 + \lambda v\Phi_H^3 + \frac{\tilde{\lambda}}{4}\Phi_H^4$$

with $m_H = 125\,\text{GeV}$ and $v = 246\,\text{GeV}$. The tri-linear higgs self-coupling (λv) can be measured via di-higgs production [2]. The leading order (LO) feynman diagrams for SM di-higgs production are shown in Fig. 123.1a, b. Due to the destructive interference of the two diagrams, di-higgs cross section within SM is small [3](10.16 fb at $\sqrt{s} = 8\,\text{TeV}$). This type of di-higgs production will be referred to as **Non resonant** production. Many BSM theories predict enhanced di-higgs production rate (up to 1 pb) if a heavy resonance X decays to a pair of higgs bosons [4] e.g. spin 0 (radion) or spin 2 (graviton). LO feynman diagram for such **Resonant** di-higgs production is shown in Fig. 123.1c.

On behalf of the CMS Collaboration.

R. K. Dewanjee (✉)
Laboratory of High Energy and Computational Physics, KBFI, Rävala pst 10, 10143 Tallinn, Estonia
e-mail: ram.krishna.dewanjee@cern.ch

Md. Naimuddin (ed.), *XXII DAE High Energy Physics Symposium*, Springer Proceedings in Physics 203, https://doi.org/10.1007/978-3-319-73171-1_123

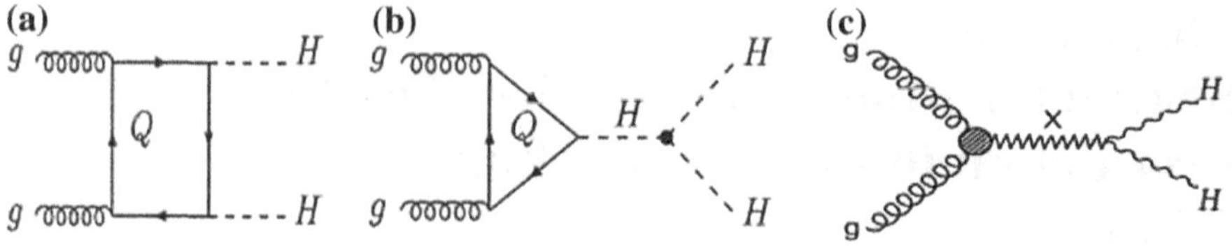

Fig. 123.1 LO Feynman diagrams for Non resonant (**a** and **b**) and Resonant (**c**) di-Higgs production. [6]

123.2 Analysis Strategy and Event Categorization

Due to the presence of b-quarks in the final state, the analysis is binned in b-tagged jet multiplicity. Three event categories: 2Jet0tag, 2Jet1tag and 2Jet2tag are defined depending on the number of jets passing b-tagging criteria out of the two (non τ_h) jets in the event that have the highest b-tagging score ($p_T > 20\,\text{GeV}$, $|\eta| < 2.4$[1]). Besides, two opposite sign τ_h jets are also required ($p_T > 45\,\text{GeV}$, $|\eta| < 2.1$) in the event. Additional selections (e.g. mass window cuts on di-jet and di-tau invariant masses, kinematic fitting of the di-jet system and vetoing any event containing an electron/muon) are applied to reject backgrounds.

123.3 Backgrounds

Important backgrounds for both resonant and non resonant di-Higgs searches are:

1. **Drell-Yan+Jets/**$Z \rightarrow \tau_h\tau_h$: This is the dominant irreducible background in both searches and is estimated from $Z \rightarrow \mu\mu$ events in data from which the muons are removed and replaced by τ_h with similar kinematics from simulation (Embedding).
2. **W+Jets**: It is caused by a jet/μ faking a τ_h. It is modelled in shape from simulation and in normalization from a high m_T[2] region ($m_T > 70$ GeV) in data having high purity of W+Jets events.
3. **QCD MultiJets**: This is the dominant reducible background in both analyses and is estimated from data using control regions defined by di-tau charge and isolation.
4. $t\bar{t}$**+Jets**: It is estimated in shape from simulation and in yield from a $t\bar{t}$ enriched control region.
5. **Drell-Yan+Jets/**$Z \rightarrow ll(l = e/\mu)$: This background primarily arises from e$\rightarrow$ τ_h and $\mu\rightarrow$ τ_h fakes and is estimated from simulation.

[1] p_T: transverse momentum of jet, η: jet pseudo-rapidity.

[2] For example, for a W$\rightarrow$ $l\nu$ decay, Transverse mass is defined as: $\sqrt{2p_T^l \not{E}_T(1-\cos\phi)}$, where p^l = lepton p_T, $\not{E}_T$ = missing transverse energy (MET) and ϕ is the angle between them.

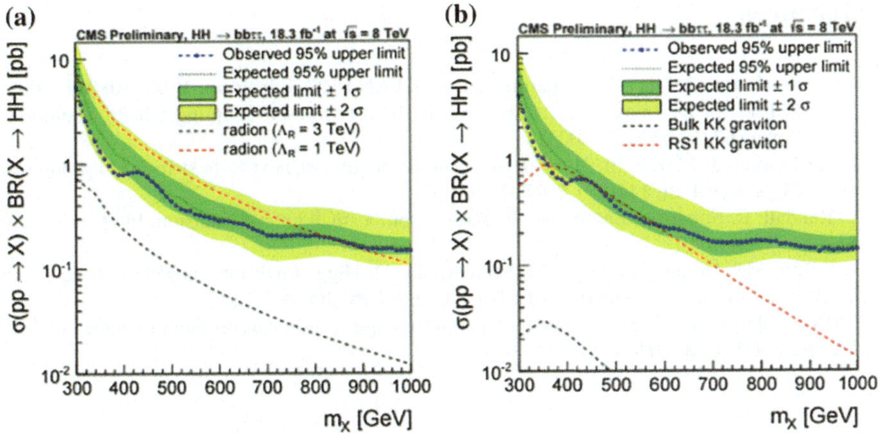

Fig. 123.2 Upper limit at 95% CL_s on $\sigma(gg \rightarrow X) \times BR(X \rightarrow HH)$ for Radion/Spin-0 (**a**) and Graviton/Spin-2 (**b**) hypotheses at 8 TeV center-of-mass energy as function of resonance mass (m_X). [6]

6. **WW/WZ/ZZ+Jets**: This is a small background in both searches. It is estimated from simulation normalised to Next to LO (NLO) cross sections at 8 TeV.
7. **Single Top**: It is estimated from simulation and is normalised to NLO cross sections at 8 TeV.
8. **SM Higgs Background**: This is the sum of all processes involving decay of SM Higgs boson which pass the event selections. It is also completely estimated from simulation.

123.4 Results

In both these searches, no signal was observed. Hence exclusion limits were set on the cross-section for di-Higgs production via a binned maximum likelihood fit to the data [6]. While resonant di-Higgs search used the di-Higgs mass (m_{HH}), a new observable called the stransverse mass (m_{T2}) [5] was used for the non-resonant search to suppress the $t\bar{t}$+jets background. The limits for the resonant search are shown in Fig. 123.2a, b. The non-resonant search limit stands at 0.50 pb ($0.87^{+0.56}_{-0.33}$ pb) observed (expected) corresponding to 50 times (86 times) the SM di-Higgs cross-section.

References

1. CMS collaboration, The CMS experiment at the CERN LHC. J. Instrum. **3**(08), S08004 (2008)
2. U. Baur, T. Plehn, D. Rainwater, Determining the Higgs boson self-coupling at hadron colliders. Phys. Rev. D **67**(3), 033003 (2003)
3. D. de Florian, J. Mazzitelli, Higgs boson pair production at next-to-next-to-leading order in QCD. Phys. Rev. Lett. **111**(20), 201801 (2013)
4. L. Randall, R. Sundrum, Large mass hierarchy from a small extra dimension. Phys. Rev. Lett. **83**(17), 3370 (1999)
5. A.J. Barr, M.J. Dolan, C. Englert, M. Spannowsky, Di-Higgs final states augMT2ed−selecting hh events at the high luminosity LHC. Phys. Lett. B **728**, 308–313 (2014)
6. CMS collaboration, Model independent search for Higgs boson pair production in the $bb\tau^+\tau^-$ final state. *CMS-PAS-HIG-15-013 (2015).*

Chapter 124
Double Parton Scattering Studies at the Large Hadron Collider Using the CMS Detector

R. Kumar, A. Mehta, R. Gupta, S. Bansal, V. Bhatnagar, K. Mazumdar and J. B. Singh

124.1 Introduction

The hard scattering between the partons from the colliding hadrons is characterized by the presence of a particle or a cluster of particles with large mass or large transverse momentum (p_T). At high energy collisions, the parton densities are large enough to cause a significant probability for two or more parton-parton scatterings within the same hadron-hadron collision, known as multiple parton interactions (MPI). Exactly two hard parton-parton scatterings in a proton-proton (pp) collision leads to DPS, a subset of MPI.

A number of DPS measurements have been performed using events from pp and $p\bar{p}$ collisions in different final states and at different centre-of-mass energies, e.g., measurements with multi-jet final-states at $\sqrt{s} = 63$ GeV, 630 GeV, 1.8 TeV and 7 TeV [1–4]; in $\gamma + 3$ jets events at $\sqrt{s} = 1.8$, 1.96 and 7 TeV [5–7]; in $W +$ jets at $\sqrt{s} = 7$ TeV [8, 9]; and in same-sign WW events at $\sqrt{s} = 8$ TeV [10].

The effective cross section (σ_{eff}) for a hadron-hadron collision is defined as the measure of the transverse distribution of partons inside the colliding hadrons and their overlap in a collision. This paper presents the measurement of σ_{eff} using the study of double parton scatterings being done with data collected by the CMS detector [12] using the W + 2-jet process at $\sqrt{s} = 7$ TeV, and same-sign WW process at $\sqrt{s} = 8$ TeV.

---

on behalf of CMS collaboration.

R. Kumar (✉) · A. Mehta · R. Gupta · S. Bansal · V. Bhatnagar · J. B. Singh
Panjab University, Chandigarh, India
e-mail: kumardeepraman@gmail.com

R. Kumar
Akal University, Talwandi Sabo, India

K. Mazumdar
Tata Institute of Fundamental Research, Mumbai, India

Md. Naimuddin (ed.), *XXII DAE High Energy Physics Symposium*, Springer Proceedings in Physics 203, https://doi.org/10.1007/978-3-319-73171-1_124

124.2 DPS Using Same-Sign W Boson Pairs

A useful feature of the same-sign W boson pair production is the cross section, which is comparable for DPS and SPS, whereas in opposite-sign W pair production the cross section for SPS events dominates over the DPS one by a large factor. The decay of two same-sign W bosons in muonic final state is considered for this analysis [10]. The data sample corresponds to an integrated luminosity of 19.7 fb^{-1} at 8 TeV. In case of DPS the two W bosons are produced in first approximation independent of each other, and they are expected to be randomly distributed in the azimuthal plane and with less p_T as compared to SPS production. Hence, the muons produced from DPS are less boosted as compared to the leptons produced from SPS and there would not be any correlation between the two muons in the azimuthal plane. After several control and validation studies, a Multi-Variate Analysis (MVA) using a Boosted Decision Tree (BDT) has been performed in order to improve sensitivity to DPS events with respect to a single observable study. The idea is to use the BDT estimator to get a response shape with the highest possible DPS sensitivity; therefore many sensitive kinematic observables have been put into the BDT training process. The BDT response is studied on top of the same-sign offline base selection and results.

The upper limit on the WW production cross section by DPS is observed to be 1.12 pb, which is calculated using the ratio of the measured DPS yield with respect to the yield expected from Monte Carlo (computed using CLs method based on modified frequentist approach). The lower limit on σ_{eff} is calculated to be 5.91 mb, which is consistent with the previous measurements.

124.3 DPS Using W + 2-Jet Events

The study of DPS using W + 2-jet events at a centre-of-mass energy of 7 TeV, collected by the CMS detector in 2011, has also been performed [9]. The data sample corresponds to an integrated luminosity of 5 fb^{-1}. Events with a W boson, reconstructed from the muon and missing transverse energy information, are required to have exactly two jets. The determination of the effective cross section (σ_{eff}) requires the measurement of the fraction of DPS events in the selected sample of W + 2-jet events (f_{DPS}). For the measurement of f_{DPS} the correlation observables, sensitive to DPS, are investigated. It is observed that the simulations of W + jets events with MADGRAPH5 + PYTHIA8 (or PYTHIA6) and NLO predictions of POWHEG2 + PYTHIA6 (or HERWIG6) provide a good description of the observables, and describe the data only if multiple parton interactions are included. The standalone PYTHIA8, due to missing higher order processes, is not able to provide a good description of the selected data events.

The fraction of DPS in selected W + 2-jet events is extracted using DPS signal, and single parton scattering (SPS) background templates produced by simulations. The obtained value of the DPS fraction is 0.055 ± 0.002 (stat.) $\pm$ 0.014 (syst.). The

σ_{eff} is calculated to be 20.7 ± 0.8 (stat.) ± 6.6 (syst.) mb. This measured value of σ_{eff} is also found to be consistent with the PYTHIA8 model predictions.

124.4 Conclusions

A firm decision of the invariance of the effective cross section with respect to centre-of-mass energy can not be drawn due to large systematic uncertainties. The low sensitivity of the correlation observables, used to extract the fraction of DPS in the selected sample, is one of the reasons behind the large systematic uncertainty. Therefore, it is required to look for new channels as well as new observables with enhanced sensitivity to DPS [11]. For the proper understanding of invariance of the effective cross section, it is required to perform the measurement with same final state at different centre-of-mass energies. The ongoing LHC run at a center-of-mass energy of 13 TeV will be important for the study of invariance of the effective cross section.

References

1. UA2 Collaboration, A study of multi-jet events at the CERN $\bar{p}$p collider and a search for double parton scattering. Phys. Lett. B **268**, 145 (1991)
2. A.F.S. Collaboration, Double parton scattering in pp collisions at $\sqrt{s}$ = 63 GeV. Z. Phys. C **34**, 163 (1987)
3. CDF Collaboration, Study of four jet events and evidence for double parton interactions in $p\bar{p}$ collisions at $\sqrt{s}$ = 1.8 TeV. Phys. Rev. D **47**, 4857 (1993)
4. C.M.S. Collaboration, Measurement of four-jet production in proton-proton collisions at $\sqrt{s}$ = 7 TeV. Phys. Rev. D. **89**(9), 092010 (2014)
5. CDF Collaboration, Double parton scattering in $p\bar{p}$ collisions at $\sqrt{s}$ = 1.8 TeV. Phys. Rev. D **56**, 3811 (1997)
6. D0 Collaboration, Double parton interactions in photon + 3 jet events $p\bar{p}$ collisions at $\sqrt{s}$ = 1.96 TeV. Phys. Rev. D **81**, 052012 (2010)
7. CMS Collaboration, Study of double parton scattering in photon + 3 jets final state in proton-proton collisions at 7 TeV. CMS-PAS-FSQ-12-017 (2015)
8. ATLAS Collaboration, Measurement of hard double-parton interactions in W(→lv) + 2-jet events at $\sqrt{s}$ = 7 TeV with the ATLAS detector. New J. Phys. **15**, 033038 (2013)
9. C.M.S. Collaboration, Study of double parton scattering using W + 2-jet events in proton-proton collisions at $\sqrt{s}$ = 7 TeV. JHEP **03**, 032 (2014)
10. CMS Collaboration, Double Parton Scattering cross section limit from same-sign W bosons pair production in di-muon final state at LHC. CMS-PAS-FSQ-13-001 (2015)
11. R. Kumar et al., New observables for multiple-parton interactions measurements using Z+jets processes at the LHC. Phys. Rev. D **93**(5), 054019 (2016)
12. CMS Collaboration, The CMS experiment at the CERN LHC. JINST (3), S08004 (2008)

Chapter 125
Supernova Neutrinos: Fast Flavor Conversions Near the Core

Manibrata Sen

125.1 Introduction

Core collapse supernovae (SN) offer a fascinating environment to study neutrino flavor evolution in dense environments. Very recently, it was predicted that fast flavor conversions, occurring with a rate $\mu \sim \sqrt{2}G_F n_\nu$ (n_ν is the neutrino density), can happen very near the SN core, as opposed to the well-known collective effects, occurring at $r \sim \mathcal{O}(10^2)$ km [1, 2] or the MSW effect at $r \sim \mathcal{O}(10^3)$ km [3, 4]. A necessary condition for this seemed to be a non-trivial angular distribution in the neutrino emission spectrum [5, 6]. Based on these claims, we make a detailed study of these fast flavor conversions. We focus on regions *close* to the SN core and hence model the source as a flat geometry as shown in the left panel of Fig. 125.1. We redo the linear stability analysis (LSA) with a physically well motivated angular emission spectrum, where the ν_e, which decouple later than the $\bar{\nu}_e$, ν_x, have a larger flux and wider angular distribution than the latter. We study possible instabilities for neutrino flavor evolution in space as well as time, and for the first time, include backward going modes also. Finally, we verify our LSA results with numerical results from the fully non-linear evolution.

125.2 Set-Up of the Problem

Neutrino flavor evolution in a dense media is explained with space-time dependent Wigner functions $\varrho_{\mathbf{p},\mathbf{x},t}$ with momentum $\mathbf{p}$ at position $\mathbf{x}$ and time t. The equation of motion (EoM) is

M. Sen (✉)
Tata Institute of Fundamental Research, Homi Bhabha Road, Mumbai 400005, India
e-mail: manibrata.sen@gmail.com

Md. Naimuddin (ed.), *XXII DAE High Energy Physics Symposium*, Springer Proceedings in Physics 203, https://doi.org/10.1007/978-3-319-73171-1_125

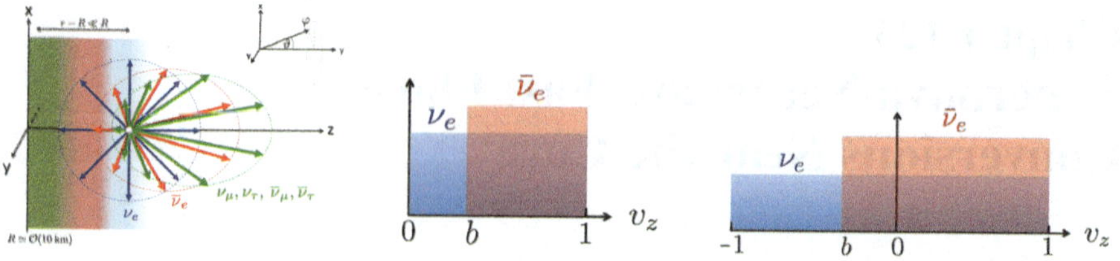

Fig. 125.1 Left panel shows schematic polar plots of the angular distributions of the ν_e (blue), $\bar{\nu}_e$ (red), and ν_x (green) emission fluxes. Middle panel shows a spectrum with no ingoing ν_e or $\bar{\nu}_e$, while the right panel shows ingoing ν_e and $\bar{\nu}_e$

$$\partial_t \varrho_{\mathbf{p},\mathbf{x},t} + \mathbf{v_p} \cdot \nabla_{\mathbf{x}} \varrho_{\mathbf{p},\mathbf{x},t} = -i[\Omega_{\mathbf{p},\mathbf{x},t}, \varrho_{\mathbf{p},\mathbf{x},t}] \ ,$$

Here we have neglected external forces acting on the system as well as collisions. The Hamiltonian matrix is $\Omega_{\mathbf{p}} = \Omega_{\text{vac}} + \Omega_{\text{MSW}} + \Omega_{\nu\nu}$, where $\Omega_{\text{vac}} = \text{diag}(-\omega/2, +\omega/2)$ is the vacuum term with $\omega = \Delta m^2/2E$; $\Omega_{\text{MSW}} = \lambda\ \text{diag}(1, 0)$ is the matter term with $\lambda = \sqrt{2}G_F n_e$, where n_e is the electron density; and $\Omega_{\nu\nu} = \sqrt{2}G_F \int \frac{d^3\mathbf{q}}{(2\pi)^3}(\varrho_{\mathbf{q}} - \bar{\varrho}_{\mathbf{q}})(1 - \mathbf{v_p} \cdot \mathbf{v_q})$ is the multi-angle neutrino-neutrino interaction term. Since the total number of neutrinos is always conserved, we can write

$$\varrho_{\omega,v_z,\varphi} = \frac{1}{2}\text{Tr}(\varrho_{\omega,v_z,\varphi})\,\mathbb{I} + \Phi_\nu \frac{g_{\omega,v_z,\varphi}}{2} \begin{pmatrix} s_{\omega,v_z,\varphi} & S_{\omega,v_z,\varphi} \\ S^*_{\omega,v_z,\varphi} & -s_{\omega,v_z,\varphi} \end{pmatrix} ,$$

and drop the trace term. Here $S_{\omega,v_z,\varphi} \ll 1$ and $s^2_{\omega,v_z,\varphi} + |S_{\omega,v_z,\varphi}|^2 = 1$. Also Φ_ν is the normalization of the differential spectrum $g_{\omega,v_z,\varphi}$, chosen accordingly. As neutrinos are produced as flavor eigenstates, no oscillation occurs as long as $S_{\omega,v_z,\varphi} = 0$. We linearize the equations in this small parameter $S_{\omega,v_z,\varphi}$ to get an eigenvalue equation [7]

$$\begin{aligned} i(\partial_t + v_z\partial_z + \mathbf{v}_T \cdot \partial_T)S_{\omega,v_z,\varphi} = & \left[\omega + \lambda + \mu \int d\Gamma' \left(1 - v_z v_z' - \mathbf{v}_T.\mathbf{v}_T'\right) g_{\omega',v_z',\varphi'}\right] S_{\omega,v_z,\varphi} \\ & -\mu \int d\Gamma' \left(1 - v_z v_z' - \mathbf{v}_T.\mathbf{v}_T'\right) g_{\omega',v_z',\varphi'}\, S_{\omega',v_z',\varphi'} \ , \end{aligned}$$

where $\mathbf{v}_T$ is the velocity vector of the neutrino projected on the x-y–plane. The important quantity here is the difference in the diiferential spectrum, given by $g_{\omega,v_z,\varphi} \propto d\phi_{\nu_e}/d\Gamma - d\phi_{\nu_x}/d\Gamma$ for neutrinos and $\propto d\phi_{\nu_x}/d\Gamma - d\phi_{\bar{\nu}_e}/d\Gamma$ for antineutrinos. Keeping in mind that the angular spectrum of emission should be different for neutrinos and antineutrinos, we choose the following schematic spectrum $g_{\omega,v_z,\varphi} = \frac{1}{2\pi}\left[(1+a)\delta(\omega)\Theta(v_z)\Theta(1-v_z) - \delta(\omega)\frac{1}{(1-b)}\Theta(v_z - b)\Theta(1-v_z)\right]$ as shown in Fig. 125.1 (middle panel). Here a denotes the neutrino-antineutrino asymmetry whereas b controls the difference in zenith angle distribution. Such a "non-trivial" angular emission spectrum seems to be crucial for fast conversion.

To solve the eigenvalue equation, we take $S = Qe^{-i(\Omega_t t+\Omega_z z)}$, where $\Omega_{(t,z)} = \gamma_{(t,z)} + i\kappa_{(t,z)}$ can take complex values. A non-zero positive $\kappa_{(t,z)}$ causes an exponential growth in S, thereby signalling an instability. It is important to mention that since we are looking for fast conversions, we can integrate out ω from the spectrum and effectively set $\omega/\mu \to 0$.

125.3 Results

Armed with this formalism, we look for instabilities for evolution in time ($\Omega_z \to 0$) and in space ($\Omega_t \to 0$). We show a contour plot κ_z for different values of a and b in Fig. 125.2. A common feature of all these plots is that no fast conversion takes place if $b = 0$, thereby indicating that a non-trivial spectrum might be necessary for fast conversions. Also, matter suppresses these growths.

Similarly, in Fig. 125.3, we plot the growth rates for κ_t. We also show similar instabilities for a spectrum including backward travelling modes as $g_{\omega,v_z,\varphi} = \frac{1}{2\pi}\left[\frac{1+a}{2}\delta(\omega)\Theta(1+v_z)\Theta(1-v_z) - \delta(\omega)\frac{1}{(1-b)}\Theta(v_z-b)\Theta(1-v_z)\right]$ as shown in Fig. 125.1 (right panel). We note that inclusion of backward going modes increases fast conversions. Interestingly, matter suppression does not occur for time evolution, as suggested in [8]. To verify our results from LSA, we have also solved the fully non-linear EoMs for specific values of a and b. The results shown in Fig. 125.4 indicate that indeed complete flavor averaging takes place within a few nanoseconds. One is also led to speculate whether it is important to have a crossing in the angular spectrum for development of fast conversions, as is evident from all the above cases. Further details are worked out in [9].

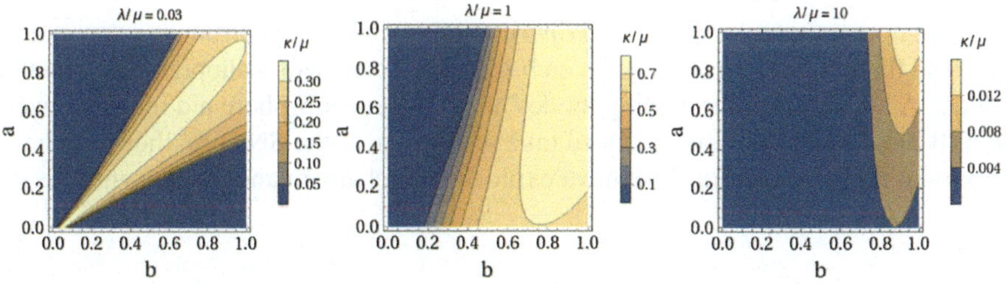

Fig. 125.2 Instability growth rates for evolution in space for three different values of $\lambda/\mu = 0.03$, 1, and 10, respectively from left to right

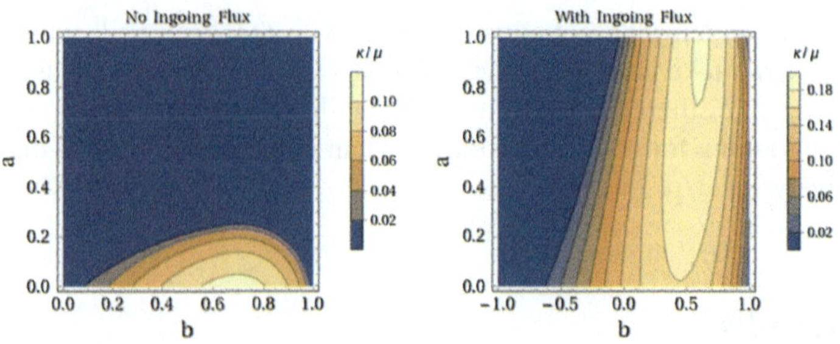

Fig. 125.3 Instability growth rates for evolution in time. Left: without inward going modes. Right: with inward going modes

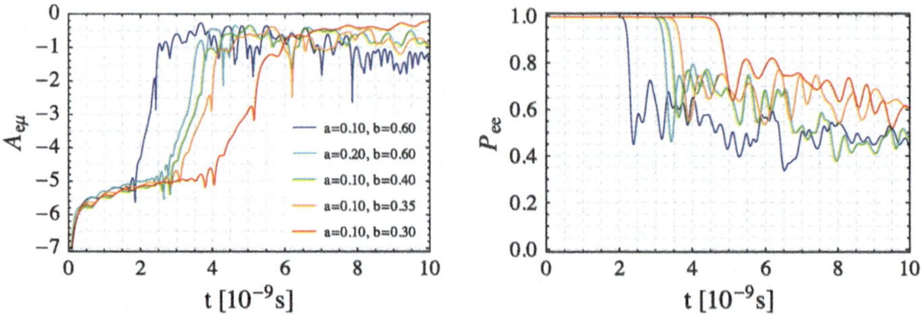

Fig. 125.4 Instability growth for evolution in time. Left panel shows the quantity $A_{e\mu} = \log_{10}|S|$ giving the extent of flavor conversion. Right panel shows the electron neutrino survival probabilities P_{ee} for same values of a and b

125.4 Conclusion

Thus we notice that for a non-trivial distribution of angular spectrum, we can indeed get rapid flavor turn-overs in time, even if in space the growth is suppressed by matter effects. Also, backward travelling modes near the neutrinosphere aid in fast conversion. If these fast conversions indeed take place so near the SN core, they may have important implications for supernova explosion mechanism and nucleosynthesis.

References

1. H. Duan, G.M. Fuller, J. Carlson, Y.Z. Qian, Phys. Rev. D **74**, 105014 (2006), https://doi.org/10.1103/PhysRevD.74.105014 [astro-ph/0606616]
2. S. Hannestad, G.G. Raffelt, G. Sigl, Y.Y.Y. Wong, Phys. Rev. D **74**, 105010 (2006) Erratum: [Phys. Rev. D **76**, 029901 (2007)], https://doi.org/10.1103/PhysRevD.74.105010, https://doi.org/10.1103/PhysRevD.76.029901 [astro-ph/0608695]
3. L. Wolfenstein, Phys. Rev. D **17**, 2369 (1978), https://doi.org/10.1103/PhysRevD.17.2369
4. S.P. Mikheev, A.Y. Smirnov, Sov. J. Nucl. Phys. **42**, 913 (1985). [Yad. Fiz. **42**, 1441 (1985)]

5. R.F. Sawyer, Phys. Rev. Lett. **116** (8), 081101 (2016), https://doi.org/10.1103/PhysRevLett.116.081101, [arXiv:1509.03323 [astro-ph.HE]]
6. S. Chakraborty, R.S. Hansen, I. Izaguirre, G. Raffelt, JCAP **1603** (03), 042 (2016), https://doi.org/10.1088/1475-7516/2016/03/042, arXiv:1602.00698 [hep-ph]
7. A. Banerjee, A. Dighe, G. Raffelt, Phys. Rev. D **84**, 053013 (2011), https://doi.org/10.1103/PhysRevD.84.053013, arXiv:1107.2308 [hep-ph]
8. B. Dasgupta, A. Mirizzi, Phys. Rev. D **92** (12), 125030 (2015), https://doi.org/10.1103/PhysRevD.92.125030, arXiv:1509.03171 [hep-ph]
9. B. Dasgupta, A. Mirizzi, M. Sen, arXiv:1609.00528 [hep-ph]

Chapter 126
Exploring Compressed Top Squark Region with Kinematic Variables

Partha Konar, Tanmoy Mondal and Abhaya Kumar Swain

126.1 Introduction

After the Higgs boson discovery supersymmetry (SUSY) is undoubtedly the most appealing theory waiting to be unveiled at the LHC. SUSY not only unifies the gauge couplings, it also explains the stabilization of the Higgs boson mass against large quantum correction with light stop which can be probed at the LHC. Direct searches exclude $m_{\tilde{t}}$ below 800–900 GeV when the stop ($\tilde{t}$) decays to a top quark and a neutralino [1, 2].

In the region of SUSY spectrum, where the mass difference between the $\tilde{t}$ and the neutralino (χ) is small, the decay of the stop produce a bunch of soft particles along with neutralino which are very difficult to identify in the detector. Naturally direct search bounds are relatively poor exploring this region. If the mass difference is such that $\Delta M \equiv (m_{\tilde{t}} - m_{\chi}) < m_W + m_b$, the stop can decay in the four-body mode $\tilde{t} \to b\, f\, f'\, \chi$. For the leptonic four body stop decay, CMS collaborations excluded stop masses below 360 GeV at 95% confidence level for $\Delta M \sim 30$ GeV [3]. However, the mass limit weaken drastically with larger mass gap and for $\Delta M \simeq m_W$ the limit slips down to 270 GeV.

We focus on this $\Delta M < m_W + m_b$ mass gap region and exploit suitable kinematic variables to constrain the parameter space. We illustrate [4], considering the clean

P. Konar · T. Mondal (✉) · A. K. Swain
Physical Research Laboratory (PRL), Ahmedabad 380009, Gujarat, India
e-mail: tanmoym@prl.res.in; tanmoymondal@hri.res.in

P. Konar
e-mail: konar@prl.res.in

A. K. Swain
e-mail: abhaya@prl.res.in

T. Mondal
Regional Centre for Accelerator-based Particle Physics, Harish-Chandra Research Institute, HBNI, Chhatnag Road, Jhusi 211019, Allahabad, India

Md. Naimuddin (ed.), *XXII DAE High Energy Physics Symposium*, Springer Proceedings in Physics 203, https://doi.org/10.1007/978-3-319-73171-1_126

and reliable leptonic decay channel, that our proposed kinematic variables can extend the state-of-the-art limits on the stop mass provided by the CMS collaboration.

126.2 Kinematic and Invariant Mass Variables

We demonstrate the effectiveness of our proposed variable in four body dileptonic channel of stop along with ISR jet(s). We demand that the signal events should consist of two opposite sign isolated leptons, ≥ 1 b-jet, ≥ 1 high P_T ISR jet(s) and large $\not{E}_T$. The main background for this signal region is the top pair production in the dileptonic channel.

Since we consider $\Delta M \leq m_W$, both the top and subsequently the $W-$boson will be produced off-shell for the signal resulting the four-body decay, unlike the background top pair case which is sequential two-body decay. These distinct kinematic topologies between the signal and the background make it possible to look for different kinematic variables which show observable singularities in phase space to discover/exclude the light stop at the LHC.

Invariant mass of the b-tagged jet and the lepton, $M_{b\ell}$ can be utilized for maximizing the signal to background ratio. The distribution of the variable $M_{b\ell}$ has an endpoint which depends on the decay topology as, $M_{b\ell}^{max} = \sqrt{m_t^2 - m_W^2} \simeq 153\,\text{GeV}$ for $t\bar{t}$ and for signal $M_{b\ell}^{max} = \Delta M$.

We also propose two new ratios [4], $R_{bE} = \sum P_T^{b_i}/\not{E}_T$, $R_{\ell E} = \sum P_T^{\ell_i}/\not{E}_T$. It is easy to follow that for the signal region R_{bE} peaks at the mass ratio (m_b/m_χ), whereas $R_{\ell E}$ peaks at $(m_\ell/m_\chi) \approx 0$. These two interesting ratios are better suited for the dileptonic decay channel exploring the stop four-body decay scenario.

126.3 Event Simulation and Basic Cuts

We simulate the signal events using `MadGraph5` and those events were passed to `Pythia8` for hadronization and parton showering and finally detector level simulation is done in `Delphes3` using the available ATLAS card.

Since most of the $p_T(\tilde{t})$ is carried away by neutralino, we choose $\not{E}_T > 200$ GeV to reduce significant amount of background events including the QCD multijet backgrounds. In the same spirit we have included another variable $\not{E}_T/M_{eff} > 0.3$ where M_{eff} is the scalar sum of all momentum including ISR.

Exploiting the fact that the ISR will be approximately in the opposite direction to the $\not{E}_T$, we introduce additional cut that $|\phi(ISR) - \phi(\not{E}_T) - \pi| < 0.4$. This will also significantly diminish the enormous QCD background. To minimize the effect of jet mis-measurement contributing into $\not{E}_T$ we also demand that $|\phi(j) - \phi(\not{E}_T)| > 0.2$ for all jets other than the ISR.

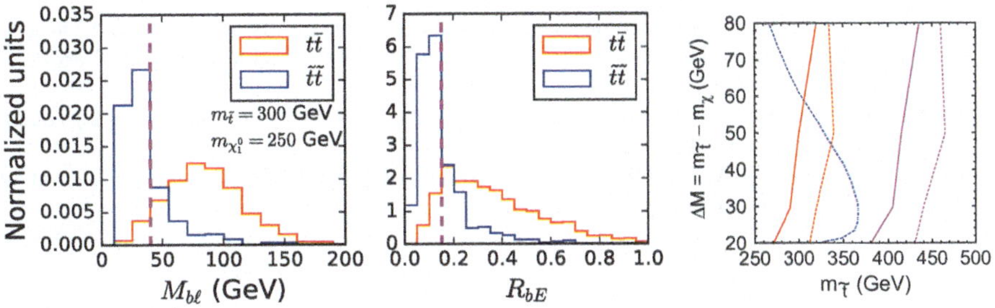

Fig. 126.1 The distribution of variables $M_{b\ell}$ and R_{bE} in the left and middle panel. Exclusion limit in right panel for 13 and 300 fb^{-1} data (red and magenta, respectively) shown with solid and dashed line for soft and hard leptons. The blue line shows present CMS limit on stop mass

126.4 Results

Using the simulated events which passed all the basic selection cuts described above, we have plotted two pivotal variables for both signal and background $t\bar{t}$ events in Fig. 126.1. In order to maximize signal to background ratio we optimized the event selection cuts as: (i) $M_{b\ell} < 40$ (60) GeV, (ii) $R_{bE} < 0.15$ (0.1), and (iii) $R_{\ell E} < 0.15$ (0.3). The numbers in the parentheses are the cut used for $\Delta M = 80\,\text{GeV}$ which is different due to the fact that at this ΔM the $W-$boson is on-shell and effectively the lepton coming from $W-$boson becomes more energetic whereas the $b-$quark becomes softer.

We plotted the 2 σ exclusion plot for 13 TeV LHC with 13 and 300 fb^{-1} data in Fig. 126.1. The blue (dash-dot) curve shows present limit on $\tilde{t}$ mass as presented by the CMS collaboration which is poor for larger mass difference. The red(magenta) solid and red(magenta) dashed curve show our exclusion limit at 13 TeV LHC with 13(300) fb^{-1} data where for solid(dashed) curve we have $p_T > 10(5)$ GeV. It is clear that the larger mass gap region can effectively be probed using the new variables proposed here. Also, in magenta curves we exhibit the 2σ exclusion limits for 13 TeV LHC with an integrated luminosity of 300 fb^{-1}. Evidently, in spite of small branching ratio leptonic channel with the help of the proposed variables [4] is capable of delivering limits comparable to that of hadronic modes [5].

References

1. Search for top squarks in final states with one isolated lepton, jets, and missing transverse momentum in $\sqrt{s} = 13$ TeV pp collisions with the ATLAS detector. Technical Report ATLAS-CONF-2016-050, CERN, Geneva (2016)
2. Search for direct top squark pair production in the single lepton final state at $\sqrt{s}$=13 TeV. Technical Report CMS-PAS-SUS-16-028, CERN, Geneva (2016)

3. Search for new physics in the compressed mass spectra scenario using events with two soft opposite-sign leptons and missing momentum energy at 13 TeV. Technical Report CMS-PAS-SUS-16-025, CERN, Geneva (2016)
4. Partha Konar, Tanmoy Mondal, and Abhaya Kumar Swain. Demystifying compressed top squark region with kinematic variables. 2016
5. Haipeng An, Lian-Tao Wang, Opening up the compressed region of top squark searches at 13 TeV LHC. Phys. Rev. Lett. **115**, 181602 (2015)

Chapter 127
Results and Future Prospects of Exclusive Vector Meson Production with pPb Collisions at CMS

Ruchi Chudasama

127.1 Introduction

Exclusive vector meson (VM) photoproduction is defined by the reaction $\gamma + \mathrm{p} \rightarrow$ VM+p, with the characteristic features that, apart from the vector meson in the final state, no other particles are produced and the vector meson has a low transverse momentum. This process can be studied at the Large Hadron Collider in ultra-peripheral collisions (UPCs) of ions and protons occurring at impact parameters much larger than the sum of their radii. Recently, CMS [1], ALICE [2] and LHCb [3] presented their measurements of exclusive heavy vector meson photoproduction at the LHC. Since the process occurs through γp or γPb interactions via the exchange of two-gluons with no net color transfer, the cross section at leading order (LO) is proportional to the square of the gluon density in the target proton or ion. It provides a valuable probe of the gluon density at a small momentum fraction x, which is kinematically related to $W_{\gamma p}$ ($x = (M_{\Upsilon} / W_{\gamma p})^2$). The exclusive photoproduction of Υ(1S, 2S, 3S) has been measured in their dimuon decay channel in ultraperipheral collisions of protons and heavy ions (pPb) with the CMS experiment [5] at $\sqrt{s_{\mathrm{NN}}}$=5.02 TeV for an integrated luminosity of $\mathcal{L}_{\mathrm{int}}$=33 nb^{-1}. The photoproduction cross section for $\Upsilon(nS)$ is measured as a function of $W_{\gamma p}$ in the range $91 < W_{\gamma p} < 826$ GeV, which corresponds to the rapidity of the Υ meson in the range $|y| < 2.2$, and x values are of the order $x \sim 10^{-4}$ to $x \sim 1.3 \cdot 10^{-2}$. The differential cross section $d\sigma/dt$, has been measured in the range $|t| < 1.0$ (GeV/c)^2 and the b-slope parameter is estimated.

On behalf of CMS collaboration.

R. Chudasama (✉)
Nuclear Physics Division, Bhabha Atomic Research Center, Mumbai, India
e-mail: ruchi.chudasama@cern.ch

Md. Naimuddin (ed.), *XXII DAE High Energy Physics Symposium*, Springer Proceedings in Physics 203, https://doi.org/10.1007/978-3-319-73171-1_127

127.2 Event Selection and Background Estimation

The UPC events are selected by applying a dedicated trigger, which selects at least one muon in each event and at least one to six tracks. Two muon tracks originating from the same primary vertex with $p_T > 3.3$ GeV, and pseudorapidity $|\eta| < 2.2$ are used to select exclusive events. The p_T of the muon pair is restricted from 0.1–1 GeV to reduce the contamination from elastic QED and inelastic background. The dominant background contribution to exclusive Υ signal comes from QED, $\gamma\gamma \rightarrow \mu^+\mu^-$, which is estimated by STARLIGHT. The contribution of non-exclusive background is estimated by a data-driven method by selecting events with more than 2 tracks. This template was normalized to a two muon track sample in the region of dimuon $p_T > 1.5$ GeV. Additional small background contributions of exclusive γPb $\rightarrow \Upsilon$ Pb events is estimated using a reweighted STARLIGHT Υ MC sample. These backgrounds were subtracted from data to obtain the exclusive signal. The background subtracted $|t|$ and y distributions were used to measure the b parameter, and estimate the exclusive Υ photoproduction cross-section as a function of $W_{\gamma p}$, respectively. The distributions were first unfolded to the region $0.01 < |t| < 1$ GeV^2, $|y| < 2.2$, and muon $p_T^{\mu} > 3.3$ GeV, using the D'Agostini method for unfolding and it's further extrapolated to transverse momenta of 0 GeV by acceptance correction factors.

127.3 Results and Future Prospects

The differential $d\sigma/dt$ cross section is extracted for the combined three Υ(nS) states as shown in Fig. 127.1. The cross section is fitted with an exponential function $Ne^{-b|t|}$ in the region $0.01 < |t| < 1.0$ GeV^2, using an unbinned χ^2 minimization method. A

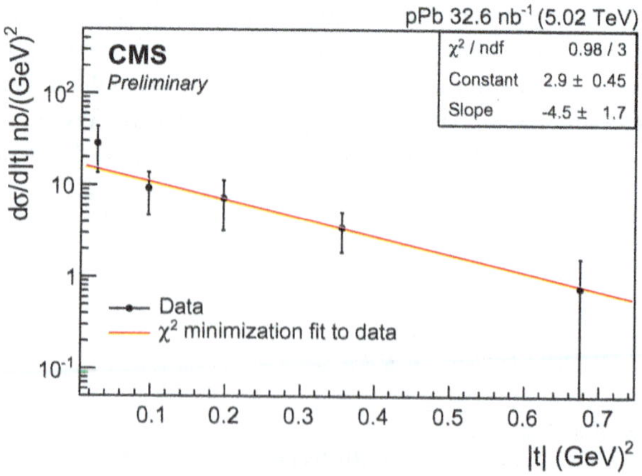

Fig. 127.1 Differential Υ photoproduction cross section as a function of $|t|$ [14]

value of $b = 4.5 \pm 1.7$ (stat) $\pm\, 0.6$ (syst) GeV^{-2} is extracted from the fit. This result is in agreement with the value $b = 4.3^{+2.0}_{-1.3}$ (stat.)$^{+0.5}_{-0.6}$(syst.) measured by the ZEUS experiment [4] for the photon-proton centre-of-mass energy $60 < W_{\gamma\mathrm{p}} < 220$ GeV. The differential $\Upsilon(1\mathrm{S})$ photoproduction cross section $\mathrm{d}\sigma/\mathrm{d}y$ is extracted in four bins of dimuon rapidity using the background-subtracted, unfolded, and acceptance-corrected number of signal events in each rapidity bin . Due to very limited statistics, the number of $\Upsilon(1\mathrm{S})$ events can not be extracted separately in each rapidity bin. Therefore, they are estimated for three states together, and corrected for the fraction of $\Upsilon(1\mathrm{S})$ to $\Upsilon(\mathrm{nS})$ events and the feed-down contribution of $\Upsilon(2\mathrm{S})$ decaying to $\Upsilon(1S) + \pi^+\pi^-$ and $\Upsilon(1S)$+$\pi^0\pi^0$.

Figure 127.2 shows the CMS data together with the previous measurements from H1 [6], ZEUS [7] and LHCb [2] data. It is also compared with different theoretical predictions of the JMRT model [8], factorized IPsat model [9, 10], IIM [11, 12] and bCGC model [13]. As $\sigma(W_{\gamma p})$ is proportional to the square of the gluon PDF of the proton, and the gluon distribution at low Bjorken x is well described by a power law, the cross section will also follow a power law. Any deviation from such trend would indicate a different behavior of the gluon density function. We fit a power law $A \times (W/400)^{\delta}$ with CMS data alone, which gives $\delta = 0.96 \pm 0.43$ and $A = 655 \pm 196$, and is shown by the black solid line. The extracted δ value is comparable to the value $\delta = 1.2 \pm 0.8$, obtained by ZEUS [4]. Our data are compatible with a power law dependence of $\sigma(W_{\gamma p})$, disfavouring faster rising predictions of LO pQCD. Using pPb collisions at 5.02 TeV, we could probe $x \sim 10^{-4}$ to $x \sim 1.3 \cdot 10^{-2}$ values. It would be possible to probe higher values of $W_{\gamma p}$, and lower $x \sim 10^{-6}$ with new pPb data at 8.16 TeV. The increase of statistics will allow us to perform differential studies in p_{T} and y for coherent and incoherent J/ψ and ψ(2S).The t distribution is expected to present a pronounced diffractive peak which can then be used to clearly discriminate between saturation and nonsaturation models [15, 16].

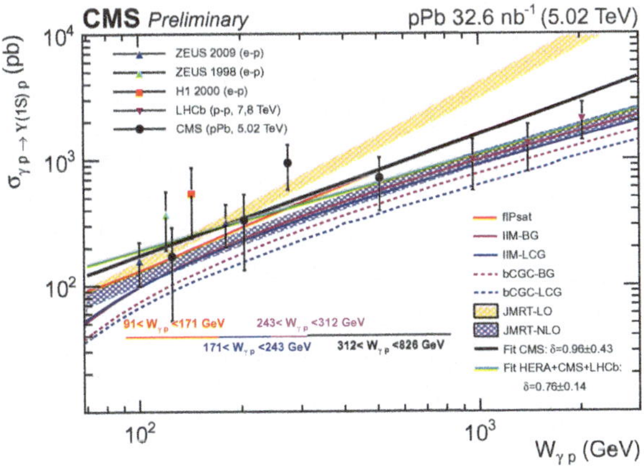

Fig. 127.2 Cross section for exclusive $\Upsilon(1\mathrm{S})$ photoproduction, $\gamma p \to \Upsilon(1S)p$ as a function of photon-proton center-of-mass energy, $W_{\gamma p}$ [14]

References

1. CMS Collaboration, arXiv:1605.06966
2. ALICE Collaboration, Phys. Rev. Lett. **113**, 232504 (2014)
3. LHCb Collaboration, JHEP **09**, 084 (2015), arXiv:1505.08139
4. ZEUS Collaboration, Phys. Lett. B **708**, 14 (2012)
5. CMS Collaboration, JINST **3**, S08004 (2008)
6. H1 Collaboration, Eur. Phys. B **46**, 585 (2006)
7. ZEUS Collaboration, Phys. Lett. B **680**, 4 (2009)
8. P. Jones, D. Martin, M.G. Ryskin, T. Teubner, JHEP **11**, 085 (2013)
9. T. Lappi, H. Mantysaari, Phys. Rev. C **83**, 065202 (2011)
10. T. Lappi, H. Mantysaari, Phys. Rev. C **87**
11. G. Sampaio dos Santos, M. V. T. Machado, Phys. Rev. C **89**, 025201 (2014), 032201 (2013)
12. G. Sampaio dos Santos, M.V.T. Machado, J. Phys. **G42**, 105001 (2015)
13. V.P. Goncalves, B.D. Moreira, F.S. Navarra, Phys. Lett. B **742**, 172 (2015)
14. CMS Collaboration, FSQ-13-009
15. T. Toll, T. Ullrich, Phys.Rev. **C87** (02), 024913 (2013)
16. N. Armesto, A. H. Rezaeian, Phys.Rev. **D90** (5), 054003 (2014)

Chapter 128
Towards Efficient Reconstruction of Semi-invisible Events from Higgs at the LHC

Akanksha Bhardwaj, Partha Konar and Abhaya Kumar Swain

128.1 Introduction

The Large Hadron Collider (LHC) is currently focused in measuring the properties of the Higgs boson after discovering it, which is a great achievement in itself. The precise measurement of different Higgs couplings, with a significant amount of data after run-II of LHC, is crucial in exploring any effect from BSM origin. With the large Yukawa coupling, the Higgs boson decays abundantly to tau leptons holding the paramount importance for examining the properties of Higgs in this channel. The measurement of the properties of the Higgs in tau channel is extremely challenging primarily because of presence of multiple neutrinos in the final state. In addition, the jet rich hadronic environment and identification efficiency contaminate the observable utilized for measuring the properties of Higgs.

In this study, we follow some of the dedicated mass reconstruction methods for reconstructing the neutrino momenta in a scenario where $h \to \tau^+\tau^- \to \pi^+\pi^- + \nu_\tau\bar{\nu}_\tau$. Collinear Approximation [1] is an frequently used technique to reconstruct the invariant mass of $m_{\tau\tau}$ in which the Higgs is produced along with an energetic ISR jet and decaying into τ pair. This technique is build upon two main assumptions that the visible decay product and the neutrino associated with it are almost collinear and the missing transverse energy $\not{E}_T$ in the semi-invisible decay is only due to neutrinos.

A. Bhardwaj (✉) · P. Konar · A. K. Swain
Physical Research Laboratory (PRL), Ahmedabad 380009, Gujarat, India
e-mail: akanksha@prl.res.in; akansha.bhardwaj@iitgn.ac.in

P. Konar
e-mail: konar@prl.res.in

A. K. Swain
e-mail: abhaya@prl.res.in

A. Bhardwaj
IIT Gandhinagar, Palaj 382 355, Gandhinagar, India

Md. Naimuddin (ed.), *XXII DAE High Energy Physics Symposium*, Springer Proceedings in Physics 203, https://doi.org/10.1007/978-3-319-73171-1_128

Under these assumptions two equations can be written from $\not{E}_T$ constraints

$$\not{E}_{Tx} = q_{m1} \sin\theta_{v1} \cos\Phi_{v1} + q_{m2} \sin\theta_{v2} \cos\Phi_{v2} \tag{128.1}$$

$$\not{E}_{Ty} = q_{m1} sin\theta_{v1} sin\Phi_{v1} + q_{m2} sin\theta_{v2} sin\Phi_{v2} \tag{128.2}$$

where $\not{E}_{Tx,y}$ are the x and y component of $\not{E}_T$ vector, $q_{m1,2}$ are the magnitude of invisible momentum of each neutrinos from τ decay, and $\theta_{v1,2}$ and $\phi_{v1,2}$ are the polar and azimuthal angels of each visible decay products. The fraction of momentum taken by visible decay product of τ can be calculated as $x_{1,2} = \frac{p_{v1,2}}{p_{v1,2}+p_{m1,2}}$. Using this, the invariant mass of $\tau\tau$ system is given by $M_{\tau\tau} = \frac{m_v}{\sqrt{x_1 x_2}}$, where m_v is the total invariant mass of visible decay products. This method has certain drawbacks, it breaks down when both the tau are produced back-to-back.

Missing Mass Calculator(MMC) was proposed [2] to reconstruct the invariant mass of $h \to \tau\tau$ in all possible decay modes of τ (1-Prong, 3-Prong and leptonic decay). This technique significantly improves over collinear approximation in reconstruction of the Higgs mass. MMC assumes that there is no other source for missing energy except neutrinos such as mismeasurement of jet momenta. In the present study, we consider only 1 prong decay of τ. In this case, there are 6 unknowns, 3 components of momentum associated with each neutrino. One can write only 4 equation which connect these unknowns.

$$\not{E}_{Tx} = q_{m1} \sin\theta_{m1} \cos\Phi_{m1} + q_{m2} \sin\theta_{m2} \cos\Phi_{m2} \tag{128.3}$$

$$\not{E}_{Ty} = q_{m1} \sin\theta_{m1} \sin\Phi_{m1} + q_{m2} \sin\theta_{m2} \sin\Phi_{m2} \tag{128.4}$$

$$M^2_{\tau 1,2} = m^2_{m1,2} + m^2_{v1,2} + 2E_{v1,2} 2E_{m1,2} - 2q_{v1,2} q_{m1,2} \cos\Delta\theta_{vm_{1,2}} \tag{128.5}$$

Hence, the number of unknowns exceeds the number of constraints. $q_{v1,2}$, $m_{v1,2}$ are the momenta, invariant masses of visible decay products of τ. $\theta_{m1,2}$, $\phi_{m1,2}$ polar and azimuthal angel of neutrinos, and $\cos\Delta\theta_{vm_{1,2}}$ are the angles between momenta of the visible and invisible decay product from each τ. Neutrino mass is considered to be zero. To solve the remaining two unknown degree of freedoms $\phi_{m1,2}$, scan is done over all possible values. Additional knowledge of τ decay kinematics is used which gives higher weight to the more preferable solutions over the less preferable ones. The probability distribution $P(\Delta R, p_\tau)$ of angular separation between the neutinos and the visible decay products of τ leptons is used as an additional information of a particular tau decay topology. For a given event, we can obtain the $M_{\tau\tau}$ distribution by scanning the parameter space for $\phi_{1,2}$ and weighted by the corresponding probability $P(\Delta R, p_\tau)$. Peak position of each $M_{\tau\tau}$ distribution of the particular event is set as the final estimator of the Higgs mass.

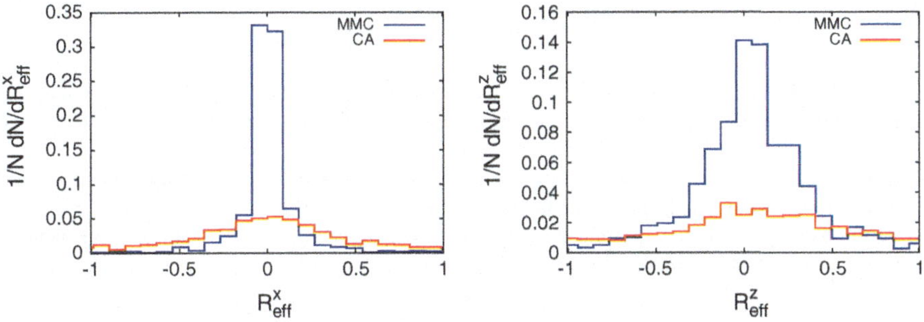

Fig. 128.1 Efficiency of momentum reconstruction both in transverse(left panel) and longitudinal (right panel) directions

128.2 Reconstructing Neutrinos Using MMC

We further extend MMC to know the invisible momenta which can be useful to explore the Higgs property. Since Higgs mass is known, that can be used as an additional constraint along with $m_{\tau\tau}$ distribution of a single event obtained from MMC to reconstruct the momentum of invisible particles, which reduces the parameter space for Φ_{m1} and Φ_{m2} and constraint them around the $\Phi_{\upsilon 1}$ and $\Phi_{\upsilon 2}$. There are many choices for Φ_{m1} and Φ_{m2} which closely satisfies the Higgs mass shell condition, from those choices we can randomly choose one Φ_{m1} and Φ_{m2}. All these choices are closed enough to true value of Φ_{m1} and Φ_{m2} and efficiency of such reconstruction can be calculated. The deviation in the component of the momentum is defined as $R^t_{ef} = \dfrac{q_t^{Reco} - q_t^{True}}{|q_t^{Reco}|}$ and $R^z_{ef} = \dfrac{q_z^{Reco} - q_z^{True}}{|q_z^{Reco}|}$ where q_t^{Reco} and q_z^{Reco} is the transverse and the longitudinal component of the reconstructed momentum and q_t^{True} and q_z^{True} are the transverse and longitudinal component of the true momentum [3].

The efficiency of reconstruction the momentum of invisible particle using MMC has improved with significant amount as compare to momentum reconstructed using collinear approximation and demonstrated in Fig. 128.1.

128.3 Conclusions

Studying physics beyond the Standard Model at the LHC is the main focus in its current and future runs. Measuring the properties of the Higgs is vital, any deviation from the SM value would unambiguously indicate BSM signature. The Higgs decays to tau leptons is a prominent channel for studying its properties for which one requires full reconstruction of the semi-invisible event. We reconstruct the neutrinos momenta by modifying the MMC using Higgs mass shell condition. The reconstructed momenta using MMC are in good agreement with true momenta and can be used for CP and polarization studies.

References

1. R.K. Ellis, I. Hinchliffe, M. Soldate, J.J. Van Der Bij, Higgs decay to $\tau^+\tau^-$ a possible signature of intermediate mass higgs bosons at high energy hadron colliders. Nucl. Phys. B **297**(2), 221–243 (1988)
2. A. Elagin, P. Murat, A. Pranko, A. Safonov, A new mass reconstruction technique for resonances decaying to di-tau. Nucl. Instrum. Methods **A654**, 481–489 (2011)
3. Partha Konar and Abhaya Kumar Swain, Reconstructing semi-invisible events in resonant tau pair production from Higgs. Phys. Lett. B **757**, 211–215 (2016)

Chapter 129
Gravitino Production in a Thermal Universe

Richa Arya, Namit Mahajan and Raghavan Rangarajan

129.1 Gravitino Problem

Gravitino is the supersymmetric partner of the graviton, the spin 2 mediator of gravitational interactions. It is massless and has spin states $\pm 3/2$. When SUSY breaks, the gravitino gains mass and spin $\pm 1/2$ states via a super-Higgs mechanism. The spin $\pm 1/2$ states are also referred to as goldstino modes.

The abundance of gravitinos directly affects the cosmology of our Universe. An excess of stable gravitinos of mass greater than 1 keV can contribute to the energy density of the Universe and can overclose the Universe. The unstable gravitinos of mass between $100\,\text{GeV} \lesssim m_{\tilde{G}} \lesssim 10\,\text{TeV}$ decay into energetic particles after primordial nucleosynthesis which destroy light nuclei created during nucleosynthesis.

129.2 Gravitino Production Rate and Finite Temperature SUSY Breaking

We study the gravitinos produced in the early Universe in the radiation dominated Universe after reheating [1–3], by the scattering of thermalised inflaton decay products. The production rate is given by [4]

$$\Gamma_s = \frac{3T^3}{16\pi M_P^2} \sum_{i=1}^{3} \left[1 + \frac{m_i^2}{3m_{\tilde{G}}^2}\right] c_i g_i^2 \ln\left(\frac{k_i}{g_i}\right) \tag{129.1}$$

where $M_{\rm P} \simeq 2.4 \times 10^{18}$ GeV is the reduced Planck mass, m_i represents the difference in gaugino and gauge boson masses squared, $m_{\tilde{G}}$ is the gravitino mass, $i = 1, 2, 3$

R. Arya (✉) · N. Mahajan · R. Rangarajan
Physical Research Laboratory, Navrangpura, Ahmedabad 380 009, India
e-mail: richaarya@prl.res.in

Md. Naimuddin (ed.), *XXII DAE High Energy Physics Symposium*, Springer Proceedings in Physics 203, https://doi.org/10.1007/978-3-319-73171-1_129

refers to the three SM gauge groups, $g_i(T)$ are the gauge coupling constants, and $c_{1,2,3}$ and $k_{1,2,3}$ are constants associated with the gauge groups. Taking all $m_i = m_{\tilde{g}}$, we get $\Gamma_s \propto \frac{1}{M_P^2}\left(1 + \frac{m_{\tilde{g}}^2}{3m_{\tilde{G}}^2}\right)$. The first term within the parentheses is associated with spin 3/2 gravitinos while the second is associated with spin 1/2 gravitino production.

SUSY is broken by non-zero temperature due to the finite thermal energy density of the Universe which splits the boson and fermion masses. The soft mass generation due to finite temperature effects is given by $m_{\tilde{g}}^2 \rightarrow m_{\tilde{g}}^2 - m_g^2 \sim \delta_3 T^2 + m_0^2$, where m_g is the gluon mass, δ_3 is some parameter and $m_0 \sim 100$ GeV represents the zero temperature mass splitting, while $m_{\tilde{G}} \sim \sqrt{\rho}/(\sqrt{3}M_P) + m_{\tilde{G}0} = \delta' T^2/(\sqrt{3}M_P) + m_{\tilde{G}0}$ where $m_{\tilde{G}0}$ is the zero temperature gravitino mass (which depends on the supersymmetry breaking mechanism), and δ' is another parameter. Then the factor in the scattering rate

$$\gamma_3 \equiv \frac{m_{\text{soft}}^2}{3m_{\tilde{G}}^2} = \frac{\delta_3 T^2 + m_0^2}{3[\delta' T^2/(\sqrt{3}M_P) + m_{\tilde{G}0}]^2} \qquad (129.2)$$

Region I: $\delta_3 T^2 > m_0^2$ and $\delta' T^2/(\sqrt{3}M_P) > m_{\tilde{G}0}$ which gives $\gamma_3 \approx \frac{\delta_3 M_P^2}{\delta'^2 T^2}$.

Region II: $\delta_3 T^2 > m_0^2$ and $\delta' T^2/(\sqrt{3}M_P) < m_{\tilde{G}0}$ which gives $\gamma_3 \approx \frac{\delta_3 T^2}{3m_{\tilde{G}0}^2}$.

Region III: $\delta_3 T^2 < m_0^2$ and $\delta' T^2/(\sqrt{3}M_P) < m_{\tilde{G}0}$ which gives $\gamma_3 \approx \frac{m_0^2}{3m_{\tilde{G}0}^2}$.

In Region I and II, this factor can be much larger than 1 and also larger than the zero temperature limit $m_0^2/(3m_{\tilde{G}0}^2)$.

129.3 Results of Gravitino Abundance Calculations

In the standard scenario, $\Gamma_s < H$ and gravitinos are out of equilibrium. But we find that because of an enhanced production rate, gravitinos are in thermal equilibrium initially ($\Gamma_s > H$) with a thermal abundance $n_{\tilde{G}} = 3\zeta(3)/2\pi^2 T^3$. The freeze out condition is $\Gamma_s(T_f) = H(T_f) = 5T_f^2/M_P$, and T_f determines the frozen abundance.

129.3.1 Very Light Gravitino $m_{\tilde{G}0} = 0.1$ eV

The freeze out temperature is calculated to be $T_f = 100$ GeV. The lifetime $t = M_P^2/m_{\tilde{G}}^3 = 1.2 \times 10^{35}$ yr, which is much larger than the age of Universe and the abundance today is calculated to be $Y_{\tilde{G}0} = Y_{\tilde{G}}(T_f) = 1.8 \times 10^{-3}$. The density parameter $\Omega_{\tilde{G}} = \rho_{\tilde{G}}/\rho_c = m_{\tilde{G}} Y_{\tilde{G}0} s(T_0)/\rho_c \approx 0.92 \times 10^{-4}$, which implies that the gravitinos will not overclose the Universe.

129.3.2 Light Gravitino $m_{\tilde{G}0} = 1\,\text{keV}$

The freeze out temperature $T_f = 600\,\text{GeV}$ and the abundance today $Y_{\tilde{G}0} = 1.8 \times 10^{-3}$. The lifetime $t = 1.2 \times 10^{23}$ yr is much larger than the age of Universe. The density parameter $\Omega_{\tilde{G}} \approx 0.92$, which is in conflict with observations.

129.3.3 Heavy Gravitino $m_{\tilde{G}0} = 100\,\text{GeV}$

The freeze out temperature $T_f = 1.2 \times 10^8\,\text{GeV}$ and the abundance before the gravitino decays is $Y_{\tilde{G}}(T_f) = 1.8 \times 10^{-3}$ which is much larger than the cosmological upper bound on the gravitino abundance of 10^{-14} [5].

129.3.4 Very Heavy Gravitino $m_{\tilde{G}0} = 30\,\text{TeV}$

The freeze out temperature $T_f = 5.5 \times 10^9\,\text{GeV}$ and the lifetime of gravitinos $t = 0.1\,\text{s}$ which implies that the gravitinos would have decayed before nucleosynthesis and not lead to any cosmological problem.

129.4 Out of Equilibrium Production of Gravitinos

Till now we presumed that $T_{\text{reh}} > T_f$ which allowed the gravitinos to be in thermal equilibrium after inflation. In order to suppress the high abundance of the gravitinos, we consider scenarios with $T \ll T_{\text{reh}} < T_f$ for the 1 keV and 100 GeV gravitinos. The gravitino production rate is given by the integrated Boltzmann equation

$$\frac{dY_{\tilde{G}}}{dT} = -\frac{\beta\gamma_3}{M_{\text{P}}}, \tag{129.3}$$

where β is some constant independent of temperature. For $m_{\tilde{G}0} = 100\,\text{GeV}$, the region of our interest is primarily *Region II*. On integrating equation (129.3) for *Region II* from T_{reh} to $T \ll T_{\text{reh}}$, we get

$$Y_{\tilde{G}}(T) \approx \beta \frac{\delta_3}{9 M_{\text{P}} m_{\tilde{G}0}^2} T_{\text{reh}}^3.$$

This implies that T_{reh} should be less than $4 \times 10^4\,\text{GeV}$ so that the abundance is within cosmological bounds. For $m_{\tilde{G}0} = 1\,\text{keV}$, on integrating equation (129.3) for *Region*

III from $T_{\text{reh}} = 300\,\text{GeV}$ to $T \sim m_0 = 100\,\text{GeV}$, the gravitino abundance obtained is $Y_{\tilde{G}} = \gamma_3\beta/2M_P m_0 = 1 \times 10^{-4}$ which is large and inconsistent with current observations. This implies that T_{reh} must be less than $m_0 = 100\,\text{GeV}$ to shut off this mode of gravitino production. Such a low reheat temperature will be consistent only with low scale baryogenesis and leptogenesis models [6–8].

References

1. J.R. Ellis, J.E. Kim, D.V. Nanopoulos, Phys. Lett. B **145**, 181 (1984)
2. M. Kawasaki, T. Moroi, Prog. Theor. Phys. **93**, 879 (1995)
3. M. Bolz, A. Brandenburg, W. Buchmuller, Nucl. Phys. B **606**, 518 (2001)
4. J. Pradler, F.D. Steffen, Phys. Rev. D **75**, 023509 (2007)
5. R.H. Cyburt, J. Ellis, B.D. Fields, F. Luo, K.A. Olive, V.C. Spanos, JCAP **0910**, 021 (2009)
6. L. Boubekeur, T. Hambye, G. Senjanovic, Phys. Rev. Lett. **93**, 111601 (2004)
7. P.S.B. Dev, R.N. Mohapatra, Phys. Rev. D **92**, 016007 (2015)
8. K. Kohri, A. Mazumdar, N. Sahu, Phys. Rev. D **80**, 103504 (2009)

Chapter 130
Design and Characterization of Discrete Analog Front-End for Resistive Plate Chamber Detector

Purnendu Kumar, Sankaran Aniruddhan and Anil Prabhakar

130.1 Introduction

The outputs of Resistive plate chamber (RPC) detectors are nanosecond electrical pulses with a few milli-volts of amplitude at 50 Ω termination impedance [1]. For accurate timing data abstraction using a precision time to digital converter (TDC), it is required to have a broadband front-end with lowest possible time-walk and jitter in the output signal.

130.1.1 Signal Characterization

We started with characterizing the signals of SG503C RPC available in the Indian Institute of Technology Madras (IITM) Physics lab with respect to plastic scintillator based muon trigger. For amplification we have used Hybrid micro-circuit pre-amplifier [2] designed by BARC. On receiving of trigger from the muon telescope, RPC signals were recorded using a 1 GHz DSO (Keysight DSOX-3104A). We analyzed a 1000 such recorded signals from the RPC operating in avalanche mode. As we already know from design parameters that, HMC pre-amplifiers have a 2 ns rise-time, we verified parameters using higher bandwidth 2 stage AC coupled MMIC pre-amplifier (gain ~25 dB, −3 dB bandwidth ~530 MHz and rise-time <1 ns). As expected, we observed that gain, rise-time, fall time and FWHM were reduced (Table 130.1).

P. Kumar (✉) · S. Aniruddhan · A. Prabhakar
Indian Institute of Technology Madras, Chennai 600036, India
e-mail: purnenduk90@gmail.com

Md. Naimuddin (ed.), *XXII DAE High Energy Physics Symposium*, Springer Proceedings in Physics 203, https://doi.org/10.1007/978-3-319-73171-1_130

Table 130.1 Signal characteristics comparison

Characteristics	HMCPAN	2 stage MMIC
Amplitude(V)	0.952 ± 0.323	0.356 ± 0.124
FWHM(ns)	5.824 ± 0.434	4.403 ± 0.448
Rise-time(ns)	3.591 ± 0.436	3.224 ± 0.462
Fall-time(ns)	5.723 ± 0.444	3.838 ± 0.621

130.2 Theory

In case of ICAL detector [3] due to ~2 m long strips, signal transmission time may go as high as 10–12 ns. Hence it is better to consider the detector strips as transmission lines rather than point capacitors. It is more suitable to use a voltage amplifier rather than charge amplifier in this case.

Higher amplification factor without clipping the bandwidth is desired to increase the slope of the signal, resulting in low jitter operation of comparator. First the input signal is converted to differential one and one of the lines is delayed while the other is attenuated. If delay is represented by t_d, attenuation factor is f, rise time of the signal is t_r, then, $t_d = t_r(1 - f)$. This yields the zero crossing at time free of signal slope. Architecture of front end triggering system was inspired from NINO [4] architecture.

130.3 Implementation and Testing

Three stage MMIC amplifier resulting in total gain of 38 dB and bandwidth of 460 MHz was used with 180° hybrid passive splitter, which was used to split the signal. External cable and attenuator assembly was used to provide 1.5 ns delay and 50% attenuation before being added and compared to ground reference. To avoid noise, a parallel fixed reference comparator with one-shot multivibrator was added, and when ANDed with CFD path, this gives a precise trigger for valid signals above the noise band (Fig. 130.1). Output of analog channel was given to a PS706 discriminator with −20 mV threshold. Efficiency and time resolution were measured using standard VME modules CAEN V830 and CAEN V1190B. These results were found to be comparable to those obtained from HMC pre-amplifier (95.68% efficiency, and 1.46 ns σ for time distribution) Considering the RMS sigma of time distribution for each scintillator detector being ~0.736 ns [5], time resolution (FWHM of Gaussian fit for timing distribution) can be calculated as follows.

$$TR = 2\sqrt{2ln2}(\sqrt{\sigma_{RPC}^2 - \sigma_{Scintillator}^2}) \approx 2.355(\sqrt{\sigma_{RPC}^2 - \sigma_{Scintillator}^2}) \quad (130.1)$$

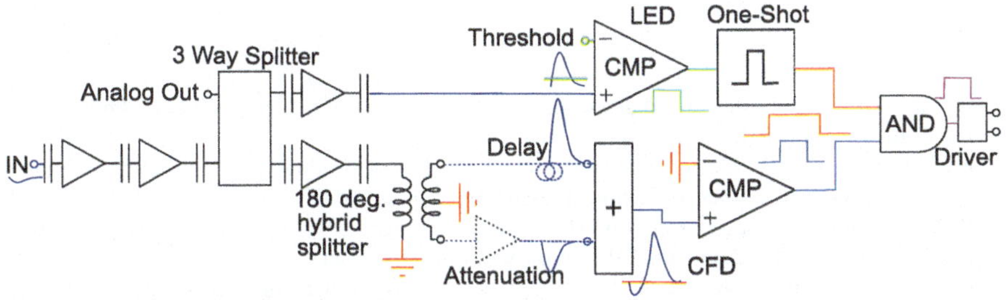

Fig. 130.1 Design of 3 stage MMIC amplifier with parallel CFD and LED for RPC (dotted lines represents the external connections)

For the purpose of verification of CFD functionality we have captured 2000 signals using DSO with 10 GHz sampling. Efficiency of 3 Stage MMIC amplifier was found to be 94.28%, and timing resolution was 2.698 and 2.350 ns when measured with LED and CFD set-up.

130.4 Conclusions

We have successfully designed and tested a discrete analog front end-for resistive plate chamber detector operating in avalanche mode. As gain and delay can be set by external components, this front-end can be used with other detectors by simply varying the component values as per rise-time of signal. Although we have seen an improvement of only ~0.35 ns in time resolution (Fig. 130.2) further improvement can be achieved by compensating for rise-time too.

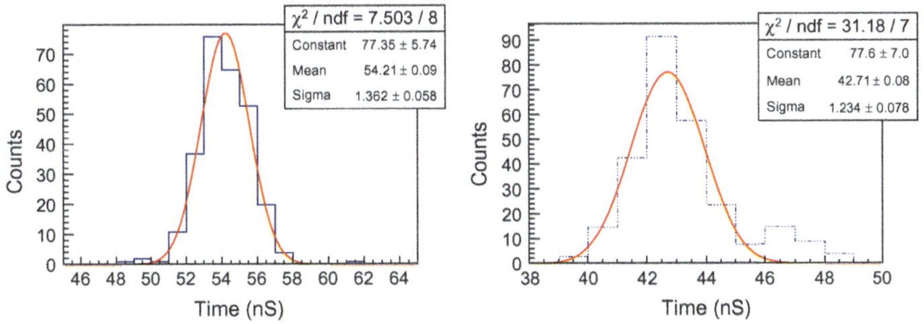

(a) LED timing distribution (b) CFD timing distribution

Fig. 130.2 Arrival time distribution from 3 stage MMIC pre-amplifier measured with DSOX-3104A

References

1. B. Satyanarayana, Design and characterisation studies of resistive plate chambers, Ph.D. thesis, IIT Bombay 04412701, 2009
2. R.S. Shastrakar et. al., Design and development of a family of ultra fast amplifier hybrid micro circuits for INO experiments. Technical report, ED, BARC, 2008
3. ICAL-Collabration, Design report ICAL electronics, trigger, control, data acquisition and monitoring systems. TIFR Mumbai, 2013. http://www.ino.tifr.res.in
4. F. Anghinolfi et. al., NINO: An ultra-fast and low-power front-end amplifier/discriminator ASIC designed for the multigap resistive plate chamber. NIMA 0018-9499. 183–187 (2004). https://doi.org/10.1016/j.nima.2004.07.024
5. K. Raveendrababu, Study of glass resistive plate chambers for the INO-ICAL detector: Ph.D. thesis, Bhabha Atomic Research Centre, Mumbai PHYS01201004019, 2016. http://www.ino.tifr.res.in/ino/theses/Raveendrababu_Ph.D_Thesis.pdf

Chapter 131
Stabilizing Electroweak Vacuum Through Modified Chaotic Inflation

Abhijit Kumar Saha and Arunansu Sil

Stability of the Standard Model (SM) higgs potential at high energy is a long lasting question. This means that at some energy scale ($\Lambda_I^{SM} \sim 10^9$ GeV with $m_h = 125.6$ GeV and $m_t = 173.3$ GeV), SM higgs quartic coupling (λ_h) turns negative. At early universe, the H field fluctuates with amplitude $\sim H_{\rm Inf}$ during inflation. So for large field inflationary scenario (e.g. chaotic inflation) ($H_{\rm Inf} \sim 10^{14}$ GeV), this may be dangerous for stability of universe. In this paper we modify the chaotic inflation by introducing an extra SM singlet scalar (apart from the inflation field) such that inflationary predictions are satisfied by current Planck data [2]. At the same time, our aim is to make the Higgs vacuum absolutely stable through the threshold effect as shown in [1].

We consider the inflation to be governed by two SM singlet scalars ϕ and χ having potential V,

$$V_{\rm I} = \frac{1}{2}m^2\phi^2 - \frac{c_1}{2}\phi^2(\chi^2 - v_\chi^2) + \frac{\lambda_\chi}{4}(\chi^2 - v_\chi^2)^2, \qquad (131.1)$$

where m, c_1, λ_χ and v_χ are real and positive parameters. At global minimum, $\langle\chi\rangle = v_\chi$ and the ϕ field settles at zero. During inflation the ϕ takes super-Planckian value and χ field receives a negative inflation dependent mass-squared term ($m_\chi^2(\phi) = 2c_1\phi^2 + 2\lambda_\chi v_\chi^2$). Therefore χ acquires a large field value from its coupling with the ϕ field $\langle\chi\rangle \simeq \sqrt{\frac{c_1}{\lambda_\chi}}\phi = \chi_{\rm I}$, $\chi_{\rm I}$ is considered to be sub-Planckian which implies $\frac{c_1}{\lambda_\chi} \ll 1$. We have assumed $v_\chi \ll m$ so that it does not play any role in predictions of the modified chaotic inflation model. Due to the super-Planckian field value of ϕ at

A. K. Saha (✉) · A. Sil
Department of Physics, Indian Institute of Technology Guwahati, Guwahati 781039, Assam, India
e-mail: abhijit.saha@iitg.ernet.in

A. Sil
e-mail: asil@iitg.ernet.in

Md. Naimuddin (ed.), *XXII DAE High Energy Physics Symposium*, Springer Proceedings in Physics 203, https://doi.org/10.1007/978-3-319-73171-1_131

the beginning and during inflation, $m_\chi^2(\phi)$ turns out to be greater than $H_{\rm Inf}^2 \simeq \frac{m^2\phi^2}{6M_P^2}$. Hence χ is expected to be stabilized at χ_I quickly. Therefore the effective inflationary potential after integrating out the heavy field χ is

$$V_{\rm Inf} \simeq M_P^4\left[\frac{1}{2}\tilde{m}^2\tilde{\phi}^2\left(1-\alpha\tilde{\phi}^2\right)\right], \tag{131.2}$$

where $\tilde{m} = m/M_P$ and $\tilde{\phi} = \phi/M_P$ and $\alpha = \frac{c_1^2}{2\lambda_\chi \tilde{m}^2}$. We assume $\alpha\tilde{\phi}^2 \ll 1$ so that this correction term does not deform the standard chaotic inflation model much. Spectral index n_s and tensor to scalar ratio are defined as $n_s = 1 - 6\epsilon + 2\eta$ and $r = 16\epsilon$ where $\epsilon = \frac{1}{2}\left(\frac{V'_{\rm Inf}}{V_{\rm Inf}}\right)^2$ and $\eta = \left(\frac{V''_{\rm Inf}}{V_{\rm Inf}}\right)$. The mass parameter $\tilde{m}$ can be fixed from curvature perturbation spectrum (P_S).

We perform a scan over the parameters $\tilde{m}$ and α involved in (131.2) so as to obtain r and n_s within the allowed range of Planck 2015 [2] for number of e-fold $N_e = 58$. We have found with the increase of α, r could be decreased with the correct order of n_s. We presented our findings in the $n_s - r$ plot (Fig. 131.1) and compared it with the allowed range of Planck 2015 [2]. There are some important conditions that must be satisfied to have a successful inflationary scenario. (i) During inflation: $m_\chi^2 > H_{\rm Inf}^2$, (ii) χ field is sub-Planckian at the onset of inflation and afterwards (iii) $\alpha\tilde{\phi}^2 < 1$, (iv) minimum 20% deviation in the modified version correspond to $\alpha\phi^2 > 0.05$, (v) universe during the inflation is expected to be dominated by the ϕ field and hence $\lambda_\chi\chi^4/4 \ll (1/2)m^2\phi^2$. Taking care of all these points we find a contour plot in Fig. 131.2 between c_1 and λ_χ.

From Fig. 131.2, we choose $c_1 = 2.5 \times 10^{-11}$, $\lambda_\chi = 1.3 \times 10^{-8}$ for studying Higgs vacuum stability. Note that H was at origin during inflation due to its large effective mass. The relevant potential for studying vacuum stability issue is given by

$$V_0 = \lambda_H\left(H^\dagger H - \frac{v^2}{2}\right)^2 + \frac{\lambda_{\chi H}}{2}(\chi^2 - v_\chi^2)\left(H^\dagger H - \frac{v^2}{2}\right) + \frac{\lambda_\chi}{4}(\chi^2 - v_\chi^2)^2. \tag{131.3}$$

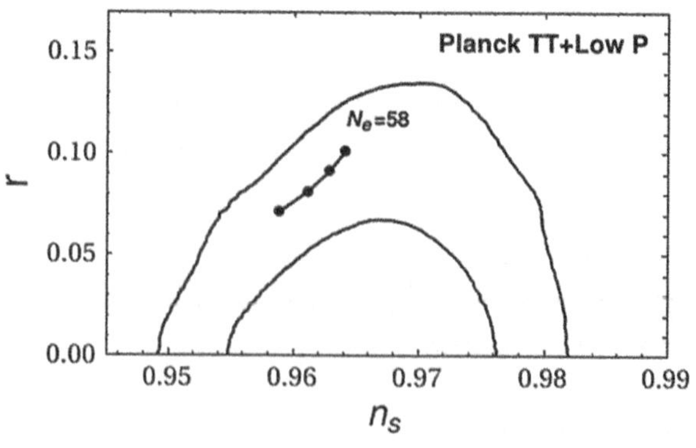

Fig. 131.1 We show the predictions for n_s and r as obtained from the modified version of standard chaotic inflation with dark dots for $N_e = 58$. Along the solid line joining them, we vary α to find n_s and r in our set up. We have also compared our results with 1σ and 2σ contours of $n_s - r$ as obtained from Planck 2015 data [2]

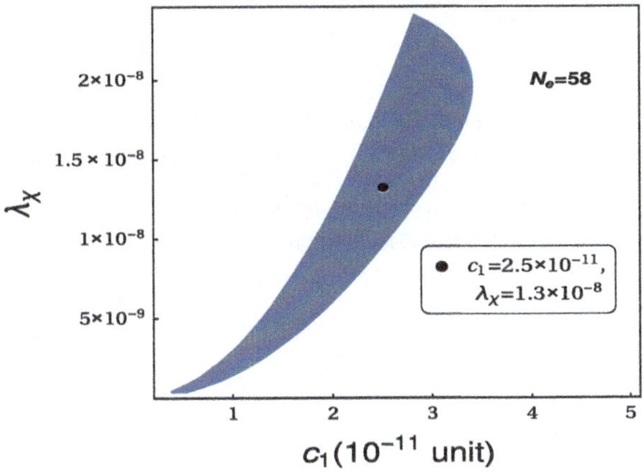

Fig. 131.2 We show the allowed region of c_1 (in unit of 10^{-11}) and λ_χ is indicated by the shadowed region by using the constraints [(i)–(v)]. The dark black dot represents the reference value of c_1 and λ_χ

After inflation the minimum of V_0 is given by $\langle H^\dagger H\rangle = \frac{v^2}{2}$, $\langle\chi\rangle = v_\chi$, with λ_H, $\lambda_\chi > 0$. To ensure the vacuum stability till M_P, these conditions must hold at any scale μ. There is an extra stability condition $4\lambda_\chi\lambda_H > \lambda_{\chi H}^2$ with $\lambda_{\chi H} > 0$. The χ field affects the Higgs vacuum stability in two ways. One is that it modifies the RG equation of λ_h from its SM form. The second contribution comes at energy scale below m_χ, where χ field is integrated out. This phenomenon gives a positive shift to SM higgs quartic coupling λ_h, parametrized as $\delta\lambda = \frac{\lambda_{\chi H}^2}{4\lambda_\chi}$. As χ field has nonzero vev, after spontaneous breaking of electroweak symmetry it will mix with H field to give new heavy and light mass eigenstates. However in the approximation $\lambda_\chi v_\chi^2 \gg \lambda_H v^2$, their mixing turns out to be very small. Hence below the scale m_χ, the effective potential of V_0 becomes

$$V_0^{\text{eff}} \simeq \lambda_h\Big(H^\dagger H - \frac{v^2}{2}\Big)^2, \text{ with } \lambda_h(m_\chi) = \Big[\lambda_H - \frac{\lambda_{\chi H}^2}{4\lambda_\chi}\Big]_{m_\chi}, \tag{131.4}$$

The positive shift on λ_h at m_χ helps in delaying the Higgs quartic coupling(λ_H) to become negative provided $m_\chi < \Lambda_I^{\text{SM}}$. We try to find that particular value of m_χ, for which λ_H stays positive for all energy scales below M_P, and takes a positive negligible value ~ 0 at M_P ($\lambda_H(M_P) \sim 0$). We identify that particular value of m_χ as Λ_I^C. To do so, we have considered $\delta\lambda \sim \lambda_h(m_\chi)$ to avoid unnaturalness. This implies $\lambda_{\chi H} \simeq 2.55 \times 10^{-5}$ and $v_\chi = 4.92 \times 10^{11}$ GeV once we use the reference value of c_1 and λ_χ (black dot in Fig. 131.2) for $N_e = 58$. So if we take $m_\chi < \Lambda_I^C$, it is expected that λ_H will be absolutely positive till M_P. On the other hand for $m_\chi > \Lambda_I^C$, λ_H should be negative at some energy scales below M_P. We found these consistent with $\Lambda_I^C \sim 8 \times 10^7$ GeV in our scenario [3].

References

1. J. Elias-Miro, J.R. Espinosa, G.F. Giudice, H.M. Lee, A. Strumia, JHEP **1206**, 031 (2012). https://doi.org/10.1007/JHEP06(2012)031, arXiv:1203.0237 [hep-ph]
2. P.A.R. Ade et al., Planck Collaboration, Astron. Astrophys. **594**, A13 (2016). https://doi.org/10.1051/0004-6361/201525830, arXiv:1502.01589 [astro-ph.CO]
3. A.K. Saha, A. Sil, Phys. Lett. B **765**, 244 (2017). https://doi.org/10.1016/j.physletb.2016.12.031, arXiv:1608.04919 [hep-ph]

Chapter 132
Particle Identification with the TOP and ARICH Detectors at Belle II

S. Sandilya

132.1 Introduction

A reliable particle identification (PID) is important for any high energy physics experiment, and in case of B-factories it is inevitable [1]. In the B-factories, PID is required to tag B-meson flavour for CP violation studies in the neutral B-meson system, and to suppress backgrounds in precision measurements of rare B and D decays. In Belle II detector [2], PID will be performed by the Time-Of-Propagation (TOP) counter in the central region and the Aerogel Ring Imaging Cherenkov (ARICH) counter in the forward endcap region. The working principle of both the TOP and ARICH counters is based on imaging the Cherenkov rings. In these proceedings, we report the design, method and present status of the PID systems in both the regions of the Belle II detector.

132.2 TOP Counters

A TOP counter module primarily consists of a quartz radiator bar, micro-channel plate photomultipliers (MCPPMTs) and a front-end readout. Two quartz bars each having dimension $(1250 \times 450 \times 20)\ \mathrm{mm}^3$, a mirror with dimension $(100 \times 450 \times 20)\ \mathrm{mm}^3$, and a small expansion prism of dimension $(100 \times 456 \times 20 - 51)\ \mathrm{mm}^3$ are epoxied to form the radiator. The radiator is enclosed in a box made of aluminum honeycomb panels and is supported by PEEK polymer buttons. The mirror focuses parallel rays in the Cherenkov cone to a single point and thus removes the effect of

(On behalf of the Belle II Collaboration).

S. Sandilya (✉)
University of Cincinnati, Cincinnati, OH 45221, USA
e-mail: saurabhsandilya@gmail.com

Md. Naimuddin (ed.), *XXII DAE High Energy Physics Symposium*, Springer Proceedings in Physics 203, https://doi.org/10.1007/978-3-319-73171-1_132

the bar thickness and also allows to correct for chromatic dispersion. The quality of the optical components of the radiator is ensured by several quality acceptance tests [3].

Cherenkov photons emitted by a charged track in the radiator go through total internal reflections and registered at the expansion volume end by an array of MCPPMTs. The arrival time, including the time of flight of the charged particle and the time of propagation of emitted photons, and position in the detection plane of each Cherenkov photon are used to compute a likelihood for a given particle mass hypothesis [4].

A square-shaped MCPPMT has been developed in collaboration with Hamamatsu Photonics KK for the TOP counter [5, 6]. It has a multi-alkali photocathode whose average quantum efficiency is about 28% for wavelengths around 380 nm. By applying the nominal operating bias, a gain of 10^6 is achieved. Each module contains two rows of 16 MCPPMTs, giving in total 32 MCPPMTs per module. The MCPPMTs are physically mounted on the front-end electronics modules. The main components of the front-end electronics are: front board to host the MCPPMT array; high voltage board to provide high voltages to the MCPPMTs; Application-Specific Integrated Circuits (ASICs); and Standard Control, Read-Out, and Data (SCROD) board [7, 8].

A small prototype of the TOP counter was tested at the 1.2 GeV/c positron beam at LEPS (Laser Electron Photon beam line at SPring-8) in June 2013 and results were found in agreement with the MC expectations.

132.3 ARICH Counters

The ARICH counter is basically a proximity focusing RICH. Its main components are: aerogel tiles as a radiator, an array of position sensitive photon detectors, and a readout system [9–11]. The aerogel radiator consists of two layers with increasing refractive indices along the particle path so that the Cherenkov photons emitted by each layer overlap at the photo-detection plane. The two aerogel tiles are of thickness 20 mm each and have refractive indices of 1.045 and 1.055 for upstream and downstream tiles, respectively. This arrangement of two layers with different refractive indices gives better performance than a single aerogel layer for the entire thickness [12]. The photo-detector for the ARICH should be sensitive to single photon detection, able to provide position information, be immune to the 1.5 T magnetic field perpendicular to the photon detection plane, and be tolerant of the high radiation environment. A Hybrid Avalanche Photo-Detector (HAPD) was developed in a joint effort with Hamamatsu which has a peak quantum efficiency of about 28% at 400 nm. The bombardment gain is of about 1800 and an additional gain of about 40 is achieved from the avalanche process of the APDs, resulting an overall gain of 70,000. The ARICH covers 3.5 m^2 in the forward endcap of the Belle II detector; the aerogel radiator plane and the photo-detection plane are separated by a distance of 200 mm. The radiator plane consists of 124 pairs of aerogel tiles while the photo-detection plane contains 420 HAPDs. At the outermost edge of these two planes, planar mirrors are placed to redirect the outside-going photons towards the detection plane. Dedicated high gain and low noise electronics were developed for the readout. To each HAPD a front-end board with four ASICs and a field programmable gate array is attached. The digitized hit information is collected by a merger board from front-end boards, and then communicated to further stages of the data acquisition system [13].

A prototype ARICH was tested at the DESY test beam in May 2013. The Cherenkov angle resolution is found to be 15.8 mrad, and on average 9 photons per track are detected.

132.4 Summary and Status

The TOP and ARICH detectors will provide PID information in the barrel and forward endcap region, respectively, of the Belle II detector. The working principle of both the detectors is based on imaging the Cherenkov ring created by the passage of charged particles. The TOP utilizes the impact position and time of arrival of the Cherenkov photons at the detection plane after total internal reflections in the quartz radiator bar to obtain PID information, whereas the ARICH is a proximity focusing RICH detector. Prototypes of both the detectors demonstrated their expected performance in test-beams. All 16 TOP counters have been successfully installed in the Belle II detector. ARICH detector is under construction now and will be finished by the spring of 2017 and its installation to the Belle II detector is expected in the summer. Detailed simulations have been performed with the Belle II software framework, based on which we expect an excellent charged kaon efficiency (>90%) with a very small (<10%) charged pion misidentification probability.

References

1. P. Krizan, J. Instrum. **4**, P11015 (2009)
2. T. Abe et al., (Belle II Collaboration), arXiv:1011.0352
3. B. Wang, Nucl. Instrum. Methods Phys. Res. Sect. A **766**, 204–207 (2013)
4. B.N. Ratcliff, Nucl. Instrum. Methods Phys. Res. Sect. A **502**, 211–221 (2003)
5. K. Inami, Nucl. Instrum. Methods Phys. Res. Sect. A **595**, 96–99 (2008)
6. M. Akatsu, Nucl. Instrum. Methods Phys. Res. Sect. A **528**, 763–775 (2004)
7. M. Andrew, in *IEEE Realtime Conf. Rec.* (2012), pp. 1–5
8. M. Andrew, in *PoS, (TIPP2014)* (2014), p. 171
9. T. Matsumoto et al., Nucl. Instrum. Methods Phys. Res. Sect. A **521**, 367–377 (2004)
10. T. Iijima et al., Nucl. Instrum. Methods Phys. Res. Sect. A **548**, 383–390 (2005)
11. S. Nishida et al., Nucl. Instrum. Methods Phys. Res. Sect. A **766**, 28–31 (2014)
12. S. Korpar et al., Nucl. Instrum. Methods Phys. Res. Sect. A **572**, 429–431 (2007)
13. S. Nishida et al., Phys. Procedia **37**, 1730–1735 (2012)

Chapter 133
Feasibility Study for Development of a PET Device Based on Multigap Resistive Plate Chambers

M. Nizam, B. Satyanarayana and R. R. Shinde

133.1 Introduction

The multigap resistive plate chamber (MRPC) is a modified version of RPC detector wherein the gas gap between the electrodes is further divided into multiple sub-gaps by introducing electrically floating highly resistive plates. The MRPC was first conceptualized and developed in 1996 [1]. These detectors consists of many highly resistive plates (e.g. glass) and very thin gas gaps between them. The high voltage is applied only on the outermost electrodes and the inner electrodes are all electrically floating.The signal is read from X plane and Y plane pickup panels placed at outermost anode and cathode. The time resolution of these detectors improves with narrower sub-gaps. We attempt to use these detectors in medical imaging techniques such as Positron Emission Tomography (PET) due to their excellent time resolution capability. A detailed description of the MRPC detectors, their fabrication as well as performance has been discussed in [2].

133.2 MRPC for Positron Emission Tomography (PET)

We have developed several MRPCs as a part of the detector R&D program of the India-based Neutrino Observatory (INO) project [3]. These detectors have been tested for their long time operation and we have achieved a time resolution of about 220 ps for cosmic muons [2]. We attempted to use these detectors for possible application in medical imaging. We have mounted two MRPCs vertically and a radioactive source

M. Nizam (✉)
Homi Bhabha National Institute, Mumbai 400094, India
e-mail: mohammad.nizam@tifr.res.in

M. Nizam · B. Satyanarayana · R. R. Shinde
Tata Institute of Fundamental Research, Mumbai 40005, India

Md. Naimuddin (ed.), *XXII DAE High Energy Physics Symposium*, Springer Proceedings in Physics 203, https://doi.org/10.1007/978-3-319-73171-1_133

(^{22}Na) is placed asymmetrically between the two detectors. ^{22}Na emits a positron which annihilates with an electron almost at rest and two gammas of 511 keV are produced with opposite momenta. Placement of the two detectors and the source is shown in Fig. 133.2a. The photons are detected by two detectors on either sides in coincidence with each other. The trigger scheme of the setup and vertically mounted MRPCs ensure that the cosmic background is minimised.

133.3 Data Acquisition System

We need both digital as well analog outputs from the preamplifiers for correcting time walk of Time to Digital Converter (TDC) data. Anusparsh boards are used as preamplifiers to obtain time and charge information simultaneously. Anusparsh is a front end ASIC designed for the ICAL experiment. It is an 8-channel amplifier and discriminator ASIC. It also provides analog output of the amplifier stage for a selected channel. The amplifier comprises of a regulated cascode transcendence amplifier, followed by two stages of differential amplifier. The threshold is common to all discriminator channels. Outputs of the discriminators are Low Voltage Differential Signals (LVDS).The charge information is obtained from the analog output of the Anusparsh and used for the calibration of the TDC data. A picture of the Anusparsh boards designed and fabricated at TIFR is shown in Fig. 133.1a. We use RPC-DAQ module to acquire the data. RPC-DAQ is an FPGA based data acquisition system designed for INO-ICAL experiment. It uses Altera Cyclone IV CE115 FPGA and an NIOS-II soft core processor from Altera. It has 128 Low Voltage Differential Signalling (LVDS) inputs to cater for 64X and 64Y strips of a 2 m × 2 m RPC. We are using only 32 channels for our MRPC-PET setup right now. It has interface to 20 channels of HPTDC and Ethernet connectivity for data and command interface to the back-end server. A picture of RPC-DAQ is shown in Fig. 133.1b. To make the trigger for the TDC and ADC, we sum signals from all the eight channels of each X- and Y- planes of each MRPC. The sum output of X- and Y- planes are ANDed and these signals from both the detectors are further ANDed to produce the final trigger, this ensures the simultaneous detection of oppositely directed photons.

(a) Anusparsh Board

(b) RPC-DAQ Module

Fig. 133.1 Anusparsh Board and RPC-DAQ Module

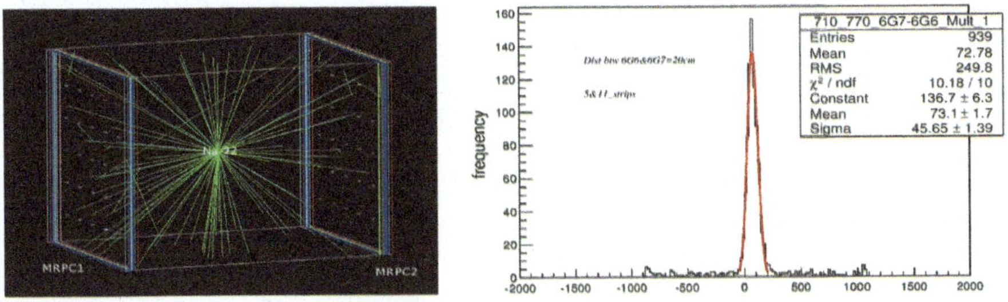

(a) Detector setup and source position

(b) Differential timing distribution

Fig. 133.2 Detector setup and TDC data

133.4 Results and Discussion

The X and Y coordinates of hits are recorded along with the time of arrival of the photon at the detector. We obtain lines of response by joining the (X,Y) of MRPC1 and (X,Y) of MRPC2 only for events with single multiplicity(with one strip hit per plane per detector). The timing Information infers the exact position of the source on the line of response. The difference between timings of two opposite photons is calculated as $\triangle t = t_{MRPC1} - t_{MRPC2}$, The position accuracy is given by $\triangle L[mm] \approx \sigma_t[ps]/2$ [4]. $\triangle L =$ FWHM of the source position, $\sigma_t =$ Resolution of time difference between MRPC1 and MRPC2. Figure 133.2b. shows distribution of $\triangle t = t_{MRPC1} - t_{MRPC2}$ of two opposite strips of MRPC1 and MRPC2 with the position of source equidistant from both MRPCs. The time resolution $\triangle t$ obtained from this study is 1.141 ± 0.035 ns.

We plan to use two detectors on both sides of the source to improve the efficiency of our system. Work is in progress with detectors in horizontal position and two scintillator paddles which covers the setup from top and bottom for cosmic veto. Simulation studies are also in progress to estimate the improvement in efficiency as a function of the number of gas gaps and to compare the time resolution with our experimental results.

References

1. E. Cerron Zeballos, I. Crotty, D. Hatzifotiadou, J. Lamas Valverde, S. Neupane, M.C.S. Williams, A. Zichichi, Nucl. Instrum. Methods A **374**, 132 (1996)
2. M.M. Devi, N.K. Mondal, B. Satyanarayana, R.R. Shinde, Eur. Phys. J. C **76**, 711 (2016). https://doi.org/10.1140/epjc/s10052-016-4570-2
3. M.S. Athar et al., [INO Collaboration], India-based Neutrino Observatory: Project Report. vol. I, INO-2006-01
4. A. Blanco et al., Nucl. Instrum. Methods A **508**, 88 (2003)

Chapter 134
Development of Fast, Low Power 8-Channel Amplifier-Discriminator Board for the RPCs

Puneet Kanwar Kaur, Pathaleswar, M. N. Saraf, B. Satyanarayana and R. R. Shinde

134.1 Introduction

ICAL is a magnetized Iron-Calorimeter to be set up by the India based Neutrino Observatory (INO) collaboration for studying neutrinos. ICAL will study atmospheric neutrinos using magnetized iron as its target mass and glass Resistive Plate Chambers (RPCs) as the active detector elements [1]. RPC detector signals are of 1–5 mV in amplitude, and with rise time of the order of 1 ns. Hence the pulse profiles are extremely narrow, viz., 5–10 ns. The preamplifier should accept single ended input current signals, should have suitable gain with peaking time of 1 ns or less and there should also be a discriminator with threshold in the range of fC to pC. ICAL will have about 3.6 million detector channels and its engineering module will have about 50,000 channels. So, the power consumption per channel should be very low, besides ensuring small form-factor of the preamplifier board in order to minimize the dead area in the detector.

A few ASICs exploiting the transimpedance as well as classical voltage amplifier architecture were designed and fabricated by the INO collaboration. The analog front-end boards developed using these ASICs were extensively tested and order for limited production of these ASICs is being processed. In the meantime, it was decided to design and fabricate analog front-end boards using NINO ASIC so that they can be deployed in the small RPC detector stacks.

P. K. Kaur—Now at Pohang University of Science and Technology, S.Korea.

P. K. Kaur · Pathaleswar · M. N. Saraf · B. Satyanarayana (✉) · R. R. Shinde
Tata Institute of Fundamental Research, Homi Bhabha Road, Mumbai 400005, India
e-mail: bsn@tifr.res.in

Md. Naimuddin (ed.), *XXII DAE High Energy Physics Symposium*, Springer Proceedings in Physics 203, https://doi.org/10.1007/978-3-319-73171-1_134

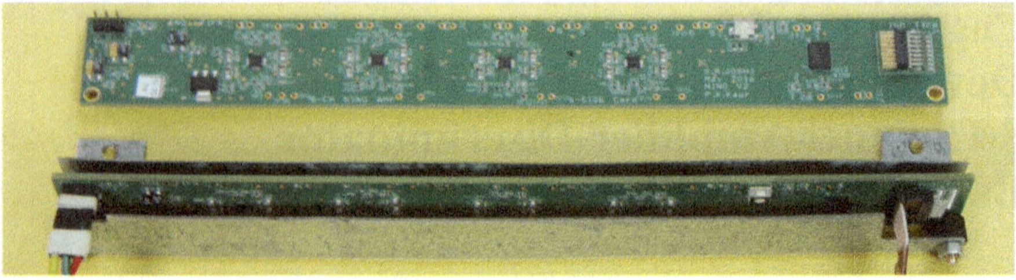

Fig. 134.1 Analog front-end board with mounting chassis

134.2 NINO ASIC

NINO is an ultrafast, low noise, 8-channel front-end preamplifier-discriminator chip developed at CERN to be used mainly in the ALICE time-of-flight detector [2]. It is a fully differential chip featuring four differential amplifier stages with built in hysteresis control.

134.3 Board Design

A charged particle passing through an RPC induces single-ended signals on the pick-up strips while NINO ASIC requires differential inputs. This conversion could be done in two ways: using passive circuits, which cause signal attenuation or by using active circuits, using which a desired gain may be added and common voltages can also be matched. THS4520, a single channel wideband (620 MHz) and fully differential operational amplifier is used for this purpose.

A 6-layer analog front-end board using NINO ASIC and eight differential drivers of unity gain and with threshold control and other accessory circuits was designed with a track separation of 5mil and board dimensions of 200 mm × 23 mm. Excluding the loss in the power supply regulator, the board consumes about 560mW of power. A board along with its mounting chassis is shown in Fig. 134.1.

134.4 Characterisation and Performance

The NINO based analog front-end board is extensively tested and characterised on the RPC detectors. Its performance is compared with those of the boards built using HMCs and ASICs developed by the collaboration. Noise rate and efficiency of the test RPC were studied using various types of boards. The NINO ASIC based board has shown similar or better performance compared to the other solutions. In particular,

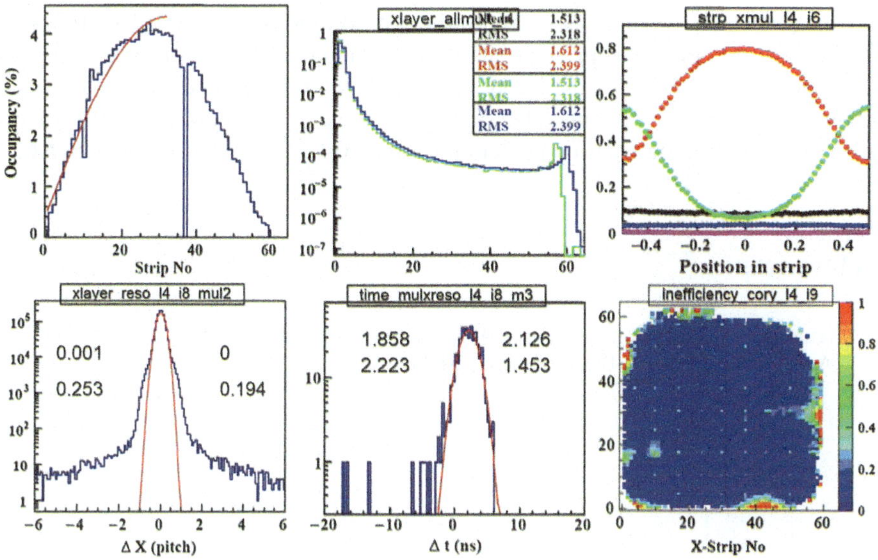

Fig. 134.2 Cosmic ray test results using NINO preamp

we obtained an efficiency of 92% and stable noise rate of 45 Hz for an RPC, readout by a NINO preamp set with a threshold of 80 fC.

Finally, performance tests of NINO prepamps boards were carried out by populating a 2 m × 2 m RPC detector in the stack at IICHEP, Madurai. The stack was triggered by cosmic ray muons and the performance obtained by the RPC readout by NINO preamps was compared with those of rest of the RPCs in the stack and readout by other preamp boards. A collage of these results is shown in Fig. 134.2.

The top-left panel of the Fig. 134.2 shows strip occupancy distribution of the RPC detector. While from the top-middle panel, we obtain a strip multiplity of about 1.6 using NINO preamp, in the top-right panel we plotted track position histogram on a strip as function of the RPC strip multiplity. Two important performance characteristics - position and timing resolutions are shown in the bottom-left and bottom-middle panels respectively. The values obtained from these plots are about 6 mm and about 1.5 ns respectively. Finally the planel shown on the bottom-right is the inefficiency scan of the RPC detector area.

134.5 Summary and Outlook

NINO ASIC based fast, low power 8-channel amplifier-discriminator board meant for INO-ICAL's single gap RPCs was successfully designed, fabricated and characterised. The board satisfies the form-factor requirements of RPC front-end electronics. 20 boards, produced as pilot production were tested on one of the RPCs of the $2\,\text{m} \times 2\,\text{m}$ RPC detector stack at IICHEP, Madurai along with ICALs digital front-end board and rest of the backend system. 200 boards produced as a limited production order, are being installed in the rest of the RPCs in the stack. This board will also be the analog front-end solution for the 20-RPC magnetised mini-ICAL.

References

1. S. Ahmed et al., (INO Collaboration) (2015), arXiv:1505.07380
2. F. Anghinolfi et al., IEEE Trans. Nucl. Sci. **51**(5) (2004)

Chapter 135
Development of a Resistive Plate Chamber with Heat Strengthened Glass

G. Majumder, V. M. Datar, S. D. Kalmani, N. K. Mondal, S. Mondal, B. Satyanarayana and R. R. Shinde

135.1 Introduction

The Resistive Plate Chamber (RPC) [1, 2] is chosen as the active detector element in the ICAL detector [3] due to its low cost and excellent time resolution. We are trying to develop RPCs using tempered glass of 3 mm thickness manufactured at the St. Gobain glass factory near Chennai, India. The hardness of this glass was tested using external pressure. A surface test of the normal glass shows a strength of up to 20–25 MPascal. Toughened or tempered glass is a type of safety glass processed by controlled thermal or chemical treatment to increase its strength compared with normal glass. This certainly changes the surface quality of glass but the long term stability of RPCs made from this glass needs to be tested. On the other hand the cooling process is slower in heat strengthened (HS) glass resulting in a lower compressive strength and a smaller change in the surface quality and hence strength in comparison with tempered glass. Thus, we first used HS glass, which is twice as hard as normal glass to make RPCs and studied their properties viz. noise rate, efficiency of detecting cosmic muons etc.

135.2 Physical Properties of Glass

An atomic force microscope was used to look for any change or damage on the surface but did not find any change with respect to normal glass. We have also looked at the composition of HS glass using electron mass spectroscopy and compared it with the composition of normal glass. We did not find any difference. But, the bulk resistivity

G. Majumder (✉) · V. M. Datar · S. D. Kalmani · N. K. Mondal · S. Mondal ·
B. Satyanarayana · R. R. Shinde
Tata Institute of Fundamental Research, Homi Bhabha Road, Mumbai 400005, India
e-mail: gobinda@tifr.res.in

Md. Naimuddin (ed.), *XXII DAE High Energy Physics Symposium*, Springer Proceedings in Physics 203, https://doi.org/10.1007/978-3-319-73171-1_135

of normal glass is found to be $2.6 \times 10^{12}\Omega - \text{cm}$, whereas it is roughly 40% in HS glasses, which could be a source of larger dark current.

135.3 RPC with HS Glass

The HS glasses of 3.2 mm thickness were used to make two RPCs of size $100 \times 100\,\text{cm}^2$ with a gap of 2 mm and compared their performance with that of an identical sized normal glass RPC. One surface of the glass was painted with graphite coating to provide high voltage on the top- and the bottom glass. The surface resistivity was about $1\text{M}\Omega/square$ across different segments of the coating. Honeycomb panels, composed of polypropylene and of 5 mm thickness, with copper strips on one side and an aluminum ground plane on the other side, are used to read out the induced signal. The width of the copper pickup strips is 2.8 cm with a pitch of 3 cm. These detectors were filled with the mixture of Freon, iso-butane and SF_6 in the proportion of 95.3:4.5:0.3 and with a flow rate 2 SCCM in all three chambers. All detectors are operated at 2 mbar above atmospheric pressure. HV is applied on both surfaces, +ve on the top glass and equal amount of −ve HV on the bottom glass. In this paper we will always quote the difference of these as the applied HV. RPC signals are amplified by two stage voltage amplifiers of combined gain of ~80. Typical signal shapes from the HS and normal glass RPCs are similar.

135.4 Dark Current and Noise Rate

We have tested three RPCs in parallel, (i) RPC with HS glass (HS01), (ii) RPC with HS glass (HS02) and (iii) RPC with normal glass (NORM). As expected the dark current in RPCs of HS glass is about twice the dark current of RPC of normal glass. The amplified signals of the RPCs were fed to a discriminator with settable threshold (usually $V_{th} = -20\,\text{mV}$). During this study, we have read out eight amplified signals from each side of the HS01 and HS02 RPCs, but readout only the top eight amplified signals from the NORM RPC. The average count rate in the strips of NORM is about $1\,\text{Hz/cm}^2$ at 10 KV, whereas those numbers are 3–4 times larger for HS.

135.5 Muon detection efficiency

Four scintillator detectors of size $30\,\text{cm}\,(L) \times 2\,\text{cm}(W) \times 1\,\text{cm}(D)$ were used as cosmic muon telescope to study the muon detection efficiency and timing performance for the three RPCs. A fourfold coincidence of the scintillator signals within 30 ns ensured that a cosmic ray muon passed through the $30 \times 2\,\text{cm}^2$ surface area.

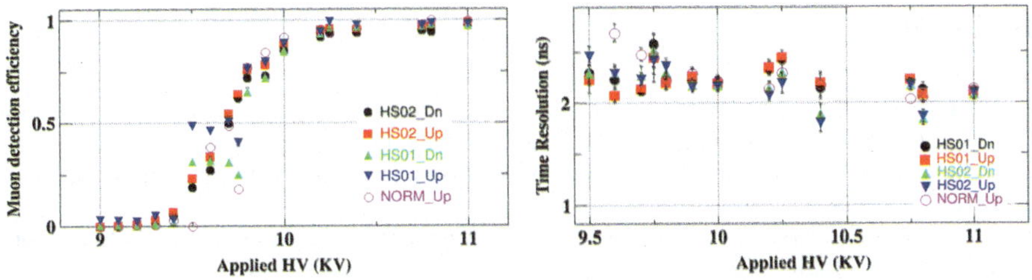

Fig. 135.1 (left) Single muon detection efficiency and (right) observed time resolutions as a function of the applied operating voltage. Only the statistical error is shown here

All RPC were aligned in such a way that this area overlapped with only one pickup strip in either surface of each RPC. So, here _Dn and _Up indicate only one strip in a RPC under the muon telescope. The muon detection efficiencies at different applied voltages are shown in left of Fig. 135.1; all strips show nearly the same efficiency, close to 95%, at the plateau.

135.6 Timing Performance

The same discriminator output signals were used for the timing measurement of the RPC signals with respect to the cosmic muon trigger. The width of the distributions includes the uncertainty of the time measurement in the scintillator telescope, which is about 1.5 ns. The observed time resolution of all these RPC strips is shown in right of Fig. 135.1. The time resolution of all RPCs is more or less the similar, between 2 and 2.5 ns.

135.7 Conclusion

The cosmic muons signals of both timing and detection efficiency of RPCs produced with heat strengthened glass show the same characteristics as of a RPC made with normal glass, except for the larger dark current in the former due to low bulk resistivity. We are in process of putting these RPC in the existing cosmic muon stack to determine the efficiency and timing of the whole detector. We have also produced RPCs with this HS glass of size 2 m × 2 m to test large scale RPC with HS glass.

References

1. Yu.N. Pestov, G.V. Fedotovich, A picosecond time-of-flight spectrometer for the vepp-2m based on local - discharge spark counter, SLAC-TRANS-0184, IYF-77-78 (1978)
2. R. Santonico, R. Cardarelli, Development of resistive plate counters. Nucl. Instrum. Methods **187**, 377 (1981)
3. S. Ahmed et al., Physics potential of the ICAL detector at the India-based Neutrino Observatory (INO), arXiv:1505.07380

Chapter 136
A Proposed Method for Extracting the Proton Geometry in the bSat Dipole Model from HERA Data

Tobias Toll

The bSat Dipole Model, developed by Kowalski and Teaney [1], models electron proton interactions. It is believed to be a universal model in the sense that it can describe all Deeply Inelastic Scattering (DIS) processes at small gluon momentum fraction x. In the dipole model, the incoming virtual photon fluctuates into a quark antiquark dipole, which interacts with gluons in the proton.

The measured charge radius (at low energies) of the proton is found to be 0.88 fm [2]. Different models of QCD give different predictions to weather the gluons are distributed similarly to the electric charge (closely surrounding the valence quarks), are predominantly in a region inside the valence quarks, resulting in a gluon radius smaller than the charge radius, or are not strongly correlated with the valence position, resulting in a gluon radius larger than the charge radius. Independent results from lattice gauge theory and string theory in the Ads/CFT frame-work both seem to favour a smaller gluon radius.

136.1 Inclusive Deeply Inelastic Scattering

For inclusive DIS the total cross-section for (virtual) photon-proton scattering is as a function of x and photon virtuality Q^2:

$$\sigma^{\gamma^* p}_{\mathrm{DIS}\ T,L}(x, Q^2) = \sum_f \int \mathrm{d}^2\mathbf{r} \int_0^1 \frac{\mathrm{d}z}{4\pi} \int \mathrm{d}^2\mathbf{b} \left(\Psi^*\Psi\right)^f_{T,L} \frac{\mathrm{d}\sigma_{q\bar{q}}}{\mathrm{d}^2\mathbf{b}} \tag{136.1}$$

Here, the wave-overlap of the virtual photon is given $(\Psi^*\Psi)^f_{T,L}$, for transversely and longitudinally polarised photons respectively. The quark flavour of the dipole is

T. Toll (✉)
Shiv Nadar University, Greater Noida, UP, India
e-mail: tobias.toll@snu.edu.in

Md. Naimuddin (ed.), *XXII DAE High Energy Physics Symposium*, Springer Proceedings in Physics 203, https://doi.org/10.1007/978-3-319-73171-1_136

denoted f, and z is the photon momentum fraction taken by the quark in the dipole. The dipole cross-section is given by $\frac{d\sigma_{q\bar{q}}}{d^2\mathbf{b}} = 2\left(1 - e^{-\Omega/2}\right)$, where the opacity $\Omega = \frac{\pi r^2}{N_C}\alpha_S(\mu^2) \cdot xg(x, \mu^2)T(b)$. The exponentiation of the dipole-cross section comes from the possibility of many dipole-gluon interactions inside the proton, and this form ensures unitarity. Here, $xg(x, \mu^2)$ is the density of gluons in the longitudinal direction (in momentum space) and $T(b)$ is the gluon-thickness in the transverse direction (in impact parameter coordinate space). The thickness function is taken to be normalised to unity. Usually it has been taken to be a Gaussian $T(b) = \frac{1}{2\pi B_G}e^{-\frac{b^2}{2B_G}}$, where B_G is a model parameter symbolising the variance of the transverse gluon disOtribution, with the root mean square impact-parameter given by $b_{\text{rms}} = \sqrt{2B_G}$. The longitudinal part is modeled with a DGLAP evolution from a starting distribution $xg(x, \mu_0^2) = A_g x^{\lambda_g}(1 - x)^{5.6}$

By Taylor expanding the dipole cross-section for small opacities one gets: $\frac{d\sigma_{q\bar{q}}}{d^2\mathbf{b}} = \Omega - \frac{\Omega^2}{2} + \mathcal{O}(\Omega^3)$. Integrating this over impact parameter, as in (136.1), it is clear that the normalised thickness only contributes to the Ω^2 term, with a factor $1/(4\pi B_G)$. This means that the total DIS cross-section, while not being very sensitive to the B_G parameter, has the following dependence upon it:

$$\frac{\partial\sigma^{\gamma^* p}_{\text{DIS } T,L}}{\partial B_G} > 0 \tag{136.2}$$

136.2 Inclusive Diffractive DIS

Now, the Inclusive Diffraction transversely and longitudinally polarised structure functions are given by [3]:

$$\sigma^{\gamma^* p}_{\text{Diffraction } T(n=1),L(n=0)} \propto \int d^2\mathbf{b}\left(\int dr\, r K_n(\epsilon r) J_n(kr)\frac{d\sigma_{q\bar{q}}}{d^2\mathbf{b}}\right)^2 \tag{136.3}$$

where $k^2 = z(1 - z)M_X^2 - m_f^2$, with M_X being the invariant mass of the dipole, and $\epsilon^2 = z(1 - z)Q^2 + m_f^2$. This expression holds for large β where the dipole is unlikely to fluctuate into higher Fock-states. Doing the same Taylor expansion of the dipole cross-section in this expression, one sees that already the leading term is an integral over the square of the thickness function, yielding the following dependence on the proton size parameter for the inclusive diffraction cross-section:

$$\frac{\partial\sigma^{\gamma^* p}_{\text{Diffraction } T,L}}{\partial B_G} < 0 \tag{136.4}$$

The dependence on B_G in the inclusive diffraction case is stronger than in the inclusive DIS case, but it is still expected to be important in the latter, given the precision of the existing measurements. The fact that the cross-section dependence on B_G is opposite for diffractive and inclusive DIS will provide a strong test of the universality of the bSat Dipole Model frame work.

136.3 Exclusive Diffractive DIS

In inclusive DIS, the impact parameter is not an observable, which it is in theory for diffraction. In practice however, it is not possible to measure the impact-parameter of an interaction. Instead one may measure the momentum transfer in the pomeron vertex $t = (p - p')^2$, where p and p' are the four momenta of the incoming and scattered proton respectively. t is a Fourier conjugate to the impact parameter. To measure t is only possible in exclusive diffraction, which is the only process in which the four-momentum of the scattered proton can be reliably reconstructed.

136.4 Discussion

Historically, the B_G parameter has been extracted from the data solely by comparisons with the t spectrum measured in exclusive diffraction, using a Gaussian distribution. This has yielded a value $B_G \sim 4\,\mathrm{GeV}^{-2}$, corresponding to $b_{\mathrm{rms}} \sim 0.6\,\mathrm{fm}$, which is considerably smaller than the measured charge radius. I propose the following procedure for pin-pointing the proton geometry in the dipole model: First fit all parameters of the model: A_g, λ_g, μ_0^2, and B_G (as well as quark masses), to *both* inclusive DIS and inclusive diffractive DIS. I propose that one does this for several different shapes of the proton, not only the Gaussian of the original. Thus one effectively only fits the size of the proton (not shape). Once this is done, the different shapes should be confronted with the measured t-spectrum of exclusive diffraction, which can to some degree discriminate against the shape. However, this discriminating power is not great. Therefore, one should choose shapes which are considerably qualitatively different, using Gaussian as a mean, I propose using a Laplacian distribution for a proton with relatively more gluons in the center, and a hard sphere (which is a special case of a Woods–Saxon) to model a proton with a wider gluon spatial distribution.

We will within shortly give a detailed description of the results following the scheme laid out here.

References

1. H. Kowalski, D. Teaney, Phys. Rev. D **68**, 114005 (2003). https://doi.org/10.1103/PhysRevD.68.114005 [hep-ph/0304189]
2. P.J. Mohr, D.B. Newell, B.N. Taylor, Rev. Mod. Phys. **88**(3), 035009 (2016). https://doi.org/10.1103/RevModPhys.88.035009, arXiv:1507.07956 [physics.atom-ph]
3. H. Kowalski, T. Lappi, C. Marquet, R. Venugopalan, Phys. Rev. C **78**, 045201 (2008). https://doi.org/10.1103/PhysRevC.78.045201, arXiv:0805.4071 [hep-ph]

Chapter 137
Elemental Analysis of Glass and Bakelite Electrodes Using PIXE Facility

Manisha, V. Bhatnagar, J. S. Shahi, S. Verma, B. P. Mohanty and A. Kumar

137.1 Introduction

An RPC is a gaseous detector utilising a constant and uniform electric field produced by two highly resistive electrode plates like glass and bakelite [1]. These characteristics make both, glass and bakelite RPCs active candidates in various running as well as future HEP experiments for different applications [2]. STAR experiment at RHIC [3]; ALICE, ATLAS and CMS experiments at LHC [4–6]; Belle experiment at KEK are using glass and bakelite RPC based detector systems [7]. The proposed Iron CALorimeter (ICAL) detector at the underground India-based Neutrino Observatory (INO) is also planning to use glass RPCs as the active detector elements [8]. For long life time of experiment, it is necessary to do elaborate study to characterize electrode materials. In reported study, elemental analysis of glass and bakelite samples which was not reported in literature so far, is done using PIXE facility available at Cyclotron laboratory [9], Chandigarh. Elemental analysis gives an insight in to the elemental composition of glass and bakelite electrodes; glass and bakelite electrode samples procured from local and international market are found to be having similar elemental composition. Therefore, locally available electrode samples could be used.

Manisha (✉) · V. Bhatnagar · J. S. Shahi · S. Verma · A. Kumar
Department of Physics, Panjab University, Chandigarh, India
e-mail: manisha.lohan@cern.ch

B. P. Mohanty
Department of Biophysics, Panjab University, Chandigarh, India

Md. Naimuddin (ed.), *XXII DAE High Energy Physics Symposium*, Springer Proceedings in Physics 203, https://doi.org/10.1007/978-3-319-73171-1_137

137.2 PIXE

PIXE is a well established analytical technique of X-ray spectroscopy, which is used for rapid and simultaneous multielement analysis. To perform PIXE of a sample, it is bombarded with high energy charged particles, which causes ionization (due to coulomb interaction) of inner shell electrons. The electrons from higher shell fill that vacancy in inner shell and difference of binding energies of two shells is emitted in the form of X-rays. Identification of constituent elements of analyzed sample is done on the basis of wavelength of emitted characteristic X-rays and concentration is predicted from intensities of characteristic X-rays. These characteristic X-rays are detected by high purity germanium (HPGe) X-ray detector-GUL0035 [9].

137.2.1 Experimental Setup and Data Analysis

PIXE measurements are done using 2.7 MeV proton beam, incident on target (sample to be analyzed) using a graphite collimator having 1 mm diameter. Al absorbers are used to improve count rate of high Z elements ($Z > 34$). Emitted X-rays are detected by HPGe X-ray detector, positioned at an angle of 45° to the incident proton beam axis. PIXE spectrum is obtained with the help of MAESTRO (for windows OS) program and analyzed with GUPIX software [10].

137.3 Results

PIXE spectrum of glass and bakelite samples are shown in Fig. 137.1. Nineteen elements Si, P, S, K, Ca, Sc, Ti, Mn, Fe, Co, Ni, Cu, Zn, Sn, Sr, Cl, Cr, Nb and Sb are predicted from PIXE measurements of glass and bakelite. Concentration of predicted elements is shown in Table 137.1.

Sample codes: A: Asahi Glass; B: Modi Glass; C: Saint Gobain Glass; D: IEL Bakelite; E: Hylam Bakelite; F: Italian Bakelite; G: FormicaW Bakelite; H: FormicaB Bakelite.

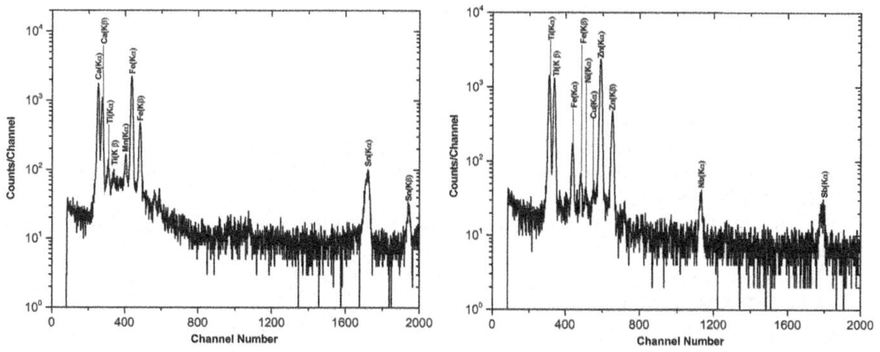

Fig. 137.1 PIXE spectrum of Modi glass (left) and Hylam bakelite (right)

Table 137.1 Elemental concentrations of glass and bakelite samples predicted from PIXE measurements

Element	A(PPM)	B(PPM)	C(PPM)	D(PPM)	E(PPM)	F(PPM)	G(PPM)	H(PPM)
Si	449221	465556	743768	914.3	2635	344	–	14883
P	698	–	940.6	195.6	111.7	448	1402	–
Ca	51499	52938.7	83120	10587	303	771	663.6	3966
Sc	228.2	648.9	1056.4	3312	29.5	8.6	1037.1	99.2
K	1292.7	1444.2	6116.9	248.3	1120	129	87.1	190
S	985	875.6	1542.3	735.2	180	2051	426.5	53051
Cl	–	–	–	2137	3163	299	285.4	2995
Fe	976.3	610.9	1751.1	321.5	282	84.1	268.4	283.4
Ti	121.4	150.5	193	258	68432.5	1.1	121531	1897.2
Mn	37.7	63.1	60.5	26.7	–	30.3	–	841.3
Cr	–	–	–	43.5	131.2	21.8	593.3	–
Zn	–	7.2	6.7	1968.7	1119	5.7	34.6	11.4
Sn	–	3413.3	2848.9	–	–	–	–	–
Co	13.2	10.8	42.7	109.6	15.4	1.6	4.6	–
Ni	3.8	–	6.3	997.4	11	3.9	10.2	6.3
Cu	–	–	–	65.6	91	2.3	66.3	46.1
Sr	25.3	13.3	95.6	–	–	–	–	–
Nb	–	–	–	1.3	51.5	2.2	1.6	1.1
Sb	–	–	–	–	310	–	–	–

References

1. R. Santonico et al., Nucl. Instrum. Methods A **187**, 377 (1981)
2. G. Bruno, Eur. Phys. J. C **33**, s01 (2004)
3. B. Bonner et al., Nucl. Instrum. Methods A **508**, 181–184 (2003)
4. ALICE Collaboration, ALICE Technical design report of the time-of-flight system (TOF), CERN- LHCC 2000-12, 2000
5. ATLAS Collaboration, ATLAS muon spectrometer: Technical design report, CERN- LHCC 2000-12, 2000
6. C.M.S. Collaboration, Technical proposal, CERN-LHCC-94-38, 1994; CMS Collaboration JINST **187**, S08004 (2008)
7. A. Abashian et al., Nucl. Instrum. Methods A **479**, 117–232 (2002)
8. V.M. Datar et al., Nucl. Instrum. Methods A **602**, 744–748 (2009)
9. N.K. Puri et al., Int. J. PIXE **16**, 7–20 (2006)
10. A. Shariff et al., Nucl. Instrum. Methods B **189**, 131–137 (2002)

Chapter 138
Calibration and Auxiliary Unit for INO's ICAL Data Acquisition System

P. Kaur, A. Lokapure, Pathaleswar, M. N. Saraf, B. Satyanarayana, D. Sil, S. S. Upadhya and E. Yuvaraj

138.1 Necessity and Overview of CAU

138.1.1 Segmented Event Data Acquisition

The segmented data of an event is recorded over participating RPCs in the detector and transferred to back-end for event building by data-collation using time stamps. The event time stamp in each RPC is the latched data of RTC on a trigger. So, RTCs in all the connected data nodes like RPCs, Global Trigger system and CAU needs to be time synchronized.

138.1.2 Trajectory Reconstruction

The X-Y position of interaction in the RPCs of consecutive layers will be used to reconstruct the tracks in the detector. The TOF of interaction points in the RPCs across layers will give the time profile of the tracks in the detector. The locally recorded TOF measurements in DAQs are to be translated to a single global reference point to reconstruct the time profile of tracks. The global trigger path delays to RPCs are different due to different trigger cable lengths from the CAU. CAU computes these

P. Kaur—Now at Pohang University of Science and Technology, S. Korea
A. Lokapure—Now at National Institute of Oceanography, Goa, India.

P. . Kaur · A. Lokapure · Pathaleswar (✉) · M. N. Saraf · B. Satyanarayana ·
D. Sil · S. S. Upadhya · E. Yuvaraj
Tata Institute of Fundamental Research, Homi Bhabha Road, Mumbai 400005, India
e-mail: pathaleswar@tifr.res.in

D. Sil
e-mail: dipankar.sil@tifr.res.in

Md. Naimuddin (ed.), *XXII DAE High Energy Physics Symposium*, Springer Proceedings in Physics 203, https://doi.org/10.1007/978-3-319-73171-1_138

trigger path delays to RPCs by measuring round path delays of trigger path to all the RPCs using TDC during calibration. The global TOF data are obtained by adding respective trigger path delay offset to local TOF data of RPCs. These global TOFs data are used to re-construct the trajectories in the detector.

138.1.3 Control and Management

The controls such as enabling/disabling of trigger, loading and synchronizing RTCs, setting operation parameters etc. are handled in CAU via command interface.

138.2 CAU Architecture and Subsystems

Figure 138.1 represents the CAU detailed architecture with the various Signal and Data interfaces to various units.

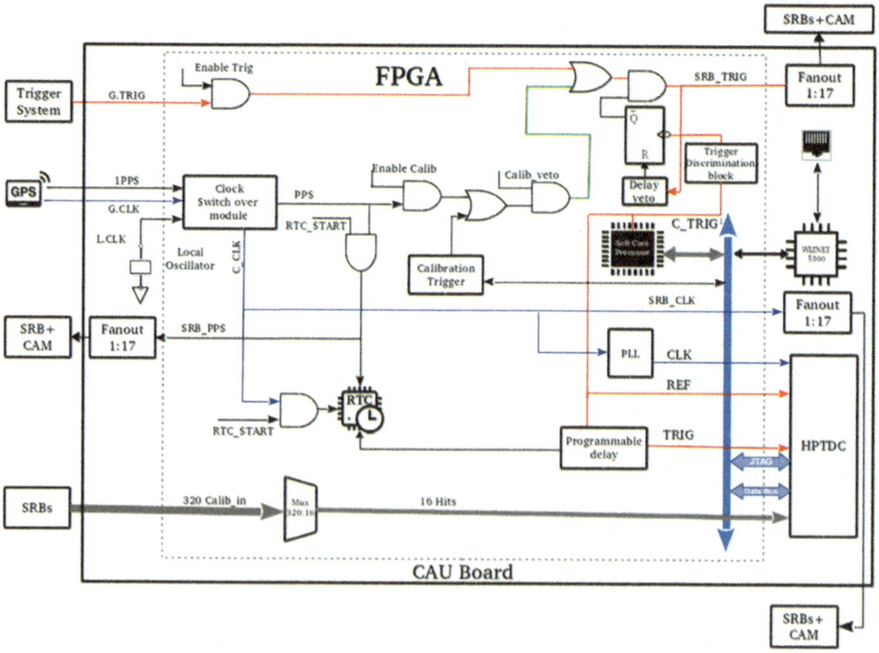

Fig. 138.1 CAU architecture

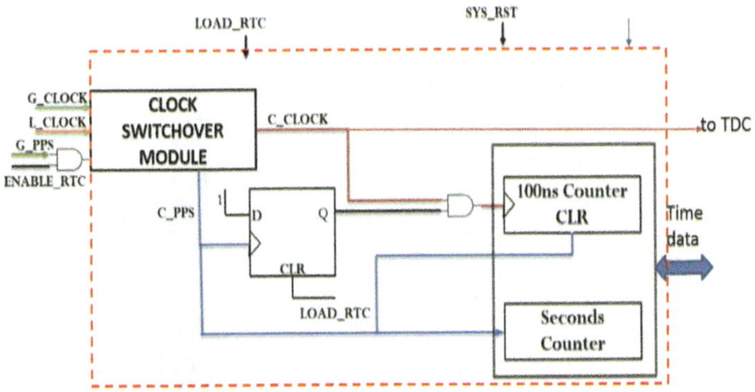

Fig. 138.2 RTC block diagram

138.2.1 Main Subsystems

1. Calibration Trigger Generator: Calibration Trigger is a periodic pulse to all the RPC nodes and Trigger system to invoke calibration event in all nodes and recording segmented calibration data. Optional selection of PPS or a pulse with a programmable period from ms to few hours is supported.
2. Caliban Multiplexer: The return path signals of calibration trigger from all nodes to CAU are sent to TDC via a sequential auto select line updating Multiplexer. The multiplexer selects a set of 16 Calib signals and feed it to TDC Hits for offset measurement.
3. RTC and Synchronization (Fig. 138.2): RTC in CAU is comprised of two 32bit counters one for 100 ns and other for Seconds. The 100 ns counter is driven by the common 10 MHz clock and cleared by common PPS signal which drives the Seconds counter. The common Clock and PPS is mainly derived from GPS and by the local oscillator in the absence of GPS signals. RTC Synchronization in all other nodes is achieved by following steps: (i) Load the seconds counter of RTCs in all the nodes with future time by remote command. (ii) Enable PPS fan-outs to all nodes via the command server, at pre-loaded time which enables common clock and PPS to all the RTCs. The subsequent PPS synchronizes RTCs by clearing 100 ns counters.
4. Time to Digital Converter: The HPTDC by CERN is used to measure round trip delays with a resolution of 100 ps [1].
5. Clock Switchover Module: Switch-overs to Internal Clock or PPS in the absence of their global counterpart signal.
6. NIOS II Microprocessor and Network Interface: The processor is used for local program control, data acquisition and remote command interface, and Wiznet 5300 is interfaced with it for remote network interface.
7. Trigger Discrimination: Identifies the trigger as Normal or calibration type based on width and acts accordingly.

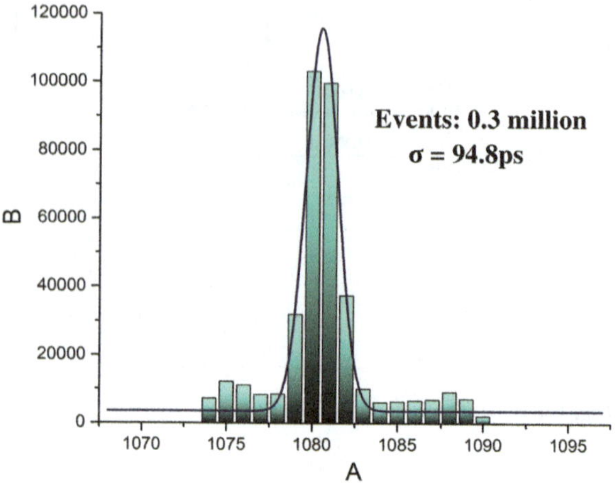

Fig. 138.3 Delay computed - 7.5 m cable

138.3 Results and Conclusions

The Fig. 138.3. shows measured round path delay 7.5 m cable with $\sigma = 94.8$ ps. Evaluation tests with lengthier cables also show the desired results [2].The observed results show good stability of RTC time synchronization among 3 DAQs over a period 24 h.

References

1. J. Christiansen et al., High performance time to digital converter, Version 2.2, 2004. (CERN)
2. P. Kaur et al., Delay and offset calibration schemes for the INO ICAL electronics, in *XXI DAE-BRNS High Energy Physics Symposium* (2014)

Chapter 139
Muon Chamber Endcap Upgrade of the CMS Experiment with Gas Electron Multiplier (GEM) Detectors and Their Performance

Mohit Gola

139.1 Introduction

The Large Hadron Collider (LHC) is the most powerful particle accelerator till date, built by the European Organization for Nuclear Research (CERN). From time to time, LHC needs upgrade so that the discovery potential is increased. This involves the upgrade of the accelerator systems of LHC and its detectors. Compact Muon Solenoid (CMS) [1] is one of the multi-purpose detectors of LHC having onion-like structure which is capable of studying many aspects of proton collisions at TeV scale. Due to the LHC Upgrade, there will be increase in the centre of mass energy upto 14 TeV and luminosity to $5 - 7 \times 10^{34}$ which will help to study the high energy range physics and rare decays that could be beyond the SM. By increasing the collision rate, the detection environment in CMS gets affected as the background rate of forward region of CMS muon end-caps increases abruptly. Therefore, CMS GEM Collaboration [2] decided to introduce additional gaseous detectors in the forward end- caps, which can operate at very high rates with good performance, which are known as Gas Electron Multiplier (GEM). The upgrade project is named as GE1/1, where "G" stands for GEM, "E" for End-cap and "1/1" corresponds to first muon station and its first ring, respectively. The full ϕ coordinate and the pseudo-rapidity region $1.55 < \eta < 2.18$ will be covered by trapezoidal super- chambers (layer of 2 triple-GEM detectors).

Few detectors are going to be installed inside CMS cavern for testing and commissioning purpose called Slice Test [3] in which 10 detectors with 2 different geometries (Short and Long) are used. These Slice Test detectors should pass few Quality controls which are described below.

M. Gola, On behalf of CMS Collaboration.

M. Gola (✉)
University of Delhi, Delhi 110007, India
e-mail: mohit.gola@cern.ch

Md. Naimuddin (ed.), *XXII DAE High Energy Physics Symposium*, Springer Proceedings in Physics 203, https://doi.org/10.1007/978-3-319-73171-1_139

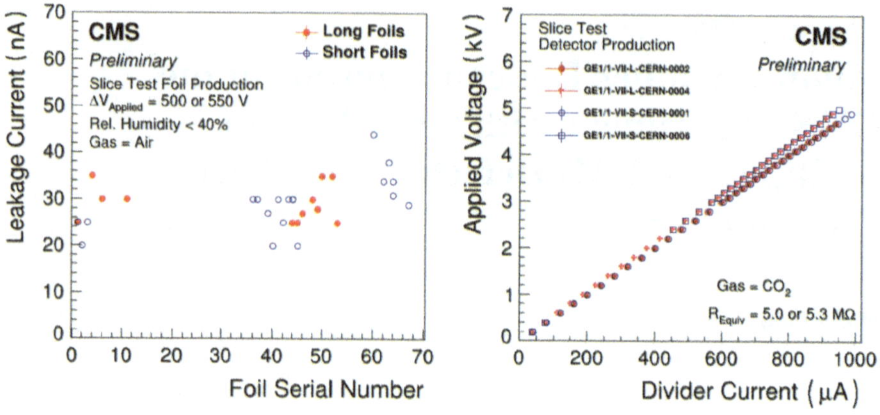

Fig. 139.1 Left: showing the leakage current of 4 long and 6 short GEM detector prototypes. Right: showing the HV test of 2 long and 2 short GEM detector prototypes

139.2 Quality Controls

CMS GEM Collaboration has decided to perform following tests before installing the detectors for Slice Test and this paper consists of results with 4 long and 6 short GEM detector prototypes (generation VII).

139.2.1 QC2: Leakage Current Test

When a voltage is applied across the GEM Foil, a current flows from top to bottom due to the surface conductivity of the polyimide (Kapton) which is known as leakage current. Applying voltage across the foil also burns the dust inside the holes and blow it away. The results are shown in Fig. 139.1(left).

139.2.2 QC3: Gas Leak Test

To test the gas tightness after assembling a detector, it is first over-pressurized under the safe limit of 25 mbar. The input and output valves of the detector are then closed to detect any gas leak. Ideally, the detector is said to be gas tight if the over-pressure in its volume remains constant with time. The results of this test are not yet approved.

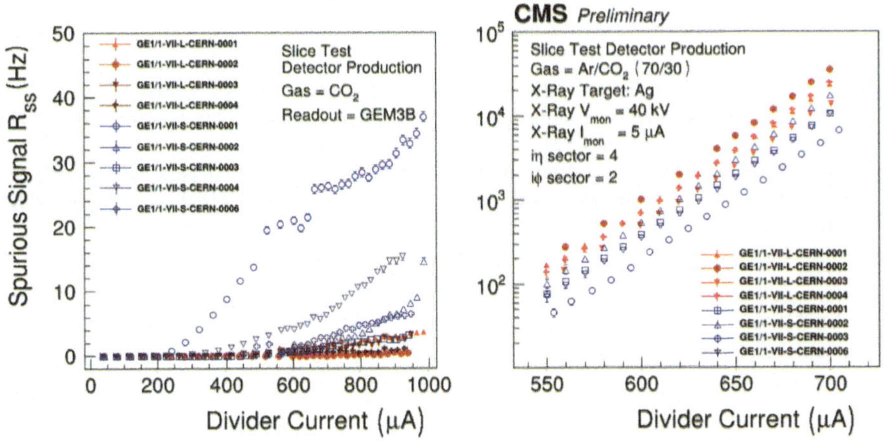

Fig. 139.2 Left: showing the spurious signal rate of 4 long and 5 short GEM detector prototypes. Right: showing the gain of 4 long and 4 short GEM detector prototypes

139.2.3 QC4: High Voltage Test and Spurious Signal

This test is performed under the high voltage (HV) environment to check the behaviour of the HV distribution circuit of the detector. The result is shown in Fig. 139.1(right). If the rate of particles is observed in cool gas at high voltage, such signals are known as Spurious Signals. We measure this rate from the bottom of the third GEM Foil as a function of the current across the HV divider. The result is shown in Fig. 139.2(left).

139.2.4 QC5: Gain Test

Gain is defined as the ratio of the output current to the input current. The current and rate are measured across the $(i\eta, i\phi) = (4, 2)$ readout sector under the radiation of X-rays as shown in Fig. 139.2(right).

139.3 Conclusions

Two prototypes of GEM detectors will be used i.e. Short ($106.1 \times 23.1 \times 42.0$) and Long ($120.6 \times 23.1 \times 44.6$) in the upcoming slice test. We have shown the performance, specifically: GEM foil leakage currents, chamber gas volume integrity, high voltage circuit performance, and effective gain of the GE1/1 chambers which are going to be installed during the slice test.

References

1. C.M.S. Collaboration, The CMS experiment at the CERN LHC. JINST **3**, S08004 (2008)
2. A. Colaleo, A. Safonov, A. Sharma, M. Tytgat, CMS Technical design report for the Muon Endcap GEM upgrade. CERN-LHCC-2015-012; CMS-TDR-013
3. CMS GEM Collaboration, Updates on production status and performance of GE1/1 slice test detectors for the GEM upgrade of the CMS forward Muon system. CMS detector performance note

Chapter 140
Numerical Investigation on RPC Time Response

Abhik Jash, Sridhar Tripathy, Nayana Majumdar, Supratik Mukhopadhyay, Satyajit Saha and Subhasis Chattopadhyay

140.1 Introduction

The proposed ICAL [1] detector at INO will be used to study the atmospheric muon neutrinos to deliver improved precision of neutrino oscillation parameters, measure the deviation of mixing angle from its maximal value, determine its correct octant along with the neutrino mass hierarchy. ICAL is a magnetized calorimeter equipped with 50 kTon target mass of iron plates interleaved with Resistive Plate Chambers (RPCs) acting as tracking devices stacked in 150 layers. Position and timing information as delivered by the RPCs are two important parameters needed for studying the neutrino properties. So, understanding the physics behind the performance of RPC is necessary in optimizing the calorimeter design, predicting and interpreting the measurements.

Attempts have been made to simulate the timing performance of a RPC using a 30 cm $\times$ 30 cm Bakelite RPC with 2 mm gas gap as a model. The effects of different operating conditions like applied voltage, gas mixture and geometrical component like edge spacer on the timing performance have been studied. The numerical results have been compared to the available analytic values as well.

A. Jash (✉) · S. Tripathy · N. Majumdar · S. Mukhopadhyay · S. Saha
Saha Institute of Nuclear Physics, 1/AF Bidhannagar, Kolkata 700064, India
e-mail: abhik.jash@saha.ac.in

S. Chattopadhyay
Variable Energy Cyclotron Center, 1/AF Bidhannagar, Kolkata 700064, India

A. Jash
HBNI, Training School Complex, Anushakti Nagar, Mumbai 400094, India

Md. Naimuddin (ed.), *XXII DAE High Energy Physics Symposium*, Springer Proceedings in Physics 203, https://doi.org/10.1007/978-3-319-73171-1_140

140.2 Numerical Calculations

Garfield simulation framework [2] has been used to calculate the signals induced on the RPC read-out strip due to passage of muons through the detector following Shockley–Ramo theorem [3, 4]. Computation of other parameters relevant for simulating the signal, such as primary ionization, electric and weighting field maps, electron transport properties, has been done using the toolkits of Garfield, namely, HEED [5], neBEM [6] and Magboltz [7] respectively.

Muons of energy 2 GeV have been passed through the RPC gas chamber in randomly varying directions spanning a range of zenith angle, $\theta = 0 - 10°$ and azimuthal angle, $\phi = 0 - 360°$. The time corresponding to the crossing of 20% of the amplitude of current signal, induced due to the passage of a muon has been considered as the signal arrival time. A histogram has been filled in for the arrival of 5000 events using ROOT [8], whose mean and RMS give the average signal arrival time and intrinsic time resolution respectively.

140.3 Results

The variation of average signal arrival time and time resolution as a function of the applied voltage and for different SF_6 percentage in the gas mixture has been shown in Fig. 140.1. The proposed method of calculation is valid for signals generated after a certain threshold voltage, V_{th} (~9.8 kV for 0.5% SF_6 and ~9.4 kV for the others in Fig. 140.1) below which the avalanche process is too small to produce proper signals. For voltages above V_{th}, the detectable signal is produced earlier as the drift velocity increases with the rise in voltage. SF_6 plays the role of limiting the electron production, which in turn causes a delay in producing a detectable signal.

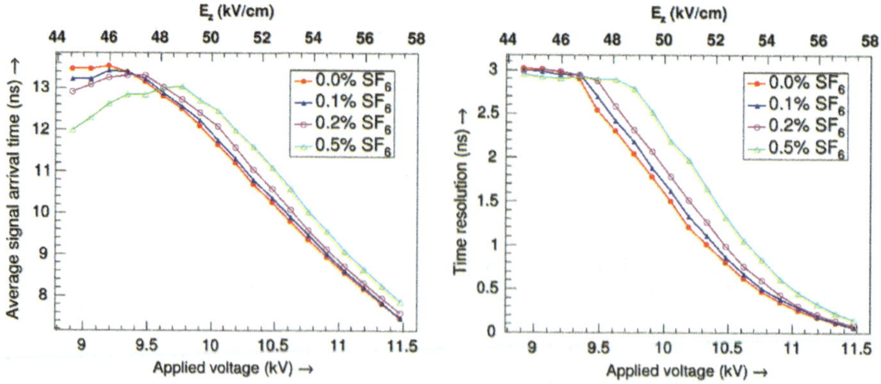

Fig. 140.1 Variation of average signal arrival time and time resolution of RPC with the applied voltage and the corresponding electric field for different amount of SF_6 in gas

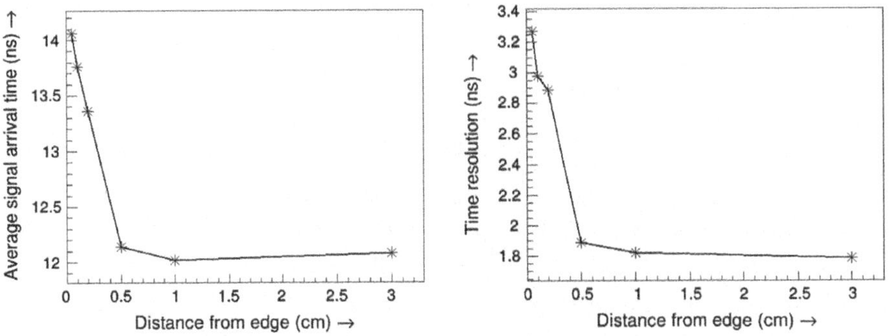

Fig. 140.2 Variation of average signal arrival time and time resolution of RPC with the distance from the edge spacer of RPC

This explains the slower arrival of signal for larger amount of SF_6 [9]. The analytic value of time resolution as calculated under certain assumptions [10], shows that it varies inversely with the effective Townsend co-efficient (α_{eff}) and the drift velocity (V_z). The increase of both the parameters with the voltage explains the improvement of time resolution at higher field values. With the increase in SF_6 amount, the value of α_{eff} decreases, whereas V_z remains almost unchanged which in turn degrade the time resolution. Our earlier calculations [11] have shown a reduced electric field near the edge spacer of RPC, which is expected to affect the timing properties at that region. The timing parameters as calculated at different distances away from the edge spacer have been depicted in Fig. 140.2 which shows that the effect of the component stretches up to 0.5 cm in this case.

140.4 Conclusion

The numerical calculations show that the average signal arrival time and time resolution improve with the increase in applied voltage. This trend is supported by the analytic formulation of time resolution [10]. Increase in the proportion of streamer quencher SF_6 has been found to deteriorate the time resolution. The presence of geometrical components affects the time resolution in comparison to the usual value as the electrostatic field gets distorted due to their presence, as has been observed in case of edge spacer.

Acknowledgements The financial support from INO collaboration is gratefully acknowledged.

References

1. A. Kumar et al., Pramana - J. Phys. **88**, 79 (2017)
2. R. Veenhoff, Nucl. Instrum. Methods A **419**, 726–730 (1998)
3. W. Shockley, J. Appl. Phys. **9**(10), p635 (1938)
4. S. Ramo, Proc. IRE **27**(9), p584 (1939)
5. I.B. Smirnov, Nucl. Instrum. Methods A **554**, 474–493 (2005)
6. N. Majumdar, S. Mukhopadhyay, J. Instrum. **2**, P09006 (2007)
7. S.F. Biagi, Nucl. Instrum. Methods A **273**, 533 (1988)
8. ROOT - Data analysis framework, https://root.cern.ch
9. M. Salim et al., J. Instrum. **10**, C04033 (2015)
10. W. Riegler, C. Lippmann, Nucl. Instrum. Methods A **508**, 14–18 (2003)
11. A. Jash et al., J. Instrum. **10**, P11009 (2015)

Chapter 141
Composite Inert Doublet Dark Matter

Siddhartha Karmakar

141.1 Introduction

In composite Higgs models the hierarchy problem of Higgs mass gets alleviated due to the presence of a new strong interaction scale around a few TeV. Here the Higgs appears as a pseudo-Nambu–Goldstone boson (pNGB) of a broken global symmetry of the strong sector. Numerous studies have been performed for the coset $SO(5)/SO(4)$, also known as the Minimal composite Higgs model. Beyond this minimal choice, models with extended composite scalar sectors can also be constructed [1–5].

We consider a Composite Inert Doublet dark matter Model (CIDM), where two pNGB scalar doublets span the $SO(6)/SO(4) \times SO(2)$ coset. One of these doublets, being odd under a remnant Z_2-symmetry of the coset, can provide a dark matter candidate. We explore some aspects of the relevant parameter space of this model allowed by relic density (RD). Such an endeavour was undertaken in [5] earlier. But we focus on the role of the quartic scalar couplings and its interplay with the compositeness scale (f) and mass differences of the new scalars in more detail.

141.2 The Model and Results

A general study of composite (extended) scalar sectors can be performed in terms of higher dimensional operators [6] and these operators can be classified according to their relative weightage following the prescriptions introduced in [7]. Some of the least suppressed operators are of type $\phi^4 D^2$ and ϕ^6, where ϕ stands for scalars and D for derivatives. For the $SO(6)/SO(4) \times SO(2)$ coset, two operators of type

S. Karmakar (✉)
Discipline of Physics, Indian Institute of Technology Indore, Khandwa Road,
Simrol, Indore 453552, India
e-mail: phd1401251010@iiti.ac.in

Md. Naimuddin (ed.), *XXII DAE High Energy Physics Symposium*, Springer
Proceedings in Physics 203, https://doi.org/10.1007/978-3-319-73171-1_141

$\phi^4 D^2$ with two inert fields survive [4] and one can do a little algebra to write them as $O_1 = (\Phi_1^\dagger D_\mu \Phi_2 + h.c)(\Phi_2^\dagger D^\mu \Phi_1 + h.c)$. We use the notations introduced in [8] for the scalars and their masses. We refrain from considering $\lambda'_{3,4,5}$ (in (141.1)) whose values depend on the representation of SM quarks under $SO(4) \times SO(2)$. As it was pointed out in [5], the low DM-mass region for a general CIDM is almost entirely ruled out by direct detection experiments or theoretically disfavoured, we only focus on $M_H \gtrsim 500\,\text{GeV}$. In this region the key annihilation channel for DM pair is into a vector boson pair. So, in this case, the relevant Lagrangian for CIDM is, $\mathcal{L}_{CIDM} = \mathcal{L}_{SM} + \mathcal{L}'$, where,

$$\mathcal{L}' \supset |D_\mu \Phi_2|^2 - \mu_2^2 |\Phi_2|^2 - \lambda_3 \left(1 + \frac{\lambda'_3}{f^2}|\Phi_1|^2\right)|\Phi_1|^2|\Phi_2|^2 - \lambda_4 \left(1 + \frac{\lambda'_4}{f^2}|\Phi_1|^2\right)|\Phi_1^\dagger \Phi_2|^2$$
$$-\frac{\lambda_5}{2}\left(1 + \frac{\lambda'_5}{f^2}|\Phi_1|^2\right)((\Phi_1^\dagger \Phi_2)^2 + h.c) + \frac{1}{f^2}(\Phi_1^\dagger D_\mu \Phi_2 + h.c)(\Phi_2^\dagger D^\mu \Phi_1 + h.c). \quad (141.1)$$

The set of parameters for studying the relic abundance is $\{M_H, \Delta M_A, \Delta M_{H^+}, \lambda_L^{eff}, \lambda_2, \lambda'_3, \lambda'_4, \lambda'_5, f\}$, where, $\lambda_L^{eff} = \sum_{i=3,4,5} \lambda_i (1 + \lambda'_i \frac{v^2}{f^2})$. We have used software package micrOMEGAs4.1 to investigate the DM properties [9]. The HHh coupling relevant for RD in CIDM case is λ_L^{eff} plus an additional contribution coming from the gauge-Higgs operator O_1. To properly disentangle the effect of compositeness coming from gauge-Higgs operator we take λ_L^{eff} as a parameter, not the HHh coupling, unlike [5]. λ_2 has almost no impact on RD and we keep $\lambda_2 = 0.9$ for all cases to satisfy the vacuum stability criteria. In the pure gauge (PG) limit (all quartic couplings set to zero) RD can be obtained for $f \gtrsim 2377\,\text{GeV}$. The RD first increases and then decreases as one increases DM mass; unlike the non-composite inert doublet model (IDM), where it keeps increasing. The impact of λ_L^{eff} in the $HH \to VV$ channel can not be numerically seen if the effect of coannihilation is too large (corresponding to $\Delta M_{A,H^+} \sim \mathcal{O}(0.1)\,\text{GeV}$). On the other hand, achieving correct relic abundance with $\Delta M_{A,H^+} \sim \mathcal{O}(1)\,\text{GeV}$ comes at the expense of relatively higher values of f compared to the PG or almost-degenerate limit. We take a few combinations of the parameters to illustrate these effects and in each case we have varied one parameter at a time. In Fig. 141.1a, C1, C2 and C3 correspond to $\lambda_L^{eff} = -0.01, 0$ and 0.01 respectively, with $\{\Delta M_A = \Delta M_{H^+} = 4\,\text{GeV}, f = 3.5\,\text{TeV}\}$. We have kept $\lambda'_{3,4,5} = 0$ to avoid unnecessary complications. The destructive interference for the negative sign of λ_L^{eff} is clearly visible in Fig. 141.1a.

In Fig. 141.1b, we demonstrate the dependence of relic density on λ_L^{eff} for scenarios with $\Delta M_{A,H^+} \sim \mathcal{O}(1)\,\text{GeV}$. C4 = $\{M_H = 800\,\text{GeV}, \Delta M_A = \Delta M_{H^+} = 7\,\text{GeV}, f = 8\,\text{TeV}\}$, C5 = $\{M_H = 800\,\text{GeV}, \Delta M_A = \Delta M_{H^+} = 7\,\text{GeV}, f = 7\,\text{TeV}\}$, C6 = $\{M_H = 600\,\text{GeV}, \Delta M_A = \Delta M_{H^+} = 3\,\text{GeV}, f = 10\,\text{TeV}\}$, and C7 = $\{M_H = 600\,\text{GeV}, \Delta M_A = \Delta M_{H^+} = 2.8\,\text{GeV}, f = 10\,\text{TeV}\}$. If λ_L^{eff} is kept decreasing from its null value, it will increase the rate of $HH \to hh$, and thus, after a certain point,

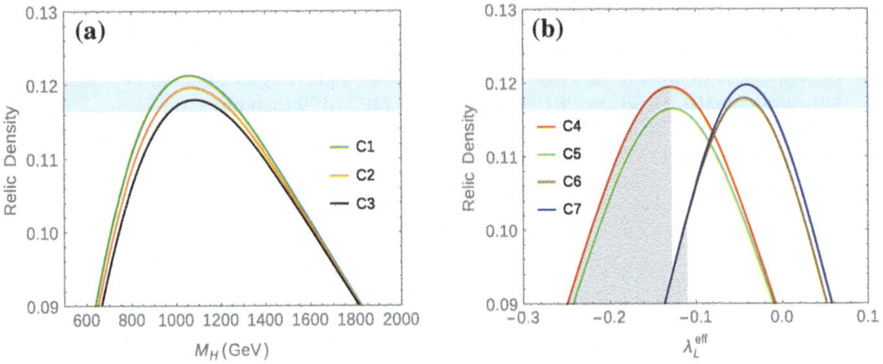

Fig. 141.1 **a** Relic density versus M_H, **b** Relic density versus λ_L^{eff}. Horizontal blue band stands for 3σ range for relic density, $\Omega h^2 = 0.1186 \pm 0.0020$ [10]. Gray shaded regions are excluded by LUX-2016 data

it will lead to the decrease of relic abundance. For example, in C4, relic abundance starts decreasing for $\lambda_L^{eff} \lesssim -0.129$. C6 and C7 show that making the scalars more degenerate leads to higher DM abundance, which happens for IDM too. C4 and C5 show that increasing the compositeness scale has similar impact on the relic density. Unlike [5], we properly disentangle the effect of the gauge-Higgs operator O_1 on relic density. In addition, we have numerically illustrated the destructive interference in $H\,H \to V\,V$ for negative values of λ_L^{eff}. We also illustrate the effect of mass splitting of the Z_2 odd scalars on the relic density and its interplay with f in order to satisfy relic density constraint.

Acknowledgements Author thanks S. Rakshit and N. Khan for discussions. This work is partially supported by DST via Grant No. EMR/2014/001177.

References

1. B. Gripaios, A. Pomarol, F. Riva, J. Serra, JHEP **0904**, 070 (2009). https://doi.org/10.1088/1126-6708/2009/04/070, arXiv:0902.1483 [hep-ph]
2. M. Frigerio, A. Pomarol, F. Riva, A. Urbano, JHEP **1207**, 015 (2012). https://doi.org/10.1007/JHEP07(2012)015, arXiv:1204.2808 [hep-ph]
3. A. Carmona, M. Chala, JHEP **1506**, 105 (2015). https://doi.org/10.1007/JHEP06(2015)105, arXiv:1504.00332 [hep-ph]
4. J. Mrazek, A. Pomarol, R. Rattazzi, M. Redi, J. Serra, A. Wulzer, Nucl. Phys. B **853**, 1 (2011).https://doi.org/10.1016/j.nuclphysb.2011.07.008, arXiv:1105.5403 [hep-ph]
5. N. Fonseca, R. Zukanovich Funchal, A. Lessa, L. Lopez-Honorez, JHEP **1506**, 154 (2015). https://doi.org/10.1007/JHEP06(2015)154, arXiv:1501.05957 [hep-ph]
6. S. Karmakar, S. Rakshit, arXiv:1707.00716 [hep-ph]
7. G.F. Giudice, C. Grojean, A. Pomarol, R. Rattazzi, JHEP **0706**, 045 (2007). https://doi.org/10.1088/1126-6708/2007/06/045 [hep-ph/0703164]

8. N. Khan, S. Rakshit, Phys. Rev. D **92**, 055006 (2015). https://doi.org/10.1103/PhysRevD.92.055006, arXiv:1503.03085 [hep-ph]
9. G. Blanger, F. Boudjema, A. Pukhov, A. Semenov, Comput. Phys. Commun. **192**, 322 (2015). https://doi.org/10.1016/j.cpc.2015.03.003, arXiv:1407.6129 [hep-ph]
10. C. Patrignani et al. [Particle Data Group], Chin. Phys. C **40**(10), 100001 (2016). https://doi.org/10.1088/1674-1137/40/10/100001

Chapter 142
Belle II Silicon Vertex Detector: Design Requirements and Construction Status

D. Dutta and G. B. Mohanty

142.1 Introduction

The Belle II experiment [1] will be the next generation flavor factory operating at an instantaneous luminosity 40 times that of Belle [2]. This will be achieved by decreasing the beam size by a factor of 20 while increasing the beam currents by a factor of 2. The increased luminosity also increases the background level.

142.2 The Belle II Silicon Vertex Detector

To operate at such a high luminosity environment, the silicon vertex detector (SVD) [3] along with the pixel detector [4] is designed to provide excellent spatial resolution ($\sim$20 μm) and low momentum tracking, be radiation tolerant, mechanically stable and immune to background hits. It is also required to have low material budget and fast readout electronics ($\mathcal{O}$(50 ns)) [1].

142.3 The SVD Design

The SVD consists of four layers of DSSD sensors (layers 3–6) at radii 38, 80, 115 and 140 mm with an angular acceptance ranging from 17° to 150°. The innermost layer of the SVD is cylindrical, while the outer three have a lantern shape. This is due to the presence of trapezoidal sensors in the forward region. The four layers are composed of 7, 10, 12 and 16 modules ('ladders') with 2, 3, 4 and 5 sensors per

D. Dutta (✉) · G. B. Mohanty
Tata Institute of Fundamental Research, Mumbai 400005, India
e-mail: deepanwita.dutta@tifr.res.in

Md. Naimuddin (ed.), *XXII DAE High Energy Physics Symposium*, Springer Proceedings in Physics 203, https://doi.org/10.1007/978-3-319-73171-1_142

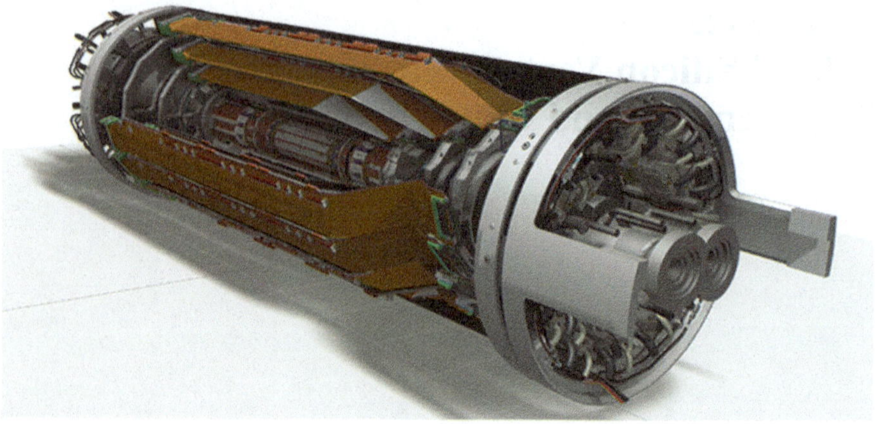

Fig. 142.1 Computer aided design of Belle II VXD showing different SVD ladders

ladder, respectively. These sensors are supported by lightweight carbon fibre ribs having high tensile strength [1]. The APV25 [5] chips are used for readout due to their short shaping time and high radiation tolerance. These chips consist of low-noise preamplifiers followed by a shaper. To retain a high signal-to-noise ratio (SNR), the capacitive load of the amplification system should be low. This requires placing the chips very close to the sensor. At Belle II, a flexible circuit called Origami containing the APVs are directly placed over the n-side of the inner DSSDs (with a thermal and electrical insulator [6] in between) to maintain high SNR while minimizing the material budget. This novel concept is called the Origami chip-on-sensor design. Another highlight of the design has been the use of dual phase CO_2 (liquid and gas mixture) for cooling the DSSDs, which enables a common cooling pipe to be used for two adjacent ladders (Fig. 142.1).

142.4 Assembly Procedure, Mechanical and Electrical Tests

The SVD ladders are assembled at various sites. TIFR is involved in the assembly of Layer 4 of the SVD. In general, these assembly procedures require a high level of precision and are accomplished with the help of several dedicated support structures, jigs. Gluing is done for mechanical connection. To make electrical connections, wire bonding is done using an ultrasonic wedge bonder. Details can be found in [7]. The gluing and wirebonding parameters need to be optimized. Precise coordinate measurements and electrical testing are done at various stages during the assembly procedure. Before the ladder assembly, the sensors are electrically tested. Current-voltage (IV) measurements are done to determine the breakdown voltage and overall sensor quality. After full ladder assembly, various coordinate measurements are done to check whether the ladder is aligned with the desired precision or not. We measure

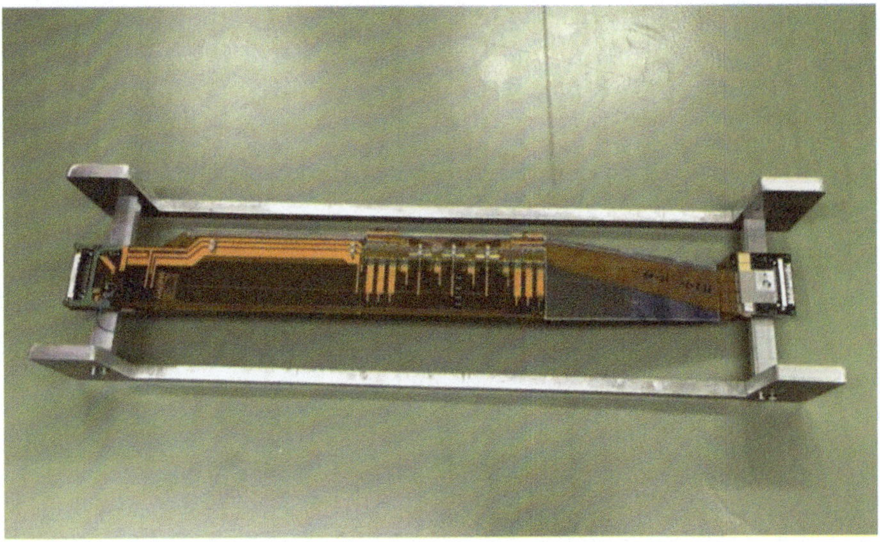

Fig. 142.2 A fully assembled layer 4 ladder

coordinates of some randomly chosen points and check the precision of various alignment parameters. Electrical tests are then performed to determine the overall electrical performance of the APV chips and fully assembled ladder. We send some electrical or radioactive signals and try to read the output via the APVDAQ [8] system. The defective strips are detected with the help of pulse height, noise, SNR and APV response curve. They are then recorded to an online database system.

142.5 Current Status

After building several working prototypes, all assembly sites are now onto the production grade ladders. Figure 142.2 shows a production grade L4 ladder. Ladder production is expected to complete by November 2017. Ladder installation in the endrings is planned to start from May 2017.

Acknowledgements The authors thank the conference organizers for organizing such a nice conference.

References

1. T. Abe et al. (Belle II Collaboration), arXiv:1011.0352
2. A. Abashian et al. (Belle Collaboration) Nucl. Instrum. Methods Phys. Res. Sect. A **479**, 117 (2002)
3. M. Friedl et al., Nucl. Instrum. Methods Phys. Res. Sect. A **732**, 83 (2013)
4. C. Marinas et al., Nucl. Instrum. Meth. Phys. Res. Sect. A **731**, 31 (2013)
5. M. J. French et al., Nucl. Instrum. Methods Phys. Res. Sect. A **466**, 359 (2001)
6. http://www.airexag.ch
7. C. Irmler et al., JINST **11**, C01087 (2016)
8. http://www.hephy.at/user/friedl/belle/apvdaq_nov2004.pdf

Chapter 143
A Holographic Description of BCS Instability at Finite Baryon Density

Swarnendu Sarkar, Sudipto Paul Chowdhury, S. Kalyana Rama, Balachandran Sathiapalan and Nilanjan Sircar

143.1 Introduction

In the recent years there has been a lot of work devoted towards the study of strongly coupled systems using AdS/CFT duality [1]. The applications of this duality range from attempts in understanding Quantum Chromodynamics (QCD) to various condensed matter systems. Holographic superconductivity is one such promising field of research. In the Ginzburg–Landau type of approach, one studies the dynamics of a charged scalar field in the background of an Reissner Nordström-AdS (RN-AdS) black hole. By the AdS/CFT dictionary the boundary value normalizable mode of the charged scalar field gives the expectation value of the scalar field (condensate) it corresponds to. The appearance of an expectation value then signals a condensation that breaks the boundary $U(1)$ symmetry. The bulk dual of the boundary instability is the instability of the RN-AdS black hole. Our main aim however is to study models that begin with a string theory. In this case the field content will be known precisely. We can then shed some light on the formation of Cooper pairs. Holography works

S. Sarkar (✉)
Department of Physics and Astrophysics, University of Delhi, Delhi 110007, India
e-mail: ssarkar@physics.du.ac.in

S. P. Chowdhury
Institute of Physics, Sachivalaya Marg, Bhubaneswar 751005, India
e-mail: sudipto@iopb.res.in

S. K. Rama · B. Sathiapalan
The Institute of Mathematical Sciences, Taramani, Chennai 600113, India
e-mail: krama@imsc.res.in

B. Sathiapalan
e-mail: bala@imsc.res.in

N. Sircar
School of Physics and Centre for Theoretical Physics, National Institute for Theoretical Physics, University of the Witwatersrand, Wits, Johannesburg 2050, South Africa
e-mail: nilanjan.tifr@gmail.com

Md. Naimuddin (ed.), *XXII DAE High Energy Physics Symposium*, Springer Proceedings in Physics 203, https://doi.org/10.1007/978-3-319-73171-1_143

for non-Abelian gauge theories. So we look at QCD models with a large number of colors, N_c.

143.2 A Microscopic Model

Holographic examples of large N_c QCD models were first studied by Witten [2], and more recent versions of it were constructed by Sakai and Sugimoto [3]. The model that we propose [4] consists of $SU(N_c)$ Yang–Mills with large N_c and two flavors of quarks. At low temperatures the theory is in the confined phase. When we have a finite density of baryons we expect the phase to be superconducting. We thus analyze this model in the following two steps: (i) Introduce finite baryon density at zero temperature and check that the phase with this finite baryon density is favourable over the vacuum phase. (ii) We then look for solutions that probe the existence of a baryon number violating condensate. The existence of these solutions will then prove the appearance of the superconducting phase.

D-brane configuration: In the Sakai–Sugimoto model $SU(N_c)$ Yang–Mills is realized on a stack of N_c $D4$ branes. With one direction along the $D4$ compactified, the low energy theory is pure $SU(N_c)$ Yang–Mills in $3+1$ dimensions. Flavor quarks are introduced through probe $D8$ branes. The $U(1)$ gauge symmetry on the probe $D8$ brane corresponds to the $U(1)$ baryonic symmetry of the theory on the $D4$. The corresponding charge d gives the baryon density. We need fields charged under this $U(1)$ which will act as probes for the formation of the superconducting condensate. However the low energy spectrum on a single $D8$ brane contains only one real scalar. We thus consider two $D8$ branes. The gauge symmetry on the pair of $D8$'s is $U(2) \cong U(1)_B \times SU(2)$. If τ^i are the generators of $SU(2)$ we can then decompose the scalar and field as $\Phi = \Phi^3 \frac{\tau^3}{2} + \Phi^+\tau^+ + \Phi^-\tau^-$ and similarly for the gauge field A. The scalar and the gauge fields $\Phi^\pm$ and $A^\pm$ are charged under the $U(1)_3$ generated by τ^3. So if we turn on sources corresponding to this $U(1)_3$, these fields can then act as probes for the condensate. The baryon charge d_3 due to the two branes thus come with opposite sign and the net charges corresponding to the branes, (d_1, d_2) are different, where $d_{1,2} = d_0 \pm \frac{d_3}{2}$, d_0 being the charge due to $U(1)_B$. In the presence of sources the $D8$ branes form a cusp at the lowest point [5]. If there are a set of $D8$'s they come at different angles at the lowest point and thus locally they intersect. This breaks the $U(1)_B \times SU(2)$ symmetry to $U(1)_B \times U(1)_3$.[1] A study of the phases shows that the phase with nonzero d_3 is favored over the vacuum phase [4]. We thus have a phase with finite baryon density. When two branes meet at an angle, in general there is a tachyonic mode. This mode drives the process of smoothening out the intersecting configuration. We identify this with the Cooper pairing instability

[1]The $U(1)$'s on the branes are sourced by instantons. So strictly speaking we should start with a configuration of $2n$ $(n>1)$ $D8$ branes thus having $U(2n)$ symmetry. This symmetry is then broken down to $U(n) \times U(n) \cong U(1)_B \times SU(n) \times U(1)_3 \times SU(n)$. However since the $SU(n)$ is not relevant for our analysis we only work with $U(1)_B \times U(1)_3$.

of the strongly coupled theory. There should be a condensate of charged fields that breaks the $U(1)_3$ symmetry. This is demonstrated numerically in the Yang–Mills approximation in [4].

Finite temperature: In a simplified setup of intersecting $D1$ branes in flat space, the tachyonic instability was studied at finite temperature in [6]. This computation is done with the aim of capturing the essential physics of the instability of intersecting $D8$ branes in $D4$ brane background. Tree-level analysis shows that the mass-squared for the tachyon is given by $-\theta/(2\pi\alpha^{'})$, where θ is the intersecting angle and the constant $\alpha^{'}$ is the square of the *string length*. At zero temperature this instability was studied in the Yang–Mills approximation in [7]. The Yang–Mills approximation is valid as long as the intersecting angle is small. Thus if the intersecting configuration is defined by $\Phi_1^3 = qx$ then $(\theta \to 0, \alpha^{'} \to 0)$, $\theta/(2\pi\alpha^{'}) = q$ (fixed). At finite temperature we expect the instability to vanish above a critical temperature T_c. This was demonstrated at the one-loop level by computing the finite temperature correction to the tachyon mass for intersecting $D1$ branes. For the Yang–Mills coupling, $g^2 = 0.01$ and for $q = 0.1, 0.2$ and 0.3 the critical temperatures are $T_c = 3.34, 9.48$ and 16.73 respectively. This analysis was further generalized to that case of intersecting $D2$ and $D3$ branes in [8] (see also [9]).

References

1. J.M. Maldacena, The Large N limit of superconformal field theories and supergravity, Int. J. Theor. Phys. **38**, 1113 (1999) [Adv. Theor. Math. Phys. **2**, 231 (1998)]. https://doi.org/10.1023/A:1026654312961 [hep-th/9711200]
2. E. Witten, Anti-de Sitter space, thermal phase transition, and confinement in gauge theories. Adv. Theor. Math. Phys. **2**, 505 (1998), arXiv:hep-th/9803131
3. T. Sakai, S. Sugimoto, Low energy hadron physics in holographic QCD, Prog. Theor. Phys. **113**, 843 (2005). https://doi.org/10.1143/PTP.113.843, arXiv:hep-th/0412141
4. S. Kalyana Rama, S. Sarkar, B. Sathiapalan, N. Sircar, Strong coupling BCS superconductivity and holography, Nucl. Phys. B **852**, 634 (2011). https://doi.org/10.1016/j.nuclphysb.2011.07.011, arXiv:1104.2843 [hep-th]
5. O. Bergman, G. Lifschytz, M. Lippert, Holographic nuclear physics, JHEP **11**, 056 (2007). https://doi.org/10.1088/1126-6708/2007/11/056, arXiv:0708.0326 [hep-th]
6. S.P. Chowdhury, S. Sarkar, B. Sathiapalan, BCS instability and finite temperature corrections to Tachyon mass in intersecting D1-Branes, JHEP **1409**, 063 (2014). https://doi.org/10.1007/JHEP09(2014)063, arXiv:1403.0389 [hep-th]
7. K. Hashimoto, S. Nagaoka, Recombination of intersecting D-branes by local tachyon condensation, JHEP **0306**, 034 (2003). https://doi.org/10.1088/1126-6708/2003/06/034 [hep-th/0303204]
8. V. Sethi S.P. Chowdhury, S. Sarkar, Finite temperature corrections to Tachyon mass in intersecting D-Branes, https://doi.org/10.1007/JHEP04(2017)109, arXiv:1610.07140 [hep-th]
9. V. Sethi, S.P. Chowdhury, S. Sarkar, Intersecting $D3$ branes at finite temperature, in *Proceedings report in this volume* (Chapter 108)

Chapter 144
Generalized Degeneracies and Their Resolution in Neutrino Oscillation Experiments

Newton Nath, Srubabati Goswami and K. N. Deepthi

144.1 Introduction

Standard three-flavor neutrino oscillation paradigm consists of six oscillation parameters, these are; (i) 3-mixing angles ($\theta_{ij}, j > i = 1, 2, 3$), (ii) 2-mass squared differences ($\Delta m^2_{i1}, i = 2, 3$) and (iii) the Dirac CP phase δ_{CP}. Almost two decades of neutrino oscillation experiments have measured or given hints about these parameters. Currently, the major three unknowns in neutrino oscillation physics are, (i) neutrino mass hierarchy, i.e. the sign of $|\Delta m^2_{31}|(\Delta m^2_{31} > 0$ is known as the normal hierarchy (NH) or $\Delta m^2_{31} < 0$ is known as the inverted hierarchy (IH)), (ii) the octant of θ_{23} ($\theta_{23} < \pi/4$ is known as the lower octant (LO) or $\theta_{23} > \pi/4$ is known as the higher octant (HO)) and (iii) the CP phase δ_{CP}, recently T2K results hint towards the maximal δ_{CP} value [1]. The (LBL) oscillation experiments like, T2K [2] and NOνA [3] which are currently taking data, can provide information on these unknowns. The major obstacles which these LBL experiments have to overcome are the issues of parameter degeneracies i.e. at least two different sets of parameters giving rise to the same oscillation probability.

In this work, we show in a comprehensive way the parameter degeneracies in the test ($\theta_{23} \times \delta_{CP}$)-plane for a given set of representative true values for both the hierarchies. Depending on right (R) or wrong (W) values of (hierarchy$-$octant$-\delta_{CP}$), there can be 8-possible solutions. We show all the possible observed degeneracies, by considering NOνA neutrino runs and then we discuss the role of antineutrinos ($\overline{\nu}$ s) to resolve these degeneracies. We then demonstrate how the addition of T2K and ICAL@INO can help in further constraining the degenerate solutions. Subleading effects originating from new physics beyond Standard Model may affect

N. Nath (✉) · S. Goswami · K. N. Deepthi
Physical Research Laboratory, Navrangpura, Ahmedabad 380009, India
e-mail: newton@prl.res.in

N. Nath
Indian Institute of Technology, Gandhinagar, Ahmedabad 382424, India

Md. Naimuddin (ed.), *XXII DAE High Energy Physics Symposium*, Springer Proceedings in Physics 203, https://doi.org/10.1007/978-3-319-73171-1_144

the determination of various unknowns in neutrino oscillation physics. In the near future, this can be probed in the neutrino oscillation experiments. In this respect, we also present a possible new physics scenario, namely NSI and discuss its effect on the determination of neutrino mass hierarchy. The oscillation probabilities which are relevant in our study are considered from [4]. The simulation details and experimental specifications that we considered are given in [5, 6] and the references there in. The current best fit values and 3σ ranges, that we considered in our study are consistent with [7, 8].

144.2 Results

In this section, we present the degeneracies present in both probability and χ^2 level and the role of the $\overline{\nu}$ s to resolve these degeneracies. We also discuss the role of T2K and ICAL@INO. The first column of Fig. 144.1 describes the degeneracies in the appearance channel. The descriptions of the various bands are given in the figure. We see here that the overlapping regions between navy-blue and green bands show the degeneracy for the same values of δ_{CP}. Whereas, by drawing a horizontal line for a given probability one can identify various other degeneracies present at the probability level. Second column shows the degeneracies at the χ^2 level. In this figure, the true point (39°, −90°) is marked with the black dot and the contours around it show the true solutions. Whereas, contours around $\theta_{23} \sim 52°$ show the degeneracies with wrong octant. In the third column we show the allowed region using NOνA+T2K+ICAL@INO. In this case the degenerate solutions are removed by $\overline{\nu}$-run and the allowed area is further constrained by T2K+ICAL@INO. The detailed analysis for other sets of parameter values are presented in [5].

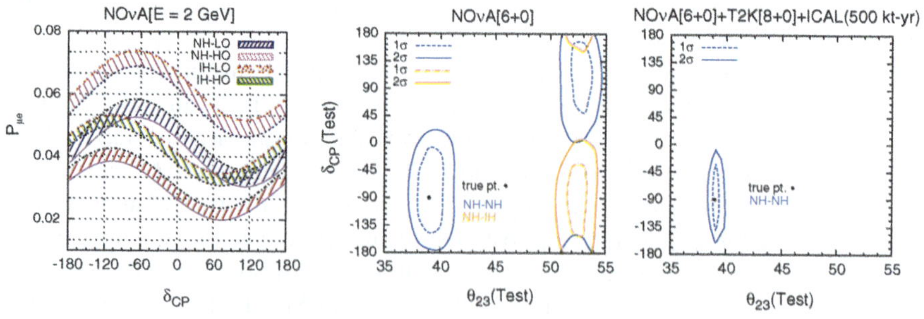

Fig. 144.1 Here, first (second) column shows the degeneracies in probability (event) level. Whereas third column shows the removal of degeneracies and the precision of the parameters due to the addition of $(\nu + \overline{\nu})$ run from T2K and ICAL@INO. Note that the plots are based on [5]

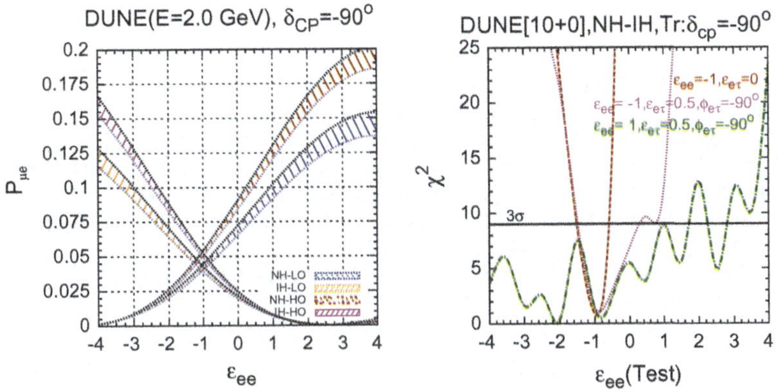

Fig. 144.2 Here, first column shows the appearance channel probability in presence of NSI parameter, ϵ_{ee} for DUNE. Whereas second column shows the hierarchy sensitivity in presence of NSI parameters ϵ_{ee}, $\epsilon_{e\tau}$. Note that the plots are based on [6]

In the Fig. 144.2, we describe the role of NSI on the determination of hierarchy for DUNE. We focus on the effects of propagation NSI for which an extra contribution to the Lagrangian can come from dimension-six four-fermion operators:

$$-\mathcal{L}^{NC}_{\text{NSI}} = 2\sqrt{2}G_F\epsilon^{fC}_{\alpha\beta}(\overline{\nu}_\alpha\gamma^\rho P_L\nu_\beta)(\bar{f}\gamma_\rho P_C f) + \text{H.c.} \tag{144.1}$$

where $\epsilon^{fC}_{\alpha\beta}$ are NSI parameters $\alpha, \beta = e, \mu, \tau$, $f = u, d, e$, C denotes the chirality and G_F is the Fermi constant. In [6], we discussed the role of the diagonal NSI parameter ϵ_{ee}. In the first column of Fig. 144.2, we present the $P_{\mu e}$ vs ϵ_{ee} (model-independent range) for fixed energy, δ_{CP}. The width of the bands are over octant for a given hierarchy as described in the figure. We mainly focus on a special point, $\epsilon_{ee} = -1$ for which, the NSI effect gets nullified by the usual matter term. Hence, in absence of off-diagonal NSI parameters any LBL experiments will not be able to lift this degeneracy. In the second column, we discuss this degeneracy at the χ^2 level and also describe the role of off-diagonal NSI parameter $\epsilon_{e\tau}$. Here, brown curve shows the degeneracy for $\epsilon_{ee} = -1$ at χ^2 level and the pink curve shows that addition of $\epsilon_{e\tau}$ is not able to lift the degeneracy once $\epsilon_{ee} = -1$. Whereas, the green curve shows that if $\epsilon_{ee} \neq -1$ then DUNE can have hierarchy sensitivity if certain ranges of ϵ_{ee} are not allowed.

In conclusion, we describe the (hierarchy−octant−δ_{CP}) generalized degeneracy and their resolution using neutrino oscillation experiments. We also discuss the impact of NSI on the mass hierarchy determination in case of DUNE.

References

1. K. Abe et al., [T2K Collaboration], Phys. Rev. Lett. **112**, 061802 (2014)
2. K. Abe et al., [T2K Collaboration], Phys. Rev. D **91**, 072010 (2015)
3. P. Adamson et al., [NOνA Collaboration], Phys. Rev. Lett. **116**, 151806 (2016)
4. E.K. Akhmedov, R. Johansson, M. Lindner, T. Ohlsson, T. Schwetz, JHEP **0404**, 078 (2004)
5. M. Ghosh, P. Ghoshal, S. Goswami, N. Nath, S.K. Raut, Phys. Rev. D **93**, 013013 (2016)
6. N. Nath, M. Ghosh, S. Goswami, The physics of antineutrinos in DUNE and determination of octant and δ_{CP}, Nucl. Phys. **B913**, 381–404 (2016)
7. M.C. Gonzalez-Garcia, M. Maltoni, T. Schwetz, JHEP **1411**, 052 (2014)
8. D.V. Forero, M. Tortola, J. Valle, Phys. Rev. D **90**, 093006 (2014)

Chapter 145
Studies on Gain of Triple GEM Using Different Argon Based Gas Mixtures

Prasant Kumar Rout, Deb Sankar Bhattacharya, Purba Bhattacharya, A. R. Sahasransu, Supratik Mukhopadhyay and Nayana Majumdar

145.1 Introduction

A typical GEM [1] foil consists of two copper layers of thickness 5 μm separated by a thin insulating polyimide material (usually Kapton) of thickness 50 μm. It typically contains a matrix of bi-conical holes of outer diameter 70 μm and inner diameter 50 μm etched with a pitch of 140 μm. Gas amplification occurs inside the holes of the GEM foils in the presence of high electric field.

145.2 Laboratory Experimental Setup

One triple GEM detector has been assembled and tested in the test bench at SINP. It is a stack of three GEM foils as shown in Fig. 145.1. In the present work we have studied the detector gain and electron transport properties of the gas mixture. The experiment has been performed using the gas mixture Argon + CO_2 with mixing ratio 80 : 20 and 70 : 30 respectively with a ^{55}Fe source which emits 5.91 keV K_α X-ray photons. In the present work, 3 : 1 : 2 : 1 (Drift: Transfer 1 : Transfer 2 : Induction) gap configuration has been used (Fig. 145.2).

P. K. Rout (✉) · S. Mukhopadhyay · N. Majumdar
Saha Institute of Nuclear Physics, HBNI, Kolkata 700064, India
e-mail: prasant.kumar.rout@cern.ch

D. S. Bhattacharya
Institute of Physics, Bhubaneswar 751005, India

P. Bhattacharya
Weizmann Institute of Science, Herzl St. 234, Rehovot 7610001, Israel

A. R. Sahasransu
IISER, Mohanpur, West Bengal 741246, India

Md. Naimuddin (ed.), *XXII DAE High Energy Physics Symposium*, Springer Proceedings in Physics 203, https://doi.org/10.1007/978-3-319-73171-1_145

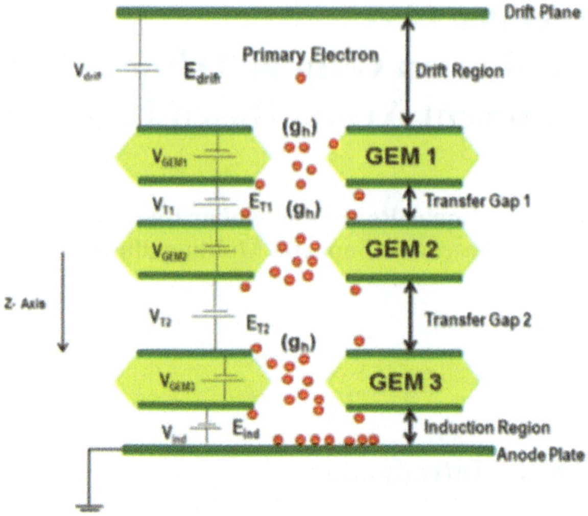

Fig. 145.1 Schematic of experimental setup

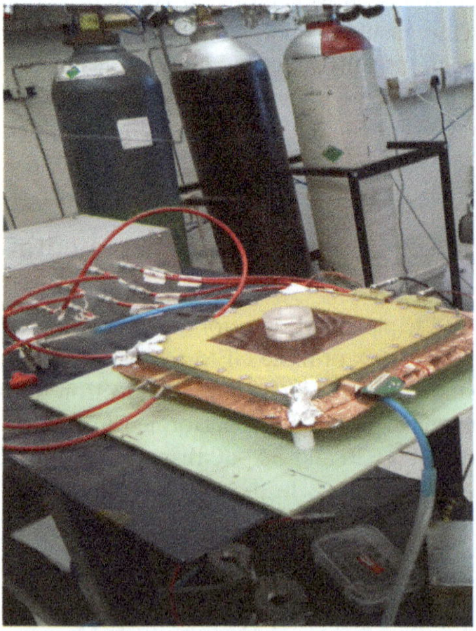

Fig. 145.2 SINP test bench

145.3 Experimental and Numerical Results

Experimental Results: The variation of gain with high voltage is shown in Fig. 145.3. As expected, the gain increases exponentially as the high voltage is increased. At very high values, however, the increase is no longer exponential (or the exponent is modified). This may be because of the effects of space charge accumulation within the triple GEM detector. A typical ^{55}Fe spectrum obtained during the experiment is shown in Fig. 145.4.

Numerical Results: The gas transport properties, like Townsend coefficient and drift velocity of electrons have been calculated using Magboltz [2] through Garfield [3] interface. The variations have been presented in Figs. 145.5 and 145.6. Numerical calculations shows that, Argon + CO_2 (80 : 20) gas mixture has higher Townsend coefficient than the Argon + CO_2 (70 : 30) gas mixture as shown in Fig. 145.5. This explains the higher gain in the Argon + CO_2 (80 : 20) gas mixture experimentally

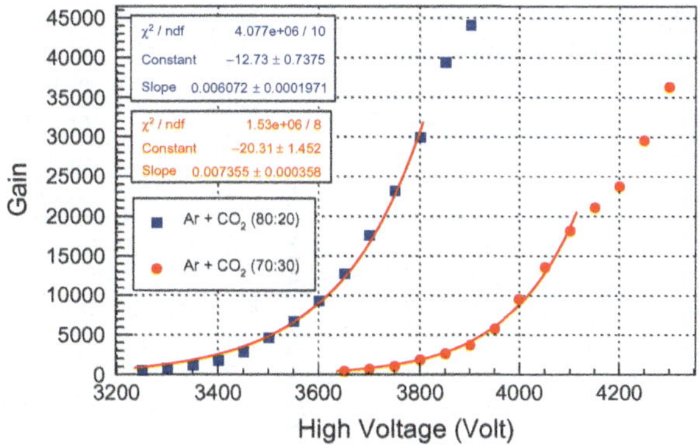

Fig. 145.3 Variation of gain with high voltage

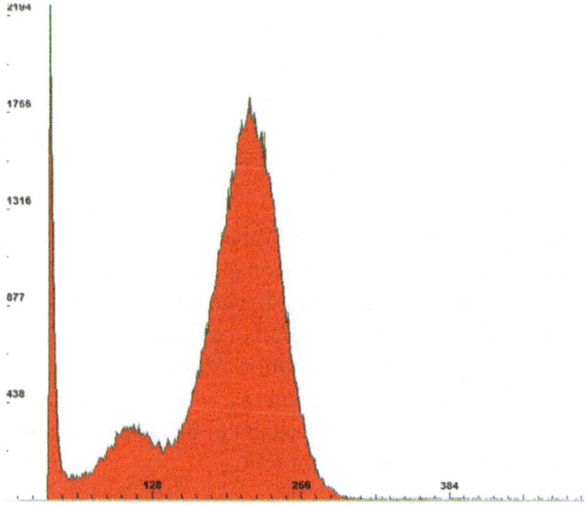

Fig. 145.4 ^{55}Fe Spectrum with Triple GEM

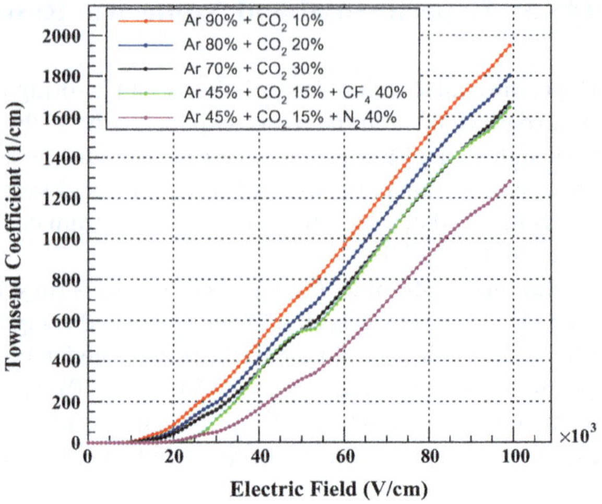

Fig. 145.5 Variation of Townsend coefficient with electric field

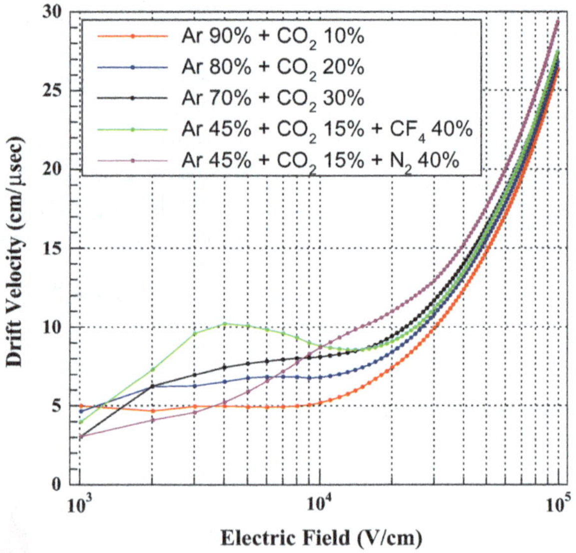

Fig. 145.6 Variation of Drift velocity with electric field

as shown in Fig. 145.3. At higher voltages, the probability of space charge effect and spark dominates. As a result, the gain drops and begins to saturate in both compositions of Argon + CO_2 gas mixture. In the present results, the onset of this regime is found to be earlier than expected. As shown in Fig. 145.6, Argon + CO_2 + CF_4 mixture is the fastest, while Argon + CO_2 (70:30) gas mixture is the next good option in the region of our interest (drift, transfer and induction fields of 1-5 kV/cm). The N_2 based gas mixture seems to be faster than other gas mixtures at higher fields and could be interesting for RPCs.

145.4 Conclusion

Preliminary steps have been taken to characterize a small triple GEM prototype detector. Experimental and numerical studies have been carried out to understand the influence of different components and proportions of gaseous mixtures on the detector performance. Choice of gas mixture, as expected, has a significant impact and, depending on the projected application, experimental and numerical studies need to be conducted.

Acknowledgements We thank Mr. Pradipta Kumar Das, Mr. Sridhar Tripathy, Mr. Abhik Jash and Prof. Satyajit Saha for their help and support. We are thankful to our institutions for providing us with the necessary facilities and financial support.

References

1. F. Sauli, GEM: a new concept for electron amplification in gas detectors. Nucl. Inst. Method A **386**(2), 531–534 (1997)
2. S.F. Biagi, Nucl. Instr. Meth. A **273**, 533 (1988)
3. R. Veenhof, NIM A **419**, 726–730 (1998)

Chapter 146
Simulation Studies of the Lambda Disks Detector for the $\overline{\text{P}}$ANDA Experiment

Ajay Kumar and Ankhi Roy

146.1 Introduction

$\overline{\text{P}}$ANDA (antiProton ANnihilations at DArmstadt) was conceived to study proton antiproton annihilations at the FAIR (Facility for Antiproton and Ion Research) [1]. The main physics motivation of $\overline{\text{P}}$ANDA is to shed light on the low energy regime of Quantum Chromodynamics (QCD) and to explore the energy region between perturbative and non-perturbative QCD. This presents the possibility to include hyperon studies in the $\overline{\text{P}}$ANDA physics program. Hyperons decay weakly and hence have a long decay length of several centimeters. This leads to a decay of hyperons in an outer part or even outside the Micro Vertex Detector (MVD). In order to improve the reconstruction of hyperons, it was proposed to include an additional Lambda Disks Detector (LDD) in forward direction in front of the MVD. It consists two layer of double sided silicon strip sensors located in a large free volume between the MVD and the Gas Electron Multipliers (GEM) detector.

146.2 Simulation Studies of Lambda Disks Detector

The reaction $\bar{p}p \rightarrow \bar{\Lambda}\Lambda \rightarrow p\bar{p}\pi^{+}\pi^{-}$ was simulated and reconstructed at incident beam momenta of 1.8 and 4.0 GeV/c to perform feasibility studies of the Lambda Disks Detector, using FairRoot and PandaRoot. The selected 1.8 GeV/c

Ajay Kumar and Ankhi Roy for the $\overline{\text{P}}$ANDA collaboration.

A. Kumar (✉) · A. Roy
Indian Institute of Technology Indore, Indore 453552, India
e-mail: ajay.kumar.phys@gmail.com

A. Roy
e-mail: ankhi@iiti.ac.in

Md. Naimuddin (ed.), *XXII DAE High Energy Physics Symposium*, Springer Proceedings in Physics 203, https://doi.org/10.1007/978-3-319-73171-1_146

beam momentum was 4.0 GeV/c chosen near to production threshold (1.436 GeV/c) of the above reaction, while 4 GeV/c was selected to assess the performance of the detector at higher energies. The EvtGenDirect [2] event generator is used for the generation of $\Lambda\bar{\Lambda}$ events. EvtGenDirect provides two different decay models for this decay, LambdaLambdaBar and LambdaLambdaBarHE. The LambdaLambdaBar model is based on an experimental angular distribution described by 8th order Legendre polynomials, for 1.8 GeV/c beam momentum. The LambdaLambdaBarHE decay model is used for the simulation at 4 GeV/c and also based on an angular distribution parametrization with a momentum dependent exponential function.

We have estimated the reconstruction efficiency and mass resolution of Λ and $\bar{\Lambda}$ hyperons with and without the Lambda Disks Detector. A double Gaussian function is used to fit the Λ peak from invariant mass distributions, since the combinatorial background has approximately a Gaussian shape. The mean values and the σ of the fitted distributions are tabulated in Table 146.1 without and including the Lambda Disks Detector at both the tested beam momenta. The reconstruction efficiencies of Λ hyperons is shown in Fig. 146.1 and estimated values are tabulated in Table 146.1. The overall $\Lambda\bar{\Lambda}$ reconstruction efficiency is found to be 28.5 and 20% at 1.8 and 4.0 GeV/c, respectively. Simulations for $\bar{p}p \rightarrow D^{*+}D^{*-} \rightarrow D^0\pi^+\bar{D}^0\pi^-$ and $\bar{p}p \rightarrow J/\psi\pi^+\pi^-$ were performed as well to study the effect of the Lambda Disks Detector on their reconstruction efficiencies. The reconstruction performance of these two channels should not be affected by the addition of Lambda Disks to the $\overline{\text{P}}$ANDA detector setup. The mean and σ values of the fitted mass distributions from both the channels and the average reconstruction efficiency for each produced particle are tabulated in Table 146.1.

It is concluded from these simulation studies that the reconstruction efficiency and resolution of produced Λ hyperons are not affected after the addition of the LDD. However, the number of hit points per track are increasing with the addition of these two disks to the central detector system of $\overline{\text{P}}$ANDA, which is a positive sign

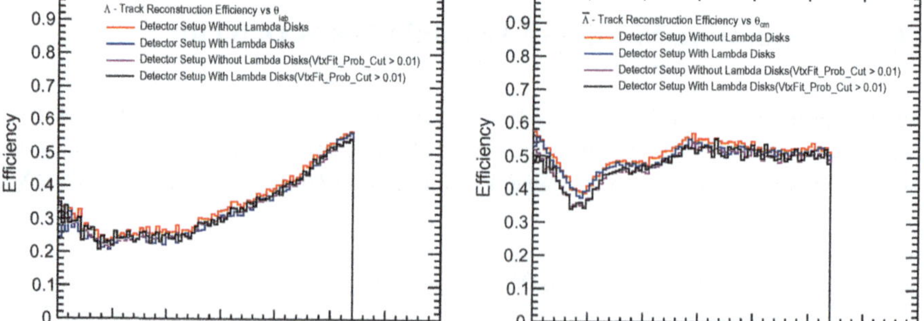

Fig. 146.1 Reconstruction efficiency of Λ (left) and $\bar{\Lambda}$ (right) at 1.8 GeV/c incident beam momentum with and without the Lambda Disks Detector

Table 146.1 Reconstructed invariant mass and efficiency without (with) the Lambda Disks Detector of Λ ($\bar{\Lambda}$) hyperons at two different beam momenta and for produced mesons from both selected channels. The statistical errors for the mean and σ values are less than 1 MeV/c^2

Detector status	P_{beam} [GeV/c]	Particle	Mean [GeV/c^2]	σ [GeV/c^2]	Avg. Reco. Effi. [%]
w/o LDD	1.8	Λ	1.114	0.004	42
		$\bar{\Lambda}$	1.115	0.003	48
	4.0	Λ	1.116	0.003	25
		$\bar{\Lambda}$	1.116	0.002	52
w/ LDD	1.8	Λ	1.115	0.004	42
		$\bar{\Lambda}$	1.115	0.003	48
	4.0	Λ	1.116	0.003	25
		$\bar{\Lambda}$	1.116	0.002	52
w/o LDD	8.0	D^0	1.86	0.019	75
		$\bar{D^0}$	1.86	0.019	75
	8.0	D^{*+}	2.01	0.017	40
		D^{*-}	2.01	0.017	40
w/ LDD	8.0	D^0	1.86	0.019	75
		$\bar{D^0}$	1.86	0.019	75
	8.0	D^{*+}	2.01	0.017	40
		D^{*-}	2.01	0.017	40
w/o LDD	6.0	J/ψ	3.086	0.052	84
w/ LDD	6.0	J/ψ	3.085	0.053	84

towards the development of the LDD. In these feasibility studies, the LDD are kept at fixed positions from the interaction point and an ideal track finder algorithm is used to find the tracks from the secondary decay vertices, considering them decayed at origin, due to the unavailability of a secondary track finder algorithm with the present PandaRoot software. One could repeat the same study with the secondary tracking algorithm while it is implemented within PandaRoot software and by changing the locations of the LDD in order to optimize the position of the detector. Finally, if one considers the improvement in the reconstruction efficiency of hyperons with LDD, then $\overline{\text{P}}$ANDA should add the LDD. This will help to improve the hyperon studies.

References

1. PANDA Collaboration, Technical Design Report for the PANDA Micro Vertex Detector, http://arxiv.org/abs/1207.6581v2 and references therein
2. S. Spataro, [PANDA Collaboration], Simulation and event reconstruction inside the PandaRoot framework. J. Phys. Conf. Ser. **119**, 032035 (2008). and references therein

Chapter 147
Flux Tubes, Field Configuration and Non-Perturbative Dynamics of QCD

Deependra Singh Rawat, H. C. Chandola, H. C. Pandey and D. Yadav

The color confinement remains to be one of the major challenging issue for many years as it is directly linked with the physical spectrum of the theory. There are many proposed conjectures made in this direction among which dual QCD based on magnetic symmetry [1–4] has been used here as a viable theory to explain the confining features of quarks inside the hadronic cage. In such a dual theory, the flux tube solutions [5] appear as a result of the topological degrees of freedom with the appearance of vector and scalar glueballs in the magnetically dominated regime of QCD vacuum. This novel picture of dual QCD vacuum is expected to systematically explore the complicated phase structure of QCD with the possibility of the formation of QGP [6, 7] as a state of nuclear matter under the extreme conditions of temperature and density [8, 9]. In view of these facts, in the present article we have investigated the field configurations of color electric field at different hadronic scale to uncover the dynamics of phase transition in the QCD vacuum.

The original version of this chapter was revised: Incorrect author name and affiliations have been corrected and a new acknowledgement text has been included. The erratum to this chapter is available at https://doi.org/10.1007/978-3-319-73171-1_227

D. S. Rawat (✉) · H. C. Chandola · D. Yadav
Department of Physics (UGC-Centre of Advanced Study),
DSB Campus, Kumaun University, Nainital, India
e-mail: dsrawatphysics@gmail.com

H. C. Chandola
e-mail: chandolaharish@gmail.com

D. Yadav
e-mail: dinoyadav02@rediffmail.com

H. C. Pandey
Department of Physics, Birla Institute of Applied Sciences, Bhimtal, India
e-mail: hempandey@birlainstitute.co.in

Md. Naimuddin (ed.), *XXII DAE High Energy Physics Symposium*, Springer Proceedings in Physics 203, https://doi.org/10.1007/978-3-319-73171-1_147

In a (4+n)-dimensional metric manifold, magnetic symmetry [10, 11] based field decomposition formulation provides a gauge invariant investigation and topological features to the QCD confinement for which the Lagrangian in quenched approximation may be expressed as [1–4],

$$\mathcal{L}_m^{(d)} = -\frac{1}{4}B_{\mu\nu}^2 + |[\partial_\mu + i\frac{4\pi}{g}B_\mu^{(d)}]\phi|^2 - \frac{24\pi^2}{g^4}[\phi_0^4 + (\phi^*\phi)^2\{2ln\frac{\phi^*\phi}{\phi_0^2} - 1\}]. \tag{147.1}$$

The one loop effective potential introduced here for inducing the dynamical breaking of magnetic symmetry which leads to the monopole condensation resulting in the QCD confinement. The field equation for the topological part ($B_{\mu\nu}$) from the above Lagrangian may be derived in the following form,

$$B_{\mu\nu}^{\nu} + i4\pi g^{-1}(\phi\partial_\mu\phi^* - \phi^*\partial_\mu\phi) - 32\pi^2 g^{-2} B_\mu^{(d)}\phi\phi^* = 0. \tag{147.2}$$

In view of the Nielsen and Olesen [5] interpretation of vortex like solutions, let us consider the single flux tube solution using the cylindrical symmetry and the flux tube orientation along the z-axis for which, the dual gauge field and the monopole field are $\mathbf{B}_\mu^{(d)} = g^{-1}cos\alpha(\partial_\mu\beta)\hat{m}$, and $\phi(x) = exp(in\varphi)\chi(\rho)$, $(n = 0, \pm1, ...)$ respectively. With the uniqueness of the function $\phi(x)$ and the Nelson-Olesen ansatz [5] for the dual gauge field and monopole field in static limit $\mathbf{B}(\rho) = -\hat{\phi}B(\rho) and\ B_t = B_\rho = B_z = 0$ leads to the color electric field in the cylindrically symmetric configuration and is given by [1],

$$E_m(\rho) = -\frac{1}{\rho}\frac{d}{d\rho}(\rho B(\rho)). \tag{147.3}$$

The associated field equations governing the flux tube structure may then be expressed in the following form,

$$\frac{d}{d\rho}[\frac{1}{\rho}\frac{d}{d\rho}(\rho B(\rho)] - \frac{8\pi}{g}(\frac{n}{\rho} + \frac{4\pi}{g}B(\rho))\chi^2(\rho) = 0 \tag{147.4}$$

$$\frac{1}{\rho}\frac{d}{d\rho}(\rho\frac{d\chi}{d\rho}) - [(\frac{n}{\rho} + \frac{4\pi}{g}B(\rho))^2 + \frac{24\pi^2}{g^4}(4\chi^2 ln\frac{\chi^2(\rho)}{\phi_0^2})]\chi(\rho) = 0. \tag{147.5}$$

The asymptotic behaviour of the monopole field $\phi \to \phi_0$ as $\rho \to \infty$ leading to the appropriate asymptotic solution for the dual gauge potential that ensures the formation of color flux tubes, is given by,

$$B(\rho) = -ng(4\pi\rho)^{-1}[1 + F(\rho)]\ and\ F(\rho) \overset{\rho\to\infty}{\longrightarrow} C\rho^{\frac{1}{2}}exp(-m_B\rho), \tag{147.6}$$

where C is a constant and $m_B(= 4\pi g^{-1}\sqrt{2}\phi_0)$ is the glueball mass. The color flux tubes formation and the confinement of color charges can be visualized more effectively on the energetic grounds by investigating the energy per unit length of the flux tube (string tension)[1, 2] and introducing a new variable R as $\rho = Rsin\theta$ that leads to the critical radius and critical density [2] of phase transition as, $R_c = 2(\frac{\pi n^2}{3}\alpha_s)^{\frac{1}{4}} m_B^{-1}$, $d_c = (\frac{64}{3}\pi^3 n^2 \alpha_s)^{-\frac{1}{2}} m_B^2$ respectively. Using the prescription for the color electric field and potential, the total electric flux penetrating the area (S) surrounded by a closed loop around the sphere of radius R is then $\phi = \int_S E.dS \equiv \oint B.dl = -2\pi \int_0^\infty \rho E_m(\rho) d\rho = -2\pi R^2 \int_0^{\frac{\pi}{2}} E_m(\theta) sin2\theta d\theta$. This alongwith the energy minimization condition, then leads to the critical color electric field at the boundary of phase transition ($R = R_c$), in the following form,

$$E_m^c = \frac{ng}{4\pi R_c^2} = \frac{ng}{2}\left(\frac{64}{3}\pi^3 n^2 \alpha_s\right)^{-\frac{1}{2}} m_B^2, \tag{147.7}$$

which leads to its numerical value as $E_m^c(fm^{-2}) = 4.97, 2.76, 0.02$ for $\alpha_s = 0.22$, 0.47, 0.96 respectively. The general form of color electric field evaluated by using (147.3) alongwith its asymptotic solution (147.6) in present spherically symmetric case is then reduced to the following form,

$$E_m(\theta) = \sqrt{\frac{\alpha_s}{16\pi}} \frac{C}{(Rsin\theta)^{\frac{3}{2}}}(1 - 2m_B Rsin\theta) exp(-m_B Rsin\theta). \tag{147.8}$$

The profiles of such color electric field as a function of the polar angle θ and for different radius (R) at optimal value of α_s have been presented by 2-d graphics (Fig. 147.1) which shows that, for a large sphere enclosing the flux tubes, the color electric flux gets localized or spreaded towards the poles ($\theta = 0$ or π) while its gets uniformly distributed for the small sphere and acquires a constant value at the critical radius R_c. With the increasing the radius in infrared sector demonstrated the

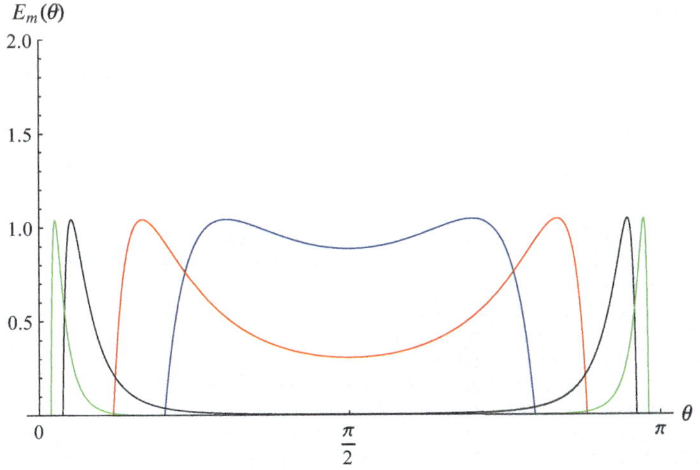

Fig. 147.1 Profiles of color electric field for R = 0.1 fm (Blue), 0.163 fm (Red), 0.5 fm (Black) and 1.0 fm (Green)

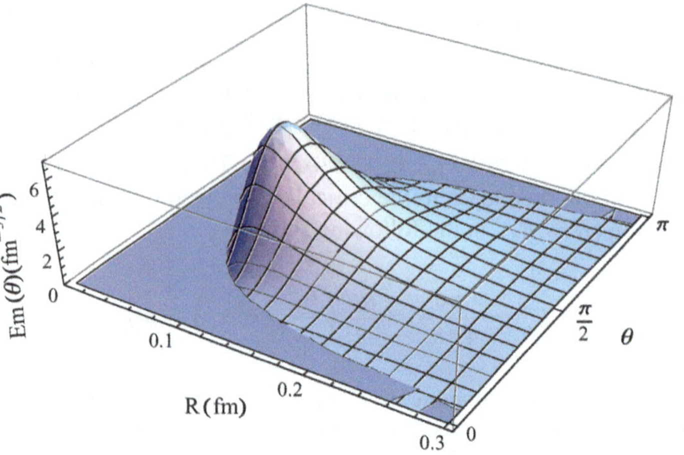

Fig. 147.2 3-D profiles of Color Electric field for $\alpha_s = 0.22$

reduction and drifting of maxima of color electric fluxes towards the higher sphere radii and consequently a reduction in flux tube density in the far-infrared sector of QCD. In the near-infrared sector, however, the increase in flux tube density may lead to the annihilation of the neighbouring flux tubes and a large homogeneous QGP formation in the central region is expected [12]. In an identical way, a more general 3-d representation (Fig. 147.2) of (147.8) for near infrared sector of QCD, also confirms similar critical behavior of color electric field near critical radius of phase transition and the possibility of QGP formation due to increased flux tube density in central region. The equally important thermal evolution of such configurations for QCD phase structure will be dealt in our forthcoming communication.

Acknowledgements One of the author DSR is thankful to University Grants Commission, New Delhi for the financial assistance in the form of UGC (BSR) Research Fellowship in Sciences for Meritorious Students. Thanks are also due to referee for his valuable comments and suggestion.

References

1. H.C. Pandey, H.C. Chandola, Phys. Lett. B **476**, 193 (2000)
2. H.C. Chandola, D. Yadav, H.C. Pandey, H. Dehnan, Int. J. Mod. Phys. A **20**, 2743 (2005)
3. H.C. Chandola, D. Yadav, Nucl. Phys. A **829**, 151 (2009)
4. H.C. Chandola, G. Punetha, H. Dehnan, Nucl. Phys. A **945**, 226 (2016)
5. H. Neilsen, B. Olesen, Nucl. Phys. B **61**, 45 (1973)
6. E.V. Shuryak, Nucl. Phys. A **750**, 64 (2005)
7. J. Liao, E.V. Shuryak, Phys. Rev. C **75**, 054907 (2007)
8. M.A. Lampert, E. Sevetitsky, Phys. Rev. D **61**, 034011 (2000)
9. H.C.M. Lizardo, E.V.L. Mello, Physica A **305**, 340 (2002)
10. Y. M. Cho, Phys. Rev D 21 (1980) 1080; ibid 2415 (1981)
11. Y.M. Cho, F.H. Cho, J.H. Yoon, Phys. Rev D **87**, 085025 (2013)
12. H. Ichi, S. Suganuma, H. Toki, Phys. Rev. D **54**, 3382 (1996)

Chapter 148
Long Range Correlations in P–Pb Collisions at $\sqrt{s_{NN}} = 5.02$ TeV with ALICE

Greeshma Koyithatta Meethaleveedu

148.1 Introduction

Two-particle angular correlations measurement is an excellent tool to probe the different mechanisms that occur in heavy-ion collisions. Angular correlations are constructed by measuring the relative angular differences in azimuth ($\Delta\varphi$) and pseudorapidity ($\Delta\eta$) between the trigger and associated particles. The correlation function in small systems like pp typically consists of a 'near side' jet peak at ($\Delta\varphi \approx 0$, $\Delta\eta \approx 0$) and a recoil 'away side' peak at ($\Delta\varphi \approx \pi$) extended in $\Delta\eta$. In addition to these features, ridge-like structures extended in $\Delta\eta$ were observed in nucleus-nucleus collisions at both near and away sides which were understood to be a characteristic of hydrodynamic flow in the Quark Gluon Plasma (QGP).

The subsequent discovery of a near side ridge in high multiplicity collisions in pp [1] and p–Pb [2] by CMS came as a surprise to the high-energy community. ALICE [3] followed by ATLAS [4] revealed symmetric near and away side ridges in p–Pb collisions when the correlations obtained in low-multiplicity events was subtracted from that in high multiplicity events. The ridge was further quantified using Fourier coefficients (v_2 and v_3). Studies using identified particles exhibited a mass ordering of v_2 as observed in Pb–Pb collisions [5, 6]. The origin of the ridge is nevertheless a question of debate and there are several theories explaining them in pp and p–Pb collisions, such as hydrodynamics and Color Glass Condensate [7–9] to mention a few. To further explore these structures in larger $\Delta\eta$, correlations between forward trigger muons ($-4 < \eta < -2.5$) and mid-rapidity associated hadrons ($-1 < \eta < -1$) was measured [10]. This is discussed in this proceeding.

Greeshma Koyithatta Meethaleveedu for the ALICE Collaboration.

G. K. Meethaleveedu (✉)
Indian Institute of Technology Bombay, Mumbai, India
e-mail: greeshmakm@iitb.ac.in

Md. Naimuddin (ed.), *XXII DAE High Energy Physics Symposium*, Springer Proceedings in Physics 203, https://doi.org/10.1007/978-3-319-73171-1_148

148.2 Analysis

The analysis was performed using the proton-lead collisions at a centre of mass energy of $\sqrt{s_{NN}} = 5.02$ TeV recorded by the ALICE detector. Muon-track correlations were studied only in p-Pb collisions while muon-tracklet correlations were studied in p-Pb and Pb-p collisions.[1] Here, tracks refer to particles fully reconstructed in Time Projection Chamber (TPC) and Inner Tracking System (ITS) while tracklets refer to short track segments reconstructed in the innermost layers of ITS.

The correlations were measured for two event classes: high-multiplicity (0–20% central events) and low-multiplicity (60–100% peripheral events). The low multiplicity correlations were subtracted from the high-multiplicity to remove the contribution from jets. The remaining correlation function was projected onto $\Delta\varphi$ and fit with a Fourier cosine series, $a_0 + 2a_1 cos\Delta\varphi + 2a_2 cos2\Delta\varphi + 2a_3 cos3\Delta\varphi + \ldots$. From the pair azimuthal anisotropy $V_{2\Delta}^{\mu-h}\{\text{2PC,sub}\} = a_2^{\mu-h}/(a_0^{\mu-h} + b)$, where $a_2^{\mu-h}$ is the a_2 extracted from $\mu - h$ correlations and b is the combinatorial baseline of the low-multiplicity correlation, the v_2 of the trigger muon was obtained (assuming the factorisation of v_2 of the trigger and associated particles) using $v_2^{\mu}\{2PC, sub\} = V_{2\Delta}^{\mu-h}\{2PC, sub\}/v_2^h = V_{2\Delta}^{\mu-h}\{2PC, sub\}/\sqrt{V_{2\Delta}^{h-h}\{2PC, sub\}}$. Figure 148.1 shows a set of muon-tracklet correlation functions in high- and low- multiplicity events along with the subtracted correlation function.

148.3 Results and Conclusions

Figure 148.2a shows the variation of $v_2^{\mu}\{\text{2PC,sub}\}$ as a function of p_T in both p-going and Pb-going directions. Consistently, a higher value of v_2 was observed in the Pb-

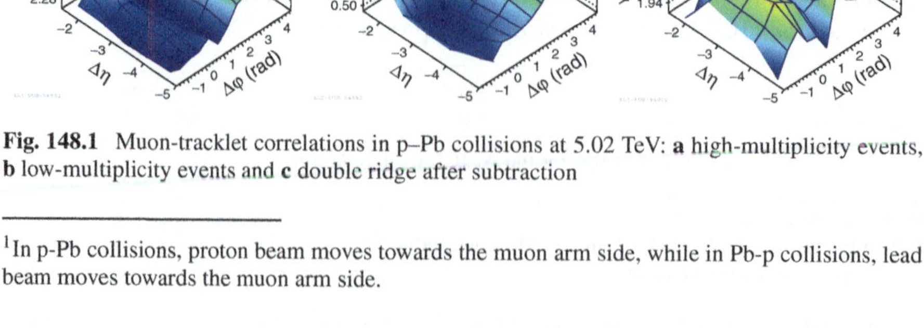

Fig. 148.1 Muon-tracklet correlations in p–Pb collisions at 5.02 TeV: **a** high-multiplicity events, **b** low-multiplicity events and **c** double ridge after subtraction

[1] In p-Pb collisions, proton beam moves towards the muon arm side, while in Pb-p collisions, lead beam moves towards the muon arm side.

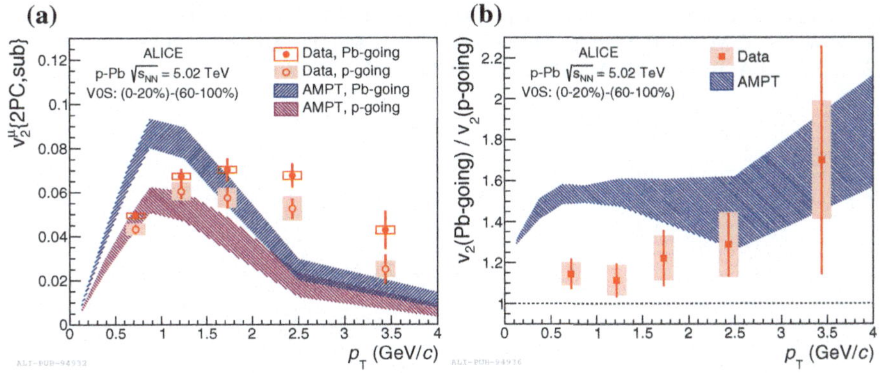

Fig. 148.2 **a** $v_2^{\mu}\{2\text{PC,sub}\}$ in p-going and Pb-going directions and **b** ratio of $v_2^{\mu}\{2\text{PC,sub}\}$ in Pb-going to p-going, as a function of p_T and comparison to AMPT

going as compared to p-going. Figure 148.2b shows the ratio of Pb-going to p-going as a function of p_T. Taking into account the large uncertainties, no dependence on p_T was observed. A constant fit to the ratio showed that the v_2 of the Pb-going is $16 \pm 6\%$ higher than the v_2 of p-going data. The figures also show the comparison of data with AMPT [11] simulation performed taking into account the efficiency of the absorber in the Forward Muon Spectrometer of ALICE. Qualitatively, AMPT reproduced the trend in data at lower p_T, though the quantitative difference is significant. While at high p_T (> 2 GeV/ c), where the source of muon production is predominantly heavy-flavor decay, AMPT did not reproduce the data. This might be because the heavy flavor muons have a non-zero v_2 while the model assumed a zero v_2. Moreover, there is the possibility of the muon parent particle composition or v_2 values in AMPT being different from that in data.

As the effects of the absorber are included in the data, a direct comparison with existing models [12] is not possible currently. Further measurements coupled with model comparisons and theoretical studies are required to understand the origin of long range correlations in small systems.

References

1. V. Khachatryan et al., (CMS Collaboration), JHEP **09**, 091 (2010)
2. S. Chatrchyan et al., (CMS Collaboration), Phys. Lett. B **718**, 795–814 (2013)
3. B. Abelev et al., (ALICE Collaboration), Phys. Lett. B **719**, 29–41 (2013)
4. G. Aad et al., (ATLAS Collaboration), Phys. Rev. Lett. **110**, 182302 (2013)
5. B. Abelev et al., (ALICE Collaboration), Phys. Lett. B **726**, 164–177 (2013)
6. V. Khachatryan et al., (CMS Collaboration), Phys. Lett. B **742**, 200–224 (2015)
7. P. Bozek, W. Broniowski, G. Torrieri, Phys. Rev. Lett. **111**, 172303 (2013)
8. P. Bozek, W. Broniowski, Phys. Lett. B **718**, 1557–1561 (2013)
9. K. Dusling, R. Venugopalan, Phys. Rev. D **87**, 054014 (2013)

10. J. Adam et al., (ALICE Collaboration), Phys. Lett. B **753**, 126–139 (2016)
11. Z.W. Lin, C.M. Ko, B.A. Li, B. Zhang, S. Pal, Phys. Rev. C **72**, 064901 (2005)
12. P. Bozek, A. Bzdak, G.L. Ma, Phys. Lett. B **748**, 301–305 (2015)

Chapter 149
Noise Rejection from the Hadron Showers in the INO-ICAL Detector

Jafar Sadiq and Prafulla Kumar Behera

149.1 Introduction

The India-based Neutrino Observatory (INO) will construct a large magnetised iron calorimeter (ICAL) detector to determine $|\Delta m^2_{32}|$ and to accurately measure θ_{23} by detecting the atmospheric neutrinos [1]. The ICAL will have 151 layers of 5.6 cm thick iron plates interleaved by a 4 cm air gap, where the Resistive Plate Chambers (RPCs) will be inserted. Three identical modules, each of size 16 m $\times$ 16 m $\times$ 14.5 m, will be constructed [1]. The charged current (CC) $\nu_\mu(\bar{\nu}_\mu)$ interactions in the iron target produce a muon ($\mu^\pm$) and a set of hadrons. The hadron interactions in the materials generate a shower of secondary particles which leaves a shower of hits in the RPCs. Inclusion of hadron energy significantly enhances the precision in the physics results of INO-ICAL detector [2, 3]. Therefore, reconstruction of energy of hadrons is crucial, and it is achieved by calibrating it with respect to the number of hits in the hadron shower [4]. The noise hits in the ICAL RPCs contribute to the number of hits, and hence it leads to an over-estimation of hadron energy. Therefore, it is important to identify the hadron showers and get the number of real hadron hits. A clustering algorithm was developed to extract the hadron showers from the noise background.

149.2 Hadron Clustering

The clustering algorithm separates the hits in the RPCs into different clusters depending on their positions. Figure 149.1 shows the schematic of clustering. A collection of hits which are close to each other was considered as a cluster. After removing the

J. Sadiq (✉) · P. K. Behera
Department of Physics, Indian Institute of Technology Madras, Chennai, India
e-mail: sadiqvengad@gmail.com

Md. Naimuddin (ed.), *XXII DAE High Energy Physics Symposium*, Springer Proceedings in Physics 203, https://doi.org/10.1007/978-3-319-73171-1_149

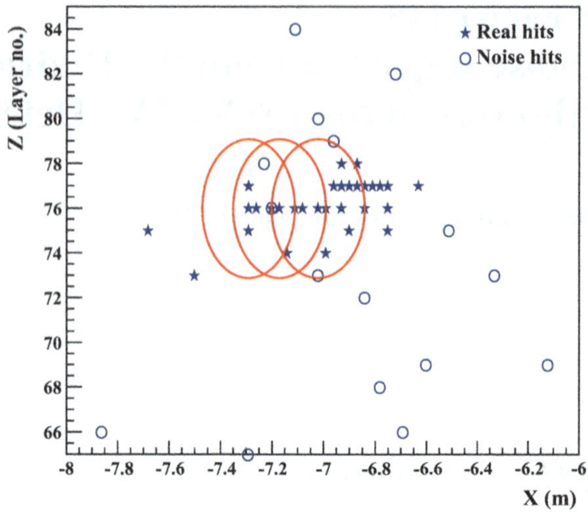

Fig. 149.1 Schematic of clustering

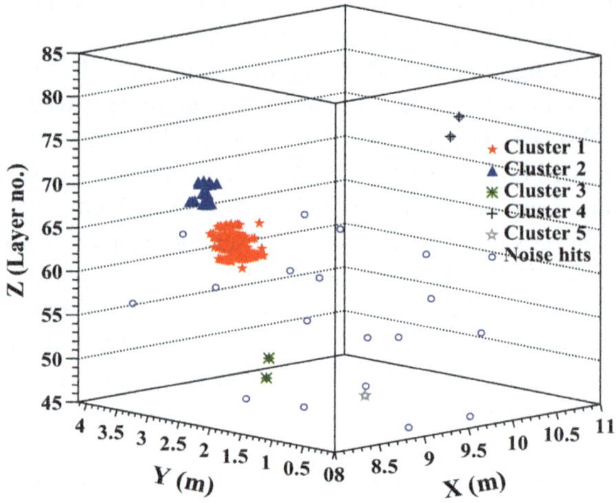

Fig. 149.2 Sorted clusters in a CC ν_μ event

muon track hits, the distance between each pair of the remaining hits was checked. More distance was considered in Z-direction, since the spacing between two consecutive layers is higher than the spacing between two consecutive strips. Starting a cluster with an arbitrary hit and considering it as the center of an ellipsoid, all the hits inside the ellipsoid were added to the cluster. Each of these hits also was taken as the center of a new ellipsoid to add the remaining hits. The clusters were sorted in the decreasing order of the number of hits. The clusters in an atmospheric CC ν_μ event is shown in Fig. 149.2. The clusters with an increased number of hits were considered as hadron showers.

The clusters with only noise hits become largest cluster in the events with very less number of hadron hits. Such events were rejected by applying a selection criteria of

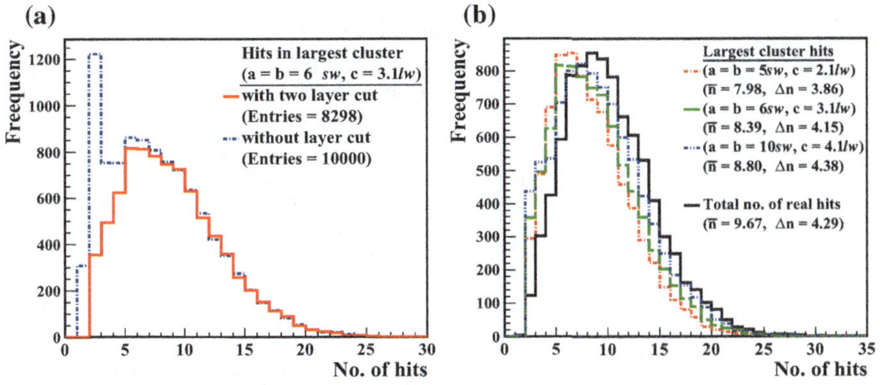

Fig. 149.3 **a** Hit distribution with and without the selection criteria. **b** Hit distributions for a set of ellipsoid parameters. sw and lw represent the strip width (3 cm) and layer width (9.6 cm) respectively. $\bar{n}$ and Δn represent the mean and RMS of the hit distributions respectively

"minimum two layers in the cluster" as shown in Fig. 149.3a. A cluster with highest number of hits was considered as the hadron shower in single pion Monte-Carlo (MC) events. This shower might miss a few number of hits, if size of the ellipsoid is small. Therefore, the ellipsoid parameters have to be optimized in such a way that most of the hadron hits were included in the largest cluster, and fraction of the noise hits in the largest cluster was reduced. The hit distribution produced by single charged pion of energy 5 GeV for a set of ellipsoid parameters is shown in Fig. 149.3b. The ellipsoid parameter values $a = b = 6 \times 3$ cm, and $c = 3.1 \times 9.6$ cm were considered for further study on the calibration of hadron energy.

149.3 Hadron Energy Calibration

Atmospheric neutrino (ν_μ) and anti-neutrino ($\bar{\nu}_\mu$) events in ICAL, generated using the neutrino event generator NUANCE, were used to study the energy response of hadrons produced by atmospheric neutrinos. These NUANCE data was passed through the GEANT4, and the hit information was stored. 100 random correlated noise hits were also generated in the digitization step of ICAL simulation. The hits in two clusters with higher number of hits was used for the calibration of hadron energy, since there are multiple hadrons in the final products of NUANCE events. The mean number of hits $\bar{n}(E)$ increases with energy and gets saturated at higher energies.

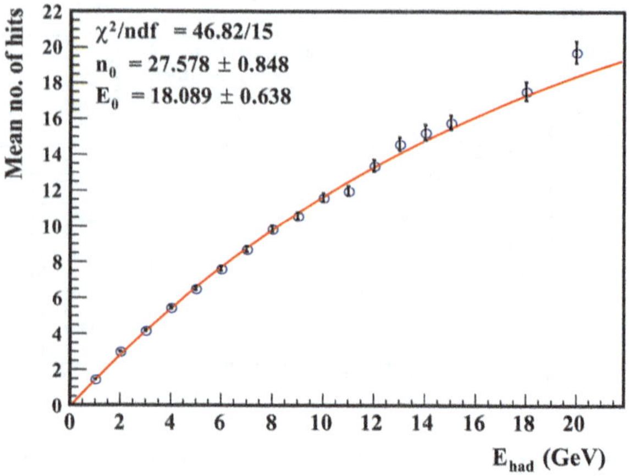

Fig. 149.4 The mean number of hits as a function of hadron energy

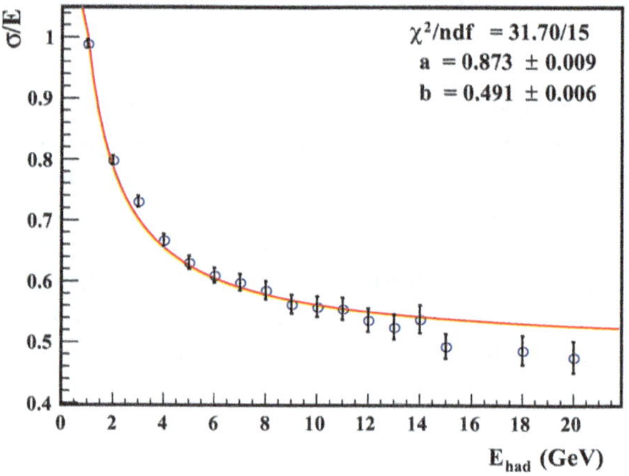

Fig. 149.5 Energy resolution as a function of hadron energy

The mean number of hits and energy resolution ($\sigma/E = \Delta n/n$) were fitted with functions (149.1) and (149.2), and are shown in Figs. 149.4 and 149.5 respectively.

$$\bar{n}(E) = n_0[1 - \exp(-E/E_0)] \tag{149.1}$$

$$\sigma/E = \sqrt{a^2/E + b^2} \tag{149.2}$$

149.4 Conclusions

Reconstruction of hadron energy is crucial to enhancing the accuracy in the physics results of INO-ICAL detector, and it is obtained by calibrating it with respect to the number of hits in the RPCs. Contribution of noise hits to the hadron shower is reduced by a clustering algorithm. The energy of hadrons, produced in atmospheric neutrino events, was calibrated with the mean number of hits in the hadron showers obtained using the algorithm. The energy resolution (σ/E) obtained from the hit distributions was found to be in the range 80−48% for the hadrons of energies in the range 2−20 GeV. The parameters in the algorithm will be optimized to improve the energy resolution, and will be used to reconstruct the energy of hadrons.

References

1. A. Kumar et al., INO Collaboration, Pramana **88**, 79 (2017)
2. M.M. Devi et al., JHEP **2017**, 1–26 (2017)
3. A. Ghosh et al., JHEP **2013**, 1–33 (2013)
4. M.M. Devi et al., JINST **8**, P11003 (2013)

Chapter 150
Review of (Anti-)Nuclei Production from High Energy Experiments

Natasha Sharma

150.1 Introduction

The Universe started with a big-bang with nearly equal abundance of matter and anti-matter. However, this symmetry got lost in the evolution of the universe with no visible amounts of anti-matter being present. The production of light (anti-)nuclei including (anti-)hypernuclei in high energy heavy-ion collisions may be related to the matter-antimatter symmetry. The production of light (anti-)nuclei has been studied in the vast energy range in heavy-ion collisions experiments at Bevalac, AGS, SPS, RHIC, and LHC. The two proposed production mechanisms for the (anti-)nuclei are the coalescence model [1] and the statistical thermal model [2]. The simple coalescence model assumes that (anti-)nuclei are formed if the constituents baryons are close in the coordinate and the momentum phase space. The thermal model predicts the dependence of the particle yields of mass m on the baryon chemical potential (μ_B) and chemical freeze-out temperature (T_{chem}) by the relation $\mathrm{d}N/\mathrm{d}y \propto \exp((\mu_B - m)/T_{\text{chem}})$.

150.2 Results and Discussions

In coalescence mechanism, the spectral distribution of the deuterons is related to that of primordial protons via $E_d \frac{\mathrm{d}^3 N_d}{\mathrm{d}p_d^3} = B_2 \left(E_{\mathrm{p}} \frac{\mathrm{d}^3 N_{\mathrm{p}}}{\mathrm{d}p_{\mathrm{p}}^3}\right)^2$, assuming that protons and neutrons have the same momentum distribution. B_2 is the coalescence parameter for deuteron with momentum of $p_d = 2\, p_{\mathrm{p}}$. Figure 150.1 (left) shows the energy dependence of the coalescence parameter B_2. It decreases from Bevalac to SPS energies, then remains almost constant up to RHIC energy [3]. From RHIC top energy to

N. Sharma (✉)
Department of Physics, Panjab University, 160014 Chandigarh, India
e-mail: natasha.sharma@cern.ch; nsharma@cern.ch

Md. Naimuddin (ed.), *XXII DAE High Energy Physics Symposium*, Springer Proceedings in Physics 203, https://doi.org/10.1007/978-3-319-73171-1_150

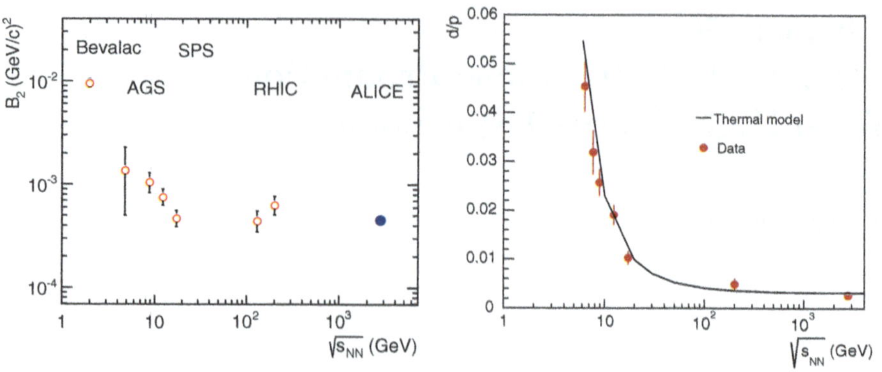

Fig. 150.1 Left: B_2 measured from various experiments as a function of $\sqrt{s_{NN}}$; Right: d/p ratio from data and thermal model prediction as a function of $\sqrt{s_{NN}}$

LHC energy B_2 shows slight decrease [4]. In the coalescence picture, this behavior is explained by an increase in the source volume i.e. the larger the distance between protons and neutrons which are created in the collision, the less likely is that they coalesce. At RHIC energies, it is observed that elliptic flow (v_2) of light (anti-)nuclei when scaled with mass number follows v_2 of (anti-)protons suggesting that light (anti-)nuclei are produced via coalescence mechanism [5]. However, the preliminary results from ALICE suggest that simple coalescence model fail to describe (anti-)deuteron v_2 at LHC energy [6].

The right plot of Fig. 150.1 shows the energy dependence of experimentally measured deuteron to proton ratio in heavy-ion collisions [4, 7]. The comparison is made with the thermal model predictions which seems to explain the data well. It is observed that for a given collision energy, T_{chem} and μ_B remain same for heavy-ion and small collisions system. At LHC, the d/p ratio for p–Pb and pp collisions is smaller than in Pb–Pb system [8]. This observation could not be explained by the thermal model.

For a given system and energy, the thermal model suggests that the nuclei yields follow an exponential decrease with the mass which is consistent with the experimental results. One has to pay a "penalty factor" to produce the heavier nuclei or each added nucleon [8]. Figure 150.2 (left) shows the penalty factor as a function of energy for heavy-ion collisions [7]. The markers show the experimental results (solid symbols for nuclei and open symbols for anti-nuclei) while the curves represent the thermal model predictions. The thermal model predictions are consistent with the heavy-ion experimental results within the uncertainties. However, this is not true for the smaller systems. For example, it is observed that penalty factor is about 600 for p–Pb collisions at LHC [8] which is well above the thermal model expectation at this energy. This deviation from thermal model expectation in smaller collision systems needs theoretical explanation.

The right plot of Fig. 150.2 shows the experimental measurements of hyper-triton ($^{3}_{\Lambda}$H) lifetime from various experiments [9]. The figure also shows the lifetime of the

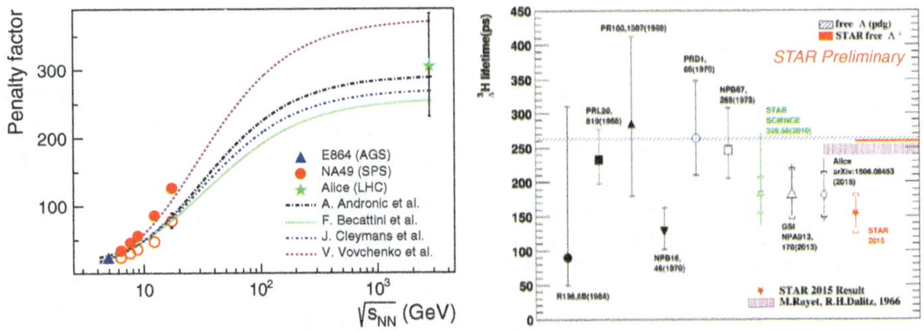

Fig. 150.2 Left: Penalty factor as a function of energy for heavy-ion collisions. Right: Lifetime measurements of hypertriton from different experiments

free Λ from the PDG and measurement from STAR. The recent experimental results from STAR [9] and ALICE [10] suggest that the $^{3}_{\Lambda}$H lifetime is less than that of free Λ. It will be interested to understand the theoretical explanation of the observed difference in lifetime of $^{3}_{\Lambda}$H and free Λ.

Recently, the ALICE experiment has measured the mass and binding energies of nuclei and anti-nuclei and found it to be compatible within uncertainties. This confirm the CPT invariance of light nuclei [11]. The Quantum Chromodynamics (QCD) predicts the existence of exotic bound states of baryons [12]. The thermal and coalescence models explain well the experimentally measured yields of light nuclei and hypertriton [4]. These models may predict the yields of exotic bound states and hence may provide baseline to test the existence of these states. Various experiments attempt to search for these weakly decaying bound state of baryons. Recently, ALICE has tried to search for the $\Lambda\Lambda$ and $\overline{\Lambda n}$ bound states [12]. No evidence for these bound states is observed. Theoretical explanation is needed to understand the non-observation of these states.

150.3 Summary

Light (anti-)nuclei measurements have been performed by various experiments. The thermal model seems to describe (anti-)(hyper-) nuclei production well in heavy-ion collisions. However, deviation from thermal model is seen for small collision systems, the simple coalescence model could explain (anti-)nuclei production in small systems. More work is needed on theoretical part to understand the production mechanism of nuclei and anti-nuclei in high energy collisions.

Acknowledgements This work is supported by DST-SERB Ramanujan Fellowship grant no. SB/S2/RJN-084/2015. NS thanks Dr. L. Kumar for the discussions.

References

1. S. Butler, C. Pearson, Phys. Rev. **129**, 836–842 (1963); J.I. Kapusta, Phys. Rev. C **21**, 1301–1310 (1980)
2. A. Andronic et al., Phys. Lett. B **697**, 203–207 (2011); J. Cleymans et al., Phys. Rev. C **84**, 054916 (2011)
3. S. Adler et al., (PHENIX Collaboration), Phys. Rev. Lett. **94**, 122302 (2005)
4. J. Adam et al., (ALICE Collaboration), Phys. Rev. C **93**, 024917 (2016)
5. L. Adamczyk et al., (STAR Collaboration), Phys. Rev. C **94**, 034908 (2016)
6. R. Lea, Nucl. Phys. A **956**, 264–267 (2016)
7. T. Anticic et al., (NA49 Collaboration), Phys. Rev. C **94**, 044906 (2016)
8. N. Sharma, Nucl. Phys. A **956**, 461–464 (2016)
9. Y. Xu, in *12th International Conference on Hypernuclear and Strange Particle Physics (HYP2015) Sendai, Japan 7–12 September (2015)* (2015)
10. J. Adam et al., (ALICE Collaboration), Phys. Lett. B **754**, 360–372 (2016)
11. J. Adam et al., (ALICE Collaboration), Nat. Phys. **11**, 811–814 (2015)
12. J. Adam et al., (ALICE Collaboration), Phys. Lett. B **752**, 267–277 (2016)

Chapter 151
Search for the Pair-Production of First Generation Leptoquarks in pp Collisions at $\sqrt{s}$ = 13 TeV

M. A. Bhat, B. Mahakud, S. I. Cooper, G. B. Mohanty, T. Aziz and P. Rumerio

151.1 Introduction

A general effective model for LQs was proposed in 1987 by Buchmuller, Ruckl and Wyler [1]. The four underlying assumptions of the model are: (1) LQs have renormalizable interactions, (2) their interactions are invariant under the SM gauge groups: $SU(3)_C \times SU(2)_L \times U(1)_Y$, (3) they couple only to SM fermions and gauge bosons and, (4) they conserve lepton and baryon number separately in order to protect against a rapid proton decay. An LQ will decay into a lepton and a quark, giving rise to a final state containing high-momentum leptons and jets. The decay into a charged lepton and quark has branching fraction β, consequently the decay into a neutrino and a quark has branching fraction $1 - \beta$, where β is a free parameter. Previous searches for rare processes mediated by lepton number violation and flavor-changing neutral currents prefer leptoquarks at LHC-accessible mass ranges, which couple to leptons and quarks of same generation (no mixing) [2, 3]. Thus in this search for first generation LQs, we assume their decay to electrons or electron neutrinos along with jets. The process can give rise to the following three final states:

- Two electrons and two jets: each LQs decays into an electron and a quark,
- One electron, missing transverse energy, and two jets: one LQ decays into an electron and a quark, while the other into a neutrino and a quark, and
- No electrons, missing transverse energy, and two jets: each LQ decays into a neutrino and a quark.

The corresponding branching fractions are β^2, $2 \times \beta(1 - \beta)$, and $(1 - \beta)^2$, respectively. We consider two values of β: 1, which corresponds to LQs always decaying to

M. A. Bhat (✉) · B. Mahakud · G. B. Mohanty · T. Aziz
Tata Institute of Fundamental Research, Mumbai 400005, India
e-mail: mbhat@cern.ch

S. I. Cooper · P. Rumerio
University of Alabama, Tuscaloosa, AL, USA

Md. Naimuddin (ed.), *XXII DAE High Energy Physics Symposium*, Springer Proceedings in Physics 203, https://doi.org/10.1007/978-3-319-73171-1_151

the first final state; and 0.5, where 50% of LQs decay to the second final state, with 25% decaying to the first final states. The first and second final states are denoted as $eejj$ and $e\nu jj$. In this analysis, we search for LQ decays in the *eejj* final state that has the largest sensitivity due to the presence of two isolated high p_T leptons. So far, earlier searches have not detected any LQs. CMS has published the result of a search using about 20 fb^{-1} of the 2012 LHC data at $\sqrt{s} = 8$ TeV, which gave first generation LQ mass limits of less than 1010 (850) GeV for $\beta = 1(0.5)$ [1]. Similarly, ATLAS used the 2012 LHC data [4] to set first generation LQ mass limits of less than 1050 (900) GeV for $\beta = 1(0.5)$; it also looked at the possibility of $\beta = 0.2$, which gave a mass limit of 900 GeV. All of these limits are set at the 95% confidence level.

151.2 Event Selection for the eejj Analysis

The $eejj$ final state is the result of two pair-produced LQs each decaying to an electron and a jet, where the branching fraction β as defined in Sect. 1 is equal to 1. Events containing at least two electrons and two jets are thus selected, where the two leading p_T electrons and jets are used in the analysis.

151.2.1 Preselection

At the preselection level, we go for a loose set of criteria in order to compare Monte Carlo (MC) simulated background predictions with data. The event selection criteria are as follow:

- Event should pass the online trigger of at least one loose electron with $p_T >$ 27 GeV.

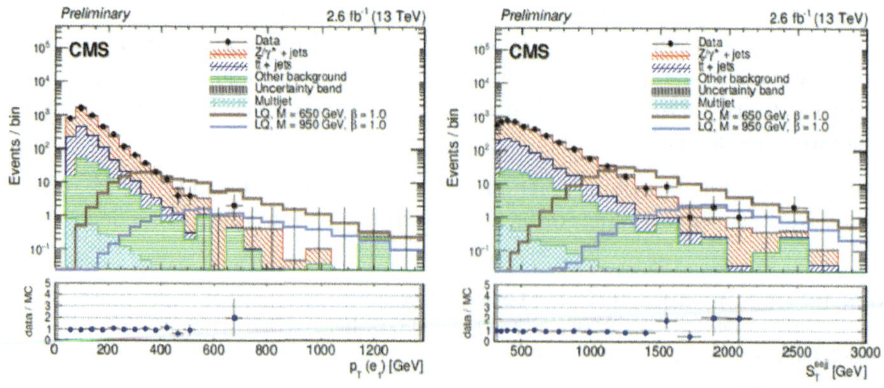

Fig. 151.1 p_T (left) of the highest-p_T electron and S_T (right) after applying the $eejj$ preselection

- It should contain exactly two highly energetic well-identified electrons having the associated ECAL supercluster $E_T > 50\,\text{GeV}$ and pseudorapidity $|\eta_{SC}| < 2.5$.
- It should not contain any muons, where muons are selected with $p_T > 45\,\text{GeV}$ and $|\eta| < 2.1$.
- It should have at least two jets with $p_T > 50\,\text{GeV}$ and $|\eta| < 2.4$.
- The dielectron invariant mass $m_{ee} > 50\,\text{GeV}$.
- The scalar sum of the p_T of the two electrons and two leading jets $S_T > 300\,\text{GeV}$.

After applying these criteria, the agreement of expected background with the data is examined in a series of plots (see Fig. 151.1).

151.3 Backgrounds

The major backgrounds from SM processes are Z+jets and $t\bar{t}$, while single top, W+jets, diboson, and γ+jets contribute at a lower level. There is also an instrumental background from QCD multijet events with jets faking electrons. Below we describe how various background contributions are determined in this analysis.

- The Z+jets and $t\bar{t}$ background shapes are taken from MC samples that are normalized to data using the *eejj* preselection.
- Single top, W+jets, diboson, and γ+jets are derived completely from MC.
- QCD "fake" jet background is determined using a data-driven method.

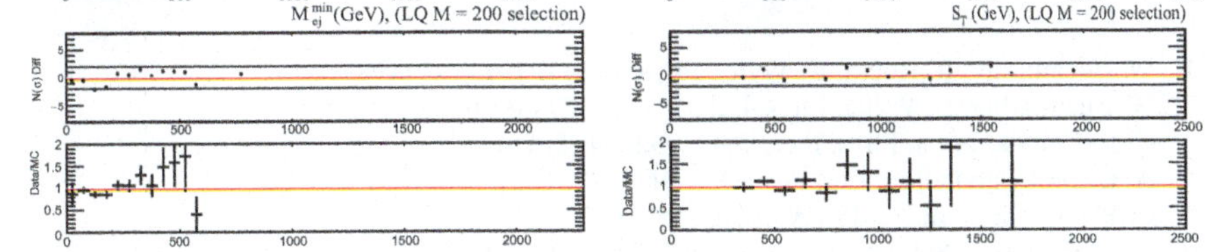

Fig. 151.2 Distributions of M_{ej}^{min} (left) and S_T (right) for the final selection optimized for a leptoquark of mass 200 GeV

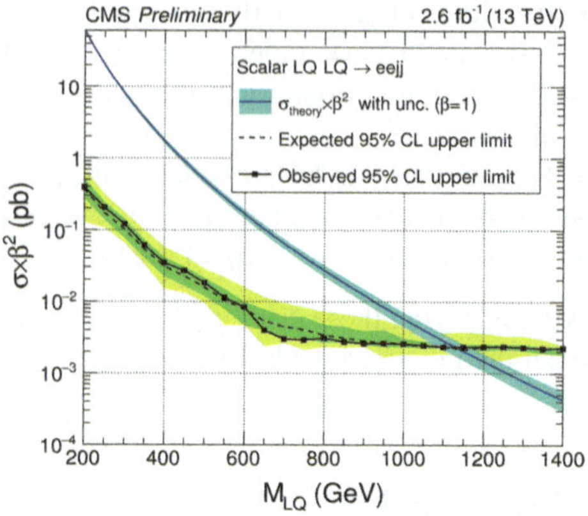

Fig. 151.3 Observed and expected upper limits for scalar LQ pair-production cross section times β^2 at the 95% confidence level.The median (dashed line), 1σ (green), and 2σ (yellow) expected limits are shown. The theoretical prediction is varied by the renormalization scale and PDF uncertainty to produce the shaded band

151.4 Results

As shown in Fig. 151.2, we do not find any evidence for an LQ signal in the data. Therefore, we proceed to set upper limits on the scalar LQ production cross section using a full frequentist CLs method [9], using the event yields. The systematic uncertainties are modeled as nuisance parameters with a log normal distribution, while the statistical uncertainties are parameterized as log-normal functions, or in the case of small yields, gamma functions with widths determined from the number of MC or data control region events. The upper limits at 95% confidence level on $\sigma \times \beta^2$ along with the predicted scalar LQ pair-production cross section at NLO are shown in Fig. 151.3 The intersection of the central value of the theoretical prediction with the upper limit on the cross section yields a limit on the LQ mass. The limit for the $\beta = 1$ case is 1130 GeV [5].

References

1. W. Buchmuller, R. Ruckl, D. Wyler, Phys. Lett. B **191**, 442 (1987)
2. W. Buchmuller, D. Wyler, Phys. Lett. B **177**, 377 (1986)
3. CMS Collaboration, Phys. Rev. D **93**, 032004 (2016)
4. ATLAS Collaboration, Eur. Phys. J. C **76**(5) (2016)
5. CMS Collaboration, CMS PAS EXO-16-043
6. B. Schrempp, F. Schrempp, Phys. Lett B **153**, 101 (1985)
7. J.C. Pati, A. Salam, Phys. Rev D **10**, 275 (1974)
8. CMS Collaboration, JINST **3**, S08004 (2008)
9. A.L. Read, Modified frequentist analysis of search results (the CLs method), CERN-OPEN 2000-205. 1st Workshop on Confidence Limits, CERN (Jan 2000)

Chapter 152
Signatures of T and CPT Violation in Presence of Sterile Neutrino

Jyotsna Singh and R. B. Singh

152.1 Introduction

Three active neutrino framework is well established still few set of anomalous data observed by LSND [1], MiniBooNE [2, 3], Gallium source experiments [4, 5] and reactor experiments [6], indicated towards neutrino oscillations with substantially different frequency. A possible explanation to these anomalous or high frequency oscillations can be addressed by adding more neutrinos to the three active neutrino framework. While the Z^0 decay width measured by LEP is consistent with, only three low mass active neutrinos [7]. The available parameter space can be restricted to the minimal 3+1 neutrino model (3 active + 1 sterile) by several null results [8]. The present and future neutrino experiments aim towards placing an stringent upper limit on such violations [9–11]. In this paper the effects of sterile neutrinos on capturing the signatures of T and CPT violation in 3+1 neutrino scheme is checked with neutrino factory muon beam of energy, 50 GeV.

152.2 Parametrization Scheme and Signatures of T and CPT Violation in 3+1 Neutrino Framework

The selection for parametrization of unitary matrix is

$$U = U_{34}(\theta_{34}, 0)U_{24}(\theta_{24}, 0)U_{14}(\theta_{14}, 0)U_{23}(\theta_{23}, \delta_3)U_{13}(\theta_{13}, \delta_2)U_{12}(\theta_{12}, \delta_1) \tag{152.1}$$

J. Singh (✉) · R. B. Singh
Department of Physics, University of Lucknow, Lucknow, India
e-mail: jyo2210@yahoo.co.in

R. B. Singh
e-mail: rajendrasinghrb@gmail.com

Md. Naimuddin (ed.), *XXII DAE High Energy Physics Symposium*, Springer Proceedings in Physics 203, https://doi.org/10.1007/978-3-319-73171-1_152

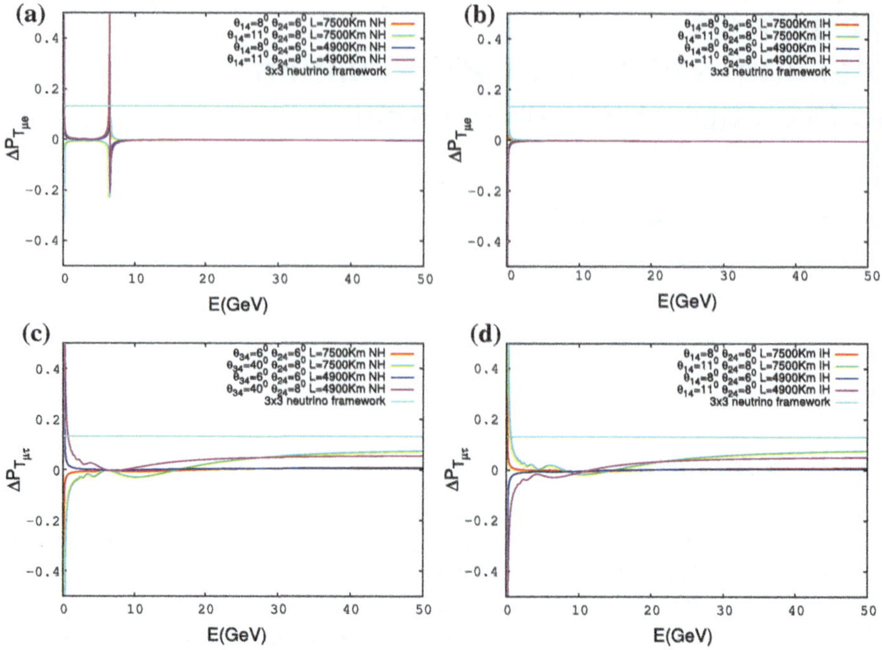

Fig. 152.1 The plot **a** and **b** show $\Delta P_{T_{\mu e}}$ Vs E_ν while plot **c** and **d** show $\Delta P_{T_{\mu\tau}}$ Vs E_ν at baselines 4900 Km and 7500 Km. The left and right plots displays the effect of varying θ_{34} and θ_{24} for normal mass hierarchy and inverted mass hierarchy

If CPT is conserved then $\Delta P_{CP} = \Delta P_T$.
Values of different neutrino oscillation parameters considered in this work are $\Delta m^2_{21} = 8\times 10^{-5} eV^2$, $\Delta m^2_{31} = 2.5\times 10^{-3} eV^2$, $\theta_{12} = 34.4°$, $\theta_{23} = 45°$, $\theta_{13} = 8.5°$

The oscillation probability equation in constant matter density are developed by the formalism of Kimura, Takamura and Yokomakura for golden and discovery channel [11, 12]

$$(\Delta P_T)_{\mu e} = 4c_{13}c^2_{14}c_{24}s_{13}s_{23}s_{14}s_{24}\sin(\delta_2 - \delta_3)\left[\frac{\Delta_e}{(\Delta_e - \Delta_{31})}\sin \Delta m^2_{31}L/2E + \frac{\Delta^2_{31}}{\Delta_e(\Delta_e - \Delta_{31})}\sin 2\Delta_e\right] \tag{152.2}$$

$$(\Delta P_T)_{\mu\tau} = 4c_{13}c^2_{14}c_{24}s_{13}s_{23}s_{14}s_{24}\sin(\delta_3)\left[\frac{\Delta_e}{(\Delta_{31})}\sin \Delta m^2_{31}L/2E - \frac{\Delta_{31}}{\Delta_e}\sin 2\Delta_e\right] \tag{152.3}$$

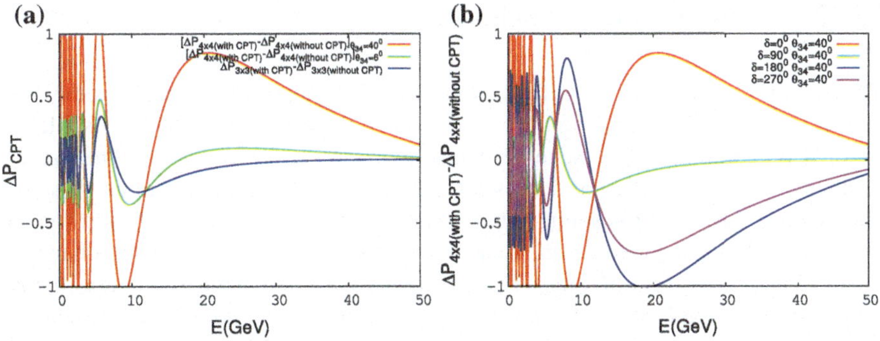

Fig. 152.2 ΔP_{CPT} Vs E_ν at baseline 7500 Km. In left plot the effect of varying θ_{34} in the allowed range is shown while all CP values are kept equal to zero. This variation is compared with ΔP_{CPT} with energy in three neutrino framework. In right plot the effect of sterile phase δ_3 on $[\Delta P_{4\times4(\delta a_{ij}=0)} - \Delta P_{4\times4(\delta a_{21}=3\times10^{-23},\delta a_{31}=4.5\times10^{-23})}]$ with energy is shown

The CPT violating hamiltonian in the flavor basis can be written as

$$H_f = \frac{1}{2E}.U_0 D(0, \Delta m_{21}^2, \Delta m_{31}^2, \Delta m_{41}^2)U_0^\dagger + U_b.D_b(0, \delta a_{21}, \delta a_{31}, \delta a_{41})U_b^\dagger$$
$$+ D_m(A_e, 0, 0, A_n) \qquad (152.4)$$

The unitary matrix U_b is checked for different angular values [10]. In our work we have considered U_0=U_b.

$$P_{\mu\mu} = 1 - 2s_{24}^2 \cos^2(\Delta_{31} + \delta a_{31}L/2) - (1 - 8\hat{s}_{23}^2)\sin^2(\Delta_{31} + \delta a_{31}L/2)$$
$$+ (c_{12}^2\Delta_{12} - 2s_{24}s_{34}\cos\delta_3\Delta_n)\sin 2(\Delta_{31} + \delta a_{31}L/2) + \frac{s_{13}^2(\Delta_{31} + \delta a_{31}L/2)}{(\Delta_{31} + \delta a_{31}L/2 - \Delta_e)^2}$$
$$\times [2(\Delta_{31} + \delta a_{31}L/2)\sin\Delta_e \cos(\Delta_{31} + \delta a_{31}L/2)\sin(\Delta_{31} + \delta a_{31}L/2 - \Delta_e) -$$
$$(\Delta_{31} + \delta a_{31}L/2 - \Delta_e)\Delta_e \sin 2(\Delta_{31} + \delta a_{31}L/2)] \qquad (152.5)$$

Latest constraints on sterile mixing angles [13] given below, are considered in this work.

$$\theta_{14} \in [7.5^\circ - 11.5^\circ];\ \theta_{24} \in [5.2^\circ - 8.6^\circ];\ \theta_{34} \in [0^\circ - 44.4^\circ] \qquad (152.6)$$

The signatures of T and CPT violation with neutrino factory muon beam and baseline are illustrated at probability level in this work.

152.3 Summary and Conclusion

The presence of sterile sector will affect the precision measurement made by the three active neutrino sector hence untill the possibility of an short wavelength oscillation is ruled out, minimal 3+1 scheme demands extensive investigation. The present available neutrino oscillation data is not able to impose tight constraints on sterile parameters, hence long baseline future experiment muon beam is selected for this work. T violation signatures are looked with two different appearance channels $\nu_\mu \rightarrow \nu_e$ and $\nu_\mu \rightarrow \nu_\tau$. Looking at Fig. 152.1 we can comment that discovery channel seems to be a better choice for constaringing leptoinc T violation. The presence of sterile neutrino with large mixing angles will hide the CPT violation in the three neutino sector, as can be seen in Fig. 152.2. Hence untill the presence of sterile neutrino is ruled out, our work demands neutrino oscillation experiments, which can impose better constraints on sterile neutrino oscillation parameters.

References

1. A. Aguilar-Arevalo et al., Phys. Rev. D **64**, 112007 (2001)
2. A. Aguilar-Arevalo et al., Phys. Rev. Lett. **102**, 101802 (2009)
3. A. Aguilar-Arevalo et al., Phys. Rev. Lett. **110**, 161801 (2013), arxiv:1207.4409
4. J. Abdurashitov et al., (SAGE Collaboration), Phys. Rev. C **80**, 015807 (2009)
5. F. Kaether et al., Phys. Lett. B **685**, 47 (2010)
6. Y. Declais et al., Nucl. Phys. B **434**, 503 (1995)
7. L3 Collaboration, SLD Electroweak and Heavy Flavour Groups. Phys. Rep. **427**, 257 (2006)
8. K. Abe et al., (super-Kamiokande), Phys. Rev. D **65**, 112001 (2002)
9. R. Gandhi et al., arXiv:1508.06275v2 [hep-ph]
10. A. Chatterjee et al., arXiv:1402.6265v1 [hep-ph]
11. Y. Pant et al., arXiv:1509:04096 [hep-ph]
12. K. Kimura et al., arXiv:hep-ph/0205295
13. C.H. Collins, C.A. Arguelles, J.M. Condrad, M.H. Shaevitz, arXiv:[hep-ph]

Chapter 153
Comparison of Silicon, Germanium and Diamond Sensors as Trackers in Collider Experiments

Shyam Kumar and Raghava Varma

153.1 Signal and Noise Levels

When an Ionizing radiation passes through a material, it creates e-h pairs which, when suitably collected create an electrical signal. We have simulated the number of e-h pairs created by the MIP particle in 300 μm thick Si, Ge and Diamond sensors in PANDARoot framework using GEANT 3 [1] (Fig. 153.1). The intrinsic noise can be calculated by using the formula $\sigma = q(n_e\mu_e + n_h\mu_h)$. We have estimated the intrinsic noise in all three materials. The intrinsic Diamond has larger signal to noise ratio than Si and Ge, because of the large band gap and hence simple ohmic contact makes it suitable for detection of radiation, while for Si and Ge we need p-n junction to reduce the intrinsic carrier concentration (Table 153.1).

153.2 Radiation Damage

There are two types of radiation damage, namely, surface damage and bulk damage. Surface damage is due to the ionizing energy loss. Ionizing energy loss is responsible for the degradation of oxide layer. In Si and Ge based detector, we have oxide layer, while in Diamond we have just ohmic contact, so this type of damage will not be present in Diamond. Bulk damage is due to non-ionizing energy loss. The threshold atomic displacement energy for Si, Ge and Diamond are 25, 20 and 43.6 eV respectively. Bulk damage will also be smaller in Diamond than Si and Ge [2].

S. Kumar (✉) · R. Varma
Department of Physics, Indian Institute of Technology Bombay, Mumbai, India
e-mail: shyam055119@gmail.com

R. Varma
e-mail: raghava.varma@gmail.com

Md. Naimuddin (ed.), *XXII DAE High Energy Physics Symposium*, Springer Proceedings in Physics 203, https://doi.org/10.1007/978-3-319-73171-1_153

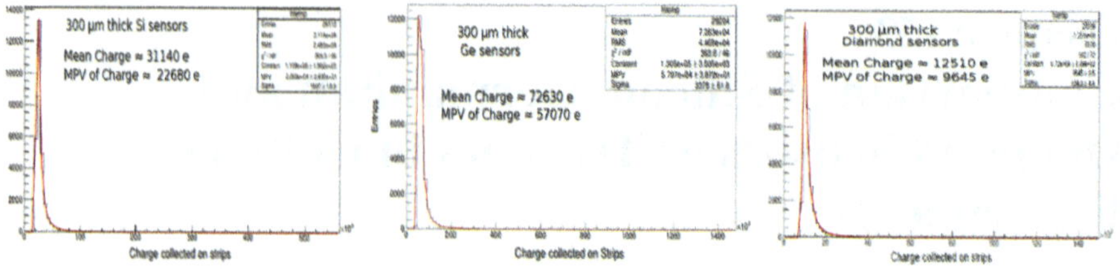

Fig. 153.1 Charge created by MIP in 300 μm thick Si, Ge and Diamond sensors [3]. The X-axis represents the charge collected and Y-axis represents the entries

Table 153.1 Signal and noise levels for $5.0 \times 5.0 \times 0.3\ \text{mm}^3$ Si, Ge and diamond sensors

Material	MPV signal(e)	$\rho\ (\Omega\text{m})$	$\mu_e(\frac{\text{cm}^2}{\text{Vs}})$	$\mu_h(\frac{\text{cm}^2}{\text{Vs}})$	Noise(e)
Silicon	22680	640	1450	505	3.8×10^8
Germanium	57070	0.46	3900	1800	1.5×10^{11}
Diamond	9645	10^{12}	1800	1600	0.14

153.3 Conclusion

Diamond has low intrinsic noise, low multiple scattering ($X_0 = 9.37$ cm for Si, $X_0 = 2.3$ cm for Ge and $X_0 = 12.14$ cm for Diamond), low radiation damage and fast timing ($\approx$3 ns) makes it excellent material for tracking in HEP experiments. In addition the possibility of creating delta electrons (The e-h pair creation energy is 13.6 eV larger in Diamond, 3.6 eV in Si and 2.96 eV in Ge), is lower in Diamond than in Si and Ge because of the high e-h pair creation energy, which disturbs the charge center of gravity. The only issue with the Diamond which limits its use in HEP experiments is small size and large cost for high quality Diamond (Charge Collection Efficiency = 100%). We are also working on the growth of Diamond by Microwave Plasma Chemical Vapor Deposition (MPCVD) process in the laboratory to produce device grade Diamond [4].

References

1. Technical Design Report for the $\bar{P}$ANDA Micro Vertex Detector, PANDA Collaboration (2011)
2. S. Kumar, Comparative study of radiation damage in Si, Ge and Diamond used as detector. Proc. DAE Symp. Nucl. Phys. **61** (2016)
3. F. Hartmann, *Evolution of Silicon Sensor Technology in Particle Physics* (Springer, Berlin, 2009), pp. 19–21
4. S. Kumar et al., Comparison of Silicon, Germanium and Diamond sensors for using it in HEP detector applications, *Advanced Detectors for Nuclear, High Energy and Astroparticle Physics* (Bose Institute Kolkata, 2017)

Chapter 154
Timing Studies of a Large Size Oil-Free Bakelite Resistive Plate Chamber

Rajesh Ganai, Mehulkumar Shiroya, Zubayer Ahammed and Subhasis Chattopadhyay

154.1 Introduction

RPC [1] was first introduced in the year 1980 by R.Santonico and R. Cardarelli. It is a gaseous detector made up of two high resistive parallel electrode plates e.g. glass, bakelite. RPCs are known to have good time resolution and efficiency to detect charged particles specially muons. INO-ICAL will be using ~30,000 RPCs with a time resolution of ~1 ns in order to study the oscillations in the atmospheric neutrino sector [2]. As a part of R&D, we have developed a large size (240 cm × 120 cm) single gap (0.2 cm) oil free RPC from bakelite sheets available in local market [3, 4].

R. Ganai (✉) · Z. Ahammed · S. Chattopadhyay
EHEP & A Group, Variable Energy Cyclotron Centre, Kolkata 700064, India
e-mail: rajesh.ganai.physics@gmail.com

R. Ganai
Bose Institute, Centre for Astroparticle Physics and Space Sciences, P-1/12,CIT Road, Scheme VII-M, Kolkata 700054, West Bengal, India

R. Ganai
GSI Helmholtzzentrum für Schwerionenforschung GmbH, Planckstrasse 1, 64291 Darmstadt, Germany

R. Ganai · Z. Ahammed · S. Chattopadhyay
Training School Complex, Homi Bhabha National Institute, Anushakti Nagar, Mumbai 400 085, India

M. Shiroya
Sardar Vallabhbhai National Institute of Technology, Surat 395007, Gujarat, India

Md. Naimuddin (ed.), *XXII DAE High Energy Physics Symposium*, Springer Proceedings in Physics 203, https://doi.org/10.1007/978-3-319-73171-1_154

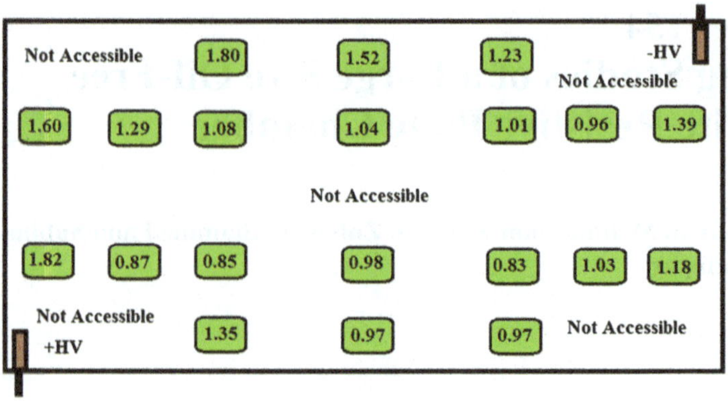

Fig. 154.1 Time resolution of the RPC in different locations

Fig. 154.2 Time resolution distribution of the RPC

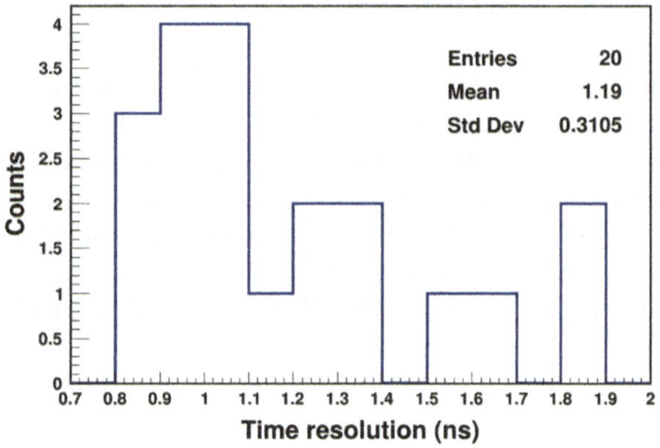

154.2 Timing Studies of the Large Size Oil-Free Bakelite RPC

The time resolution of the RPC was measured with cosmic rays in a standard cosmic ray test bench using three scintillators in 20 different locations. Figure 154.1 shows the measured time resolution after correcting for the scintillators at different accessible locations on the detector. It can be seen that the central regions of the detector has a resolution of <1 ns. The inaccessible regions are due to supports given to the platform where RPC has been kept as discussed in [3]. A distribution of the measured time resolution values have been shown in Fig. 154.2 with a mean of 1.19 ns.

154.3 Conclusion and Outlook

The time resolution of the RPC was measured at 20 different locations with an average of 1.19 ns which is close to the requirement of the ICAL experiment. The time resolution around the edges of the RPC deviates from the central region. The reasons for such a deviation need to be studied.

References

1. R. Santonico, R. Cardarelli, Nucl. Instrum. Method **187**, 377 (1981)
2. INO Project Report, INO/2006/01. http://www.ino.tifr.res.in/ino//OpenReports/INOReport.pdf
3. R. Ganai et al., JINST **11**, P04026 (2016)
4. R. Ganai et al., JINST **11**, C09010 (2016)

Chapter 155
D and *B* Mesons in Hot and Dense Symmetric Nuclear Medium

Rahul Chhabra and Arvind Kumar

155.1 Introduction

Study of in-medium $D(B)$ mesons is important to clearly understand about $J/\psi(\Upsilon)$ suppression and formation of mesic nuclei in the HIC experiments. If the masses of $D(B)$ mesons decrease in the medium then, higher (bottomonium) states produced in the HIC experiments may decay to these mesons, instead of $J/\psi(\Upsilon)$ state. The results of the present work, may be used to understand the various outcomes of the HIC experiments like CBM and Panda under the FAIR Facility, at Darmstadt Germany.

155.2 Methodology

Chiral SU(3) model includes an effective Lagrangian density, which we solve by using mean field theory to find the coupled equations of motion for the scalar fields σ, ζ and scalar dilation field χ. We solve these equations to investigate the medium modified scalar σ, ζ and scalar dialaton field χ. By using explicit symmetry breaking term and broken scale invariance property of QCD we express the quark and gluon condensate in terms of these medium modified scalar and dialaton fields, respectively [1]. In the QCD sum rules, we divide the two correlation function $\Pi_{\mu\nu}(q)$, into vacuum part, nucleon depended part [2]. We neglect the pion bath term as we consider the temperature dependence through the Chiral SU(3) model. In QCD sum rules, we equate the Borel transformed equation of the scattering matrix on the

R. Chhabra (✉) · A. Kumar
National Institute of Technology, Jalandhar, India
e-mail: rahulchhabra@ymail.com

Md. Naimuddin (ed.), *XXII DAE High Energy Physics Symposium*, Springer Proceedings in Physics 203, https://doi.org/10.1007/978-3-319-73171-1_155

phenomenological and OPE side [2]. In this equation we use the above mentioned medium depended quark and gluon condensates to calculate the in-medium masses and decay constants of the D and B mesons.

155.3 Results and Discussion

The values of the various parameters are taken from ref. [3]. We observe a decrease (decrease) in the masses (decay constants) of D and B mesons in the nuclear medium (Fig. 155.1). This drop is more in the high dense medium than in the low dense medium, whereas the values of masses of D and B mesons increase as we move from zero temperature to finite temperature medium (Fig. 155.1). This drop in the mass of above D and B mesons indicate possibility of formation of bound states of D and B mesons with nucleons. In addition, this may also cause suppression of J/ψ and Υ states in the HIC experiments respectively.

Using QCD sum rules, the shift in masses and decay constants of pseudoscalar $D(B)$ mesons had also been observed to be $-72(-473)$, and $-6(-71)$ MeV, respectively, whereas, by neglecting leading order term the observed shift was $-47(-329)$,

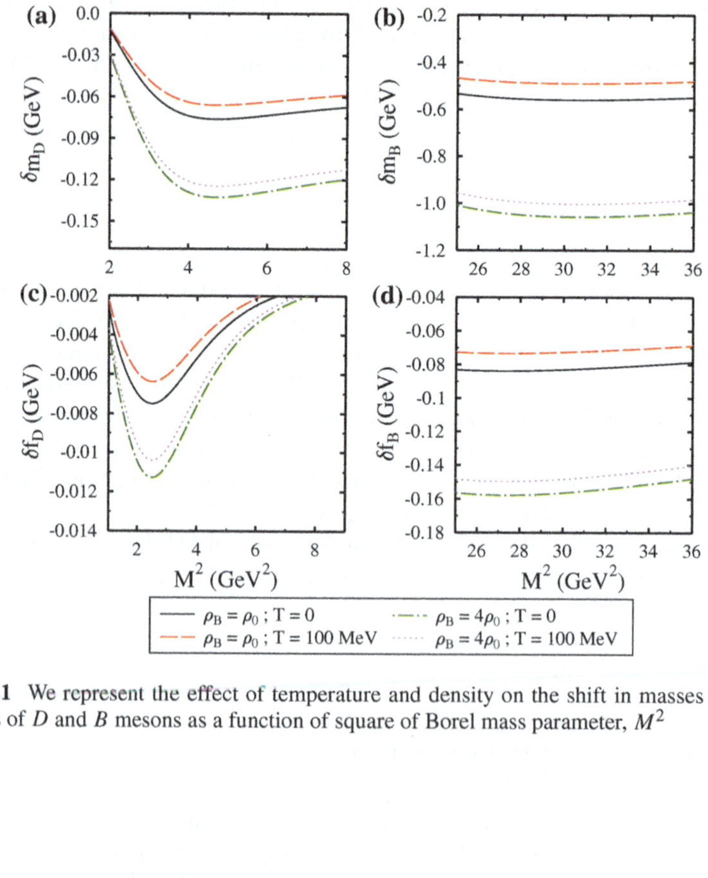

Fig. 155.1 We represent the effect of temperature and density on the shift in masses and decay constants of D and B mesons as a function of square of Borel mass parameter, M^2

Table 155.1 We enlist the numerical data (in unit of MeV) of the shift in masses and decay constants of *D* and *B* mesons for two values of densities $\rho_B = \rho_0$, $\rho_B = 4\rho_0$, and for each of these two values we represent the data for two values of temperature, T = 0 and 100 MeV

	T = 0		T = 100 (MeV)	
	ρ_0	$4\rho_0$	ρ_0	$4\rho_0$
δm_D	−76	−65	−132	−124
δm_B	−560	−488	−1056	−1001
δf_D	−3.9	−3.3	−5.6	−5.1
δf_B	−83	−73	−157	−149

and −4(−48) MeV respectively, at nuclear saturation density, zero temperature symmetric nuclear medium [2]. We can compare these results with our result of mass modification of *D* meson as 72 MeV (Table. 155.1).

155.4 Conclusions

Drop of mass and decay constant of $D(B)$ mesons in hot and dense symmetric nuclear medium may cause the $J/\psi(\Upsilon)$ suppression in the future HIC experiments like CBM and Panda.

Acknowledgements Authors greatfully acknowledge the Department of Science and Technology(Government of India) for providing the financial help under the project(SR/FTP/PS-209/2012).

References

1. A. Kumar, R. Chhabra, Phys. Rev. C **92**, 035208 (2015)
2. Z. Wang, Phys. Rev. C **92**, 065205 (2015)
3. R. Chhabra, A. Kumar, arXiv:1611.09549

Chapter 156
Coulomb Modified Glauber Model Analysis for Interaction of $^{56}Fe_{26}$, $^{84}Kr_{36}$, $^{132}Xe_{54}$, $^{197}Au_{79}$ and $^{238}U_{92}$ Projectiles

N. Marimuthu, V. Singh and S. S. R. Inbanathan

156.1 Introduction

The Relativistic heavy ion collision has been extensively studied both theoretically and experimentally during last couple of decades [1–4]. The photographic nuclear emulsion technique is used to investigate nucleus-nucleus (AA) and hadron-nucleus (hA) interactions at high energy [5]. Glauber Multiple scattering modal (GM) is used to calculate the total reaction cross section, which is important parameters for the collision processes. This model has been modified in accordance of coulomb field, that is colliding nuclei deviated from the straight-line trajectory. This approach is called the Coulomb modified Glauber model (CMGM) [2]. In the present work, the total reaction cross section has been calculated using the CMGM with and without medium effect for various projectiles such as $^{56}Fe_{26}$, $^{84}Kr_{36}$, ^{132} Xe_{54}, $^{197}Au_{79}$ and $^{238}U_{92}$ with nuclear emulsion nuclei at KE ~1 GeV/n, and compared with the experimental data.

156.2 CMGM Formalism

According to the Glauber model [2], the total reaction cross section are given by

$$\sigma_R\ (mb) = 2\pi \int [1 - T\ (b)]b.db. \tag{156.1}$$

where, T(b) = exp [$-\chi$(b)] and nucleus-nucleus interaction is given by [4].

V. Singh (✉) · N. Marimuthu
Physics Department, Institute of Science, Banaras Hindu University, Varanasi 221005, India
e-mail: venkaz@yahoo.com

N. Marimuthu · S. S. R. Inbanathan
Department of physics, The American College, Madurai 625002, Tamilnadu, India

Md. Naimuddin (ed.), *XXII DAE High Energy Physics Symposium*, Springer Proceedings in Physics 203, https://doi.org/10.1007/978-3-319-73171-1_156

$$\chi_{PT}(b) = \frac{\pi^2 \rho P(0) \rho T(0) a_P^3 a_T^3}{10(a_P^2 + a_T^2 + r_0^2)} \bar{\sigma}_{NN} \exp(-\frac{b^2}{a_P^2 + a_T^2 + r_0^2}). \tag{156.2}$$

Now consider effects of coulomb field between projectile and target according to the Ref. [2], the impact parameter b is replaced by b' as

$$b' = \frac{\eta + (\sqrt{\eta + k^2 b^2})}{k}, \eta = \frac{Z_P Z_T e^2}{\hbar \nu}. \tag{156.3}$$

156.3 Results and Discussion

Figure 156.1 (left) shows the comparison between the calculated average values of total reaction cross section without nuclear medium effect (σ^{theory}) using nuclear matter density $\rho = 0$ and corresponding experimental values of $^{56}Fe_{26}$, $^{84}Kr_{36}$, $^{132}Xe_{54}$, $^{197}Au_{79}$ and $^{238}U_{92}$ - Em. σ^{theory} shows slightly higher than experimental value for projectiles $^{56}Fe_{26}$, $^{84}Kr_{36}$, $^{132}Xe_{54}$ withal it shows disagreement for projectiles $^{197}Au_{79}$, $^{238}U_{92}$. The calculated average values of reaction cross section with nuclear medium effect and corresponding experimental values are shown in Fig. 156.1 (right). These total reaction cross sections calculation using $\rho = 0.15$, 0.17 and 0.19 fm^{-3} and it shows good agreement with projectiles $^{56}Fe_{26}$, $^{84}Kr_{36}$, $^{132}Xe_{54}$. It concludes the average value of projectile, target nuclear emulsion reaction cross section increases with increasing projectile mass number.

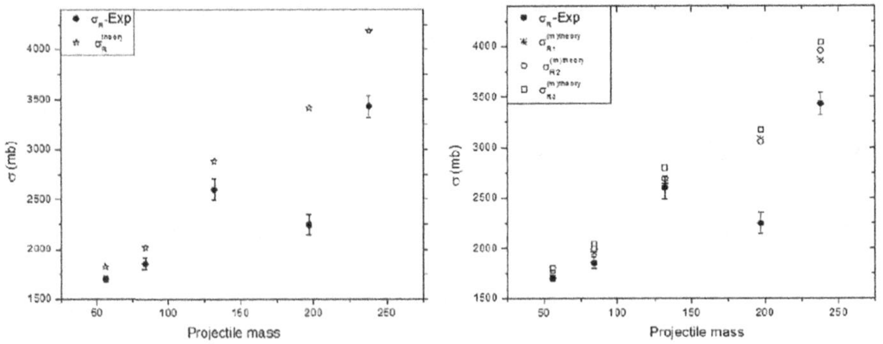

Fig. 156.1 Total reaction cross section without (left) and with (right) nuclear medium effect in the CMGM model for different projectiles at ~1 GeV/n

156.4 Conclusion

The calculated reaction cross section values with nuclear medium effect shows good agreement with experimental results compared to without nuclear medium effect, which indicates that the nuclear medium effect is essential for calculation of reaction cross section in CMGM model. The calculation of reaction cross section depends not only on the projectile and target radius but also depend on the mass number of the projectile.

References

1. S.K. Charagi, S.K. Gupta, Phys. Rev. C **41**, 1171 (1997)
2. A.Y. Abul-magd et al., Nuovo cimento **110**, 1281–1288 (1997)
3. X.Z. Cai et al., Phys. Rev. C **58**, 572 (1998)
4. M.S.M. Nour El-Din, M. E. Solite, Can. J. Phys. **86**, 1209–1217 (2008)
5. N. Marimuthu et al., Indian J. Phys. **91**, 431–438 (2017)

Chapter 157
Discriminating Hybrid Textures of Neutrino Mass Matrix in the Light of Latest Neutrino Data and Baryon Asymmetry

Rupam Kalita

157.1 Introduction

Standard Model (SM) of particle physics become the most successful model after the discovery of the Higgs boson and can explain all the fundamental interactions except gravity. But still Standard Model of particle physics can not explain many of the observed phenomena including neutrino masses and mixing [1] and matter-antimatter asymmetry [2] of the Universe. So far neutrino physics represents the particle physics beyond the Standard Model (BSM). To explain the mass of the neutrino we have to consider some famous BSM mechanism known as seesaw mechanism [3] where some extra heavy fermions are added to the SM particle content. Seesaw mechanism can be divided into three categories Type I, Type II and Type III respectively and in this work we consider only type I seesaw. In our present work we try to study type I seesaw with some texture zero neutrino mass matrices known as hybrid texture neutrino mass matrix. Hybrid texture [4] neutrino matrices are symmetric matrix where two elements are are equal and one element is zero. There are 60 possible structure of hybrid texture matrix is possible but out of these 60 only 39 are compatible with latest neutrino oscillation data. In this work we consider 39 of hybrid texture neutrino mass matrix and try to study baryon asymmetry of the Universe incorporating different neutrino parameters. Baryon asymmetry can be studied very well by a mechanism known as leptogenesis [5] and leptogenesis can be generated with the decay of right handed neutrinos. It will be more useful and predictive if we can connect right handed neutrino to generate neutrino mass and leptogenesis simultaneously using seesaw mechanism.

R. Kalita (✉)
Department of Physics, Tezpur University, Tezpur 784028, India
e-mail: rup@tezu.ernet.in

Md. Naimuddin (ed.), *XXII DAE High Energy Physics Symposium*, Springer Proceedings in Physics 203, https://doi.org/10.1007/978-3-319-73171-1_157

157.2 Numerical Analysis

In this work we consider type I seesaw to generate neutrino mass and type I seesaw mass formula can be written as $M_\nu = m_D M_{RR}^{-1} m_D^T$, where m_D, M_{RR} are Dirac and right handed neutrino mass matrices respectively. Since the diagonalizing matrix of M_ν is U_{PMNS} (PMNS matrix) so we can write above equation [6] as

$$M_\nu = U_{\text{PMNS}} M_\nu^{\text{diag}} U_{\text{PMNS}}^T \tag{157.1}$$

For normal hierarchy, the diagonal mass matrix of the light neutrinos [7] can be written as

$$M_\nu^{\text{diag}} = \text{diag}(m_1, \sqrt{m_1^2 + \Delta m_{21}^2}, \sqrt{m_1^2 + \Delta m_{31}^2})$$

whereas for inverted hierarchy it can be written as

$$M_\nu^{\text{diag}} = \text{diag}(\sqrt{m_3^2 + \Delta m_{23}^2 - \Delta m_{21}^2}, \sqrt{m_3^2 + \Delta m_{23}^2}, m_3).$$

Now if we consider neutrino mass matrix as Majorana type there are four independent parameters are present in the neutrino mass matrix in (157.1), namely one Dirac CP phase, two majorana phases and lightest neutrino mass. In this work we consider neutrino mass matrix as hybrid texture mass matrix and we compare this hybrid texture matrix with above (157.1) to observed correct baryon asymmetry of the

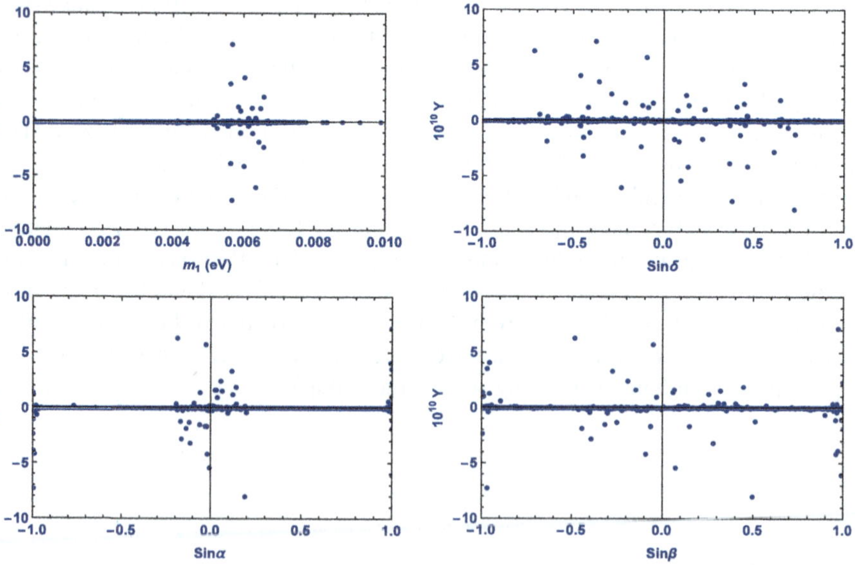

Fig. 157.1 Correct baryon asymmetry results of B4 hybrid texture model

universe given by $Y_B = (6.06 \pm 0.090) \times 10^{-10}$ [2]. Now we arrive at the two constrains one which sets one the light neutrino matrix element to zero and the other which equates two elements of the mass matrix. From these two complex equations we can write four real equations and these equations can be solved numerically to determine four independent neutrino parameters. Finally using all these independent neutrino parameters we can compute baryon asymmetry of the Universe for different hybrid texture neutrino mass matrices (Fig. 157.1).

157.3 Results and Conclusion

In this work we try to study the baryon asymmetry of the Universe by considering hybrid texture neutrino mass matrices. Here we consider type I seesaw mechanism to generate the mass of the neutrino and leptogenesis. The expressions for the leptogenesis can be found in [8, 9]. Leptogenesis can be converted to baryogenesis using sphaleron process. Here we study only two flavor leptogenesis and leave the other flavor for future study. We study the leptogenesis for all 39 hybrid texture model [4] considering both normal and inverted hierarchy of neutrino mass. We have not observed correct baryon asymmetry for all inverted hierarchy neutrino case but observed correct baryon asymmetry only for few models in normal hierarchy case. We observed correct baryon asymmetry only for models A2, B4, B5, C5, D5, F7 [4] in normal hierarchy case. We have shown one figure how baryon to photon ratio varies with lightest neutrino mass, delta, alpha and beta for B4 hybrid texture model. The other disfavored hybrid texture models may however, be saved by suitable model building works incorporating different sources of baryon asymmetry. We can constrain these hybrid texture models with observed correct baryon asymmetry of the Universe. In our present work we not only constrain hybrid texture mass model but also constrain experimentally undetermined neutrino parameters. Experimentally Still we do not have the correct values of neutrino mass, Dirac CP phase and Majorana phases and therefor our analysis may give some phenomenological scenario to constrain all these undetermined neutrino parameters by observing correct baryon asymmetry of the Universe. Also our analysis disfavored inverted hierarchy of neutrino mass and gives a hint to solve the hierarchy problem in neutrino mass. Baryon asymmetry of the universe is one of the major unresolved issuc in these days and our work is trying to connect neutrino mass and mixing with baryon asymmetry of the Universe considering the same source. Neutrinoless double beta decay, renormalization group equation also can be studied with the help of these models and we leave this work for future studies.

Acknowledgements I would like to thank Dr Debasish Borah from IIT guwahati for useful discussions and comments regarding this work.

References

1. S. Fukuda et al., (Super-Kamiokande), Phys. Rev. Lett. **86**, 5656 (2001), arXiv:hep-ex/0103033; Q.R. Ahmad et al., (SNO), Phys. Rev. Lett. **89**, 011301 (2002)
2. P.A.R. Ade et al., [Planck Collaboration], Astron. Astrophys. **571**, A16 (2014)
3. P. Minkowski, Phys. Lett. **B67**, 421 (1977); S. Antusch, S.F. King, Phys. Lett. **B597**(2), 199 (2004); R. Foot, H. Lew, X.G. He, G.C. Joshi, Z. Phys. **C44**, 441 (1989)
4. J.-Y. Liu, S. Zhou, Phys. Rev. D **87**, 093010 (2013)
5. S. Davidson, E. Nardi, Y. Nir, Phys. Rep. **466**, 105 (2008)
6. M. Borah, D. Borah, M.K. Das, Phys. Rev. D **91**, 113008 (2015)
7. M.C. Gonzalez-Garcia, M. Maltoni, T. Schwetz, JHEP **1411**, 052 (2014); D.V. Forero, M. Tortola, J.W.F. Valle, Phys. Rev. D **90**, 093006 (2014)
8. R. Barbieri, P. Creminelli, A. Strumia, N. Tetradis, Nucl. Phys. **B575**, 61 (2000); A. Abada, S. Davidson, F. -X. Josse-Michaux, M. Losada, A. Riotto, JCAP **0604**, 004 (2006)
9. R. Kalita, D. Borah, Phys.Rev. **D92**(5), 055012 (2015)

Chapter 158
Lepton Flavor Violating Higgs Decay

Priya Maji, Debika Banerjee and Sukadev Sahoo

158.1 Introduction

There are many decay modes for Higgs boson in standard model (SM) background which have been probed in LHC already. But from those decay channels of Higgs boson, there are few lepton flavour violating (LFV) decays ($h \rightarrow \mu\tau, h \rightarrow e\tau, h \rightarrow e\mu$) which are not allowed within the SM theory. The coupling of Higgs boson to quarks and leptons are suppressed due to small Yuakwa couplings in SM, which can be easily dominated by new physics contributions. Some scientists have analyzed LFV decays ($h \rightarrow l_i^+ l_j^-$) in several extension of SM including both effective lagrangian approach and also some specific models. Previously, LHC predicted [1] the branching ratio for ($h \rightarrow \mu\tau, h \rightarrow e\tau$) which can reach up to $O(10\%)$. Recently, CMS [2] and ATLAS [3] collaboration have found the best fit of branching ration for the decay channel $h \rightarrow \mu\tau$ which is 0.84% and 0.77% with 2.4σ and 1.3σ respectively and the upper limit is constrained as $B(h \rightarrow \mu\tau) < 1.51\%$ [2] by CMS and $B(h \rightarrow \mu\tau) < 1.43\%$ [4] by ATLAS.

158.2 Theoretical Framework

In our work, we are interested to follow the model independent calculations which are carried out by effective lagrangian approach. Yukawa interactions are mostly occurring interactions between Higgs and fermions. The Yukawa lagrangian is given as [5]

$$L_Y = -m_i \bar{f}_L^i f_R^i - Y_{ij} \bar{f}_L^i f_R^j h + h.c., \tag{158.1}$$

P. Maji (✉) · D. Banerjee · S. Sahoo
Department of Physics, National Institute of Technology,
Durgapur 713209, West Bengal, India
e-mail: majipriya@gmail.com

Md. Naimuddin (ed.), *XXII DAE High Energy Physics Symposium*, Springer Proceedings in Physics 203, https://doi.org/10.1007/978-3-319-73171-1_158

where Y_{ij} is Yukawa coupling constant [1] and f_L, f_R are the charged leptons. The partial decay width and branching ratio of Higgs particle can be written as

$$(h \to l_\alpha l_\beta) = \frac{m_h}{8\pi}(|Y_{l_\alpha l_\beta}|^2 + |Y_{l_\beta l_\alpha}|^2), \tag{158.2}$$

$$BR(h \to l_\alpha l_\beta) = \frac{\Gamma(h \to l_\alpha l_\beta)}{\Gamma_{SM} + \Gamma(h \to l_\alpha l_\beta)}, \tag{158.3}$$

where, $l_\alpha = l_\beta = e, \mu, \tau$ and $\alpha \neq \beta$ and $\Gamma_{SM} = 4.1\,\text{MeV}$.

158.3 Result and Discussion

From [5] we get the values of Yukawa coupling constants which are shown in the Table 158.1. Putting the values of $Y_{\mu\tau}$ in (158.2), we have calculated the decay widths and branching ratios for the decay $h \to \mu\tau$. Varying the range of Yukawa constants within 0 to 2.55×10^{-3}, we plot a graph of decay width and branching ratio vs. Yukawa constants. From Fig. 158.1, we observe that decay width increases exponentially and at branching ratio increases slowly with Yukawa constants. The best fit for Yukawa constants give almost the same value of experimental result of branching ratios. The decay width for $Y_{\mu\tau} = 1.87 \times 10^{-3}$ (CMS) and $Y_{\mu\tau} = 1.79 \times 10^{-3}$ (ATLAS) are found to be $\Gamma_{\mu\tau} = 0.0348$ and $\Gamma_{\mu\tau} = 0.03187$ respectively. We hope

Table 158.1 Upper limit and experimental values of Yukawa coupling constant

Searches	Limits on Yukawa
$h \to \mu\tau(CMS)$	$Y_{\mu\tau} < 2.55 \times 10^{-3}$
	$Y_{\mu\tau} = 1.87 \times 10^{-3}$
$h \to \mu\tau(ATLAS)$	$Y_{\mu\tau} < 2.45 \times 10^{-3}$
	$Y_{\mu\tau} = 1.79 \times 10^{-3}$

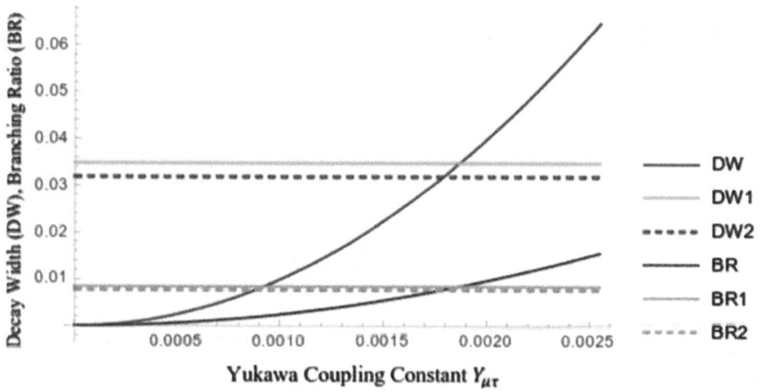

Fig. 158.1 Variation of decay width and branching ratio with Yukawa coupling constant

that the experimental confirmation of such processes could be helpful to enlighten the new physics.

References

1. R. Harnik et al., Flavor violating Higgs decays. JHEP **1303**, 026 (2013)
2. V. Khachatryan et al., (CMS Collaboration): Search for lepton-flavour-violating decays of the Higgs boson. Phys. Lett. B **749**, 337 (2015)
3. G. Aad et al., (ATLAS Collaboration): Search for lepton-flavour-violating $h \to \mu\tau$ decays of the Higgs boson with ATLAS detector. JHEP **1511**, 211 (2015)
4. Dell' Asta, L.:BEH fermionic decays and combinaion. Rencontres de Moriond - Electroweak Session (2016). https://indico.in2p3.fr/event/12279/session/5/contribution/202/material/slides/0.pdf
5. S. Banerjee et al., The Lepton flavour violating Higgs decays at the HL-LHC and the ILC. JHEP **1607**, 026 (2013)

Chapter 159
Excited States Searches for Light and Heavy Flavour Quarks with CMS Data at $\sqrt{s} = 13\,\text{TeV}$

Rocky Bala Garg, Varun Sharma and Brajesh C. Choudhary

159.1 Introduction

Proton-proton collisions at high energy produces events containing particles with high transverse momenta, thereby probing the interacting partons at the shortest distance scales. These tests help to search for signals predicted by new physics models, such as substructure of quarks (quark compositeness). A common signature for quark compositeness is to search for its excited state. In this letter, coupling between excited quarks (q^*), ordinary quarks and gauge bosons is uniquely fixed to be of magnetic moment type by gauge invariance. These can be described by effective Lagrangian [1]:

$$\mathcal{L}_{eff} = \frac{1}{2\Lambda}\bar{q}^*_R\,\sigma^{\mu\nu}\left[g_s f_s \frac{\lambda_a}{2} G^a_{\mu\nu} + g f \frac{\tau}{2} W_{\mu\nu} + g' f' \frac{Y}{2} B_{\mu\nu}\right] q_L + h.c., \quad (159.1)$$

In proton-proton collisions, excited quarks would be produced predominantly by quark-gluon annihilation, and would then decay into a quark and a gauge boson (g, W, Z, γ). This analysis searches for the processes qg $\rightarrow$ q* $\rightarrow$ qγ (where q = u, d) and bg $\rightarrow$ b* $\rightarrow$ bγ.

159.2 Event Selection

Events from Single photon data are required to pass the single photon trigger HLT_Photon165_HE10 and primary vertex selection. A high P_T photon (jet) with $P_T^{\gamma} > 190$ GeV and $\eta^{\gamma} < 1.4442$ ($\eta^{Jet} < 2.4$) are selected by utilizing the properties

R. B. Garg (✉) · V. Sharma · B. C. Choudhary
Department of Physics and Astrophysics, University of Delhi, Delhi, India
e-mail: rocky.bala.garg@cern.ch

Md. Naimuddin (ed.), *XXII DAE High Energy Physics Symposium*, Springer Proceedings in Physics 203, https://doi.org/10.1007/978-3-319-73171-1_159

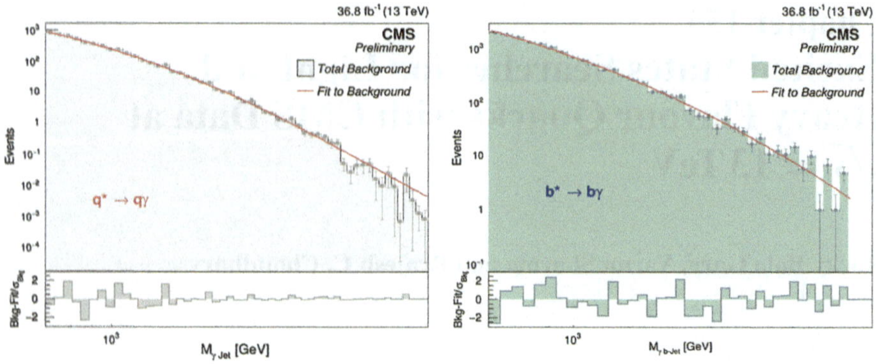

Fig. 159.1 Invariant mass fitting with background polynomial function for q* (left) and b* (right)

of these particles to identify [2]. The jets originating from decay of b-quarks are identified using CSVv2 algorithm [3]. The invariant mass of the γ+jet (γ+b-jet) system is required to be greater than 695 GeV. The $M_{\gamma+Jet}$/$M_{\gamma+b-Jet}$ fitted with polynomial background function is presented in Fig. 159.1.

159.3 Results and Conclusion

The limits on the $\sigma \times$ BR $\times\ \epsilon$ is computed by constructing a poissonion likelihood. The expected limits obtained for q* and b* are shown in Fig. 159.2.

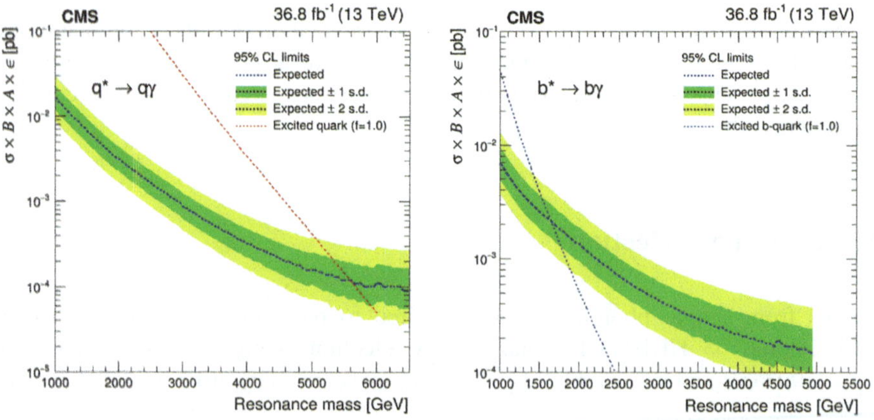

Fig. 159.2 Mass limit on $\sigma \times$ BR $\times\ \epsilon$ for q* (left) and b* (right)

Acknowledgements We would like to thank CMS collaboration for providing opportunity for carrying out this reasearch. We also thank CSIR & DST for supporting this work.

References

1. Excited quark production at Hadron colliders, Int. J. Mod. Phys. **A2**, 1285 (1987)
2. Jet Performance in pp Collisions at 7 TeV, CMS Physics Analysis Summary CMS-PAS-JME-10-003 (2010)
3. Identification of b-quark jets with the CMS experiment, JINST **8** (2013)

Chapter 160
Bianchi III Cosmologies for the Form $f(R) = R - \beta/R^n$ Using the Palatini Formalism

Debika K. Banik, Sebika K. Banik and Kalyan Bhuyan

160.1 Introduction

$f(R)$ gravity theory [1–3] have attracted considerable interest for explaining the late time acceleration of the universe. Among the various version of $f(R)$ gravity, we have considered the Palatini $f(R)$ gravity [1–3] and study the dynamics of the model $f(R) = R - \beta/R^n$ using Dynamical System Approach (DSA). Here we have extended the work of our recent paper [3] and analyse the phase space of the spatial curvature for this particular form of $f(R)$.

160.2 Palatini $f(R)$ gravity in Bianchi III metric

The Bianchi type III metric is described by the line element [3]

$$ds^2 = -dt^2 + A^2(t)dx^2 + e^{-2px}B^2(t)dy^2 + C^2(t)dz^2 \tag{160.1}$$

where $A(t)$, $B(t)$, $C(t)$ are the expansion scale factors and p is a constant. x, y, z are the comoving coordinates and t is the cosmic time.

The resultant field equation for this metric is given by [3]

D. K. Banik (✉) · S. K. Banik · K. Bhuyan
Department of Physics, Dibrugarh University, Dibrugarh 786004, Assam, India
e-mail: debikabanik05@gmail.com

S. K. Banik
e-mail: sebikabanik063@gmail.com

K. Bhuyan
e-mail: kalyanbhuyan@gmail.com

Md. Naimuddin (ed.), *XXII DAE High Energy Physics Symposium*, Springer Proceedings in Physics 203, https://doi.org/10.1007/978-3-319-73171-1_160

$$\theta^2 = \frac{6\kappa(\rho_m + \rho_r) + 3(f'R - f) + 6f'\sigma^2 + 6f'p^2(^2K)}{2f'\xi^2} \tag{160.2}$$

where θ, σ, ^2K, ρ_m and ρ_r are the volume expansion scalar, shear scalar, spatial curvature, matter energy density and radiation energy density, respectively. The expression for ξ is included in (160.5)

The dimensionless variables we have introduced are as follows:

$$\Sigma = \frac{\sqrt{3}\sigma}{\xi\theta},\ K = -\frac{3p^2(^2K)}{\xi^2\theta^2},\ x = \frac{3}{2}\frac{(f'R - f)}{f'\xi^2\theta^2},\ \Omega_r = \frac{3\kappa\rho_r}{f'\xi^2\theta^2},\ \Omega_m = \frac{3\kappa\rho_m}{f'\xi^2\theta^2} \tag{160.3}$$

Using the variables (160.3), the following evolution equations are obtained

$$\begin{aligned}
\frac{d\Sigma}{d\tau} &= \Sigma[-3 + 3\Sigma^2 + K - 3x + \Omega_r - 9C(R)x + 9D(R)(\Sigma^2 + K - 1)] - \xi K,\\
\frac{dK}{d\tau} &= K[1 + 3\Sigma^2 + K - 3x + \Omega_r - 9C(R)x + 9D(R)(\Sigma^2 + K + 1)],\\
\frac{dx}{d\tau} &= x[3 + 3\Sigma^2 + K - 3x + \Omega_r + 9C(R)(1 - x) + 9D(R)(\Sigma^2 + K)],\\
\frac{d\Omega_r}{d\tau} &= \Omega_r[-1 + 3\Sigma^2 + K - 3x + \Omega_r - 9C(R)x + 9D(R)(\Sigma^2 + K)],\\
1 &= \Sigma^2 - K + x + \Omega_r + \Omega_m
\end{aligned} \tag{160.4}$$

where τ stands for ln a (a is the average scale factor) and

$$C(R) = \frac{1}{3\theta}\frac{R\dot{f}'}{(f'R - f)},\ D(R) = \frac{1}{3\theta}\frac{f''\dot{R}}{f'},\ \xi = 1 + \frac{3}{2}\frac{f''\dot{R}}{f'\theta} \tag{160.5}$$

Setting (160.4) equal to zero we get the following fixed points: $P_r : (\Sigma, K, x, \Omega_r) = (0, 0, 0, 1)$, $P_m : (\Sigma, K, x, \Omega_r) = (0, 0, 0, 0)$, $P_d : (\Sigma, K, x, \Omega_r) = (0, 0, 1, 0)$, $P_s^{\pm} : (\Sigma, K, x, \Omega_r) = (\pm 1, 0, 0, 0)$ $P_k : (\Sigma, K, x, \Omega_r) = (1/2, -3/4, 0, 0)$, $P_a : (\Sigma, K, x, \Omega_r) = (1, -1, 0, -1)$

160.3 Conclusion

In this analysis, we plot the phase space for $f(R) = R - \beta/R^n$ in Fig. 160.1. We plot the phase portrait for $n = 1$ and 0.1 (β arbitrary) and observed the heteroclinic sequence of the type $P_k \to P_s \to P_d$ for both the values of n. This curve depicts the evolution of universe from non flat anisotropic point to the flat anisotropic point and finally towards the flat isotropic de-sitter point.

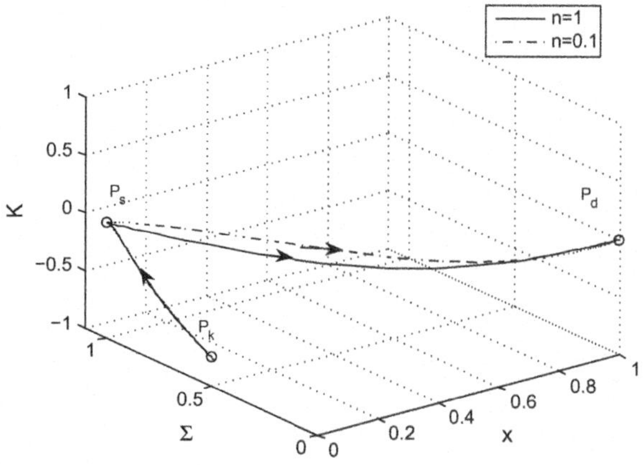

Fig. 160.1 Phase portrait for $n = 1, 0.1$

References

1. T.P. Sotiriou, V. Faraoni, f(R) theories of gravity. Rev. Mod. Phys. **82**, 451–497 (2010)
2. S. Fay, R. Tavakol, f(R) gravity theories in palatini formalism: cosmological dynamics and observational constraints. Phys. Rev. D. **75**, 063509 (2007)
3. D.K. Banik, S.K. Banik, K. Bhuyan, Anisotropic Bianchi type-III model in palatini $f(R)$ gravity. Astrophys. Space Sci. **362**, 51 (2017). https://doi.org/10.1007/s10509-017-3031-1

Chapter 161
Explaining the Observed Deviation in R($D^{(*)}$) in an Anomalous 2HDM

Lobsang Dhargyal

161.1 Introduction

It has been reported first by Babar and Belle, a possible deviation from the lepton flavor universality in $R(D^{(*)}) = \frac{Br(B \to D^{(*)}\tau\nu)}{Br(B \to D^{(*)}l\nu)}$. The present world average gives a deviation of 1.8σ for the R(D) and 3.3σ for the $R(D^*)$ and the taken together corresponds close to a 3.8σ deviation from SM prediction. There is also deviation reported in Br($B \to \tau\nu$) which is 1.3σ above the SM prediction. When the errors in $R(D^*)$ and $B(B \to \tau\nu)$ are added in quadrature, deviation from SM is about 4σ. In this work we present a Higgs mediated flavor universality violation in a Flipped or Lepto-Specific 2HDM with anomalous charged Higgs coupling to τ lepton or b quark respectively. This work is base on the materials contain in the reference [1].

161.2 Results

Due to limited space given, we will only show the main results from [1]. In Table 161.1 we have shown for two different values of the parameters from the fits to the data:

161.3 Conclusions

By adding theoretical and experimental errors in quadrature [1], we conclude that our model agrees within 1σ for the combination of $R(D^{(*)})$ and $\mathcal{B}r(B \to \tau\nu_\tau)$ data compare to about 4σ deviation from SM predictions. The same results can be

L. Dhargyal (✉)
Institute of Mathematical Sciences, Chennai 600113, India
e-mail: dhargyal2011@gmail.com

Md. Naimuddin (ed.), *XXII DAE High Energy Physics Symposium*, Springer Proceedings in Physics 203, https://doi.org/10.1007/978-3-319-73171-1_161

Table 161.1 $\chi^2_{min} = 10.95$ and we have taken the $\tan\beta$ as $100 > \tan\beta > 1$

S.no	$\tan\beta$	$M_{\pm}$ GeV	$R(D)_{Th}$	$R(D^*)_{Th}$	$Br_{Th}(B \to \tau\nu)$
1	69.97	700	0.348	0.255	1.29×10^{-4}
2	99.95	1000	0.348	0.255	1.29×10^{-4}

achieved if b quark replaces the τ lepton in a Lepton Specific 2HDM. From the form of the Yukawa couplings it is expected that if we require $\eta = -1$ for the b quark or τ lepton in the 2HDM-II will also work, an anomalous SUSY?

Acknowledgements Author would like to thank Nita Sinha and Rahul Srivastava, Institute of Mathematical Sciences for helpful discussions and comments. Author would also like to thank Shrihari Gopalakrishna, Institute of Mathematical Sciences for helpful comments. This work is supported and funded by the Department of Atomic Energy of the Government of India and by the Government of Tamil Nadu.

Reference

1. L. Dhargyal, Phys. Rev. D **93**(11), 115009 (2016)

Chapter 162
Study of Intermittency in Pb-Pb Collisions at 158 A GeV

Sunil Dutt

162.1 Introduction

The word "intermittency" originally came from the field of fluid-dynamics [1]. In an isotropic turbulent fluid of a high Reynold number, an intermittent structure appears as tube-like regions of a high vorticity isosurface. This stochastic, irregular behavior leads to a similar power-law variations of the moments as the size of region is decreased. The properties of such fluctuations have been extensively measured and discussed in the turbulence of fluids.

162.2 Mathematical Formulation

Bialas and Peschanski [2, 3] suggested a means of suppressing the fluctuations due to finite multiplicity by calculating the mean scaled factorial moments $<F_q>$ of the multiplicity distribution. Given a total interval of rapidity divided into Δy divided M equal bins of size $\delta y = \frac{\Delta y}{M}$, the mean scaled factorial moments $<F_q>$ of order q is defined as

$$<F_q> = M^{q-1} \sum_{m=1}^{M} \frac{K_m (K_m - 1) .. (K_m - q + 1)}{N (N - 1) \ldots (N - q + 1)} \tag{162.1}$$

where K_m denotes the number of particles in mth bin. The behavior of F_q as a function of is an indicator of the correlation length for fluctuations. $<\ldots>$ indicates an average over events, and $<n>$ is the mean multiplicity within δy. In particular,

S. Dutt (✉)
Department of Physics, Govt. College for Women, Gandhi Nagar, Jammu, J&K, India
e-mail: sunil_dutt_22@yahoo.in

Md. Naimuddin (ed.), *XXII DAE High Energy Physics Symposium*, Springer Proceedings in Physics 203, https://doi.org/10.1007/978-3-319-73171-1_162

Table 162.1 The values $b_{q,0}$, $b_{q,1}$ and $b_{q,2}$for q = 2, 3 and 4

$b_{q,k}$	2	3	4
$b_{q,0}$	0.419	1.241	1.147
$b_{q,1}$	1.803	−0.385	0.021
$b_{q,2}$	−2.117	−0.811	−1.040

Table 162.2 Experimental data of WA98 experiment

Data sets	Type	E_T(GeV)	No. of events
Cent-I	Most central	> 347.6	342
Cent-II	Central	225.5–298.6	1000
Cent-III	Peripheral	89.9–124.3	748

a mechanism with a self-similar ("branching") structure would exhibit a power-law dependence

$$< F_{q>\alpha}\delta y^{-\emptyset_q} \tag{162.2}$$

This power-law dependence is known as intermittency. The moments can either be normalized to the whole event sample or to the individual events. The moments are studied as a function of the chosen η-bin and the variation of the moments with varying $\delta\eta$ may indicate an intermittent behaviour. Horizontal Analysis

$$<F_q>_H = \frac{1}{N_{evts}} \sum\nolimits_{i=1}^{N_{evts}} M^{q-1} \sum\nolimits_{m=1}^{M} \frac{K_{m,i}\left(K_{m,i}-1\right)\ldots\left(K_{m,i}\ -q+1\right)}{< N>^q} \tag{162.3}$$

where N_{evts} is the number of events in the sample, $K_{m,i}$ is the content of bin m in event i and $<N>$ is the average multiplicity in pseudorapidity window $\Delta\eta$. A correction factor for horizontal moments has been proposed by Fialkowski et al. [4] to compensate for the non uniform shape of the rapidity distributions. The horizontal moments are corrected by dividing the factor (Tables 162.1 and 162.2)

$$R_F = \frac{1}{M} \sum\nolimits_{m=1}^{M} M^q \frac{< K_m>^q}{< N>^q}$$

The results from the analysis of data on photon multiplicities distributions.

162.3 Details of Calculations

The analysis is made on central collisions only which are characterized by ET cut of 348.8 GeV. From the sample of the interactions, we have collected 1000 central events. The quality of the events is tested for various plots like η- distribution and

Table 162.3 The values of coefficients of A, B and C obtained from horizontal scaled factorial moment corrected for Experiment central data

$b_{q,k}$	2	3	4
$b_{q,0}$	0.022 ± 0.01	0.025 ± 0.01	0.237 ± 0.03
$b_{q,1}$	0.024 ± 0.02	0.045 ± 0.11	0.443 ± 0.03
$b_{q,2}$	0.058 ± 0.01	0.068 ± 0.10	0.142 ± 0.02

Table 162.4 The values of coefficients of A, B and C obtained from horizontal scaled factorial moment corrected for simulated central data

$b_{q,k}$	2	3	4
$b_{q,0}$	0.011 ± 0.01	0.022 ± 0.01	0.147 ± 0.02
$b_{q,1}$	0.041 ± 0.02	0.041 ± 0.03	0.442 ± 0.01
$b_{q,2}$	0.065 ± 0.01	0.066 ± 0.02	0.852 ± 0.04

Ø- distribution which are in line with the published results. For this purpose the VENUS 4.12 and GEANT (GWA98) package is used and the number of central collisions Simulated was about 1342.The data is analysed in restricted Pseudorapidity window 3.2 $<\eta<$ 4.0 with full azimuthal coverage. These events were also subjected to all the procedures laid down for the experimental data. A relative comparison of the above studies was made for the relevant parameters and attempt has been made to understand the mechanism of the particle production and the observation of the fluctuations in the photon multiplicities (Tables 162.3 and 162.4).

162.4 Conclusion

We find poor evidence for an intermittent pattern of rapidity density fluctuations. The origin of such intermittent fluctuations is still unclear and remains an interesting area for both theoretical and experimental studies at LHC energies. The study of multiplicity fluctuations of hadrons in final state enables one to find some new quantities to reflect the features of different kinds of phase transitions. The experimental results are compared with predictions of simulation studies using Monte Carlo (MC) code, VENUS, for particle production in relativistic nucleus-nucleus collisions.

References

1. T.H. Burnett et al., Extremely high multiplicities in high-energy nucleus-nucleus collisions. Phys. Rev. Lett. **50**, 2062 (1983)

2. A. Bialas, R. Peschanski, Moments of rapidity distributions as a measure of short-range fluctuations in high-energy collisions. Nucl. Phys. B **273**, 703 (1986)
3. A. Bialas, R. Peschanski, Intermittency in Multiparticle production at high energy. Nucl. Phys. B **308**, 857 (1988)
4. K. Fialkowski et al., Intermittency and QCD jets. Acta Physica Polonica. B **20**, 639 (1989)

Chapter 163
Lepton Flavor Violation in a Model with Softly Broken $\mathbb{Z}_4$ Symmetry

Roopam Sinha

163.1 Introduction

The Lagrangian of our model

$$\mathcal{L}_Y = f_{ij}^{(e)} \bar{l}_{iL} e_{jR} \Phi + f_{ij}^{(\nu)} \bar{l}_{iL} N_{jR} \tilde{\Phi} + + g_{ij} \overline{(l_{iL})^c} i\tau_2 l_{jL} \chi_1^+ + h_{ij} \bar{S}_{iL} e_{jR} \chi_2^+$$
$$+ h'_{ij} \overline{(N_{iR})^c} e_{jR} \chi_1^+ + m^{ij} \overline{S_{iL}} N_{jR} \qquad (163.1)$$

Particle content	$SU(2)_L \times U(1)_Y$	$\mathbb{Z}_4$
$l_{iL} = (\nu_{iL}, e_{iL})^T$	$(2, -1)$	i
e_{iR}	$(1, -2)$	i
N_{iR}	$(1, 0)$	i
S_{iL}	$(1, 0)$	i
Φ	$(2, 1)$	1
χ_1^+	$(1, 2)$	-1
χ_2^+	$(1, 2)$	1

A mass term $M\bar{S}_L(\nu_L)^c$ is induced at the one-loop level. M_L, M_R and μ_S are all $\mathbb{Z}_4$ violating. If we take $m_\nu \sim 0.1$ eV, $M \sim 1$ eV, $m \sim 1$ TeV. The Majorana mass matrix M_ν in the seesaw basis $X_L = \left(\nu_L \; (N_R)^c \; S_L\right)^T$ is given by

R. Sinha (✉)
Saha Institute of Nuclear Physics, HBNI, 1/AF Bidhannagar,
Kolkata 700064, India
e-mail: roopam.sinha@saha.ac.in

Md. Naimuddin (ed.), *XXII DAE High Energy Physics Symposium*, Springer Proceedings in Physics 203, https://doi.org/10.1007/978-3-319-73171-1_163

$$\mathcal{L}_{mass} = \left(\overline{\nu_L}\ \overline{(N_R)^c}\ \overline{S_L}\right)\begin{pmatrix} 0 & m_D & M \\ m_D^T & 0 & m^T \\ M^T & m & 0 \end{pmatrix}\begin{pmatrix} (\nu_L)^c \\ N_R \\ (S_L)^c \end{pmatrix} = \overline{X_L} M_\nu (X_L)^c. \quad (163.2)$$

The scalar potential of the model is

$$V(\chi_1, \chi_2, \Phi) = \sum_{i=1}^{2} m_i^2 \chi_i^\dagger \chi_i + m_{12}^2 \chi_1^\dagger \chi_2 + m_\phi^2 (\Phi^\dagger \Phi) + \lambda_\phi (\Phi^\dagger \Phi)^2$$

$$+ \sum_{i=1}^{2} \lambda_i (\chi_i^\dagger \chi_i)^2 + \lambda_{12} (\chi_1^\dagger \chi_1)(\chi_2^\dagger \chi_2) + \sum_{i=1}^{2} \lambda_i' (\Phi^\dagger \Phi)(\chi_i^\dagger \chi_i) \quad (163.3)$$

Diagonalizing the matrix is $\begin{pmatrix} m_1 + \lambda_1' v^2 & m_{12}^2 \\ m_{12}^{*2} & \lambda_2' v^2 + m_2^2 \end{pmatrix}$ through the basis transformation $\begin{pmatrix} \chi_1 \\ \chi_2 \end{pmatrix} = \begin{pmatrix} \cos\theta & \sin\theta \\ -\sin\theta & \cos\theta \end{pmatrix}\begin{pmatrix} h_1 \\ h_2 \end{pmatrix}$, the masses of h_1, h_2 turns out to be,

$$M_{1,2}^2 = \frac{1}{2}[(m_1^2 + m_2^2) + (\lambda_1' + \lambda_2')v^2] \pm \frac{1}{2}[(m_1^2 - m_2^2) + (\lambda_1' - \lambda_2')v^2]\cos 2\theta \mp m_{12}^2 \sin 2\theta \quad (163.4)$$

where $\tan 2\theta = -2m_{12}^2 / \left((m_1^2 - m_2^2) + (\lambda_1' - \lambda_2')v^2\right)$.

163.2 Radiative Mass M of M_ν

Taking charged lepton mixings into account the radiative mass is given by

$$M_{ik} \overline{S_{iL}} (\nu_{kL})^c = \overline{S_{iL}} m_j h_{ij}' g_{kj}^{*\prime} \sin\theta\cos\theta \frac{1}{16\pi^2}\left[\frac{M_2^2}{m_j^2 - M_2^2}\ln\frac{m_j^2}{M_2^2} - \frac{M_1^2}{m_j^2 - M_1^2}\ln\frac{m_j^2}{M_1^2}\right](\nu_{kL})^c \quad (163.5)$$

163.3 LFV Decays $e_i \to e_j \gamma$ and $Z \to e_i^\pm e_j^\mp$

Apart from the W-boson mediated amplitude, the amplitude mediated by h_1, h_2:

$$\Gamma_{h_1,h_2}(e_i \to e_j \gamma) = \frac{\alpha}{48\pi} \sum_k \frac{m_i^5}{192\pi^3} (U^{\nu\dagger} g U_L^e)_{ik} (U^{\nu\dagger} g U_L^e)_{kj}^* / \bar{M}^2)^2 \quad (163.6)$$

where $1/\bar{M}^2 = \cos^2\theta/M_1^2 + \sin^2\theta/M_2^2$. Using $\Gamma(\mu \to e\bar{\nu}_e\nu_\mu) = \frac{m_\mu^5 G_F^2}{192\pi^3}$, and the bounds on the branching ratios such as $B(\mu \to e\gamma) = \frac{\Gamma(\mu \to e\gamma)}{\Gamma(\mu \to e\bar{\nu}_e\nu_\mu)} < 1.2 \times 10^{-11}$ the parameters of the model is constrained. Similar constraints are also obtained from various LFV Z decays and the decay $\mu \to e\bar{\nu}_e\nu_\mu$.

Chapter 164
Lepton Number Violation and Lepton Flavour Violation in Left-Right Symmetric Model

Happy Borgohain and Mrinal Kumar Das

164.1 Introduction

The landmark discovery of neutrino oscillation has provided clear evidence that these tiny weakly interacting particles are massive, which has been an enthralling manifestation of physics beyond the Standard Model (SM). We have done a phenomenological study of two important BSM phenomenon, viz. NDBD and LFV in the framework of LRSM which can explain the origin of small neutrino mass via seesaw mechanism. We have studied for two $\mu - \tau$ symmetric neutrino mass models, TBM and BM for both the hierarchies, normal and inverted.

164.1.1 NDBD and LFV in LRSM

In the context of LRSM, several new contributions to NDBD arises due to the presence several heavy fields. Many of the earlier works have explained it in details (like [1–3]). We have considered the new physics contributions to NDBD coming from $\mathrm{W_R}^-$ and Δ_R respectively. For simple approximations, we assumed similar mass scales for the heavy particles i.e., $\mathrm{M_R} \approx \mathrm{M_{W_R}} \approx \mathrm{M_{\Delta_L^{++}}} \approx \mathrm{M_{\Delta_R^{++}}} \approx$ TeV. For new physics contribution to NDBD considering perturbation in type II (I) seesaw, effective mass governing NDBD is [1, 2],

$$\mathrm{m_{N+\Delta_R}}^{\mathrm{eff}} = \mathrm{p}^2 \frac{\mathrm{M_{W_L}}^4}{\mathrm{M_{W_R}}^4} \frac{\mathrm{U_{Rei}}^{*2}}{\mathrm{M_i}} + \mathrm{p}^2 \frac{\mathrm{M_{W_L}}^4}{\mathrm{M_{W_R}}^4} \frac{\mathrm{U_{Rei}}^2 \mathrm{M_i}}{\mathrm{M_{\Delta_R}}^2} \qquad (164.1)$$

H. Borgohain (✉) · M. Kumar Das
Department of Physics, Tezpur University, Tezpur 784028, India
e-mail: happy@tezu.ernet.in

M. Kumar Das
e-mail: mkdas@tezu.ernet.in

Md. Naimuddin (ed.), *XXII DAE High Energy Physics Symposium*, Springer Proceedings in Physics 203, https://doi.org/10.1007/978-3-319-73171-1_164

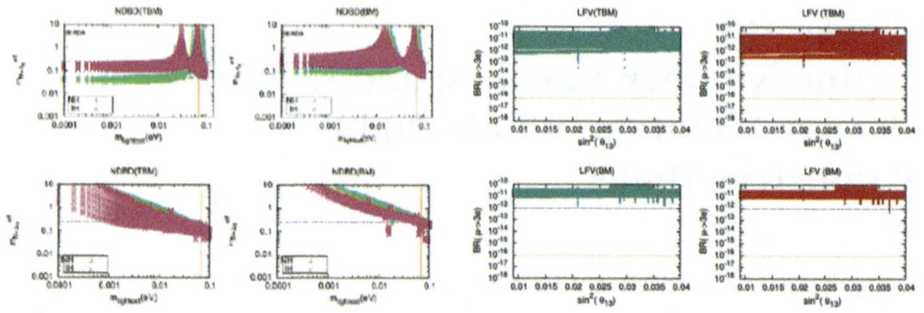

Fig. 164.1 In figure left, new physics contribution to effective mass governing NDBD considering perturbation in type II (top) and type I (bottom) seesaw respectively, the dashed and the solid lines represents the GERDA [4] bound on the effective mass and the PLANCK bound on the sum of the absolute neutrino mass respectively and in figure right, total contribution to LFV shown as a function of θ_{13} for the TBM and BM for both NH and IH

The relevant branching ratios for the LFV process $\mu \to 3e$ is [1],

$$\mathrm{BR}\,(^{-} \to 3\mathrm{e}) = \frac{1}{2}\left|\mathrm{h}_{-\mathrm{e}}\mathrm{h}^{*}_{\mathrm{ee}}\right|^{2}\left(\frac{\mathrm{m_{W_L}}^{4}}{\mathrm{M}^{++4}_{\Delta_\mathrm{L}}} + \frac{\mathrm{m_{W_R}}^{4}}{\mathrm{M}^{++4}_{\Delta_\mathrm{R}}}\right),\ \mathrm{h_{ij}} = \sum_{\mathrm{n}=1}^{3}\mathrm{V_{in}V_{jn}}\left(\frac{\mathrm{M_n}}{\mathrm{M_{W_R}}}\right) \tag{164.2}$$

164.2 Results and Discussions

While considering type II as perturbation, TBM pattern shows consistent results of $\mathrm{m_{eff}}$ for NH and IH. Whereas considering type I as perturbation,the values that are consistent with experimental bound imposed by GERDA are found for lightest mass (0.001-0.1)eV for TBM and about 0.1 eV for BM. The BR for the process $\mu \to 3e$ in the LRSM remains consistent in the 3σ range for θ_{13} for TBM mixing pattern whereas for BM, BR lies out of the limit propounded by the SINDRUM experiment (Fig. 164.1).

References

1. H. Borgohain, M.K. Das, Neutrinoless double beta decay and lepton flavour violation in broken $\mu - \tau$ symmetric neutrino mass models. Int. J. Theor. Phys. https://doi.org/10.1007/s10773-017-3458-8
2. D. Borah, A. Dasgupta, *Neutrinoless Double Beta Decay in Type I+II Seesaw Models*, https://doi.org/10.1007/JHEP11(2015)208
3. J. Barry, W. Rodejohann, Lepton number and flavour violation in TeV-scale left-right symmetric theories with large left-right mixing, https://doi.org/10.1007/JHEP09(2013)153
4. M. Agostini, et al., (GERDA Collaboration), https://doi.org/10.1103/PhysRevLett.111.122503

Chapter 165
Search for CP Violation in $D^0 \rightarrow K_S^0 K_S^0$ Decay at Belle and Belle II

N. Dash

165.1 Introduction

Charge-conjugation Parity Violation (CPV) in charmed meson decays has not yet been observed and is predicted to be (10^{-3}) in the Standard Model (SM). An evidence for CPV in 2012 by LHCb ($-0.82 \pm 0.21 \pm 0.11$)% [1] suggested a 3.5 standard deviation (σ), confirmed by CDF [2], took people by surprise and revived the field. Measurements of ΔA_{CP} ($A_{CP}^{D^0 \rightarrow \pi^+\pi^-} - A_{CP}^{D^0 \rightarrow K^+K^-}$) have been performed by LHCb, CDF, BaBar and Belle collaborations [3, 4]. Recently LHCb updated ΔA_{CP} result with a larger data sample and there is no evidence of non-zero asymmetry [4]. The combined result of ΔA_{CP} by HFAG [5] gives an agreement with no CPV at 9.3% confidence level (CL). Though there is no current evidence of non-zero asymmetry, CPV in charm decays is investigated in other channels. Singly Cabibbo-suppressed (SCS) decays are of special interest as the possibility of interference with NP amplitudes could lead to larger CPV than predicted by SM, The $D^0 \rightarrow K_S^0 K_S^0$ decay is one such channel [6].

The D^0 candidates are selected as coming from the decay $D^{*+} \rightarrow D^0 \pi_s^+$, where π_s^+ denotes the low-momentum "slow" pion. The charge of this slow pion reveals the flavor content of neutral D meson (whether it is a D^0 or $\bar{D}^0$) at its production vertex. A stringent selection criterion is applied on the momentum of the D^{*+} candidate in the e^+e^- center-of-mass frame, $p^*(D^*)$, to suppress D^{*+} coming from B decays as well as to reduce the combinatorial background. Assuming the total decay width to be same for particles and antiparticles, the time-integrated A_{CP} is given as:

$$A_{CP} = \frac{\Gamma(D^0 \rightarrow f) - \Gamma(\bar{D}^0 \rightarrow \bar{f})}{\Gamma(D^0 \rightarrow f) + \Gamma(\bar{D}^0 \rightarrow \bar{f})} \tag{165.1}$$

(On behalf of the Belle Collaboration).

N. Dash (✉)
Indian Institute of Technology, Bhubaneswar, India
e-mail: nd11@iitbbs.ac.in

Md. Naimuddin (ed.), *XXII DAE High Energy Physics Symposium*, Springer Proceedings in Physics 203, https://doi.org/10.1007/978-3-319-73171-1_165

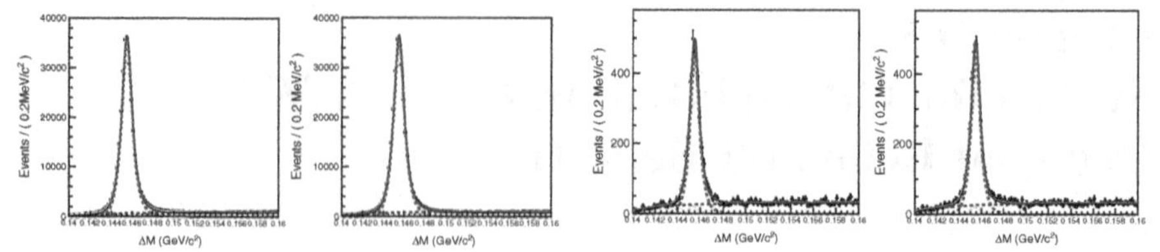

Fig. 165.1 Distributions of the mass difference ΔM for the $K_S^0\pi^0$ (left two) and $K_S^0K_S^0$ (right two) final states. Points with error bars are the data, the solid curves show the results of the fit, dashed (blue) curves are the background predictions and solid (cyan) is for the peaking background

where, Γ represents the partial decay width and f is specific final state. The extracted raw asymmetry is given by:

$$A_{\text{raw}} = \frac{N(D^0 \to f) - N(\bar{D}^0 \to \bar{f})}{N(D^0 \to f) + N(\bar{D}^0 \to \bar{f})} = A_{CP} + A_{FB} + A_{\epsilon}^{\pm} \tag{165.2}$$

Here, A_{FB} is the forward-backward production asymmetry, and $A_{\epsilon}^{\pm}$ is the asymmetry due to different detection efficiencies for positively and negatively charged pions. Both can be eliminated through a relative measurement of A_{CP} if the charged final-state particles are identical. The A_{CP} of signal mode (S) is measured relative to other well measured normalization decay mode (N). Such an approach enables the cancellation of several sources of systematic uncertainties that are common to both the signal and normalisation mode. The *CP* asymmetry of the signal mode can then be expressed as: $A_{CP}(S) = A_{\text{raw}}(S) - A_{\text{raw}}(N) + A_{CP}(N)$. For $A_{CP}(N)$, the world average value [7] is used. The result of the analysis presented here is based on data sample corresponding to 921 fb^{-1} integrated luminosity collected with the Belle detector at the KEKB asymmetric energy e^+e^- collider at center of mass energy $\sqrt{s}$ $\approx$ 10.58 GeV [8].

165.2 Search for $D^0 \to K_S^0K_S^0$ Decay

The $D^0 \to K_S^0K_S^0$ decay is Single Cabibbo Suppressed channel [6]. The most recent SM-based analysis obtained a 95% CL upper limit of 1.1% for direct *CP* violation in this decay [9]. The search for *CP* asymmetry in $D^0 \to K_S^0K_S^0$ has been performed first by the CLEO Collaboration [10] using a data sample of 13.7 fb^{-1} as $(-23 \pm 19)\%$. Recently, LHCb measured time-integrated *CP* asymmetry in $D^0 \to K_S^0K_S^0$ as $(-2.9 \pm 5.2 \pm 2.2)\%$, where the first uncertainty is statistical and the second is systematic [11]. The LHCb result is consistent with no CPV, in agreement with SM expectations. We present here the preliminary result of the measurement of the CP asymmetry in $D^0 \to K_S^0K_S^0$ decays using 921 fb^{-1} data collected at the Belle detector.

165.2.1 Signal Extraction and Systematic Uncertainties

A simultaneous fit of the ΔM for D^{*+} and D^{*-} is used (Fig. 165.1) to estimate the asymmetry. Here the normalization mode is $D^0 \to K_S^0\pi^0$. The ΔM (the mass difference between reconstructed D^* and D^0) distributions for the signal mode $D^0 \to K_S^0 K_S^0$ and normalization mode are shown in Fig. 165.1. The signal yield for $D^0 \to K_S^0 K_S^0$ is $5,399 \pm 87$ events and for $D^0 \to K_S^0\pi^0$ is $531,807 \pm 796$ events. The A_{raw} observed in data for signal and normalization modes are $(+0.45 \pm 1.53)\%$ and $(+0.16 \pm 0.14)\%$, respectively. The total systematic uncertainty for the time integrated CP violating asymmetry A_{CP} in the $D^0 \to K_S^0 K_S^0$ decay is $\pm 0.17\%$ [12]. A nonvanishing asymmetry originating from the different nuclear interaction of K^0 and $\bar{K}^0$ mesons with the detector material estimated in Ref. [13] in the presence of neutral kaon in the final states.

165.2.2 Result and Conclusion

The measured time-integrated CP-violating asymmetry A_{CP} in the $D^0 \to K_S^0 K_S^0$ decay is found to be $A_{CP} = (-0.02 \pm 1.53 \pm 0.17)$ % using a data sample of 921 fb^{-1} integrated luminosity. The dominant systematic uncertainty comes from the A_{CP} error of the normalisation channel. The result is consistent with SM expectations and is a significant improvement compared to the previous results from CLEO [10] and LHCb Collaborations [11], already probing the region of interest. The Belle II experiment is designed to record data at SuperKEKB, a major upgrade of KEKB, expect A_{CP} with a precision of 0.2% and should be able to test the SM.

Acknowledgements We thank the KEKB group, all institutes and agencies that have supported the work of the members of the Belle Collaboration.

References

1. R. Aaij et al., LHCb Collaboration. Phys. Rev. Lett. **108**, 111602 (2012)
2. T. Aaltonen et al., CDF Collaboration. Phys. Rev. Lett. **109**, 111801 (2012)
3. B. Aubert et al., BABAR Collaboration. Phys. Rev. Lett. **100**, 061803 (2008)
4. R. Aaij et al., (LHCb Collaboration), LHCb PAPER 2015 055, arXiv:1602.03160v1 (2016)
5. Y. Amhis et al., arXiv:1612.07233 and online update at http://www.slac.stanford.edu/xorg/hfag
6. G. Hiller et al., Phys. Rev. D **87**, 014024 (2013). arXiv:1211.3734
7. K.A. Olive et al., Particle Data Group. Chin. Phys. C **38**, 090001 (2014)
8. A. Abashian et al., Nuclear instruments and methods. Phys. Res. A **479**, 117–232 (2002)
9. U. Nierste, A. Schacht, Phys. Rev. D **92**, 054036 (2015). arXiv:1508.00074
10. G. Bonvicini et al., CLEO Collaboration. Phys. Rev. D **63**, 071101(R) (2001). arXiv:hep-ex/0012054
11. R. Aaij et al., LHCb Collaboration. JHEP **10**, 055 (2015). arXiv:1508.06087
12. A. Abdesselam et al., BELLE-CONF-1609, arXiv:1609.06393 [hep-ex]
13. B.R. Ko, E. Won, B. Golob, P. Pakhlov, Phys. Rev. D **84**, 111501 (2011)

Chapter 166
Charged and Neutral Current Pion Production in Neutrino-Nucleus Scattering

Kapil Saraswat, Prashant Shukla, Vineet Kumar and Venktesh Singh

166.1 Introduction

The neutrinos in the range between 1 and 3 GeV can interact with matter by many processes such as quasi elastic scattering (QES), interaction via resonance pion production (RES) and deep inelastic scattering (DIS). One of the most important processes in the RES region is coherent pion production. Coherent pion production is a process where the nucleus interact as a whole with neutrino and remain in the same quantum state as it was initially when the neutrino arrived.

166.2 PCAC Based Model

The PCAC based approach is used to calculate the differential cross section for the charged and neutral coherent pion production process which s given in [1, 2]. The used kinematic cuts are given in [2]. ξ is a parameter which is used for the pion dominating picture in the scattering process. The pion-nucleus cross section is based on the Glauber model approach which is described in [2].

K. Saraswat (✉) · V. Singh
Department of Physics, Banaras Hindu University, Varanasi 221005, India
e-mail: kapilsaraswat76@gmail.com

P. Shukla · V. Kumar
Nuclear Physics Division, Bhabha Atomic Research Centre, Mumbai 400085, India

P. Shukla
Homi Bhabha National Institute, Anushakti Nagar, Mumbai 400094, India

Md. Naimuddin (ed.), *XXII DAE High Energy Physics Symposium*, Springer Proceedings in Physics 203, https://doi.org/10.1007/978-3-319-73171-1_166

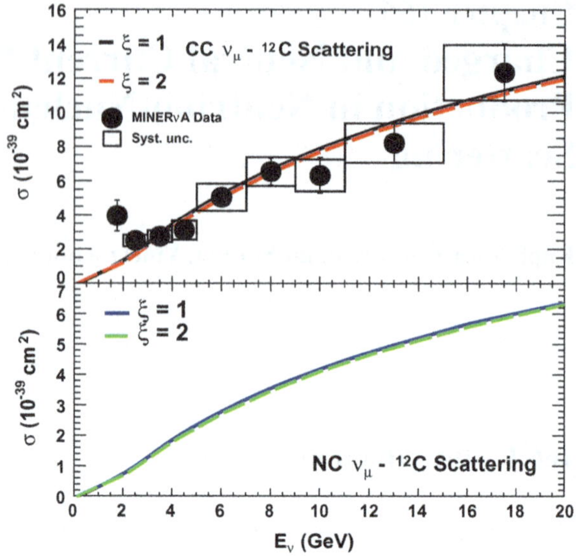

Fig. 166.1 σ of $\nu_\mu -^{12} C$ scattering as a function of E_ν

166.3 Results and Discussions

Figure 166.1 shows the total cross section σ for the neutrino-carbon scattering as a function of the neutrino energy E_ν obtained using the PCAC model for ξ = 1 and 2. The upper panel shows the charged current (CC) case while the lower panel shows the neutral current (NC). In the upper panel, the calculations are compared with the MINERνA data [3]. The calculations are compatible with the data. In both panels, the cross section is reduced with increasing the value of ξ. Figure 166.2 shows the σ for the neutrino-iron scattering as a function of E_ν obtained using the PCAC model for ξ = 1 and 2. Figure 166.3 shows the σ for the neutrino-oxygen scattering as a function of E_ν obtained using the PCAC model for ξ = 1 and 2. In the upper panel, the calculations are compared with the T2K data [4]. The calculations are compatible with the data at lower neutrino energy. With increasing the value of ξ, the total cross section is reduced.

Fig. 166.2 σ of $\nu_\mu -^{56} Fe$ scattering as a function of E_ν

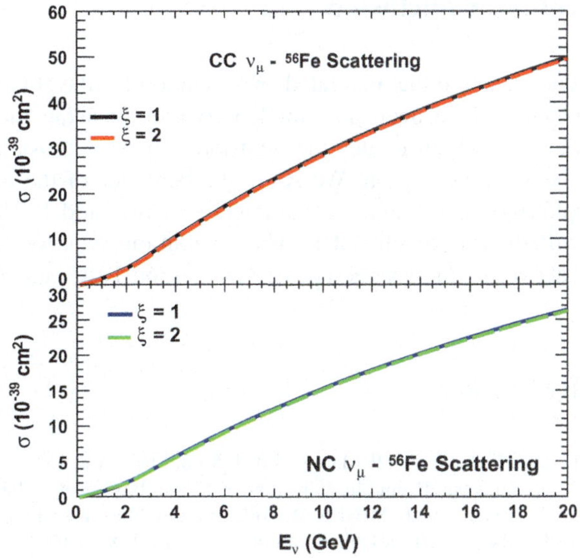

Fig. 166.3 σ of $\nu_\mu -^{16} O$ scattering as a function of E_ν.

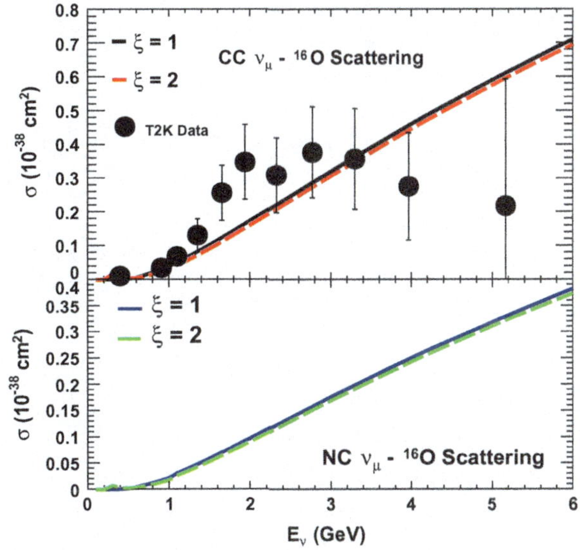

166.4 Conclusion

We presented the integrated cross section for the charged and neutral coherent pion production in neutrino - nucleus scattering using the formalism based on PCAC theorem which relates the neutrino - nucleus cross section to the pion - nucleus elastic cross section. We study the behavior of the cross section as a function of neutrino energy and the parameters of the model. The calculations are compared with the experimental data. The calculation gives good description of the MINERνA data while it gives good description of the T2K data at low neutrino energy.

References

1. B.Z. Kopeliovich, P. Marage, Int. J. Mod. Phys. A **8**, 1513 (1993)
2. K. Saraswat, P. Shukla, V. Kumar, V. Singh, Phys. Rev. C **93**, 035504 (2016)
3. A. Higuera et al., MINERvA Collaboration. Phys. Rev. Lett. **113**, 261802 (2014)
4. K. Abe et al., T2K Collaboration. Phys. Rev. D **95**, 012010 (2017)

Chapter 167
Freeze-Out of Strange Hadron in pp, pPb and PbPb Collisions at LHC Energies

Kapil Saraswat, Prashant Shukla, Vineet Kumar and Venktesh Singh

167.1 Introduction

Heavy ion collisions at Large Hadron Collider (LHC) and Relativistic Heavy Ion Collider (RHIC) are performed to create and study the properties of Quark Gluon Plasma (QGP). The p_{T} spectra of the hadrons is useful tools to study the particle production mechanisms, thermalization and collective effects and at high p_{T} probe jet quenching effects. Three types of the collision processes such as pp, pPb and PbPb are performed to study about the different aspects of the particle production at different center of mass energies at LHC.

167.2 Modified Tsallis Distribution Function

In [1], the Tsallis function with the transverse flow is given as

$$E\frac{d^3N}{dp^3} = C_n\left[\exp\left(\frac{-\gamma\,\beta\,p_{\mathrm{T}}}{n\,T}\right) + \frac{\gamma\,m_{\mathrm{T}}}{n\,T}\right]^{-n}. \tag{167.1}$$

Here C_n is the normalization constant, m_{T} is the transverse mass and $\gamma = 1/\sqrt{1-\beta^2}$. The parameter n is also called the degree of the thermalization. It is related to the non-extensivity parameter q by the relation $n = 1/(q-1)$.

K. Saraswat (✉) · V. Singh
Department of Physics, Banaras Hindu University, Varanasi 221005, India
e-mail: kapilsaraswat76@gmail.com

P. Shukla · V. Kumar
Nuclear Physics Division, Bhabha Atomic Research Centre, Mumbai 400085, India

P. Shukla
Homi Bhabha National Institute, Anushakti Nagar, Mumbai 400094, India

Md. Naimuddin (ed.), *XXII DAE High Energy Physics Symposium*, Springer Proceedings in Physics 203, https://doi.org/10.1007/978-3-319-73171-1_167

167.3 Results and Discussions

Figure 167.1 shows the invariant yields of K_s^0 for different multiplicity events in pp collision at $\sqrt{s} = 7$ TeV (panel(a)), pPb collision at $\sqrt{s_{NN}} = 5.02$ TeV (panel(b)) and PbPb collision at $\sqrt{s_{NN}} = 2.76$ TeV (panel(c)) as a function of the p_T measured by the CMS experiment [2] in the mid rapidity region $|y_{CM}| < 1$. The solid curves are the modified Tsallis distribution fitted to K_s^0 p_T spectra. The modified Tsallis gives good description of the data.

Figure 167.2 shows the Tsallis parameters n $\big(\text{panel(a)}\big)$, T $\big(\text{panel(b)}\big)$ and β $\big(\text{panel(c)}\big)$ as a function of the efficiency corrected average track multiplicity $< N_{trk} >$ in the pp collision at $\sqrt{s} = 7$ TeV, pPb collision at $\sqrt{s_{NN}} = 5.02$ TeV and

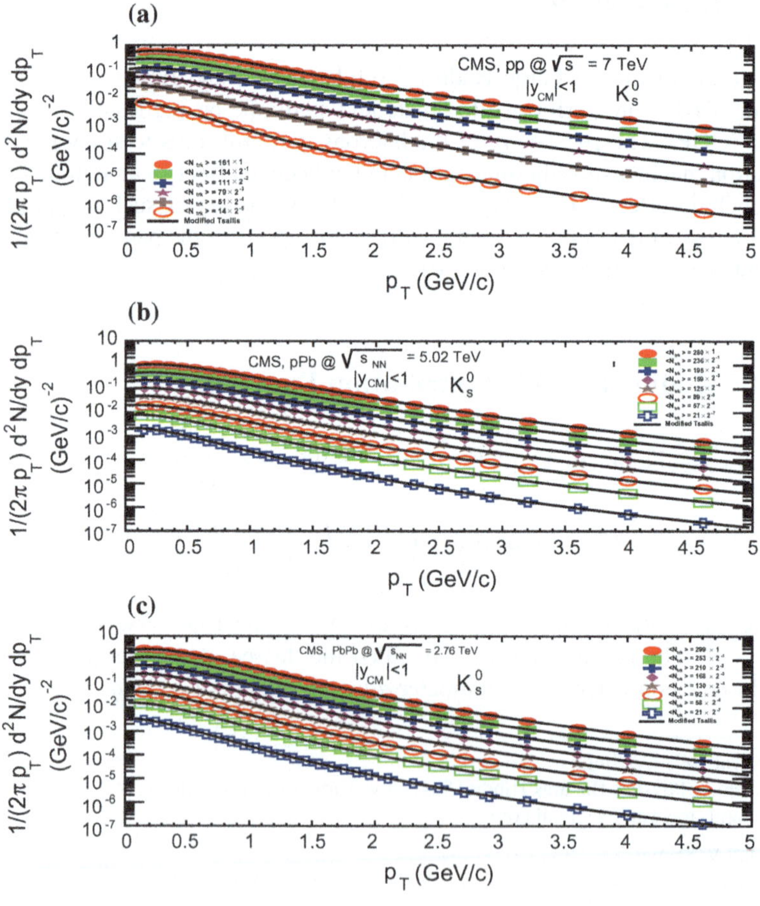

Fig. 167.1 The invariant yield of K_s^0 as a function of p_T for pp, pPb and PbPb collisions

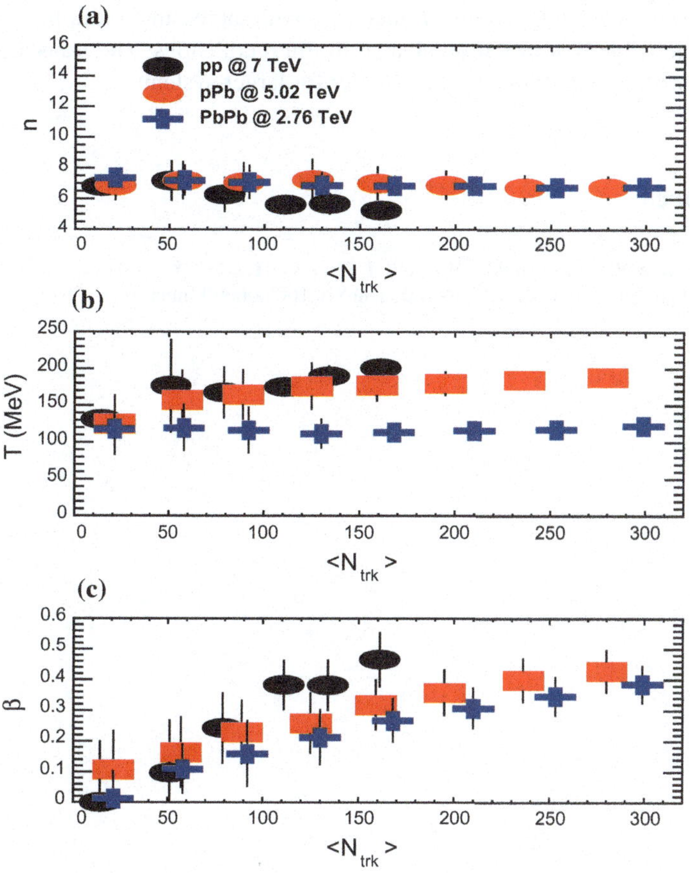

Fig. 167.2 n, T and β for K_s^0 as a function of $< N_{\text{trk}} >$ in pp, pPb and PbPb collisions

PbPb collision at $\sqrt{s_{\text{NN}}} = 2.76$ TeV. In the panel (a), It is found that the parameter n decreases with increasing multiplicity in pp collisions but its behaviour remains similar in other collision systems. In the panel (b), the temperature T increases with increasing multiplicity in pp and pPb collisions but it varies slowly in PbPb collisions. In the panel (c), The transverse flow β increases with increasing multiplicity for all systems.

167.4 Conclusion

The analysis of p_{T} spectra of K_s^0 for different multiplicity classes in pp, pPb and PbPb collisions is performed using the TSallis distribution which includes the transverse flow. The parameter n remains similar for different multiplicity classes in all systems. The Tsallis temperature T increases with increasing multiplicity in pp and pPb collisions but it shows little change in PbPb collisions. The transverse flow β increases with multiplicity and it is smaller for larger systems.

References

1. P.K. Khandai, P. Sett, P. Shukla, V. Singh, J. Phys. G **41**, 025105 (2014)
2. V. Khachatryan et al., CMS Collaboration, arXiv:1605.06699 [nucl-ex]

Chapter 168
Study of $B_c^+ \to D^+\nu\bar{\nu}$ Decay in the Light-Cone Quark Model

Nisha Dhiman and Harleen Dahiya

168.1 Introduction

The study of an exclusive semileptonic rare $B_c^+ \to D^+\nu\bar{\nu}$ decay is prominent among all the B_c meson decay modes as it plays a significant role for precision tests of the flavor sector in the SM and its possible new physics extensions. At quark level, the decay $B_c^+ \to D^+\nu\bar{\nu}$ proceed via $b \to d$ FCNC transition with the intermediate u, c and t quarks and most of the contribution comes from the intermediate t quark. Also, due to the neutral and massless final states ($\nu\bar{\nu}$), it opens an unique opportunity to study the Z penguin effects [1]. The theoretical analysis of CP violating effects in rare semileptonic decays requires knowledge of the transition form factors that are model dependent quantities and are scalar functions of the square of momentum transfer. As a theoretical input, hadronic matrix elements of quark currents will be required to calculate the transition form factors [2] in order to study the decay rates and branching ratios of the decay process. In the present work, we have evaluated the form factors for $B_c^+ \to D^+\nu\bar{\nu}$ decay in the light-cone framework.

168.2 Form Factors for $B_c^+ \to D^+\nu\bar{\nu}$ Decay in LCQM

The form factor $f_+(q^2)$ and $f_T(q^2)$ can be obtained in $q^+ = 0$ frame with the "good" component of current, i.e. $\mu = +$, from the hadronic matrix elements [2]. The form factors $f_+(q^2)$ and $f_T(q^2)$ can be expressed in explicit form as [3]

N. Dhiman (✉) · H. Dahiya
Department of Physics, Dr. B. R. Ambedkar National Institute of Technology, Jalandhar 144011, India
e-mail: nishdhiman1292@gmail.com

Md. Naimuddin (ed.), *XXII DAE High Energy Physics Symposium*, Springer Proceedings in Physics 203, https://doi.org/10.1007/978-3-319-73171-1_168

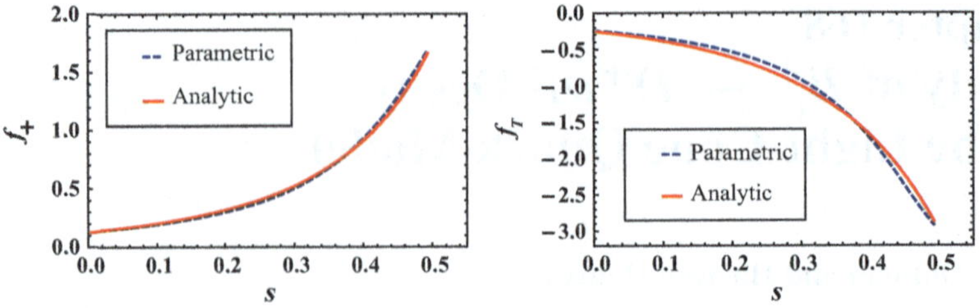

Fig. 168.1 Analytic solutions of f_+ and f_T (thick solid curve) compared with the parametric results (dashed curve), with definition $s = q^2/M^2_{B_c^+}$

$$f_+(q^2) = \int_0^1 dx \int d^2\mathbf{k}_\perp \sqrt{\frac{\partial k'_z}{\partial x}} \sqrt{\frac{\partial k_z}{\partial x}} \phi_d(x, \mathbf{k}'_\perp)\phi_b(x, \mathbf{k}_\perp) \frac{A_b A_d + \mathbf{k}_\perp \cdot \mathbf{k}'_\perp}{\sqrt{A_b^2 + \mathbf{k}_\perp^2}\sqrt{A_d^2 + \mathbf{k}'^2_\perp}},$$

and

$$f_T(q^2) = -\int_0^1 dx \int d^2\mathbf{k}_\perp \sqrt{\frac{\partial k'_z}{\partial x}} \sqrt{\frac{\partial k_z}{\partial x}} \phi_d(x, \mathbf{k}'_\perp)\phi_b(x, \mathbf{k}_\perp) \times \frac{x(M_{B_c^+} + M_{D^+})\left[(m_d - m_b)\frac{\mathbf{k}_\perp\cdot\mathbf{q}_\perp}{\mathbf{q}_\perp^2} + A_b\right]}{\sqrt{A_b^2 + \mathbf{k}_\perp^2}\sqrt{A_d^2 + \mathbf{k}'^2_\perp}}.$$

We choose a double pole parametric form to compare our analytic solutions [4]

$$f(q^2) = \frac{f(0)}{1 + \mathcal{A}\, s + \mathcal{B}\, s^2}.$$

The constituent quark masses and the scale parameter β (in units of GeV) that we have used in our work are given as

$$m_b = 4.8, m_d = 0.25, m_c = 1.4, \beta_{B_c^+} = 0.81 \text{ and } \beta_{D^+} = 0.46.$$

Using these parameters, we present the analytic solutions of the form factors f_+ and f_T (thick solid curve) for $0 \leq q^2 \leq (M_{B_c^+} - M_{D^+})^2$ in the Fig. 168.1.

References

1. H.-M. Choi, Light-front quark model analysis of the exclusive rare $B_c \to D_{(s)}(\ell^+\ell^-, \nu_\ell\bar{\nu}_\ell)$ decays. Phys. Rev. D **81**, 054003 (2010). https://doi.org/10.1103/PhysRevD.81.054003
2. C. -D. Lu, W. Wang, Z.-T. Wei, Heavy-to-light form factors on the light cone. Phys. Rev. D **76**, 014013 (2007). https://doi.org/10.1103/PhysRevD.76.014013

3. T. Wang, T. Liu, D. Zhang, B.-Q. Ma, B_c Meson rare decays in the light-cone Quark Model. Eur. Phys. J. C **71**, 1758 (2011). https://doi.org/10.1140/epjc/s10052-011-1758-3
4. C.Q. Geng, C.W. Hwang, C.C. Liu, Study of Rare $B_c^+ \to D_{d,s}^{(*)+} l\,\bar{l}$ Decays. Phys. Rev. D **65**, 094037 (2002). https://doi.org/10.1103/PhysRevD.65.094037

Chapter 169
Event Selection and Background Estimation for Decay $B^0 \to J/\psi\gamma$

Rajeev Kumar

169.1 Introduction

The decay $B^0 \to J/\psi\gamma$ is very rare decay with a theoretical predicted branching fracton of 5.398×10^{-8} in naive factorization hypothesis [1] and is 2.435×10^{-9} in the framework of QCD factorization at the order of α_s [2]. The decay is also sensitive to New Physics (NP), for example right handed currents [2] or non-spectator charm in the B^0 meson [3]. An upper limit of $1.6 \times 10^{-6}(1.5 \times 10^{-6})$ on the branching fraction was set at 90% confidence level by the BaBar collaboration (LHCb collaboration) [4, 5].

169.2 Reconstruction of J/ψ, γ And B-Meson

The identification of the leptons is very vital for the reconstruction of J/ψ. The J/ψ mesons are reconstructed from the decay mode $J/\psi \to \ell^+\ell^-$, where ℓ is muon or electron. A charged track is said to be as muon track if its muon likelihood ratio $\mathcal{R}(\mu)$ is greater than 0.1, where $\mathcal{R}(\mu) = \mathcal{L}_\mu/(\mathcal{L}_\mu + \mathcal{L}_\pi + \mathcal{L}_K)$. The likelihoods $\mathcal{L}_\mu$, $\mathcal{L}_\pi$ and $\mathcal{L}_K$ for a track to be identified muon, pion and kaon, respectively are calculated using the method explained in reference [6]. The electron tracks are identified by combining dE/dx from the CDC, E/p and shower shape in the ECL. The four-momenta of soft photons within 50 milliradian of the e^+ or e^- directions are included in the invariant mass calculation of J/ψ to recover di-electron events which are radiating a photon. The invariant mass

Rajeev Kumar—(Belle collaboration).

R. Kumar (✉)
Punjab Agricultural University, Ludhiana 141004, Punjab, India
e-mail: rajeevsharma@pau.edu

Md. Naimuddin (ed.), *XXII DAE High Energy Physics Symposium*, Springer Proceedings in Physics 203, https://doi.org/10.1007/978-3-319-73171-1_169

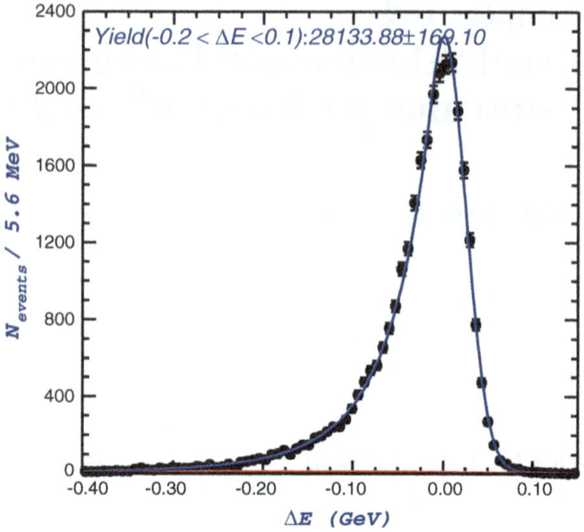

Fig. 169.1 The ΔE distribution from signal data

windows are -0.07 (-0.14) GeV$/c^2$ $\leq M_{\ell^+\ell^-} - M_{J/\psi} \leq 0.040$ GeV$/c^2$ to select J/ψ candidates in the $\mu^+\mu^-(e^+e^-)$ channels where $M_{J/\psi}$ denotes the world average of J/ψ mass [7]. Photon candidates are selected as ECL energy clusters which are not associated with a charged track. The photon candidate is rejected if the energy ratio in the central array of 3×3 ECL cells to that of 5×5 cells (E_9/E_{25}) is less than 0.95. A $\pi^0 \to \gamma\gamma$ $(\eta \to \gamma\gamma)$ veto is applied to reject the photons when combined with another and satisfy 0.110 GeV$/c^2$ $\leq M_{\gamma\gamma} \leq$ 0.150 GeV$/c^2$ (0.500 GeV$/c^2$ $\leq M_{\gamma\gamma} \leq$ 0.575 GeV$/c^2$). The B-candidate is reconstructed by combining a selected photon with a J/ψ candidate. The $\Delta E \equiv E_B^* - E_{\text{beam}}^*$ called, energy difference, is used to separate signal from background, where E_{beam}^* is the run dependent beam energy and E_B^* is the reconstructed energy. The 100,000 signal Monte Carlo (MC) $B\bar{B}$ events generated using EvtGen and the response of Belle detector is simulated by using GEANT4 software package [8], which accommodates the geometry of each detector component. The fitted ΔE distribution is shown in Fig. 169.1, where signal shape is parametrized by CBline shaped function and combinatorial background by Chebyshev polynomial of order 2. The obtained reconstruction efficiency is $(28.1 \pm 0.2)\%$. The background in this analysis include background from decays of B and from continuum quark production $(e^+e^- \to q\bar{q})$. The Belle simulated sample contains all these types of the background. The comparative studies of both on-resonance simulated continuum background and off-resonance experimental data indicate that there is negligible fraction of these events satisfying the selection criteria. The background from hadronic decays of B mesons is estimated by using a large MC sample having size 100 times that of data. The asymmetric ΔE signal region ensures that majority of these background events lies outside the signal region. The major fraction of the background in the signal region is primarily due to $B^0 \to J/\psi\pi^0(\eta)$ where a photon from $\pi^0(\eta) \to \gamma\gamma$ is misidentified as a $B^0 \to J/\psi\gamma$ photon. Also, the background from $B^0 \to J/\psi K_L$ decays have peaking structure in the signal ΔE region because $K_L \to 3\pi^0$ in the electromagnetic calorimeter and mimic as a shower from a single photon. The obtained background estimates of 3.7(0.6) events for $B^0 \to J/\psi\pi^0(\eta)$ mode and 0.8 events for $B^0 \to J/\psi K_L$ mode, which result in a total background of $N_{\text{bkg}} = 4.5 \pm 1.1$ events in the signal region. The peaking shape from $B^0 \to J/\psi\pi^0(\eta)$ is modelled by sum of two bifurcated gaussian and that of $B^0 \to J/\psi K_L(K_S)$ by that of bifurcated gaussian. The remaining components of the background have flat structure in the signal ΔE region.

On summarizing, the Belle has started the analysis $B^0 \to J/\psi\gamma$ using 771 million $B\bar{B}$ events. The obtained reconstruction efficiency is $(28.1 \pm 0.2)\%$ and background parametrization has been completed. Using this large data sample, we are expecting a first evidence or better upper limit on this analysis from the previous measurements by BaBar and LHCb [4, 5].

References

1. M. Neubert, B. Stech, Non-leptonic weak decays of b mesons, in *Heavy Flavours II*, vol. 15, ed. by A. Buras, M. Linder (World Scientific Publishing, Singapore, 1988), pp. 294–344
2. Y. Yang, L. Gongru, R. Wang, The rare annihilation decays $\bar{b}^0_{s,d} \to j/\psi\gamma$. Eur. Phys. J. C **34**, 291–296 (2004). https://doi.org/10.1140/epjc/s2004-01711-y
3. S. Brodsky, S. Gardner, Evading the ckm hierarchy: intrinsic charm in b decays. Phys. Rev. D **65**, 054016 (2002). https://doi.org/10.1103/PhysRevD.65.054016
4. B. Aubert, R. Barate et al., Search for the decay $b^0 \to j/\psi\gamma$. Phys. Rev. D **70**, 091104 (2004). https://doi.org/10.1103/PhysRevD.70.091104
5. R. Aaij, C. Beteta et al., Search for rare decays $b^0 \to j/\psi\gamma$ and $b^0_s \to j/\psi\gamma$. Phys. Rev. D **92**, 112,002 (2015). https://doi.org/10.1103/PhysRevD.92.112002
6. A. Abashian, K. Abe et al., Muon identification in the belle experiment at kekb. Nucl. Instrum. Meth. A **491**, 69–82 (2002)
7. C. Partignani et al., The review of particle physics. Chin. Phys. C **40**, 100,001 (2016). https://doi.org/10.1088/1674-1137/40/10/100001
8. S. Agostinelli, J. Allison et al., Geant4 -a simulation toolkit. Nucl. Instrum. Meth. A **506**, 250–303 (2003). https://doi.org/10.1016/S0168-9002(03)01368-8

Chapter 170
Bounds on Sterile Neutrino Component in the Solar Neutrino Flux

Govind Singh, Ashish Sharma, Gazal Sharma, Shankita Bhardwaj, Surender Verma and B. C. Chauhan

170.1 Introduction

The Sun glows due to fusion reactions such as hydrogen burning occuring in the core of the Sun. These fusion reactions generate enormous flux of neutrinos. The hunt for solar neutrinos began with the Homestake Solar neutrino experiment during 1960s. At present, we have solar neutrino data from numerous neutrino experiments. In our current work, we are using solar neutrino data from Salt phase of Sudbury Neutrino Observatory (SNO-II), Neutral Current Detector (NCDs) phase of SNO, also known as SNO-III phase, SuperK-II, SuperK-III, Borexino and KamLAND [1, 2]. In this paper, we have done model independent analysis of the available solar neutrino data from above experiments and derived the constraints on the sterile neutrinos, which might be present in the solar neutrino flux.

G. Singh · A. Sharma · G. Sharma · S. Bhardwaj · S. Verma · B. C. Chauhan (✉)
Department of Physics and Astronomical Science, School of Physical and Material Sciences, Central University of Himachal Pradesh (CUHP), Dharamshala, Kangra 176215, Himachal Pradesh, India
e-mail: chauhan@associate.iucaa.in

G. Singh
e-mail: gsgovind22@gmail.com

A. Sharma
e-mail: sharma.ashish2615@gmail.com

G. Sharma
e-mail: gazzal.sharma555@gmail.com

S. Bhardwaj
e-mail: shankita.bhardwaj982@gmail.com

S. Verma
e-mail: s_7verma@yahoo.co.in

Md. Naimuddin (ed.), *XXII DAE High Energy Physics Symposium*, Springer Proceedings in Physics 203, https://doi.org/10.1007/978-3-319-73171-1_170

170.2 Theory of Model Independent Analysis

The model independent equations in terms of solar neutrino flux, neglecting electronic antineutrino component [3, 4], are given as

$$\phi^{CC} = \phi_{\nu_e}, \qquad \phi^{NC} = \phi_{\nu_e} + \phi_{\nu_x} + \bar{r}_d \phi_{\bar{\nu}_x}, \qquad \phi^{ES} = \phi_{\nu_e} + r\phi_{\nu_x} + \bar{r}_x \phi_{\bar{\nu}_x}. \tag{170.1}$$

The parameters $r, \bar{r}_x$ are the ratios of the NC neutrino and non-electronic antineutrino event rates to the NC+CC neutrino event rate, respectively. However, $\bar{r}_d$ is the ratio of the antineutrino deuteron fission event rate to neutrino deuteron fission event rate. Here the subscript 'x' in ν_x and $\bar{\nu}_x$ represents the non-electronic (μ/τ) components.
The active neutrino $(\nu_e + \nu_x + \bar{\nu}_x)$ and sterile neutrino $(\nu_{sterile})$ fluxes are given by

$$\phi_{active} = \frac{[(r - \bar{r}_x)\phi^{NC} + (1 - \bar{r}_d)((1 - r)\phi^{CC} - \phi^{ES})]}{r\bar{r}_d - \bar{r}_x}, \tag{170.2}$$

We get the flux for sterile neutrinos by subtracting the active neutrino flux from the SSM predictions

$$\phi_{sterile} = \phi^{B}_{SSM} - \phi_{active} \tag{170.3}$$

The fraction of active neutrinos present in the solar neutrino flux can be estimated by

$$sin^2\alpha = \frac{\phi_{active} - \phi^{CC}}{\phi_{SSM} - \phi^{CC}}, \tag{170.4}$$

By model-independent way, Elastic Scattering (ES) rate for Borexino and KamLAND experiment for no-antineutrino component $(sin^2\psi = 1)$ is given as

$$R_{Bor/KL} = P^{M}_{ee} + (1 - P^{M}_{ee})r sin^2\alpha, \tag{170.5}$$

where P^{M}_{ee} is the survival probability for medium energy neutrinos and r is the cross-sectional ratio for Borexino and KamLAND Solar phase.

170.3 Results and Discussion

We have used SSM predictions for ^{8}B solar neutrino flux, ϕ_{SSM}= 5.88±0.65 × 10^6cm^{-2} s^{-1} [5].

Table 170.1 Limits on active and sterile neutrino fluxes for Case-I and Case-II in units of $10^6 cm^{-2} s^{-1}$

	Case-I			Case-II		
	SNO-II	SK-II	SK-II	SNO-III	SK-II	SK-III
ϕ_{active}	4.64 ± 0.71	4.59 ± 0.62	4.69 ± 0.57	6.31 ± 0.74	5.34 ± 0.67	5.44 ± 0.62
$\phi_{sterile}$	1.24 ± 0.96	1.29 ± 0.90	1.19 ± 0.87	−0.43 ± 0.18	0.54 ± 0.93	0.44 ± 0.90
$sin^2\alpha$	0.71 ± 0.15	0.69 ± 0.13	0.72 ± 0.13	1.10 ± 0.20	0.87 ± 0.16	0.90 ± 0.16

Table 170.2 Cross sectional ratios

	r	$\bar{r}_x$	$\bar{r}_d$
SNO-II	0.150	0.115	0.954
SNO-III	0.151	0.116	0.955
SK-II	0.149	0.114	–
SK-III	0.151	0.116	–
Bor	0.213	0.181	–
KL	0.210	–	–

Table 170.3 Bounds on R_{Bor} for Case-I and Case-II

	R_{Bor}(sterile)	R_{Bor}(no-sterile)
	Case-I	
SNO-II	0.524 ± 0.18	0.561 ± 0.17
SK-II	0.523 ± 0.18	0.561 ± 0.17
SK-III	0.527 ± 0.18	0.561 ± 0.17
	Case-II	
SNO-III	–	–
SK-II	0.552 ± 0.18	0.565 ± 0.17
SK-III	0.554 ± 0.18	0.565 ± 0.17

The second row of Table 170.1 represents the constraints on $\phi_{sterile}$. In Case-I the bounds on sterile flux at 1 σ range is $\phi_{sterile} \leq 2.20 \times 10^6 cm^{-2} s^{-1}$, $\phi_{sterile} \leq 2.19 \times 10^6 cm^{-2} s^{-1}$ and $\phi_{sterile} \leq 2.06 \times 10^6 cm^{-2} s^{-1}$ for SNO-II with SK-II and SK-III, respectively which is more constrained as predicted by the work in literature [4], which is $\phi_{sterile} \leq 2.24 \times 10^6 cm^{-2} s^{-1}$ at 1 σ. For Case-II, the bounds on sterile flux at 1 σ range is $\phi_{sterile} \leq -0.25 \times 10^6 cm^{-2} s^{-1}$, $\phi_{sterile} \leq 1.47 \times 10^6 cm^{-2} s^{-1}$ and $\phi_{sterile} \leq 1.34 \times 10^6 cm^{-2} s^{-1}$ for SNO-III with SK-II and SK-III, respectively. There is also a probability of no-sterile solar neutrino flux in the lower side of 1 σ band for this case (Tables 170.2, 170.3 and 170.4).

The ES rates obtained for Borexino and KamLAND are:

Table 170.4 Bounds on R_{KL} for Case-I and Case-II

	R_{KL}(sterile)	R_{KL}(no-sterile)
	Case-I	
SNO-II	0.523 ± 0.18	0.556 ± 0.17
SK-II	0.522 ± 0.18	0.556 ± 0.17
SK-III	0.524 ± 0.18	0.556 ± 0.17
	Case-II	
SNO-III	–	–
SK-II	0.550 ± 0.18	0.564 ± 0.17
SK-III	0.551 ± 0.18	0.564 ± 0.17

170.4 Conclusion

In this work we have obtained, in a model independent way, the active solar neutrino flux bounds for the sub-cases of Case-I (SK-II and SK-III ES). This clearly indicates the presence of sterile neutrino component in solar neutrino flux data. The sub-cases of Case-II (SK-II and SK-III ES) show that there may or may not be presence of sterile component in solar neutrino data. The upper bounds on $\phi_{sterile}$ are more constrained than previously existing in literature. The constraints on sterile neutrino component in KamLAND and Borexino are also presented.

Acknowledgements B.C. Chauhan and all the other authors acknowledge the financial support provided by University Grant Commission (UGC), Government of India vide Grant No. MRP-MAJOR-PHYS-2013-12281 and CUHP.

References

1. B. Aharmim et al., (SNO Collaboration), Phys. Rev. Lett. **101**, 111301 (2008)
2. J.P. Cravens et al., (Super-Kamiokande Collaboration), Phys. Rev. D **78**, 032002 (2008); K. Abe et al., (Super-Kamiokande Collaboration), Phys. Rev. D **83**(5), 052010 (2011); G. Bellini et al., (Borexino Collaboration), Phys. Rev. D **82**, 033006 (2010); S. Abe et al., (Borexino Collaboration), Phys. Rev. D **82**, 033006 (2010); S. Abe et al., (KamLAND Collaboration), Phys. Rev. Lett. **100**, 221803 (2008)
3. B.C. Chauhan, Joao Pulido, JHEP **12**, 040 (2004)
4. S. Dev, S. Kumar, S. Verma, Modern Phys. Lett. A **21**, 1761 (2006)
5. Aldo M. Serenalli, Astrophys. Space Sci. **328**, 13 (2010)

Chapter 171
Study of Bulk Properties in Cu+Au, Cu+Cu and Au+Au Collision at $\sqrt{s_{NN}} = 200$ GeV Using AMPT and UrQMD Models

Sumit Kumar, Lokesh Kumar and Natasha Sharma

171.1 Introduction

One of the main aim of high energy heavy-ion collisions is to explore the QCD phase diagram, search for possible QCD critical point and phase boundary. To explore the phase diagram we need to obtain two axis-temperature and baryon-chemical potential [1]. These quantities are obtained by measuring particle yields and ratios. We have studied particle yields and ratios for Au+Au, Cu+Cu and Cu+Au collisions at $\sqrt{s_{NN}} = 200$ GeV using the AMPT [2] and UrQMD [3] models. The models results are compared with the available experimental data.

171.2 Results and Discussions

171.2.1 Energy Dependence of Yields and Ratios

The experimental data [4, 5] as well as both models (AMPT and UrQMD) show increase in π^+, π^-, K^+, K^-, $\bar{p}$ yield and decrease in proton yield with the increasing energy for Au+Au and Cu+Cu collisions. The experimental yields of $\pi^{\pm}$, $K^{\pm}$, p and $\bar{p}$ is higher for Au+Au collisions than Cu+Cu at $\sqrt{s_{NN}} = 200$ GeV. The AMPT and UrQMD both models predict that the yields of $\pi^{\pm}$, $K^{\pm}$, p and $\bar{p}$ is higher for Au+Au collisions than for Cu+Au than for Cu+Cu collisions at $\sqrt{s_{NN}} = 200$ GeV. Both the models explain the experimental data qualitatively. The AMPT model seems to describe particle yields better in most of the cases.

S. Kumar · L. Kumar · N. Sharma
Department of Physics, Panjab University, Chandigarh 160014, India
e-mail: sumit@rcf.rhic.bnl.gov

Md. Naimuddin (ed.), *XXII DAE High Energy Physics Symposium*, Springer Proceedings in Physics 203, https://doi.org/10.1007/978-3-319-73171-1_171

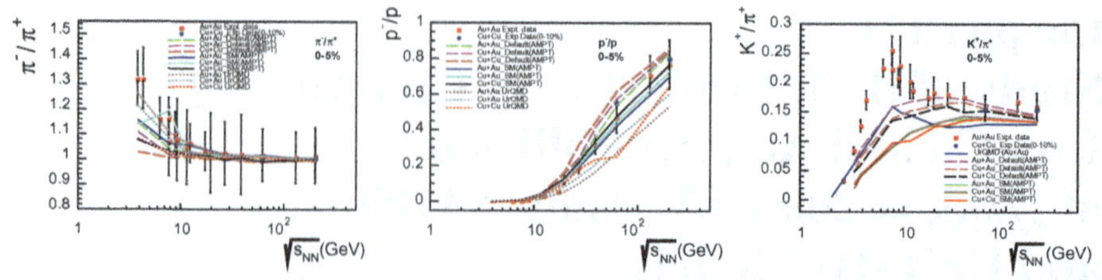

Fig. 171.1 Comparison of experimental anti-particle to particle ratios with models for different collision systems as a function of energy for 0–5% centrality class

The experimental measurements of K^-/K^+ and $\bar{p}/p$ ratios increase with increasing energy from 3.8 up to 200 GeV. However, π^-/π^+ ratio decreases with increasing energies and approaches to unity around 10 GeV and remains constant with further increase in energy. The mixed anti-particle ratios K^-/π^- and $\bar{p}/\pi^-$ show increasing trend with increasing energy. The p/π^+ ratio decrease with increase in energy then remains constant for higher energies. The K^+/π^+ ratio increases with increasing energies, show a peak/horn around 7.7 GeV and then decreases and become constant with further increase in energy.

We have compared the experimental measured ratios with both AMPT and UrQMD models. Both models qualitatively explains the ratios, the AMPT seems to describe the ratios better in most cases. However, the UrQMD model reproduce the peak/horn structure qualitatively in K^+/π^+ ratio whereas, AMPT default version shows smooth trend with no horn but explains the ratio quite well above 10 GeV. Figure 171.1 show the comparison of experimental ratios with both models for different collision systems as a function of energy for 0–5% centrality class.

171.2.2 Centrality Dependence of Yields and Ratios

The experimental yields increase from peripheral to central collisions. Both AMPT and UrQMD models qualitatively explains the measured yields and ratios. Figure 171.2 show the comparison of experimental yields with models for different collision systems at $\sqrt{s_{NN}} = 200$ GeV as a function of centrality.

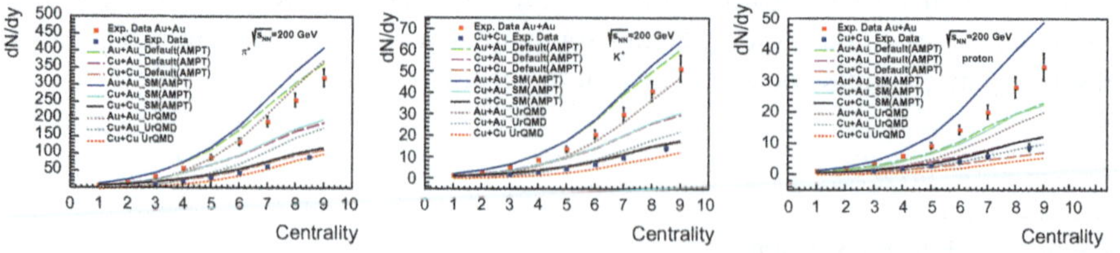

Fig. 171.2 Comparison of experimental yields with models for different collision systems at $\sqrt{s_{NN}} = 200$ GeV as a function of centrality

Acknowledgements This work is supported by DST-SERB project of LK (ECR/2016/000109). NS acknowledges the support of Ramanujan Fellowship (D.O. No. SB/S2/RJN- 084/2015).

References

1. M.M. Aggarwal et al., An Experimental Exploration of the QCD Phase Diagram: The Search for the Critical Point and the Onset of De-confinement, arXiv:1007.2613
2. Z.-W. Lin, S. Pal, C.M. Ko, B.-A. Li, B. Zhang, Charged particle rapidity distributions at relativistic energies. Phys. Rev. C **64**, 011902 (2001)
3. S.A. Bass et al., Microscopic models for ultrarelativistic heavy ion collisions. Prog. Part. Nucl. Phys. **41**, 255–369 (1998)
4. L. Kumar, Review of recent results from the RHIC beam energy scan. Mod. Phys. Lett. A **28**, 1330033 (2013)
5. L. Kumar, Systematics of kinetic freeze-out properties in high energy collisions from STAR. Nucl. Phys. A **931**, 1114–1119 (2014)

Chapter 172
Mass Spectra of Bottonia Using Linear Potential in Relativistic Frame Work

Tanvi Bhavsar, Manan Shah, Bhavin Patel and P. C. Vinodkumar

172.1 Introduction

Observations made in the last few years have generated great interest in heavy quark sector. Decays of heavy quarkonium(Charmonium and Bottomonium) with emission of light quarks are important to study new exotic states and help us to understand the quark dynamics and the light hadron formation. Bottomonium is considered to be a best tool to understand Quantum Chromodynamics (QCD) at low energies [1]. The Belle detector and KEKB e^-e^+ collider have collected data for bottomonium sector above threshold i.e. for higher excite states $\Upsilon(4S, 5S, 6S)$ [2, 3]. In this study we have computed mass spectra of bottomonium states in relativistic frame work by using linear confinement potential. This formalism has been successfully applied to describe mass spectra of the heavy-light meson and their decay properties [4].

172.2 Theoretical Formalism

We assume here that quark and anti quark inside the mesonic bound state is independently confined by a potential [4],

$$V(r) = \frac{1}{2}(1 + \gamma_0)(\lambda r + V_0)$$

T. Bhavsar (✉) · P. C. Vinodkumar
Department of Physics, Sardar Patel University,
Vallabh Vidyanagar 388120, India
e-mail: tanvibhavsar1992@yahoo.com

M. Shah
P. D. Patel Institute of Applied Sciences, Charusat, Changa 388421, India

B. Patel
20 Tuxedo Crt, M1G 3S5 Scarborough, Canada

Md. Naimuddin (ed.), *XXII DAE High Energy Physics Symposium*, Springer Proceedings in Physics 203, https://doi.org/10.1007/978-3-319-73171-1_172

Table 172.1 S-wave and P-wave mass spectrum for $(b\bar{b})$ (in MeV)

nL	State	Present	Experimental [6]	[7]	nL	State	Present	Experimental [6]	[7]
1S	1^3S_1	9460.78	9460.30 ± 0.26	9460.38	1P	1^3P_2	9902.5	9912.21 ± 0.26	9910.63
	1^1S_0	9394.36	9390.7 ± 2.9	9392.91		1^3P_1	9903.4	9892.78 ± 0.26	9891.33
						1^3P_0	9895.7	9859.44 ± 0.42	9861.39
						1^1P_1	9815.2	9900.00	9899.93
2S	2^3S_1	10023.8	10023.26 ± 0.31	10023.3	2P	2^3P_2	10227.8	10268.65 ± 0.22	10271.2
	2^1S_0	9948.39	–	9987.42		2^3P_1	10228.9	10255.46 ± 0.22	10254.8
						2^3P_0	10221.3	10232.50 ± 0.40	10230.5
						2^1P_1	10154.2	–	10261.8
3S	3^3S_1	10337.6	10355.2 ± 0.5	10364.2	3P	3^3P_2	10481.4	–	–
	3^1S_0	10272.5	–	10333.9		3^3P_1	10482.4	–	–
						3^3P_0	10475.1	–	–
						3^1P_1	10415.9	10551 ± 14 ± 17	–

where λ is the strength of the confinement part of the potential. V_0 is a constant negative potential depth. To get binding energy we have solved the two component (positive and negative energy) Dirac equation [5]. After solving these equations we express the energy eigen value as

$$\epsilon = (E_D - m_q - V_0)(m_q + E_D)^{\frac{2}{3}}\lambda^{\frac{-2}{3}}$$

The parameters are fixed to get the ground state masses of $b\bar{b}$. The confined one gluon exchange interactions used in our earlier study of open flavor meson are employed for the present study [4]. Sample results for S and P states are given in Table 172.1.

172.3 Result and Discussion

The predicted masses of bottomonia are found to be in accordance with experimental results and also with other available theoretical results. The predicted results of S-wave bottomonium states agree well up to 3S states and the result for P-wave bottomonium agree well up to 2P states.

Acknowledgements We acknowledge the financial support from DST-SERB, India (research Project number: SERB/F/8749/2015-16).

References

1. W.-J. Deng, H. Liu, L.-C. Gui, X.-H. Zhong, arXiv:1607.04696v1
2. A. Abashian et al., Nucl. Instrum. Meth. A **479**, 117 (2002)
3. S. Kurokawa, E. Kikutani, Nucl. Instrum. Meth. A **499**, 1 (2003)
4. M. Shah, B. Patel, P.C. Vinodkumar, Phys. Rev. D **93**, 0940 (2016)
5. W. Greiner, *Relativistic Quantum Mechanics Wave Equations*
6. K.A. Olive et al., (Particle data group) Chin. Phys. C **38**, 090001 (2014)
7. S.F. Radford, W.W. Repko, Nucl. Phys. A **865**, 69–75 (2011)

Chapter 173
Mass Spectra of Doubly Heavy Ξ_{bc} Baryons

Zalak Shah and Ajay Kumar Rai

173.1 Introduction and Theoretical Framework

During the last few years, the significant experimental progress (LHCb, BELLE, BABAR, FOCUS) has been developed on the spectroscopy of the heavy baryons. Although, no experimental evidences of doubly heavy Ξ_{bc} baryon is observed yet [1]. Therefore, it would be interesting to present reliable theoretical predictions of this baryon [2–5]. The Hamiltonian of three body baryonic system in terms of reduced mass and six dimensional hyperradial Schrödinger equation can be written (in terms of radial hyper central coordinate x) in the hCQM as [6, 7]

$$\left[\frac{d^2}{dx^2} + \frac{5}{x}\frac{d}{dx} - \frac{\gamma(\gamma+4)}{x^2}\right]\Psi_\gamma(x) = -2m[E - V(x)]\Psi_\gamma(x) \tag{173.1}$$

$$V(x) = V^0(x) + \left(\frac{1}{m_\rho} + \frac{1}{m_\lambda}\right) V^{(1)}(x) + V_{SD}(x) \tag{173.2}$$

$$V^{(0)}(x) = \frac{\tau}{x} + \beta x \qquad \& \qquad V^{(1)}(x) = -C_F C_A \frac{\alpha_s^2}{4x^2} \tag{173.3}$$

Z. Shah (✉) · A. Kumar Rai
Department of Applied Physics, S.V.N.I.T., Surat 395007, Gujarat, India
e-mail: zalak.physics@gmail.com

A. Kumar Rai
e-mail: raiajayk@gmail.com

Md. Naimuddin (ed.), *XXII DAE High Energy Physics Symposium*, Springer Proceedings in Physics 203, https://doi.org/10.1007/978-3-319-73171-1_173

We consider the hypercentral potential $V(x)$ as the color coulomb plus linear potential with first order correction along with the spin dependent part to calculate the mass spectra [details can be found in [7]].

173.2 Results and Discussions

The obtained mass spectra upto L = 2 (1S-3S, 1P-3P, 1D-2D) are listed as well as compared with other theoretical predictions in Table 173.1. We observe that ground state masses are in good agreement with each others [2–4] except [5]. Since the spectra are unknown experimentally, the regge trajectory ($M^2 \rightarrow n$) is plotted in Fig. 173.1 using the obtained values. The plot can be seen linear, parallel and equidistant. Using $n = \beta M^2 + \beta_0$, we can find slope (β) and intercept (β_0) values. This study will be useful to identify the resonances and also to assign the J^P values of the Ξ_{bc} baryon in future experiments.

Table 173.1 S, P and D state masses of Ξ_{bc} baryon (in GeV)

State	Our	[2]	[3]	[4]	[5]	State	Our	[5]	State	Our	[5]
$(1^2S_{1/2})$	6.920	6.919	6.943	6.904	7.014	$(1^2P_{1/2})$	7.160	7.390	$(1^4D_{3/2})$	7.326	7.706
$(1^2S_{3/2})$	6.986	6.986	6.985	6.936	7.064	$(1^2P_{3/2})$	7.149	7.394	$(1^4D_{5/2})$	7.313	
$(2^2S_{1/2})$	7.244			7.478	7.321	$(2^2P_{1/2})$	7.425		$(2^4D_{3/2})$	7.571	
$(2^2S_{3/2})$	7.267			7.495	7.353	$(2^2P_{3/2})$	7.415		$(2^4D_{5/2})$	7.559	
$(3^2S_{1/2})$	7.509				7.904	$(3^2P_{1/2})$	7.664				
$(3^2S_{3/2})$	7.521				7.917	$(3^2P_{3/2})$	7.655				

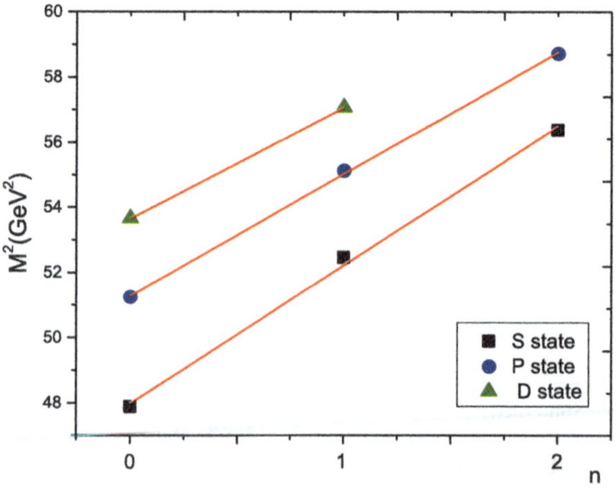

Fig. 173.1 Regge Trajectories of Ξ_{bc} baryon in (n, M^2) plane (n = principal quantum number) for S ($\frac{1}{2}^+$), P ($\frac{1}{2}^-$) and D ($\frac{5}{2}^+$) state

Acknowledgements A. K. Rai acknowledges the financial support extended by DST, India under SERB fast track scheme SR/FTP/PS-152/2012.

References

1. C. Patrignani et al., Chin. Phys. C **40**, 100001 (2016)
2. C. Albertus, E. Hernandez et al., Eur. Phys. J. A **32**, 183 (2007)
3. Z.S. Brown, W. Detmold, S. Meinel, K. Orginos, Phys. Rev. D **90**, 094507 (2014)
4. F. Giannuzzi, Phys. Rev. D **79**, 094002 (2009)
5. B. Eakins, W. Roberts, Int. J. Mod. Phys. A **27**, 1250039 (2012)
6. E. Santopinto, F. Iachello, M.M. Giannini, Eur. Phys. J. A **1**, 307–315 (1998)
7. Z. Shah, K. Thakkar, A.K. Rai, P.C. Vinodkumar, Eur. Phys. J A **52**, 513 (2016); Chin. Phys. C **40**, 123102 (2016); Eur. Phys. J. C **76**, 530 (2016)

Chapter 174
Mass and Hadronic Decay Widths of Z States as Di-meson Molecule

N. R. Soni , R. R. Chaturvedi , A. K. Rai and J. N. Pandya

After the discovery of Z_c^+ and Z_b^+ hadronic states by BESIII [1] and BELLE [2] collaborations respectively, there have been many attempts to describe these states either as tetra quark states or as hadronic molecules. The charged states and the masses of these exotic states are nearer to the threshold of $D^+\bar{D}^*$ and $B\bar{B}^*$ and respectively suggesting them to be di-mesonic molecular states. We treat them as hadronic molecules of $D^+\bar{D}^*$ mesons and $B\bar{B}^*$ mesons. We consider the interaction between the constituent mesons of the type modified Woods-Saxon plus Coulomb repulsive terms of the form [3]

$$V(r) = \frac{V_0}{1 + Exp\left[\frac{r-R_0}{a}\right]} + \frac{C\,Exp\left[\frac{r-R_0}{a}\right]}{\left(1 + Exp\left[\frac{r-R_0}{a}\right]\right)^2} - \frac{b}{r} \tag{174.1}$$

where the potential parameters are the strength of the potential $V_0 = 15\ MeV$, $b = 0.08$, size of the hadron $R_0 = 1.75$ fm, diffuseness of the surface of the molecule $a = -0.51$ fm. C determines the depth of the potential ($0 < C < 150$ MeV) [3]. For computation of binding energy, we solve the Schrödinger equation numerically for the interaction potential defined in Eq. 174.1. Binding energy and thus the di-mesonic molecular masses are given in Table 174.1.

N. R. Soni · J. N. Pandya (✉)
Applied Physics Department, Faculty of Technology and Engineering,
The Maharaja Sayajirao University of Baroda,Vadodara, Gujarat 390001, India
e-mail: jnpandya-apphy@msubaroda.ac.in

N. R. Soni
e-mail: nrsoni-apphy@msubaroda.ac.in

R. R. Chaturvedi · A. K. Rai
Applied Physics Department, Sardar Vallabhbhai National Institute of Technology,
Surat, Gujarat 395007, India
e-mail: raghavr.chaturvedi@gmail.com

A. K. Rai
e-mail: raiajayk@gmail.com

Md. Naimuddin (ed.), *XXII DAE High Energy Physics Symposium*, Springer Proceedings in Physics 203, https://doi.org/10.1007/978-3-319-73171-1_174

Table 174.1 Masses of $Z_c^+(D^+\bar{D}^*)$ and $Z_b^+(B\bar{B}^*)$ molecular states (in MeV) with the variation in potential depth C (in MeV)

C	$D^+\bar{D}^*$		$B\bar{B}^*$	
	Binding energy	Mass	Binding energy	Mass
0	11.82	3864.74	5.58	10598.9
50	11.96	3864.61	7.05	10597.4
100	12.07	3864.5	8.04	10596.4
150	12.15	3864.42	8.72	10595.7
PDG [5]		3883.9±4.5		10607.2±2.0

Table 174.2 Decay widths of Z_c^+ and Z_b^+ molecular states (in MeV)

Decay mode	Decay width					
	C = 0	C = 50	C = 100	C = 50	Exp [6]	[4]
$Z_c^+ \to \psi(1s) + \pi$	11.7202	11.7553	11.7849	11.8064	–	10.43–23.89
$Z_c^+ \to \psi(2s) + \pi$	2.1166	2.1146	2.1127	2.1114	–	1.28–2.94
$Z_b^+ \to \Upsilon(1s) + \pi$	22.8443	22.9280	22.9998	23.0567	22.9±7.3	13.3–30.8
$Z_b^+ \to \Upsilon(2s) + \pi$	26.9257	26.9858	27.0443	27.0930	21.1±4.0	15.4–35.7

We employ the method of Phenomenological Lagrangian mechanism developed by Y. Dong et al. [4] to compute hadronic decay widths. The corresponding two body decay widths for Z_c^+ and Z_b^+ can be written as [4]

$$\Gamma_{Z_c^+ \to \Psi(ns)\pi^+} \simeq \frac{g^2_{Z_c\Psi(ns)\pi}}{96\pi M^3_{Z_c}} \lambda^{3/2}(M^2_{Z_c}, M^2_{\psi(ns)}, M^2_\pi)\left(1 + \frac{M^2_{\psi(ns)}}{2M^2_{Z_c}}\right) \quad (174.2)$$

$$\Gamma_{Z_b^+ \to \Upsilon(ns)\pi} \simeq \frac{g^2_{Z_b\Upsilon(ns)\pi}}{16\pi M_{Z_b}} \lambda^{1/2}(M^2_{Z_b}, M^2_{\Upsilon(ns)}, M^2_\pi) \quad (174.3)$$

where $\lambda(x, y, z) = x^2 + y^2 + z^2 - 2xy - 2yz - 2zx$ is the Källen function, $g's$ are the coupling constant corresponds to the coupling between hadron to its constituent mesons. The computed decay widths are given in Table 174.2.

Result and Conclusion

The masses of Z_c^+ and Z_b^+ considering them as molecular states of $D^+\bar{D}^*$ and $B\bar{B}^*$ respectively are found to be below the mass of their resonance. We have analysed the nature of potential with the parameters such as diffuseness and depth of the potential. We have also computed the hadronic decay widths of these states in formalism of interaction Lagrangian mechanism and compare with the experiments. Our predictions of decay widths are in good agreement with the data from experiments.

Acknowledgements This work is supported by the University Grants Commission of India under Major Research Project F.No.42-775/2013(SR)

References

1. M. Ablikim et al., BESIII collaboration. Phys. Rev. Lett. **110**, 252001 (2013)
2. Z.Q. Liu et al., Belle collaboration. Phys. Rev. Lett. **110**, 252002 (2013)
3. H. Fakhri, J. Sadeghi, Mod. Phys. Lett. A 29, 615 (2004); C. Berkdemir, A. Berkdemir, R. Sever, Phys. Rev. C **72**, 027001 (2005)
4. Y. Dong et al., Phys. Rev. D **88**, 014030 (2013); J. Phys. G **40**, 015002 (2013)
5. C. Patrignani et al., Particle data group. Chin. Phys. C **40**, 100001 (2016)
6. I. Adachi et al., (Belle Collaboration), arXiv:hep-ex/1105.4583

Chapter 175
Implications of Fermionic Dark Matter on Recent Neutrino Oscillation Data

Shivaramakrishna Singirala

175.1 Model Description and Neutrino Phenomenonology

Models that can shed light on the origin of neutrino masses and the nature of dark matter (DM) are captivating. Scotogenic model [1] is one such model with an additional scalar doublet η and three right-handed neutrinos N_i $(i = 1, 2, 3)$. A Z_2 symmetry forbids tree level neutrino mass. The Yukawa lagrangian is [2]

$$\mathcal{L}_N = \overline{N_i} i \partial\!\!\!/ P_R N_i + \left(D_\mu \eta\right)^\dagger (D^\mu \eta) - \frac{M_i}{2} \overline{N_i{}^c} P_R N_i + h_{\alpha i} \overline{\ell_\alpha} \eta^\dagger P_R N_i + \text{h.c.} \tag{175.1}$$

In this model, neutrinos get their mass at loop level given by

$$(\mathcal{M}_\nu)_{\alpha\beta} = \sum_{i=1}^{3} h_{\alpha i} h_{\beta i} \Lambda_i, \quad \Lambda_i = \frac{\lambda_5 v^2}{8\pi^2 M_i} \left[\frac{r_i{}^2}{1 - r_i{}^2} \left(1 + \frac{r_i{}^2}{1 - r_i{}^2} \ln r_i{}^2 \right) \right]. \tag{175.2}$$

Here $r_i = \frac{M_i}{m_0}$, $m_0{}^2 = \frac{(m_R{}^2 + m_I{}^2)}{2}$. Various experiments confirmed that neutrinos oscillate in flavor as they propagate. We consider a mixing matrix of tri-bimaximal (TBM) form with a perturbation to obtain non-zero value for θ_{13} [3].

$$U_{PMNS} = \begin{pmatrix} \cos\theta & \sin\theta & 0 \\ -\frac{\sin\theta}{\sqrt{2}} & \frac{\cos\theta}{\sqrt{2}} & \frac{1}{\sqrt{2}} \\ \frac{\sin\theta}{\sqrt{2}} & -\frac{\cos\theta}{\sqrt{2}} & \frac{1}{\sqrt{2}} \end{pmatrix} U_\nu^{\text{corr}} \quad \text{where } U_\nu^{\text{corr}} = \begin{pmatrix} \cos\varphi & 0 & e^{-i\zeta}\sin\varphi \\ 0 & 1 & 0 \\ -e^{i\zeta}\sin\varphi & 0 & \cos\varphi \end{pmatrix}, \tag{175.3}$$

where $\theta \simeq 35^\circ$ and $\varphi = 12^\circ$ [4]. Now, diagonalizing the mass matrix (175.2) gives the flavor structure written in terms of $h_{ei} (= h_i)$, and the mass eigenstates as

S. Singirala (✉)
School of Physics, University of Hyderabad, Hyderabad 500046, India
e-mail: krishnas542@gmail.com

Md. Naimuddin (ed.), *XXII DAE High Energy Physics Symposium*, Springer Proceedings in Physics 203, https://doi.org/10.1007/978-3-319-73171-1_175

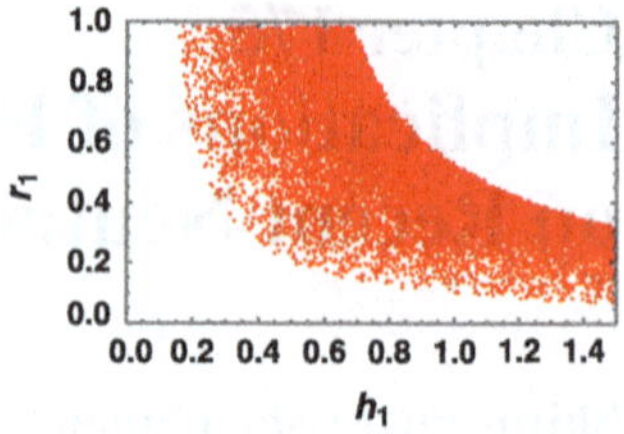

Fig. 175.1 h_1 v/s r_1 region consistent with current bound on Ωh^2

$$h_{\alpha i} = \begin{pmatrix} h_1 & h_2 & h_3 \\ -0.68\, h_1 & h_2 & 3.56\, h_3 \\ 0.31\, h_1 & -h_2 & 4.55\, h_3 \end{pmatrix}, \quad \begin{aligned} m_1 &= 1.55\,(h_1^2 \Lambda_1), \\ m_2 &= 3.04\,(h_2^2 \Lambda_1), \\ m_3 &= 34.44\,(h_3^2 \Lambda_3). \end{aligned} \tag{175.4}$$

We have considered N_1, N_2 are degenerate in mass. Now we constrain the parameter space $h_{1,2,3}$, $r_{1,3}$, m_0 with DM and lepton flavor violating (LFV) decays.

175.2 Relic Abundance

We compute relic abundance of fermionic DM N_1 to constrain h_1 and r_1 (Fig. 175.1). The effective annihilation cross-section including co-annihilation effects is defined as $\sigma_{\rm eff}|v_{\rm rel}| = a_{\rm eff} + b_{\rm eff} v_{\rm rel}^2$, where $a_{\rm eff}$ and $b_{\rm eff}$ are [5]

$$\begin{aligned} a_{\rm eff} &= \frac{1}{16\pi} \frac{r_1^2}{m_0{}^2 (r_1^2+1)^2} (s_{12} h_1^2 h_2^2), \\ b_{\rm eff} &= \frac{1}{48\pi} \frac{r_1^2}{m_0{}^2 (r_1^2+1)^4} [(r_1^4+1)(s_1 h_1^4 + s_2 h_2^4) \\ &\quad + (1 - 3r_1^2 - r_1^4)(s_{12}\, h_1^2 h_2^2)], \\ \text{where } s_1 &= 2.42,\ s_2 = 9.24 \text{ and } s_{12} = 9.47. \end{aligned}$$

175.3 Lepton Flavour Violating Decays

The evidence of neutrino oscillations and the violation of family lepton number could in principle allow flavor changing neutral current (FCNC) transitions in the charged lepton sector as well, such as $\ell_\alpha \to \ell_\beta \gamma$ and $\ell_\alpha \to \ell_\beta \overline{\ell_\beta} \ell_\beta$. The branching ratios are given as [5, 6]

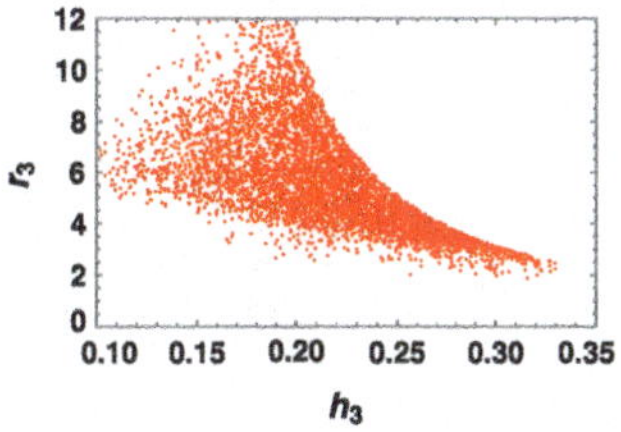

Fig. 175.2 h_3 v/s r_3

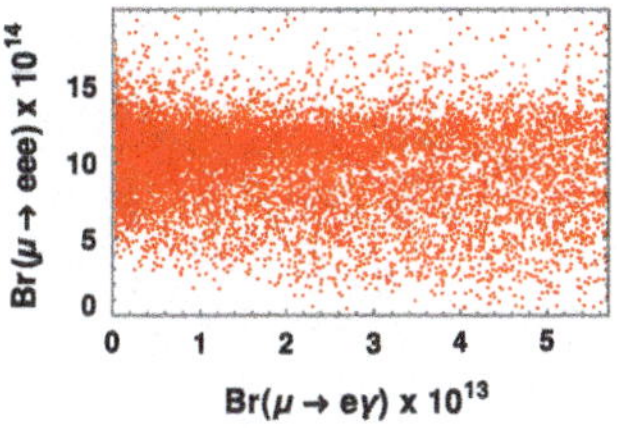

Fig. 175.3 Br($\mu \to e\gamma$) v/s Br($\mu \to eee$)

$$\mathrm{Br}(\ell_\alpha \to \ell_\beta \gamma) = \frac{3(4\pi)^3 \alpha_{\mathrm{em}}}{4G_F^2} |A_D|^2 \mathrm{Br}\left(\ell_\alpha \to \ell_\beta \nu_\alpha \overline{\nu_\beta}\right),$$

$$\frac{\mathrm{Br}\left(\ell_\alpha \to \ell_\beta \overline{\ell_\beta} \ell_\beta\right)}{\mathrm{Br}\left(\ell_\alpha \to \ell_\beta \nu_\alpha \overline{\nu_\beta}\right)} = \frac{3(4\pi)^2 \alpha_{\mathrm{em}}^2}{8G_F^2} \left[|A_D|^2 \left(\frac{16}{3} \log\left(\frac{m_\alpha}{m_\beta}\right) - \frac{22}{3} \right) \right.$$
$$\left. |A_{ND}|^2 + \frac{1}{6}|B|^2 + \left(-2A_{ND}A_D^* + \frac{1}{3}A_{ND}B^* - \frac{2}{3}A_D B^* + \mathrm{h.c.} \right) \right].$$

Here α_{em}, G_F denote fine structure and Fermi constants, dipole A_D, non-dipole A_{ND} and the box B contributions are given in [5].
$\mu \to e\gamma$ channel gives a stringent constraint on h_3, r_3 parameter space (Fig. 175.2). Figure 175.3 shows the correlation plot between $\mu \to e\gamma$ and $\mu \to eee$.

To conclude, we have investigated the neutrino radiative mass matrix using a mixing matrix of TBM type with a perturbation matrix to invoke non-zero θ_{13}. We have shown the allowed region of parameters in the model consistent with neutrino oscillation data, relic abundance of fermionic DM and LFV decays.

I would like to thank Prof. Rukmani Mohanta for helpful suggestions and DST-Inspire Fellowship division - IF130927 for the financial support.

References

1. E. Ma, Phys. Rev. D **73**, 077301 (2006). arXiv:hep-ph/0601225
2. D. Suematsu, T. Toma, T. Yoshida, Phys. Rev. D **79**, 093004 (2009)
3. F.P. An et al., DAYA-BAY collaboration. Phys. Rev. Lett. **108**, 171803 (2012)
4. M. Sruthilaya, C. Soumya, K.N. Deepthi et al., New J. Phys. **17**, 083028 (2015)
5. S. Singirala, Chin. Phys. C **41**, 043102 (2017). arXiv:1607.03309
6. T. Toma, A. Vicente, JHEP **01**, 160 (2014). arXiv:1312.2840

Chapter 176
Study of D^* Polarization to Discriminate New Physics in $\bar{B} \to D^* \tau \bar{\nu}$

Suman Kumbhakar, Ashutosh Kumar Alok, Dinesh Kumar and S. Uma Sankar

176.1 Introduction

In recent years the most striking anomaly is the deviation of 3.5σ from the SM expectation of the ratio $R_{D^*} = \Gamma(B \to D^* \tau \bar{\nu})/\Gamma(B \to D^* l \bar{\nu})$ $(l = e,\ \mu)$[1, 2]. The four fermion interaction $b \to c\, \tau\, \bar{\nu}$, which induces the decays $B \to (D,\ D^*)\, \tau \bar{\nu}$, occurs at the tree level within the SM. To distinguish various NP operators which can explain the excess in R_{D^*}, we study various angular observables in $B \to D^*\, \tau \bar{\nu}$. In this semileptonic decay the differential decay distribution can be written with respect to three angles - (a) θ_D, the angle between B and D where the D meson comes from D^* decay, (b) θ_τ, the angle between τ and B and (c) ϕ, the angle between D^* decay plane and the plane defined by the lepton momenta [3]. In the present data which measured R_{D^*}, it is possible to measure θ_D but not θ_τ or ϕ. We study the capability of $f_L(q^2)$ to discriminate between various NP solutions which account for R_{D^*} excess. We find that this quantity provides good discrimination when the NP operators have scalar or tensor form.

S. Kumbhakar (✉) · D. Kumar · S. Uma Sankar
Indian Institute of Technology, Bombay, India
e-mail: suman@phy.iitb.ac.in

A. K. Alok
Indian Institute of Technology, Jodhpur, India

Md. Naimuddin (ed.), *XXII DAE High Energy Physics Symposium*, Springer Proceedings in Physics 203, https://doi.org/10.1007/978-3-319-73171-1_176

176.2 Discrimination of New Physics Using D^* Polarization

The effective Hamiltonian for the quark level transition $b \to c\,\tau\,\bar{\nu}$ is given by

$$H_{eff} = \frac{4G_F}{\sqrt{2}} V_{cb} \left[\mathcal{O}_{V_L} + \frac{\sqrt{2}}{4G_F V_{cb}} \frac{1}{\Lambda^2} \left\{ \sum_i \left(C_i \mathcal{O}_i + C_i^{'} \mathcal{O}_i^{'} + C_i^{''} \mathcal{O}_i^{''} \right) \right\} \right], \tag{176.1}$$

where the scale Λ is assumed to be 1 TeV. The Lorentz structures of the unprimed $\mathcal{O}_i$ and primed $\mathcal{O}_i^{'}$ and $\mathcal{O}_i^{''}$ operators are given in [4]. The values of all operator coefficients which provide a good fit to the data are also given in [4]. The solutions with operators $\mathcal{O}_{S_L}$ are excluded because the predict the leptonic partial decay width of B_c meson is larger than the measured total decay width. The solutions which obey this constraint have either tensor operators or $\mathcal{O}_{S_L}^{''}$ operators.

The D^* polarization fraction $f_L(q^2)$ is defined to be $f_L(q^2) = A_L/(A_L + A_T)$, where the quantities A_L and A_T are defined in [3]. From this formula we compute $f_L(q^2)$ and $\langle f_L(q^2) \rangle$ for the allowed NP couplings listed in [4] and display them in Fig. 176.1. We discuss the form of $f_L(q^2)$ for each plot: Only C_T present: Inspite of having a different Lorentz structure from SM, it is difficult to distinguish the small $C_T(-0.07 \pm 0.02)$ case from SM prediction. But for the large $C_T(0.52 \pm 0.02)$ case, $\langle f_L(q^2) \rangle$ is 0.14 as opposed to 0.46 for SM. $C_{S_R}^{''}$ and $C_{S_L}^{''}$ present: The solutions with large $C_{S_L}^{''}$ values have $\langle f_L(q^2) \rangle = 0.26$ which again is significantly lower than the SM prediction. So the measurement of the average D^* polarization, $\langle f_L(q^2) \rangle$, is capable of distinguishing three of the allowed twelve NP solutions. The Belle collaboration is in process of measuring $\langle f_L(q^2) \rangle$. It is expected that the uncertainty in this measurement will be about ± 0.1. Hence the above three solutions to the R_{D^*} problem can be distinguished at 95% C.L. or better.

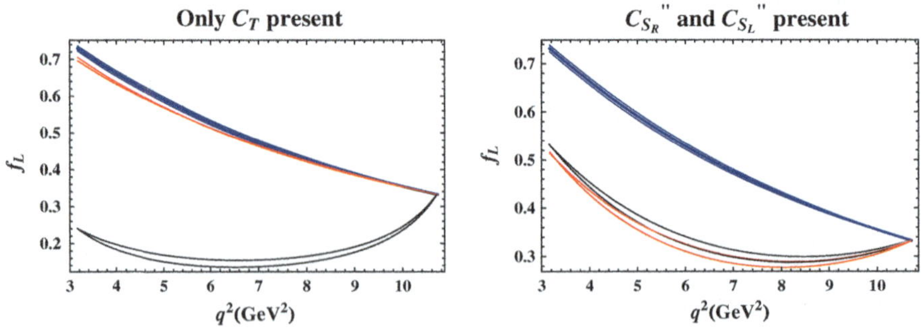

Fig. 176.1 Plots of $f_L(q^2)$ as a function of the dilepton invariant mass q^2. The blue curve represents the SM and the red and the black curves represent NP operators. The black and red bands in left panel are for $f_L(q^2)$ with NP couplings $C_T = (0.52 \pm 0.02)$ and $C_T = (-0.07 \pm 0.02)$, respectively. The right panel shows the plots of solutions with NP coefficients $(C_{S_R}^{''}, C_{S_L}^{''}) = (0.96, 2.41)$ (black band) and $(-6.34, -2.39)$ (red band)

References

1. R. Aaij et al., [LHCb Collaboration], Phys. Rev. Lett. **115**(11), 111803 (2015); Phys. Rev. Lett. **115**(15), 159901 (2015)
2. M. Huschle et al., [Belle Collaboration], Phys. Rev. D **92**(7), 072014 (2015)
3. A.K. Alok et al., JHEP **1111**, 121 (2011)
4. M. Freytsis, Z. Ligeti, J.T. Ruderman, Phys. Rev. D **92**(5), 054018 (2015)

Chapter 177
Resonance Measurements in p–Pb Collisions at $\sqrt{s_{NN}}$ = 5.02 TeV with ALICE

Sarita Sahoo

Resonance production data in pp and p–Pb collisions provides input for tuning QCD inspired event generators and serve as a baseline for results obtained in heavy-ion collisions. Resonances with lifetimes of a few fm/c may be sensitive to the hadronic phase produced either in elementary or in heavy-ion collisions. In addition, multiplicity-differential measurements of resonance production in pp and p–Pb collisions contribute to the search for the onset of collective effects.

The main components of the ALICE detector [1] used in the analyses described here are the Inner Tracking System (tracking and vertex finding), the Time Projection Chamber (tracking and particle identification), the Time-Of-Flight detector (particle identification) and the VZERO detector (event multiplicity and centrality classes definition).

Figures 177.1 and 177.2 show yield ratios of resonances to long-lived particles. The ratio K^{*0}/K shows that K^{*0} ($\tau \sim 4.16$ fm/c) [2] is significantly suppressed in central Pb–Pb collisions with respect to peripheral Pb–Pb and pp collisions. This behaviour is commonly attributed to the dominance of rescattering of its decay daughters over regeneration processes in the hadronic phase. The ϕ ($\tau \sim 46.3$ fm/c) [2], due to its longer lifetime, shows no such suppression. A similar suppression pattern for K^{*0} is observed in p–Pb collisions though not significant within uncertainties. In p–Pb collisions the K^{*0} and ϕ have higher $\langle p_T \rangle$ values than the proton and Λ [4]. As in pp collisions, this violates the mass ordering, which is observed in central Pb–Pb collisions. The Ξ^{*0}/π and $\Sigma^{*\pm}/\pi$ ratios show an increasing trend with multiplicity which approaches values predicted by the Grand Canonical thermal model [5]. The amount of increase is the same as for the ground state baryons with the same valence quarks content, thus Ξ^{*0}/Ξ^{-} and $\Sigma^{*\pm}/\Lambda$ are flat versus multiplicity. The hyperon to pion ratio increases with multiplicity and strangeness content.

Sarita Sahoo (for the ALICE collaboration).

S. Sahoo (✉)
Institute of Physics, HBNI, Bhubaneswar, India
e-mail: sarita.sahoo@cern.ch

Md. Naimuddin (ed.), *XXII DAE High Energy Physics Symposium*, Springer Proceedings in Physics 203, https://doi.org/10.1007/978-3-319-73171-1_177

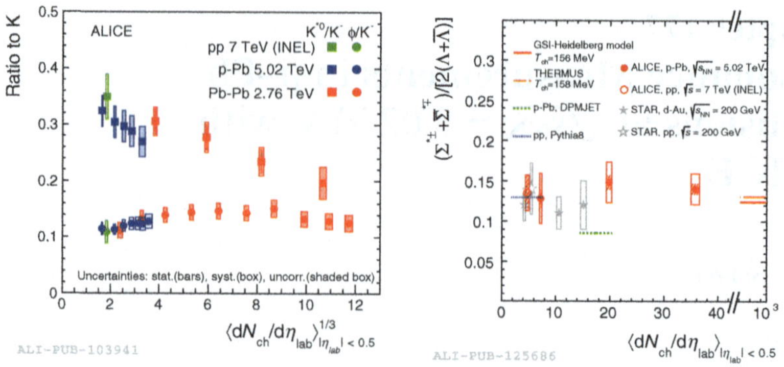

Fig. 177.1 Yield ratios of K^{*0} and ϕ to K as a function of $\langle dN_{ch}/d\eta_{lab}\rangle^{1/3}$ at mid-rapidity in pp, p–Pb and Pb–Pb collisions (left panel) [4]. Yield ratios of $\Sigma^{*\pm}$ to Λ as a function of $\langle dN_{ch}/d\eta_{lab}\rangle$ in pp collisions at $\sqrt{s} = 7$ TeV and in p–Pb collisions at $\sqrt{s_{NN}} = 5.02$ TeV, compared to lower energy data from RHIC (right panel) [3]. Different model predictions are shown as horizontal lines

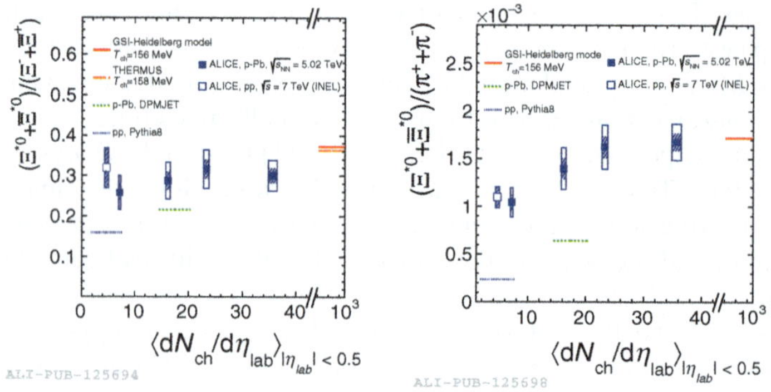

Fig. 177.2 Yield ratios of Ξ^{*0} over Ξ (left) and π (right) as a function of $\langle dN_{ch}/d\eta_{lab}\rangle$ in pp and p–Pb collisions. Different model predictions are shown as horizontal lines [3]

The measured $\Sigma^{*\pm}/\Lambda$ ratio is consistent (within systematic uncertainties) with the values predicted by PYTHIA8 [6] and thermal models. DPMJET [7] is found to underpredict the measured values for p–Pb.

References

1. K. Aamodt et al., (ALICE Collaboration), J. Instrum. **3**, S08002 (2008)
2. C. Patrignani et al., (Particle Data Group), Chin. Phys. C **40**, 100001 (2016)
3. D. Adamova et al., (ALICE Collaboration), arXiv:1701.07797
4. J. Adam et al., (ALICE Collaboration), Eur. Phys. J. C **76**, 245 (2016); B. Abelev et al., (ALICE Collaboration), Phys. Rev. C **91**, 024609 (2015)

5. A. Andronic et al., Phys. Lett. B **728**, 216–227 (2014); S. Wheaton et al., Comput. Phys. Commun. **180**, 84–106 (2009)
6. T. Sjöstrand et al., Comput. Phys. Commun. **178**, 852–867 (2008)
7. S. Roesler et al., arXiv:hep-ph/0012252

Chapter 178
Large Extra Dimensions Search in the Photon + MET Final State in pp Collisions at $\sqrt{s} = 13$ TeV at CMS in LHC

Shamik Ghosh

178.1 Introduction

One of the proposed solutions to the hierarchy problem, is ADD extra dimensions which postulates that there exist n compactified extra dimensions, in which gravitons can propagate freely. In LHC they can be created in proton-proton collisions [1–4].

178.2 Event Selection

The analyzed data sample corresponds to an integrated luminosity of 12.9fb^{-1}. A single-photon trigger is used that requires at least one photon candidate with $p_T > 165$ GeV.The photon candidate is identified by requiring H/E(Hadronic over Electromagnetic calorimeter energy deposit) to be below 0.1 to reject jets. Events require at least one Barrel photon ($|\eta| < 1.44$) with $p_T > 175$ GeV and Missing E^T (MET) > 170 GeV. Photon and MET must be separated by minimum 2 rad in ϕ.

Events are rejected if the minimum opening angle between MET and up to four leading jets (min$\Delta\phi$ (p_T miss,jet)) is less than 0.5. This requirement suppresses the background where jet energy mismeasurement gives rise to large MET. Event is rejected if it contains an electron or a muon with $p_T > 10$GeV that is away from the photon by ΔR > 0.5.

On behalf of CMS Collaboration.

S. Ghosh (✉)
Saha Institute of Nuclear Physics, HBNI, Kolkata 700064, India
e-mail: shamik.ghosh@cern.ch

Md. Naimuddin (ed.), *XXII DAE High Energy Physics Symposium*, Springer Proceedings in Physics 203, https://doi.org/10.1007/978-3-319-73171-1_178

178.3 Background Estimation

Bulk of the background(~70%) comes from Z($\rightarrow \nu\nu$)+γ and W($\rightarrow l\nu$)+γ processes leading to large MET and γ when neutrino escapes detection and lepton may be outside detector acceptance region or not identified. Electrons fake photons when the pixel seed is missing. Jets fake photons when a π_0 decaying to two photons is misidentified as a single photon.Beam halo(when a muon travelling with the beam emits a photon due to bremsstrahlung) and ECAL spikes(anomalous ECAL signals) are also significant contributors.

178.4 Results

400 events are observed in data, which is in agreement with the total expected background of 414.6 ± 38 events. No excess with respect to the SM prediction is observed. Limits are set on the ADD extra dimension scenario (Fig. 178.1).

Process	Estimate
$Z(\rightarrow \nu\bar{\nu}) + \gamma$	215 ± 32
$W(\rightarrow \ell\nu) + \gamma$	57.2 ± 8.0
Electron misidentification	54.1 ± 3.2
ECAL spikes	27 ± 13
Beam halo	15 ± 11
γ + jets	10.1 ± 5.7
$W \rightarrow \mu\nu$	8.5 ± 3.0
$t\bar{t}\gamma$	8.2 ± 0.6
Jet misidentification	5.9 ± 1.7
$W \rightarrow \tau\nu$	5.2 ± 2.3
WZ	3.5 ± 1.7
$Z(\rightarrow \ell\ell) + \gamma$	2.9 ± 0.2
ZZ	1.3 ± 0.7
$WW\gamma$	0.7 ± 0.1
Total background	414.6 ± 38
Data	400

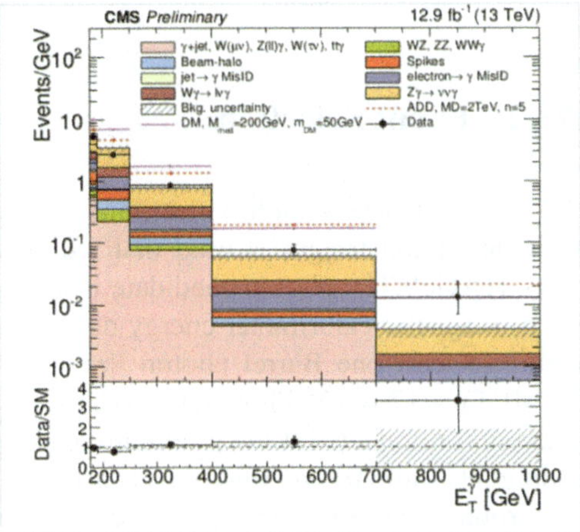

95% CL observed and expected lower limits on M_D as a function of n.

n	Obs. limit (TeV)	Exp. limit (TeV)
3	2.44	2.33
4	2.48	2.38
5	2.54	2.45
6	2.60	2.50

Fig. 178.1 Upper left table compares data and MC. Upper right figure shows number of events as a function of photon pT. Bottom table lists limits for ADD model [5]

References

1. N. Arkani-Hamed, S. Dimopoulos, G. Dvali, The hierarchy problem and new dimensions at a millimeter. Phys. Lett. B **429**, 263 (1998)
2. CMS Collaboration, The CMS experiment at the CERN LHC. JINST **3**, S08004 (2008)
3. Gian F. Giudice, Riccardo Rattazzi, James D. Wells, Quantum gravity and extra dimensions at high-energy colliders. Nucl. Phys. B **544**, 338 (1999)
4. CMS Collaboration, Search for Dark Matter and Large Extra Dimensions in the gamma + MET final state in pp Collisions at sqrt(s) = 13 TeV, CMS Physics Analysis Summary CMS-PAS-EXO-16-014, CERN, 2016
5. Search for dark matter and graviton produced in association with a photon in pp collisions at $\sqrt{s}$ = 13 TeV with an integrated luminosity of 12.9 fb^{-1} . CMS-PAS-EXO-16-039 . The CMS Collaboration

Chapter 179
Effects of Z' Mediated FCNC on $B \to K\mu^+\mu^-$ in the 331 Model

Soram Robertson Singh and Barilang Mawlong

179.1 Standard Model Calculation

The effective Hamiltonian for the $B \to Kl^+l^-$ transition is given by [1],

$$\mathcal{H}_{\text{eff}} = -4\frac{G_F}{\sqrt{2}} V_{tb} V_{tq}^* \sum_{i=1}^{10} C_i(\mu) O_i(\mu), \tag{179.1}$$

where $q = s$, $C_i(\mu)$ are the Wilson coefficients and $O_i(\mu)$ are the local operators. The form of the differential decay width for $B \to Kl^+l^-$ can be found in [2].

179.2 New Physics - 331 Model

It is a model based on $SU(3)_C \otimes SU(3)_L \otimes U(1)_X$ which breaks down into $SU(3)_C \otimes U(1)_Q$ in two steps. This symmetry breaking gives rise to a neutral Z' boson and four other extra gauge bosons [3].

The contribution of Z' to the $b \to s\mu^+\mu^-$ transition affects only the coefficients C_9 and C_{10}.

$$\sin^2\theta_W C_9^{NP} = -\frac{1}{g_{SM}^2 M_{Z'}^2} \frac{\Delta_L^{sb}(Z')\Delta_V^{\mu^+\mu^-}(Z')}{V_{ts}^* V_{tb}} \tag{179.2}$$

S. R. Singh (✉) · B. Mawlong
School of Physics, University of Hyderabad, Hyderabad 500046, India
e-mail: robsoram@gmail.com

B. Mawlong
e-mail: barilang05@gmail.com

Md. Naimuddin (ed.), *XXII DAE High Energy Physics Symposium*, Springer Proceedings in Physics 203, https://doi.org/10.1007/978-3-319-73171-1_179

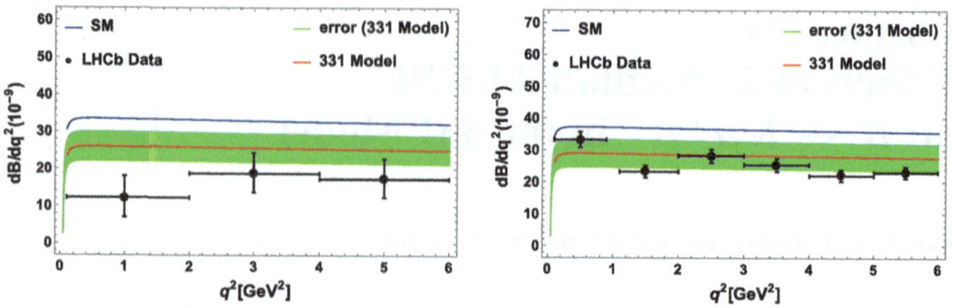

(a) $B^0 \to K^0\mu^+\mu^-$ (b) $B^+ \to K^+\mu^+\mu^-$

Fig. 179.1 Differential branching ratio

$$\sin^2\theta_W C_{10}^{NP} = -\frac{1}{g_{SM}^2 M_{Z'}^2}\frac{\Delta_L^{sb}(Z')\Delta_A^{\mu^+\mu^-}(Z')}{V_{ts}^* V_{tb}} \tag{179.3}$$

where $g_{SM}^2 = 4\frac{G_F}{\sqrt{2}}\frac{\alpha}{2\pi sin^2\theta_W}$. $\Delta_L^{sb}(Z')$ and $\Delta_{V(A)}^{\mu^+\mu^-}(Z')$ are the Z' quark and lepton couplings, respectively. In this paper, we have shown results for model parameter values $\beta = -1/\sqrt{3}$ and $M_Z' = 1$ TeV only. The values of $\Delta_{V(A)}^{\mu^+\mu^-}(Z')$ at $M_{Z'} = 3$ TeV can be found in [3]. We have taken the same values of $\Delta_{V(A)}^{\mu^+\mu^-}(Z')$ at $M_{Z'} = 1$ TeV as there is negligible difference for the two cases [3].

$\Delta_L^{sb}(Z')$ is parameterised as $\Delta_L^{sb}(Z') = -s_{23}e^{i\delta_{23}}$ and constraints on (s_{23}, δ_{23}) are obtained by comparing ΔM_s , ϕ_s and branching ratio of $B_s \to \mu^+\mu^-$ with their experimental observations taken from [4]. Two scenarios are obtained for (s_{23}, δ_{23}). Out of two, only one scenario gives favourable results. We have shown the results for the scenario (0.01, 5.3) only in this paper.

Conclusion: From Fig. 179.1, it can be seen that for $\beta = -1/\sqrt{3}$, there is a reasonable enhancement of the differential branching ratio for $B^0 \to K^0\mu^+\mu^-$ in the 331 model. The enhancement of the differential branching ratio for $B^+ \to K^+\mu^+\mu^-$ in the 331 model is significant especially when it is compared with the LHCb data [5]. For the isospin asymmetry, there is negligible deviation of the 331 results from the SM one and the same are not shown here.

Acknowledgements S. R. Singh would like to acknowledge DST Inspire for financial support.

References

1. A. Ali, P. Ball, L.T. Handoko, G. Hiller, Phys. Rev. D **61**, 074024 (2000)
2. W.-F. Wang, Z.-J. Xiao, Phys. Rev. D **86**, 114025 (2012)
3. A.J. Burasa, F. De Fazioc, J. Girrbacha, JHEP **02**, 112 (2014)
4. Y. Amhis et al., arXiv:1612.07233; R.Aaij et al., [The LHCb Collaboration], Phys. Lett. B **762**, 253 (2016); CMS Collaboration and LHCb Collaboration, Nature **522**, 68 (2015)
5. R. Aaij et al., [The LHCb collaboration]. JHEP **06**, 133 (2014)

Chapter 180
Measurement of Underground Cosmic Muons Charge Ratio at INO-ICAL Detector

Jaydip Singh and Jyotsna Singh

Introduction: The planned India-based Neutrino Observatory(INO) is a 52 ktons Iron CALorimeter(ICAL) detector [1] to be built at Theni district of Tamilnadu in Southern India. It is designed primarily to study the atmospheric neutrino flavor oscillations through interaction of atmospheric neutrinos in the detector and the main goal of the ICAL detector is to make precise measurement of neutrino oscillation parameters and determine the neutrino mass hierarchy. At a depth of 1.2 Km below the Earth's surface, the INO-ICAL detector will be the biggest magnetized detector to measure cosmic ray muon flux with capability to distinguish μ+ from μ- with large statistics. The average height of the peak around the tunnel region is 1587.32 m and the latitude and longitude at the cavern location are as follows:

- Latitude: $77°15'0''E - 77°30'0''E$
- Longitude: $9°57'30''N - 10°0'0''N$

We add this geometry to the INO-ICAL code and estimated the threshold energy for muon to hit the detector, standard rock with density of 2.650 gm/cm^3 is used for rock composition in Geant4 code (Fig. 180.1).

Charge Identification efficiency of the detector: Details of the detector simulation for low energy muon is discussed in the [1], in this work we have followed the same approach for detector simulation in the high energy region. In our analysis right plot of Fig. 180.2, charge identification efficiency decreases with increasing energy and falls down to 70% at 400 GeV/c at the detector and the cutoff for surface muon to reach at the detector is around 1600 GeV from the INO cavern peak (Fig. 180.3).

J. Singh (✉) · J. Singh
Department of Physics, Lucknow University, Lucknow 226007, India
e-mail: jaydip.singh@gmail.com

Md. Naimuddin (ed.), *XXII DAE High Energy Physics Symposium*, Springer Proceedings in Physics 203, https://doi.org/10.1007/978-3-319-73171-1_180

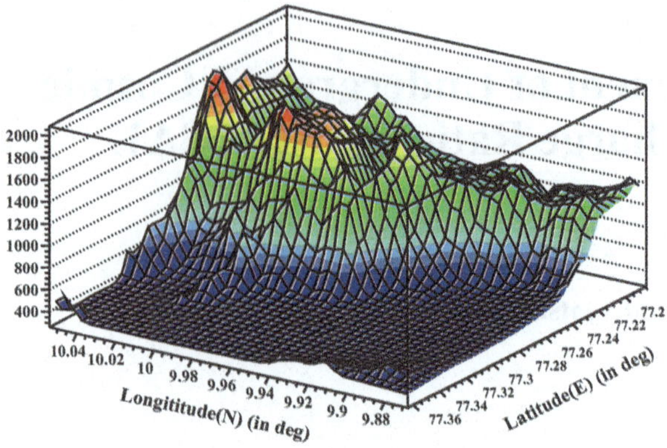

Fig. 180.1 Mountain profile at Pottipuram site, 10 km distance around the INO peak

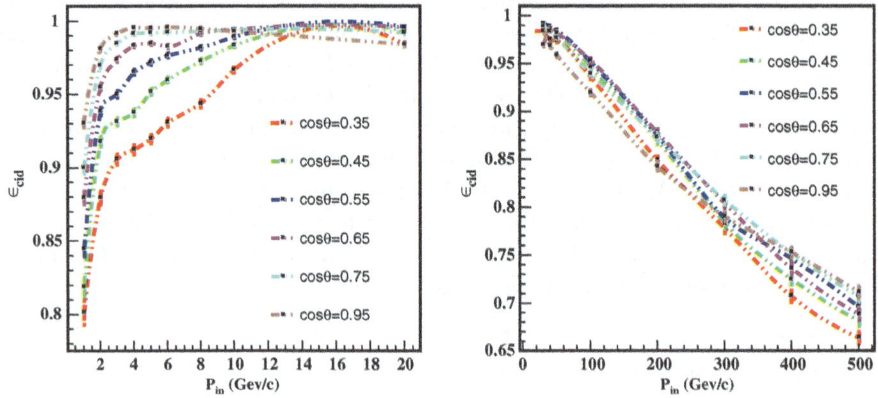

Fig. 180.2 The relative charge identification efficiency as a function of the input momentum for different cosθ values at low and high energy

Charge Ratio: Cosmic muons were generated using Corsika package, all particles were generated at 2 km above the sea level and only muon flux is used for our analysis. Muon flux generated using Corsika will be used as an input for the INO-ICAL code to estimate the charge ratio.

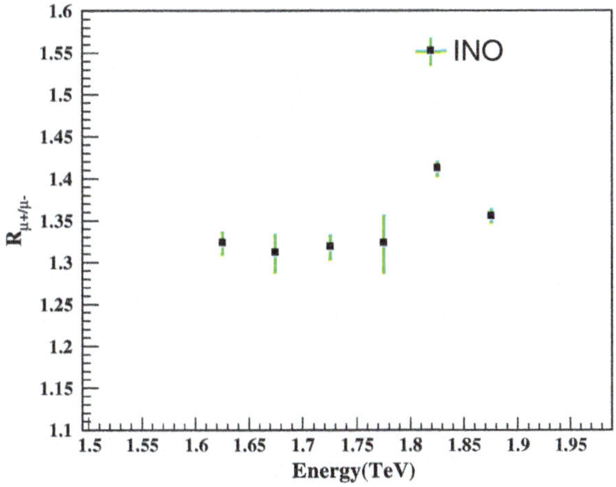

Fig. 180.3 Muon charge ratio at INO-ICAL as a function of muon energy

References

1. A. Chatterjee et al., (INO Collab). JINST **9**(2014), PO7001 (2014)
2. D. Heck, J. Knapp, J.N. Capdevielle, G. Schatz, T. Thouw, Report FZKA 6019 (1998), Forschungszentrum Karlsruhe, https://www.ikp.kit.edu/corsika/70.php
3. GEANT4 collaboration, S. Agostinelli et al., GEANT4: a simulation toolkit, Nucl. Instrum. Methods A **506**, 250 (2003), http://geant4.cern.ch/
4. P. Adamson et al., (MINOS Collab.), Phys. Rev. D **76**, 052003 (2007)
5. N. Agafonova et al., (OPERA Collab.), Eur. Phys. J. **C67**, 25 (2010)
6. V. Khachatryan et al., (CMS Collab.), Phys. Lett. **B692**, 83 (2010)

Chapter 181
Performance Study of Large Size RPC Detector for INO-ICAL Experiment

Aman Phogat, Ankit Gaur, Ashok Kumar, Moh. Rafik and Md. Naimuddin

181.1 Introduction

The India-based Neutrino Observatory (INO) is a planned underground research laboratory at Theni in Southern India. The ICAL detector at INO will consist of three modules, each having the dimension's of 16.2 m×16 m×14.5 m made out of total 150 layers of 5.6 cm Iron plates which are interleaved by glass resistive plate chambers as active detector elements. The detector is magnetized with a field of nearly 1.5 Tesla in order to identify the charge of the muons produced in the charged current interaction of neutrinos or anti-neutrinos with the target iron nuclei, which is important in order to study matter effects.

181.2 Resistive Plate Chamber Detectors

Resistive Plate Chambers (RPCs) [1–3] are charged particle gaseous detectors made of a pair of high resistivity electrodes such as glass or bakelite. We have fabricated 1 m×1 m in size glass RPCs using 3 mm thick Saint Gobain glass. A uniform gas gap of 2 mm thickness is maintained between the pates using regularly spaced highly insulated polycarbonate spacers. In order to distribute the operating voltage, over the large detector area, a thin layer of semi-conductive graphite paint is coated on the outer surface of the glass. The passage of a charged particle will induce signals on the external copper pickup strips of 3 mm strip widths. The fabricated RPCs are tested for its efficiency and counting rate for three gas compositions.

1. R134a (95.0%), Isobutane (5.0%)
2. R134a (94.7%), Isobutane (5.0%), Sulfur Hexaflouride (0.3%)

A. Phogat (✉) · A. Gaur · A. Kumar · Moh. Rafik · Md. Naimuddin
Department of Physics and Astrophysics, University of Delhi, Delhi, India
e-mail: amanphogat.phogat@gmail.com

Md. Naimuddin (ed.), *XXII DAE High Energy Physics Symposium*, Springer Proceedings in Physics 203, https://doi.org/10.1007/978-3-319-73171-1_181

3. R134a (94.5%), Isobutane (5.0%), Sulfur Hexaflouride (0.5%).

181.3 Experimental Setup

The experimental set-up consists of a cosmic ray hodoscope using prototype RPCs sandwiched between the three plastic scintillators as shown in Fig. 181.1. The pre-amplified analog output signal of RPCs and each scintillator are fed into a CAEN V814 discriminator for noise rejection and digital conversion. The logical ANDing of the triple coincidence of the scintillators pulses and scintillators plus RPC (4-fold) is done by CAEN V976 logic unit. The ratio of the counts of 4-fold to 3-fold gives RPC detection efficiency.

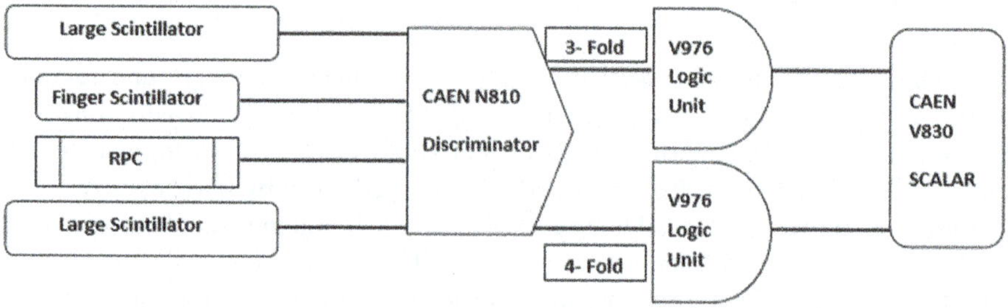

Fig. 181.1 Schematic diagram of the experimental setup used for the efficiency measurements

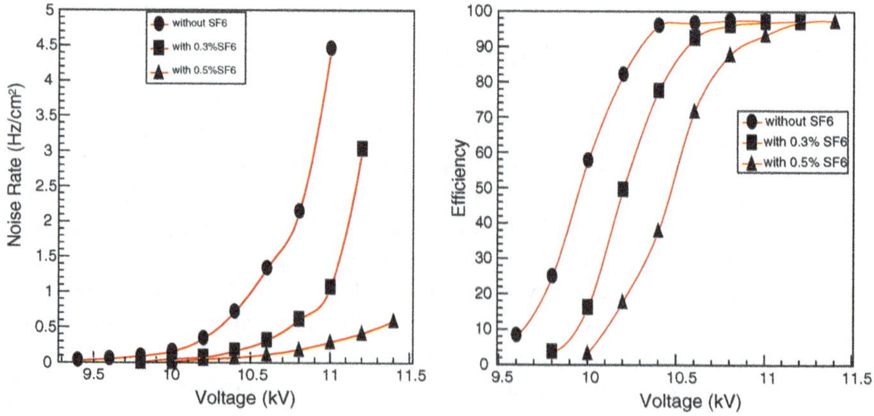

Fig. 181.2 Counting rate (left) and Efficiency (right) of the Saint Gobain RPC detector as a function of applied bias voltage, under different SF_6 concentrations

181.4 Results

The performance study of the fabricated RPC shows that the efficiency of the detector is above 95% under all the gas composition, but with the increasing concentration of SF_6 shifts the beginning of the plateau towards the higher applied bias voltage. The count rate decreases with the addition of electronegative SF_6 gas (Fig. 181.2).

References

1. R. Santonico, R. Cardarelli, Development of resistive plate counters. Nucl. Instrum. Methods A **187**, 377 (1981)
2. A. Kumar et al., Study of RPC bakelite electrodes and detector performance for INO-ICAL. JINST **9**, C10042 (2014)
3. D. Kaur et al., Characterization of 3 mm glass electrodes and development of RPC detectors for INO-ICAL experiment. Nucl. Instrum. Methods A **774**, 74 (2015)

Chapter 182
Cobimaximal Neutrino Mixing with A_4 Symmetry

M. Sruthilaya and R. Mohanta

182.1 Introduction

Neutrinos exist in three flavours ν_e, ν_μ and ν_τ and mix with each other as they propagate which is known as neutrino oscillation. Neutrino oscillation occurs since neutrinos are massive and the flavour eigenstates are mixture of mass eigenstates. The mixing is described by PMNS matrix [1], which can be parametrized in terms of three mixing angles (θ_{12}, θ_{13} and θ_{23}) and three phases (one Dirac phase δ_{CP} and two Majorana phases). Neutrino oscillation experiments explore mass-square differences ($\Delta m_{ij}^2 = m_i^2 - m_j^2$), mixing angles ($\theta_{ij}$) and Dirac phase ($\delta_{CP}$). Results from various neutrino oscillation experiments show that cobimaximal mixing ($\theta_{13} \neq 0$, $\theta_{12} \neq 0$, $\theta_{23} = \frac{\pi}{4}$, $\delta_{CP} = \pm\frac{\pi}{2}$) is a good approximation to lepton mixing. Here we consider a model based on A_4 symmetry which gives cobimaximal mixing in neutrino sector and $Z_2 \times Z_2$ invariant perturbation in charged lepton sector to accommodate the results from recent neutrino oscillation experiments. The details of model and perturbation is given in Sects. 182.2 and 182.3 respectively and we conclude our discussion in Sect. 182.4.

182.2 The Model

The particle content of the model includes, in addition to standard model fermions (i.e., the lepton doublets l_{iL} and charged lepton singlets l_{iR}), three right-handed neutrinos (ν_{iR}), four Higgs doublets (ϕ_i, ϕ_0) and three Higgs singlets (χ_i). They belong to four irreducible representations of A_4 [2] as in [3]. Here A_4 symmetry is accompanied

M. Sruthilaya (✉) · R. Mohanta
University of Hyderabad, Hyderabad 500046, India
e-mail: msruthi28@gmail.com

Md. Naimuddin (ed.), *XXII DAE High Energy Physics Symposium*, Springer Proceedings in Physics 203, https://doi.org/10.1007/978-3-319-73171-1_182

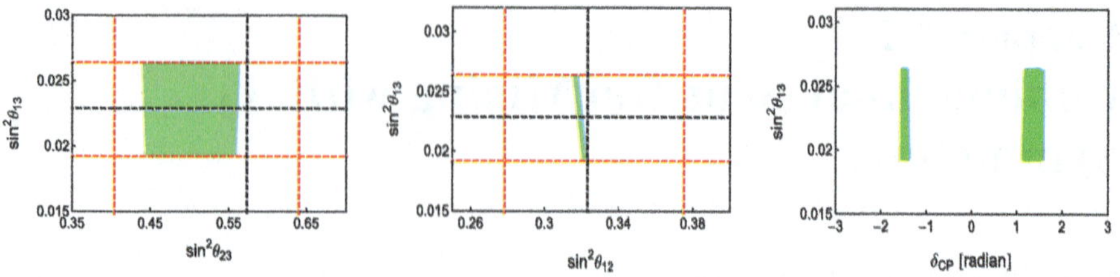

Fig. 182.1 Black and red doted lines represent best-fit values and 3σ ranges of mixing angles taken from [6]

by an additional $U(1)_X$ symmetry as in [4]. Thus, the Yukawa Lagrangian for the leptonic sector is given as [5],

$$\mathcal{L} = -\left\{\left[\lambda_1\left(\bar{l}_{iL}\phi_i\right)l_{1R}\right] + \left[\lambda_2\left(\bar{l}_{iL}\phi_i\right)'' l_{2R}\right] + \left[\lambda_3\left(\bar{l}_{iL}\phi_i\right)' l_{3R}\right]\right\} \quad (182.1)$$
$$-\left\{\lambda_0\left[\left(\bar{l}_{iL}\nu_{iR}\right)\phi_0\right] + \frac{1}{2}\left[M\left(\bar{\nu}_{iR}\hat{\nu}_{iR}\right)\right] + \lambda_\chi\left[\left(\bar{\nu}_{iR}\hat{\nu}_{iR}\right)_3\chi_i\right]\right\} + \text{h.c.}$$

The vacuum alignment $\langle\phi_i\rangle = v(1, 1, 1)$, $\langle\phi_0\rangle = u$ and $\langle\chi_i\rangle = (\omega_1, \omega_2, -\omega_2)$ will leads to cobimaximal mixing in leptonic sector and a correlation between θ_{12} and θ_{13} given as $\sin^2\theta_{12} = \frac{1-3\sin^2\theta_{13}}{3(1-\sin^2\theta_{13})}$.

182.3 Perturbation in Charged Lepton Sector

To accommodate the possible deviation of θ_{23} from $\frac{\pi}{4}$ and the possibility of δ_{CP} to be anywhere between 0 to 2π as indicated by recent experimental results, we consider a perturbation in charged lepton sector which is invariant under the group $Z_2 \times Z_2$. As discussed in [3], such perturbation can have the form

$$\lambda_1\bar{l}_{iL}M_1\phi_i l_{1R} + \lambda_2\bar{l}_{iL}M_2\phi_i l_{2R} + \lambda_3\bar{l}_{iL}M_3\phi_i l_{2R}\ , \quad (182.2)$$

which modifies the charged lepton mass matrix hence the mixing angles also get modified. We obtain the correlation plots between the mixing angles and δ_{CP} as given in Fig. 182.1 to compare with the experimental results.

182.4 Conclusions

We consider a model based on A_4 symmetry, which gives cobimaximal form for the leading order neutrino mixing matrix. We found that with a $Z_2 \times Z_2$ invariant perturbation in charged lepton sector the mixing angles obtained are within the 3σ ranges of their experimental values and δ_{CP} is around $\pm\frac{\pi}{2}$.

References

1. B. Pontecorvo, Sov. Phys. JETP **7**, 172 (1958)
2. E. Ma, G. Rajasekaran, Phys. Rev. D **64**, 113012 (2001)
3. A. Dev, P. Ramadevi, S. Uma Sankar, arXiv: 1504.04034
4. X.-G. He, Y.-Y. Keum, R. Volkas, JHEP **0604**, 39 (2006), arXiv:hep-ph/0601001
5. W. Grimus, Phys. Part. Nucl. **42**, 566 (2011)
6. D.V. Forero, M. Tortola, J.W.F. Valle, Phys. Rev. D **90**, 093006 (2014), arXiv: 1405.7540

References

Chapter 183
Optimisation and Characterisation of Glass RPC

R. Kanishka, Vipin Bhatnagar and D. Indumathi

183.1 Introduction

India-based Neutrino Observatory (INO) [1], a proposed underground facility will look for atmospheric neutrinos. The magnetized Iron CALorimeter (ICAL) detector at INO with its charge identification capability will study the oscillation pattern of atmospheric neutrinos. It aims at precise measurement of oscillation parameters [1], probing neutrino mass hierarchy as well as new physics. In the next sections we describe Resistive Plate Chambers, methodology, results and conclusions.

183.1.1 Resistive Plate Chambers

The RPCs [2] are the parallel plate gas detectors made of electrodes of high bulk resistivity such as glass or bakelite. These RPCs are fabricated using standardized procedure and operated in **Avalanche** or **Streamer Mode**.

The quality of glass plays a crucial role in long-term stability of Resistive Plate Chambers. Factors like aging can deteriorate the glass quality and hence the performance of the RPCs. Aging can be due to the possible contamination of the detector gas with impurities and moisture causing damage to the electrodes due to formation of Hydrogen Fluoride (HF) inside the gas gap. So, its necessary to study the

R. Kanishka (✉) · V. Bhatnagar
Physics Department, Panjab University, Sector 14, Chandigarh 160 014, India
e-mail: kanishka.rawat.phy@gmail.com

V. Bhatnagar
e-mail: vipin@pu.ac.in

D. Indumathi
The Institute of Mathematical Sciences, CIT Campus, Chennai 600 113, India
e-mail: indu@imsc.res.in

Md. Naimuddin (ed.), *XXII DAE High Energy Physics Symposium*, Springer Proceedings in Physics 203, https://doi.org/10.1007/978-3-319-73171-1_183

different properties of glasses; for this we characterised various glass electrodes of Saint Gobain, Asahi, Modi procured from local market. Based on the different test of glass samples we measured efficiency of all the fabricated RPCs. We describe methodology and results in next section.

183.2 Methodology and Results

We have done detailed characterisation studies based on various techniques, for Saint Gobain, Asahi and Modi glass samples procured from different manufacturers from the local market. To determine the surface properties, AFM of all the glass samples were done. Asahi and Saint Gobain were better than Modi glass sample as shown in Fig. 183.1. We investigated it further by fabrication and characterisation of RPCs of the same samples.

We fabricated 30 cm × 30 cm RPCs of Asahi, Saint Gobain and Modi glass using standardized procedure. We tested the RPCs in avalanche (Freon(R134a): Isobutane: SF_6 :: 95.15 : 4.51 : 0.34) and streamer mode (Freon(R134a): Isobutane: Argon ::

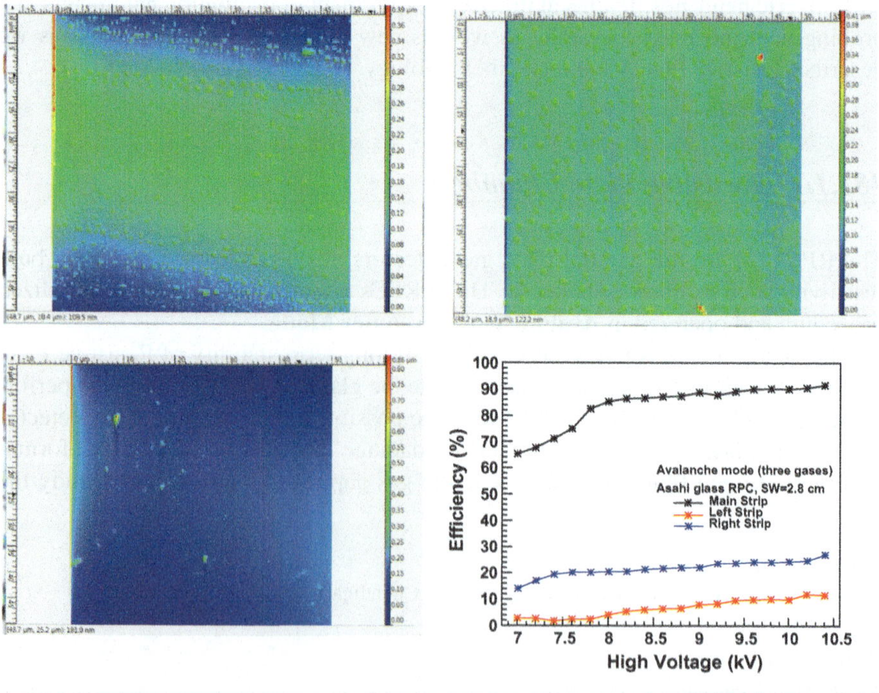

Fig. 183.1 AFM scans of Asahi (UL), Saint Gobain (UR) and Modi (BL) glass samples respectively. Cross-talk and efficiency of Asahi glass RPC with the strip width size = 2.8 cm in avalanche mode (three gases) (BR)

62 : 8 : 30). For efficiency, scintillator paddles were ANDed $P_1.P_2.P_3.P_4$ to give a 4-fold signal, that acted as the trigger pulse. The trigger signal and main strip signal from the RPC was then ANDed together to form the signal. An improper alignment of the RPC was considered as cross-talk and was measured from left and right strips of RPC. The efficiency and cross-talk of Asahi glass RPC for strip width = 2.8 cm in avalanche mode are shown in Fig. 183.1. We conclude by saying that on the basis of their surface properties, efficiency and cross-talk Asahi glass gave better results than others.

References

1. M.S. Athar et al., *India-based Neutrino Observatory: Project Report*, vol. I, http://www.ino.tifr.res.in/ino/OpenReports/INOReport.pdf (2006)
2. B. Satyanarayana, *Design and Characterisation Studies of Resistive Plate Chambers*, Ph.D. thesis, Department of Physics, IIT Bombay, PHY-PHD-10-701, (2009)

Chapter 184
Calculating Neutral Current Events at the Proposed ICAL Detector in INO

R. Thiru Senthil and D. Indumathi

184.1 Introduction

One of the primary goals of the magnetized Iron Calorimeter (ICAL) detector which will be located at the India-based Neutrino Observatory (INO) [1] is precise determination of neutrino oscillation parameters. In the absence of oscillations, the atmospheric neutrino fluxes are in the ratio $\nu_\mu : \nu_e$ as $2 : 1$. Due to large θ_{23} which drives $\nu_\mu \leftrightarrow \nu_\tau$ oscillations, the up-going flux ratio is $\nu_\mu : \nu_e : \nu_\tau = 1 : 1 : 1$; hence there are large number of ν_τ in the (up-going) atmospheric neutrino flux. They will produce τ leptons as a result of charged-current (CC) interactions of ν_τ with the iron in ICAL detector [2],

$$\nu_\tau N \to \tau^- \mathbf{X} \, , \tag{184.1}$$

where X are hadrons. Although these events are kinematically suppressed due to the large mass of the τ, they are significant [3] due to the large fluxes.

The produced τ leptons will predominantly decay into hadrons ($\sim$ 67% of the time):

$$\tau^- \to \nu_\tau \Sigma_i h_i \, , \tag{184.2}$$

where h_i are hadrons. The semi-leptonic decay into μ and e are comparatively smaller, with a branching fraction of about 17% each:

$$\tau^- \to \nu_\tau \ell \, \nu_\ell \, , \tag{184.3}$$

R. Thiru Senthil (✉) · D. Indumathi
The Institute of Mathematical Sciences, Chennai, India
e-mail: rtsenthil@imsc.res.in

D. Indumathi
e-mail: indu@imsc.res.in

Md. Naimuddin (ed.), *XXII DAE High Energy Physics Symposium*, Springer Proceedings in Physics 203, https://doi.org/10.1007/978-3-319-73171-1_184

where $\ell = e, \mu$. Hence there will be huge hadron energy ($\mathbf{X} + \Sigma_i h_i$), in CC τ interactions. These events will therefore imitate neutral current (NC) events with no charged leptons in the final state. Hence to precisely determine the number of tau events, the NC background must be calculated precisely.

Due to the large mass of the τ, the threshold for the CC interaction is $E_{\nu_\tau} > 3.5$ GeV, i.e., predominantly via deep inelastic scattering (DIS).

184.2 Methodology and Results

We have used a C-based program to calculate the deep inelastic cross section and generate NC events using Honda atmospheric neutrino fluxes at ICAL. The program generates Monte Carlo type events by using a random number generator weighted by the fluxes and cross sections. At low energy ($E < 3$ GeV), there are additional contributions from quasi-elastic and resonance NC cross-sections which we have ignored; hence we only consider the region where the neutrino energy, $E_\nu > 3$ GeV.

The generator has been validated against the NUANCE neutrino generator [4]. There is reasonable agreement between the two generators as can be seen from Fig. 184.1. Hence, this generator can be used to study τ CC events over the NC background.

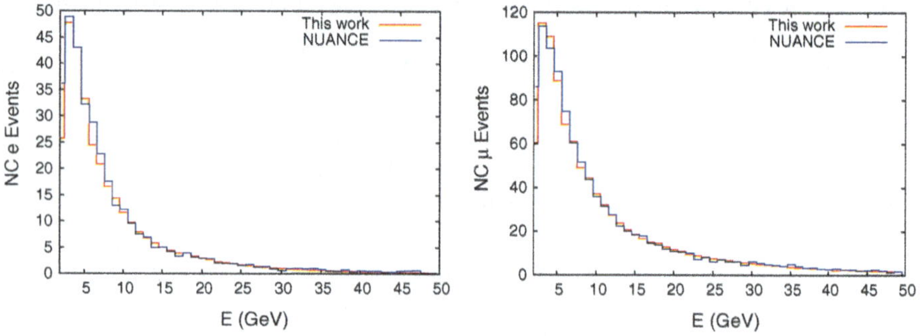

Fig. 184.1 Electron (L) and Muon (R) neutral current event rates in ICAL using NUANCE and neutrino event generator from present work

References

1. India Based Neutrino Observatory, http://www.ino.tifr.res.in/ino
2. See Particle Data Group Review on Structure Functions
3. Sumanta Pal, Ph. D. thesis, HBNI University (2014)
4. D. Casper, The Nuance neutrino physics simulation. Nucl. Phys. Proc. Suppl. **112**, 161 (2002), arXiv:0208030[hep-ph/0208030]

Chapter 185
Statistical Hadronisation in High Energy Particle Collisions

S. Sharma, S. Thakur and M. Kaur

185.1 Introduction

Collisions at relativistic high energies result in producing multitude of new particles whose number can be predicted from the simple rules of statistical mechanics. Several models have been proposed which combine the statistics with thermodynamics to describe the production of particles in a systematic way. Extensive (standard) statistical approaches were used in nearly all the previous models to describe the particle multiplicity. In the present work, Tsallis [1] model (non extensive approach) and Gamma distribution [2] are used to study the charged particle multiplicity on the proton-Emulsion data as well as on p Au and p W data at P_{lab} from 27 GeV to 800 GeV. The dependence of entropic index parameter q, of Tsallis statistics on the energy, P_{lab}, is also calculated.

185.2 Distributions for Charged Multiplicity

185.2.1 Tsallis Distribution

Tsallis q statistics provides the distribution in which entropy is non extensive in nature i.e.

$$S_q(A, B) = S_A + S_B + (1 - q)S_A S_B \quad (185.1)$$

S. Sharma (✉) · S. Thakur · M. Kaur
Department of Physics, Panjab University, Chandigarh, India
e-mail: raisandeep6dec@gmail.com

Md. Naimuddin (ed.), *XXII DAE High Energy Physics Symposium*, Springer Proceedings in Physics 203, https://doi.org/10.1007/978-3-319-73171-1_185

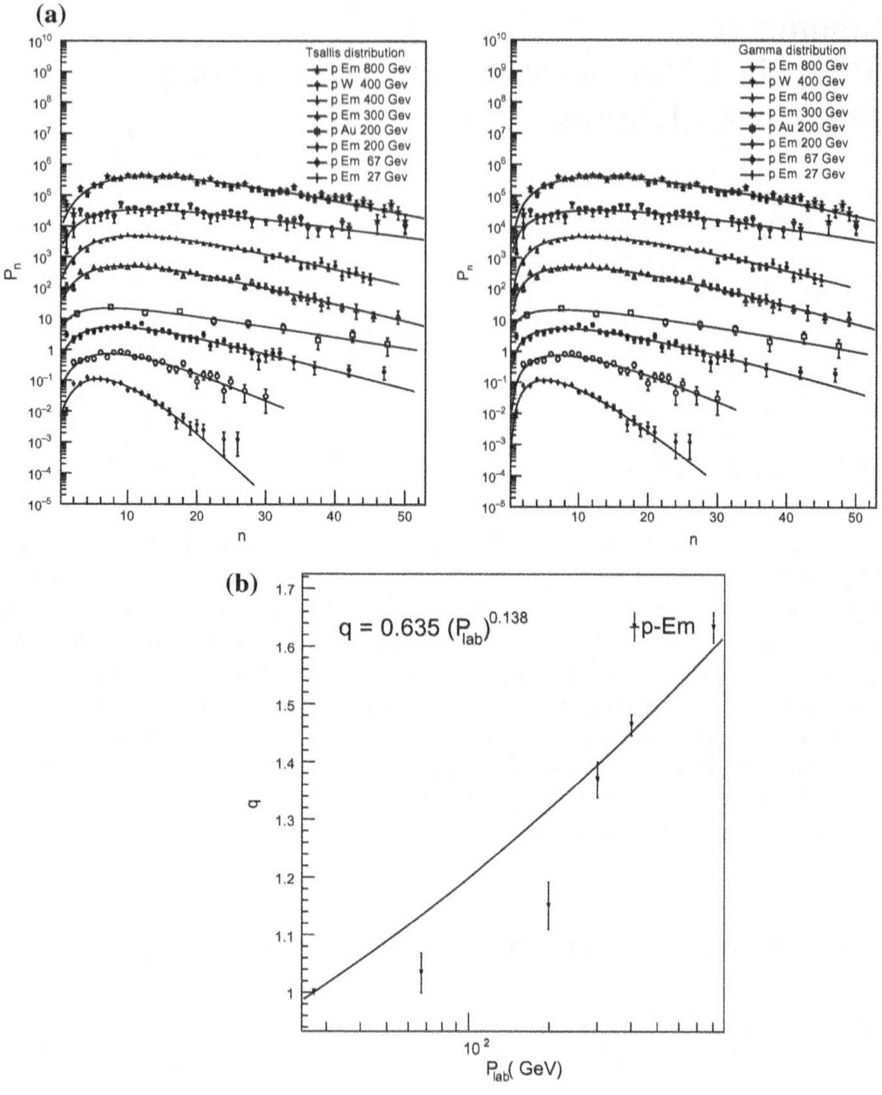

Fig. 185.1 **a** Fitting of Tsallis and Gamma distribution at various energy for proton-Emulsion. Each curve in both the figures is multiplied by the factor of 10. **b** Dependence of entropic index, q on the energy, P_{lab} for the proton-Emulsion data

where q is entropic index with value, q >1 and factor (1-q) gives the departure of entropy from its additive behaviour. In Tsallis q statistics probability is calculated using the partition function Z [3].

$$P_N = \frac{Z_q^N}{\sum Z}, \tag{185.2}$$

where Z represents the total partition function at particular charged multiplicity. For N particles, partition function can be written as

$$Z(\beta, \mu, V) = \sum (\frac{1}{N!})(nV - nv_0 N)^N \tag{185.3}$$

n is gas density, V and v0 are the volume and excluded volume of the system, with $\bar{N}$ average number of particles

$$\bar{N} = Vn[1 + (q-1)\lambda(Vn\lambda - 1) - 2v_0 n], \tag{185.4}$$

185.2.2 Gamma Distribution

Probability [2] in this case is $P_n = Ax^{(\alpha-1)}e^{-\beta x}$, with α, β as its parameters.

185.3 Results and Discussion

Figure 185.1a clearly indicates both the distributions describe the experimental data very elegantly. Entropic index q in every case comes out to be greater than 1, exemplify the non extensive nature of the entropy and increases with increase in P_{lab} exhibiting power law relation $q = a(P_{lab})^b$,with a=0.635 $\pm$ 0.007 and b=0.138$\pm$0.003 as depicted in Fig. 185.1b.

References

1. C. Tsallis, J. Stat. Phys. **52**, 479 (1998)
2. K. Urmossy, G.G. Barnafoldi, T.S. Brio, Phys. Lett. B **701**, 11 (2011)
3. C.E. Aguair, T. Kodama, Physica A **320**, 271 (2003)

Chapter 186
A Novel Assembly Procedure of GE1/1 Detectors for CMS High Luminosity Phase of the LHC

Aashaq Shah, Ashok Kumar, Md. Naimuddin, Mohit Gola and Shivali Malhotra

186.1 Introduction

The CMS Collaboration has proposed several muon system upgrades during the future High Luminosity phase of the LHC in order to maintain its high level of performance. The CMS muon system [1] is designed to provide robust, redundant and fast identification of the muons traversing the system, in addition to trigger capabilities and momentum measurement. CMS-Muon Collaboration has therefore proposed to install two layers of triple-GEM [2] chambers known as GE1/1 and GE2/1 in the endcap regions $1.6<|\eta|<2.18$ and $1.55<|\eta|<2.45$ respectively. Integration of these new detectors with existing muon system can highly improve the muon trigger, momentum resolution.

186.2 The Full Size Prototype Construction

The GE1/1 prototype has a trapezoidal shape with active area of approximately ~220 mm and ~456 mm on narrower and wider side and length of about ~990 mm. Pre-assembly of such detectors involves optical and electrical tests to determine geometrical defects and leakage current measurements. The variation in the leakage current should not exceed 30 nA and 50 nA corresponding to the humidity values of 30% and 45% respectivelly when a potential of 550 V is applied across it using an insulation testing Megger. This test can lead the observation of initial discharges which can be heard or seen at the beginning but eventually should disappear. The

On behalf of the CMS-Muon Collaboration.

A. Shah (✉) · A. Kumar · Md. Naimuddin · M. Gola · S. Malhotra
Department of Physics and Astrophysics, University of Delhi, Delhi, India
e-mail: aashaq.shah@cern.ch

Md. Naimuddin (ed.), *XXII DAE High Energy Physics Symposium*, Springer Proceedings in Physics 203, https://doi.org/10.1007/978-3-319-73171-1_186

detector is then built by piling up various components as shown in the Fig. 186.1a over a precision machined reference table called as assembly table. To assemble a single GE1/1 detector with gap configuration of 3/1/2/1, firstly a GEM stack is formed by putting alignment pins positioned on the Plexiglass board with 3 mm side spacers inserted over these guidening pins which are lying on the top of the Plexiglass board before putting the first GEM foil. Similarly a stack of three foils is formed with 3/1/2/1 respective spacing spacing. The Stack is then removed and placed over the drift board and is stretched using the mechanism as shown in Fig. 186.1b and on the other side is closed using a readout board. The whole mechanism involved is shown in Fig. 186.2.

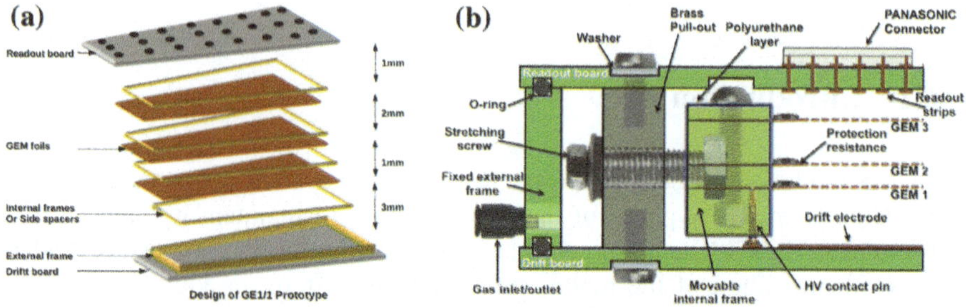

Fig. 186.1 **a** Basic sketch of GEM assembly and its components **b** Foil stretching mechanism [3]

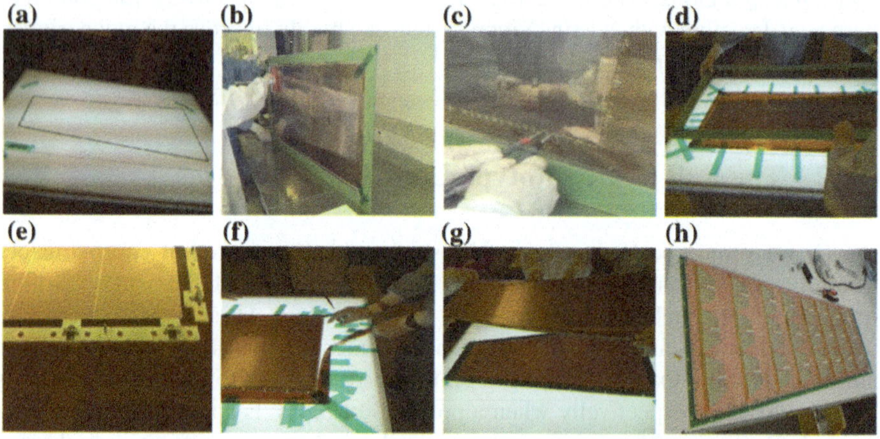

Fig. 186.2 **a** Placement of side spacers over the guidening pins to form the fine trapezium **b** Cleaning of the GEM foil using adhesive roller **c** Leakage current check using insulation megger **d** Placement of the GEM foils on assembly table **e** Placement of stretching nuts for foil stretching **f** Removing the dead area from the GEM stack **g** Placement of triple-GEM stack over the drift board and **h** Final GE1/1 detector

186.3 Conclusion

The GEM GE1/1 assembly for the CMS has been described and such detector's would be installed in the CMS during the phase II (LS2) upgrade scheduled to be in 2019. The detectors will provide fast triggering, precise tracking, will improve overall muon trigger and also can sustain the high radiation environment for long-term.

References

1. The CMS Collaboration, JINST **3**, S08004 (2008)
2. F. Sauli, NIM **A386**(1), 531–534 (1997)
3. The CMS Collaboration: CMS-TDR-013 (2015)

Chapter 187
Charged Particle Multiplicity and Transverse Energy Distribution Using Weibull–Glauber Approach in Heavy-Ion Collisions

Bharati Naik, Nirbhay K. Behera, Sadhana Dash, Basanta K. Nandi and Tanmay Pani

187.1 Introduction

The charged particle multiplicity distribution and the transverse energy distribution of the produced particles are regarded as the basic global observables in the ultra-relativistic nucleus-nucleus collision [1]. In this we have used the two component approach based on Glauber model, combined with the Weibull model of multi-particle production in the hadronic interaction to simulate the multiplicity distribution in heavy-ion collisions [2, 3]. To determine the particle multiplicity for a single event, one defines the number of independent sources emitting particles known as 'ancestors' [4]. The number of ancestors have been parametrized in terms of both N_{part} and N_{coll} as

$$N_{ancestors} = xN_{part} + (1-x)N_{coll} \tag{187.1}$$

where, N_{part} - number of participant, N_{coll} - number of binary collisions, x is the relative contribution of both hard and soft process in the final multiplicity. The probability of producing n particles per ancestor is given by the two parameter Weibull distribution

$$P(n;\lambda,k) = \frac{k}{\lambda}\left(\frac{n}{\lambda}\right)^{k-1} e^{-(\frac{n}{\lambda})^k} \tag{187.2}$$

where λ is related to the mean multiplicity per ancestor and k is related to the production dynamics.

B. Naik (✉) · N. K. Behera · S. Dash · B. K. Nandi · T. Pani
Indian Institute of Technology Bombay, Mumbai 400076, India
e-mail: bharati.naik@cern.ch

Md. Naimuddin (ed.), *XXII DAE High Energy Physics Symposium*, Springer Proceedings in Physics 203, https://doi.org/10.1007/978-3-319-73171-1_187

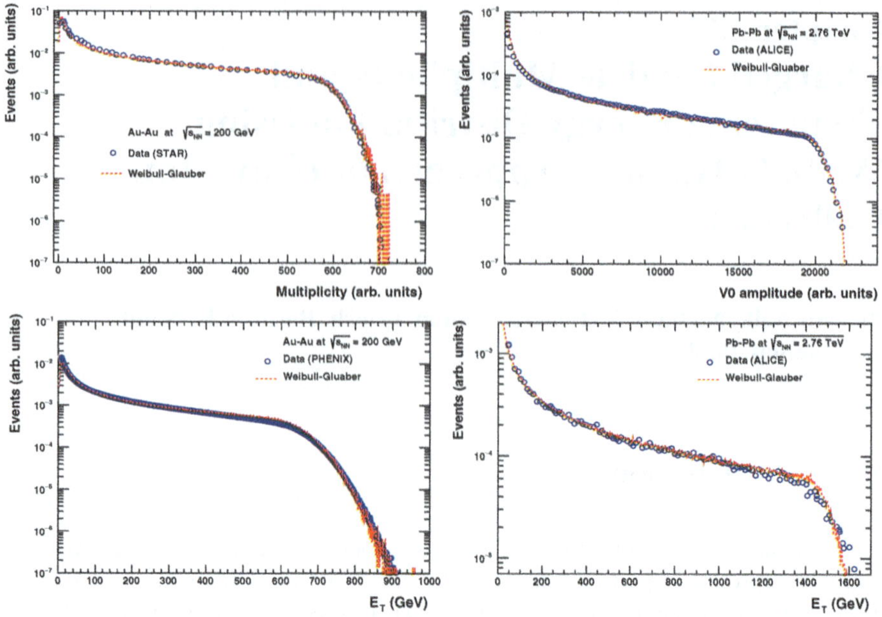

Fig. 187.1 **Upper**: (Left) The charged particle multiplicity distribution measured by the STAR TPC detector for Au–Au collisions at $\sqrt{sNN}$ = 200 GeV, (Right) VZERO scintillators amplitude distribution in Pb–Pb collisions at $\sqrt{sNN}$ = 2.76 TeV in ALICE [4] **Lower**: (Left) Transverse energy recorded by PHENIX lead-scintillator detector in Au–Au collision at $\sqrt{sNN}$ = 200 GeV (Right)transverse energy distribution in Pb–Pb collisions at $\sqrt{sNN}$ = 2.76 TeV in ALICE [5]

187.2 Results and Summary

The model is successfully fit to the charged particle multiplicity distribution and transverse energy distribution in different system and different energies which is shown in Fig. 187.1.

The charged particle multiplicity and transverse energy distributions in heavy ion collisions at RHIC and LHC energies are well described by the two component Glauber–Weibull model. Weibull–Glauber approach can be used to determine the centrality classes of heavy-ion collision at a given energy.

References

1. X.N. Wang, M. Gyulassy, Phys. Rev. Lett. **86**, 3496 (2001)
2. S. Dash, B.K. Nandi, P. Sett, Phys. Rev. D **93**, 114022 (2016)
3. R.J. Glauber, Phys. Rev. **100**, 242 (1955); R.J. Glauber, Nucl. Phys. A **774**, 3 (2006)
4. B. Abelev et al., [ALICE Collaboration], Phys. Rev. C **88**(4), 044909 (2013)
5. J. Adam et al., [ALICE Collaboration], arXiv:1603.04775 [nucl-ex]

Chapter 188
Transverse Momentum Distributions in Spin-1 Diquark Model

Navdeep Kaur, Narinder Kumar and Harleen Dahiya

188.1 Introduction

The electron is a structureless object. But an electron can fluctuate into a virtual electron-photon pair, carrying the same quantum number as an electron, i.e. $e \to e\gamma \to e$ because of the Heisenberg uncertainty principle. Further, these virtual photons can themselves break up into pairs of virtual electrons and positrons and, as a result, an isolated electron is treated as a composite system of virtual photons, electrons and positrons. Therefore, the bare electron effectively becomes a dressed electron. Photons, electrons, and positrons can be interpreted in this case as partons contained in the original electron [1].

188.2 Methodology

We calculate the TMDs using the LFWFs given in [2]. We have used the generalized form of QED by assigning a mass M to the external electrons and a different mass m to the internal electron and a mass λ to the internal photon. To improve the convergence of the wave functions at the end points of x as well as the $\mathbf{k}_\perp^2$ behaviour, we differentiate $\varphi(x, \mathbf{k}_\perp)$ with respect to M^2 and have

$$\varphi(x, \mathbf{k}_\perp) = -\frac{e}{\sqrt{1-x}} \frac{1}{\left(M^2 - \frac{\mathbf{k}_\perp^2 + m^2}{x} - \frac{\mathbf{k}_\perp^2 + \lambda^2}{1-x})^2\right)}.$$

N. Kaur (✉) · N. Kumar · H. Dahiya
Department of Physics, Dr. B. R. Ambedkar National Institute of Technology, Jalandhar 144011, India
e-mail: nkmangat91@gmail.com

Md. Naimuddin (ed.), *XXII DAE High Energy Physics Symposium*, Springer Proceedings in Physics 203, https://doi.org/10.1007/978-3-319-73171-1_188

The following definition of electron TMDs in terms of LFWFs are used to calculate the analytic results for TMDs [1]:

$$g_{1L}^{e}(x,\mathbf{k}_\perp) = \frac{1}{2(2\pi)^3}[\mid \Psi^{\uparrow}_{+,+1}(x,\mathbf{k}_\perp)\mid^2 + \mid \Psi^{\uparrow}_{+,-1}(x,\mathbf{k}_\perp)\mid^2 - \mid \Psi^{\uparrow}_{-,+1}(x,\mathbf{k}_\perp)\mid^2],$$

$$g_{1T}^{e}(x,\mathbf{k}_\perp) = \frac{M}{4(2\pi)^3 k_\perp^2}\sum_{\lambda}[(k_x + \iota k_y)\Psi^{\uparrow *}_{+,\lambda}(x,\mathbf{k}_\perp)\Psi^{\downarrow}_{+,\lambda}(x,\mathbf{k}_\perp)$$
$$+ (k_x - \iota k_y)\Psi^{\downarrow *}_{+,\lambda}(x,\mathbf{k}_\perp)\Psi^{\uparrow}_{+,\lambda}(x,\mathbf{k}_\perp)],$$

$$h_{1L}^{\perp e}(x,\mathbf{k}_\perp) = \frac{M}{2(2\pi)^3 k_\perp^2}\sum_{\lambda}[(k_x + \iota k_y)\Psi^{\uparrow *}_{-,\lambda}(x,\mathbf{k}_\perp)\Psi^{\uparrow}_{+,\lambda}(x,\mathbf{k}_\perp)$$
$$+ (k_x - \iota k_y)\Psi^{\uparrow *}_{+,\lambda}(x,\mathbf{k}_\perp)\Psi^{\uparrow}_{-,\lambda}(x,\mathbf{k}_\perp)].$$

The analytic results, found by inserting the explicit expressions of the LFWFs, are

$$g_{1L}^{e}(x,\mathbf{k}_\perp) = \frac{e^2}{(2\pi)^3}\Big[\frac{x^2(1-x)(k_\perp^2(1+x)^2 - (1-x)^2(Mx-m)^2)}{(M^2x(1-x) - (k_\perp^2)(1-x) - (k_\perp^2+m^2) - (k_\perp^2+\lambda^2)x)^4}\Big],$$

$$g_{1T}^{e}(x,\mathbf{k}_\perp) = \frac{e^2}{(2\pi)^3}\Big[\frac{x^3(1-x)^2M(Mx-m)}{(M^2x(1-x) - (k_\perp^2)(1-x) - (k_\perp^2+m^2) - (k_\perp^2+\lambda^2)x)^4}\Big],$$

$$h_{1L}^{\perp e}(x,\mathbf{k}_\perp) = \frac{-2e^2}{(2\pi)^3}\Big[\frac{x^2(1-x)^2M(Mx-m)}{(M^2x(1-x) - (k_\perp^2)(1-x) - (k_\perp^2+m^2) - (k_\perp^2+\lambda^2)x)^4}\Big].$$

188.3 Results

In Fig. 188.1, we plotted the result of T-even TMDs in transverse momentum plane at fixed value of x = 0.3.

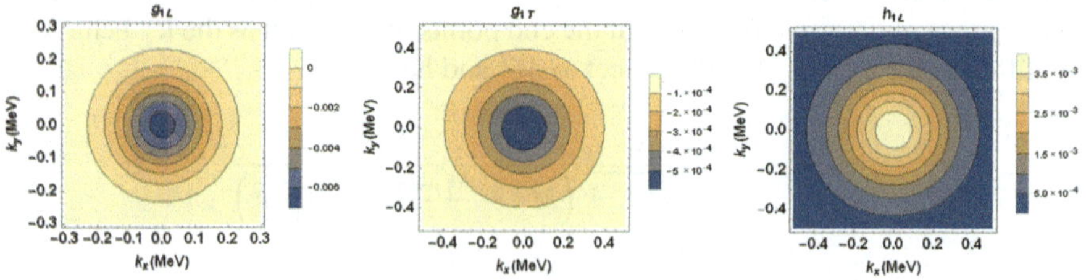

Fig. 188.1 Density plots in transverse momentum plane $k_x - k_y$ at fixed value of x = 0.3

References

1. A. Bacchetta, L. Mantovani, B. Pasquini: Electron in three-dimensional momentum space. Phys. Rev. D **93**, 013005 (2015). https://doi.org/10.1103/PhysRevD.93.013005
2. S.J. Brodsky, D.S. Hwang, B.Q. ma, I. Schmidt: Light-cone representation of the spin and orbital angular momentum of relativistic composite systems. Nucl. Phys. B **593**, 311–335 (2001). https://doi.org/10.1016/S0550-3213(00)00626-X

Chapter 189
Event-by-Event Charge Separation in Pb-Pb Collisions at $\sqrt{s_{NN}} = 2.76$ TeV with ALICE at the LHC

Sonia Parmar

189.1 Introduction

ALICE, A Large Ion Collider Experiment, is specially optimized to study the properties of the deconfined state of quarks and gluons known as the "Quark Gluon Plasma (QGP)". In non-central heavy-ion collisions, strong parity violation coupled with a strong magnetic field leads to a charge separation along the system's angular momentum direction resulting in the "Chiral Magnetic Effect (CME)" [1]. This effect has been studied by the STAR experiment at RHIC for different beam energies $\sqrt{s_{NN}} = 7.7 - 200\,\text{GeV}$ [2] and by the ALICE at LHC at $\sqrt{s_{NN}} = 2.76\,\text{TeV}$ [3]. In view of this, Voloshin [4] proposed a multi-particle correlator $\langle cos(\phi_\alpha + \phi_\beta - 2\Psi_{RP})\rangle$ to observe CME, where ϕ_α, ϕ_β are the azimuthal angles of the particles α, β, respectively and ψ_{RP} is the reaction plane angle. The ALICE collaboration investigated the charge separation effect by studying the centrality dependence of this multi-particle correlator for the different charge combinations such as opposite sign charge pairs $(+-)$ and same sign charge pairs $(++, --)$, as shown in Fig. 189.1 [3]. This indicates that the correlation for same sign charge pairs increases as one moves from the central to peripheral collisions. Figure 189.1 [3] also displays the STAR data points, which exhibits the similar trend.

For this analysis, the ALICE time projection chamber (TPC) covering the pseudorapidity region, $\eta < |0.9|$ is used to reconstruct the charged particle tracks and the VZERO detector is used to measure the collision centrality. The accepted events are divided into various centrality classes from 20–30% to 60–70%. The analysis is performed in pseudorapidity coverage $\eta < |0.8|$ and $0.2 < p_T < 5$ GeV/c.

Sonia Parmar (for the ALICE Collaboration).

S. Parmar (✉)
Panjab University, Chandigarh 160014, India
e-mail: sonia.parmar@cern.ch

Md. Naimuddin (ed.), *XXII DAE High Energy Physics Symposium*, Springer Proceedings in Physics 203, https://doi.org/10.1007/978-3-319-73171-1_189

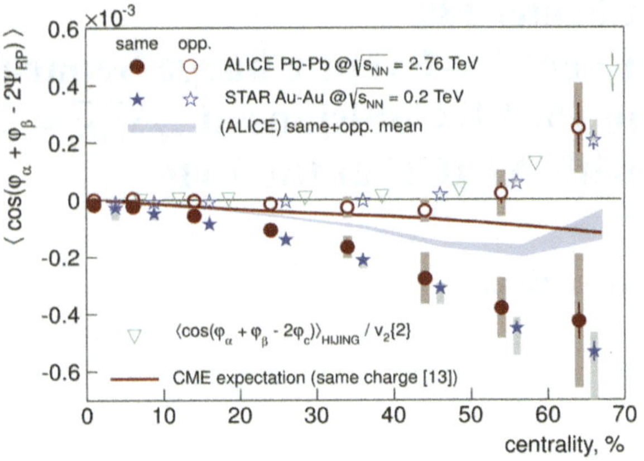

Fig. 189.1 Centrality dependence of the correlator for same sign charged pairs (red colour) and opposite sign charged pairs (blue colour) [3]

189.2 Sliding Dumbbell Method

To investigate the localized charge separation on event-by-event basis, the SDM is used, which is similar to the Sliding Window Method [5] used for charged-neutral fluctuations in Pb-Pb collisions at 158 A GeV at SPS [6]. In this method, the observable Db_{+-} is measured and defined as:

$$Db_{+-} = \frac{N_+^L}{(N_+^L + N_-^L)} + \frac{N_-^R}{(N_+^R + N_-^R)}, \tag{189.1}$$

where, N_+^L and N_-^L are the numbers of positively and negatively charged particles on the left side of the dumbbell respectively, whereas N_+^R and N_-^R are the numbers of positively and negatively charged particles on the right side of the dumbbell respectively. The whole azimuthal plane is scanned by sliding the dumbbell of $\Delta\phi = 90°$ in steps of $\delta\phi = 1°$ and calculating the fraction Db_{+-} in each event to extract the maximum value of Db_{+-}. The Db_{+-} can also be calculated for different centrality intervals ranging from 20–30% to 60–70%.

189.3 Results and Discussion

A sample of about 1 million minimum bias events of Pb-Pb collision data at $\sqrt{s_{NN}} = 2.76\,\text{TeV}$ were analysed. The Db_{+-} distributions are obtained for different centralities using the SDM and also obtained by placing the dumbbell at randomly chosen azimuth in each event for the similar centrality intervals. Db_{+-} distribution obtained from randomly placed dumbbells shows a gaussian distribution, which peaks ~ 1 but the one obtained using SDM shifts towards higher Db_{+-} values and this shift is more for semi-central events as compared to those for central events. Some of the semi-central events have been observed with much larger Db_{+-} values, which indicates that the positive charged particles are on one side of the dumbbell and the negative charged particles are on the other side of the dumbbell. The percentage of such events increases with decreasing collision centrality. Further studies on Monte Carlo Generators without the CME (e.g. HIJING) are needed to investigate any biases of this method.

References

1. K. Fukushima, et al., Phys. Rev. D **78**, 074033 (2008)
2. L. Adamczyk, et al., (STAR Collaboration), Phys. Rev. Lett. **113**, 052302 (2014)
3. B. Abelev, et al., (ALICE Collaboration), Phys. Rev. Lett. **110**, 012301 (2013)
4. S. Voloshin, Phys. Rev. C **70**, 057901 (2004)
5. M.M. Aggarwal, G. Sood, Y.P. Viyogi, Phys. Lett. B **638**, 39 (2006)
6. M.M. Aggarwal, et al., (WA98 Collaboration), Phys. Lett. B **701**, 300 (2011)

Chapter 190
Rotating Boson Star Under Weak Gravity Potential

Bharti Jarwal and S. Somorendro Singh

190.1 Introduction

Boson stars are hypothetical and astronomical objects formed of self gravitating field of bosonic particles. These particles may be detected by the gravitational radiation emitted by a pair of co-orbiting boson stars. Boson stars may have been formed through gravitational collapse during the primordial stages of the big bang process. According to the earliest theory, a super massive boson star could exist at the core of a galaxy, which might explain many of the observed properties of active galactic cores. Bosonic objects were first studied by considering second quantized scalar field satisfying Klein-Gordon equation [1, 2] in non interacting systems. In this equation the semi-classical Einstein equation had been solved and gravitationally bound state of field was obtained. The maximum mass for boson stars was found to be of the same order as that of neutron star given by $M_{max} \approx 0.633 M_P^2/m$, where $M_P = 1.2 \times 10^{19} GeV$ is the Planck mass and m is the mass of the individual bosons. Phenomenological changes in these system were investigated by [3] while study under weak gravity limit [4].

In this work weak potential is added as perturbation [5] and oscillation of BS is studied under this correction.

B. Jarwal · S. S. Singh (✉)
University of Delhi, Delhi 110007, India
e-mail: sssingh@physics.du.ac.in

B. Jarwal
e-mail: jarwalbharti22@gmail.com

Md. Naimuddin (ed.), *XXII DAE High Energy Physics Symposium*, Springer Proceedings in Physics 203, https://doi.org/10.1007/978-3-319-73171-1_190

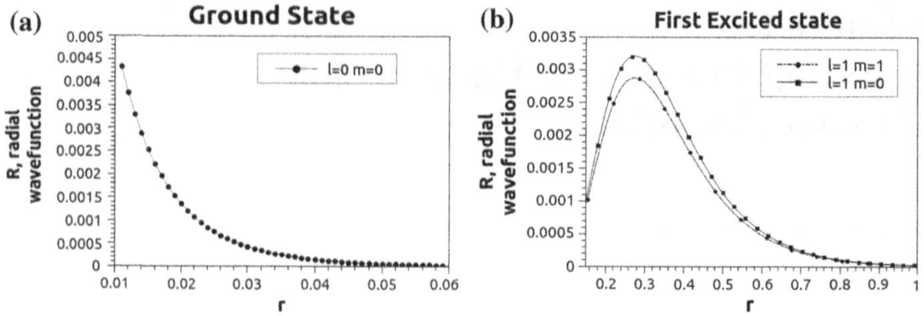

Fig. 190.1 **a** Radial wave function of ground state (l = 0). **b** Radial wave function of first excited state (l = 1)

190.1.1 BS Under Weak Potential

The potential is chosen in such a way that it exists due to the interaction of quark, anti quark, gluon etc. The weak gravity potential used as perturbation is defined as

$$V_{perturb}(r) = \sigma r_D(1 - e^{-r/r_D}) - (\alpha_s e^{-r/r_D}/r) \tag{190.1}$$

where α_s, σ and r_D are constant having values 0.471, 0.192 GeV^2 and 2.13 GeV^{-1} respectively.

190.1.2 Result and Discussion

The radial wave function for ground and excited states are obtained for different values of energy. It's results are shown in figures given below. The energy of the ground state and first excited state are obtained as 6.4 GeV and 1.38 GeV while mass of the BS used is around 30 GeV. It is clear from the graph that wave function for ground state attain maximum value around 0.012 fm. In the case of first excited state it is reached around the peak value of 0.28 fm. The amplitude of ground state wave function is higher than that of first excited state. This shifting of the peaks shows the rotating behavior of the BS under the correction of this weak gravity potential (Fig. 190.1).

Acknowledgements I am very much thankful to Dr.Naimuddin, faculty, University of Delhi, Delhi-7, for his support in publishing the manuscript in this reputed journal series. I thank University of Delhi, Delhi-110007 for providing the R and D research grant, 2016 as financial support.

References

1. R. Ruffini, S. Bonazolla, Phys. Rev. **187**, 1767 (1969)
2. Kaup, Phys. Rev. **172**(5), 1333–1341 (1968)
3. M. Colpi, S.L. Shapiro, I. Wasserman, Phys. Rev. Lett. **57**, 2485 (1986)
4. S. Weinberg, *Gravitation and Cosmology* (Wiley, NewYork, 1972)
5. S. Kumar, J. Prasanth, V. Bannur, Physica A **432**, 71 (2015)

Chapter 191
A Model Study of D-h Azimuthal Correlation at LHC Energies

S. K. Tripathy, M. Younus, Z. Naik and P. K. Sahu

Introduction

Charm-hadron azimuthal correlation is able to infer properties like charm production and fragmentation process while that of charm-anti-charm can discern the contribution of NLO-pQCD and LO-pQCD into two particle correlation spectra. $p + p$ being the baseline, correlation study in $p + Pb$ will clearly indicate the effect due to initial cold nuclear matter effect, while $p + Pb$ study of D-h and D-$\bar{D}$ correlation will show the early effect due to quark gluon plasma.

Analysis Method

Charm cross-section in HIJING [1] is given by the following equation:

$$\frac{d\sigma_{cc}^{pp}}{dp_T^2 dy_1 dy_2} = K \sum_{a,b} x_1 f_a(x_1, p_T^2) x_2 f_b(x_2, p_T^2) \times \frac{d\hat{\sigma}_{ab}}{d\hat{t}}, \qquad (191.1)$$

here a, b are the parton species, y_1, y_2 are the rapidities of the scattered partons, and x_1, x_2 are the fraction of momentum carried by the initial partons. Value of K = 2.0 has been used to account for the higher order corrections. AMPT [2] also uses leading order two body partonic interaction as in HIJING, followed by secondary interaction of partons after hadronisation. In NLO [3], p_T differential spectrum of heavy quarks produced in $p + p$ collisions is defined in general as:

S. K. Tripathy (✉) · M. Younus · P. K. Sahu
Institute of Physics, Bhubaneswar, India
e-mail: srikanta2012@iopb.res.in

S. K. Tripathy · Z. Naik
Sambalpur University, Burla, India

Md. Naimuddin (ed.), *XXII DAE High Energy Physics Symposium*, Springer Proceedings in Physics 203, https://doi.org/10.1007/978-3-319-73171-1_191

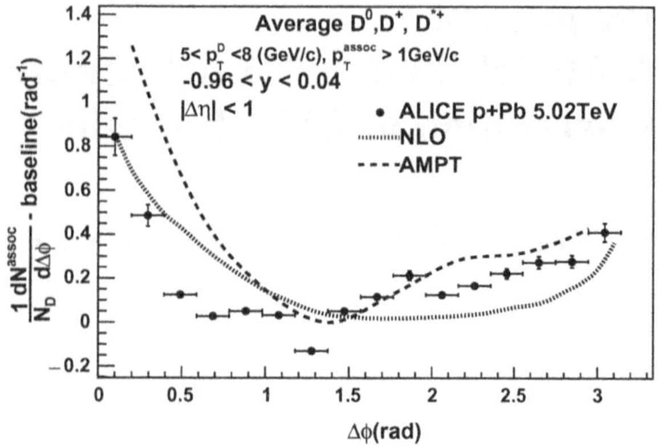

Fig. 191.1 Azimuthal correlation of D-mesons in $p + Pb$ collisions at $\sqrt{s_{NN}} = 5.02\,\text{TeV}$ after baseline subtraction. Uncertainties in ALICE data [4] are putted by hand to be 10%

$$E_1E_2\frac{d\sigma}{d^3p_1d^3p_2} = x_ax_b\sum_{ij}\left[f_i^{(a)}(x_a, Q^2)f_j^{(b)}(x_b, Q^2)\frac{d\hat{\sigma}_{ij}(\hat{s},\hat{t},\hat{u})}{d\hat{t}}\right.$$
$$\left.+f_j^{(a)}(x_a, Q^2)f_i^{(b)}(x_b, Q^2)\frac{d\hat{\sigma}_{ij}(\hat{s},\hat{u},\hat{t})}{d\hat{t}}\right]/(1+\delta_{ij}), \qquad (191.2)$$

where x_a and x_b are the fractions of the momenta carried by the partons from their interacting parent hadrons.

Results and Conclusion

In Fig. 191.1 we have presented azimuthal correlation of D^0, D^+ and D^{*+} mesons in $p + Pb$ data at $\sqrt{s_{NN}} = 5.02$ TeV. Here p_T of D-mesons taken from $5 < p_T < 8\ GeV/c$ and that of charged particles taken to be $p_T > 1\ GeV/c$, both particles falling under pseudo-rapidity window $|\eta| < 1$.

AMPT results give a better correlation pattern comparable to data while NLO is almost flat for $\Delta\phi > \pi/2$, showing near-side correlation is more prominent than away side pattern. In future, we will study the parametric dependence of differences between these two models.

References

1. X.N. Wang, M. Gyulassy, Phys. Rev. D **44**, 3501 (1991)
2. Z.-W. Lin, C.M. Ko, B.-A. Li, B. Zhang, S. Pal, Phys. Rev. C **72**, 064901 (2005)
3. M.L. Mangano, P. Nason, G. Ridolfi, Nucl. Phys. B **373**, 295 (1992); S. Frixione, M.L. Mangano, P. Nason, G. Ridolfi, Adv. Ser. Direct. High Energy Phys. **15**, 609 (1998)
4. J. Adam et al., arXiv:1605.06963v1

Chapter 192
Impact of New Physics on CP-Asymmetries at Long Baselines

Jogesh Rout, Mehedi Masud and Poonam Mehta

CP-Asymmetries in the Appearance Channels

The CP-asymmetry is given by $\Delta P^{CP}_{\alpha\beta} = P_{\alpha\beta} - \bar{P}_{\alpha\beta}$, where $\alpha, \beta = e, \mu, \tau$. In vacuum, it is given by

$$\begin{aligned}\Delta P^{CP}_{\alpha\beta} &= 8\,J\left[\sin(r_\lambda \lambda L)\sin^2\frac{\lambda L}{2} - \sin(\lambda L)\sin^2\frac{r_\lambda \lambda L}{2}\right] \\ &= 4\,\sin\delta_{13}\,J_r\,[\sin\lambda L/2\,\sin r_\lambda\lambda L/2\,\sin(1-r_\lambda)\lambda L/2] \qquad (192.1)\end{aligned}$$

where $J = s_{12}c_{12}s_{23}c_{23}s_{13}c^2_{13}\sin\delta_{13}$ is the Jarlskog invariant, $J_r = J/\sin\delta_{13}$, $\lambda \equiv \frac{\delta m^2_{31}}{2E}$ and $r_\lambda \equiv \frac{\delta m^2_{21}}{\delta m^2_{31}}$. In matter, $\Delta P^{CP}_{\alpha\beta}$ gets modified due to the fact that matter is CP-asymmetric. Matter induced CP-effect is referred to as extrinsic in order to distinguish it from vacuum induced effect (intrinsic). The two contributions (intrinsic and extrinsic) to the CP-violation signal can be separated using the following observable

$$\delta[\Delta P^{CP}_{\alpha\beta}] = [P_{\alpha\beta} - \bar{P}_{\alpha\beta}](\delta_{13} = \pi/2) - [P_{\alpha\beta} - \bar{P}_{\alpha\beta}](\delta_{13} = 0)\,. \qquad (192.2)$$

J. Rout (✉) · P. Mehta
School of Physical Sciences, Jawaharlal Nehru University,
New Delhi 110067, India
e-mail: jogesh.rout1@gmail.com

P. Mehta
e-mail: pm@jnu.ac.in

M. Masud
AHEP Group, IFIC –C.S.I.C./Universitat de València, Parc Cientific de Paterna,
C/Catedratico José Beltrán, 2, 46980 Paterna, València, Spain
e-mail: masud@hri.res.in

M. Masud
Harish-Chandra Research Institute, Chhatnag Road, Allahabad 211019, India

Md. Naimuddin (ed.), *XXII DAE High Energy Physics Symposium*, Springer Proceedings in Physics 203, https://doi.org/10.1007/978-3-319-73171-1_192

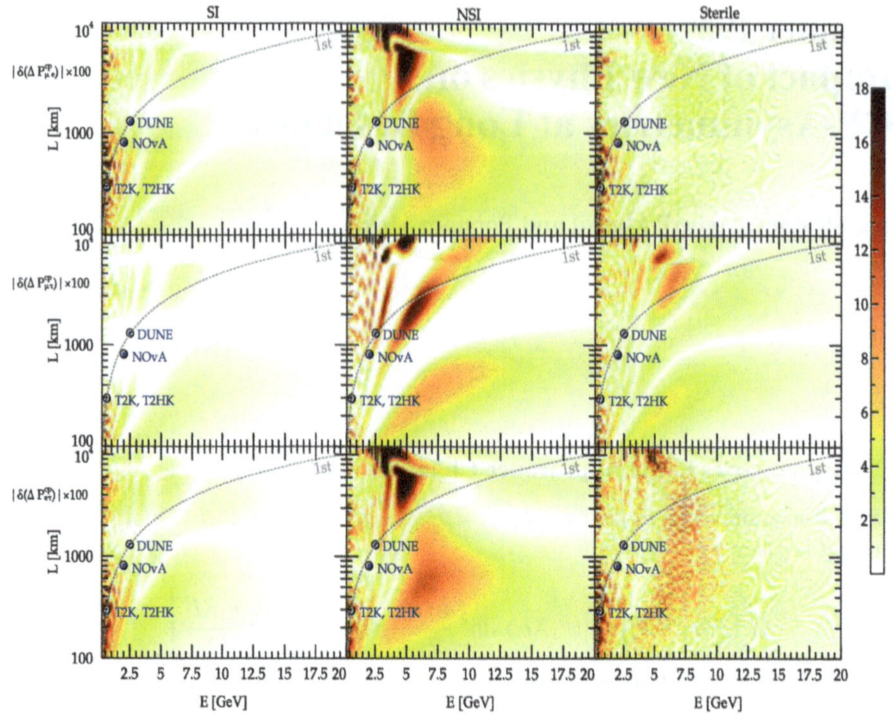

Fig. 192.1 Oscillogram of absolute relative CP asymmetry for the appearance channels. The location of first oscillation maximum (for $P_{\mu e}$) is depicted by dashed grey curve. Darker regions imply larger amount absolute relative CP asymmetry for those values of E and L

which is difference between the CP asymmetries computed at two different values of δ_{13}. For a discussion on the usefulness and limitation of using this observable, see [1].

Results and Discussion

In Fig. 192.1, we show the oscillograms of $\delta[\Delta P^{CP}_{\alpha\beta}]$ in the plane of E and L for the appearance channels for the standard interaction(SI) and two new physics scenarios. The large asymmetry regions are found at relatively lower values of energies (≤ 2 GeV). We take two new physics scenarios - non-standard interactions (NSI) and sterile neutrinos which represent good examples of going beyond the standard framework [1]. In case of NSI, the three flavour unitarity is preserved while in the sterile case, it is destroyed.

Conclusion

By means of oscillogram plots, we show which regions will be well-suited in order to extract the intrinsic CP-violating contribution. In the case of SI, the dark region implies better distinguishability. However, in presence of new physics, one can get darker patches at some values but that basically implies that new physics obstructs the clean determination of Dirac delta CP-phase δ_{13}. The wiggles in the plots arise due to the fast oscillations induced by the larger mass-squared difference, δm^2_{41}. For a detailed study, see [1].

Reference

1. J. Rout, M. Masud, P. Mehta, Phys. Rev. D **95**, 075035 (2017)

Chapter 193
Multiplicity Distributions in Hadronic and Leptonic Collisions Using Weibull Model

Priyanka Sett, Sadhana Dash and Basanta Kumar Nandi

193.1 Introduction

The charged hadron multiplicity distributions are one of the most important observables in high energy collisions, especially as they provide an insight on particle production. The hard processes are nicely explained by pQCD and the soft processes are described by incorporating the fragmentation and cascade processes. The systems which have fractal nature or systems evolving with fragmentation or sequential branching are nicely described by Weibull distribution [1]. The functional form of Weibull distribution for any random variable n is given by,

$$P(n, \lambda, k) = \frac{k}{\lambda}\left(\frac{n}{\lambda}\right)^{k-1} e^{(-(n/\lambda)^k)}; \quad \langle n \rangle = \lambda\Gamma(1 + 1/k)$$

where, k is known as the shape parameter and λ is known as scale parameter. In this work the $p + p\ (p\bar{p})$ and e^+e^-data are described by the Weibull distribution and the parameters are physically interpreted.

193.2 Results

For this study the $p + p\ (p\bar{p})$ data are taken from UA5, ISR, CMS and LHCb collaborations for various available center of mass energies and pseudo-rapidity (η) ranges. The e^+e^-data are used from ALEPH and TASSO collaborations for different collision energies and different rapidity (y) ranges. The details can be found in [2, 3]. Figure 193.1 shows the plots for both $p + p\ (p\bar{p})$ (left) and e^+e^-(right) data. The solid lines are the Weibull fit and describes the data very nicely. The Weibull

P. Sett (✉) · S. Dash · B. K. Nandi
Indian Institute of Technology Bombay, Mumbai, Powai, India
e-mail: sett.priya@phy.iitb.ac.in

Md. Naimuddin (ed.), *XXII DAE High Energy Physics Symposium*, Springer Proceedings in Physics 203, https://doi.org/10.1007/978-3-319-73171-1_193

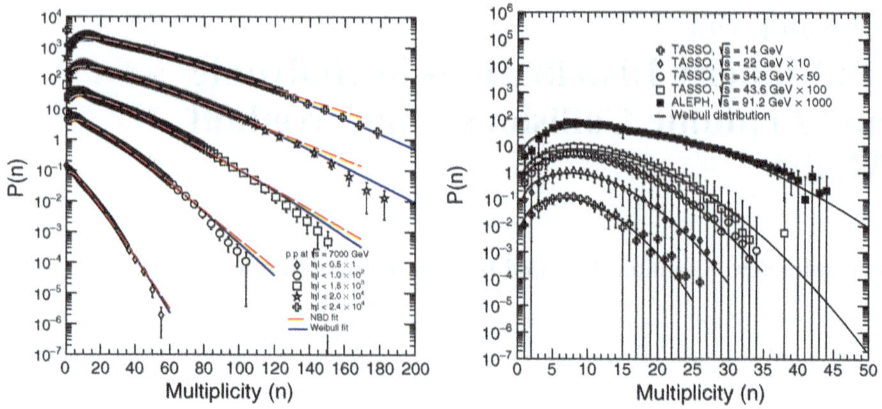

Fig. 193.1 Left Plot: The multiplicity distributions in p+p collisions at 7 TeV described by both the Weibull (solid lines) and NBD distributions (dashed lines) for different η ranges. Right Plot: The multiplicity distributions in e^+e^-collisions for different $\sqrt{s}$ fitted with Weibull distribution

parameters λ and k has been studied as a function of both collision energies and η (y in case of e^+e^-collisions) intervals. It is observed that values of λ increases both the cases and hence it is associated with the mean multiplicity. The parameter k increases with increasing η and y intervals but does not show any significant collision energy dependence. The parameter k is associated with soft processes.

193.3 Summary

The Weibull distribution describes the multiplicity distribution for leptonic and hadronic collisions very nicely. The parameter λ can be associated with mean multiplicity $\langle n \rangle$ and the parameter k can be associated to the mechanisms involved with soft processes.

References

1. W. Weibull, A statistical distribution function of wide applicability. J. Appl. Mech. **18**, 293 (1951)
2. S. Dash, B.K. Nandi, P. Sett, Weibull model of multiplicity distribution in hadron-hadron collisions. Phys. Rev. D **93**(11), 114022 (2016)
3. D. Dash, B.K. Nandi, P. Sett, Multiplicity distributions in e^+e^- collisions using Weibull distribution. Phys. Rev. D **94**(7), 074044 (2016)

Chapter 194
K_s^0 Production at the Main Injector Particle Production Experiment at Fermilab

Amandeep Singh, Ashok Kumar and Lalit Saini

194.1 Introduction

The Main Injector Particle Production (MIPP) experiment [1] is a large acceptance spectrometer to measure the particle production on variety of targets using protons beams with momentum ranging from 5–120 GeV/c. The MIPP spectrometer has been used to study the K_s^0 particles production in the interactions of 120GeV/c protons beam with Beryllium, Carbon and Bismuth targets.

194.2 MIPP Spectrometer

The MIPP spectrometer consists of "Jolly Green Giant" and "Rossie" magnets. The magnetic fields were set so that the kick angles are about equal and opposite. The track reconstruction of the incoming beam particles were done using three Beam Chambers (BC). The trajectories of the secondary charged particles were done using the hits from four Drift Chambers (DC) and two Multi-wire Proportional Chambers (PWC). The scintillator counter (SCINT) was placed immediately after the targets. The physical properties of the targets are mentioned in [2].

A. Singh (✉) · A. Kumar
Panjab University, Chandigarh 160014, India
e-mail: asingh.phy@gmail.com

L. Saini
DAV University, Jalandhar, India

Md. Naimuddin (ed.), *XXII DAE High Energy Physics Symposium*, Springer Proceedings in Physics 203, https://doi.org/10.1007/978-3-319-73171-1_194

194.3 Analysis Technique

Event selection cuts were applied to reject events in which beam particles interacted with detector material instead of target and, to determine the incident beam flux. We have applied conditions in analysis that events with properly reconstructed single beam track where retained so that initial state was defined and The reconstructed beam track positions on upstream face of the target had to be consistent to the target′s dimension.

In K_s^0 hypothesis, the positive (negative) charged particle assumed to be $\pi^+\pi^-$ meson. Any pair of tracks, with opposite charge and distance of closest approach between them smaller than 1 cm, was taken as a possible V^0s candidate. To select the candidates which correspond to K_s^0 with high probability additional cuts were applied in which the longitudinal distance along the beam direction between primary vertex and V^0 decay vertex had to be greater than 2 cm. The impact parameter of the track on the primary vertex had to be greater than 0.2 cm.

The hypothetical $\pi^+\pi^-$ invariant mass distributions were plotted in different momentum bins covering the range from 0.4 to 30 GeV/c. The raw number of K_s^0 in each momentum bin was estimated using the fit [3] for the signal and background. The corrections were applied to the raw number of K_s^0 to obtain the K_s^0 yield in the primary p-A interactions. These corrections were obtained from FLUKA hadron production model.

194.4 K_s^0 Cross-Section Measurements

The K_s^0 inclusive differential cross-sections in bins of momentum is calculated as follow

$$\frac{d\sigma}{dp} = \frac{N_{K_S^0}}{N_{beam}} \times \frac{1}{n_t} \times \frac{1}{\Delta p} \times 10^4, [mb/GeV/c] \tag{194.1}$$

Here $N_{K_S^0}$ is the corrected number of K_s^0; N_{beam} is the incident beam flux; n_t is the number of nuclei per cm^2 in the target; Δ p is the width of the momentum bin and factor 10^4 to bring the results to mb. The overall systematic uncertainties arising from different sources were found 15–17% depending upon target. Table 194.1 presents the K_s^0 production cross-sections from targets in 120 GeV/c protons beam interactions.

Table 194.1 K_s^0 production cross-sections using 120 GeV/c protons beam interactions

Interaction	σ (mb) FLUKA	σ (mb) data $\pm$ stat $\pm$ syst
p + Be/120 GeV/c	44.35	39.54±1.46±6.97
p + C/120 GeV/c	66.34	57.49±2.69±9.54
p + Bi/120 GeV/c	664.40	520.33±16.87±80.77

References

1. R. Raja: Nucl. Instrum. Methods Phys. Res. Sect. A **553**, 225 (2005)
2. T.S. Nigmanov et al., Phys. Rev. D **83**, 012002 (2011)
3. N. Abgrall et al., arXiv:physics.acc-ph/1309.1997v1

Chapter 195
Front-End Readout for INO-ICAL GRPC

Moh. Rafik, Aman Phogat, Ankit Gaur, Ashok Kumar and Md. Naimuddin

195.1 Introduction

The Indian Based Neutrino Observatory will going to use about the 28000 RPCs for the operation of the ICAL detector [1–3]. The complete opeartion requires commissioning of huge channels. For this purpose, we are proposing very compact front-end ASIC HARDROC, which can read 64 channels simultaneously. It consists of fast low impedance preamplifier with 8 bits variable gain per channel, one slow shaper (50–150 ns), 3 variable CRRC fast shapers followed by 3 discriminators with a range of 10 fC- 10 pC.

195.2 Analog Cross Talk Study of Hardroc2

We have injected the 1 pC charge in one channel. The crosstalk is measured from the neighbor channels by looking directly into the channel output. The 3% crosstalk is well differentiated and located on the input. It has also been confirmed that there is no long distance cross talk, as shown in the Fig. 195.1.

195.3 Jitter Measurement of Discriminated Pulse

Jitter is measured by using EZJIT Jitter analysis software. TIE (time interval error) is measured here. We have injected different charges (pC) in one particular channel and discriminated pulse jitter is measured (Fig. 195.2).

Moh. Rafik (✉) · A. Phogat · A. Gaur · A. Kumar · Md. Naimuddin
Department of Physics & Astrophysics, University of Delhi, Delhi, India
e-mail: mohdrafik.ece12@iitbhu.ac.in

Md. Naimuddin (ed.), *XXII DAE High Energy Physics Symposium*, Springer Proceedings in Physics 203, https://doi.org/10.1007/978-3-319-73171-1_195

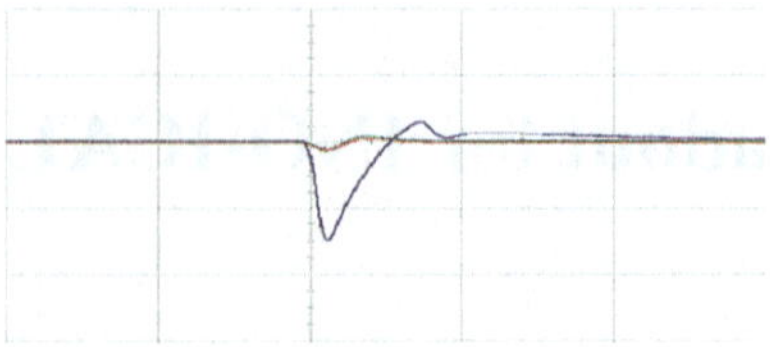

Fig. 195.1 Cross talk Measurement

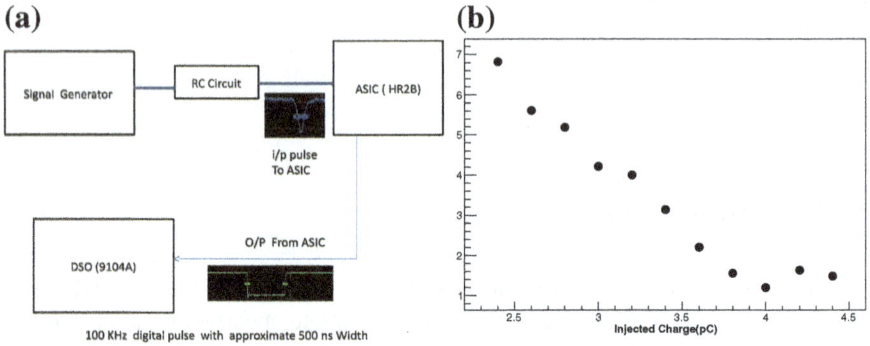

Fig. 195.2 **a** Block Diagram for Jitter Measurement setup and **b** Jitter measurement

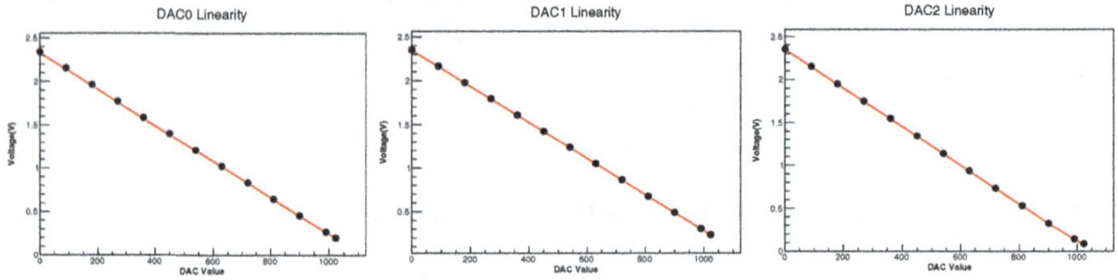

Fig. 195.3 DAC0 (Left most), DAC1 (Middle), DAC2 (Right most) Linearity Curve

195.4 DACs Linearity

The threshold of each discriminator is decided by a 10 bit-DAC. These three DACs are similar, except that the reference current for DAC0 equal or half of the I(ref) of DAC1 and DAC2. In this way we can set the fine tuning for a small charge dynamic range of charge input (fC). DAC0 sets the smallest thresholds, from 10 fC up to 100 fC (Fig. 195.3).

195.5 Conclusion

Approximately 3% Cross talk is noted and there is no long distance cross talk. Time interval error of HARDROC2 Discriminated pulse is measured which varies from few Pico second to 7 ns and very less compare to the discriminated pulse width 500 ns. Linearity of DACs is studied and relationship between DAC and voltage is derived, which is very useful for the further study.

References

1. *Development and Commissioning of the HARDROC based Readout for the INO-ICAL Experiment*, https://doi.org/10.1088/1748-0221/11/10/C10004
2. Md. Naimuddin et al., Characterization of glass electrodes and RPC detectors for INO-ICAL experiment. JINST **9**, C10039 (2014)
3. A. Kumar et al., RPC detector characteristics and performance for INO-ICAL experiment. J. Instrum. **11** C03034 (2016)

Chapter 196
Production of D-Mesons in p+p and p+Pb Collisions at LHC Energies

R. C. Baral, S. K. Tripathy, M. Younus, Z. Naik and P. K. Sahu

The modification in spectra of the observed particles in the heavy ion collision has effect of cold nuclear matter (CNM) [2] before formation of QGP which are often masked by hot and dense matter effects. The p+Pb collisions give us a unique opportunity to study these initial nuclear effects. A heavy quark owing to its large mass is produced much before the formation of QGP and remains free to probe thermalized medium without carrying any prior effects due to nucleus [3, 4]. From the recent result of p+Pb data [5] on particle production, the value of nuclear modification factor (R_{pPb}) deviates from unity by a significant amount mostly in low and mid-p_{T} regions, which shows a considerable CNM effect on heavy quark. Different models used for this analysis are HIJING [6], AMPT [7], NLO [8] and FONLL [9].

ALICE has recently published results on D-meson in p+p and p+Pb collisions [10]. For p+p system, the study is based on the mid-rapidity region, i.e., $|y_{cms}| < 0.5$, whereas for p+Pb system the rapidity range is $-0.96 < y_{cms} < 0.04$. We have ensured that no D-meson is coming from B-meson in our models. Normalized p+p yield was divided by $T_{pp} = 1.39 \times 10^{-5}\,\mu\mathrm{b}^{-1}$ to obtain cross-section, while for p+Pb, $T_{pPb} = 9.8334 \times 10^{-5}\,\mu\mathrm{b}^{-1}$ [12]. R_{pPb} can be defined as:

$$R_{\mathrm{pPb}} = \frac{(\frac{d\sigma}{dp_T})_{pPb}}{A \times (\frac{d\sigma}{dp_T})_{pp}}, \qquad (196.1)$$

where A is the mass number of a nucleus, $d\sigma$ is the p_{T} differential cross-section. Figure 196.1 shows p_{T}-dependence of average R_{pPb} of D^0, D^+ and D^{*+} mesons in p+Pb data at $\sqrt{s_{\mathrm{NN}}} = 5.02\,\mathrm{TeV}$. The calculations from HIJING and AMPT in p+Pb data are showing prominent CNM effects such as shadowing, EMC and multi-parton

R. C. Baral (✉) · S. K. Tripathy · M. Younus · P. K. Sahu
Institute of Physics, HBNI, Bhubaneswar, India
e-mail: rcbaral@iopb.res.in

S. K. Tripathy · Z. Naik
Sambalpur University, Burla, India

Md. Naimuddin (ed.), *XXII DAE High Energy Physics Symposium*, Springer Proceedings in Physics 203, https://doi.org/10.1007/978-3-319-73171-1_196

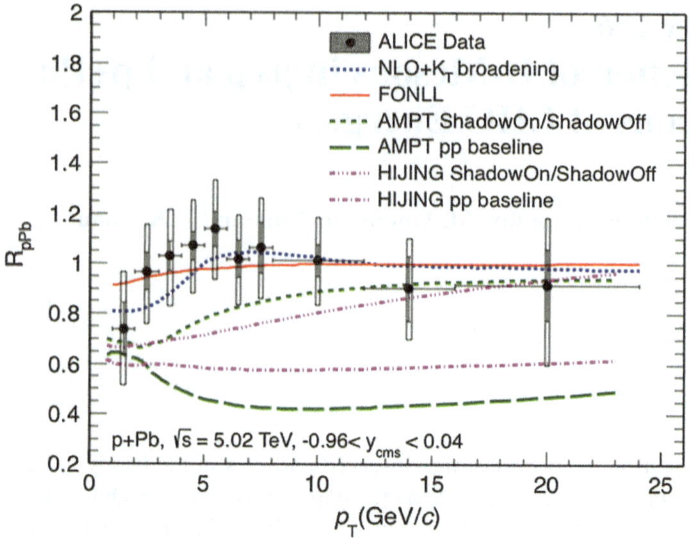

Fig. 196.1 Nuclear modification factor for D-meson in p+Pb at $\sqrt{s_{NN}} = 5.02\,\text{TeV}$ [11]

scattering effects, for the entire p_T range. The results underestimate the experimental data. The reason that R_{pPb} from AMPT differs from that of HIJING may be due to additional partonic and hadronic transport parts in AMPT. In the case of NLO having shadowing feature and momentum broadening effect (Cronin) due to re-scattering is closer to the trend of the data within error bars. FONLL gives only very small shadowing effect for $p_T < 10\,\text{GeV}/c$. To study explicitly the shadowing effect, we have taken the ratio of p_T-differential cross-section with shadow on and same with shadow off in p+Pb data. We see significant shadowing effect particularly at the lower and intermediate p_T regions, while any other effects due to Pb nucleus are canceled both from numerator and denominator of the ratio. In conclusion, there is an initial CNM effect playing an important role in all models. Further improvements are required in our parameter dependent models, to explain the experimental data.

References

1. EM Collab., J. Ashman et al., Phys. Lett. B **202**, 603 (1988); EM Collab., M. Arneodo et al., Phys. Lett. B **211**, 493 (1988)
2. PHENIX Collab. (M. G. Wysocki), Nucl. Phys. A **904–905**, 67c–74c (2013)
3. Z. W. Lin, M. Gyulassy, Phys. Rev. C **51** 2177 (1995); M. Younus, D.K. Srivastava, J. Phys. G, Nucl. Part. Phys. **37** 115006 (2010)
4. E. Eichten, I. Hinchliffe, K. Lane, C. Quigg, Rev. Mod. Phys. **56**, 579 (1984)
5. ALICE Collab. (B. Abelev et al.), Phys. Rev. Lett. **110**, 082302 (2013)
6. X.N. Wang, M. Gyulassy, Phys. Rev. D **44**, 3501 (1991)
7. Z.-W. Lin, C.M. Ko, B.-A. Li, B. Zhang, S. Pal, Phys. Rev. C **72**, 064901 (2005)

8. M.L. Mangano, P. Nason, G. Ridolfi, Nucl. Phys. B **373**, 295 (1992)
9. M. Cacciari, M. Greco, P. Nason, J. High Energy Phys. **9805**, 007 (1998); M. Cacciari, PoS High Energy Phys. **2005**, 137(2006), arXiv:hep-ph/0512251
10. ALICE Collab. (B. Abelev et al.), J. High Energy. Phys. **01**, 128 (2012); ALICE Collab. (B. Abelev et al.), Phys. Rev. Lett. **113**, 232301 (2014)
11. R.C. Baral, S.K. Tripathy, M. Younus, Z. Naik, P.K. Sahu, Int. J. Mod. Phys. E **25**(11), 1650092 (2016)
12. ALICE Collab. (J. Adam et al.), Phys. Rev. C **91**, 064905 (2015)

Chapter 197
Study of Discrete Symmetries for B_s Meson System

Bharti Kindra and Namit Mahajan

197.1 Introduction

CP violation has been well studied in $K^0 - \bar{K}^0$, $B^0 - \bar{B}^0$ and $D^0 - \bar{D}^0$ meson system [1, 2]. CPT theorem states that lorentz covariant theories are invariant under CPT transformation which is consistent with experimental results. This implies T violation in neutral meson systems.

T-violation has been studied in K meson system by observing a difference in rates of $K^0 \to \bar{K}^0$ and $\bar{K}^0 \to K^0$. However, this observable is also CP violating, it can not guarantee T violation [3]. So, a genuine T-asymmetry was suggested by J. Bernabeu et al. in context of B meson system [4]. The idea was based on Einstein–Podolsky–Rosen (EPR) entanglement [5], which was used to form asymmetries by measuring time reversed processes. These asymmetries were tested at PEP-II at SLAC and the results were in agreement with standard model [6].

We extend the same formalism to study discrete symmetries in $B_s^0 - \bar{B}_s^0$ meson states which are obtained from decays of $\Upsilon(5s)$ at B factories. This is relatively complicated both phenomenologically and analytically. This is because unlike B_d system, $\Delta\Gamma \neq 0$ for B_s system. Also, in B factories, $B_s - \bar{B}_s$ pairs are produced in both symmetrically and anti-symmetrically correlated state while $B - \bar{B}$ pairs are produced in anti-symmetrically correlated state only.

We follow the same conventions and definitions as given in [4] with two changes. First, initial state is different, as it is a superposition of $C = -$ and $C = +$ state. It is written as

B. Kindra (✉) · N. Mahajan
Physical Research Laboratory, Ahmedabad, India
e-mail: bharti@prl.res.in

B. Kindra
Indian Institute of Technology, Gandhinagar, India

Md. Naimuddin (ed.), *XXII DAE High Energy Physics Symposium*, Springer Proceedings in Physics 203, https://doi.org/10.1007/978-3-319-73171-1_197

$$|i\rangle = \frac{1}{\sqrt{2}}[\alpha(|B^0\bar{B}^0\rangle - |\bar{B}^0 B^0\rangle) + \beta(|B^0\bar{B}^0\rangle + |\bar{B}^0 B^0\rangle)]$$

Second, we are not neglecting $\Delta\Gamma$.

197.2 Asymmetries

Using the double tagging method, we have derived five asymmetries which are genuine signs of CPT/CP/T violation [7]. One of them is, Equal semileptonic asymmetry which is a CP and T asymmetry, and is defined as

$$A_1 = \frac{P(l^+, l^+) - P(l^-, l^-)}{P(l^+, l^+) + P(l^-, l^-)}$$

where $P(a, b)$ is the transition probability of observing a at time t_1 and b at time t_2. Expression for $C = -$ state:

$$A_1 = 4\kappa\Re(\epsilon)$$

For $C = +$ state,

$$A_1 = \frac{\begin{pmatrix} 2\kappa\big[-\Re(\delta)\left(x^3\sinh(\Delta\Gamma\text{tm}) + x^3 y\cosh(\Delta\Gamma\text{tm}) + 2xy\cos(\Delta\text{mtm}) + 8y\sin(\Delta\text{mtm})\right) + \\ \Im(\delta)\left(2x^2\cosh(\Delta\Gamma\text{tm}) + x^2\cos(\Delta\text{mtm}) + 3x\sin(\Delta\text{mtm}) + 8y\sinh(\Delta\Gamma\text{tm})\right) + \\ 2x\Re(\epsilon)\left(-\cos(\Delta\text{mtm}) + x^2\cosh(\Delta\Gamma\text{tm}) + x^2 y\sinh(\Delta\Gamma\text{tm}) + x\sin(\Delta\text{mtm})\right)\big] \end{pmatrix}}{x\left(-\cos(\Delta\text{mtm}) + x^2\cosh(\Delta\Gamma\text{tm}) + x^2 y\sinh(\Delta\Gamma\text{tm}) + x\sin(\Delta\text{mtm})\right)}$$

Here, $x = \frac{\Delta m}{\Gamma}$ and $y = \frac{\Delta\Gamma}{\Gamma}$.

197.3 Conclusion

Five such asymmetries have been calculated which can be measured using data for $\Upsilon(5s)$ at B factories. The expressions are functions of four parameters. Hence, these asymmetries will constrain these CPT-,T- and CP- violating parameters in B_s meson system.

References

1. J.H. Christenson, J.W. Cronin, V.L. Fitch, R. Turlay, Evidence for the 2π decay of the K_2^0 meson. Phys. Rev. Lett. **13**, 138–140 (1964)
2. Bernard Aubert et al., Observation of CP violation in the B^0 meson system. Phys. Rev. Lett. **87**, 091801 (2001)
3. L. Alvarez-Gaume, C. Kounnas, S. Lola, Panagiotis Pavlopoulos, Violation of time reversal invariance and CPLEAR measurements. Phys. Lett. B **458**, 347–354 (1999)
4. M.C. Banuls, J. Bernabeu, Studying indirect violation of CP, T and CPT in a B factory. Nucl. Phys. B **590**, 19–36 (2000)
5. A. Einstein, B. Podolsky, N. Rosen, Can quantum-mechanical description of physical reality be considered complete? Phys. Rev. **47**, 777–780 (1935)
6. J.P. Lees et al., Observation of Time Reversal Violation in the B^0 Meson System. Phys. Rev. Lett. **109**, 211801 (2012)
7. N. Mahajan, B. Kindra, in preperation

Chapter 198
Texture Zeros of the Lepton Mass Matrices

Neelu Mahajan

The discovery of neutrino oscillations have provided us convincing evidence for massive neutrinos and a signal to explore New Physics (NP). The neutrino experiments [1] give us a very precise, finite and large value of reactor angle θ_{13} and nearly maximal atmospheric angle θ_{23}. But, the data provide us no clue regarding the absolute neutrino mass scale and the related issue of the neutrino mass hierarchy. Also, the nature of neutrinos whether Dirac or Majorana is still unknown. From the theoretical point of view, the elegant way to explain this is the seesaw mechanism that gives us an explanation for the smallness of the neutrino mass as compared to their charged counterparts which requires the neutrinos to be Majorana particles. In this paper, we are comparing the two phenomenological models of the Majorana neutrino mass matrix and confront them with the recent neutrino oscillation data to check their compatibility.

To begin with, we present hermitian texture 3 zero mass matrices, e.g.,

$$M_l = \begin{pmatrix} B_l & A_l & 0 \\ A_l^* & 0 & 0 \\ 0 & 0 & C_l \end{pmatrix}, \qquad M_{\nu D} = \begin{pmatrix} C_\nu & 0 & A_\nu \\ 0 & 0 & B_\nu \\ A_\nu^* & B_\nu^* & 0 \end{pmatrix}, \tag{198.1}$$

M_l and $M_{\nu D}$ respectively corresponding to Dirac-like charged lepton and neutrino mass matrices. Details of diagonalizing transformation can be looked in our previous paper [2]. In case of Majorana neutrinos, the neutrino mass matrix M_ν is given by seesaw mechanism, for example,

$$M_\nu = -M_{\nu D}^T \, (M_R)^{-1} \, M_{\nu D}, \tag{198.2}$$

where $M_{\nu D}$ and M_R are respectively, the Dirac neutrino mass matrix and the right-handed Majorana neutrino mass matrix. In Model I, following Fukugita et al. [3] we

N. Mahajan (✉)
Department of Physics, G.G.D.S.D College, Chandigarh, India
e-mail: neeluggdsd@gmail.com

Md. Naimuddin (ed.), *XXII DAE High Energy Physics Symposium*, Springer Proceedings in Physics 203, https://doi.org/10.1007/978-3-319-73171-1_198

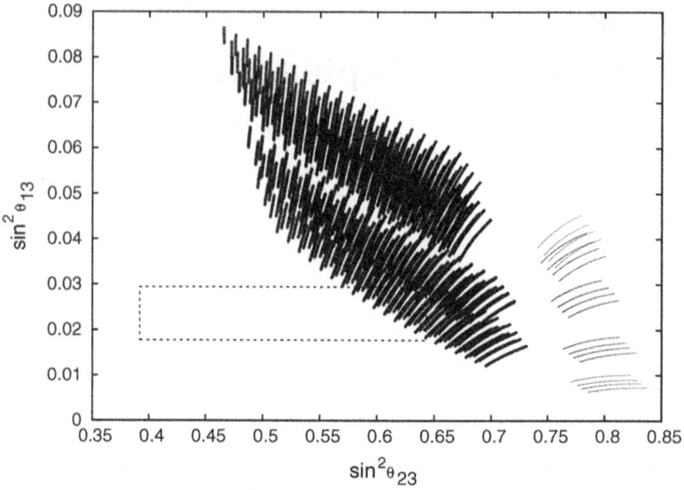

Fig. 198.1 Plot showing the parameter space for mixing angles $sin^2\theta_{13}$ and $sin^2\theta_{23}$ as shown by dots and crosses for model I and II respectively. The blank rectangular region shows the experimentally allowed 3σ region for $sin^2\theta_{13}$ and $sin^2\theta_{23}$

consider the structure of $M_R = m_R I$ as simple as possible to keep the number of parameters under control, where I is the unity matrix and m_R denotes a very large mass scale. In Model II, following Fritzsch et al. [4] we can introduce the same texture structure on M_R as on M_{ν_D} given by (198.1). Using seesaw equation (198.2), we can calculate M_ν and it comes out to be texture 2 zero type. Its matrix elements are given by

$$\mathcal{A}_\nu = \frac{\mathcal{A}_D^2}{\mathcal{A}_R}, \mathcal{B}_\nu = \frac{\mathcal{B}_D\mathcal{C}_D}{\mathcal{C}_R} + \frac{\mathcal{C}_D\mathcal{A}_D}{\mathcal{A}_R}\left(\frac{\mathcal{B}_D}{\mathcal{C}_D} - \frac{\mathcal{B}_R}{\mathcal{C}_R}\right), \mathcal{C}_\nu = \frac{\mathcal{C}_D^2}{\mathcal{C}_R}, \mathcal{D}_\nu = \frac{\mathcal{C}_D^2}{\mathcal{A}_R}\left(\frac{\mathcal{B}_D}{\mathcal{C}_D} - \frac{\mathcal{B}_R}{\mathcal{C}_R}\right)^2. \tag{198.3}$$

To check the viability of these two models, we have presented a plot showing the parameter space by constraining $sin^2\theta_{12}$ by its 3σ experimental bound [5]. Interestingly, plot shown by dots does not show any overlap with the allowed rectangular region, concludes that the model with M_R diagonal is not able to reproduce the oscillation data. Model II shown by crosses have a significant overlap with the rectangular region, survives current mixing data having θ_{23} maximal and also reveals the normal hierarchy of Majorana neutrino masses. Thus, texture zero model for M_R enhances the predictive power of the model (Fig. 198.1).

Acknowledgements N.M. would like to thank Principal, GGDSD College, Chandigarh for providing facilities to work.

References

1. P. Adamson, [MINOS Collaboration], Phys. Rev. Lett. **107**, 181802 (2011), K. Abe et al. (T2K), Phys. Rev. D. **91**, 072010 (2015)
2. N. Mahajan, M. Randhawa, M. Gupta, P.S. Gill, Prog. Theor. Exp. Phys. **2013**, 083B02 (2013)
3. M. Fukugita, Y. Shimizu, M. Tanimoto, T. Yanagida, Phys. Lett. B **716**, 294 (2012)
4. H. Fritzsch, S. Zhou, Phys. Lett. B **718**, 1457 (2013)
5. M.C. Gonzalez-Garcia, M. Maltoni, J. Salvado, JHEP 1212 (2016)

Chapter 199
Study of a Triple GEM Detector Operated with Different Argon Based Gas

Rajendra Nath Patra, R. N Singaraju, S. Biswas, Z. Ahammed, T. K. Nayak and Y. P. Viyogi

199.1 Introduction

The study of GEM detectors over the last two decades has mainly focused on its rate capability, position and timing resolution and ion suppression [1]. A large number of studies have used the gas mixture of Ar:CO_2 (70:30) but very few studies are available for other gas mixtures [1], that too mainly for single GEMs. Having more Argon may be sometimes advantageous if high count rate is not the issue but running the detector at lower voltage becomes more important. Our study with Ar:CO_2 (70:30) gas mixture has been presented earlier [2]. Here we report our results for the Ar:CO_2 (90:10) mixture and compare the two results.

The construction of $10 \times 10\,\text{cm}^2$ triple GEM detector is described in [3]. A negative high voltage (HV) is applied using a resistor chain such that the voltages across the GEM foils are in the range of 280–350 V.

RNP supported by UGC-NET fellowship and YPV thanks to INSA Senior Scientist.

R. N. Patra (✉) · R. N. Singaraju · Z. Ahammed · T. K. Nayak · Y. P. Viyogi
Variable Energy Cyclotron Centre, HBNI, Kolkata 700064, India
e-mail: rajendrapatra07@gmail.com

S. Biswas
Bose Institute, Department of Physics and CAPSS, Sector V, Kolkata 700091, India

T. K. Nayak
CERN, Geneva 23, Switzerland

Md. Naimuddin (ed.), *XXII DAE High Energy Physics Symposium*, Springer Proceedings in Physics 203, https://doi.org/10.1007/978-3-319-73171-1_199

199.2 Experimental Measurements and Results

To measure the efficiency as a function of applied HV we used ^{106}Ru-Rh β-source, with triggers provided by the coincidence signal of two crossed scintillator detectors kept above and a third scintillator detector kept below the GEM detector under study. The result is shown in Fig. 199.1a. The efficiency at the plateau is found to be ~95%.

The gain and energy resolution are measured with ^{55}Fe X-ray source. The energy spectrum of ^{55}Fe is shown in Fig. 199.1b. The gas gain is calculated assuming that the main peak in Fig. 199.1b corresponds to full energy deposition of 5.9 keV X-ray. Variation of gas gain with applied HV is shown in the Fig. 199.2a along with the earlier results of Ar:CO_2 (70:30) gas mixture [2].

Energy resolution calculated from the Gaussian fit parameters of ^{55}Fe main peak is shown in Fig. 199.2(b) for various HV values along with the results of Ar:CO_2

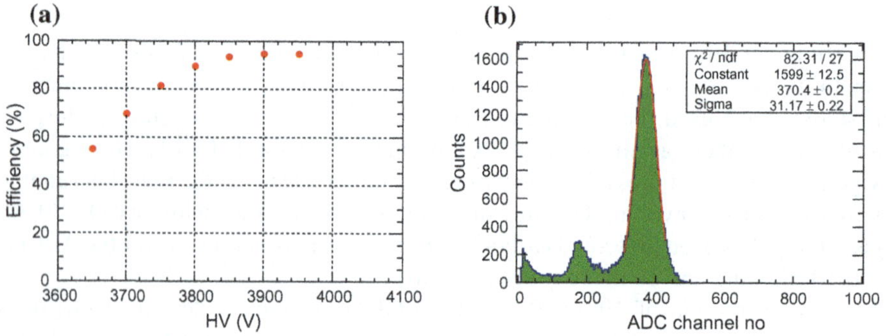

Fig. 199.1 **a** Efficiency as a function of HV obtained by using ^{106}Ru-Rh source, **b** ^{55}Fe spectrum at 3800 V for the GEM detector operated with 90:10 gas mixture

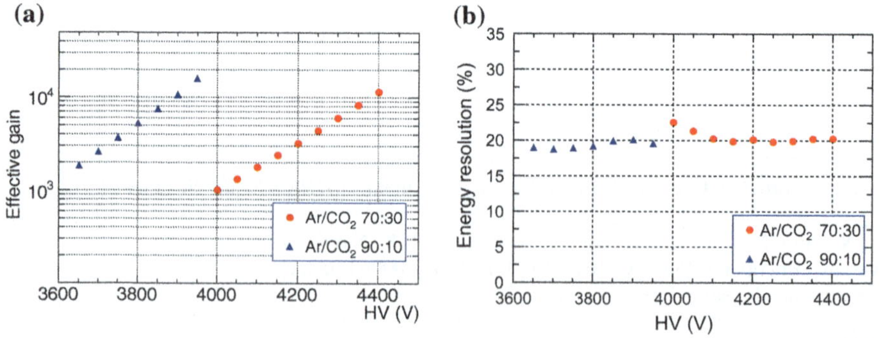

Fig. 199.2 **a** Gain as a function of HV for both gas mixtures, **b** Energy resolution (FWHM) as a function of HV for both gas mixtures

(70:30) gas mixture [2]. It is found that for Ar:CO_2 (90:10) gas mixture the gain is much higher at relatively much lower applied voltage and the energy resolution is also slightly better.

199.3 Summary

The triple GEM detector operated with Ar:CO_2 (90:10) gas mixture is found to provide same efficiency at much lower operating voltage compared to the one using Ar:CO_2 (70:30) gas mixture, the gas gain is also higher and energy resolution is slightly better.

References

1. F. Sauli, Nucl. Instrum. Methods **A805**, 2 (2016)
2. R.N. Patra et al., in Proceedings of the DAE-BRNS Symposium on Nuclear Physics, vol. 61 (2016), p. 1050
3. R.N. Patra et al., Nucl. Instrum. Methods A. **824**, 501–503 (2016)

Chapter 200
Masses of P-Wave Charmonia in Isospin Asymmetric Strange Hadronic Matter Using QCD Sum Rules

Sushil Kumar

200.1 Introduction

Many different efforts have been made in many different directions for understanding and clarifying the origin of mass of the matter that surrounds us. Thus in-medium properties of hadrons containing charm and strange mesons is of considerable interest. At sufficiently high temperature, the QCD medium will undergo a phase transition from confined phase(Hadronic matter)to deconfined phase (QGP). The upcoming Compressed Baryonic Matter (CBM) experiment at Facility for Antiproton and Ion Research (FAIR) at Institute for Heavy Ion Research (GSI), Germany intends to study the properties of the hadrons containing charm and strange quarks extensively, while the PANDA Collaboration will study the modifications in the mass and width of charmed hadrons in the nuclear matter and the charm spectroscopy. Observation of new states with exotic quantum numbers of charm and strangeness may be possible outcome of these experiments in future. In the chiral $SU(3)$ model, the modifications in the P-wave charmonia masses are due to interactions with the hadrons and scalar mesons (σ, ζ, δ and χ). In the [1] it is suggested that within QCD sum rule calculations, interactions of heavy quarkonium states with the hadronic matter are through the gluon condensates. In the QCD sum rules, modifications of the scalar gluon condensates, $\left\langle \frac{\alpha_s}{\pi} G^a_{\mu\nu} G^{a\mu\nu} \right\rangle$ and twist-2 tensorial gluon condensates, $\left\langle \frac{\alpha_s}{\pi} G^a_{\mu\sigma} G^{a}{}_{\nu}{}^{\sigma} \right\rangle$ give rise to the mass modifications of $^3P_0(\chi_{c0})$ and $^3P_1(\chi_{c1})$ states.

S. Kumar (✉)
Department of Physics and Electronics, Hans Raj College,
University of Delhi, Delhi 110007, India
e-mail: sushil8207@gmail.com

Md. Naimuddin (ed.), *XXII DAE High Energy Physics Symposium*, Springer Proceedings in Physics 203, https://doi.org/10.1007/978-3-319-73171-1_200

200.2 Chiral $SU(3)$ Model

The Chiral $SU(3)$ model [2] is a field theoretical model based on chiral symmetry [3–5] and broken scale invariance [2, 6]. To include the charm meson interactions with the light hadrons, Chiral $SU(3)$ model is generalized to $SU(4)$. The Chiral $SU(4)$ model has been used to study charmed strange meson in the dense hadronic matter [7] and to study the in-medium modification of vector meson properties [2]. The general form of Lagrangian in Chiral $SU(3)$ model is as follows:

$$\mathcal{L} = \mathcal{L}_{kinetic} + \sum_{W=X,Y,V,A,u} \mathcal{L}_{BaryW} + \mathcal{L}_{vect} + \mathcal{L}_{MM} + \mathcal{L}_{SymBreak}. \tag{200.1}$$

The spin-0, spin-1 meson multiplets and baryon octet are elementary degrees of freedom in the lagrangian. In (200.1), $\mathcal{L}_{kinetic}$ is the kinetic energy term, $\mathcal{L}_{BaryW}$ includes the interaction term of the baryons with the spin-0 and spin-1 mesons. The $\mathcal{L}_{MM}$ gives the meson-meson interaction terms and $\mathcal{L}_{SymBreak}$ introduces an explicit symmetry breaking of the chiral symmetry.

The mean field approximation have been used in the present investigation to study the in medium properties of P-wave charmonia. In the mean field approximation, all the meson fields are treated as classical fields and only the scalar and vector fields contribute to the baryon-meson interactions [8]. The effective mass of the baryon of species j is given as,

$$m_j^* = -g_{\sigma j}\sigma - g_{\zeta j}\zeta - g_{\delta j}\tau_3\delta, \tag{200.2}$$

where m_j^* is the effective mass of the baryon of type, j ($j = \Lambda, \Xi^{-,0}\Sigma^{\pm,0}, p, n$), $g_{\sigma j}$, $g_{\zeta j}$ and $g_{\delta j}$, are the coupling constants for the interaction of baryon species j with the σ, ζ and δ fields respectively and τ_3 is the third component of isospin. The coupled equations of motion for the σ, ζ,δ and χ fields are solved at different values of baryon density and temperature. The parameters η and f_s give the isospin asymmetry and strangeness of the hadronic medium and are defined as

$$\eta = \frac{(\rho_n - \rho_p)}{(2\rho_B)} + \frac{(\rho_{\Sigma^-} - \rho_{\Sigma^+})}{(\rho_B)} + \frac{(\rho_{\Xi^-} - \rho_{\Xi^+})}{(2\rho_B)},\ f_s = \frac{\Sigma_j \rho_j S_j}{(\rho_B)} \tag{200.3}$$

where ρ_j is the number density of the baryon of type, $j(j = \Lambda, \Xi^{-,0}\Sigma^{\pm,0}, p, n)$ and ρ_B is the baryon density and S_j is the number of strange quarks of baryon 'j'.

200.3 QCD Sum Rules

In QCD sum Rules we approach the bound state problem(large distances) from the asymptotic freedom side (small distances). The asymptotic freedom starts to break down and confinement effects become important. Breakdown of the asymptotic freedom is signaled by the emergence of power corrections due to the non perturbative effects of QCD vacuum. In QCD sum rule [1], the in-medium masses of the lowest charmonium states can be written as

$$m^2 \simeq \frac{M_{n-1}^J(k)}{M_n^J(k)} - 4m_c^2 k \tag{200.4}$$

where M_n^J is the nth moment of the meson and k is the normalization scale. Using operator product expansion, the moment M_n^J can be written as [1]

$$M_n^J(k) = A_n^J(k)\left[1 + a_n^J(k)\alpha_s + b_n^J(k)\phi_b + c_n^J(k)\phi_c\right], \tag{200.5}$$

where $A_n^J(k)$, $a_n^J(k)$, $b_n^J(k)$ and $c_n^J(k)$ are the Wilson coefficients. The coefficients a_n^J and b_n^J are associated with perturbative radiative corrections and scalar gluon condensate term respectively.

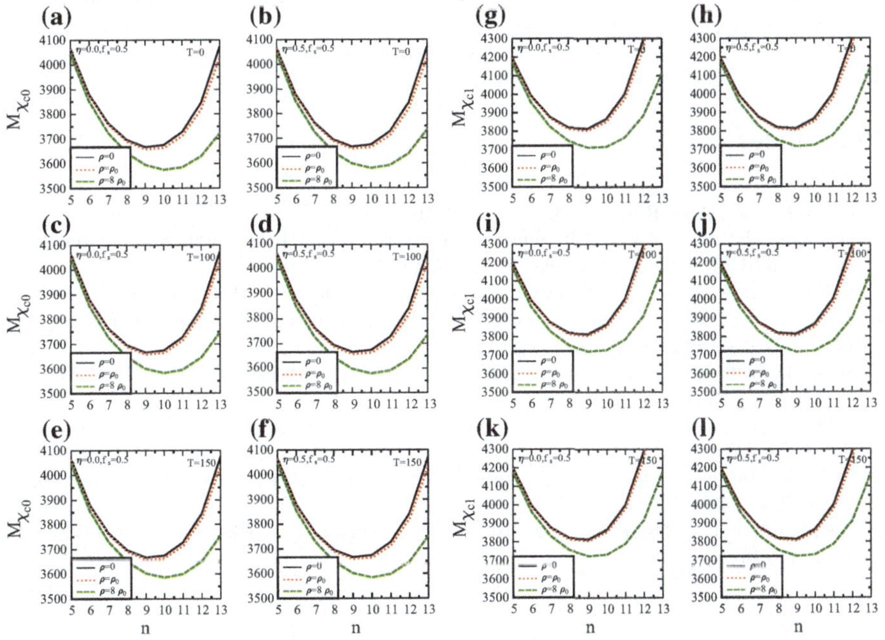

Fig. 200.1 The masses of the χ_{c0} and χ_{c1} are plotted as functions of n, for baryon density, $\rho_B = 0, \rho_0, 8\rho_0$ at different temperatures for isospin asymmetry parameter, η as 0 and 0.5 in the isospin asymmetric strange hadronic matter

200.4 Result and Discussions

Figure 200.1 shows the masses of χ_{c0} and χ_{c1} states in isospin asymmetric strange hadronic matter for baryon densities $\rho_B = \rho_0$ and $8\rho_0$ respectively, at temperatures T = 0.100 and 150 MeV for $\eta = 0.0$ and 0.5. As dilaton field changes very feebly with isospin asymmetry of the medium, so the effect of isospin asymmetry on the mass modifications of the χ_{c0} and χ_{c1} mesons are almost negligible.

200.5 Summary

In the present investigation, the masses of the P- Wave charmonia in isospin asymmetric strange hadronic matter is calculated using modification of a dilaton field. The effect of the isospin asymmetry of the medium on masses of P-wave charmonia is almost negligible. The present study of in medium properties of P-wave charmonia can be particularly relevant to the experiments at the Facility for Antiproton and Ion Research (FAIR) at GSI, Germany.

References

1. F. Klingl et al., Phys. Rev. Lett. **82**, 3396 (1999)
2. D. Zschiesche, A. Mishra, S. Schramm, H. Stocker, W. Greiner, In-medium vector meson masses in a chiral SU(3) model. Phys. Rev. **70**, 045202 (2004) (01–13)
3. S. Weinberg, Nonlinear realizations of chiral symmetry. Phys. Rev. **166**, 1568–1577 (1968)
4. S. Coleman, J. Wess, B. Zumino, Structure of phenomenological lagrangians-I. Phys. Rev. **177**(5), 2239–46 (1969)
5. W.A. Bardeen, B.W. Lee, Some considerations on nonlinear realizations of Chiral $SU(3) \times SU(3)$. Phys. Rev. **177**, 2389–2397 (1969)
6. P. Papazoglou, D. Zschiesche, S. Schramm, J. Schaffner-Bielich, H. Stöcker, W. Greiner, Phys. Rev. C **59**, 411 (1999)
7. S. Kumar, In medium properties of charmed strange mesons in dense hadronic matter. Int. J. Theor. Phys **54**(5), 1644–53 (2015)
8. B.D. Serot, J.D. Waleska, The relativistic nuclear many-body problem. Adv. Nucl. Phys. **16**, 1–327 (1986)

Chapter 201
HO Weight Factor in Particle Flow Algorithm in CMS Experiment

Suman Chatterjee and Gobinda Majumder

On behalf of the CMS Collaboration

201.1 Introduction

The Compact Muon Solenoid (CMS) [1], one of the two general purpose detectors at LHC, contains, as we go radially outward from the beam line, a tracker, an electromagnetic calorimeter, a hadron calorimeter inside 3.8 T solenoidal magnetic field and a muon spectrometer outside the magnetic field. CMS also has a Outer Hadron calorimeter (HO) outside the magnetic field, before the muon chamber, to sample the tail of the highly energetic hadron showers as hadron calorimeter inside the magnet is not sufficiently thick to capture completely the shower initiated by the highly energetic jets. HO consists of five rings in total. The one at central region with two layers, about $\eta = 0$, at different depths is known as Ring 0. The other four rings are Ring $1(-1)$, Ring $2(-2)$ as one goes along positive (negative) z direction. Each ring covers the full angular space (2π) in ϕ direction and they are subdivided to form towers of $\Delta\eta \times \Delta\phi \simeq 0.87 \times 0.87$ mapping the hadron calorimeter layers inside the magnet.

In Particle Flow algorithm, total energy measured in calorimeter is taken as

$$E_{total} = E_{ECAL} + 1.3 \times (E_{HB} + w \times E_{HO})$$

S. Chatterjee (✉) · G. Majumder (✉)
Tata Institute of Fundamental Research, Mumbai 400005, India
e-mail: suman.chatterjee@tifr.res.in

G. Majumder
e-mail: gobinda@tifr.res.in

Md. Naimuddin (ed.), *XXII DAE High Energy Physics Symposium*, Springer Proceedings in Physics 203, https://doi.org/10.1007/978-3-319-73171-1_201

where w is the weight factor HO w.r.t. HB (portion of hadron calorimeter inside the magnet up to $\eta = 1.4$). Aim of this project is to measure w and its energy and direction dependence at a function of the energy deposit the jets in HO.

201.2 Event Selection

The events with at least one jet with $P_T > 200\,\text{GeV}$ in barrel (HB) calorimeter inside the magnet and energy deposit in HO more than 5 GeV are selected. Roughly 7% of the total hadronic events satisfy these selection criteria.

201.3 Method of Approach and Experimental Results

HO weight can be optimized to counterbalance the energy leakage which gives rise to Missing Transverse Energy (MET). HO weight is also calculated from the best resolution of P_T difference of two jets in the dijet events. Figure 201.1 depicts the performance of HO weight to reduce MET (left) and to make the resolution of dijet balance better (right).

From Fig. 201.1, it can be concluded that HO weight factor increases with the energy recorded in HO cluster; to describe this pattern a functional form was tried for HO weight factor where HO weight has a linear dependence on HO cluster energy, $w = 0.5 \times (1 + \beta \times E_{HO})$ and checked how MET and dijet resolution varies with β. We find that almost same value of β ($\sim$0.012) corresponds to minimum MET/best resolution for dijet balance for all the ranges of HO cluster energy. Photon+jet balance, also used in this study, exhibit similar results.

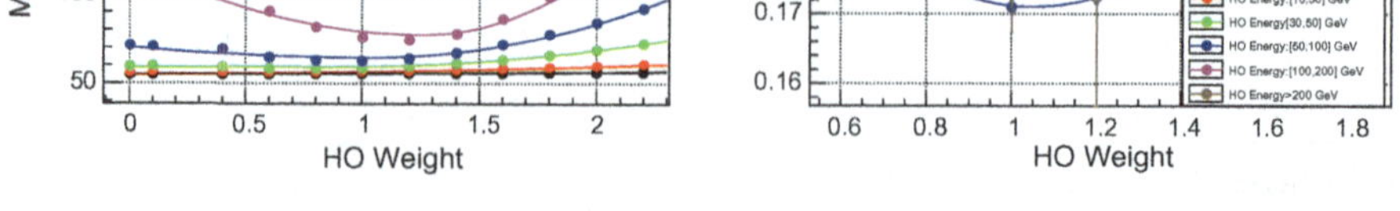

Fig. 201.1 Variation of the mean of MET distribution and variation of Gaussian width of relative P_T difference of two jets with HO weight factor for different ranges of HO cluster energy [2]

201.4 Conclusion

We observe that HO weight, with a clear energy dependence, has an appreciable impact on physical observables and can be optimized for different ranges of the energy measured in HO. We have also shown that it is possible to optimize the measurement of different observables by approximating HO weight as a linear function of the energy recorded in HO cluster.

References

1. CMS Collaboration, JINST **3**, S08004 (2008)
2. Details of this study are documented in CMS note DN-2016/023

Chapter 202
Measurement of Missing Transverse Energy in CMS Experiment

Pallabi Das and Kajari Mazumdar

202.1 Introduction

Particles that go undetected inside a detector are the main cause of missing transverse energy or MET or E_T^{miss}. By definition,

$$\vec{E}_T^{miss} = - \sum_{i \in visible} \vec{p}_T^{\,i} \tag{202.1}$$

where $\vec{p}_T^{\,i}$ is the transverse momentum of the ith detected particle, like muons, electrons or hadrons produced by hadronization of quarks and gluons. The momentum imbalance due to detector resolution and sensitivity of the detector to various identified particles must be scrutinised thoroughly before relevant physics interpretation. In practice, the MET is corrected for the potential effects of the detector measurements and details of event reconstruction algorithms.

202.2 CMS Detector

CMS detector [1] consists of four detector subsystems: tracker, electromagnetic calorimeter (ECAL), hadronic calorimeter (HCAL) and muon system. The Tracker reconstructs the tracks of charged particles and measures the transverse momentum (p_T). The ECAL and HCAL measure the energy deposits from charged and neutral particles via electromagnetic and hadronic interactions. The Muon system is used

on behalf of the CMS Collaboration

P. Das (✉) · K. Mazumdar
Tata Institute of Fundamental Research, Mumbai, India
e-mail: pallabi.das@cern.ch

Md. Naimuddin (ed.), *XXII DAE High Energy Physics Symposium*, Springer Proceedings in Physics 203, https://doi.org/10.1007/978-3-319-73171-1_202

for the detection of muons and measuring their momenta, it is the outermost part of the detector.

202.3 E_T^{miss} Corrections and Uncertainties

The simulated and reconstructed top pair production events are studied, as the inclusive final state consists of leptons, jets and genuine E_T^{miss}: $t \rightarrow Wb$, $W \rightarrow qq'$ or $W \rightarrow l\nu$. The estimate of E_T^{miss} is affected by energy thresholds for calorimeter response and the non-linearity of calorimeters. For example, the jet energy corrections applied to each of the measured jet momentum in an event are incorporated in the calculations via the following relation:

$$\vec{E}_T^{miss} = - \sum_{j \in jet} \vec{p}_T^{\,j,JEC} - \sum_{i \in uncl.} \vec{p}_T^{\,i} \tag{202.2}$$

All the particles in a particular event can be divided into two parts: some are clustered into jets while the others remain unclustered. The p_T of the unclustered particles are not corrected by the jet energy correction factor.

The E_T^{miss} uncertainty is the resultant of the uncertainties in the measurement of four-momenta of particles in various detector components. For example: the unclustered energy part of E_T^{miss}, which is not associated with well reconstructed objects, has the following contributions: charged hadron, neutral hadron, photons and forward particles. Effect of each component uncertainty on E_T^{miss} can be seen from Fig. 202.1.

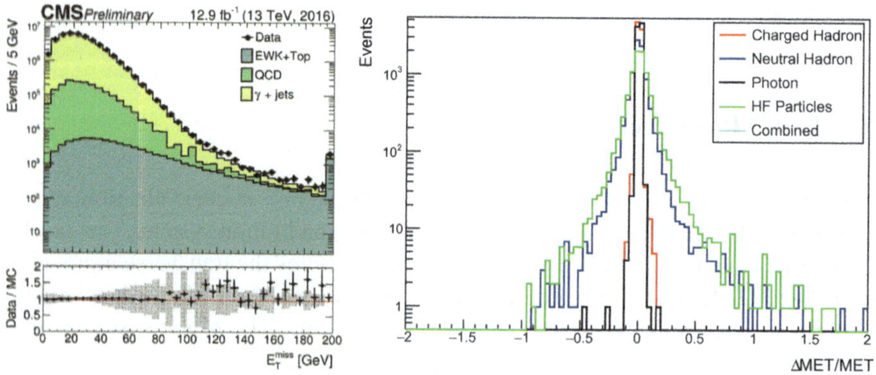

Fig. 202.1 Typical E_T^{miss} distribution from Drell–Yan events [2] and variations in fractional change in E_T^{miss} observed in $t\bar{t}$ simulation

202.4 Future Work

Identification of physics processes like W production, $t\bar{t}$ production as well as search for various beyond Standard Model physics phenomena requires presence of significant E_T^{miss} in the event. Efforts are being made to improve the E_T^{miss} measurements in the high pileup environment as foreseen in upcoming LHC runs.

References

1. The CMS Collaboration, The CMS experiment at the CERN LHC, JINST3:S08004 (2008)
2. The CMS Collaboration, CMS PAS JME-016-004

Chapter 203
Study of Medium Effects for Structure Properties of Quark Matter Star in Color Flavour Locking Phase

Suman Thakur and Shashi K. Dhiman

203.1 Introduction

It has been investigated [1] that the material in the core of compact stars may consist of quark matter. Compact stars which are absolutely made up of quark matter, additionally there maybe a small layer of a crust of nuclei are called Quark stars. Witten [1] claim that the quarks matter may be the true ground state of nuclear dense matter in the compact stars. Theoretical researchers generally admit that the ground state of QCD with three flavors is the color-flavor-locked (CFL) phase [2]. In present work, we propose to investigate the quarks matter within the framework of self consistent chemical potential dependent quarks quasiparticle model. Here, a set of EOS is employed to investigate the mass-radius relationship of quarks star.

203.2 Theoretical Framework

The effective hamiltonian of Quark quasiparticle Quark model may be expressed as [3],

$$H_{eff}(p, \mu_q) = \frac{d_q}{2\pi^2} \sum_{\nu_q} \sum_{q} \left(\sqrt{p^2 + m_q^2} - \mu_q\right) a_i^{\dagger} a_j + B^*(m_q, g), \quad (203.1)$$

where $q = u, d$ and s quarks, d_q is the color-spin dengenracy of quarks, ν_q represents the fermi momentum of the quarks. The $B^*(m_q, g)$ represents the mass and medium effects dependent Bag function as shown in Fig. 203.1, g is the coupling constant

S. Thakur (✉) · S. K. Dhiman
Department of Physics, Himachal Pradesh University, Shimla 171005, India
e-mail: sumanthakur88@gmail.com

Md. Naimuddin (ed.), *XXII DAE High Energy Physics Symposium*, Springer Proceedings in Physics 203, https://doi.org/10.1007/978-3-319-73171-1_203

of asymptotic freedom. The total thermodynamical potential of quarks matter in the CFL color superconducting phase is

$$\Omega_{CFL}(\bar{\mu}, \mu_e) = \Omega_{CFL}(\bar{\mu}) + \Omega^{GB}_{CFL}(\bar{\mu}, \mu_e) \tag{203.2}$$

The total energy density and pressure for CFL color superconducting quarks matter can be calculated as: $\mathcal{E}_{QM} = \Omega_{CFL}(\bar{\mu}, \mu_e) + 3\bar{\mu}n_q$, $P_{QP} = -\Omega_{CFL}(\bar{\mu}, \mu_e)$.

203.3 Results

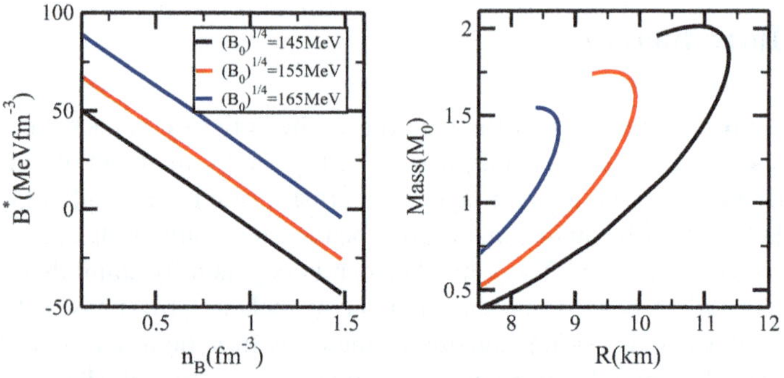

Fig. 203.1 Variation in Effective bag function as a function of density(left) and Mass-Radius Relationship(right) of Quarks Star having masses of each quarks (m_u = 5 MeV, m_d = 5MeV and m_s = 100 MeV) at CFL gap (Δ = 10 MeV) (Paired Quarks matter)

203.4 Conclusions

The range of gravitational maximum mass is found to be lie in the (1.54–2.01)$M_\odot$, corresponding radius in (8.3–10.9) km for quark stars. These theoretical results of Gravitational mass and radius are good in agreement with the recent observations for masses of $1.97 \pm 0.04 M_\odot$ and $2.01 \pm 0.04 M_\odot$ of the pulsars PSR J1614-2230 and PSR J0348+0432, respectively [4].

References

1. E. Witten, Cosmic separation of phases. Phys. Rev. D. **30**, 272–285 (1984)
2. K. Rajagopal, F. Wilczek, Enforced electrical neutrality of the color-flavor locked phase. Phys. Rev. Lett. **86**, 3492–3495 (2001)
3. K. Schertler, C. Greiner, M.H. Thoma, Medium effects in strange quark matter and strange stars. Nucl. Phys. A. **616**, 659–679 (1997)
4. J. Antoniadis et al., A massive pulsar in a compact relativistic binary. Science **340**, 1233232, 9 (2013)

Chapter 204
Simulation Study of a Possible Indium-Based Cryogenic Detector for Electron Neutrinos

Neha, G. Majumder and V. M. Datar

204.1 Introduction

An Indium based detector for studying low energy pp neutrinos from the Sun in real time was proposed by Raghavan [1] who also discussed an implementation based on a pixelated In-loaded liquid scintillator (In-LS) [2]. Another possibility based on a cryogenic Indium detector was discussed by Booth et al. [3]. The Indium detector detects electron neutrinos through the charged current interaction via an allowed Gamow-Teller transition. The prompt electron is followed by a delayed γ-ray cascade (115–498 V) from the $7/2^+$ s excited state in ^{115}Sn ($\tau \sim 4.7\,\mu$s). A triple coincidence requirement has been shown to provide a very good background rejection for the In-LS detector with a unit size of ~ 7.6 cm. In case of cryogenic detector, the smaller the size of the cube the better the discrimination from the random coincident background due to the intrinsic ^{115}In radioactivity but the larger the number of readout channels and consequent thermal load for the cryogenic system.

Neha (✉) · V. M. Datar (✉)
India-based Neutrino Observatory, Tata Institute of Fundamental Research, Mumbai 400005, India
e-mail: neha_005@tifr.res.in

V. M. Datar
e-mail: vivek.datar@tifr.res.in

Neha
Homi Bhabha National Institute, Mumbai 400094, India

G. Majumder (✉)
DHEP, Tata Institute of Fundamental Research, Mumbai 400005, India
e-mail: gobinda@tifr.res.in

Md. Naimuddin (ed.), *XXII DAE High Energy Physics Symposium*, Springer Proceedings in Physics 203, https://doi.org/10.1007/978-3-319-73171-1_204

204.2 Estimation of smallest unit of detector

The capture rate of low energy pp solar neutrinos with the ^{115}In is given by the product of cross-section of interaction (σ in units cm^2) and the estimated solar neutrino flux (ϕ in units of $cm^{-2}s^{-1}$). Estimation of [1] gives $\sigma\phi \sim 750$ SNU [1 SNU(solar neutrino units) = 10^{-36} captures/(target nucleus)(s)] for the combined capture rate due to pp, pep, ^{7}Be solar ν's. The capture rate for only pp is 693 SNU computed by taking $\sigma = 11 \times 10^{-45}\,\mathrm{cm}^2$ and $\phi = 6.3 \times 10^{10}\,\mathrm{cm}^{-2}\mathrm{s}^{-1}$ [1]. For total mass of 5 ton In, event rate is calculated as

$$(S) = (\sigma\phi)(N_A)\left(\frac{M}{A}\right) = 1.8 \times 10^{-5}s^{-1} \tag{204.1}$$

The activity of In is 0.25 Bq/gm which leads to a background event rate of 1.25 (2.5) $\times\ 10^6\ s^{-1}$ for 5 (10) ton of mass, resulting in S/N $\sim 10^{-11}$. However, a triple co-incidence technique [1] can be used to suppress this background and achieve $S/N \sim 1$, as discussed below.

For any particle detector the localization of event is a very important. In the proposed Indium-based detector the event is localized by making small cells of In whose dimensions are optimized by (a) chance co-incidence rate and (b) attenuation factor of gamma rays(μ_m). The signal in the detector comprises of (i) the prompt coincidence of e^- pulse due to ν_e - In CC interaction and 3.3 μs ($\tau_1 = 10\ \mu s$) delayed γ of E = 115.7 keV energy, (ii) the delayed coincidence of 115.7 keV gamma (from (i)) and 497.3 keV γ with $T_{1/2}$ = few ns ($\tau_2 = 10$ ns). Random coincidence rate becomes

$$N = \left(\frac{N_1^2\tau_1}{n}\right)\left(\frac{N_2\tau_2}{n}\right) = \frac{1.95 \times 10^5 s^{-1}}{n^2} \tag{204.2}$$

where $N_1 = N_2 = 1.25 \times 10^6\ s^{-1}$ for 5 ton detector and n is the number of cells in the detector.

$$\frac{S}{N} = (9 \times 10^{-11})(n^2) \tag{204.3}$$

For S/N = 1, $n = 77459$ and the size of the cell can be calculated as

$$volume = \frac{M_{In}}{\rho_{In} \times n} \approx 9 \tag{204.4}$$

Assuming the cell to be a cube, the appropriate dimension would be 2 cm $\times$ 2 cm $\times$ 2 cm.

For a γ-ray of energy E = 100 V (500 V), $\mu_m = 1.609\,\mathrm{cm}^2/\mathrm{g}$ ($9.371 \times 10^{-2}\,\mathrm{cm}^2/\mathrm{g}$) [5] which results in attenuation length = 0.085 cm(1.46 cm). This means that the 115 keV γ-ray would be detected in the same cell whereas roughly half of the 497 keV γ-rays would leave the cell for cell dimensions of 2 cm $\times$ 2 cm $\times$ 2 cm.

However the additional randoms due to a larger number of nearest neighbors would be offset by a smaller energy window due to the excellent energy resolution expected from a cryogenic detector.

References

1. R.S. Raghavan, Phys. Rev. Lett. **37**, 259 (1976)
2. C. Grieb, R.S. Raghavan, Phys. Rev. Lett. **98**, 141102 (2007)
3. N.E. Booth et al., in *Proceedings of Workshop on Low Temperature Detectors for Neutrinos and Dark Matter at Tergensee* (Springer, 1987)
4. R.S. Raghavan et al., Phys. Rev. Lett. **41**, 63 (1978)
5. XCOM, *NIST*, http://physics.nist.gov/PhysRefData/Xcom/html/xcom1.html

Chapter 205
Measurement of Angular Distribution and Integrated Flux of Cosmic Ray Muons Using 2 m × 2 m RPC Stack at IICHEP Madurai

S. Pethuraj, V. M. Datar, S. D. Kalmani, G. Majumder, N. K. Mondal, S. Mondal, P. Nagaraj, Pathaleswar, K. C. Ravindran, M. N. Saraf, B. Satyanarayana, R. R. Shinde, Dipankar Sil, S. H. Thoker, S. S. Upadhya, P. Verma and E. Yuvaraj

205.1 Analysis of Experimental Data and Monte-Carlo Generation

Events are triggered with atleast one strip hit in L1, L2, L9, L10. Muon hits are selected with multiplicity of 1, 2 and 3. So that noisy layers will be removed. Accepted Muon hits in all layers are fitted with Straight Line in both X-Z and Y-Z plane. Physical shift in the detector will be aligned by iterative method using muon data. Position offset, Multiplicity, X and Y residues, uncorrelated, correlated and trigger efficiency are calculated. Slope, intercept, deviation, χ^2 and NDF will be extracted from straight line fit in both X-Z and Y-Z plane, consequently θ and ϕ of cosmic ray muons.

In MC, the point(x, y) on the bottom trigger layer (Layer 1) is generated randomly. Theta is generated using $cos^n\theta$ over solid angle for n = 2. Phi is generated uniformly from $-\pi$ to π. If hits in top and bottom trigger layers inside the detector, then hits on the other layers also generated otherwise it will start next event. The generated hit position is smeared using resolution plots from data. The corrected multiplicity from data is used to generate the strip multiplicity depends on hit position in strip. Correlated and Uncorrelated inefficiency maps are also incorporated.

S. Pethuraj (✉)
Homi Bhabha National Institute, Mumbai 400094, India
e-mail: spethuraj135@gmail.com

V. M. Datar · S. D. Kalmani · G. Majumder · N. K. Mondal · S. Mondal · P. Nagaraj
Pathaleswar · K. C. Ravindran · M. N. Saraf · B. Satyanarayana · R. R. Shinde · D. Sil · S. S. Upadhya · P. Verma · E. Yuvaraj
Tata Institute of Fundamental Research, Mumbai 400005, India

S. H. Thoker
University of Kashmir, Srinagar 190006, India

Md. Naimuddin (ed.), *XXII DAE High Energy Physics Symposium*, Springer Proceedings in Physics 203, https://doi.org/10.1007/978-3-319-73171-1_205

205.2 Calculation of Exponent and Integrated Flux

The n value and integrated flux is calculated using (205.1) and (205.2),

$$\chi^2 = \Sigma_\theta \frac{(N^\theta_{Obs} - P_0 \sin\theta \cos^n\theta\, w(\theta))^2}{N^\theta_{Obs} + P_0 \times \sigma_w \times P_0 \times \sigma_w} \quad (205.1)$$

N^θ_{Obs} is experimentally observed theta distribution, $w(\theta)$ is weight $(N^{reco}_{MC}/N^{gen}_{MC})$ each theta bin, σ_w is weight error.

$$I_0 = \frac{I_{data}}{\epsilon_{trig} \times \epsilon_{selec} \times \epsilon_{Daq} \times \omega \times T_{tot}} \quad (205.2)$$

I_{data} - Integral of the theta distribution histogram in data. $\epsilon_{trig} = \frac{N_{trigger}}{N_{incident}}$, $\epsilon_{selec} = \frac{N_{selected}}{N_{trigger}}$, ϵ_{Daq} = Daq efficiency, $\omega = \frac{A \times N}{N'} \int_0^{\pi/3} cos^n\theta sin\theta d\theta \int_0^{2\pi} d\phi$, N - No of events inside the top and bottom trigger layer, N' - No of events generated on bottom trigger layer, T_{tot} = Total time taken to record the data (in seconds).

The calculated $n = 2.06 \pm 0.21$ and $I_0 = (5.99 \pm 0.48)\times\ 10^{-3}\,\text{cm}^{-2}\text{s}^{-1}\text{sr}^{-1}$. Errors includes the statistical error in data, as well as systematic error due to mismatch of data and MC, trigger efficiency, selection efficiency, event selection criteria(minimum number of layers for reconstructed trajectory, maximum χ^2/ndf) and Daq efficiency. Using similar 12 stack of 1 m × 1 m at TIFR (Mumbai) obtained $n = 2.15 \pm 0.01$ and $I_0 = (6.217 \pm 0.005) \times 10^{-3}\ \text{cm}^{-2}\text{s}^{-1}\text{sr}^{-1}$ [1].

References

1. S. Pal et al., Measurement of integrated flux of cosmic ray muons at sea level using the INO-ICAL prototype detector. JCAP **1207** (2012)
2. J. Crookes, B. Rastin, An investigation of the absolute intensity of muons at sea-level. Nucl. Phys. B **39**, 493 (1972). [IN SPIRE]

Chapter 206
Development of a ±6 kV Bias Supply for ICAL Detectors of INO

A. Manna, B. Satyanarayana, R. R. Shinde, M. N. Saraf, D. Sil and E. Yuvaraj

206.1 RPC HV Bias Supply

The supply for RPC biasing is designed as a ± 0–6 kV split supply [1], so as to minimize HV leak problem on the chamber surface. Figure 206.1 shows a simplified block diagram of the supply. The positive and negative HV outputs are generated by two low cost, low noise HV DC-DC converters specially developed for this application. The converter topology, based on a current fed resonant Royer circuit [2], minimizes harmonic generation and RFI due to quasi sinusoidal power delivery of the inverter. In addition to programming of output voltages, the supply has provision for remote on/off control, setting adjustable HV ramp rates and output voltage/load current read back facility. An on-board Atmel AVR microcontroller receives all the HV control commands from the local RPC Digital Front-End data acquisition (DFE) board through a 4-wire SPI interface. For debugging purposes, a separate RS-232 serial port has been provided in the card, to test HV supply functionalities in the absence of the DFE controller. The debugging commands can be issued from a PC hyper terminal directly.

206.2 Salient Features and Performance

The HV supply is constructed in a triangular shaped PCB (dimension: 10.5 cm × 10.5 cm) for mounting in one corner of the RPC as shown in Fig. 206.2. To

A. Manna (✉)
Electronics Division, Bhabha Atomic Research Centre, Trombay, Mumbai 400085, India
e-mail: abhman@barc.gov.in

B. Satyanarayana · R. R. Shinde · M. N. Saraf · D. Sil · E. Yuvaraj
Department of High Energy Physics, Tata Institute of Fundamental Research, Homi Bhabha Road, Navy Nagar, Colaba, Mumbai 400005, India

Md. Naimuddin (ed.), *XXII DAE High Energy Physics Symposium*, Springer Proceedings in Physics 203, https://doi.org/10.1007/978-3-319-73171-1_206

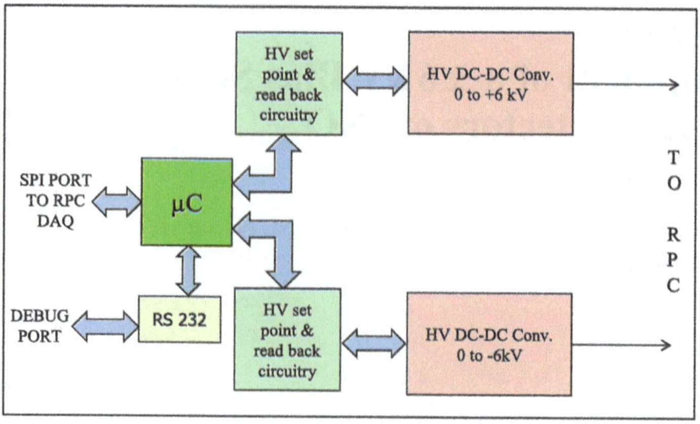

Fig. 206.1 HV Supply Schematic

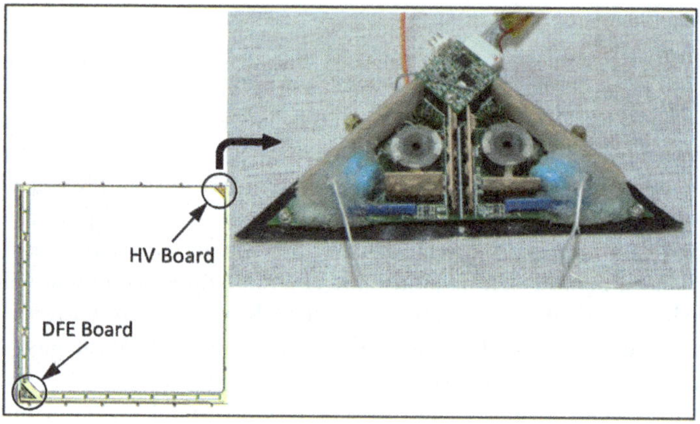

Fig. 206.2 HV Supply and its location

accommodate all the supply circuitries in the available space, the design was partitioned into several daughter boards sitting on the main triangular board. Measured performance of the bias supply is as follows: (a) O/P voltage adjustable in the range from $\pm$100 V to $\pm$6 kV. Setting resolution 2 V, accuracy $\pm$ 1% of F.S. (b) O/P ripple and noise: less than 200 mV(p-p). (c) Load current read back: 0–2000 nA (accuracy $\pm$1% of F.S.), measuring resolution 1 nA. (d) HV Ramp rate settable: 1–100 V/s.

206.3 Results and Conclusions

The Module has been tested on 2m $\times$ 2m RPC at TIFR for long term (15 days) stability of voltage, current and RPC noise rate. The performance was found to be satisfactory. The interface to DFE though the SPI link has been implemented and tested. We are ready to manufacture the HV modules for the Mini-ICAL after more long term tests for RPC noise and efficiency with the HV module.

References

1. INO-Collaboration, India-based neutrino observatory, Technical Report INO/2006/01, Tata Institute of Fundamental Research (2006), http://www.ino.tifr.res.in/ino/OpenReports/INOReport.pdf
2. J. Williams, High-voltage, low-noise dc/dc converters, EDN, 7 August 2008, pp 59-69

Chapter 207
Estimation of Leak of a Resistive Plate Chamber by Monitoring Absolute Pressure

Suryanarayan Mondal, V. M. Datar, S. D. Kalmani, G. Majumder, N. K. Mondal and B. Satyanarayana

207.1 Resistive Plate Chambers

Resistive Plate Chambers (RPC) are basically gaseous parallel-plate detectors [1, 2]. RPCs are often operated in avalanche mode with a non-flammable gas mixture of 1, 1, 1, 2-Tetrafluoroethane (95.2%), iso-C_4H_{10} (4.5%) and SF_6 (0.3%), which is continuously flushed through the gas gap. The performance of a RPC is affected by possible leaks in the detector. Hence, a proper leak test has to be performed on each RPC.

S. Mondal (✉)
Homi Bhabha National Institute, Mumbai 400094, India
e-mail: suryamondal@gmail.com

V. M. Datar · S. D. Kalmani · G. Majumder · N. K. Mondal
B. Satyanarayana
Tata Institute of Fundamental Research, Mumbai 400005, India
e-mail: vivek.datar@tifr.res.in

S. D. Kalmani
e-mail: kalmani@tifr.res.in

G. Majumder
e-mail: gobinda@tifr.res.in

N. K. Mondal
e-mail: nkm@tifr.res.in

B. Satyanarayana
e-mail: bsn@tifr.res.in

Md. Naimuddin (ed.), *XXII DAE High Energy Physics Symposium*, Springer Proceedings in Physics 203, https://doi.org/10.1007/978-3-319-73171-1_207

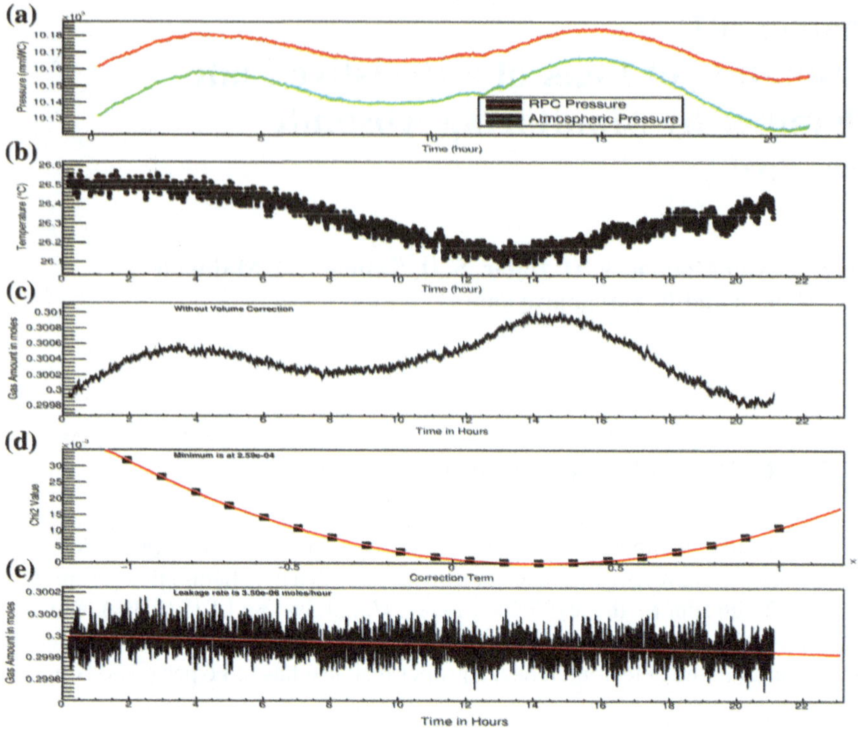

Fig. 207.1 Leak test results

207.2 Leak Rate Estimation

The chamber is first pressurised up to 45 mmWC above the atmospheric pressure (Note. 1 mmWC = 0.098 mbar). Then, the measurements of absolute pressure, both inside and outside of the RPC, are recorded continuously using two separate pressure sensors, BMP180 which is a piezo-resistive sensor, manufactured by BOSCH [3]. But, the volume of the RPC changes by a small amount due to the variation of atmospheric pressure (see Fig. 207.1a) and temperature (see Fig. 207.1b) during the day. To compensate this change in volume, a linear correction factor (x_p) is introduced. Here, it is assumed that the change in volume of RPC depends linearly on the pressure difference (ΔP) between inside and outside. The amount of gas (n), inside a chamber of volume V (6.5L for current RPC), can be calculated using the Ideal Gas equation and the correction term:

$$n_{rpc} = \left(\frac{V_{rpc}}{R} \times \frac{P_{rpc}}{T_{rpc}}\right) \times \left(1 - x_p\left(\Delta P - \Delta P_{time=0}\right)\right) \qquad (207.1)$$

For different values of x_p, the variation of n_{rpc} calculated is plotted and fitted with a straight line. For each case, χ^2 value is calculated and plotted with x_p (see Fig. 207.1c). At minimum χ^2, x_p is 2.59×10^{-4} mmWC^{-1}. The n_{rpc} plot is shown in Fig. 207.1c before volume correction. After correction, Fig. 207.1e shows a leak rate of 3.5×10^{-6} mole/hr.

Also, during the test period, the supporting button spacers inside the RPC may get detached due to any manufacturing defect. This effect can be detected clearly by observing the sudden fall of inside pressure. This method can be used for any kind of sealed chambers other than RPC where the structure of the chamber is prone to deform due to variation in atmospheric pressure.

References

1. Yu. N. Pestov, G.V. Fedotovich, A picosecond time-of-flight spectrometer for the vepp-2m based on local - discharge spark counter, SLAC-TRANS-0184, IYF-77-78 (1978)
2. R. Santonico, R. Cardarelli, Development of resistive plate counters. Nucl. Instrum. Methods Phys. Res. **187**, 377–380 (1981)
3. Bosch Sensortec, Digital pressure sensor, 7 May 2015, [Revision 2.8]
4. S. Colafranceschi et al., A study of gas contaminants and interaction with materials in RPC closed loop systems, J. Instrum. **8** (2013)

Chapter 208
Understanding Light Nuclei Production Using A Multi Phase Transport (AMPT) Model

Ashpreet Kaur, Lokesh Kumar and Natasha Sharma

208.1 Introduction

The high energy heavy-ion collision experiment aims to study the fundamental constituents of matter and their interactions with each other. Along with the study of QGP formation the high energy collisions aims to study the production mechanism of matter and anti matter. Matter(anti-matter) is composed of particles(anti-particles). It is believed that during the initial stage of universe, matter and anti matter exists in equal abundance. But it is still a mystery how this symmetry got lost in the evolution of universe with no significant amount of anti-matter being present.

To understand the production mechanism of (anti-)nuclei we use AMPT (A Multi Phase Transport) [1] model that is constructed specifically for the study of QGP in high energy heavy-ion collisions such as those at RHIC and LHC. The AMPT model includes both initial partonic and final hadronic interactions and transition between these two phases of matter.

208.2 Results

We show the results on rapidity distributions, particle ratios and coalescence parameter of proton and deuteron as well as their anti-particles. Figure 208.1 shows the basic quality assurance plot of rapidity distribution of protons (left) and of deuterons (right) for Pb–Pb collisions at $\sqrt{s_{\mathrm{NN}}}$= 2760GeV for default AMPT version. The comparison of available experimental data [2, 3] with the default and string melting versions of the AMPT model are studied. The left plot of Fig. 208.2 show the d/p ratio as function of energy for heavy-ion collisions, in addition comparison with the data

A. Kaur (✉) · L. Kumar · N. Sharma
Department of Physics, Panjab University, Chandigarh 160014, India
e-mail: ashpreet.kaur.duggal@cern.ch

Md. Naimuddin (ed.), *XXII DAE High Energy Physics Symposium*, Springer Proceedings in Physics 203, https://doi.org/10.1007/978-3-319-73171-1_208

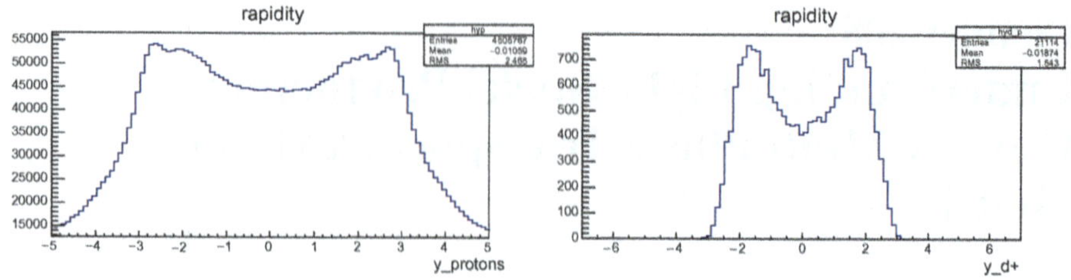

Fig. 208.1 Rapidity distribution of protons (left) and of deuterons (right) for Pb–Pb collisions at $\sqrt{s_{\rm NN}}$= 2760 GeV

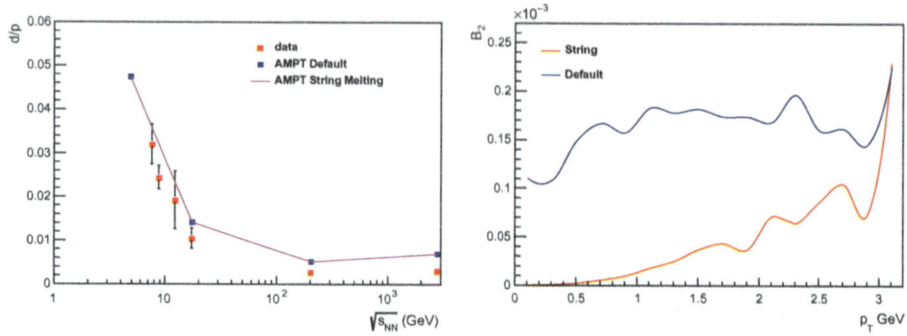

Fig. 208.2 Comparison of default and string melting version of AMPT. Left: d/p ratio as a function of energy; Right: B_2 as a function of transverse momentum for Pb–Pb collisions at $\sqrt{s_{\rm NN}}$= 2760 GeV

is also shown. Both the default and string melting version explain the experimental measurements of d/p ratio well from AGS up to LHC energies. The right plot of Fig. 208.2 shows the typical example of B_2 as a function of transverse momentum for Pb–Pb collisions at $\sqrt{s_{\rm NN}}$= 2760 GeV. B_2 increases with $p_{\rm T}$ in string melting model and remains almost constant in default version.

208.3 Summary and Outlook

We have generated events for Pb-Pb at $\sqrt{s_{\rm NN}}$= 4.8, 17.3, 2760 GeV and Au-Au at 200 GeV in AMPT model. Basic QA plots are presented. d/p is obtained for various energies and its comparison with experimental data is shown. As an outlook,we plan to obtain coalescence parameter B_2 for heavy-ion collisions from AGS, SPS, RHIC up to LHC energies using both versions of AMPT model and will compare it to experimental measurements. We will also compare elliptic flow (v_2) results from both default and string melting versions with available data to understand the production mechanism of nuclei in these collisions.

Acknowledgements This work is supported by DST-SERB project of LK (ECR/2016/000109). NS acknowledges the support of Ramanujan Fellowship (D.O. No. SB/S2/RJN- 084/2015).

References

1. L. Zhu, C.M. Ko, X. Yin, Phys. Rev. C **92**, 064911 (2015); Z.-W. Lin, S. Pal, C. M. Ko, B.-A. Li, B. Zhang, Phys. Rev. C **64**, 011902 (2001)
2. J. Adam (ALICE Collaboration), Phys. Rev. C **93**, 024917 (2016)
3. T. Anticic (NA49 Collaboration), Phys. Rev. C **94**, 044906 (2016)

Chapter 209
Orbital Dynamics Using Pseudo-Newtonian Potential

Tamal Sarkar and Arunava Bhadra

209.1 Introduction

The Pseudo-Newtonian Potentials (PNPs), which are constructed/proposed to replicate few general relativistics features approximately in Newtonian framework, are often used to study inner relativistic dynamics of the accretion flow around spacetime geometries describing black holes. For a general class of static spherically symmetric space time metrics $ds^2 = -f(r)^\beta c^2 dt^2 + \frac{1}{f(r)^\beta} dr^2 + f(r)^{1-\beta} r^2 d\Omega^2$, where $f(r)$ is the generic metric function, β is an arbitrary constant parameter, the PNP can be written as [1]

$$V_{\rm GN} = \frac{c^2(f^\beta - 1)}{2} - \left(\frac{1 - f^{2\beta-1}}{2\, f^{2\beta-1}}\right)\left[\frac{f^{2\beta} - 1}{f\,\left(f^{2\beta-1} - 1\right)}\,\dot{r}^2 + r^2\,\dot{\Omega}^2\right] . \quad (209.1)$$

The particle trajectories can be obtained by solving the Lagrangian equations for the potential given in (209.1). In the below, we studied particle trajectories for a well known naked sigularity spacetime - Janis-Newman-Winicour (JNW) metric for which $\beta = \gamma$ and $f(r) = 1 - \frac{2r_s}{\gamma r}$ where, $0 < \gamma \leq 1$. The geodesic equations are

$$\ddot{r} = -c^2\left(1 - \frac{2r_s}{\gamma r}\right)^{3\gamma-1}\frac{r_s}{r^2} + \frac{2\dot{r}^2}{\left(1 - \frac{2r_s}{\gamma r}\right)}\frac{r_s}{r^2} + \left[r - \frac{r_s}{\gamma}(1 + 2\gamma)\right]\left(\dot{\theta}^2 + \sin^2\theta\dot{\phi}^2\right) , \quad (209.2)$$

T. Sarkar (✉)
USIC, University of North Bengal, Siliguri, WB, India
e-mail: ta.sa.nbu@hotmail.com

A. Bhadra
HECRRC, University of North Bengal, Siliguri, WB, India
e-mail: aru_bhadra@yahoo.com

Md. Naimuddin (ed.), *XXII DAE High Energy Physics Symposium*, Springer Proceedings in Physics 203, https://doi.org/10.1007/978-3-319-73171-1_209

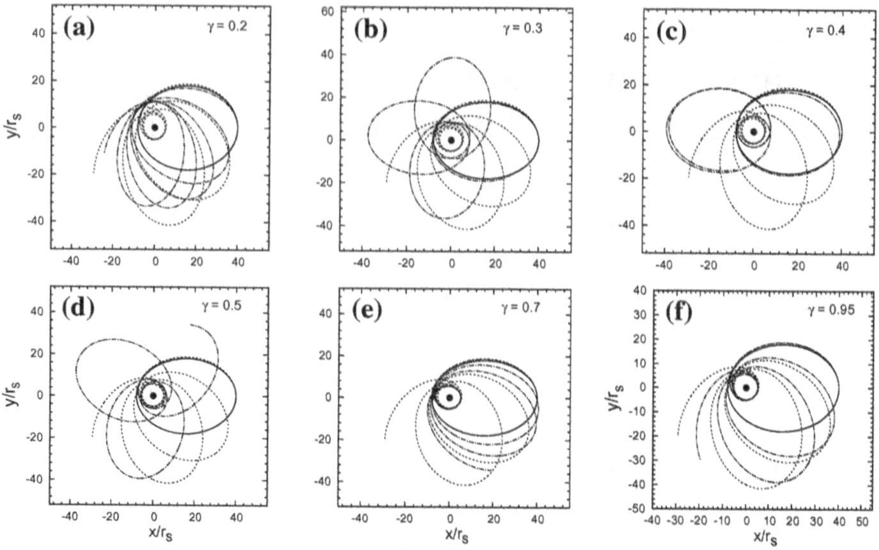

Fig. 209.1 Comparison of elliptic like trajectories of particle orbit in equatorial plane in JNW spacetime with those in Schwarzschild and Newtonian cases projected in the x-y plane. Solid and short-dashed lines corresponding to Newtonian and Schwarzschild cases, respectively. Long dotted-dashed curve in Fig. 209.1a, b, c, d, e, f are for $\gamma = 0.2, 0.3, 0.4, 0.5, 0.7, 0.95$, respectively. The particle starts from apogee with $r_a = 40r_s$ with $v_x = 0.0$ and $v_y \equiv v_{\rm in} = 0.092$ (in units of c). For $\gamma < 0.2$, no proper well defined elliptic like orbits exist with the orbital parameters chosen here

and

$$\ddot{\phi} = -\frac{2\dot{r}\,\dot{\phi}}{r}\left[\frac{\gamma r - r_s(1+2\gamma)}{\gamma r - 2r_s}\right] - 2\cot\theta\,\dot{\phi}\,\dot{\theta}\,,\;\ddot{\theta} = -\frac{2\dot{r}\,\dot{\theta}}{r}\left[\frac{\gamma r - r_s(1+2\gamma)}{\gamma r - 2r_s}\right] + \sin\theta\,\cos\theta\,\dot{\phi}^2 \tag{209.3}$$

For particle dynamics along circular orbit, $\dot{r} = 0$ and $\ddot{r} = 0$. The particle trajectories obtained by solving (209.2)–(209.3) are shown in Fig. 209.1.

209.2 Conclusions

The test particle dynamics along circular orbit in JNW space-time departs to those in Schwarzschild geometry. The stated deviation is larger for smaller γ.

References

1. S. Ghosh, T. Sarkar, A. Bhadra, Phys. Rev. D **90**, 063008 (2015)
2. E. Tejeda, S. Rosswog, MNRAS **433**, 1930 (2013)
3. E. Tejeda, S. Rosswog (2014), arXiv:1402.1171v1
4. T. Sarkar, S. Ghosh, A. Bhadra, Phys. Rev. D **92**, 083010 (2014)

Chapter 210
Ethernet Scheme for Command and Data Acquisition for the INO ICAL Detector

P. Nagaraj, M. N. Saraf, B. Satyanarayana, D. Sil, S. S. Upadhya and E. Yuvaraj

210.1 Introduction

There will be around 28,800 $2\,\text{m} \times 2\,\text{m}$ RPCs in the INO-ICAL detector [1]. Each RPC will have 64 X-side strips and 64 Y-side strips. Each RPC will have a dedicated FPGA based data acquisition unit (DFE) with Ethernet controller [2]. RPCs with On Board DFE with Ethernet facility makes the entire ICAL as LAN and helps in data transportation to server.

210.2 Event Data Acquisition

On every event an event trigger is generated by the global trigger system and is fanned out to all the RPCs (DFE modules) as shown in Fig. 210.1. On receiving the trigger signal, the participating DFEs will collect event data (Event time stamp, strip hit and time of flight) from their respective RPCs and push the event packet to transmit buffer. The integral number of event packets are sent to (one or more) Data Concentrator(s) at the back end over dedicated and pre-established TCP connections. Each DFE modules are capable of handling 38 Mbps data throughput.

P. Nagaraj · M. N. Saraf · B. Satyanarayana · D. Sil · S. S. Upadhya · E. Yuvaraj (✉)
Tata Institute of Fundamental Research, Mumbai, India
e-mail: e.yuvaraj@tifr.res.in

Md. Naimuddin (ed.), *XXII DAE High Energy Physics Symposium*, Springer Proceedings in Physics 203, https://doi.org/10.1007/978-3-319-73171-1_210

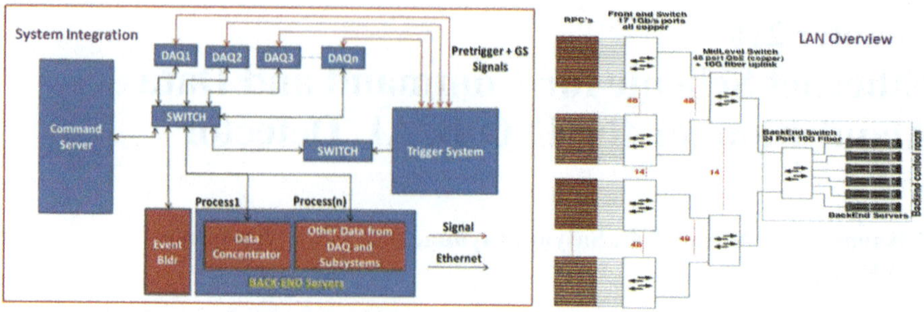

Fig. 210.1 Block diagram highlighting data and signal flow

210.3 Data Concentrator

The Data Concentrator in back end will act as a server and the DFEs will act as clients in terms of the network protocol as shown in Fig. 210.1. The Data Concentrator(s) gather the RPC data packets and attach an event number to each of the data packets belong to the event based on event time stamp comparison before transmitting the data to another server, the Event-Builder.

210.4 Command Interface

UDP based command interface was implemented in MULTICAST and UNICAST modes to control DFE functionality. Suitable handshaking and checksum schemes are implemented to ensure reliability of command and acknowledgment. Average Cycle time of a command is 1 ms.

210.5 Conclusion

Our proposed scheme will result in reduced system cost by using off the shelf Ethernet LAN components, as also the use of open source Ethernet protocols. It is very easy and flexible to configure, build and monitor the large systems. We will deploy this scheme in the proposed Mini-ICAL by 2017.

References

1. INO-Collaboration, *India-based neutrino observatory*. Technical Report INO/2006/01 (Tata Institute of Fundamental Research, 2006), http://www.ino.tifr.res.in/ino/OpenReports/INOReport.pdf
2. Soft-Core Processor Based Data Acquisition Module for ICAL RPCs with Network Interface Chapter 86, in *Proceedings XXI DAE-BRNS High Energy Physics Symposium*, (Guwahati, India, 8, 12 Dec 2014), http://www.springer.com/gp/book/9783319256177#otherversion=9783319256191

Chapter 211
Curvature Effect on QGP Equation of State

Yogesh Kumar and S. Somorendro Singh

211.1 Introduction

Quantum Chromo-Dynamics (QCD) predicts the order of phase transition from hadronic phase to the Quark–Gluon Plasma (QGP) phase.

211.2 Model Description

Earlier the free energy with effect of curvature term calculated using dynamical quark mass [1]. Further Kumar et al. [2] have modified the calculation at finite thermal quark mass. Now, we extend the previous work by using modified effective quark mass of quasiparticles and defined as [3]:

$$m_{eff}^2(T) = m_0^2 + \sqrt{2}m_0 m_q + m_q^2 \,. \tag{211.1}$$

where, m_0 is current quark mass with $N_f = 3$ and m_q is the finite quark mass [2].

Y. Kumar (✉)
Department of Physics, Deshbandhu College, University of Delhi, Kalkaji, New Delhi 110019, India
e-mail: yogesh.du81@gmail.com

S. S. Singh
Department of Physics and Astrophysics, University of Delhi, New Delhi 110007, India

Md. Naimuddin (ed.), *XXII DAE High Energy Physics Symposium*, Springer Proceedings in Physics 203, https://doi.org/10.1007/978-3-319-73171-1_211

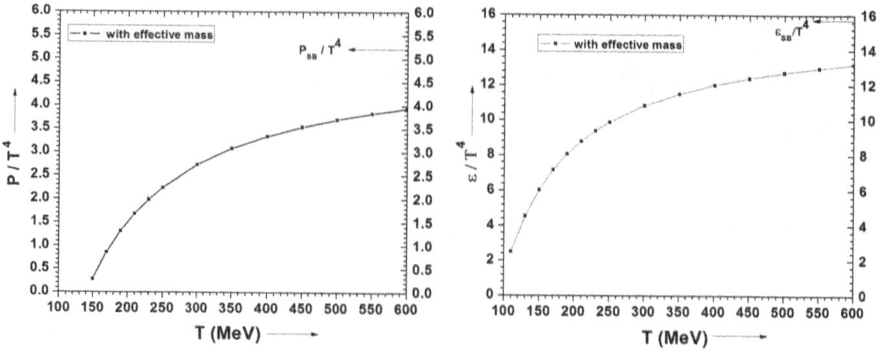

Fig. 211.1 The variation of P/T^4 (left) and ε/T^4 (right) with temperature (T) is shown

211.3 Evolution of Free Energy and EoS

The free energy, F_i for quarks and gluons is modified with the inclusion of curvature term using modified effective quark mass using [2, 3]. It is defined as:

$$F_i = \mp T g_i \int dk\rho_i(k) \ln(1 \pm e^{-(\sqrt{m_{eff}^2 + k^2})/T}) . \tag{211.2}$$

Using equation (2), we calculate pressure and energy density given as [4]:

$$P_i = -\frac{d}{dv} F_i \quad ; \quad \varepsilon = T \frac{d}{dT} P_i - P_i . \tag{211.3}$$

Using these relations, we calculate P/T^4 and ε/T^4. These relations are useful to study the quark–gluon plasma equation of state. Finally, all calculation has been done at high temperature T $\sim$ 600 MeV.

211.4 Results

Our results using effective quark mass with the effect of curvature term at such high temperature T $\sim$ 600 MeV are almost same as [4]. It shows that the theoretical model with the thermal correction in the effective quark mass including curvature term does not alter the results of Gosain et al. [4] very much, in fact the change caused in the QGP EoS is negligible as shown in Fig. 211.1. Our results are in good agreement with our earlier work [4]. Also the EoS prediction by other models are in conformity with lattice results. Overall, our model gives a negligibly small improvement so it varifies the previous results in QGP EoS. Thus, the result shows useful information to study QGP EoS in high energy heavy-ion collisions.

References

1. Singh, S.S., Gosain, D.S., Kumar, Y., Jha, A.K. Effect of curvature on a statistical model of quark-gluon-plasma fireball in the hadronic medium. Pram. J. Phys. 74, 27 (2010). http://www.ias.ac.in
2. Kumar, Y. Stable droplet formation of quark-gluon plasma in quark-hadron phase transition. J. Phys. Conf. Ser. 668, 012110 (2016). http://iopscience.iop.org
3. Srivastava, P.K., Tiwari, S.K., Singh, C.P. QCD critical point in a quasiparticle model. Phys. Rev. D **82**, 014023 (2010). https://doi.org/10.1103/PhysRevD.82.014023
4. Gosain, D.S., Singh, S.S. Equation of state of quark-gluon plasma using a simple statistical model. Int. J. Theo. Phys. **53**, 2688 (2014). https://doi.org/10.1007/s10773-014-2064-2

References

Chapter 212
Performance Metrics of a GPU Based Track Fitting Code for the INO Prototype Stack

Deepak Samuel, Varun Neelamana, S. Mishra, Anwesha Mahapatra, Chandrima Mallick, Lopamudra Nayak, Swetalin Mohapatra and M. Pooja

212.1 Introduction

The use of GPUs in the field of high energy physics is not uncommon. For example, GPUs have been used in particle transport codes, track reconstruction and trigger algorithms in experiments like the LHC [1]. We anticipate a variety of scenarios in INO where GPUs can be employed to achieve performance gains. Unlike CPU coding, GPU coding is not straightforward and requires a stringent watch over parameters that might be trivial under a CPU execution. In a first step to explore the functionality of GPU, we developed a code for the track reconstruction for the $1\,\text{m} \times 1\,\text{m}$ prototype detector at TIFR. The trigger rate of the stack is about 10 Hz [2] for fully contained events. The stack collects about a million cosmic muon events everyday from which the detector parameters like efficiency and resolution are estimated. In addition other particle physics studies are also undertaken. The data are stored in ROOT TTree structures which is not a compatible format for GPU programming. GPU functions typically operate and are efficient on data structured as arrays. To circumvent this problem and to benchmark only the linear fitting part of the routine, we tested the code using a dummy dataset (stored in array) wherein the slope and intercept are known.

212.2 Benchmarking

Datasets of different sizes (2.5, 5 and 10 lakh events) were generated for both 12 layers and 120 layers. Every event is ensured to have a different slope and intercept. The entire dataset is flattened into an one dimensional array. A simple linear regression

D. Samuel (✉) · V. Neelamana · S. Mishra · A. Mahapatra · C. Mallick ·
L. Nayak · S. Mohapatra · M. Pooja
Department of Physics, Central University of Karnataka, Kalaburagi 585367, India
e-mail: deepaksamuel@cuk.ac.in

Md. Naimuddin (ed.), *XXII DAE High Energy Physics Symposium*, Springer Proceedings in Physics 203, https://doi.org/10.1007/978-3-319-73171-1_212

function was written which calculates and returns only the slope and intercept in one dimensional arrays. The same function can be used as a GPU function (kernel) by switching to the CUDA kernel syntax [3]. The total execution times were computed using standard C timer functions for the CPU code and CUDA event timers for GPU code. For GPU computations, the array data has to be transferred to and from the GPU memory. The effect of this transfer overhead was also studied by placing the CUDA timers at appropriate locations in the code.

212.3 Results and Discussion

The results showed an expected performance gain parameterized by the speed-up factor of 5.5 for 12 layers and 1.5 for 120 layers. The drastic decline in the speed-up factor for 120 layers is, however, not clearly understood. A possible reason could be the warp size of the GPU. A warp in the CUDA model is a group of 32 threads which operate synchronously. Only after a warp execution is complete does the next warp start the execution. If one of the threads in the warp is stalled due to a memory access, for example, the execution time of a warp gets prolonged. Further, we studied the execution times with the data transfer from and to the GPU. The performance gain is only marginal for 12 layers (factor 1.5) and much worse than the CPU for 120 layers (factor 0.7). This is the impact of memory bandwidth between the GPU and CPU. We thus conclude that a performance gain on a GPU cannot be achieved by mere porting of the code. Attention is required on aspects such as memory bandwidth and thread synchronization without which the CPU might even outperform the GPU.

References

1. V. Halyo, A. Hunt, P. Jindal, P. LeGresley, P. Lujan, GPU enhancement of the trigger to extend physics reach at the LHC. J. Instrum. **8**, P10005 (2013). https://doi.org/10.1088/1748-0221/8/10/P10005
2. M. Bhuyan, V.B. Chandratre, S. Dasgupta, V.M. Datar, S.D. Kalmani, S.M. Lahamge, N.K. Mondal, P. Nagaraj, S. Pal, S.K. Rao and others, VME-based data acquisition system for the India-based Neutrino Observatory prototype detector. Nucl. Instrum. Methods Phys. Res. Sect. A Accel. Spectrom. Detect. Assoc. Equip. **661**, S73–S76 (2006). https://doi.org/10.1016/j.nima.2010.08.075
3. J. Nickolls, I. Buck, M. Garland, K. Skadron, Scalable parallel programming with CUDA. Queue **6**, 195–197 (2008). https://doi.org/10.1145/1365490.1365500

Chapter 213
Probing Wrong-Sign *hbb* Couplings in $h \to \Upsilon\gamma$

Tanmoy Modak, Jorge C. Romão, Rahul Srivastava, João P. Silva and Soumya Sadhukhan

213.1 Motivation

Couplings of the 125 GeV scalar discovered at the LHC should be probed in detail to compare it with the SM Higgs. One interesting possibility is the "wrong-sign" solution, where the hbb coupling has a sign opposite to that of the SM. Among variants of two Higgs doublet model (2HDM) with a softly broken Z_2 symmentry [1, 2], type II and Flipped can have a "wrong sign" solution. These models have BSM scalars $H^{\pm}$, A, H, apart from a scalar h, which can be identified to be the 125 GeV scalar of LHC.

A hbb sign change does not alter the SM Higgs total decay width, dominated by $h \to b\bar{b}$ rate. To probe wrong-sign effects indirectly through interference effects, one-loop contribution to $gg \to h$ and $h \to \gamma\gamma$ can be probed. For these cases there are uneven competition between bottom and top loops and top and W boson loops respectively. So even with wrong sign the values will be close to the SM, and only a very precise LHC measurement of order 5% in $pp \to h \to \gamma\gamma$ will distinguish the normal sign from the wrong-sign solutions. In contrast, the rare decay $h \to \Upsilon\gamma$ consists of two diagrams with almost the same magnitude, suppressing the rate in the SM due to an accidental precise cancellation between them [3]. Reversal of the

T. Modak
NTU, Taipei, Taiwan

J. C. Romão · J. P. Silva
CFTP, Lisboa, Portugal

R. Srivastava
IFIC, Valencia, Spain

S. Sadhukhan (✉)
PRL, Ahmedabad, India
e-mail: soumyas@prl.res.in

Md. Naimuddin (ed.), *XXII DAE High Energy Physics Symposium*, Springer Proceedings in Physics 203, https://doi.org/10.1007/978-3-319-73171-1_213

hbb sign will destroy the precise cancellation and this makes $h \to \Upsilon\gamma$ decay the best channel to probe the wrong-sign solutions. In this proceeding, we show impact of $h \to \Upsilon\gamma$ channel in probing wrong-sign solution, following our work, [4].

213.2 Theoretical Framework

The direct and indirect diagrams for $h \to \Upsilon\gamma$ decay are given in Fig. 2 of [4]. The direct diagram arises from the direct $hb\bar{b}$ coupling. The indirect diagram arises from the effective $h\gamma\gamma$ coupling with a virtual photon giving an Υ. With a CP-conserving 2HDM scalar potential with a softly broken Z_2 and the field parametrization used in [1, 2], the gauge and Yukawa couplings of the lightest 125 GeV scalar (h) are given as,

$$\begin{aligned}
\mathcal{L}_{hVV} &= \sin(\beta-\alpha)h\left[\tfrac{m_Z^2}{v}Z^\mu Z_\mu + 2\,\tfrac{m_W^2}{v}W^{+\mu}W_\mu^-\right],\\
-\mathcal{L}_{Yuk} &= \tfrac{m_t}{v}k_U h\bar{t}t + \tfrac{m_b}{v}k_D h\bar{b}b + \tfrac{m_\tau}{v}k_\tau h\tau^+\tau^-.
\end{aligned} \tag{213.1}$$

Here $k_U = \frac{\cos\alpha}{\sin\beta}$, $k_D = -\frac{\sin\alpha}{\cos\beta}$ and $k_\tau = k_D$ (Type II), $k_\tau = k_U$ (Flipped). The SM (alignment) limit corresponds to $\sin(\beta-\alpha) = 1$ which translates to $k_U = k_D = k_\tau = 1$. The wrong sign limit is defined as $\sin(\beta+\alpha) = 1$ which gives $k_U = 1, k_D = -1$. Only type II and Flipped 2HDMs are experimentally consistent with this wrong sign possibility [5–7]. Fig. 1 of [4] shows the parameter space allowing the wrong sign solution.

213.3 Results

In Fig. 213.1 results are presented where the red/dark-grey points pass all theoretical constraints. The blue/black (green/light-grey) points pass those and also μ_{VV} ($V \equiv W, Z$), $\mu_{\gamma\gamma}$, and $\mu_{\tau\tau}$ at 20% (10%). With only the theoretical constraints, a very large range of k_D gets allowed but it does not improve BR($h \to \Upsilon\gamma$) much, as it also increases the total width with k_D. After adding experimental constraints, only right-sign ($k_D \sim 1$) and wrong-sign ($k_D \sim -1$) regions get allowed. This is mostly due to μ_{VV} being very close to the SM values. In contrast to the $k_D = 1$ case, constructive interference in the wrong sign case makes BR($h \to \Upsilon\gamma$) larger by two orders of magnitude.

The possible experimental reach at 13 TeV LHC is presented in Fig. 213.1 (right), where we find a $\sigma \times BR$ value around 0.06 fb. For total integrated luminosity around 100 fb^{-1}, a measurement is becoming possible. A high-Luminosity LHC, will allow for either the detection or complete ruling out of the wrong-sign solution.

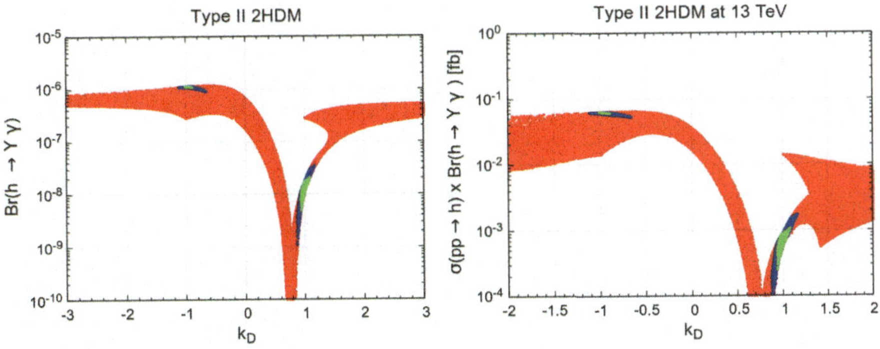

Fig. 213.1 BR($h \to \Upsilon\gamma$) (left), $\sigma \times$ BR($h \to \Upsilon\gamma$) at 13 TeV LHC (right), as a function of k_D

References

1. *The Higgs Hunter's Guide* (Westview Press, Boulder, CO, 2000)
2. Phys. Rept. **516**, 1 (2012) arXiv:1106.0034 [hep-ph]
3. Phys. Rev. D **88**(5), 053003 (2013) arXiv:1306.5770 [hep-ph]
4. Phys. Rev. D **94**(7), 075017 (2016) arXiv:1607.07876 [hep-ph]
5. JHEP **1312**, 095 (2013) arXiv:1310.7941 [hep-ph]
6. Phys. Rev. D **89**(11), 115003 (2014) arXiv:1403.4736 [hep-ph]
7. Phys. Rev. D **90**(1), 015021 (2014) arXiv:1406.6080 [hep-ph]

Chapter 214
Studying Medium Modification of Jets Using Jet Evolution with Energy Loss (JEWEL) Model

Subikash Choudhury, Rathijit Biswas, S. K. Prasad, Supriya Das and Subhasis Chattopadhyay

214.1 Introduction

The jet suppression was one of the first experimental signatures of the formation of quark gluon plasma (QGP) in HI collisions at RHIC and LHC [1]. Jets are produced from the fragmentation of hard scattered partons at the initial stage of collision. In HI collisions, these hard scattered partons propagate through hot and dense expanding medium, and lose their energy via in-medium interactions. This phenomena is referred to as "jet quenching", leading to suppression of jet yields in HI collisions relative to that in pp collisions.

In this article we present results of the modification of inclusive jet-p_T spectra in Pb-Pb collisions at $\sqrt{s_{\rm NN}} = 2.76$ and 5.02 TeV from a p-QCD inspired model of jet energy loss, JEWEL [2]. The model implements scattering between hard partons and constituents of hydrodynamically evolving medium, which leads to collisional (elastic) and radiative energy loss.

214.1.1 Jet Reconstruction

Jets are reconstructed using anti-k_T [3] jet reconstruction algorithm provided by the FASTJET package for resolution parameter R ($\sqrt{|\Delta\eta|^2 + |\Delta\varphi|^2}$) = 0.2 in the pseudo-rapidity interval $-0.5 < \eta < 0.5$.

S. Choudhury (✉) · R. Biswas · S. K. Prasad · S. Das
Bose Institute, Centre for Astroparticle Physics and Space Science (CAPSS), EN-80 Sector-V, Kolkata 700091, India
e-mail: subikash.vecc@gmail.com

S. Chattopadhyay
Variable Energy Cyclotron Centre, 1/AF Bidhan Nagar, Kolkata 700064, India

Md. Naimuddin (ed.), *XXII DAE High Energy Physics Symposium*, Springer Proceedings in Physics 203, https://doi.org/10.1007/978-3-319-73171-1_214

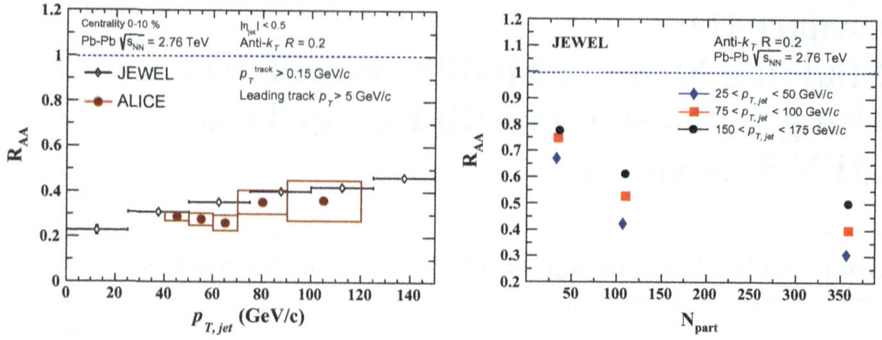

Fig. 214.1 Comparison of R_{AA}^{jet} in JEWEL and ALICE (**left**). Centrality and jet-p_T dependence of R_{AA}^{jet} (**right**)

Fig. 214.2 Jet-R_{AA} at $\sqrt{s_{NN}} = 2.76$ TeV and 5.02 TeV, calculated from JEWEL

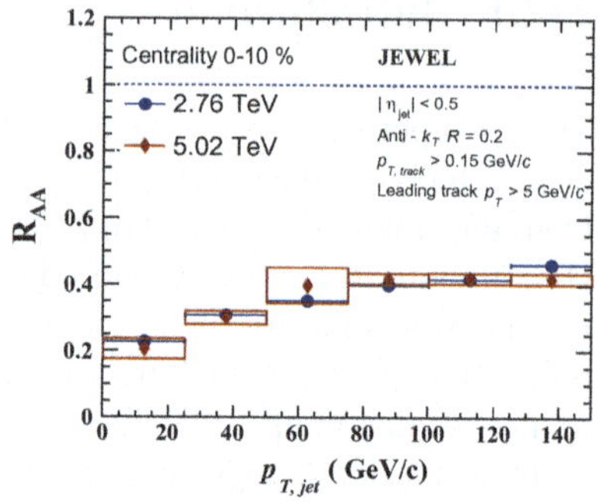

214.2 Results

The modification of jet yield in HI collisions is quantified by jet nuclear modification factor, R_{AA}^{jet}, which is defined as ratio of jet p_T spectra in HI collisions to that in pp, scaled by the number of "binary collision" (N_{coll}) at same $\sqrt{s}$. Figure 214.1 (left) shows the comparison of R_{AA}^{jet} in 0–10% central Pb–Pb collisions in JEWEL and ALICE-data. JEWEL describes jet suppression reasonably well within the experimental uncertainties. Further, the centrality dependence of jet yield modification is shown in Fig. 214.1 (right) for different jet-p_T intervals. A significant jet suppression is observed even in most peripheral (60–80%) collisions for all jet p_T bin reaching upto 175 GeV/c. Figure 214.2 shows the $\sqrt{s}$ dependence of R_{AA}^{jet} as obtained from JEWEL. R_{AA}^{jet} seems to be independent of $\sqrt{s}$ at these two energies.

Acknowledgements Subikash Choudhury acknowledges the support from DST- SERB Ramanujan Fellowship of Dr. S. Biswas (D.O. no. SR/S2/RJN-02/2012).

References

1. Phys. Lett. B (ALICE collaboration): measurement of jet suppression in central Pb-Pb collisions at $\sqrt{s_{\rm NN}}$=2.76 TeV. **746**, 1 (2015). https://doi.org/10.1016/j.physletb.2015.04.039
2. Eur. Phys. J. C. JEWEL 2.0.0 - directions for use. **74**, 2 (2014). https://doi.org/10.1140/epjc/s10052-014-2762-1
3. M. Cacciari, G.P. Salam, G. Soyez, The anti-k_t jet clustering algorithm. JHEP 0804 063 (2008). https://doi.org/10.1088/1126-6708/2008/04/063

Chapter 215
Hints on Neutrino Mixings from Flavour Data

Disha Bhatia

The flavour observable $R_K \equiv \text{BR}(B \to K\mu\mu)/\text{BR}(B \to Kee)$ presently deviates from its standard model (SM) prediction by 2.6σ, hinting at lepton flavour universality violating new physics (NP) interactions [1]. This anomaly can be addressed by introducing a $U(1)_X$ symmetry, which gets spontaneously broken by a SM singlet, S with X-charge a [2]. Neutrino oscillation data also provides strong evidences for massive neutrinos with non trivial PMNS matrix. We therefore augment the SM by three right handed neutrinos and generate neutrino masses and mixings in a Type-I seesaw framework. Note that with non-universal X-charges for neutrinos, the scalar S can also help in generating neutrino masses and could connect the observations in the flavour sector to the ones in the neutrino sector. The X-charges are considered to be vector-like, which ensures gauge anomaly cancellation in the presence of three right handed neutrinos. The charges of leptons are denoted as y_e, y_μ and y_τ. The Lagrangian describing the mass terms of neutrinos is

$$\mathcal{L}_\nu^{\text{mass}} = -\overline{\nu_L}\, m_D\, \nu_R - \frac{1}{2}\overline{\nu_R^c}\, M_R\, \nu_R - \frac{1}{2}\overline{\nu_R^c}\mathcal{Y}_R \nu_R\, S + h.c.. \qquad (215.1)$$

The mass matrix for right handed neutrinos then is

$$[M_R^S]_{\alpha\beta} = [M_R]_{\alpha\beta} + \frac{v_S}{\sqrt{2}}[\mathcal{Y}_R]_{\alpha\beta} \neq 0 \quad \text{if} \quad y_\alpha + y_\beta = 0\,, \pm a\,. \qquad (215.2)$$

Using Type-I seesaw mechanism, the mass matrix for three light neutrinos can be expressed in terms of the neutrino oscillation parameters as

$$m_{\nu_L} = -m_D^T {M_R^S}^{-1} m_D = U_{\text{PMNS}}\, m_{\nu_L}^{\text{diag}}\, U_{\text{PMNS}}^T\,. \qquad (215.3)$$

D. Bhatia (✉)
Tata Institute of Fundamental Research, Mumbai 400005, India
e-mail: disha@theory.tifr.res.in

Md. Naimuddin (ed.), *XXII DAE High Energy Physics Symposium*, Springer Proceedings in Physics 203, https://doi.org/10.1007/978-3-319-73171-1_215

Without any loss of generality, we work in the basis where m_D and m_ℓ are diagonal. Neutrino oscillations can be described in terms of nine physical parameters, i.e., mixing angles ($\theta_{12}, \theta_{13}, \theta_{23}$), phases ($\delta_{cp}, \alpha, \beta$) of the PMNS matrix, and the masses for the light neutrinos (m_1, m_2 and m_3). Experimental measurements exist for: Δm^2_{21}, Δm^2_{31}, θ_{12}, θ_{23} and θ_{13}. Therefore, we have four independent observables, which we choose to be m_{light}, δ_{cp}, α and β. These will help us determine the allowed $U(1)_X$ symmetry combinations.

The X-charges may result in texture-zeros in the M_R^S mass matrix if the condition on X-charges in (215.2) is not satisfied. These textures further result in either zero-textures or vanishing minors in the m_{ν_L} mass matrix, which can be translated as conditions on the physical parameters using (215.3). With four independent parameters, we are allowed to have four such conditions, implying a maximum of two-zero textures in the complex symmetric M_R^S mass matrix.

The allowed textures and vanishing minor conditions for m_{ν_L} have been extensively studied [3, 4]. Implications for allowed textures in M_R^S matrix can be inferred using (215.3) and have been studied in [5, 6]. We use the results for the allowed textures [5, 6] in M_R^S mass matrix and determine the plausible symmetry combinations by solving (215.1). Our solutions of two-zero textures match with the ones obtained in [6]. The symmetry combinations determined are listed in a table below,

Textures (M_R^S)	Symmetries	Textures/Minors (m_{ν_L})
2	$L_e \pm 3L_\mu - L_\tau$, $L_e - L_\mu \pm 3L_\tau$	2 texture zeros $\oplus$ 2 vanishing minors
2	$L_\mu - L_\tau$	2 vanishing minors
1	L_e, L_μ, $L_e - 3L_\mu + L_\tau$, $3L_e - L_\mu - L_\tau$	1 vanishing minor
0	$L_e - L_\mu \pm L_\tau$	—

We now determine the NP contributions required to explain the R_K anomaly. The X-charges of electron and muon should be necessarily different from each other for lepton flavour non-universality. The dominant Z' contributions to the $b \to s\ell\ell$ operator requires a tree level bsZ' vertex, which can be achieved with non-universality appearing in the third generation of quarks. The elaborate details about the model construction may be found in [2]. Following global fit results performed on the $b \to s\ell\ell$ and $b \to s\gamma$ data [7–9], NP contributions are introduced only to the operator $\mathcal{O}_9 = (\overline{s_L}\gamma^\mu b_L)(\bar{\ell}\gamma_\mu \ell)$ with Wilson coefficient of electron and muon related as $C_9^{\text{NP},e}/C_9^{\text{NP},\mu} = y_e/y_\mu$. The symmetry combinations viz. $L_\mu - L_\tau$, L_μ, $L_e - 3L_\mu \pm L_\tau$, $L_e + 3L_\mu - L_\tau$, $L_e - L_\mu \pm 3L_\tau$, and $L_e - L_\mu \pm L_\tau$ passes through the 1σ contour of the global fits as shown in the left panel of the Fig. 215.1 are hence selected [2].

The selected symmetry combinations can now be analyzed to predict the allowed regions for the four free parameters. As an example, we consider one-zero texture combinations, L_μ and $L_e - 3L_\mu + L_\tau$, both of which yields a zero texture at $[M_R^S]_{22}$ and hence result in same predictions. We show predictions for the inverted ordering in the right panel of Fig. 215.1. It is interesting to note that for higher values of m_{light}, α_1 and α_2 are severely constrained and have values around $\pi/2$ [2].

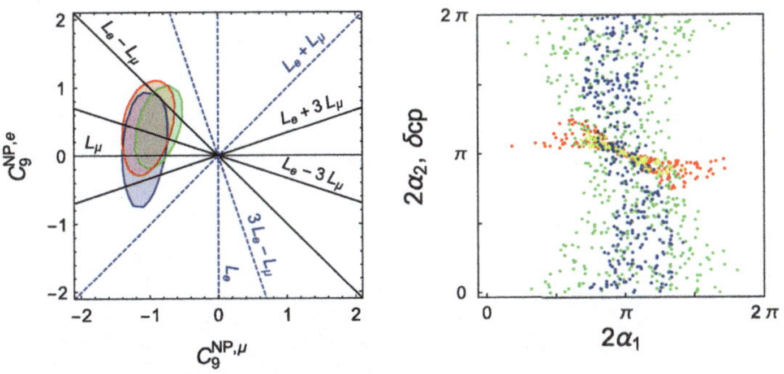

Fig. 215.1 In the left panel, predictions for the symmetry combinations are plotted (with τ charge suppressed) in the $\left(C_9^{\mathrm{NP},\mu}, C_9^{\mathrm{NP},e}\right)$ plane along with the 1σ contour obtained from global fits [7–9]. In the right panel, the predictions for α_1, α_2 and δ_{cp} for symmetries $L_\mu/L_e - 3L_\mu + L_\tau$ are plotted for inverted ordering. The points marked in yellow (red) correspond to $(2\alpha_1, 2\alpha_2)$ plane for $m_{\mathrm{light}} = 0.05(0.2)$ eV, while points in blue (green) correspond to $(2\alpha_1, \delta_{\mathrm{CP}})$ plane for $m_{\mathrm{light}} = 0.05(0.2)$ eV

Our symmetry combinations account for both R_K and neutrino mixings. They also satisfy bounds from neutral meson mixings (K–$\overline{K}$, B_d–$\overline{B_d}$, B_s–$\overline{B_s}$) and direct Z' searches at the colliders, for substantial parameter space [2]. The Z' can also be probed at 13 TeV LHC run in the di-muon channel with few fb^{-1} of Luminosity [2].

References

1. R. Aaij et al., [LHCb Collaboration], Test of lepton universality using $B^+ \to K^+\ell^+\ell^-$ decays. Phys. Rev. Lett. **113**, https://doi.org/10.1103/PhysRevLett.113.151601
2. D. Bhatia, S. Chakraborty, A. Dighe, Neutrino mixing and R_K anomaly in $U(1)_X$ models: a bottom-up approach. JHEP **1703**, 117 (2017). https://doi.org/10.1007/JHEP03(2017)117
3. E.I. Lashin, N. Chamoun, Zero minors of the neutrino mass matrix. Phys. Rev. D **78**, 073002 (2008). https://doi.org/10.1103/PhysRevD.78.073002
4. H. Fritzsch, Z.z. Xing, S. Zhou, JHEP **1109**, 083 (2011). https://doi.org/10.1007/JHEP09(2011)083
5. E.I. Lashin, N. Chamoun, One vanishing minor in the neutrino mass matrix. Phys. Rev. D **80**, 093004 (2009). https://doi.org/10.1103/PhysRevD.80.093004
6. T. Araki, J. Heeck, J. Kubo, Vanishing Minors in the Neutrino Mass Matrix from Abelian Gauge Symmetries. JHEP **1207**, 083 (2012). https://doi.org/10.1007/JHEP07(2012)083
7. W. Altmannshofer, D.M. Straub, New physics in $b \to s$ transitions after LHC run 1. Eur. Phys. J. C **75**(8), 382 (2015). https://doi.org/10.1140/epjc/s10052-015-3602-7
8. S. Descotes-Genon, L. Hofer, J. Matias, J. Virto, Global analysis of $b \to s\ell\ell$ anomalies. JHEP **1606**, 092 (2016). https://doi.org/10.1007/JHEP06(2016)092
9. T. Hurth, F. Mahmoudi, S. Neshatpour, On the anomalies in the latest LHCb data. Nucl. Phys. B **909**, 737 (2016). https://doi.org/10.1016/j.nuclphysb.2016.05.022

Chapter 216
Strong Decays and Coupling Constants of 1P and 1D Bottom Meson

Pallavi Gupta and A. Upadhyay

Heavy light mesons made from one heavy quark and one light quark are similar to hydrogen atom of hadronic physics and are ideally studied to understand the strong interactions. But, the present experimental status for bottom mesons is not well known as that of charm meson. Recently, CDF collaboration has announced a state B(5970) with $M = 5978 \pm 5 \pm 12\,\text{MeV}$ and decay width $\Gamma = 70 \pm 30\,\text{MeV}$ [1]. Considering this limitation, we studied the Okubo-Zeing-Iizuka (OZI) allowed two body strong decays of n = 1 P and D wave bottom mesons. In this work, we choose the effective Lagrangian approach to study the strong interactions of 1P and 1D bottom states with the light pseudoscalar mesons. Denoting a heavy-light meson as $Q\overline{q}$, the total angular momentum is expressed as $\overrightarrow{J} = \overrightarrow{j_l} \pm \frac{1}{2}$, where $\overrightarrow{j_l} = s_l + l$ i.e. sum of spin and orbital angular momentum of light degree of freedom. Consequently, the doublet for l = 0 (S-wave) is represented by $J^P_{j_l} = (0^-, 1^-)_{\frac{1}{2}}$, which for L = 1 (P-wave), there are two doublets represented by $(0^+, 1^+)_{\frac{1}{2}}$ and $(1^+, 2^+)_{\frac{3}{2}}$ respectively. Two doublets of L = 2 (D-wave) are represented by $(1^-, 2^-)_{\frac{3}{2}}$ and $(2^-, 3^-)_{\frac{5}{2}}$ respectively. These doublets are described by the effective super-field H_a, S_a, T_a, X_a, Y_a respectively [2, 4]. At the leading order of the heavy quark mass, we can write out the Lagrangian of decay process A→ H+M, where A = S,T,X,Y and M denotes the light pseudoscalar meson. And with the help of lagrangian, decay width formulas are derived which contains strong coupling constants like g_{sh}, g_{th}, g_{xh} and g_{yh}. We define four ratios: $R_1^{B(B_s)} = \frac{BR(B(B_s)\rightarrow B^*(B_s^*)\pi)}{BR(B(B_s)\rightarrow B(B_s)\pi)}$, $R_2^{B(B_s)} = \frac{BR(B(B_s)\rightarrow B(B_s)\eta)}{BR(B(B_s)\rightarrow B_s(B)K)}$, $R_3^{*B(B_s)} = \frac{BR(B(B_s)\rightarrow B^*(B_s^*)\eta)}{BR(B(B_s)\rightarrow B_s(B)K)}$ and $R_4^{B(B_s)} = \frac{BR(B(B_s)\rightarrow B_s^*(B^*)K)}{BR(B(B_s)\rightarrow B_s(B)K)}$. With the above fields and lagrangian, we calculated the partial decay widths and the Ratios $R_i (i = 1, \ldots, 4)$ for the 1P and 1D bottom meson family which are listed in Tables 216.1

P. Gupta (✉) · A. Upadhyay
School of Physics and Material Science, Thapar University,
Patiala 147004, Punjab, India
e-mail: pallavigupta-10gupta.pallavi@gmail.com

A. Upadhyay
e-mail: alka.iisc@gmail.com

Md. Naimuddin (ed.), *XXII DAE High Energy Physics Symposium*, Springer Proceedings in Physics 203, https://doi.org/10.1007/978-3-319-73171-1_216

Table 216.1 Tables shows the decay modes of 1P and 1D non strange and strange bottom states

Decay channel	1P				1D			
	0^+	1^+	1^+	2^+	1^-	2^-	2^-	3^-
$B_s\pi$	$147.16g_{sh}^2$	–	–	$31.62g_{th}^2$	$647.00g_{xh}^2$	–	–	$63.70g_{yh}^2$
$B_s^*\pi$	–	$160.71g_{sh}^2$	$44.45g_{th}^2$	$91.09g_{th}^2$	$237.24g_{xh}^2$	$775.18g_{xh}^2$	$86.46g_{yh}^2$	$53.50g_{yh}^2$
BK	–	–	–	$105.50g_{th}^2$	$2468.09g_{xh}^2$	–	–	$120.82g_{yh}^2$
$B_s\eta$	–	–	–	–	$320.42g_{xh}^2$	–	–	$5.64g_{yh}^2$
$B_s^*\eta$	–	–	–	–	$92.39g_{xh}^2$	$324.66g_{xh}^2$	$3.32g_{yh}^2$	$2.37g_{yh}^2$
B^*K	–	–	–	$0.66g_{th}^2$	$435.75g_{xh}^2$	$1447.54g_{xh}^2$	$66.57g_{yh}^2$	$42.80g_{yh}^2$
Total	46.15	50.39	8.22	37.97	151.65	91.96	43.92	81.14

Table 216.2 Table shows the decay modes of 1P and 1D non strange bottom states

Decay channel	1P				1D			
	0^+	1^+	1^+	2^+	1^-	2^-	2^-	3^-
$B\pi$	$875.39g_{sh}^2$	–	–	$95.47g_{th}^2$	$1756.22g_{xh}^2$	–	–	$152.58g_{yh}^2$
$B^*\pi$	–	$291.82g_{sh}^2$	$36.09g_{th}^2$	$28.03g_{th}^2$	$218.40g_{xh}^2$	$709.92g_{xh}^2$	$70.38g_{yh}^2$	$43.67g_{yh}^2$
B_sK	–	–	–	–	$207.83g_{xh}^2$	–	–	$1.51g_{yh}^2$
$B\eta$	–	–	–	–	$272.53g_{xh}^2$	–	–	$3.39g_{yh}^2$
$B^*\eta$	–	–	–	–	$79.32g_{xh}^2$	$277.72g_{xh}^2$	$1.80g_{yh}^2$	$1.33g_{yh}^2$
B_s^*K	–	–	–	–	$749.99g_{xh}^2$	$183.25g_{xh}^2$	$0.39g_{yh}^2$	$0.33g_{yh}^2$
R_4	–	–	–	–	0.24	–	–	0.22
Total	271.53	78.85	6.677	22.83	93.29	42.26	20.39	56.97

and 216.2. We take the predicted masses from the [3–5] as input, for the missing experimental information.

216.1 Conclusion

Using the theoretical couplings $g_{sh} = 0.56$, $g_{th} = 0.43$ [4], $g_{xh} = 0.19$ and $g_{yh} = 0.53$ [2], total decay widths for the states is shown in the Tables 216.1 and 216.2. Table 216.1 show that B_0^* is a board state and $B_1^{'}$ is narrow with total decay width of 271.53 and 6.67 MeV respectively. Decay width for B_3^* comes out to be 56 MeV which is comparable with the experimentally measured 55 MeV width of B(5970), thus confirming its J^P state to be 3^-.

Acknowledgements The authors gratefully acknowledge the financial support by the Department of Science and Technology (SB/FTP/PS-037/2014), New Delhi.

References

1. T. Aaltonen et al., (CDF Collaboration), Phys. Rev. D **90**, 012013 (2014)
2. Z.G. Wang, Eur. Phys. J. Plus **129**, 186 (2014)
3. M. Di Pierro et al., Phys. Rev. D **64**, 114004 (2001)
4. P. Colangelo et al., Phys. Rev. D **86**, 054024 (2012)
5. K.A. Olive et al., (Particle Data Group), Review of particle physics. Chin. Phys. C **38**, 090001 (2014)

Chapter 217
Design and Development of Gas Leakage Station for Gas Electron Multiplier (GEM) Chamber

Rizwan Ahmad, Aashaq Shah, Ashok Kumar, Md. Naimuddin, Mohit Gola and Shivali Malhotra

217.1 Introduction

A gas electron multiplier (GEM) is a type of gaseous ionization detector used in nuclear and particle physics [1]. It includes a thin, metal clad polymer foil, perforated with a high density of holes. Under application of a suitable potential difference between the two sides a high field is built up in the holes. When an ionising particle traverses the detector it releases electronion pairs by ionisation. Electrons produced in the gap between the drift cathode and GEM drift into the holes where they are multiplied by the high electrical field, and transferred into the gap between GEM and the readout board. The collection of the electrons induces a detectable current on the pick-up electrodes.

217.2 Hardware and Software Description

We used different types of pneumatic and electronic components such as flow meters, gauge pressure sensor, atmospheric pressure sensor, temperature sensor, digital display and arduino board etc. Flow meters, variable area flow meters for measuring the flow rate of gases are used to display input and output flow rate of the gas supplied in to the GEM chamber. A gauge pressure sensor which is a piezoelectric type sensor, provides constant current output for given gauge pressure used to measure pressure drop in the GEM chamber corresponding gauge pressure. Also we incorporated atmospheric pressure and temperature sensor which are piezoresistive, monolithic,signal conditioned silicon sensors for the local temperature and atmospheric pressure change in the environment. To display the readings of these sensors

R. Ahmad (✉) · A. Shah · A. Kumar · Md. Naimuddin · M. Gola · S. Malhotra
Department of Physics and Astrophysics, University of Delhi, Delhi, India
e-mail: ahmadrizwan19@gmail.com

Md. Naimuddin (ed.), *XXII DAE High Energy Physics Symposium*, Springer Proceedings in Physics 203, https://doi.org/10.1007/978-3-319-73171-1_217

separately we used four digit seven segment digital displays. Above maintained components except flow meters are configured with Arduino Mega 2560 board which is an open source physical computing platform based on a simple I/O board having several analog and digital pins to input and output pins as per requirements used to data logging and real time plotting using excel coding and interfaced through PC.

217.3 Methodology

217.3.1 Inherent Leak Test and Calibration

The inherent leakage of pneumatics components and full setup without GEM chamber is measured using U-tube manometer (water) in which height (cm) of the limbs of manometer is adjusted in turn, working with different pressure values (with time) to check leak in the components and setup itself and then calibrated.

217.3.2 Testing Results

All the equipments used in gas leak setup are leak tight (under the acceptance threshold) along with full setup without GEM chamber. Initially we set 25mbar pressure in the setup and then the leak rate is calculated after one hour and found to be 5mbar/hr. The reason could be the ambient pressure and temperature changes, drift in the pressure transducer and some leakage in the detector itself (Fig. 217.1).

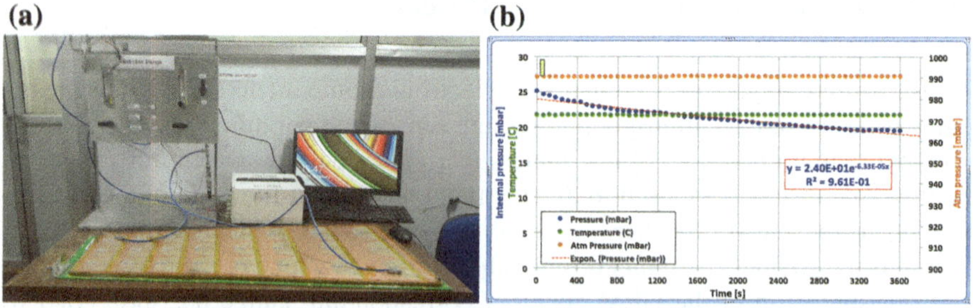

Fig. 217.1 Test result **a** Complete setup **b** Pressure versus time graph

217.4 Conclusion

We have designed and developed the setup successfully and the setup is working properly for testing gas leak in GE1/1-V chamber. We measured the leakage and ploted the graph between pressure versus time for GE1/1- chamber. It can be used with any gaseous detector to check leakage too. An actual leak in the GEM chamber is highly impossible.

Reference

1. F. Sauli, GEM: A new concept for electron amplification in gas detectors. Nucl. Instr. Meth. A **386**, 531 (1997)

Chapter 218
Design and Fabrication of a Controlled Water Based Cooling System for CBM Muon Chamber

D. Nag, S. Biswas, S. Chattopadhyay, S. Das, A. K. Dubey, C. Ghosh, A. Kumar, S. K. Prasad and J. Saini

218.1 Introduction

In CBM MUCH, Front End Boards [1, 2] are sensitive to temperature and can only operate reliably within the reliable temperature range [3] Since these FEB's are connected using wire-bonding technique, traditional air cooling is not an option,as vibrations resulting form air cooling might loosen the bonds. So water cooling is the preferable option. The preferable temperature range should be kept between 25–30 °C. Thus an automated controller is needed.

218.2 Mechanical Design and Working Flow-Chart

A copper tube of diameter 5 mm was welded on a 2 mm thick copper plate. Seven heating elements (coil resistors of 4 W each) were welded on the copper plate to simulate heat load. The temperature sensor was also fixed on this plate. In real experiment the setup is to be replaced by real-sized Aluminium plates. Also the heating elements will be replaced using original FEB's. A real-sized prototype is also built and tested recently in the lab and the Test Beam facility at CERN SPS H4 beam line.

Two different hardware approaches were followed to build the Al cooling plates. In first design, A 10 mm thick Al plate was taken, and 5 mm × 7 mm 'T' shaped grooves were made on the plate. Then the top of the groove was sealed with 5 mm × 2 mm strips of Al, resulting in a 5 mm × 5 mm water channel inside the plate.

D. Nag (✉) · S. Biswas · S. Das · S. K. Prasad
Centre for Astroparticle Physics and Space Science, Bose Institute, Kolkata, India
e-mail: dipanjannag19@gmail.com

S. Chattopadhyay · A. K. Dubey · C. Ghosh · A. Kumar · J. Saini
VECC, Kolkata, India

Md. Naimuddin (ed.), *XXII DAE High Energy Physics Symposium*, Springer Proceedings in Physics 203, https://doi.org/10.1007/978-3-319-73171-1_218

Appropriate connectors were fitted at both ends of the channel. The groove layout was chosen such that the water flow covers the maximum area, specially close to the FEB's. The Advantage of this design was that, maximum area of contact between the plate and water is available for heat transfer.

In second design, two identical Al plates having thickness of 5 mm each were taken, and grooves in mirror image were made on both. An Al pipe was press-fitted inside the groove and then the two plates were welded. This approach had an advantage of water-leak protection by it's design.

A water chiller was used as the heatsink. Cold water was circulated through the plate and poured back into the chiller using a submersible pump.

218.3 Automated Control System

The control system has been devised with a microcontroller running a negative feedback PID algorithm, taking the temperature value as the input and generating the PWM (Pulse Width Modulated) value to drive the motor as an output. The submersible pump was driven using this PWM pulse via a Power Transistor. Provision for setting various different set-points were kept. The instantaneous temperature was monitored through a computer.

218.4 Results

At first the heating elements were turned on and then the cooling system was turned on. Several different temperatures were set as the required "set-point" and the resulting temperature was recorded with time. Figures 218.1 and 218.2 shows water flow

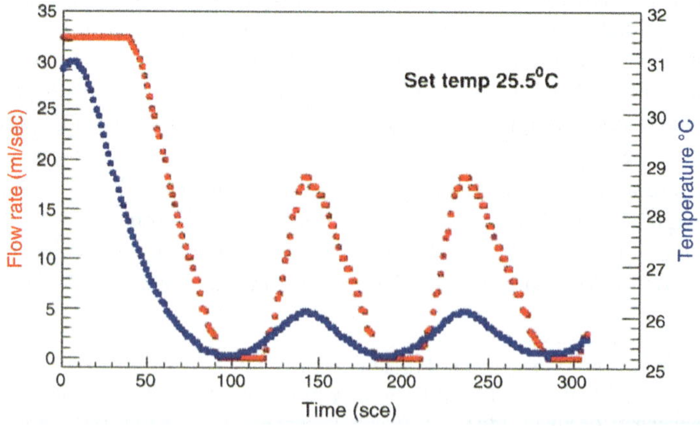

Fig. 218.1 Temp versus time, setpoint = 25 °C

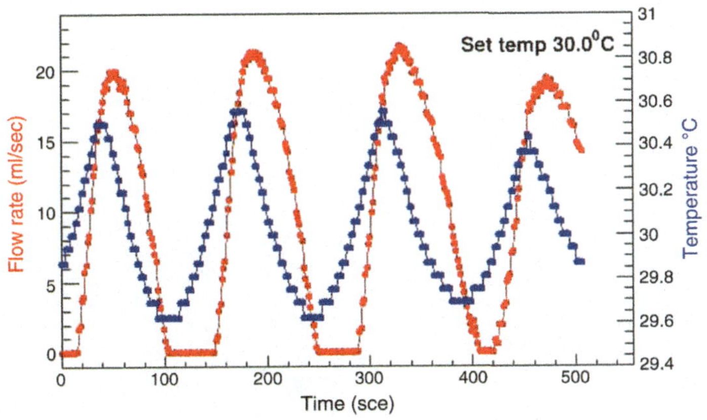

Fig. 218.2 Temp versus time, setpoint = 30 °C

rate (red marker) and plate temperature (blue marker) as a function of time while cooling is ON for set-points respectively 25.5 and 30.0 °C. The plots show a variation of ±0.5 °C around the set-point, which is within the permissible limits.

218.5 Outlook

1. The real size prototype is to be tested in detail.
2. Mechanism for water leakage detection is to be found out and implicated.
3. The controller must be scaled up to fulfil the real experimental requirements.

References

1. CBM Technical Design Report 2013 (TDR)
2. A.K. Dubey et al., DAE Symp. Nucl. Phys. **vv**, ppp (2010); A. Kiseleva, P.P. Bhaduri et al., **85**, 211–216 (2011)
3. V. Jain et al., DAE-BRNS Symp. Nucl. Phys. 60 (2015)

Chapter 219
Development and Simulation of Silicon PAD

Preeti Dhankher, Manoj Jadhav and Raghava Varma

219.1 Introduction

Silicon detector is a large surface diode with guard rings. The steps below briefly describes the techniques we have used in silicon detector fabrication [1].

1. Thermal Oxidation: The silicon wafer is first cleaned using RCA process and then oxidised by heating the wafer in oxygen environment at a temperature of $\sim$1000 °C.
2. Photolithography: We used photolithography to transfer the pattern(designed by software) from photomask to silicon wafer. Pattering of SiO_2 was done using double sided mask aligner.
3. Doping: Ion implantation technique was used to create the p-n junction.
4. Metallisation: Metallisation was done with Aluminium to create contacts and then patterning of Al was done using photolithography.

219.2 Simulation Results

In this section we present the results of TCAD simulation. Figure 219.1a shows schematic view of the detector simulated in SILVACO. We first carried out the simulation of detector developed at IIT Bombay as shown in Fig. 219.1b, where $b = 200\,\mu m$, $c = 80\,\mu m$, $d = 45\,\mu m$. Figure 219.1c shows the doping profile and potential distribution of the detector simulated. Figure 219.1b shows the distribution of electric field which is very high at the edge of junction. However, it should be uniformly distributed over the guard rings.

P. Dhankher (✉) · M. Jadhav · R. Varma
Indian Institute of Technology Bombay, Mumbai, India
e-mail: preeti.dhankher@iitb.ac.in

Md. Naimuddin (ed.), *XXII DAE High Energy Physics Symposium*, Springer Proceedings in Physics 203, https://doi.org/10.1007/978-3-319-73171-1_219

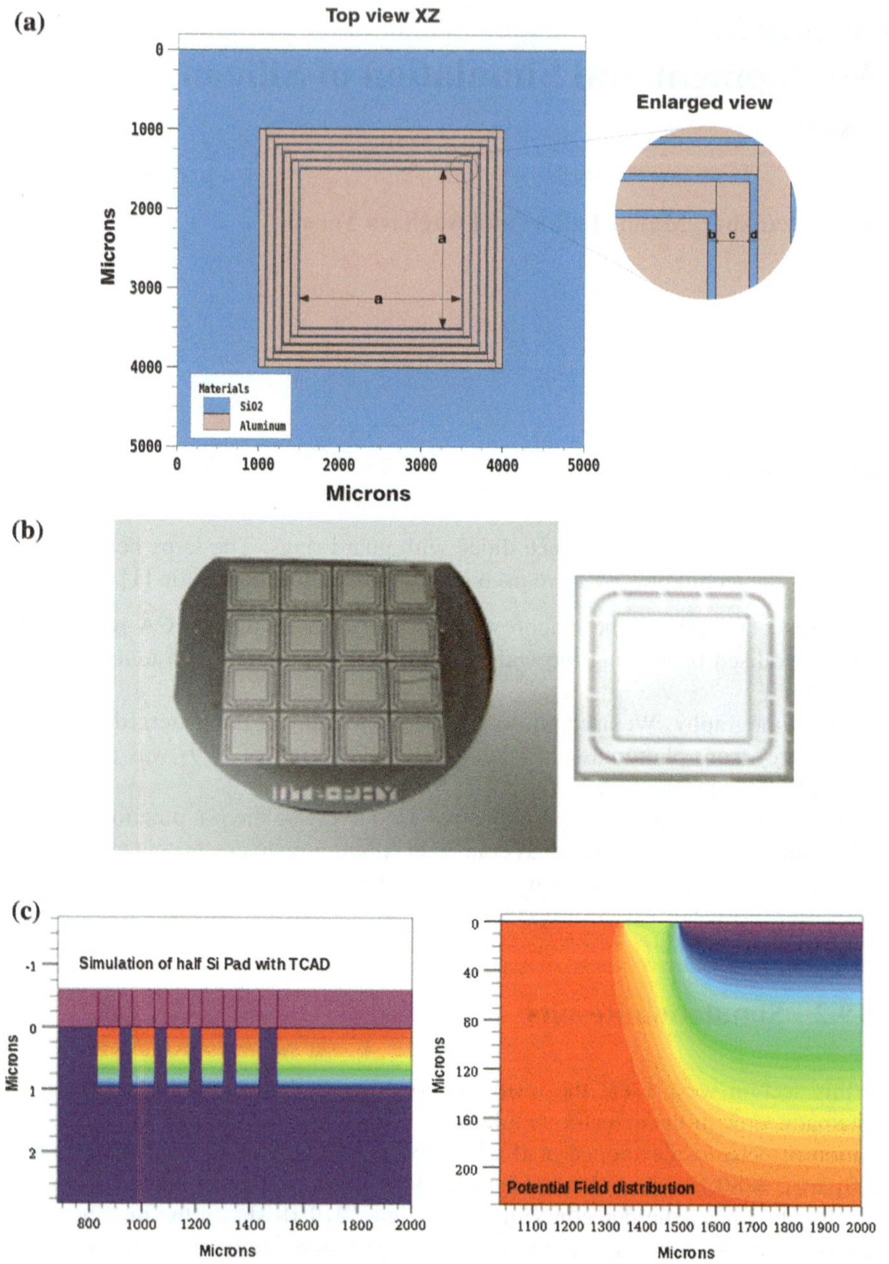

Fig. 219.1 **a** Schematic view of the detector geometry as simulated in SILVACO, **b** 1 cm × 1 cm Silicon pad developed at IIT Bombay. **c** Doping, Potential and **d** Electric field distribution

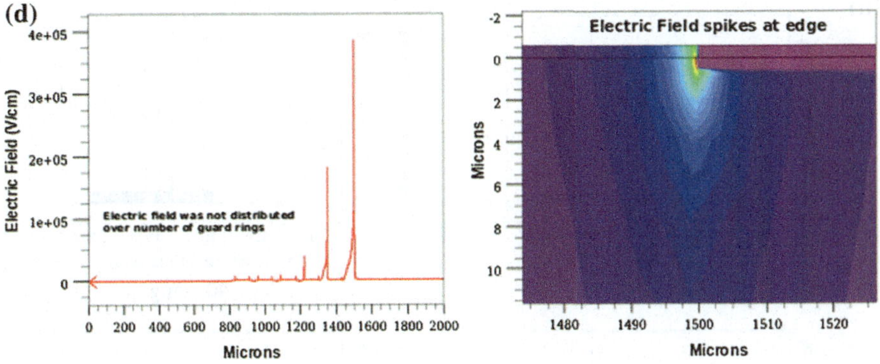

Fig. 219.1 (continued)

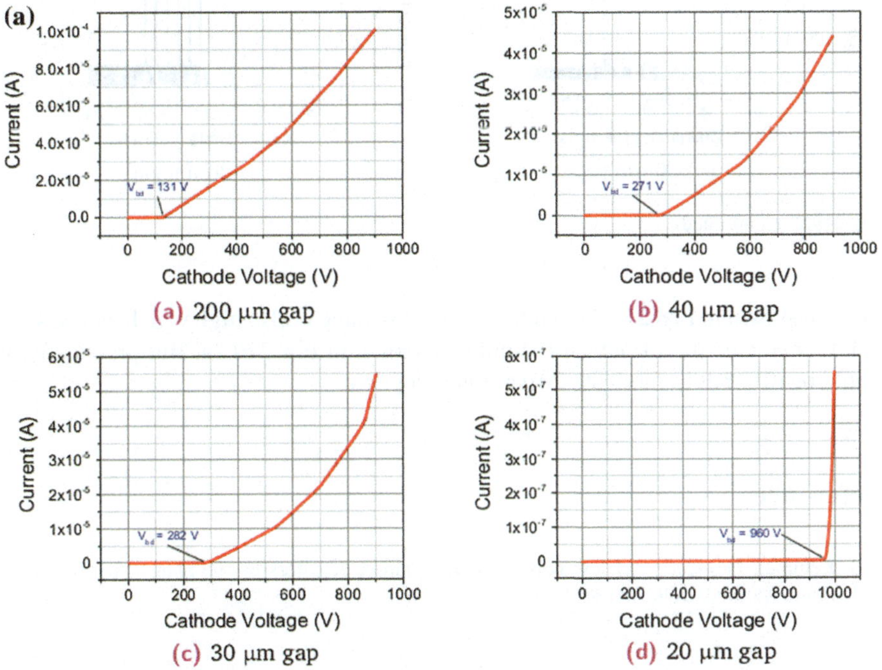

Fig. 219.2 **a** I-V characteristics of silicon pad detectors with decreasing pitch. and **b** Electric field distribution across the surface at the active-edge to the p+ implantation

To understand the effect of guard rings, we simulated three more geometries. For simplicity, we kept b = d, c = 80 μm and performed simulations for b = 20 μm, 30 μm and 40 μm and found out that, for 20 μm the electric field distribution is

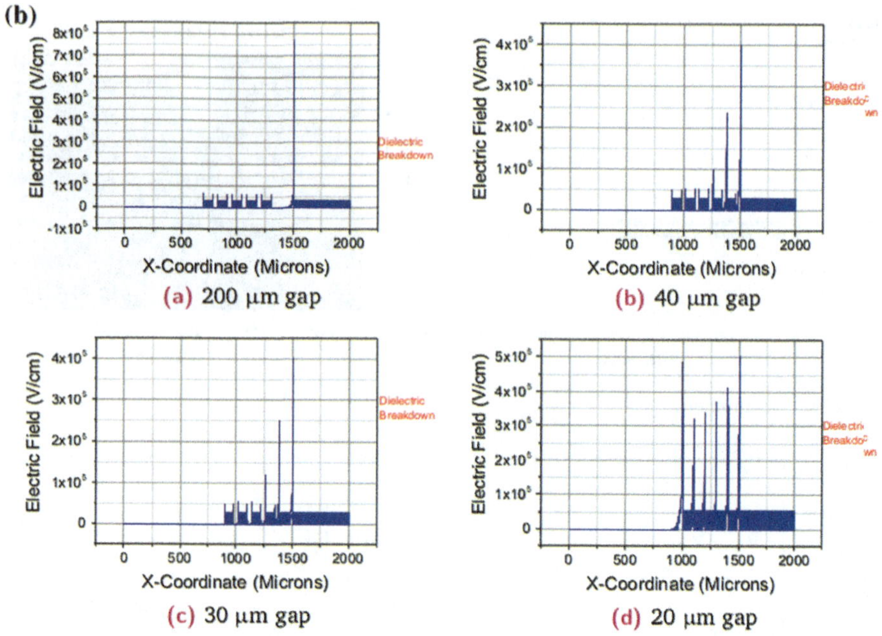

Fig. 219.2 (continued)

uniform shown in Fig. 219.2b and break down voltage is also high which can be seen in I-V characteristic of detector simulated shown in Fig. 219.2a. But practically, it would be difficult to fabricate 20 µm pitch detector.

Reference

1. J. Kemmer, Fabrication of low noise silicon radiation detectors by the planar process. Nucl. Instrum. Methods **169**(3) (1980)

Chapter 220
Photon-Scalar Oscillation in Magnetized Plasma

Manoj K. Jaiswal and Avijit K. Ganguly

220.1 Introduction

Scalar and pseudoscalar particles are possible candidates of dark matter. The scalar particle arises out of scale symmetry breaking through quantum effects while the pseudoscalar particle as axion arises due to breaking of $U(1)_{PQ}$ symmetry. Their interacting Lagrangian of scalar and pseudoscalar with photons (γ) are given by,

$$L_I = \begin{cases} \frac{1}{4} g_{\gamma\gamma\phi} \phi F^{\mu\nu} F_{\mu\nu} \text{ for scalar } \phi, \\ \frac{1}{4} g_{\gamma\gamma a} a \tilde{F}^{\mu\nu} F_{\mu\nu} \text{ for Pseudoscalar } a, \text{ axion}. \end{cases} \quad (220.1)$$

In (220.1), $g_{\gamma\gamma\phi/a}$ are the scalar (axion) photon coupling constants. These interactions, in external magnetic field $\mathbf{B}$ makes: the vacuum (i) optically active and dichroic for the photons [1, 2], (ii) possibility of superluminal velocities for photons, in a background magnetic field $\mathbf{B}$, in some energy interval [3] and (iii) finite oscillation probability between scalar photon ($P_{(\phi\rightarrow\gamma)}(\omega, z)$) or pseudoscalar photon ($P_{(a\rightarrow\gamma)}(\omega, z)$), states.

Here we have studied the photon-scalar/pseudoscalar oscillation dynamics taking the magnetized photon self energy tensor, $\Pi_{\mu\nu}(e\mathbf{B}, T, \mu)$, into account. Inclusion of this term changes the mixing dynamics. The mixing matrix (MM) in case of scalar-photon ($\phi\gamma$) system, $\phi F^{\mu\nu} F_{\mu\nu}$ turns out to be 3×3 and the same for axion-photon system $a \tilde{F}^{\mu\nu} F_{\mu\nu}$ turns out to be 4×4, [4, 5].

For high energy (ω) photons, $\omega >> \omega_p$, the MM for the second system, reduces effectively to a 3×3 one. However the role of the two transverse degrees of freedom gets interchanged (vis-a-vis, $\phi\gamma$). We have carried out the analysis for $\phi\gamma$, system and on that basis drawn some relevant conclusions, for $a\gamma$ system. The mixing dynamics under consideration follows from the solutions of,

---

M. K. Jaiswal · A. K. Ganguly (✉)
Department of Physics (MMV), Banaras Hindu University, 221005 Varanasi, India
e-mail: avijitk@hotmail.com

Md. Naimuddin (ed.), *XXII DAE High Energy Physics Symposium*, Springer Proceedings in Physics 203, https://doi.org/10.1007/978-3-319-73171-1_220

$$\left[k^2\mathbf{I} - \begin{pmatrix} \omega_p^2 & i\frac{\omega_p^2 eB_\parallel}{(\omega m_e)} & -ig_{\phi\gamma\gamma}B_\perp\omega \\ -i\frac{\omega_p^2 eB_\parallel}{(\omega m_e)} & \omega_p^2 & 0 \\ ig_{\phi\gamma\gamma}B_\perp\omega & 0 & m_\phi^2 \end{pmatrix} \right] \begin{bmatrix} A_\parallel(k) \\ A_\perp(k) \\ \phi(k) \end{bmatrix} = 0. \quad (220.2)$$

Here $\mathbf{I}$ is 3×3 identity matrix. $B\cos\theta = B_\parallel$ and $B\sin\theta = B_\perp$. The angle θ here corresponds to the angle between the photon propagation vector $\mathbf{k}$ and the magnetic field B (of O(1.0 μG), as considered here).

The Oscillation Probabilities: The oscillation probability, for ϕ to $\gamma_\parallel$ and ϕ to $\gamma_\perp$ in the limit $m_\phi = \omega_p$, upon traveling a distance z, follows from (220.2). The respective probabilities are:

$$P_{\phi\to\gamma_\parallel} = \frac{g_{\gamma\gamma\phi}^2 B_\perp^2 z^2}{4}\frac{sin^2(\beta)}{\beta^2}; \; P_{\phi\to\gamma_\perp} = \frac{g_{\gamma\gamma\phi}^2 B_\perp^2 z^2}{4}\frac{F^2}{(F^2+G^2)}\frac{sin^2(\beta)}{\beta^2}. \quad (220.3)$$

In (220.3), the undefined variables are, $\beta = \frac{\sqrt{F^2+G^2}}{2\omega}z$, $F = \frac{\omega_p^2 eB\cos\theta}{\omega m_e}$ and $G = -g_{\phi\gamma\gamma}B\sin\theta\omega$. Rest of the symbols have their usual significance. Using these probabilities, along with the gamma-ray data of supernova 1987A, one can estimate the coupling constants $g_{\gamma\gamma\phi}$ or $g_{\gamma\gamma a}$ from the relation,

$$\Phi(P_{\gamma_\parallel\to\phi} + P_{\gamma_\perp\to\phi})\delta t \leq 0.6\,\text{cm}^{-2}, \quad (220.4)$$

following, [6, 7]. Where, δt is the time duration of axion emission and Φ is the flux of axions, produced in the core of the supernova (SN), during the event. It depends on the distance between SN and earth D, radius of the SN core R_s, SN core temperature T and the proton no. density n_p, there in. The axion flux Φ, is very sensitive to the core temperature.

These limits are extracted from the non-observation of gamma rays signals– in the 25–100 MeV energy band – by the Gamma-Ray Spectrometer, on board of the Solar Maxima Mission satellite, that was operational during the event of SN1987A.

Conclusion: Considering the core temperature, T = 96 MeV $n_p = 2.8 \times 10^{38}\,\text{cm}^{-3}$ we have estimated the bound on the magnitude of the coupling constant $g_{\gamma\gamma\phi}$. The same turns out to be $g_{\gamma\gamma\phi} < 1.18 \times 10^{-11}\,\text{GeV}^{-1}$. This is a conservative bound, that is closer to the one reported in [6].

References

1. G. Raffelt, L. Stodolosky, Phys. Rev. D **37**, 1237 (1988)
2. A.K. Ganguly, P. Jain, S. Mandal, Phys. Rev. D **79**, 115014 (2009)
3. A.K. Ganguly, M.K. Jaiswal, Phys. Rev. D **90**, 026004 (2014)
4. A.K. Ganguly, P. Jain, S. Mandal. Phys. Rev. D **79**, 115014 (2009)

5. A.K. Ganguly, S. Konar, P.B. Pal, Phys. Rev. D **60**, 105014 (1999)
6. J. W. Brockway, E. D. Carlson, G.G. Raffelt, Phys. Lett. **B383**, 439 (1996). Report No. astro-ph/9605197
7. J.A. Grifols, E. masso, R. Toldra, Phys. Rev. Lett. **77**, 2372 (1996)

Chapter 221
Search for Heavy Neutral Higgs in Di-Boson Final State at $\sqrt{s} = 13$ TeV

Aashaq Shah, Ashok Kumar and Mario Pelliccioni

221.1 Introduction

A new boson with a mass of about 125 GeV was reported in 2012 by the ATLAS [1] and CMS [2, 3] collaborations at the CERN LHC which is consistent with the Higgs boson of the standard model (SM) of electroweak (EW) interactions. However, such a particle can be realised in several Beyond Standard Model (BSM) scenarios such as two-Higgs doublet model (2HDM) which predict the existence of an additional resonance at high mass with couplings similar to those of the SM Higgs boson. In this paper, we cover the search results for heavy neutral resonance decaying to WW and ZZ with fully final leptonic states using CMS.

221.2 Data and Simulated Samples

We cover the details of $H \to ZZ \to 4\ell$ ($\ell = \mathrm{e}, \mu$). Details for $H \to WW \to 2\ell 2\nu$ can be found in [4]. The data samples used for the analysis were recorded in proton-proton collisions at 13 TeV corresponding to total integrated luminosity of 12.9 fb^{-1}.

On behalf of the CMS Collaboration.

A. Shah (✉) · A. Kumar
Department of Physics and Astrophysics, University of Delhi, Delhi, India
e-mail: aashaq.shah@cern.ch

M. Pelliccioni
Istituto Nazionale di Fisica Nucleare (INFN), Turino, Italy

Md. Naimuddin (ed.), *XXII DAE High Energy Physics Symposium*, Springer Proceedings in Physics 203, https://doi.org/10.1007/978-3-319-73171-1_221

The signal samples were generated for the gluon-gluon fusion (ggH) and vector boson fusion (VBF) production modes at next-to-leading order (NLO) in QCD using POWHEG v2 for different masses.

221.3 Event Selection and Background Estimation

Each event is required to have four isolated leptons grouped into $2e2\mu$, $4e$ and 4μ final states and to separate gluon fusion and VBF production, a matrix-element based discriminant has been used. The main processes producing the backgrounds are $Z + jets$, $t\bar{t} + jets$, $Z\gamma + jets$, $WW + jets$ and $WZ + jets$ and are denoted as $Z + X$ since they are dominated by the $Z + jets$ process. The background is estimated using two independent control regions in data by requiring the presence of two leptons which satisfy the "tight" identification criteria, plus two additional opposite-sign (OS) or same-sign (SS) leptons satisfying relaxed ("loose") identification requirements. The category of events composed with two leptons which pass (P) the lepton tight identification requirements and two leptons which pass the loose identification but fail (F) the tight identification are denoted as the 2P2F region. Backgrounds which intrinsically have only two prompt leptons, such as Z + jets and $t\bar{t}$ are estimated by this control region [5].

221.4 Results

The observed p-value and the 95 percent CL upper exclusion limit are calculated for every generated mass point from 130 GeV up to 2.5 TeV for different signal width hypothesis. The expected and observed exclusion limits on the sum of ggH and VBF cross sections times branching fraction is shown in Fig. 221.1.

221.5 Conclusion

A search for a heavy Higgs boson in the $X \rightarrow WW \rightarrow 2\ell 2\nu$ and $X \rightarrow ZZ \rightarrow 4\ell$ decay channels has been carried out in various jet categories in order to maximize the exclusion limits using 2015 and 2016 data samples recorded at $\sqrt{s}$= 13 TeV corresponding to total integrated luminosity of 12.9 fb^{-1} of proton-proton collisions by the CMS. The exclusion limits on the cross section times branching ratio have been reported and no significant excess with respect to the SM background expectation has been observed.

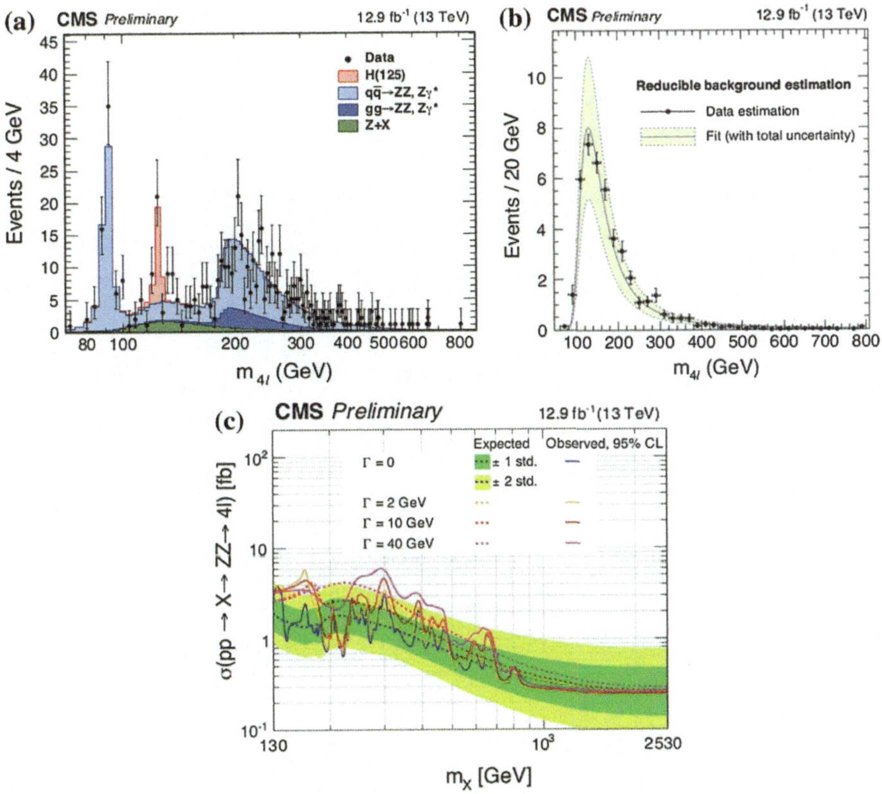

Fig. 221.1 **a** Distribution of reconstructed invariant mass m_{4l} **b** Combination of the OS and SS method predictions for the background in the signal region and the parametrized m_{4l} shape. **c** Observed and expected upper limits at the 95% CL on the $X \to ZZ \to 4\ell$ cross section as a function of resonance mass at several Γ_X values [5]

References

1. ATLAS Collaboration: Observation of a new particle in the search for the Standard Model Higgs boson with the ATLAS detector at the LHC. Phys. Lett. B **716**(1) (2012)
2. C.M.S. Collaboration, Observation of a new boson at a mass of 125 GeV with the CMS experiment at the LHC. Phys. Lett. B **716**, 30 (2012)
3. C.M.S. Collaboration, The CMS experiment at the CERN LHC. JINST **3**, S08004 (2008)
4. CMS Collaboration: Search for high mass Higgs to WW with fully leptonic decays using 2015 data. CMS PAS HIG-16-023
5. CMS Collaboration: Measurements of properties of the Higgs boson and search for an additional resonance in the four-lepton final state at $\sqrt{s} = 13$ TeV. CMS PAS HIG-16-033

Chapter 222
Transverse Single Spin Asymmetry in J/ψ Production at $\sqrt{s} = 200$ GeV

Bipin Sonawane, Rohini Godbole, Abhiram Kaushik, Anuradha Misra and Vaibhav Rawoot

222.1 Introduction and Formalism

Sivers function describes the momentum distribution of unpolarised partons inside transversely polarised hadron [1]. In the past, there have been many attempts to extract quark Sivers functions from experimental data. Recently, a phenomenological study to estimate gluon Sivers function (GSF) was peformed in [2] by fitting GSF to data on asymmetry in the process $p + p^{\uparrow} \rightarrow \pi^0 + X$ taken by PHENIX collaboration at RHIC [3].

TSSA in the process $p + p^{\uparrow(\downarrow)} \rightarrow J/\psi + X$ is defined as-

$$A_N = \frac{d\sigma^{\uparrow} - d\sigma^{\downarrow}}{d\sigma^{\uparrow} + d\sigma^{\downarrow}}$$

In Color Evaporation modal (CEM) of charmonium production, the numerator of TSSA is

B. Sonawane (✉) · A. Misra · V. Rawoot
Department of Physics, University of Mumbai, Mumbai 400098, India
e-mail: bipin.sonawane@physics.mu.ac.in

A. Misra
e-mail: misra@physics.mu.ac.in

V. Rawoot
e-mail: vaibhavrawoot@gmail.com

R. Godbole · A. Kaushik
Centre for High Energy Physics, Indian Institute of Science, Banglore 560012, India
e-mail: rohini@chep.iisc.ernet.in

A. Kaushik
e-mail: abhiramb@chep.iisc.ernet.in

Md. Naimuddin (ed.), *XXII DAE High Energy Physics Symposium*, Springer Proceedings in Physics 203, https://doi.org/10.1007/978-3-319-73171-1_222

$$\frac{d^4\sigma^\uparrow}{dydM^2d^2\mathbf{q}_T} - \frac{d^4\sigma^\downarrow}{dydM^2d^2\mathbf{q}_T} = \frac{F_{J/\psi}}{s}\int[d^2\mathbf{k}_{\perp a}d^2\mathbf{k}_{\perp b}]\Delta^N f_{g/p^\uparrow}(x_a,\mathbf{k}_{\perp a})f_{g/p}(x_b,\mathbf{k}_{\perp b})$$
$$\delta^2(\mathbf{k}_{\perp a}+\mathbf{k}_{\perp b}-\mathbf{q}_T)\hat{\sigma}_0^{gg\to c\bar{c}}(M^2) \quad (222.1)$$

and the denominator is

$$\frac{d^4\sigma^\uparrow}{dydM^2d^2\mathbf{q}_T} + \frac{d^4\sigma^\downarrow}{dydM^2d^2\mathbf{q}_T} = \frac{2F_{J/\psi}}{s}\int d^2\mathbf{k}_{\perp a}d^2\mathbf{k}_{\perp b} f_{g/p^\uparrow}(x_a,\mathbf{k}_{\perp a})f_{g/p}(x_b,\mathbf{k}_{\perp b})$$
$$\delta^2(\mathbf{k}_{\perp a}+\mathbf{k}_{\perp b}-\mathbf{q}_T)\hat{\sigma}_0^{gg\to c\bar{c}}(M^2) \quad (222.2)$$

$F_{J/\psi}$ is the fraction which gives the probability of J/ψ production below $D\bar{D}$ threshold.

We use the following Sivers function parameterization [2, 4]

$$\Delta^N f_{g/p^\uparrow}(x,k_\perp;Q) = 2\mathcal{N}_g(x)f_{g/p}(x,Q)h(k_\perp)\frac{e^{-k_\perp/\langle k_\perp^2\rangle}}{\pi\langle k_\perp^2\rangle} \quad (222.3)$$

$$\mathcal{N}_a(x) = N_a x^\alpha(1-x)^\beta\frac{(\alpha+\beta)^{\alpha+\beta}}{\alpha^\alpha\beta^\beta} \quad (222.4)$$

where, $a = q, \bar{q}$ or g, $h(k_\perp) = \sqrt{2e}\frac{k_\perp}{M_1}e^{-k_\perp^2/M_1^2}$ and N_a, α, β and M_1 are parameters determined by fits to data. e is Euler's number. SIDIS -1 and SIDIS -2 parameterizations of GSF use the best fit parameters of [2] while BV-a and BV-b refer to the models of GSF proposed by Boer and Vogelsang [5]:

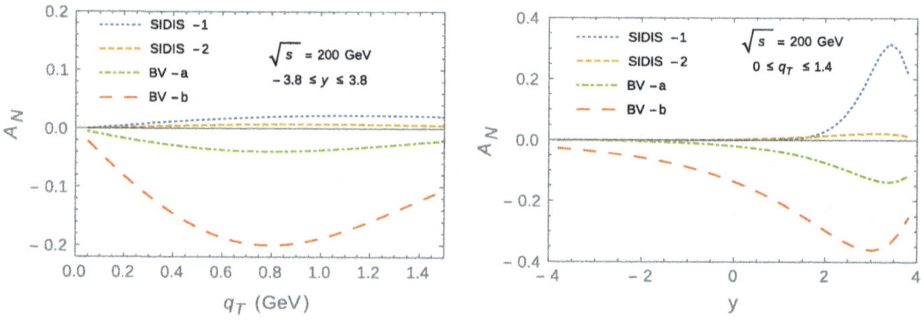

Fig. 222.1 Comparision of predictions of asymmetry in $p^\uparrow + p \to J/\psi + X$ using BV-a and BV-b parametrization with those obtain using SIDIS-1 and SIDIS-2 at $\sqrt{s} = 200\,\text{GeV}$. q_T and y integration ranges are $0 < q_T < 1.4\,\text{GeV}$ and $-3.8 < y < 3.8$ respectively. Left panel and right panel are plots of q_T and y distribution of asymmetry respectively [6].

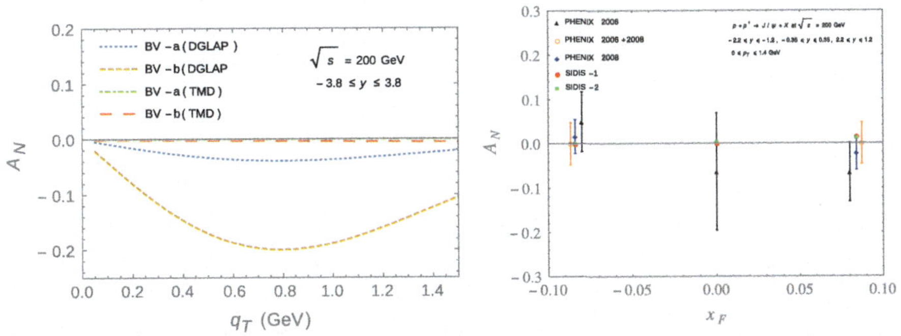

Fig. 222.2 Comparision of asymmetry with DGLAP and TMD evolution at $\sqrt{s} = 200\,\text{GeV}$. Right panel is comparision of asymmetry using SIDIS-1 and SIDIS-2 parameters with PHENIX data. Region of q_T integrations are $0 < q_T < 1.4\,\text{GeV}$ and. Asymmetry is in backward $(-2.2 < y < -1.2)$, mid$(-0.35 < y < 0.35)$ and forward $(1.2 < y < 2.2)$ rapidity regions [6].

BV-a: $\mathcal{N}_g(x) = (\mathcal{N}_u(x) + \mathcal{N}_d(x))/2$ and BV-b: $\mathcal{N}_g(x) = \mathcal{N}_d(x)$.
For making predictions with DGLAP evolved densities, we use the set of parameters given in [4] and for TMD evolved densities, we use parameter set given in [6].

222.2 Results

Summary: Here,we have presented estimates of TSSA in J/ψ production at $\sqrt{s} = 200\,\text{GeV}$ (Figs. 222.1 and 222.2). Asymmetries are found to be consistent with almost zero result of PHENIX experiment. Magnitude of asymmetry is found to be reduced when the effect of TMD evolution is taken into account.

References

1. D.W. Sivers, Single spin production asymmetries from the hard scattering of point-like constituents. Phys. Rev. D **41**, 83 (1990)
2. U. D'Alesio, F. Murgia, C. Pisano, Towards a first estimate of the gluon Sivers function from A_N data in pp collisions at RHIC. JHEP **09**, 119 (2015)
3. A. Adare et al., Measurement of transverse-single-spin asymmetries for midrapidity and forward-rapidity production of hadrons in polarized p+p collisions at $\sqrt{s}$ =200 and 62.4 GeV. Phys. Rev. D **90**(1), 012006 (2014)
4. Rohini M. Godbole, Anuradha Misra, Asmita Mukherjee, Vaibhav S. Rawoot, Sivers effect and transverse single spin asymmetry in $e + p^{\uparrow} \to e + J/\psi + X$. Phys. Rev. D **85**, 094013 (2012)
5. Daniel Boer, Werner Vogelsang, Asymmetric jet correlations in p p uparrow scattering. Phys. Rev. D **69**, 094025 (2004)
6. R.M. Godbole, A. Kaushik, A. Misra, V. Rawoot, B. Sonawane, Heavy flavour production as probe of gluon sivers function, arXiv:1612.01167 [hep-ph]

Chapter 223
Some Studies Using Capillary for Flow Control in a Closed Loop Gas Recirculation System

S. D. Kalmani, Surya Mondal, R. R. Shinde and P. V. Hunagund

223.1 Introduction

In the initial design of the CLS, as shown in Fig. 223.1, a high pressure to low pressure (HPLP) diaphragm based device was used to regulate, the high pressure (100 Kpa) from the storage tank to low pressure (200 pa) feeding to the RPCs. Due to the continuous atmospheric pressure variation cycle and the set flow rate (6 SCCM) in glass RPCs, it is necessary for the pressure controller to adjust pressure so as not to exceed the safe operating differential pressure of 3 mbar. However the HPLP regulator was found not responding as quickly as required to compensate pressure changes. Therefore, an external pressure sensor in the room was used to correct the pressure inside the CLS was implemented. A detailed study on the use of the capillary that could be used in lieu of the HPLP is described below.

The experimental setup, as shown in Fig. 223.2, consists of two MFCs (Mass Flow Controllers) and differential pressure sensors ASHCOFT XLdp. A microcontroller ARDUINO board is programmed to read data of the pressure difference across the capillary at a rate of 60 Hz. Two available MFCs are modified such that, one for controlling the flow and the other for reading the flow are used, and these are calibrated. The full scale range of the MFCs is 45 SCCM which is in the range of interest. Using Poiseuille's law is designed for the capillary length of 2.5 m of diameter of 300 microns. This tube is wound on a bobbin for safe interconnections and handling.

The studies are performed using gasses $C_2H_2F_4$ (R 134a), iC_4H_{10} (Isobutane) and SF_6 (Sulphur hexafluoride) used in the glass RPCs.

S. D. Kalmani (✉) · S. Mondal · R. R. Shinde
Tata Institute of Fundamental Research, Mumbai, India
e-mail: kalmani@tifr.res.in

P. V. Hunagund
Department of Applied Electronics and Research,
Gulbarga University, Gulbarga, India

Md. Naimuddin (ed.), *XXII DAE High Energy Physics Symposium*, Springer Proceedings in Physics 203, https://doi.org/10.1007/978-3-319-73171-1_223

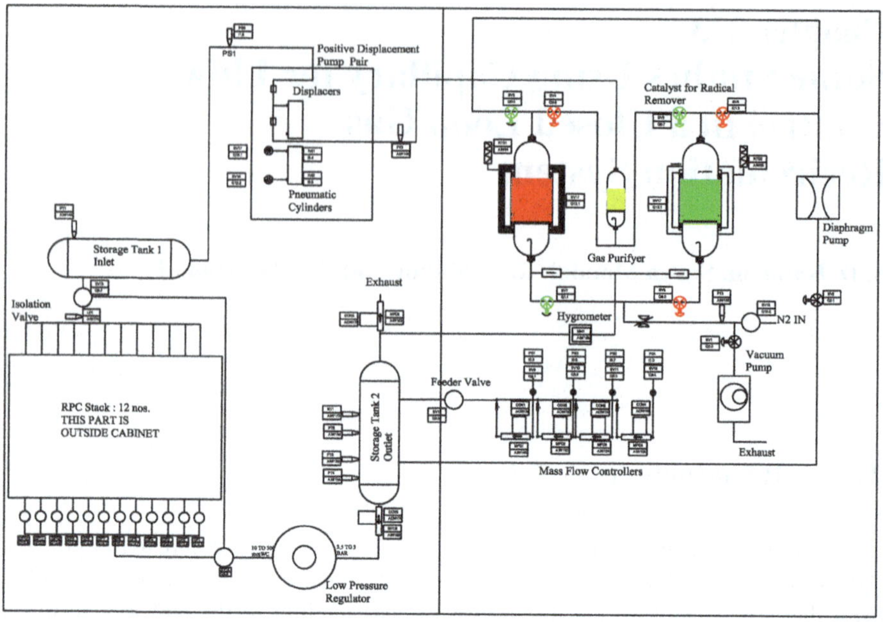

Fig. 223.1 Block Diagram of CLS

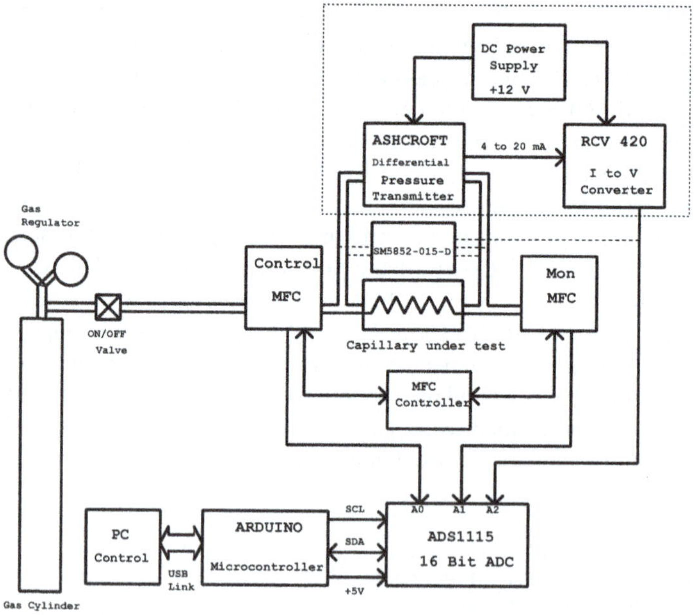

Fig. 223.2 The Experimental setup

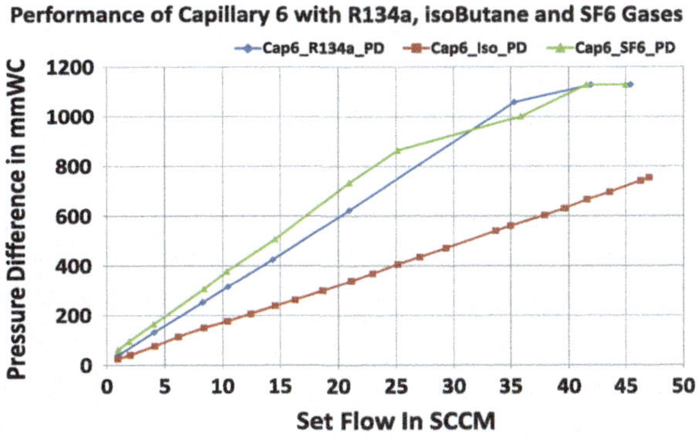

Fig. 223.3 Linear flow up to 25 SCCM with different gases of interest

223.2 Results and Conclusions

1. The flow resistance dependence on the type of the gas that flows through it, higher the viscosity, higher pressure is required to obtain a given rate. The SF_6 gas has the higher viscosity and the lowest is the R134a Fig. 223.3.
2. The capillary is functioning as expected. But is bulky and hence needed to be engineered for compactness.
3. Hagen poiseuille law holds good for the capillary designed, the flow can be considered to be laminar in the region of interest and can be used for flow rate upto 25 SCCM.

References

1. INO-Collaboration, India-based neutrino observatory, Technical Report INO/2006/01, Tata Institute of Fundamental Research (2006). http://www.ino.tifr.res.in/ino/OpenReports/INOReport.pdf
2. M. Bhuyan, et. al., Performance of the prototype gas recirculation system with built-in RGA for INO RPC system. https://doi.org/10.1016/j.nima.2010.09.169

Chapter 224
Constraints on Electromagnetic Properties of Neutrinos with Sub-keV Germanium Detectors

Lakhwinder Singh and H. T. Wong

224.1 Introduction

The neutrino oscillation experiments with various sources suggested the existence of nonzero masses and mixing between different neutrino states. Their nonzero mass has triggered intensive studies of nontrivial electromagnetic properties in both theoretical and experimental framework. The study of electromagnetic characteristics of neutrino may shed light on Dirac or Majorana nature and also as probe of new physics that might exist beyond the Standard Model(SM).

The research program of the TEXONO Collaboration [1] on neutrino electromagnetic properties, νN coherent scattering and beyond the SM is pursuing at the Kuo-Sheng Neutrino Laboratory (KSNL). A detailed description of the KSNL facilities can be found in [1]. In order to achieve a 100 eV threshold, we have studied operational characteristics and performance of several Ge detectors with different configurations under ultra-low background environment [2].

224.2 Electromagnetic Properties of Neutrino

The anomalous magnetic moment (μ_ν) automatically emerges from quantum correction when right-handed neutrinos include into SM. The value of μ_ν is proportional to the neutrino mass as required by the symmetry. The origin of neutrino millicharged (q_ν) is related to electric charge quantization which is one of the profound mystery of nature. The anomaly cancellation constraints provide the electric charge quantization within SM. However, it is no longer ensured in many extensions of SM. Both μ_ν and q_ν contributions are enhanced in sub-keV energy region due to $(1/T)$ and $(1/T^2)$-dependence, respectively, and illustrated in Fig. 224.1a. A typical measured

On the behalf of TEXONO Collaboration.

L. Singh (✉) · H. T. Wong
Institute of Physics, Academia Sinica, Taipei 11529, Taiwan
e-mail: lakhwinder@gate.sinica.edu.tw

Md. Naimuddin (ed.), *XXII DAE High Energy Physics Symposium*, Springer Proceedings in Physics 203, https://doi.org/10.1007/978-3-319-73171-1_224

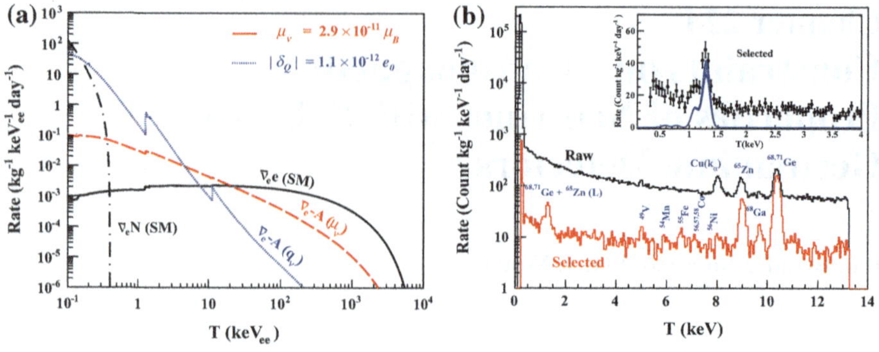

Fig. 224.1 The observable spectra due to neutrino interactions on Ge target with reactor $\bar{\nu}_e$ at $\phi_{\bar{\nu}_e} = 10^{13}$ cm^{-2}s^{-1}, with current experimental bound on neutrino magnetic moment, neutrino millicharge, together with the SM $\bar{\nu}_e - e$ and coherent scattering $\bar{\nu}_e - N$. **b** Typical spectra from point-contact Ge detectors at the KSNL [3]

energy spectra is shown in Fig. 224.1b with several identified K-L shell captured X-ray lines. The analysis threshold down to 300 eV is achieved with PCGe as depicted in Fig. 224.1b. It is observed that q_ν contribution has enhancement in cross-section when atomic effects are calculated with *ab-initio* Many-Body Theory. To include the atomic effect in the analysis, we adopted the *ab-initio* Multi-Configuration Relativistic Random-Phase Approximation theory [4]. The upper limits of $q_\nu < 2.1 \times 10^{-12}$ and $\mu_\nu < 2.6 \times 10^{-10} \mu_B$ are derived using PCGe 124.2(70.3) kg-days of ON (OFF) data at 90% C.L. The upper bounds on q_ν and μ_ν are limited due to low statistics and high background in sub-keV region. Our best experimental upper bound $\mu_\nu < 7.4 \times 10^{-11} \mu_B$ at 90% CL is derived from 570.7(127.8) kg-days of ON (OFF) with 12 keV threshold data. The transition magnetic moment ($\mu_{\nu_{sa}}$) of a sterile neutrino can rise from its conversion to an active neutrino through radiative decay. At sterile neutrino mass $m_{sa} = 7.1$ keV, the upper limit of $\mu_{\nu_{sa}} < 2.5 \times 10^{-14} \mu_B$ at 90% C.L. is derived [5]. The background in sub-keV region is not fully explained with conventional background modeling. Therefore, intensive studies on understanding of background by simulation, background measurements and new analysis techniques are still in progress.

References

1. H. B. Li et al., Phys. Rev. Lett. **90**, 131802 (2003); H. T. Wong et al., Phys. Rev. D **75**, 012001 (2007); H.T. Wong, Universe **3**, 22 (2015)
2. A.K. Soma et al., Nucl. Instrum. Methods Phys. Res. A **836**, 67–82 (2016)
3. J.-W. Chen et al., Phys. Rev. D **90**, 011301(R) (2014)
4. J.-W. Chen et al., Phys. Rev. D **91**, 013005 (2015)
5. J.-W. Chen et al., Phys. Rev. D **93**, 093012 (2016)

Chapter 225
Dark Matter, Neutrino Mass and Muon $(g-2)$ in a $U(1)_{L_\mu} - L_\tau$ Model

Anirban Biswas, Sandhya Choubey and Sarif Khan

In SM, neutrinos are massless because of the absence of right handed (RH) counterpart of the left handed (LH) neutrinos. This is in direct contradiction with the neutrino oscillation data which indicates the existence of tiny neutrino masses. Although, $U(1)_{L_\mu - L_\tau}$ extension is anomaly free, in order to generate tiny neutrino masses via Type-I seesaw mechanism we have introduced three RH neutrinos and the present scenario satisfy the neutrino oscillation parameters in 3σ range. The Yukawa interactions in generating the neutrino masses are as follows,

$$\mathcal{L}_Y = - \sum_{i,j=1,2,3} \frac{1}{2} (M_R)_{ij} \bar{N_i^c} N_j - \sum_{i=e,\mu,\tau} (M_D)_i \bar{L_i} \tilde{\phi}_h N_i + h.c. \quad (225.1)$$

where M_D is the Dirac mass matrix and M_R is the Majorana mass matrix (see [1] for detailed discussion). Another well established puzzle is the existence of DM and

A. Biswas · S. Choubey · S. Khan (✉)
Harish-Chandra Research Institute, Chhatnag Road,
Jhunsi, Allahabad 211019, India
e-mail: sarifkhan@hri.res.in

A. Biswas
e-mail: anirbanbiswas@hri.res.in

S. Choubey
e-mail: sandhya@hri.res.in

A. Biswas · S. Choubey · S. Khan
HBNI, Mumbai, India

S. Choubey
Department of Theoretical Physics, School of Engineering Sciences,
KTH Royal Institute of Technology, 106 91 Stockholm, Sweden

S. Choubey
AlbaNova University Center, Stockholm, Sweden

Md. Naimuddin (ed.), *XXII DAE High Energy Physics Symposium*, Springer Proceedings in Physics 203, https://doi.org/10.1007/978-3-319-73171-1_225

SM does not contain any candidate which can serve as the DM candidate. Since the gauge extension is local hence an additional gauge boson $Z_{\mu\tau}$ appeared. Here in this work we have extended the scalar sector by two additional singlet scalars ϕ_H (takes VEV and generated Majorana masses for the RH neutrinos and additional gauge boson $Z_{\mu\tau}$) and ϕ_{DM} (serve as the DM candidate). The interaction terms of DM with the SM like Higgs and other BSM scalar are as follows,

$$\mathcal{L}_{DM} \approx -\lambda_{Dh}(\phi^\dagger_{DM}\phi_{DM})(\phi^\dagger_h\phi_h) - \lambda_{DH}(\phi^\dagger_{DM}\phi_{DM})(\phi^\dagger_H\phi_H) - \lambda_{hH}(\phi^\dagger_h\phi_h)(\phi^\dagger_H\phi_H)\,, \tag{225.2}$$

where ϕ_h is SM like Higgs doublet which mixes with ϕ_H through the term $\lambda_{hH}(\phi^\dagger_h\phi_h)(\phi^\dagger_H\phi_H)$. The extra gauge boson $Z_{\mu\tau}$ has a significant contribution to muon $(g-2)$ anomaly by an additional one loop diagram which is given by,

$$\Delta a_\mu(Z_{\mu\tau}) = \frac{g^2_{\mu\tau}}{8\pi^2}\int_0^1 dx\frac{2x(1-x)^2}{(1-x)^2+rx}\,, \tag{225.3}$$

As shown in (225.3), the additional contribution to muon $(g-2)$ anomaly for $M_{Z_{\mu\tau}} =$ 100 MeV and $g_{\mu\tau} = 9\times 10^{-4}$ is $\Delta a_\mu = 22.6\times 10^{-10}$, which lies within the 2σ range. Next, we have calculated the neutrino oscillation parameters which satisfy the experimental data in 3σ range. In the left panel (LP) of Fig. 225.1, we have shown the variation of Dirac CP phase δ_{CP} with the atmospheric angle θ_{23}. Here we observe that for the present model δ_{CP} is very small while the values of θ_{23} lie in the both octant. On the other hand, in the right panel we have shown the variation of M_{ee} ([1, 1] element M_R matrix) with $V_{e\mu}$ ([1, 2] element of M_R matrix) and one can see that the correlation exists between the two parameter (more detail plots are given in [1]).

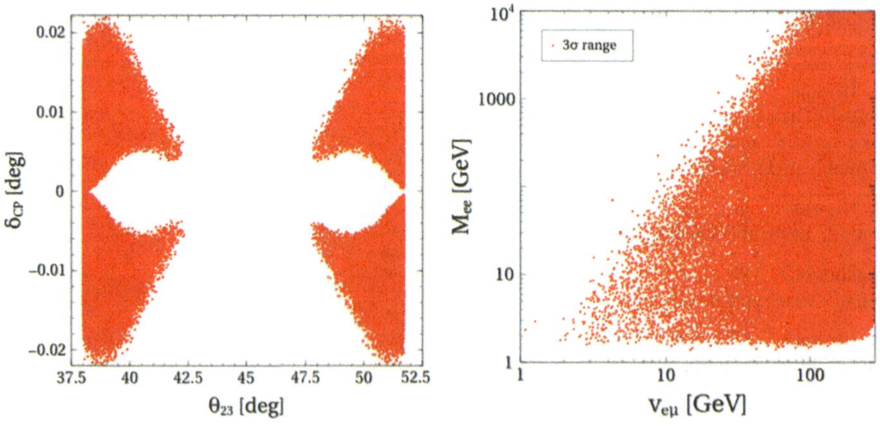

Fig. 225.1 Left panel: Variation of Dirac CP phase δ_{CP} with mixing angle θ_{23}. Right panel: Variation of M_{ee} with $V_{e\mu}$

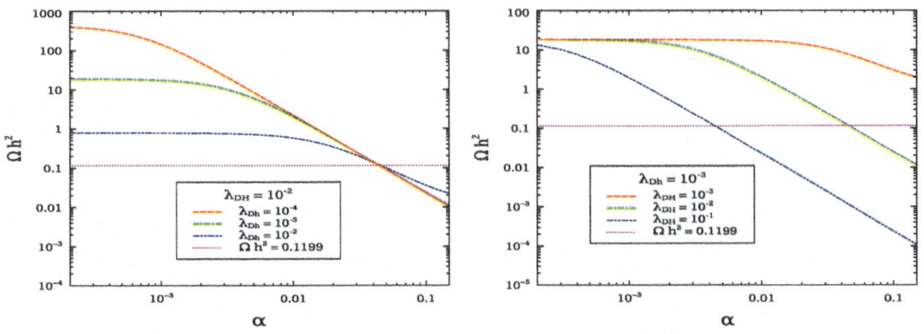

Fig. 225.2 Left (Right) panel: Variation of DM relic density with the mixing angle α for three different values of λ_{Dh} (λ_{DH})

In LP and RP of Fig. 225.2, we have shown the variation of DM relic density with the mixing angle α between the two Higgses. In the left panel, we have shown the variation of $\Omega_{DM}h^2$ for three different values of λ_{Dh} (quartic coupling between DM ϕ_{DM} and SM like Higgs ϕ_h) while in the right panel, three different values of λ_{DH} are used (quartic coupling between ϕ_{DM} and extra Higgs ϕ_H, for details see [1]). Magenta dot line in both the figures represents the central value of DM relic density measured by Planck satelite [2]. For the present model, we can also explain the Fermi-LAT [3] Galactic centre γ-ray excess by considering DM annihilation to $b\bar{b}$ final state at the Galactic centre. The produced b quarks therefore hadronise to produce γ-rays. We have done the calculation of γ-ray flux in detail in [1].

References

1. A. Biswas, S. Choubey, S. Khan, JHEP **1609**, 147 (2016)
2. P.A.R. Ade et al., [Planck Collaboration], arXiv:1502.01589 [astro-ph.CO]
3. W.B. Atwood et al., [Fermi-LAT Collaboration], Astrophys. J. **697**, 1071 (2009)

Chapter 226
Measurement of the Double-Differential Inclusive Jet Cross Section in Proton–Proton Collisions at 13 TeV Centre-of-Mass Energy with the Compact Muon Solenoid Detector

Sourav Dey

226.1 Introduction

A measurement of the double-differential inclusive jet cross section by CMS Collaborations [2] is presented as a function of the jet transverse momentum p_T and absolute jet rapidity $|y|$. A detailed description of the analysis is provided in [1]. Proton-proton collision data were recorded at a centre-of-mass energy of 13 TeV in 2015 with integrated luminosities of 71 and 44 pb^{-1} for rapidity range $|y| < 3$ and $3.2 < |y| < 4.7$, respectively.

226.2 Event Selection, Jet Reconstruction and Measurement of the Double-Differential Inclusive Jet Cross Section

The double-differential inclusive jet cross section is defined as:

$$\frac{d^2\sigma}{dp_T dy} = \frac{1}{\epsilon\mathcal{L}} \frac{N_j}{\Delta p_T \Delta y} \tag{226.1}$$

where $\mathcal{L}$ is the integrated luminosity, N_j is the number of jets in a bin of a width p_T in transverse momentum and y in rapidity, and ϵ is the product of the trigger and

Sourav Dey on behalf of the CMS Collaboration.

S. Dey (✉)
Saha Institute of Nuclear Physics, Kolkata, India
e-mail: sourav.dey@cern.ch

Md. Naimuddin (ed.), *XXII DAE High Energy Physics Symposium*, Springer Proceedings in Physics 203, https://doi.org/10.1007/978-3-319-73171-1_226

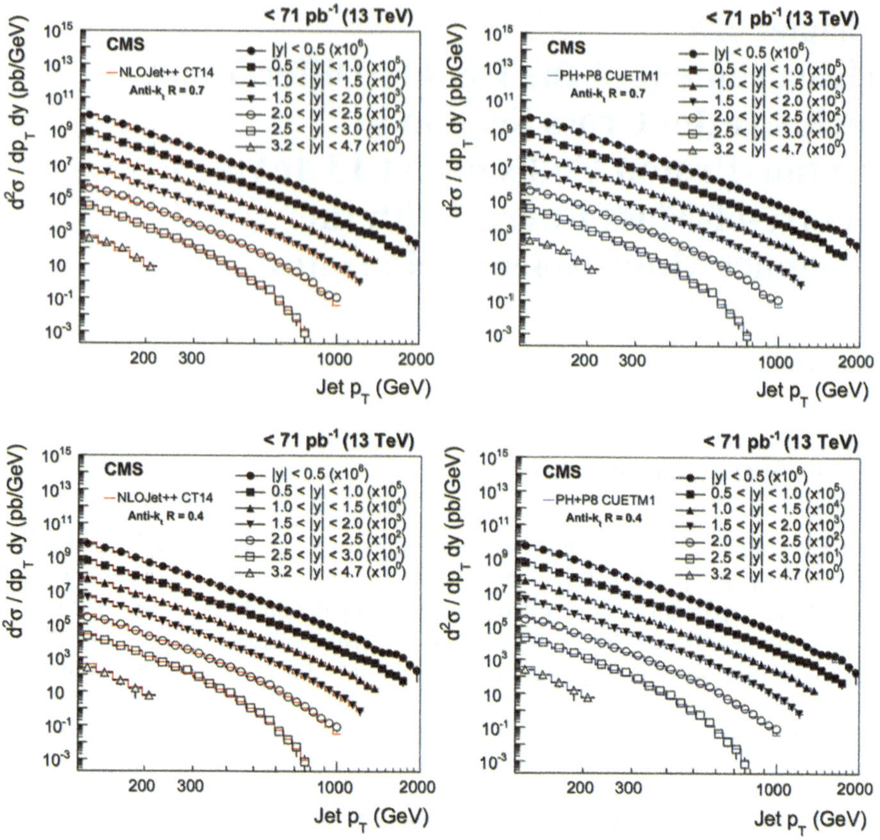

Fig. 226.1 Double-differential inclusive jet cross section as function of jet p_T [1]. On the left, data (points) and predictions from NLOJet++ based on the CT14 PDF set corrected for the non-perturbative and electroweak effects (line) are shown. On the right, data (points) and predictions from Powheg (PH) + Pythia8 (P8) with tune CUETM1 (line) are shown. Top two figures correspond to jets are clustered with the anti-k_t algorithm ($R = 0.7$), while the bottom figures are for ($R = 0.4$)

jet selection efficiencies, which is greater than 99%. Figure 226.1 show the double-differential inclusive jet cross section measurements, presented as a function of p_T for seven $|y|$ ranges, after unfolding for detector effects. The data are consistent with the predictions over a wide range of jet p_T from 114 GeV up to 2 TeV.

References

1. CMS Collaboration, Measurement of the double-differential inclusive jet cross section in protonproton collisions at $\sqrt{s} = 13$ TeV Eur. Phys. J. C **76**, 451 (2016). https://doi.org/10.1140/epjc/s10052-016-4286-3
2. CMS Collaboration, The CMS experiment at the CERN LHC. JINST **3**, S08004 (2008)

Erratum to: Flux Tubes, Field Configuration and Non-Perturbative Dynamics of QCD

Deependra Singh Rawat, H. C. Chandola, H. C. Pandey and D. Yadav

Erratum to: Chapter "Flux Tubes, Field Configuration and Non-Perturbative Dynamics of QCD" in: Md. Naimuddin (ed.), *XXII DAE High Energy Physics Symposium*, Springer Proceedings in Physics 203, https://doi.org/10.1007/978-3-319-73171-1_147

The original version of the book was inadvertently published with incorrect author name and affiliations in Chapter 147, which have to be corrected as follows:

In the first occurence, the author name "H. C. Pandey" has to be changed as "H.C. Chandola" and the corresponding email id "hempandey@birlainstitute.co.in" has to be replaced with "chandolaharish@gmail.com".

In the second occurence, the email id "chandolaharish@gmail.com" has to be replaced with "hempandey@birlainstitute.co.in" for author "H. C. Pandey".

And also, a new acknowledgement text has to be included at the end of the chapter. The erratum chapter and the book have been updated with the changes.

The updated online version of this chapter can be found at
https://doi.org/10.1007/978-3-319-73171-1_147

Md. Naimuddin (ed.), *XXII DAE High Energy Physics Symposium*, Springer Proceedings in Physics 203, https://doi.org/10.1007/978-3-319-73171-1_227

Index

Md. Naimuddin (ed.), *XXII DAE High Energy Physics Symposium*, Springer Proceedings in Physics 203, https://doi.org/10.1007/978-3-319-73171-1

The manufacturer's authorised representative in the EU is Springer Nature Customer Service Centre GmbH, Europaplatz 3, 69115 Heidelberg, Germany. If you have any concerns regarding our products, please contact ProductSafety@springernature.com

Printed and bound by CPI Group (UK) Ltd, Croydon, CR0 4YY
30/09/2026
02225663-0001